IEBC

2015

CODE AND COMMENTARY

The complete IEBC with commentary after each section

INTERNATIONAL CODE COUNCIL

2015 International Existing Building Code® and Commentary

First Printing: November 2015

ISBN: 978-1-60983-289-6 (soft-cover edition)

COPYRIGHT © 2015
by
INTERNATIONAL CODE COUNCIL, INC.

ALL RIGHTS RESERVED. This 2015 *International Existing Building Code® and Commentary* is a copyrighted work owned by the International Code Council, Inc. Without advance written permission from the copyright owner, no part of this book may be reproduced, distributed or transmitted in any form or by any means, including, without limitation, electronic, optical or mechanical means (by way of example, and not limitation, photocopying, or recording by or in an information storage retrieval system). For information on permission to copy material exceeding fair use, please contact: Publications, 4051 Flossmoor Road, Country Club Hills, IL 60478. Phone 1-888-ICC-SAFE (422-7233).

Trademarks: "International Code Council," the "International Code Council" logo and the "International Existing Building Code" are trademarks of the International Code Council, Inc.

PRINTED IN THE U.S.A.

PREFACE

The principal purpose of the Commentary is to provide a basic volume of knowledge and facts relating to building construction as it pertains to the regulations set forth in the 2015 *International Existing Building Code*. The person who is serious about effectively designing, constructing and regulating existing buildings and structures will find the Commentary to be a reliable data source and reference to almost all components in the built environment.

Throughout all of this, strenuous effort has been made to keep the vast quantity of material accessible and its method of presentation useful. With a comprehensive yet concise summary of each section, the Commentary provides a convenient reference for regulations applicable to the construction of buildings and structures. In the chapters that follow, discussions focus on the full meaning of application and the consequences of not adhering to the code text. Illustrations and examples are provided to aid understanding; they do not necessarily illustrate the only methods of achieving code compliance.

The format of the Commentary includes the full text of each section, table and figure in the code, followed immediately by the commentary applicable to that text. At the time of printing, the Commentary reflects the most up-to-date text of the 2015 *International Existing Building Code*. As stated in the preface to the *International Existing Building Code*, the content of sections in the code that begin with a letter designation (i.e., Section [B]1301.2) are maintained by another code development committee. Each section's narrative includes a statement of its objective and intent, and usually includes a discussion about why the requirement commands the conditions set forth. Code text and commentary text are easily distinguished from each other. All code text is shown as it appears in the *International Existing Building Code*, and all commentary is indented below the code text and begins with the symbol ❖.

Readers should note that the Commentary is to be used in conjunction with the *International Existing Building Code* and not as a substitute for the code. The Commentary is advisory only; the code official alone possesses the authority and responsibility for interpreting the code.

Comments and recommendations are encouraged, for through your input, we can improve future editions. Please direct your comments to the Code and Standards Department in the Chicago District Office.

The International Code Council would like to extend its thanks to Susan Gentry, Melvyn Green, Wayne Jewell, Ken Schoonover, Mark Stimac and the Structural Engineers Association of California (SEAOC) whose help as contributing authors made this document possible.

TABLE OF CONTENTS

CHAPTER 1	SCOPE AND ADMINISTRATION	1-1 – 1-24
CHAPTER 2	DEFINITIONS	2-1 – 2-6
CHAPTER 3	PROVISIONS FOR COMPLIANCE METHODS	3-1 – 3-10
CHAPTER 4	PRESCRIPTIVE COMPLIANCE METHOD	4-1 – 4-28
CHAPTER 5	CLASSIFICATION OF WORK	5-1 – 5-4
CHAPTER 6	REPAIRS	6-1 – 6-8
CHAPTER 7	ALTERATIONS—LEVEL 1	7-1 – 7-16
CHAPTER 8	ALTERATIONS—LEVEL 2	8-1 – 8-30
CHAPTER 9	ALTERATIONS—LEVEL 3	9-1 – 9-10
CHAPTER 10	CHANGE OF OCCUPANCY	10-1 – 10-14
CHAPTER 11	ADDITIONS	11-1 – 11-8
CHAPTER 12	HISTORIC BUILDINGS	12-1 – 12-8
CHAPTER 13	RELOCATED OR MOVED BUILDINGS	13-1 – 13-4
CHAPTER 14	PERFORMANCE COMPLIANCE METHODS	14-1 – 14-40
CHAPTER 15	CONSTRUCTION SAFEGUARDS	15-1 – 15-6
CHAPTER 16	REFERENCED STANDARDS	16-1 – 16-6
APPENDIX A	GUIDELINES FOR THE SEISMIC RETROFIT OF EXISTING BUILDINGS (GSREB)	A1-1
CHAPTER A1	SEISMIC STRENGTHENING PROVISIONS FOR UNREINFORCED MASONRY BEARING WALL BUILDINGS	A1-1 – A1-28
CHAPTER A2	EARTHQUAKE HAZARD REDUCTION IN EXISTING REINFORCED CONCRETE AND REINFORCED MASONRY WALL BUILDINGS WITH FLEXIBLE DIAPHRAGMS	A2-1 – A2-10
CHAPTER A3	PRESCRIPTIVE PROVISIONS FOR SEISMIC STRENGTHENING OF CRIPPLE WALLS AND SILL PLATE ANCHORAGE OF LIGHT, WOOD-FRAME RESIDENTIAL BUILDINGS	A3-1 – A3-30

TABLE OF CONTENTS

**CHAPTER A4 EARTHQUAKE RISK REDUCTION IN WOOD-FRAMED
             RESIDENTIAL BUILDINGS WITH SOFT, WEAK OR OPEN FRONT WALLS** .... A4-1 – A4-14

**CHAPTER A5 EARTHQUAKE HAZARD REDUCTION IN EXISTING
             CONCRETE BUILDINGS** .... A5-1 – A5-8

CHAPTER A6 REFERENCED STANDARDS .... A6-1 – A6-2

**APPENDIX B SUPPLEMENTARY ACCESSIBILITY REQUIREMENTS
             FOR EXISTING BUILDINGS AND FACILITIES** .... B-1 – B-4

APPENDIX C GUIDELINES FOR THE WIND RETROFIT OF EXISTING BUILDINGS C1-1

CHAPTER C1 GABLE END RETROFIT FOR HIGH-WIND AREAS .... C1-1 – C1-22

CHAPTER C2 ROOF DECK FASTENING FOR HIGH-WIND AREAS .... C2-1 – C2-2

**RESOURCE A GUIDELINES ON FIRE RATINGS OF ARCHAIC
             MATERIALS AND ASSEMBLIES** .... RESOURCE A-1 – A-140

INDEX   .... INDEX-1 – INDEX-6

Chapter 1:
Scope and Administration

General Comments

This chapter contains provisions for the scope and application (Part 1), and the enforcement and administration (Part 2) of subsequent requirements of the code. In addition to establishing the scope of the code, Chapter 1 identifies which buildings and structures come under its purview. Part 1, Scope and Application, includes Sections 101 and 102. Section 101 addresses the scope of the code as it applies to existing structures undergoing repairs, alterations, change of occupancy and additions or relocation. Section 102 establishes the applicability of the code and addresses existing structures. Part 2, Administration and Enforcement, includes the balance of the provisions of Chapter 1. Section 103 establishes the Department of Building Safety and details the appointment of department personnel. Section 104 outlines the duties and authority of the code official with regard to permits, inspections and right of entry. It also establishes the authority of the code official to approve alternative materials, used materials and modifications. Section 105 states when permits are required and establishes the procedures for reviewing applications and issuing permits. Section 106 describes the information that must be included on the documents submitted with the application. Section 107 authorizes the code official to issue permits for temporary structures and uses. Section 108 establishes requirements for a fee schedule. Section 109 includes the inspection duties of the code official or an inspection agency that has been approved by the code official. Provisions for issuing certificates of occupancy are detailed in Section 110. Section 111 gives the code official the authority to approve utility connections. Section 112 establishes the board of appeals and the criteria for making applications for appeal. Administrative provisions for violations are addressed in Section 113, including provisions for unlawful acts, violation notices, prosecution and penalties. Section 114 describes procedures for stop work orders. Section 115 establishes the criteria for unsafe structures and equipment, and the procedures to be followed by the code official to abate such conditions and notify the responsible party. Section 116 describes the emergency measures that address structures in danger of collapse. Section 117 authorizes the code official to have structures demolished that are dangerous, unsafe, insanitary or otherwise unfit for human habitation or occupancy. Each state's building-code-enabling legislation, which is grounded in the police power of the state, is the source of all authority to enact building codes. In terms of how it is used, police power is the power of the state to legislate for the general welfare of its citizens. This power enables passage of such laws as building codes. If the state legislature has limited this power in any way, the municipality may not exceed these limitations. While the municipality may not further delegate its police power (e.g., by delegating the burden of determining code compliance to the building owner, contractor or architect), it may turn over the administration of the building code to a municipal official, such as a code official, provided that sufficient criteria are given to the code official to establish clearly the basis for decisions as to whether a proposed building conforms to the code.

Chapter 1 is largely concerned with maintaining "due process of law" in enforcing the building performance criteria contained in the body of the code. Only through careful observation of the administrative provisions can the code official reasonably hope to demonstrate that "equal protection under the law" has been provided. While it is generally assumed that the administration and enforcement section of a code is geared toward a code official, this is not entirely true. The provisions also establish the rights and privileges of the design professional, contractor and building owner. The position of the code official is merely to review the proposed and completed work, and to determine if the construction conforms to the code requirements. The design professional is responsible for the design of a safe structure. The contractor is responsible for constructing the structure in conformance to the plans.

During the course of construction, the code official reviews the activity to ascertain that the spirit and intent of the law are being met and that the safety, health and welfare of the public will be protected. As a public servant, the code official enforces the code in an unbiased, proper manner. Every individual is guaranteed equal enforcement of the provisions of the code. Further, design professionals, contractors and building owners have the right of due process for any requirement in the code.

Purpose

The code, as with any other code, is intended to be adopted as a legally enforceable document to safeguard health, safety, property and public welfare. A code cannot be effective without adequate provisions for its administration and enforcement. The code official charged with the administration and enforcement of building regulations has a great responsibility and with this responsibility comes authority. No matter how detailed the code may be, the code official must, to some extent, exercise his or her own judgment in determining code compliance. The code official has the responsibility to establish that the homes in which the citizens of the community reside and the buildings in which they work are designed and constructed to be structurally stable, with adequate means of egress, light and ventilation, and to provide a minimum acceptable level of protection to life and property from fire.

A large number of existing buildings and structures do not comply with the current building code requirements for new construction. Although many of these buildings are potentially salvageable, rehabilitation is often cost prohibitive because they may not be able to comply with all the requirements for new construction. At the same time, it is necessary to regulate construction in existing buildings that undergo additions, alterations, renovations, extensive repairs or change of occupancy. Such activity represents an opportunity to ensure that new construction complies with the current building codes and that existing conditions are maintained, at a minimum, to their current level of compliance or are improved as required. To accomplish this objective, and to make the rehabilitation process easier, this chapter allows for a controlled departure from full compliance with the technical codes, without compromising the minimum standards for the fire prevention and life safety features of the rehabilitated building.

PART 1—SCOPE AND APPLICATION

SECTION 101
GENERAL

[A] 101.1 Title. These regulations shall be known as the *Existing Building Code* of **[NAME OF JURISDICTION]**, hereinafter referred to as "this code."

❖ The purpose of this section is to identify the adopted regulations by inserting the name of the adopting jurisdiction into the code.

[A] 101.2 Scope. The provisions of the *International Existing Building Code* shall apply to the *repair*, *alteration*, *change of occupancy*, *addition* to and relocation of *existing buildings*.

❖ This section establishes where the regulations contained in the code must be followed, whether all or in part. Something must happen (modification to an existing building or allowing an existing building or structure to become unsafe) for the code to be applicable. While such activity may not be as significant as for a new building, a fence is considered a structure and, therefore, its erection is within the scope of the code. The code is not a maintenance document requiring periodic inspections that will, in turn, result in an enforcement action, although periodic inspections are addressed by the *International Fire Code®* (IFC®).

[A] 101.3 Intent. The intent of this code is to provide flexibility to permit the use of alternative approaches to achieve compliance with minimum requirements to safeguard the public health, safety and welfare insofar as they are affected by the *repair*, *alteration*, *change of occupancy*, *addition* and relocation of *existing buildings*.

❖ The code is intended to set forth regulations that establish the minimum acceptable level to safeguard public health, safety and welfare. The intent becomes important in the application of sections such as Sections 102, 104.11 and 113, as well as any enforcement-oriented interpretive action or judgment. Like any code, the written text is subject to interpretation. Interpretations should not be affected by economics or the potential impact on any party. The only considerations should be the protection of public health, safety and welfare.

[A] 101.4 Applicability. This code shall apply to the *repair*, *alteration*, *change of occupancy*, *addition* and relocation of *existing buildings*, regardless of occupancy, subject to the criteria of Sections 101.4.1 and 101.4.2.

❖ All existing structures must comply with the provisions of the code when undergoing repair, alteration, change of occupancy, addition and relocation, subject to the criteria in Sections 101.4.1 and 101.4.2. Sections 101.4.1 and 101.4.2 contain provisions that are significantly different, based on whether the building has been previously occupied. Basically, if the building has not been previously occupied, it must comply with the requirements for new construction. This also applies to buildings undergoing alterations or additions.

[A] 101.4.1 Buildings not previously occupied. A building or portion of a building that has not been previously occupied or used for its intended purpose in accordance with the laws in existence at the time of its completion shall be permitted to comply with the provisions of the laws in existence at the time of its original permit unless such permit has expired. Subsequent permits shall comply with the *International Building Code* or *International Residential Code*, as applicable, for new construction.

❖ This section requires all buildings that have not been previously occupied to comply with the *International Building Code®* (IBC®) or the *International Residential Code®* (IRC®) if the permit for such building has expired. It also applies to any building that may have been completed and not occupied or used for its intended purpose. The building remains a new structure in terms of code compliance until such a time as it is occupied in whole or in part. Again this is based on the permit being expired. If the permit has not expired then the building can comply with the laws in existence at the time of construction.

[A] 101.4.2 Buildings previously occupied. The legal occupancy of any building existing on the date of adoption of this code shall be permitted to continue without change, except as is specifically covered in this code, the *International Fire Code*, or the *International Property Maintenance Code*, or as is deemed necessary by the *code official* for the general safety and welfare of the occupants and the public.

❖ This section allows for buildings that were legally occupied in part or in whole at the time the code was adopted to continue. There is a maintenance concern that is addressed by the requirement that the building comply with either the IFC or the *International Property Maintenance Code*® (IPMC®). These codes ensure that life safety systems, such as means of egress pathways and fire protection systems, are kept in place and able to protect the life and safety of the inhabitants of these existing structures.

[A] 101.5 Safeguards during construction. Construction work covered in this code, including any related demolition, shall comply with the requirements of Chapter 15.

❖ The fundamental rationale behind this section is to establish reasonable safety precautions, in accordance with Chapter 15, during all phases of construction and demolition. Chapter 15 also covers the protection of adjacent public and private properties.

[A] 101.6 Appendices. The *code official* is authorized to require rehabilitation and retrofit of buildings, structures or individual structural members in accordance with the appendices of this code if such appendices have been individually adopted.

❖ This section describes a unique aspect of the code whereby any appendix referenced in the code becomes a part of the code without the jurisdiction having to specifically adopt it. For example, in Chapter 7, voluntary alterations to lateral-force-resisting systems are allowed when conducted in accordance with Appendix A. Therefore, Appendix A, having been specifically referenced, is enforceable without having to be specifically adopted by the local jurisdiction. Any appendices not specifically referenced in the code must be individually adopted to be legally enforced.

[A] 101.7 Correction of violations of other codes. *Repairs* or *alterations* mandated by any property, housing, or fire safety maintenance code or mandated by any licensing rule or ordinance adopted pursuant to law shall conform only to the requirements of that code, rule, or ordinance and shall not be required to conform to this code unless the code requiring such *repair* or *alteration* so provides.

❖ This section is intended to keep the requirements of other codes or ordinances intact and separate from the requirements of the code.

SECTION 102
APPLICABILITY

[A] 102.1 General. Where there is a conflict between a general requirement and a specific requirement, the specific requirement shall be applicable. Where in any specific case different sections of this code specify different materials, methods of construction or other requirements, the most restrictive shall govern.

❖ In cases where the code establishes a specific requirement for a certain condition, that requirement is applicable even if it is less restrictive than a general requirement elsewhere in the code. Also, the most restrictive code requirement is to apply where there may be different requirements in the code for a specific installation.

[A] 102.2 Other laws. The provisions of this code shall not be deemed to nullify any provisions of local, state, or federal law.

❖ In some cases, other laws enacted by the jurisdiction, or the state or federal government may be applicable to a condition that is also governed by a requirement in the code. In such circumstances, the requirements of the code are in addition to that other law that is still in effect, although the code official may not be responsible for its enforcement.

[A] 102.3 Application of references. References to chapter or section numbers or to provisions not specifically identified by number shall be construed to refer to such chapter, section, or provision of this code.

❖ In a situation where the code may make reference to a chapter or section number or to another code provision without specifically identifying its location in the code, assume that the referenced section, chapter or provision is in the code and not in a referenced code or standard.

[A] 102.4 Referenced codes and standards. The codes and standards referenced in this code shall be considered part of the requirements of this code to the prescribed extent of each such reference and as further regulated in Sections 102.4.1 and 102.4.2.

Exception: Where enforcement of a code provision would violate the conditions of the listing of the equipment or appliance, the conditions of the listing shall govern.

❖ A referenced code, standard or portion thereof is an enforceable extension of the code as if the content of the standard were included in the body of the code. For example, Section 301.1.4 references ASCE 41 in its entirety for the seismic evaluation and design of an existing building. In those cases where the code references only portions of a standard, the use and application of the referenced standard is limited to those portions that are specifically identified. For example, Item 4, Section 403.5 references structural irregularity as defined in ASCE 7. Therefore, it is only these portions of ASCE 7 that are applicable to this specific code requirement. Lastly, if conflicts between the requirements of the code and a referenced standard occur, the requirements of the code govern regardless of which requirement is more restrictive. The exception deals with possible conflicts between code requirements and the conditions of an equipment listing. If the code conflicts with or deviates from the conditions of the listing, this may or may not mean that the code violated the listing. For example, the listing for an appliance might allow a particular application of an appliance that is expressly prohibited by the code. In this case, the code has not violated the listing, but instead has simply limited the application allowed by the listing. The intent is for the highest level of safety to prevail.

[A] 102.4.1 Conflicts. Where conflicts occur between provisions of this code and referenced codes and standards, the provisions of this code shall apply.

❖ This section clarifies that the code takes precedence over a referenced code or standard. As discussed in the commentary to Section 102.4, it is also important to understand the extent to which another code or standard has been referenced. This may reduce the number of possible conflicts to be addressed.

[A] 102.4.2 Conflicting provisions. Where the extent of the reference to a referenced code or standard includes subject matter that is within the scope of this code, the provisions of this code as applicable, shall take precedence over the provisions in the referenced code or standard.

❖ Often it can be confusing when a standard is referenced in its entirety and has a similar scope but very different or simply conflicting requirements. This section clarifies that, although that standard may address similar issues, the code will take precedence over those requirements as a whole.

[A] 102.5 Partial invalidity. In the event that any part or provision of this code is held to be illegal or void, this shall not have the effect of making void or illegal any of the other parts or provisions.

❖ Only invalid sections of the code (as established by the court of jurisdiction) can be set aside. This is essential to safeguard the application of the code text to situations where a provision is declared illegal or unconstitutional. This section preserves the legislative action that put the legal provisions in place.

PART 2—ADMINISTRATION AND ENFORCEMENT

SECTION 103
DEPARTMENT OF BUILDING SAFETY

[A] 103.1 Creation of enforcement agency. The Department of Building Safety is hereby created, and the official in charge thereof shall be known as the *code official*.

❖ This section creates the building department and describes its composition (see Section 109 for a discussion of the inspection duties of the department). Appendix A of the IBC contains qualifications for the employees of the building department involved in the enforcement of the code. If a jurisdiction desires to establish these qualifications for its employees, Appendix A must be specifically referenced in the adopting ordinance.

The executive official in charge of the building department is named the "code official" by this section. In actuality, the person who is in charge of the department may hold a different title, such as building commissioner, building inspector or construction official. For the purpose of the code, that person is referred to as the "code official."

[A] 103.2 Appointment. The *code official* shall be appointed by the chief appointing authority of the jurisdiction.

❖ This section establishes the code official as an appointed position of the jurisdiction.

[A] 103.3 Deputies. In accordance with the prescribed procedures of this jurisdiction and with the concurrence of the appointing authority, the *code official* shall have the authority to appoint a deputy *code official*, the related technical officers, inspectors, plan examiners, and other employees. Such employees shall have powers as delegated by the *code official*.

❖ This section provides the code official with the authority to appoint other individuals to assist with the administration and enforcement of the code. These individuals have the authority and responsibility as designated by the code official. Such appointments, however, may be exercised only with the authorization of the chief appointing authority.

SECTION 104
DUTIES AND POWERS OF CODE OFFICIAL

[A] 104.1 General. The *code official* is hereby authorized and directed to enforce the provisions of this code. The *code* *official* shall have the authority to render interpretations of this code and to adopt policies and procedures in order to clarify the application of its provisions. Such interpretations, policies, and procedures shall be in compliance with the intent and purpose of this code. Such policies and procedures shall not have the effect of waiving requirements specifically provided for in this code.

❖ The duty of the code official is to enforce the code, and he or she is the "authority having jurisdiction" for all matters relating to the code and its enforcement. It is the duty of the code official to interpret the code and to determine compliance. Code compliance will not always be easy to determine and will require judgment and expertise, particularly when enforcing the provisions of Sections 104.10 and 104.11. In exercising this authority, however, the code official cannot set aside or ignore any provision of the code.

[A] 104.2 Applications and permits. The *code official* shall receive applications, review construction documents, and issue permits for the *repair*, *alteration*, *addition*, demolition, *change of occupancy*, and relocation of buildings; inspect the premises for which such permits have been issued; and enforce compliance with the provisions of this code.

❖ The code enforcement process is normally initiated with an application for a permit. The code official is responsible for processing applications and issuing permits for the modification of buildings in accordance with the code.

[A] 104.2.1 Determination of substantially improved or substantially damaged existing buildings and structures in flood hazard areas. For applications for reconstruction, rehabilitation, repair, alteration, addition or other improvement of existing buildings or structures located in flood hazard areas, the building official shall determine where the proposed work constitutes substantial improvement or repair of substantial damage. Where the building official determines that the proposed work constitutes substantial improvement or repair of substantial damage, and where required by this code, the building official shall require the building to meet the requirements of Section 1612 of the *International Building Code*.

❖ This language is similar to IRC Section R105.3.1.1, which has the building official making a finding with regard to the value of the proposed work and market value of the building. This subsection describes what the building official does to determine whether work proposed for existing buildings meets those definitions. Some code officials have suggested to FEMA that the simple presence of the definitions is insufficient to ensure that these determinations are made and it would be helpful if the building official's responsibilities clearly specified making these determinations.

FEMA published extensive guidance on substantial improvement and substantial damage, including a number of acceptable methods to estimate market value and project costs. Most jurisdictions require the applicant to provide an estimate of costs, which is already required to be included in the application.

[A] 104.2.2 Preliminary meeting. When requested by the permit applicant or the *code official*, the *code official* shall meet with the permit applicant prior to the application for a construction permit to discuss plans for the proposed work or *change of occupancy* in order to establish the specific applicability of the provisions of this code.

Exception: *Repairs* and Level 1 *alterations*.

❖ The preliminary meeting is an important aspect of any repair, alteration, change of occupancy, addition or relocation of any building. At this phase in a project, it is considerably less expensive to make changes or corrections. Possible problem issues can be identified and solutions devised ahead of time resulting in fewer correction notices and rework being required. The exception recognizes that a preliminary meeting would not be necessary for a repair or alteration level for one project based on the fact that this level of work is not complicated or involved.

[A] 104.2.2.1 Building evaluation. The *code official* is authorized to require an *existing building* to be investigated and evaluated by a registered design professional based on the circumstances agreed upon at the preliminary meeting. The design professional shall notify the *code official* if any potential nonconformance with the provisions of this code is identified.

❖ This section authorizes the code official to have an existing structure investigated and evaluated by a design professional. Existing structures may have some structural problems that are not immediately visible. The ability to call in an experienced design professional to aid in the proper evaluation of an existing structure is invaluable.

[A] 104.3 Notices and orders. The *code official* shall issue necessary notices or orders to ensure compliance with this code.

❖ An important element of code enforcement is the necessary advisement of deficiencies and correction, which is accomplished through written notices and orders. The code official is required to issue orders to abate illegal or unsafe conditions. Section 115.3 contains additional information for these notices.

[A] 104.4 Inspections. The *code official* shall make the required inspections, or the *code official* shall have the authority to accept reports of inspection by approved agencies or individuals. Reports of such inspections shall be in writing and be certified by a responsible officer of such approved agency or by the responsible individual. The *code official* is authorized to engage such expert opinion as deemed necessary to report upon unusual technical issues that arise, subject to the approval of the appointing authority.

❖ The code official is required to make inspections as necessary to determine compliance with the code or to accept written reports of inspections by an approved agency. The inspection of the work in progress or accomplished to date is another significant element in determining code compliance. While a department does not have the resources to inspect every aspect of all work, the required inspections are those that are dictated by administrative rules and procedures based on many parameters, including available inspection resources. In order to expand the available resources for inspection purposes, the code official may approve an agency that, in his or her opinion, is objective and competent, has adequate equipment to perform any required tests, and employs experienced personnel educated in conducting, supervising and evaluating tests and inspections. When unusual, extraordinary or complex technical issues arise relative to building safety, the code official has the authority to seek the opinion and advice of experts. Since this usually involves the expenditure of funds, the approval of the jurisdiction's chief executive (or similar position) is required. A technical report from an expert requested by the code official can be used to assist in the approval process.

[A] 104.5 Identification. The *code official* shall carry proper identification when inspecting structures or premises in the performance of duties under this code.

❖ This section requires the code official (including, by definition, all authorized designees) to carry identification in the course of conducting the duties of the position. This removes any question as to the purpose and authority of the inspector.

[A] 104.6 Right of entry. Where it is necessary to make an inspection to enforce the provisions of this code, or where the *code official* has reasonable cause to believe that there exists in a structure or upon a premises a condition that is contrary to or in violation of this code that makes the structure or premises unsafe, *dangerous*, or hazardous, the *code official* is authorized to enter the structure or premises at reasonable times to inspect or to perform the duties imposed by this code, provided that if such structure or premises be occupied that credentials be presented to the occupant and entry requested. If such structure or premises be unoccupied, the *code official* shall first make a reasonable effort to locate the owner, the owner's authorized agent or other person having charge or control of the structure or premises and request entry. If entry is refused, the *code official* shall have recourse to the remedies provided by law to secure entry.

❖ The first part of this section establishes the right of the code official to enter the premises in order to make the permit inspections required by Section 109.3. Permit application forms typically include a statement in the certification signed by the applicant (who is the owner or owner's agent) granting the code official the authority to enter areas covered by the permit in order to enforce code provisions related to the permit. The right to enter other structures or premises is more limited. First, to protect the right of privacy, the owner or occupant must grant the code official permission before an interior inspection of the property can be conducted. Permission is not required for inspections that can be accomplished from the public right-of-way. Second, such access may be denied by the owner or occupant. Unless the inspector has reasonable cause to believe that a violation of the code exists, access may be unattainable. Third, code officials must present proper identification (see Section 104.5) and request admittance during reasonable hours—usually the normal business hours of the establishment—to be admitted. Fourth, inspections must be aimed at securing or determining compliance with the provisions and intent of the regulations that are specifically in the established scope of the code official's authority.

Searches to gather information for the purpose of enforcing the other codes, ordinances or regulations are considered unreasonable and are prohibited by the Fourth Amendment to the U.S. Constitution. "Reasonable cause" in the context of this section must be distinguished from "probable cause," which is required to gain access to property in criminal cases. The burden of proof establishing reasonable cause may vary among jurisdictions. Usually, an inspector must show that the property is subject to inspection under the provisions of the code; that the interests of the public's health, safety and welfare outweigh the individual's right to maintain privacy; and that such an inspection is required solely to determine compliance with the provisions of the code.

Many jurisdictions do not recognize the concept of an administrative warrant and may require the code official to prove probable cause in order to gain access upon refusal. This burden of proof is usually more substantial, often requiring the code official to stipulate in advance why access is needed (usually access is restricted to gathering evidence for seeking an indictment or making an arrest); what specific items or information is sought; its relevance to the case against the individual subject; how knowledge of the relevance of the information or items sought was obtained; and how the evidence sought will be used. In all such cases, the right to privacy must always be weighed against the right of the code official to conduct an inspection to verify that public health, safety and welfare are not in jeopardy. Such important and complex constitutional issues should be discussed with the jurisdiction's legal counsel. Jurisdictions should establish procedures for securing the necessary court orders when an inspection is deemed necessary following a refusal.

[A] 104.7 Department records. The *code official* shall keep official records of applications received, permits and certificates issued, fees collected, reports of inspections, and notices and orders issued. Such records shall be retained in the official records for the period required for retention of public records.

❖ In keeping with the need for efficient business practices, the code official must keep records pertaining to permit applications, permits, fees collected, inspections, notices and orders issued. Such documentation provides a valuable resource if questions arise regarding the department's actions with respect to a building. The code does not require that construction documents be kept after the project is complete. It requires that other documents be kept for the length of time mandated by the jurisdiction or state laws or administrative rules for retaining public records.

[A] 104.8 Liability. The *code official*, member of the Board of Appeals, or employee charged with the enforcement of this code, while acting for the jurisdiction in good faith and without malice in the discharge of the duties required by this code or other pertinent law or ordinance, shall not thereby be rendered civilly or criminally liable personally and is hereby relieved from personal liability for any damage accruing to persons or property as a result of any act or by reason of an act or omission in the discharge of official duties.

❖ The code official, other department employees and members of the appeals board are not intended to be held liable for those actions performed in accordance with the code in a reasonable and lawful manner. The responsibility of the code official in this regard is subject to local, state and federal laws that may supersede this provision.

[A] 104.8.1 Legal defense. Any suit or criminal complaint instituted against an officer or employee because of an act performed by that officer or employee in the lawful discharge of duties and under the provisions of this code shall be defended by legal representatives of the jurisdiction until the final termination of the proceedings. The code official or any subordinate shall not be liable for cost in any action, suit, or proceeding that is instituted in pursuance of the provisions of this code.

❖ This section establishes that code officials (or subordinates) must not be liable for costs in any legal action instituted in response to the performance of lawful duties. These costs are to be borne by the state, county or municipality. The best way to be certain that the code official's action is a "lawful duty" is always to cite the applicable code section on which the enforcement action is based.

[A] 104.9 Approved materials and equipment. Materials, equipment, and devices approved by the *code official* shall be constructed and installed in accordance with such approval.

❖ The code is a compilation of criteria with which materials, equipment, devices and systems must comply to be suitable for a particular application. The code official has a duty to evaluate such materials, equipment, devices and systems for code compliance and, when compliance is determined, approve the same for use. The materials, equipment, devices and systems must be constructed and installed in compliance with, and all conditions and limitations considered as a basis for, that approval. For example, the manufacturer's instructions and recommendations are to be followed if the approval of the material was based, even in part, on those instructions and recommendations. The approval authority given the code official is a significant responsibility and is a key to code compliance.

The approval process is first technical and then administrative and must be approached as such. For example, if data to determine code compliance are required, such data should be in the form of test reports or engineering analysis and not simply taken from a sales brochure.

[A] 104.9.1 Used materials and equipment. The use of used materials that meet the requirements of this code for new materials is permitted. Used equipment and devices shall be permitted to be reused subject to the approval of the *code official*.

❖ The code criteria for materials and equipment have changed over the years. Evaluation of testing and materials technology has permitted the development of new criteria that the old materials may not satisfy. As a result, used materials are required to be evaluated in the same manner as new materials. Used materials, equipment and devices must be specifically approved by the code official as being equivalent to that required by the code if they are to be used again in a new installation.

[A] 104.10 Modifications. Wherever there are practical difficulties involved in carrying out the provisions of this code, the *code official* shall have the authority to grant modifications for individual cases upon application of the owner or owner's authorized representative, provided the *code official* shall first find that special individual reason makes the strict letter of this code impractical, the modification is in compliance with the intent and purpose of this code and such modification does not lessen health, accessibility, life and fire safety, or structural requirements. The details of action granting modifications shall be recorded and entered in the files of the Department of Building Safety.

❖ The code official may amend or make exceptions to the code as needed where strict compliance is impractical. Only the code official has authority to grant modifications. Consideration of a particular difficulty is to be based on the application of the owner or owner's authorized agent and a demonstration that the intent of the code is accomplished. This section is not intended to permit setting aside or ignoring a code provision; rather, it is intended to provide for the acceptance of equivalent protection. Such modifications do not, however, extend to actions that are necessary to correct violations of the code. In other words, a code violation or the expense of correcting one cannot constitute a practical difficulty.

[A] 104.10.1 Flood hazard areas. For *existing buildings* located in *flood hazard areas* for which *repairs*, *alterations* and *additions* constitute *substantial improvement*, the *code official* shall not grant modifications to provisions related to flood resistance unless a determination is made that:

1. The applicant has presented good and sufficient cause that the unique characteristics of the size, configuration or topography of the site render compliance with the flood-resistant construction provisions inappropriate.

2. Failure to grant the modification would result in exceptional hardship.

3. The granting of the modification will not result in increased flood heights, additional threats to public safety, extraordinary public expense nor create nuisances, cause fraud on or victimization of the public or conflict with existing laws or ordinances.

4. The modification is the minimum necessary to afford relief, considering the flood hazard.

5. A written notice will be provided to the applicant specifying, if applicable, the difference between the design flood elevation and the elevation to which the building is to be built, stating that the cost of flood insurance will be commensurate with the increased risk resulting from the reduced floor elevation and that construction below the design flood elevation increases risks to life and property.

❖ This section addresses additional requirements for all buildings and structures in designated flood hazard areas. These areas are commonly referred to as "flood plains" and are shown on a community's Flood Insurance Rate Map (FIRM), prepared by the Federal Emergency Management Agency (FEMA), or other adopted flood hazard map. Through the adoption of the code, communities meet a significant portion of the flood plain management regulation requirements necessary to participate in the National Flood Insurance Program (NFIP). To participate in the NFIP, a jurisdiction must adopt, in addition to the flood-resistant requirements found in the code, Appendix G, Flood-Resistant Construction, of the IBC, or a flood plain management ordinance that contains, at a minimum, the provisions contained in Appendix G. In either case, flood plain management requirements must be applied for all existing buildings located in flood hazard areas for buildings undergoing repairs, alterations and additions that constitute substantial improvements. The code official may grant modifications to the provisions of this section for any of the five reasons listed in Section 104.10.1.

[A] 104.11 Alternative materials, design and methods of construction, and equipment. The provisions of this code are not intended to prevent the installation of any material or to prohibit any design or method of construction not specifically prescribed by this code, provided that any such alternative has been approved. An alternative material, design, or method of construction shall be approved where the *code official* finds that the proposed design is satisfactory and complies with the intent of the provisions of this code, and that the material, method, or work offered is, for the purpose intended, not less than the equivalent of that prescribed in this code in quality, strength, effectiveness, fire resistance, durability and safety. Where the alternative material, design or method of construction is not approved, the *code official* shall respond in writing, stating the reasons the alternative was not approved.

❖ The code is not intended to inhibit innovative ideas or technological advances. A comprehensive regulatory document, such as a building code, cannot envision and then address all future innovations in the industry. As a result, a performance code must be applicable to and provide a basis for the approval of an increasing number of newly developed, innovative materials, systems and methods for which no code text or referenced standards yet exist. The fact that a material, product or method of construction is not addressed in the code is not an indication that such material, product or method is intended to be prohibited. The code official is expected to apply sound technical judgment in accepting materials, systems or methods that, while not anticipated by the drafters of the current code text, can be demonstrated to offer equivalent performance. By virtue of its text, the code regulates new and innovative construction practices while addressing the relative safety of building occupants. The code official is responsible for determining if a requested alternative provides the equivalent level of protection of public health, safety and welfare as required by the code.

The last sentence is similar to that included in Section 105.3.1 regarding the rejection of a permit application. The reason for this additional level of communication is that the nonapproval of an alternative method is not the same as the nonapproval of a permit. In other words, the permit application may have been approved but an alternative method might not be approved until a later date. However, the reasons for responding to the applicant in writing are the same, as noted in the commentary to Section 105.3.1: "In order to ensure effective communication and due process of law, the reasons for denial of an application for a permit are required to be in writing." Similar language is found in all of the *International Codes*® (I-Codes®).

[A] 104.11.1 Research reports. Supporting data, where necessary to assist in the approval of materials or assemblies not specifically provided for in this code, shall consist of valid research reports from approved sources.

❖ When an alternative material or method is proposed for construction, it is incumbent upon the code official to determine whether this alternative is, in fact, an equivalent to the methods prescribed by the code. Reports providing evidence of this equivalency are required to be supplied by an approved source, meaning a source that the code official finds to be reliable and accurate. The ICC Evaluation Service is an example of an agency that provides research reports for alternative materials and methods.

[A] 104.11.2 Tests. Whenever there is insufficient evidence of compliance with the provisions of this code or evidence that a material or method does not conform to the requirements of this code, or in order to substantiate claims for alternative materials or methods, the *code official* shall have the authority to require tests as evidence of compliance to be made at no expense to the jurisdiction. Test methods shall be as specified in this code or by other recognized test standards. In the absence of recognized and accepted test methods, the *code official* shall approve the testing procedures. Tests shall be performed by an approved agency. Reports of such tests shall be retained by the *code official* for the period required for retention.

❖ To provide the basis on which the code official can make a decision regarding an alternative material or method, sufficient technical data, test reports and documentation must be provided for evaluation by the code official. If evidence satisfactory to the code official indicates that the alternative material or construction method is equivalent to that required by the code, the code official may approve it. Any such approval cannot have the effect of waiving any requirements of the code. The burden of proof of equivalence lies with the applicant who proposes the use of alternative materials or methods.

The code official must require the submission of any appropriate information and data to assist in the determination of equivalency. This information should be submitted before a permit can be issued. The type of information required includes test data in accordance with referenced standards, evidence of compliance with the referenced standard specifications and design calculations. A research report issued by an authoritative agency is particularly useful in providing the code official with the technical basis for evaluation and approval of new and innovative materials and methods of construction. The use of authoritative research reports can greatly assist the code official by reducing the time-consuming engineering analysis necessary to review these materials and methods. Failure to adequately substantiate a request for the use of an alternative is a valid reason for the code official to deny a request. Any tests submitted in support of an application must have been performed by an agency approved by the code official based on evidence that the agency has the technical expertise, test equipment and quality assurance to properly conduct and report the necessary testing. The test reports submitted to the code official must be retained in accordance with the requirements of Section 104.7.

SECTION 105
PERMITS

[A] 105.1 Required. Any owner or owner's authorized agent who intends to repair, add to, alter, relocate, demolish, or change the occupancy of a building or to repair, install, add, alter, remove, convert, or replace any electrical, gas, mechanical, or plumbing system, the installation of which is regulated by this code, or to cause any such work to be performed, shall first make application to the *code official* and obtain the required permit.

❖ This section contains the administrative rules governing the issuance, suspension, revocation or modification of building permits. It also establishes how and by whom the application for a building permit is to be made, how it is to be processed, fees and what information it must contain or have attached to it.

In general, a permit is required for all activities that are regulated by the code or its referenced codes (see Section 101.4) and these activities cannot begin until the permit is issued, unless the activity is specifically exempted by Section 105.2. Only the owner or a person authorized by the owner can apply for the permit. Note that this section indicates a need for a permit for a change in occupancy, even if no work is contemplated. Although the occupancy of a building or portion thereof may change and the new activity is still classified in the same group, different code provisions may be applicable. The means of egress, structural loads, and light and ventilation provisions are examples of requirements that are occupancy sensitive. The purpose of the permit is to cause the work to be reviewed, approved and inspected to determine compliance with the code.

[A] 105.1.1 Annual permit. Instead of an individual permit for each *alteration* to an already approved electrical, gas, mechanical, or plumbing installation, the *code official* is authorized to issue an annual permit upon application therefor to any person, firm, or corporation regularly employing one or more qualified trade persons in the building, structure, or on the premises owned or operated by the applicant for the permit.

❖ In some instances, such as large buildings or industrial facilities, the repair, replacement or alteration of electrical, gas, mechanical or plumbing systems occurs on a frequent basis. This section allows the code official to issue an annual permit for this work. This relieves both the building department and the owners of such facilities from the burden of filing and processing individual applications for this activity; however, there are restrictions on who is entitled to these permits. They can be issued only for work on a previously approved installation and only to an individual or corporation that employs persons specifically qualified in the trade for which the permit is issued. If tradespeople who perform the work involved are required to be licensed in the jurisdiction, then only those persons would be permitted to perform the work. If trade licensing is not required, then the code official needs to review and approve the qualifications of the persons who will be performing the work. The annual permit can apply only to the individual property that is owned or operated by the applicant.

[A] 105.1.2 Annual permit records. The person to whom an annual permit is issued shall keep a detailed record of *alterations* made under such annual permit. The *code official* shall have access to such records at all times, or such records shall be filed with the *code official* as designated.

❖ The work performed in accordance with an annual permit must be inspected by the code official, so it is necessary to know the location of such work and when it was performed. This can be accomplished by having records of the work available to the code official either at the premises or in his or her office, as determined by the official.

SCOPE AND ADMINISTRATION

[A] 105.2 Work exempt from permit. Exemptions from permit requirements of this code shall not be deemed to grant authorization for any work to be done in any manner in violation of the provisions of this code or any other laws or ordinances of this jurisdiction. Permits shall not be required for the following:

Building:

1. Sidewalks and driveways not more than 30 inches (762 mm) above grade and not over any basement or story below and that are not part of an accessible route.
2. Painting, papering, tiling, carpeting, cabinets, counter tops, and similar finish work.
3. Temporary motion picture, television, and theater stage sets and scenery.
4. Shade cloth structures constructed for nursery or agricultural purposes, and not including service systems.
5. Window awnings supported by an exterior wall of Group R-3 or Group U occupancies.
6. Movable cases, counters, and partitions not over 69 inches (1753 mm) in height.

Electrical:

Repairs and maintenance: Minor *repair* work, including the replacement of lamps or the connection of approved portable electrical equipment to approved permanently installed receptacles.

Radio and television transmitting stations: The provisions of this code shall not apply to electrical equipment used for radio and television transmissions, but do apply to equipment and wiring for power supply, the installations of towers, and antennas.

Temporary testing systems: A permit shall not be required for the installation of any temporary system required for the testing or servicing of electrical equipment or apparatus.

Gas:

1. Portable heating appliance.
2. Replacement of any minor part that does not alter approval of equipment or make such equipment unsafe.

Mechanical:

1. Portable heating appliance.
2. Portable ventilation equipment.
3. Portable cooling unit.
4. Steam, hot, or chilled water piping within any heating or cooling equipment regulated by this code.
5. Replacement of any part that does not alter its approval or make it unsafe.
6. Portable evaporative cooler.

7. Self-contained refrigeration system containing 10 pounds (4.54 kg) or less of refrigerant and actuated by motors of 1 horsepower (746 W) or less.

Plumbing:

1. The stopping of leaks in drains, water, soil, waste, or vent pipe; provided, however, that if any concealed trap, drainpipe, water, soil, waste, or vent pipe becomes defective and it becomes necessary to remove and replace the same with new material, such work shall be considered as new work, and a permit shall be obtained and inspection made as provided in this code.
2. The clearing of stoppages or the repairing of leaks in pipes, valves, or fixtures, and the removal and reinstallation of water closets, provided such *repairs* do not involve or require the replacement or rearrangement of valves, pipes, or fixtures.

❖ Section 105.1 essentially requires a permit for any activity involving work on a building and its systems and other structures. This section lists activities allowed without first obtaining a permit from the building department. Note that in some cases, such as Items 4 and 5, the work is exempt only for certain occupancies. Further, it is intended that even though work may be exempted for permit purposes, it must still comply with the code and the owner is responsible for proper and safe construction for all work being done. Work exempted by the codes adopted by reference in Section 101.4 is also included here.

[A] 105.2.1 Emergency repairs. Where equipment replacements and *repairs* must be performed in an emergency situation, the permit application shall be submitted within the next working business day to the *code official*.

❖ This section recognizes that, in some cases, emergency replacement and repair work must be done as quickly as possible and it is not practical to take the time necessary to apply for and obtain approval. A permit for the work must be obtained the next day that the building department is open for business. Any work performed before the permit is issued must be done in accordance with the code and corrected if not approved by the code official.

[A] 105.2.2 Repairs. Application or notice to the *code official* is not required for ordinary *repairs* to structures and items listed in Section 105.2. Such *repairs* shall not include the cutting away of any wall, partition, or portion thereof, the removal or cutting of any structural beam or load-bearing support, or the removal or change of any required means of egress or rearrangement of parts of a structure affecting the egress requirements; nor shall ordinary *repairs* include *addition* to, *alteration* of, replacement, or relocation of any standpipe, water supply, sewer, drainage, drain leader, gas, soil, waste, vent, or similar piping, electric wiring, or mechanical or other work affecting public health or general safety.

❖ This section distinguishes between what might be termed by some as repairs, but are in fact alterations that cause the code to be applicable, and ordinary repairs, which are maintenance activities and thus do not require a permit.

[A] 105.2.3 Public service agencies. A permit shall not be required for the installation, *alteration*, or *repair* of generation, transmission, distribution, or metering or other related equipment that is under the ownership and control of public service agencies by established right.

❖ Utility-owned transmission lines and metering equipment that supply electricity, gas, water, telephone, television cable, etc., may be worked on without a permit. Utilities typically are regulated by other laws that give them specific rights and authority in this area. Appliances or equipment installed or serviced but not owned or controlled by such agencies is not exempt from a permit.

[A] 105.3 Application for permit. To obtain a permit, the applicant shall first file an application therefor in writing on a form furnished by the Department of Building Safety for that purpose. Such application shall:

1. Identify and describe the work in accordance with Chapter 3 to be covered by the permit for which application is made.
2. Describe the land on which the proposed work is to be done by legal description, street address, or similar description that will readily identify and definitely locate the proposed building or work.
3. Indicate the use and occupancy for which the proposed work is intended.
4. Be accompanied by construction documents and other information as required in Section 106.3.
5. State the valuation of the proposed work.
6. Be signed by the applicant or the applicant's authorized agent.
7. Give such other data and information as required by the *code official*.

❖ This section requires that a written application for a permit be filed on forms provided by the building department, and it details the information required on the application. Permit forms will typically have sufficient space to write a very brief description of the work to be accomplished, which is sufficient only for small jobs. For larger projects, the description will be augmented by construction documents as indicated in Item 4. As required by Section 105.1, the applicant must be the owner of the property or an authorized agent of the owner, such as an engineer, architect, contractor or tenant. The applicant must sign the application, and permit forms typically include a statement that if the applicant is not the owner, he or she has permission from the owner to make the application.

[A] 105.3.1 Action on application. The *code official* shall examine or cause to be examined applications for permits and amendments thereto within a reasonable time after filing. If the application or the construction documents do not conform to the requirements of pertinent laws, the *code official* shall reject such application in writing, stating the reasons therefor. If the *code official* is satisfied that the proposed work conforms to the requirements of this code and laws and ordinances applicable thereto, the *code official* shall issue a permit therefor as soon as practicable.

❖ This section requires the code official to act with reasonable speed on a permit application. In some instances, this time period is set by state or local law. The code official must refuse to issue a permit when the application and accompanying documents do not conform to the code. In order to ensure effective communication and due process of law, the reasons for denial of an application for a permit are required to be in writing. Once the code official determines that the work described conforms to the code and other applicable laws, the permit must be issued upon payment of the fees required by Section 108.

[A] 105.3.2 Time limitation of application. An application for a permit for any proposed work shall be deemed to have been abandoned 180 days after the date of filing, unless such application has been pursued in good faith or a permit has been issued; except that the *code official* is authorized to grant one or more extensions of time for additional periods not exceeding 90 days each. The extension shall be requested in writing and justifiable cause demonstrated.

❖ Typically, an application for a permit is submitted and goes through a review process that ends with the issuance of a permit. If a permit has not been issued 180 days after the date of filing, however, the application is considered abandoned, unless the applicant was diligent in efforts to obtain the permit. The code official has the authority to extend this time limitation (in increments of 90 days), provided there is reasonable cause. This would cover delays beyond the applicant's control, such as prerequisite permits or approvals from other authorities in the jurisdiction or state. The intent of this section is to limit the time between the review process and the issuance of a permit.

[A] 105.4 Validity of permit. The issuance or granting of a permit shall not be construed to be a permit for, or an approval of, any violation of any of the provisions of this code or of any other ordinance of the jurisdiction. Permits presuming to give authority to violate or cancel the provisions of this code or other ordinances of the jurisdiction shall not be valid. The issuance of a permit based on construction documents and other data shall not prevent the *code official* from requiring the correction of errors in the construction documents and other data. The *code official* is authorized to prevent occupancy or use of a structure where in violation of this code or of any other ordinances of this jurisdiction.

❖ This section states the fundamental premise that the permit is only a license to proceed with the work. It is not a license to violate, cancel or set aside any provisions of the code. This is significant because it means that, despite any errors or oversights in the approval process, the permit applicant, not the code official, is responsible for code compliance. Also, the permit can be suspended or revoked in accordance with Section 105.6.

[A] 105.5 Expiration. Every permit issued shall become invalid unless the work on the site authorized by such permit is commenced within 180 days after its issuance, or if the work authorized on the site by such permit is suspended or abandoned for a period of 180 days after the time the work is commenced. The *code official* is authorized to grant, in writing, one or more extensions of time for periods not more than 180 days each. The extension shall be requested in writing and justifiable cause demonstrated.

❖ The permit becomes invalid under two distinct situations—both based on a 180-day period. The first situation is when no work has started 180 days from the issuance of the permit. The second situation is when the authorized work has stopped for 180 days. The person who was issued the permit should be notified, in writing, that the permit is invalid and what steps must be taken to reinstate it and restart the work. The code official has the authority to extend this time limitation (in increments of 180 days), provided the extension is requested in writing and there is reasonable cause, typically events beyond the permit holder's control.

[A] 105.6 Suspension or revocation. The *code official* is authorized to suspend or revoke a permit issued under the provisions of this code wherever the permit is issued in error or on the basis of incorrect, inaccurate, or incomplete information or in violation of any ordinance or regulation or any of the provisions of this code.

❖ A permit, is in reality, a license to proceed with the work. The code official, however, can suspend or revoke permits shown to be based, all or in part, on any false statement or misrepresentation of fact. A permit can also be suspended or revoked if it was issued in error, such as an omitted prerequisite approval or code violation indicated on the construction documents. An applicant may subsequently apply for a reinstatement of the permit with the appropriate corrections or modifications made to the application and construction documents.

[A] 105.7 Placement of permit. The building permit or copy shall be kept on the site of the work until the completion of the project.

❖ The permit, or copy thereof, is to be kept on the job site until the work is complete and made available to the code official or representative to conveniently make required entries thereon.

SECTION 106
CONSTRUCTION DOCUMENTS

[A] 106.1 General. Submittal documents consisting of construction documents, special inspection and structural observation programs, investigation and evaluation reports, and other data shall be submitted in two or more sets with each application for a permit. The construction documents shall be prepared by a registered design professional where required by the statutes of the jurisdiction in which the project is to be constructed. Where special conditions exist, the *code official* is authorized to require additional construction documents to be prepared by a registered design professional.

Exception: The *code official* is authorized to waive the submission of construction documents and other data not required to be prepared by a registered design professional if it is found that the nature of the work applied for is such that reviewing of construction documents is not necessary to obtain compliance with this code.

❖ This section establishes the requirement to provide the code official with submittal documents, including construction drawings, specifications and other documents that describe the structure or system for which a permit is sought. It describes the information that must be included in the documents, who must prepare them, and procedures for approving them.

A detailed description of the work for which the application is made must be submitted. When the work can be briefly described on the application form and the services of a registered design professional are not required, the code official may utilize judgment in determining the need for detailed documents. An example of work that may not involve the submission of detailed construction documents is the replacement of an existing 60-amp electrical service with a 200-amp service. These provisions are intended to reflect the minimum scope of information needed to determine code compliance. The code official should establish a consistent policy of the number of sets required by the jurisdiction and make this information readily available to applicants.

This section also requires the code official to determine that any state professional registration laws be complied with as they apply to the preparation of construction documents.

[A] 106.2 Construction documents. Construction documents shall be in accordance with Sections 106.2.1 through 106.2.5.

❖ This section describes, in detail, the minimum requirements for the construction documents portion of the submittal documents. Sections 106.2.1 through 106.2.5 contain the minimum information to be provided on the construction documents. The following subsections specify the detailed information that must be shown on the submitted construction documents. When specifically allowed by the code official, documents can be submitted in electronic form.

[A] 106.2.1 Construction documents. Construction documents shall be dimensioned and drawn upon suitable material. Electronic media documents are permitted to be submitted where approved by the *code official*. Construction documents shall be of sufficient clarity to indicate the location, nature and extent of the work proposed and show in detail that it will conform to the provisions of this code and relevant laws, ordinances, rules and regulations, as determined by the *code official*. The work areas shall be shown.

❖ The construction documents are required to be of a quality and detail such that the code official can determine that the work conforms to the code and other applicable laws and regulations. General statements on the documents, such as "All work must comply with the *International Existing Building Code*," are not an acceptable substitute for showing the required information.

[A] 106.2.2 Fire protection system(s) shop drawings. Shop drawings for the fire protection system(s) shall be submitted to indicate conformance with this code and the construction documents and shall be approved prior to the start of system installation. Shop drawings shall contain information as required by the referenced installation standards in Chapter 9 of the *International Building Code*.

❖ Since the fire protection contractor(s) may not have been selected at the time a permit was issued for the construction of a building, detailed shop drawings for fire protection systems are not available. Because they provide the information necessary to determine code compliance, as specified in the appropriate referenced standard in Chapter 9 of the IBC, they must be submitted and approved by the code official before the contractor can begin installing the system. For example, the professional responsible for the design of an automatic sprinkler system should determine that the water supply is adequate, but will not be able to prepare a final set of hydraulic calculations if the specific materials and pipe sizes, lengths and arrangements have not been identified. Once the installing contractor is selected, specific hydraulic calculations can be prepared. Factors such as classification of the hazard, amount of water supply available and the density or concentration to be achieved by the system are to be included with the submission of the shop drawings. Specific data sheets identifying sprinklers, pipe dimensions, power requirements for smoke detectors, etc., should also be included with the submission.

[A] 106.2.3 Means of egress. The construction documents for *Alterations*—Level 2, *Alterations*—Level 3, *additions* and *changes of occupancy* shall show in sufficient detail the location, construction, size and character of all portions of the means of egress in compliance with the provisions of this code. The construction documents shall designate the number of occupants to be accommodated in every *work area* of every floor and in all affected rooms and spaces.

❖ The complete means of egress system is required to be indicated on the plans to permit the code official to initiate a review and identify pertinent code requirements for each component. Additionally, requiring such information to be reflected in the construction documents requires the designer not only to become familiar with the code, but also to be aware of egress principles, concepts and purposes. The need to ensure that the means of egress leads to a public way is also a consideration during the plan review. Such an evaluation cannot be made without the inclusion of a site plan, as required by Section 106.2.5.

Information essential for determining the required capacity of the egress components (see Section 1005 of the IBC) and the number of egress components required from a space (see Sections 1015.1 and 1020.1 of the IBC) must be provided. The designer must be aware of the occupancy of a space and properly identify it, along with its resultant occupant load, on the construction documents. In Group I-1, R-2 and R-3 occupancies, the occupant load can be readily determined with little difference in the number; thus, designating the occupant load on the construction documents is not required.

[A] 106.2.4 Exterior wall envelope. Construction documents for work affecting the exterior wall envelope shall describe the exterior wall envelope in sufficient detail to determine compliance with this code. The construction documents shall provide details of the exterior wall envelope as required, including windows, doors, flashing, intersections with dissimilar materials, corners, end details, control joints, intersections at roof, eaves, or parapets, means of drainage, water-resistive membrane, and details around openings.

The construction documents shall include manufacturer's installation instructions that provide supporting documentation that the proposed penetration and opening details described in the construction documents maintain the wind and weather resistance of the exterior wall envelope. The supporting documentation shall fully describe the exterior wall system that was tested, where applicable, as well as the test procedure used.

❖ This section specifically identifies details of exterior wall construction that are critical to the weather resistance of the wall and requires those details to be provided on the construction documents. Where the weather resistance of the exterior wall assembly is based on tests, the submitted documentation is to describe the details of the wall envelope and the test procedure that was used. This provides the code official with enough information necessary to determine code compliance.

[A] 106.2.5 Site plan. The construction documents submitted with the application for permit shall be accompanied by a site plan showing to scale the size and location of new construction and existing structures on the site, distances from lot lines, the established street grades, and the proposed finished grades; and it shall be drawn in accordance with an accurate boundary line survey. In the case of demolition, the site plan shall show construction to be demolished and the location and size of existing structures and construction that are to remain on the site or plot. The *code official* is authorized to waive or modify the requirement for a site plan where the application for permit is for *alteration*, *repair* or *change of occupancy*.

❖ Certain code requirements are dependent on the structure's location on the lot and the topography of the site. As a result, a scaled site plan containing the data listed in this section is required to permit review for compliance. The code official can waive the requirement for a site plan when it is not required to determine code compliance, such as work involving only interior alterations or repairs.

[A] 106.3 Examination of documents. The *code official* shall examine or cause to be examined the submittal documents and shall ascertain by such examinations whether the construction or occupancy indicated and described is in accordance with the requirements of this code and other pertinent laws or ordinances.

❖ The requirements of this section are related to those found in Section 105.3.1 regarding the action of the code official in response to a permit application. The code official can delegate the review of the submittal documents to subordinates as provided for in Section 103.3.

[A] 106.3.1 Approval of construction documents. Where the *code official* issues a permit, the construction documents shall be approved in writing or by stamp as "Reviewed for Code Compliance." One set of construction documents so reviewed shall be retained by the *code official*. The other set shall be returned to the applicant, shall be kept at the site of work, and shall be open to inspection by the *code official* or a duly authorized representative.

❖ The code official must stamp or otherwise endorse as "Reviewed for Code Compliance" the construction documents on which the permit is based. One set of approved construction documents must be kept on the construction site to serve as the basis for all subsequent inspections. To avoid confusion, the construction documents on the site must be the documents that were approved and stamped. This is because inspections are to be performed with regard to the approved documents, not the code itself. Additionally, the contractor cannot determine compliance with the approved construction documents unless they are readily available. Unless the approved construction documents are available, the inspection should be postponed and work on the project halted.

[A] 106.3.2 Previous approval. This code shall not require changes in the construction documents, construction or designated occupancy of a structure for which a lawful permit has been issued and the construction of which has been pursued in good faith within 180 days after the effective date of this code and has not been abandoned.

❖ If a permit is issued and construction proceeds at a normal pace and a new edition of the code is adopted by the legislative body, requiring the building to be constructed to conform to the new code is unreasonable. This section provides for the continuity of permits issued under previous codes, as long as such permits are being "actively prosecuted" subsequent to the effective date of the ordinance adopting this edition of the code.

[A] 106.3.3 Phased approval. The *code official* is authorized to issue a permit for the construction of foundations or any other part of a building before the construction documents for the whole building or structure have been submitted, provided that adequate information and detailed statements have been filed complying with pertinent requirements of this code. The holder of such permit for the foundation or other parts of a building shall proceed at the holder's own risk with the building operation and without assurance that a permit for the entire structure will be granted.

❖ The code official has the authority to issue a partial permit to allow for the practice of "fast tracking" a job. Any construction under a partial permit is "at the holder's own risk" and "without assurance that a permit for the entire structure will be granted." The code official is under no obligation to accept work or issue a complete permit in violation of the code, ordinances or statutes, simply because a partial permit had been issued. Fast tracking places an unusual administrative and technical burden on the code official. The purpose is to proceed with construction while the design continues for other aspects of the work. Coordinating and correlating the code aspects into the project in phases requires attention to detail and project tracking so that all code issues are addressed. The coordination of these submittals is the responsibility of the registered design professional in responsible charge.

[A] 106.3.4 Deferred submittals. Deferral of any submittal items shall have the prior approval of the *code official*. The *registered design professional in responsible charge* shall list the deferred submittals on the construction documents for review by the *code official*.

Submittal documents for deferred submittal items shall be submitted to the *registered design professional* in responsible charge who shall review them and forward them to the *code official* with a notation indicating that the deferred submittal documents have been reviewed and that they have been found to be in general conformance to the design of the building. The deferred submittal items shall not be installed until their deferred submittal documents have been approved by the *code official*.

❖ Often, especially on larger projects, details of certain building parts are not available at the time of permit issuance because they have not yet been designed (e.g., exterior cladding, prefabricated items such as trusses and stairs, and the components of fire protection systems). The design professional in responsible charge must identify on the submittal documents the items to be included in any deferred submittals. Documents required for the approval of deferred items must be reviewed by the design professional in responsible charge for compatibility with the design of the building, forwarded to the code official with a notation that this is the case and approved by the code official before installation of the items. Sufficient time must be allowed for the approval process. Note that deferred submittals differ from the phased permits described in Section 106.3.3 in that they occur after the permit for the building is issued and are not for work covered by separate permits.

[A] 106.4 Amended construction documents. Work shall be installed in accordance with the reviewed construction documents, and any changes made during construction that are not in compliance with the approved construction documents shall be resubmitted for approval as an amended set of construction documents.

❖ Any amendments to the approved construction documents must be filed before constructing the amended item. In the broadest sense, amendments include all addenda, change orders, revised drawings and marked-up shop drawings. Code officials should maintain a policy that all amendments be submitted for review. Otherwise, a significant amendment may not be submitted because of misinterpretation, resulting in an activity that is not approved and that causes a needless delay in obtaining approval of the finished work.

[A] 106.5 Retention of construction documents. One set of approved construction documents shall be retained by the *code official* for a period of not less than the period required for retention of public records.

❖ A set of the approved construction documents must be kept by the code official as may be required by state or local laws, but for a period of not less than 180 days after the work is complete. Questions regarding an item shown on the approved documents may arise in the period immediately following completion of the work and the documents should be available for review. See Section 104.7 for requirements to retain other records that are generated as a result of the work.

[A] 106.6 Design professional in responsible charge. Where it is required that documents be prepared by a registered design professional, the *code official* shall be authorized to require the owner or the owner's authorized agent to engage and designate on the building permit application a registered design professional who shall act as the *registered design professional in responsible charge*. If the circumstances require, the owner or the owner's authorized agent shall designate a substitute *registered design professional in responsible charge* who shall perform the duties required of the original *registered design professional in responsible charge*. The code official shall be notified in writing by the owner or the owner's authorized agent if the *registered design professional in responsible charge* is changed or is unable to continue to perform the duties. The *registered design professional in responsible charge* shall be responsible for reviewing and coordinating submittal documents prepared by others, including phased and deferred submittal items, for compatibility with the design of the building. Where structural observation is required, the inspection program shall name the individual or firms who are to perform structural observation and describe the stages of construction at which structural observation is to occur.

❖ At the time of permit application and at various intervals during a project, the code requires detailed technical information to be submitted to the code official. This will vary depending on the complexity of the project, but typically includes the construction documents with supporting information, applications utilizing the phased approval procedure in Section 106.3.3, and reports from engineers, inspectors and testing agencies as required in Chapter 17 of the IBC. Since these documents and reports are prepared by numerous individuals, firms and agencies, it is necessary to have a single person charged with the responsibility for coordinating their submittal to the code official. This person is the point of contact for the code official for all information relating to the project. Otherwise, the code official could waste time and effort attempting to locate the source of accurate information when trying to resolve an issue, such as a discrepancy in plans submitted by different designers.

The requirement that the owner engage a person to act as the design professional in responsible charge is applicable to projects where the construction documents are required by law to be prepared by a registered design professional (see Section 106.1) and when required by the code official. The person employed by the owner to act as the design professional in responsible charge must be identified on the permit application, but the owner can change the designated person at any time during the course of the review process or work, provided the code official is so notified in writing.

SECTION 107
TEMPORARY STRUCTURES AND USES

[A] 107.1 General. The *code official* is authorized to issue a permit for temporary uses. Such permits shall be limited as to time of service but shall not be permitted for more than 180 days. The *code official* is authorized to grant extensions for demonstrated cause.

❖ In the course of construction or other activities, structures that have a limited service life are often necessary. This section contains the administrative provisions that permit such temporary structures without full compliance with the code requirements for permanently occupied structures.

This section allows the code official to issue permits for temporary structures or uses. The applicant must specify the time period desired for the temporary structure or use, but the approval period cannot exceed 180 days. Structures or uses that are "temporary" but are anticipated to be in existence for more than 180 days are required to conform to code requirements for permanent structures and uses. The code official is authorized to grant time extensions if the applicant can provide a valid reason for the extension. A typical example would be circumstances that have occurred beyond the applicant's control. This provision is not intended to be used to circumvent the 180-day limitation.

[A] 107.2 Conformance. Temporary uses shall conform to the structural strength, fire safety, means of egress, accessibility, light, ventilation and sanitary requirements of this code as necessary to ensure the public health, safety and general welfare.

❖ This section prescribes those categories of the code that must be complied with, despite the fact that the structure will be removed or the use discontinued at some time in the future. These criteria are essential for measuring the safety of any use, temporary or permanent. Therefore, the application of these criteria to a temporary structure cannot be waived.

"Structural strength" refers to the ability of the temporary structure to resist anticipated live, environmental and dead loads (see Chapter 16 of the IBC). It also applies to anticipated live and dead loads imposed by a temporary use in an existing structure.

"Fire safety" provisions are those required by Chapters 8, 9 and 10 of the IBC invoked by virtue of the structure's size, use or location on the property.

"Means of egress" refers to full compliance with Chapter 10 of the IBC.

"Accessibility" refers to full compliance with Chapter 11 of the IBC for making buildings accessible to physically disabled persons, a requirement that is repeated in Section 1103.1 of the IBC.

"Light, ventilation and sanitary" requirements are those imposed by Chapter 12 of the IBC or applicable sections of the *International Plumbing Code*® (IPC®) or *International Mechanical Code*® (IMC®).

[A] 107.3 Temporary power. The *code official* is authorized to give permission to temporarily supply and use power in part of an electric installation before such installation has been fully completed and the final certificate of completion has been issued. The part covered by the temporary certificate shall comply with the requirements specified for temporary lighting, heat or power in NFPA 70.

❖ Commonly, the electrical service on most construction sites is installed and energized long before all of the wiring is completed. This procedure allows the power supply to be increased as construction demands; however, temporary permission is not intended to waive the requirements set forth in NFPA 70. Construction power from the permanent wiring of the building does not require the installation of temporary ground-fault circuit-interrupter (GFCI) protection or the assured equipment grounding program, because the building wiring installed as required by the code should be as safe for construction use as it would be for use after the completion of the building.

[A] 107.4 Termination of approval. The *code official* is authorized to terminate such permit for a temporary use and to order the temporary use to be discontinued.

❖ This section provides the code official with the necessary authority to terminate the permit for a temporary use. The code official can order that a temporary structure be removed or a temporary use be discontinued if conditions of the permit have been violated or the structure or use poses an imminent hazard to the public, in which case the provisions of Section 116 become applicable. This text is important because it allows the code official to act quickly when time is of the essence in order to protect public health, safety and welfare.

SECTION 108
FEES

[A] 108.1 Payment of fees. A permit shall not be valid until the fees prescribed by law have been paid. Nor shall an amendment to a permit be released until the additional fee, if any, has been paid.

❖ The code anticipates that jurisdictions will establish their own fee schedules. It is the intent that the fees collected by the department for building permit issuance, plan review and inspection be adequate to cover the costs to the department in these areas.

This section requires that all fees be paid prior to permit issuance or release of an amendment to a permit. Since department operations are intended to be supported by fees paid by the user of department activities, it is important that these fees are received before incurring any expense. This philosophy has resulted in a common practice of having fees paid prior to plan review and inspection.

[A] 108.2 Schedule of permit fees. On buildings, electrical, gas, mechanical, and plumbing systems or *alterations* requiring a permit, a fee for each permit shall be paid as required in accordance with the schedule as established by the applicable governing authority.

❖ The jurisdiction inserts its desired fee schedule at this location. The fees are established by law, such as in an ordinance adopting the code (see page xi of the code for a sample), a separate ordinance or legally promulgated regulation, as required by state or local law. Fee schedules are often based on a valuation of the work to be performed. This concept is based on the proposition that the valuation of a project is related to the amount of work to be expended in the plan review, inspections and administering the permit, plus an excess to cover department overhead.

[A] 108.3 Building permit valuations. The applicant for a permit shall provide an estimated permit value at time of application. Permit valuations shall include total value of work including materials and labor for which the permit is being issued, such as electrical, gas, mechanical, plumbing equipment, and permanent systems. If, in the opinion of the *code official*, the valuation is underestimated on the application, the permit shall be denied unless the applicant can show detailed estimates to meet the approval of the *code official*. Final building permit valuation shall be set by the *code official*.

❖ As indicated in Section 108.2, jurisdictions usually base their fees on the value of the work being performed. This section, therefore, requires the applicant to provide this figure, which is to include the total value of the work, including materials and labor, for which the permit is sought. If the code official believes that the value provided by the applicant is underestimated, the permit is to be denied unless the applicant can substantiate the value by providing detailed estimates of the work to the satisfaction of the code official. For the construction of new buildings, the building valuation data referred to in Section 108.2 can be used by the code official as a yardstick against which to compare the applicant's estimate.

[A] 108.4 Work commencing before permit issuance. Any person who commences any work before obtaining the necessary permits shall be subject to an additional fee established by the *code official* that shall be in addition to the required permit fees.

❖ The code official will incur certain costs (i.e., inspection time and administrative) when investigating and citing a person who has commenced work without having obtained a permit. The code official is, therefore, entitled to recover these costs by establishing a fee imposed on the responsible party, in addition to that collected when the required permit is issued. Note that this is not a penalty, as described in Section 113.4, for which the person can also be liable.

[A] 108.5 Related fees. The payment of the fee for the construction, *alteration*, removal, or demolition of work done in connection to or concurrently with the work authorized by a building permit shall not relieve the applicant or holder of the permit from the payment of other fees that are prescribed by law.

❖ The fees for a building permit may be in addition to other fees required by the jurisdiction or others for related items such as sewer connections, water service taps, driveways and signs. It cannot be construed that the building permit fee includes these other items.

[A] 108.6 Refunds. The *code official* is authorized to establish a refund policy.

❖ This section allows for a refund of fees, which may be full or partial, typically resulting from the revocation, abandonment or discontinuance of a construction project for which a permit has been issued and fees have been collected. The refund of fees should be related to the cost of enforcement services not provided because of the termination of the project. The code official, when authorizing a fee refund, is authorizing the disbursement of public funds. Therefore, the request for a refund must be in writing and for good cause.

SECTION 109
INSPECTIONS

[A] 109.1 General. Construction or work for which a permit is required shall be subject to inspection by the *code official*, and such construction or work shall remain accessible and exposed for inspection purposes until approved. Approval as a result of an inspection shall not be construed to be an approval of a violation of the provisions of this code or of other ordinances of the jurisdiction. Inspections presuming to give authority to violate or cancel the provisions of this code or of other ordinances of the jurisdiction shall not be valid. It shall be the duty of the permit applicant to cause the work to remain accessible and exposed for inspection purposes. Neither the *code official* nor the jurisdiction shall be liable for expense entailed in the removal or replacement of any material required to allow inspection.

❖ The inspection function is one of the more important aspects of building department operations. This section authorizes the code official to inspect the work for which a permit has been issued and requires that the work to be inspected remain accessible to the code official until inspected and approved. Any expense incurred in removing or replacing material that conceals an item to be inspected is not the responsibility of the code official or the jurisdiction. As with the issuance of permits (see Section 105.4), approval as a result of an inspection is not a license to violate the code, and an approval in violation of the code does not relieve the applicant from complying with the code and is not valid.

[A] 109.2 Preliminary inspection. Before issuing a permit, the *code official* is authorized to examine or cause to be examined buildings and sites for which an application has been filed.

❖ The code official is granted authority to inspect the site before permit issuance. This may be necessary to verify existing conditions that impact the plan review and permit approval. This section provides the code official with the right-of-entry authority that otherwise does not occur until after the permit is issued (see Section 104.6).

[A] 109.3 Required inspections. The *code official*, upon notification, shall make the inspections set forth in Sections 109.3.1 through 109.3.9.

❖ The code official is required to verify that the work is completed in accordance with the approved construction documents. It is the responsibility of the permit holder to notify the code official when the item is ready for inspection. The inspections that are necessary to provide such verification are listed in the following sections, with the caveat in Section 109.3.8 that special inspections, in addition to those listed here, may be required depending on the work involved.

[A] 109.3.1 Footing or foundation inspection. Footing and foundation inspections shall be made after excavations for footings are complete and any required reinforcing steel is in place. For concrete foundations, any required forms shall be in place prior to inspection. Materials for the foundation shall be on the job, except where concrete is ready-mixed in accordance with ASTM C94, the concrete need not be on the job.

❖ It is necessary for the code official to inspect the soil on which the footing or foundation is to be placed. This inspection also includes any reinforcing steel, concrete forms and materials to be used in the foundation, except for ready-mixed concrete that is prepared off site.

[A] 109.3.2 Concrete slab or under-floor inspection. Concrete slab and under-floor inspections shall be made after in-slab or under-floor reinforcing steel and building service equipment, conduit, piping accessories, and other ancillary equipment items are in place but before any concrete is placed or floor sheathing installed, including the sub floor.

❖ The code official must be able to inspect the soil and any required under-slab drainage, waterproofing or dampproofing material, as well as reinforcing steel, conduit, piping and other service equipment embedded in or installed below a slab prior to placing the concrete. Similarly, items installed below a floor system other than concrete must be inspected before they are concealed by the floor sheathing or subfloor.

[A] 109.3.3 Lowest floor elevation. For *additions* and *substantial improvements* to *existing buildings* in *flood hazard areas*, upon placement of the lowest floor, including basement, and prior to further vertical construction, the elevation documentation required in the *International Building Code* shall be submitted to the *code official*.

❖ Where a structure is located in a flood hazard area, as established in Section 1612.3 of the IBC, the code official must be provided with certification that either the lowest floor elevation (for structures located in flood hazard areas not subject to high-velocity wave action) or the elevation of the lowest horizontal structural member (for structures located in flood hazard areas subject to high-velocity wave action) is in compliance with Section 1612 of the IBC. This certification must be submitted prior to any construction proceeding above this level.

[A] 109.3.4 Frame inspection. Framing inspections shall be made after the roof deck or sheathing, framing, fire blocking, and bracing are in place and pipes, chimneys, and vents to be concealed are complete and the rough electrical, plumbing, heating wires, pipes, and ducts are approved.

❖ This section requires that the code official be able to inspect the framing members, such as studs, joists, rafters and girders, and other items such as vents and chimneys, that will be concealed by wall construction. Rough electrical work, plumbing, heating wires, pipes and ducts must have already been approved in accordance with the applicable codes prior to this inspection.

[A] 109.3.5 Lath or gypsum board inspection. Lath and gypsum board inspections shall be made after lathing and gypsum board, interior and exterior, is in place but before any plastering is applied or before gypsum board joints and fasteners are taped and finished.

> **Exception:** Gypsum board that is not part of a fire-resistance-rated assembly or a shear assembly.

❖ In order to verify that lath and gypsum board is properly attached to framing members, it is necessary for the code official to be able to inspect before the plaster or joint finish material is applied. This is required only for gypsum board that is part of either a fire-resistant assembly or a shear wall.

[A] 109.3.6 Fire and smoke-resistant penetrations. Protection of joints and penetrations in fire-resistance-rated assemblies, smoke barriers and smoke partitions shall not be concealed from view until inspected and approved.

❖ The code official must have an opportunity to inspect joint protection and penetration protection as required by Sections 714 and 715 of the IBC for fire- and smoke-resistance-rated assemblies before it is concealed from view.

[A] 109.3.7 Other inspections. In addition to the inspections specified above, the *code official* is authorized to make or require other inspections of any construction work to ascertain compliance with the provisions of this code and other laws that are enforced by the Department of Building Safety.

❖ Any item regulated by the code is subject to inspection by the code official to determine compliance with the applicable code provision and no list can include all items in a given building. This section, therefore, gives the code official the authority to inspect any regulated items.

[A] 109.3.8 Special inspections. Special inspections shall be required in accordance with the *International Building Code*.

❖ Special inspections are to be provided by the owner for the types of work required in Section 1704 of the IBC. The code official is to approve special inspectors and verify that the required special inspections have been conducted.

[A] 109.3.9 Final inspection. The final inspection shall be made after work required by the building permit is completed.

❖ Upon completion of the work for which the permit has been issued and before the issuance of the certificate of occupancy required by Section 110.3, a final inspection is to be made. All violations of the approved construction documents and permit are to be noted and the holder of the permit is to be notified of the discrepancies.

[A] 109.4 Inspection agencies. The *code official* is authorized to accept reports of approved inspection agencies, provided such agencies satisfy the requirements as to qualifications and reliability.

❖ As an alternative to the code official conducting the inspection, he or she is permitted to accept inspections of and reports by approved inspection agencies. Appropriate criteria on which to base approval of inspection agencies can be found in Section 1703 of the IBC.

[A] 109.5 Inspection requests. It shall be the duty of the holder of the building permit or their duly authorized agent to notify the *code official* when work is ready for inspection. It shall be the duty of the permit holder to provide access to and means for any inspections of such work that are required by this code.

❖ It is the responsibility of the permit holder or other authorized person, such as the contractor performing the work, to arrange the required inspections when completed work is ready and to allow sufficient time for the code official to schedule a visit to the site to prevent work from being concealed prior to inspection. Access to the work to be inspected must be provided, including any special means such as a ladder.

[A] 109.6 Approval required. Work shall not be done beyond the point indicated in each successive inspection without first obtaining the approval of the *code official*. The *code official*, upon notification, shall make the requested inspections and shall either indicate the portion of the construction that is satisfactory as completed or shall notify the permit holder or an agent of the permit holder wherein the same fails to comply with this code. Any portions that do not comply shall be corrected and such portion shall not be covered or concealed until authorized by the *code official*.

❖ This section establishes that work cannot progress beyond the point of a required inspection without the code official's approval. Upon making the inspection, the code official must either approve the completed work or notify the permit holder or other responsible party of that which does not comply with the code. Approvals and notices of noncompliance must be in writing, as required by Section 104.4, to avoid any misunderstanding as to what is required. Any item not approved cannot be concealed until it has been corrected and approved by the code official.

SECTION 110
CERTIFICATE OF OCCUPANCY

[A] 110.1 Altered area use and occupancy classification change. Altered areas of a building and relocated buildings shall not be used or occupied, and change in the existing use or occupancy classification of a building or portion thereof shall not be made until the code official has issued a certificate of occupancy therefor as provided herein. Issuance of a certificate of occupancy shall not be construed as an approval of a violation of the provisions of this code or of other ordinances of the jurisdiction.

❖ This section establishes that a building or structure that has been repaired, altered, relocated or has experienced a change of use or occupancy cannot be occupied until a certificate of occupancy is issued by the code official, which reflects the conclusion of the work allowed by the building permit. Also, no change in occupancy of an existing building is permitted without first obtaining a certificate of occupancy for the new use.

The tool that the code official uses to control the uses and occupancies of various buildings and structures is the certificate of occupancy. It is unlawful to use or occupy a building or structure unless a certificate of occupancy has been issued for that use. Its issuance does not relieve the building owner from the responsibility for correcting any code violation that may exist.

[A] 110.2 Certificate issued. After the *code official* inspects the building and does not find violations of the provisions of this code or other laws that are enforced by the Department of Building Safety, the *code official* shall issue a certificate of occupancy that shall contain the following:

1. The building permit number.
2. The address of the structure.
3. The name and address of the owner or the owner's authorized agent.
4. A description of that portion of the structure for which the certificate is issued.
5. A statement that the described portion of the structure has been inspected for compliance with the requirements of this code for the occupancy and division of occupancy and the use for which the proposed occupancy is classified.
6. The name of the *code official*.
7. The edition of the code under which the permit was issued.
8. The use and occupancy in accordance with the provisions of the *International Building Code*.
9. The type of construction as defined in the *International Building Code*.
10. The design occupant load and any impact the *alteration* has on the design occupant load of the area not within the scope of the work.
11. If fire protection systems are provided, whether the fire protection systems are required.
12. Any special stipulations and conditions of the building permit.

❖ The code official is required to issue a certificate of occupancy after a successful final inspection has been completed and all deficiencies and violations have been resolved. This section lists the information that must be included on the certificate. This information is useful to both the code official and owner because it indicates the criteria under which the structure was evaluated and approved at the time the certificate was issued. This is especially important when later applying Chapter 13 to existing structures.

[A] 110.3 Temporary occupancy. The *code official* is authorized to issue a temporary certificate of occupancy before the completion of the entire work covered by the permit, provided that such portion or portions shall be occupied safely. The *code official* shall set a time period during which the temporary certificate of occupancy is valid.

❖ The code official is permitted to issue a temporary certificate of occupancy for all or a portion of a building prior to the completion of all work. Such certification is to be issued only when the building or portion in question can be safely occupied prior to full completion. The certification is intended to acknowledge that some building features may not be completed even though the building is safe for occupancy, or that a portion of the building can be safely occupied while work continues in another area. This provision precludes the occupancy of a building or structure that does not contain all of the required fire protection systems and means of egress. Temporary certificates should be issued only when incidental construction remains, such as site work and interior work, that is not regulated by the code and exterior decoration not necessary to the integrity of the building envelope. The code official should view the issuance of a temporary certificate of occupancy as equally significant as the issuance of the final certificate. Indeed, the issuance of a temporary certificate of occupancy offers a greater potential for conflict because once the building or structure is occupied, it is very difficult to remove the occupants through legal means. The certificate must specify the time period for which it is valid.

[A] 110.4 Revocation. The *code official* is authorized to, in writing, suspend or revoke a certificate of occupancy or completion issued under the provisions of this code wherever the certificate is issued in error or on the basis of incorrect information supplied, or where it is determined that the building or structure or portion thereof is in violation of any ordinance or regulation or any of the provisions of this code.

❖ The code official is authorized to, in writing, suspend or revoke a certificate of occupancy or completion issued under the provisions of the code wherever the certificate is issued in error, or on the basis of incorrect information supplied, or where it is determined that the building or structure, or portion thereof, is in violation of any ordinance or regulation or any of the provisions of the code.

This section is needed to give the code official the authority to revoke a certificate of occupancy for the reasons indicated in the code text. The code official may also suspend the certificate of occupancy until all of the code violations are corrected.

SECTION 111
SERVICE UTILITIES

[A] 111.1 Connection of service utilities. A person shall not make connections from a utility, source of energy, fuel, or power to any building or system that is regulated by this code for which a permit is required, until approved by the *code official*.

❖ This section establishes the authority of the code official to approve utility connections to a building for items such as water, sewer, electricity, gas and steam; and to require their disconnection when hazardous conditions or emergencies exist.

The approval of the code official is required before a connection can be made from a utility to a building system that is regulated by the applicable code, including those referenced in Section 101.4. This includes utilities supplying water, sewer, electricity, gas and steam services. For the protection of building occupants, including workers, such systems must have had final inspection approvals, except as allowed by Section 111.2 for temporary connections.

[A] 111.2 Temporary connection. The *code official* shall have the authority to authorize the temporary connection of the building or system to the utility source of energy, fuel, or power.

❖ The code official is permitted to issue temporary authorization to make connections to the public utility system prior to the completion of all work. This acknowledges that, because of seasonal limitations, time constraints, or the need for testing or partial operation of equipment, some building systems may be safely connected even though the building is not suitable for final occupancy. The temporary connection and use of connected equipment should be approved when the requesting permit holder has demonstrated to the code official's satisfaction that public health, safety and welfare will not be endangered.

[A] 111.3 Authority to disconnect service utilities. The *code official* shall have the authority to authorize disconnection of utility service to the building, structure or system regulated by this code and the referenced codes and standards in case of emergency where necessary to eliminate an immediate hazard to life or property or where such utility connection has been made without the approval required by Section 111.1 or 111.2. The *code official* shall notify the serving utility and, wherever possible, the owner or the owner's authorized agent and occupant of the building, structure or service system of the decision to disconnect prior to taking such action. If not notified prior to disconnecting, the owner or occupant of the building, structure or service system shall be notified in writing, as soon as practical thereafter.

❖ When an immediate hazard to life or property exists, the code official has the authority to order the disconnection of utility services. This can also occur when utility service has been connected without the necessary approvals required by the code. Whenever possible, the building owner or owner's authorized agent and the building occupant or occupants should be notified prior to the disconnection of the services. Then, at the first practical opportunity, the code official is to formally notify the building owner or owner's authorized agent, in writing, of the disconnection activities. As with all administrative functions, all aspects of due process must be followed.

SECTION 112
BOARD OF APPEALS

[A] 112.1 General. In order to hear and decide appeals of orders, decisions, or determinations made by the code official relative to the application and interpretation of this code, there shall be and is hereby created a board of appeals. The board of appeals shall be appointed by the governing body and shall hold office at its pleasure. The board shall adopt rules of procedure for conducting its business.

❖ This section provides an aggrieved party with a material interest in the decision of the code official a process to appeal such a decision before a board of appeals. This provides a forum, other than the court of jurisdiction, in which to review the code official's actions.

This section literally allows any person to appeal a decision of the code official. In practice, this section has been interpreted to permit appeals only by those aggrieved parties with a material or definitive interest in the decision of the code official. An aggrieved party may not appeal a code requirement per se. The intent of the appeal process is not to waive or set aside a code requirement; rather, it is intended to provide a means of reviewing a code official's decision on an interpretation or application of the code or to review the equivalency of protection to the code requirements. The members of the appeals board are appointed by the "governing body" of the jurisdiction, typically a council or administrator, such as a mayor or city manager, and remain members until removed from office. The board must establish procedures for electing a chairperson, scheduling, conducting meetings and administration. Note that Appendix B of the IBC contains complete, detailed requirements for creating an appeals board, including the number of members, qualifications and administrative procedures. Jurisdictions desiring to utilize these requirements must include Appendix B of the IBC in their adopting ordinance.

[A] 112.2 Limitations on authority. An application for appeal shall be based on a claim that the true intent of this code or the rules legally adopted thereunder have been incorrectly interpreted, the provisions of this code do not fully apply, or an equally good or better form of construction is proposed. The board shall not have authority to waive requirements of this code.

❖ This section establishes the grounds for an appeal, which claims that the code official has misinterpreted or misapplied a code provision. The board is not allowed to set aside any of the technical requirements of the code; however, it is allowed to consider alternative methods of compliance with the technical requirements (see Section 104.11).

[A] 112.3 Qualifications. The board of appeals shall consist of members who are qualified by experience and training to pass on matters pertaining to building construction and are not employees of the jurisdiction.

❖ It is important that the decisions of the appeals board are based purely on the technical merits involved in an appeal. It is not the place for policy or political deliberations. The members of the appeals board are, therefore, expected to have experience in building construction matters.

SECTION 113
VIOLATIONS

[A] 113.1 Unlawful acts. It shall be unlawful for any person, firm, or corporation to *repair*, alter, extend, add, move, remove, demolish, or change the occupancy of any building or equipment regulated by this code or cause same to be done in conflict with or in violation of any of the provisions of this code.

❖ Violations of the code are prohibited and form the basis for all citations and correction notices.

[A] 113.2 Notice of violation. The *code official* is authorized to serve a notice of violation or order on the person responsible for the *repair*, *alteration*, extension, *addition*, moving, removal, demolition, or change in the occupancy of a building in violation of the provisions of this code or in violation of a permit or certificate issued under the provisions of this code. Such order shall direct the discontinuance of the illegal action or condition and the abatement of the violation.

❖ The code official is required to notify the person responsible for the erection or use of a building found to be in violation of the code. The section that is allegedly being violated must be cited so that the responsible party can respond to the notice.

[A] 113.3 Prosecution of violation. If the notice of violation is not complied with promptly, the *code official* is authorized to request the legal counsel of the jurisdiction to institute the appropriate proceeding at law or in equity to restrain, correct, or abate such violation or to require the removal or termination of the unlawful occupancy of the building or structure in violation of the provisions of this code or of the order or direction made pursuant thereto.

❖ The code official must pursue, through the use of legal counsel of the jurisdiction, legal means to correct the violation. This is not optional.

Any extensions of time so that the violations may be corrected voluntarily must be for a reasonable, bona fide cause or the code official may be subject to criticism for "arbitrary and capricious" actions. In general, it is better to have a standard time limitation for the correction of violations. Departures from this standard must be for a clear and reasonable purpose, usually stated in writing by the violator.

[A] 113.4 Violation penalties. Any person who violates a provision of this code or fails to comply with any of the requirements thereof or who *repairs* or alters or changes the occupancy of a building or structure in violation of the approved construction documents or directive of the *code official* or of a permit or certificate issued under the provisions of this code shall be subject to penalties as prescribed by law.

❖ Penalties for violating provisions of the code are typically contained in state law, particularly if the code is adopted at that level and the building department must follow those procedures. If there is no such procedure already in effect, one must be established with the aid of legal counsel.

SECTION 114
STOP WORK ORDER

[A] 114.1 Authority. Whenever the *code official* finds any work regulated by this code being performed in a manner contrary to the provisions of this code or in a *dangerous* or unsafe manner, the *code official* is authorized to issue a stop work order.

❖ This section provides for the suspension of work for which a permit was issued, pending the removal or correction of a severe violation or unsafe condition identified by the code official.

Normally, correction notices, issued in accordance with Section 109.6, are used to inform the permit holder of code violations. Stop work orders are issued when enforcement can be accomplished no other way or when a dangerous condition exists.

[A] 114.2 Issuance. The stop work order shall be in writing and shall be given to the owner of the property involved, the owner's authorized agent or to the person doing the work. Upon issuance of a stop work order, the cited work shall immediately cease. The stop work order shall state the reason for the order and the conditions under which the cited work will be permitted to resume.

❖ Upon receipt of a violation notice from the code official, all construction activities identified in the notice must immediately cease, except as expressly permitted to correct the violation.

[A] 114.3 Unlawful continuance. Any person who shall continue any work after having been served with a stop work order, except such work as that person is directed to perform to remove a violation or unsafe condition, shall be subject to penalties as prescribed by law.

❖ This section states that the work in violation must terminate and that all work, except that which is necessary to correct the violation or unsafe condition, must cease as well. As determined by the municipality or state, a penalty may be assessed for failure to comply with this section.

SECTION 115
UNSAFE BUILDINGS AND EQUIPMENT

[A] 115.1 Conditions. Buildings, structures or equipment that are or hereafter become *unsafe*, shall be taken down, removed or made safe as the *code official* deems necessary and as provided for in this code.

❖ This section describes the responsibility of the code official to investigate reports of unsafe structures and equipment, and provides criteria for such determination.

"Unsafe structures" are defined as buildings or structures that are insanitary, deficient in light and ventilation or adequate exit facilities, constitute a fire hazard or are otherwise dangerous to human life.

All unsafe buildings must either be demolished or made safe and secure as deemed appropriate by the code official.

[A] 115.2 Record. The *code official* shall cause a report to be filed on an *unsafe* condition. The report shall state the occupancy of the structure and the nature of the *unsafe* condition.

❖ The code official must file a report on each investigation of unsafe conditions, stating the occupancy of the structure and the nature of the unsafe condition. This report provides the basis for the notice described in Section 115.3.

[A] 115.3 Notice. If an *unsafe* condition is found, the *code official* shall serve on the owner, the owner's authorized agent or person in control of the structure a written notice that describes the condition deemed *unsafe* and specifies the required *repairs* or improvements to be made to abate the *unsafe* condition, or that requires the *unsafe* building to be demolished within a stipulated time. Such notice shall require the person thus notified to declare immediately to the *code official* acceptance or rejection of the terms of the order.

❖ When a building or structure is deemed unsafe, the code official is required to notify the owner or owner's authorized agent of the building as the first step in correcting the problem. Such notice must describe the necessary repairs and improvements to correct the deficiency or must require the unsafe building or structure to be demolished in a specified time in order to provide for public health, safety and welfare. Additionally, such notice requires the immediate response of the owner or agent. If the owner or agent is not available, public notice of such declaration should suffice for the purposes of complying with this section (see Section 115.4). The code official may also determine that immediate work is necessary to correct an unsafe condition and seek a lien against the building or structure to compensate the municipality for the cost of remedial action.

[A] 115.4 Method of service. Such notice shall be deemed properly served if a copy thereof is delivered to the owner or the owner's authorized agent personally; sent by certified or registered mail addressed to the owner or the owner's authorized agent at the last known address with the return receipt requested; or delivered in any other manner as prescribed by local law. If the certified or registered letter is returned showing that the letter was not delivered, a copy thereof shall be posted in a conspicuous place in or about the structure affected by such notice. Service of such notice in the foregoing manner upon the owner's authorized agent or upon the person responsible for the structure shall constitute service of notice upon the owner.

❖ The notice must be delivered personally to the owner or owner's authorized agent. If the owner or agent cannot be located, additional procedures are established, including posting the unsafe notice on the premises in question. Such action may be considered the equivalent of personal notice; however, it may or may not be deemed by the courts as representing a "good faith" effort to notify. Therefore, in addition to complying with this section, public notice through the use of newspapers and other postings in a prominent location at the government center should be used.

[A] **115.5 Restoration.** The building or equipment determined to be *unsafe* by the *code official* is permitted to be restored to a safe condition. To the extent that *repairs*, *alterations*, or *additions* are made or a *change of occupancy* occurs during the restoration of the building, such *repairs*, *alterations*, *additions*, or *change of occupancy* shall comply with the requirements of this code.

❖ This section provides that unsafe structures may be restored to a safe condition. This means that the cause of the unsafe structure notice can be abated without the structure being required to comply fully with the provisions for new construction. Any work done to eliminate the unsafe condition, as well as any change in occupancy that may occur, must comply with the code.

SECTION 116
EMERGENCY MEASURES

[A] **116.1 Imminent danger.** Where, in the opinion of the *code official*, there is imminent danger of failure or collapse of a building that endangers life, or where any building or part of a building has fallen and life is endangered by the occupation of the building, or where there is actual or potential danger to the building occupants or those in the proximity of any structure because of explosives, explosive fumes or vapors, or the presence of toxic fumes, gases, or materials, or operation of defective or dangerous equipment, the *code official* is hereby authorized and empowered to order and require the occupants to vacate the premises forthwith. The *code official* shall cause to be posted at each entrance to such structure a notice reading as follows: "This Structure Is Unsafe and Its Occupancy Has Been Prohibited by the *Code Official*." It shall be unlawful for any person to enter such structure except for the purpose of securing the structure, making the required *repairs*, removing the hazardous condition, or of demolishing the same.

❖ If the code official has determined that failure or collapse of a building or structure is imminent, failure has occurred that results in a continued threat to the remaining structure or adjacent properties or any other unsafe condition as described in this section exists in a structure, the code official is authorized to require the occupants to vacate the premises and to post such buildings or structures as unsafe and unoccupiable. Unless authorized by the code official to make repairs, secure or demolish the structure, it is illegal for anyone to enter the building or structure. This will minimize the potential for injury.

[A] **116.2 Temporary safeguards.** Notwithstanding other provisions of this code, whenever, in the opinion of the *code official*, there is imminent danger due to an unsafe condition, the *code official* shall order the necessary work to be done, including the boarding up of openings, to render such structure temporarily safe whether or not the legal procedure herein described has been instituted; and shall cause such other action to be taken as the *code official* deems necessary to meet such emergency.

❖ This section recognizes the need for immediate and effective action in order to protect the public. This section empowers the code official to cause the necessary work to be done to minimize the imminent danger temporarily without regard for due process. This section has to be viewed critically insofar as the danger of structural failure to which the code official has responded must be "imminent"; that is, readily apparent and immediate.

[A] **116.3 Closing streets.** Where necessary for public safety, the *code official* shall temporarily close structures and close or order the authority having jurisdiction to close sidewalks, streets, public ways, and places adjacent to unsafe structures, and prohibit the same from being utilized.

❖ The code official is authorized to temporarily close sidewalks, streets and adjacent structures as needed to protect the public from the unsafe building or structure when an imminent danger exists. Since the code official may not have the direct authority to close sidewalks, streets and other public ways, the agency having such jurisdiction (e.g., the police or highway department) must be notified.

[A] **116.4 Emergency repairs.** For the purposes of this section, the *code official* shall employ the necessary labor and materials to perform the required work as expeditiously as possible.

❖ The cost of emergency work may have to be paid initially by the jurisdiction. The important principle here is that the code official must act immediately to protect the public when warranted, leaving the details of costs and owner notification for later.

[A] **116.5 Costs of emergency repairs.** Costs incurred in the performance of emergency work shall be paid by the jurisdiction. The legal counsel of the jurisdiction shall institute appropriate action against the owner of the premises or the owner's authorized agent where the unsafe structure is or was located for the recovery of such costs.

❖ The cost of emergency repairs is to be paid by the jurisdiction, with subsequent legal action against the owner or owner's authorized agent to recover such costs. This does not preclude, however, reaching an alternative agreement with the owner.

[A] **116.6 Hearing.** Any person ordered to take emergency measures shall comply with such order forthwith. Any affected person shall thereafter, upon petition directed to the appeals board, be afforded a hearing as described in this code.

❖ Anyone ordered to take an emergency measure or to vacate a structure because of an emergency condition must do so immediately.

Thereafter, any affected party has the right to appeal the action to the appeals board to determine whether the order should be continued, modified or revoked.

It is imperative that appeals to an emergency order occur after the hazard has been abated, rather than before, to minimize the risk to the occupants, employees, clients and public.

SECTION 117
DEMOLITION

[A] 117.1 General. The *code official* shall order the owner or owner's authorized agent of any premises upon which is located any structure that in the *code official's* judgment is so old or dilapidated, or has become so out of *repair* as to be *dangerous*, unsafe, insanitary or otherwise unfit for human habitation of occupancy, and such that it is unreasonable to *repair* the structure, to demolish and remove such structure; or if such structure is capable of being made safe by *repairs*, to *repair* and make safe and sanitary or to demolish and remove to the owner's or the owner's authorized agent's option; or where there has been a cessation of normal construction of any structure for a period of more than two years, to demolish and remove such structure.

❖ This section describes the conditions where the code official has the authority to order the owner to remove the structure. Conditions where the code official may give the owner or owner's authorized agent the option of repairing the structure are also in this section. The code official should carefully document the condition of the structure prior to issuing a demolition order to provide an adequate basis for ordering the owner to remove the structure.

[A] 117.2 Notices and orders. Notices and orders shall comply with Section 113.

❖ Before the code official can pursue action to demolish a building in accordance with Section 117.1 or 117.3, it is imperative that all owners and any other persons with a recorded encumbrance on the property be given proper notice of the demolition plans. See Section 113 for notice and order requirements.

[A] 117.3 Failure to comply. If the owner or the owner's authorized agent of a premises fails to comply with a demolition order within the time prescribed, the *code official* shall cause the structure to be demolished and removed, either through an available public agency or by contract or arrangement with private persons, and the cost of such demolition and removal shall be charged against the real estate upon which the structure is located and shall be a lien upon such real estate.

❖ When the owner or owner's authorized agent fails to comply with a demolition order, the code official is authorized to take action to have the building razed and removed. The costs are to be charged as a lien against the real estate. To reduce complaints regarding the validity of demolition costs, the code official should obtain competitive bids from several demolition contractors before authorizing any contractor to raze the structure.

[A] 117.4 Salvage materials. Where any structure has been ordered demolished and removed, the governing body or other designated officer under said contract or arrangement aforesaid shall have the right to sell the salvage and valuable materials at the highest price obtainable. The net proceeds of such sale, after deducting the expenses of such demolition and removal, shall be promptly remitted with a report of such sale or transaction, including the items of expense and the amounts deducted, for the person who is entitled thereto, subject to any order of a court. If such a surplus does not remain to be turned over, the report shall so state.

❖ The governing body may sell any valuables or salvageable materials for the highest price obtainable. The costs of demolition are then to be deducted from any proceeds from the sale of salvage. If a surplus of funds remains, it is to be remitted to the owner with an itemized expense and income account. If no surplus remains, this must also be reported.

Bibliography

The following resource materials were used in the preparation of the commentary for this chapter of the code.

Legal Aspects of Code Administration. Washington, DC: International Code Council, 2003.

NEHRP, *Recommended Provisions fo Seismic Regulations for New Buildings and Other Structures.* Washington, DC: National Earthquake Hazards Reduction Program, 2009.

UCBC, *Uniform Code for Building Conservation.* Washington, DC: International Code Council, 2009.

Chapter 2:
Definitions

General Comments

All terms defined in the code are located alphabetically in Chapter 2. The user should be familiar with the terms in this chapter because definitions are essential to the correct interpretation of the code, and the user might not be aware that a particular term encountered in the text has the special definition found herein.

Section 201.1 contains the scope of the chapter. Section 201.2 establishes the interchangeability of the terms in the code. Section 201.3 establishes the use of terms defined in other codes. Section 201.4 establishes the use of undefined terms, and Section 202 lists terms and their definitions according to the code.

Purpose

Codes by their very nature are technical documents. Every word, term and punctuation mark can alter a sentence's meaning and, if misused, muddy its intent.

Further, the code, with its broad scope of applicability, includes terms that have a different meaning than the generally accepted meaning of the term. These terms can have multiple meanings depending on the context or discipline.

For these reasons, maintaining a consensus on the specific meaning of terms contained in the code is essential. Chapter 2 performs this function by stating clearly what specific terms mean for the purpose of the code.

SECTION 201
GENERAL

201.1 Scope. Unless otherwise expressly stated, the following words and terms shall, for the purposes of this code, have the meanings shown in this chapter.

❖ This section contains language and provisions that are supplemental to the use of Chapter 2. It gives guidance to the use of the defined words relevant to tense, gender and plurality. Finally, this section provides direction on how to apply terms that are not defined in the code.

201.2 Interchangeability. Words used in the present tense include the future; words stated in the masculine gender include the feminine and neuter; the singular number includes the plural and the plural, the singular.

❖ While the definitions contained or referenced in Chapter 2 are to be taken literally, gender and tense are interchangeable; thus, any grammatical inconsistencies with the code text will not hinder the understanding or enforcement of the requirements.

201.3 Terms defined in other codes. Where terms are not defined in this code and are defined in the other *International Codes,* such terms shall have the meanings ascribed to them in those codes.

❖ When a word or term appears in the code and that word or term is not defined in this chapter, other references may be used to find its definition, such as the *International Building Code®* (IBC®), *International Residential Code®* (IRC®), *International Fire Code®* (IFC®), *International Plumbing Code®* (IPC®), *International Mechanical Code®* (IMC®), *International Fuel Gas Code®* (IFGC®), *International Code Council Performance Code® for Buildings and Facilities®* (ICC PC®), *International Private Sewage Disposal Code®* (IPSDC®), *International Property Maintenance Code®* (IPMC®), *International Energy Conservation Code®* (IECC®), *International Wildland-Urban Interface Code®* (IWUIC®) and *International Zoning Code®* (IZC®). These codes contain additional definitions (some parallel and duplicative), which may be used in the enforcement of the code or in the enforcement of the other codes by reference.

201.4 Terms not defined. Where terms are not defined through the methods authorized by this chapter, such terms shall have ordinarily accepted meanings such as the context implies.

❖ Words or terms not defined in the *International Codes®* series are intended to be applied based on their "ordinarily accepted meanings." The intent of this statement is that a dictionary definition may suffice, provided it is in context. Sometimes the construction terms used in the code are not specifically defined in the code or even in a dictionary. In such a case, definitions contained in the referenced standards (see Chapter 15) and published textbooks on the subject in question are good resources.

SECTION 202
GENERAL DEFINITIONS

❖ This section contains definitions of terms that are used throughout the code. It is important to emphasize that these definitions are applicable everywhere the term is used in the code. Definitions of terms can help in the understanding and application of the code requirements.

[A] ADDITION. An extension or increase in floor area, number of stories, or height of a building or structure.

❖ This term is used to describe the condition when the floor area or height of an existing building or structure is increased. This term is only applicable to existing buildings, never new ones.

[A] ALTERATION. Any construction or renovation to an existing structure other than a *repair* or *addition*. Alterations are classified as Level 1, Level 2 and Level 3.

❖ The code utilizes this term to reflect construction operations intended for an existing building, but not in the scope of an addition or repair (see the definitions of "Addition" and "Repair").

[A] APPROVED. Acceptable to the *code official* or authority having jurisdiction.

❖ As related to the acceptance process for building installations, including materials, equipment and construction systems, this definition identifies where the ultimate authority rests. Where this term is used, it intends that only the enforcing authority can accept a specific installation or component as complying with the code. For the IBC and IRC, the "building official" is identified as the person responsible for administering code provisions. For the IFC, the "fire code official" is identified as the person responsible for administering IFC provisions. For the IECC, *International Existing Building Code® (IEBC®)*, IFGC, *International Green Construction Code® (IgCC®)*, IMC, IPC, IPMC, *International Swimming Pool and Spa Code® (ISPSC®)* and IWUIC, the "code official" is identified as the person responsible.

[A] CHANGE OF OCCUPANCY. A change in the use of the building or a portion of a building. A change of occupancy shall include any change of occupancy classification, any change from one group to another group within an occupancy classification or any change in use within a group for a specific occupancy classification.

❖ Changing the occupancy classification in an existing structure may change the level of inherent hazards addressed by the code. For example, a change from a mercantile occupancy to a business occupancy renders all Group B provisions applicable to all portions of the structure where the occupancy was changed. In addition, change of occupancy by way of change of use can change the level of inherent hazards (e.g., a Group A-2 restaurant changed to a Group A-2 nightclub). The classification may be the same, but the risk level has changed. Change of occupancy, whether change of occupancy classification or simply use, is specifically addressed in Chapter 10 of the code.

[A] CODE OFFICIAL. The officer or other designated authority charged with the administration and enforcement of this code.

❖ Regardless of title, the individual designated by the jurisdiction as the person who administers and enforces the code is the code official. In addition, the code official may appoint various other individuals to assist in the activities of the Department of Building Safety. In many jurisdictions, the authority of the code official is extended, to some degree, to plans examiners and inspectors. Section 104 sets forth the duties and responsibilities of the code official.

[BS] DANGEROUS. Any building, structure or portion thereof that meets any of the conditions described below shall be deemed dangerous:

1. The building or structure has collapsed, has partially collapsed, has moved off its foundation, or lacks the necessary support of the ground.

2. There exists a significant risk of collapse, detachment or dislodgement of any portion, member, appurtenance or ornamentation of the building or structure under service loads.

❖ This definition describes what is considered to be dangerous in building construction. There is a list of two specific criteria that will cause a structure to be classified as dangerous. These criteria take into consideration the collapse or likelihood of collapse of a structure, or a portion or element of a structure. This would also include any building element that could become detached and potentially result in the failure of one or more of the building's inherent systems.

[A] DEFERRED SUBMITTAL. Those portions of the design that are not submitted at the time of the application and that are to be submitted to the *code official* within a specified period.

❖ Submittal documents are required at the time of permit application (see Section 107). If a building is "fast-track" or is a shell building, a complete set of construction documents for the building may not be available at the time of the initial permit application. So that a complete set of documents is available for the building department review and records, some of the construction drawings may be provided at a later time. What documents and when they will be provided must be included in the initial submittal and approved by the building official.

EQUIPMENT OR FIXTURE. Any plumbing, heating, electrical, ventilating, air conditioning, refrigerating, and fire protection equipment, and elevators, dumb waiters, escalators, boilers, pressure vessels and other mechanical facilities or installations that are related to building services. Equipment or fixture shall not include manufacturing, production, or process equipment, but shall include connections from building service to process equipment.

❖ This definition outlines the type of building systems and devices that are considered to be categorized as equipment or fixtures. It is important to note that while the list is quite extensive, the definition does not specifically exclude the terms "manufacturing," "production" or "process equipment."

[A] EXISTING BUILDING. A building erected prior to the date of adoption of the appropriate code, or one for which a legal building permit has been issued.

❖ This term is used to identify those structures or buildings that were constructed before the current edition of the code was adopted by the jurisdiction. Often erected under the provisions of an earlier edition of the code, the buildings are exempt from compliance with current code provisions unless otherwise stated or a hazardous condition is present, or when alterations or changes in building height and areas are made.

[A] FACILITY. All or any portion of buildings, structures, site improvements, elements and pedestrian or vehicular routes located on a site.

❖ This term is intentionally broad and includes all portions within a site and all aspects of that site containing features required to be accessible. This includes parking areas, exterior walkways leading to accessible features, recreational facilities such as playgrounds and picnic areas and any structures on the site (see also the commentary to the definition of "Site" in the IBC).

[BS] FLOOD HAZARD AREA. The greater of the following two areas:

1. The area within a flood plain subject to a 1-percent or greater chance of flooding in any year.

2. The area designated as a *flood hazard area* on a community's flood hazard map, or otherwise legally designated.

❖ The Federal Emergency Management Agency (FEMA) prepares Flood Insurance Rate Maps (FIRMs) that delineate the land area subject to inundation by the 1-percent annual chance flood. Some states and local jurisdictions develop and adopt maps of flood hazard areas that are more extensive than the areas shown on FEMA's maps. For the purpose of the code, the flood hazard area in which the requirements are to be applied is the greater of the two delineated areas.

[A] HISTORIC BUILDING. Any building or structure that is one or more of the following:

1. Listed, or certified as eligible for listing, by the State Historic Preservation Officer or the Keeper of the National Register of Historic Places, in the National Register of Historic Places.

2. Designated as historic under an applicable state or local law.

3. Certified as a contributing resource within a National Register, state designated or locally designated historic district.

❖ This definition specifies the criteria for consideration as a historic building. Chapter 11 contains the provisions for buildings that qualify as historic buildings.

LOAD-BEARING ELEMENT. Any column, girder, beam, joist, truss, rafter, wall, floor or roof sheathing that supports any vertical load in addition to its own weight or any lateral load.

❖ This term relates to all of the load-bearing elements in a structure. It is important to identify all such load-bearing elements to ensure that they can continue to accomplish their designed function of being able to transfer any vertical load, any lateral load and their own weight effectively to the earth.

NONCOMBUSTIBLE MATERIAL. A material that, under the conditions anticipated, will not ignite or burn when subjected to fire or heat. Materials that pass ASTM E 136 are considered noncombustible materials.

❖ A material that will not ignite or burn when subjected to fire or heat or that successfully passes the ASTM E136 test is considered to be noncombustible. The test determines whether a building material will act to aid combustion or add appreciable heat to a fire. A material may have a limited amount of combustible content but not contribute appreciably to a fire; thus, it may still qualify as noncombustible.

PRIMARY FUNCTION. A *primary function* is a major activity for which the facility is intended. Areas that contain a *primary function* include, but are not limited to, the customer services lobby of a bank, the dining area of a cafeteria, the meeting rooms in a conference center, as well as offices and other work areas in which the activities of the public accommodation or other private entity using the facility are carried out. Mechanical rooms, boiler rooms, supply storage rooms, employee lounges or locker rooms, janitorial closets, entrances, corridors and restrooms are not areas containing a *primary function*.

❖ Primary function areas contain the major activities for the building or space. Determination of what constitutes a primary function space can be somewhat subjective. There can be multiple areas containing a primary function in a single building. Primary function areas are not limited to public use areas. For example, both a bank lobby and the bank's employee areas, such as the teller areas and walk-in safe, are primary function areas. Areas that are not primary function spaces are support and circulation spaces. Determination of the primary function areas for a building will also determine when the route to that area and associated toilet rooms and drinking fountains must be evaluated for accessibility. If these items are not accessible, additional alternatives may be necessary (see commentary, Section 705.2).

[A] REGISTERED DESIGN PROFESSIONAL IN RESPONSIBLE CHARGE. A registered design professional engaged by the owner to review and coordinate certain aspects of the project, as determined by the *code official*, for compatibility with the design of the building or structure, including submittal documents prepared by others, deferred submittal documents and phased submittal documents.

❖ A registered design professional in responsible charge is a person typically in charge of the review and coordination of submittal documents prepared by others, deferred submittal documents and phased submittal documents for compatibility with the design of the building or structure. Refer to Section 106.3.4 for specific language dealing with this term.

REHABILITATION. Any work, as described by the categories of work defined herein, undertaken in an *existing building*.

❖ This process of returning a property to a state of utility through repair or alteration makes it possible to effect a positive contemporary use while preserving those portions and features of the property that are significant to its historic, architectural and cultural values.

REHABILITATION, SEISMIC. Work conducted to improve the seismic lateral force resistance of an *existing building*.

❖ This definition relates specifically to the efforts designed to help improve and perhaps reestablish seismic-lateral-force resistance of a property to an appropriate level.

RELOCATABLE BUILDING. A partially or completely assembled building constructed and designed to be reused multiple times and transported to different building sites.

[A] REPAIR. The reconstruction or renewal of any part of an *existing building* for the purpose of its maintenance or to correct damage.

❖ As indicated in Section 105.2.2, the repair of an item typically does not require a permit. This definition makes it clear: repair is limited to work on the item and does not include complete or substantial replacement or other new work. Note that the definition deals with both repair as it relates to maintenance and repairs as they relate to fixing damage inflicted on a building for various reasons. More specifically, the replacement of stairs due to daily wear and tear is related to the maintenance of a building; whereas a wall hit by a forklift or damage as a result of an earthquake would be considered damage as it relates to the definition.

[BS] REROOFING. The process of recovering or replacing an existing roof covering. See "Roof recover" and "Roof replacement."

❖ This term refers to the process of covering or replacing an existing roof system with a new roofing system (see Section 706).

[BS] ROOF RECOVER. The process of installing an additional roof covering over a prepared existing roof covering without removing the existing roof covering.

❖ This term refers to the process of covering an existing roof system with a new roofing system.

[BS] ROOF REPAIR. Reconstruction or renewal of any part of an existing roof for the purposes of its maintenance.

❖ Roofs should be maintained for the purpose of protection. If a section is damaged, then it should be repaired immediately.

[BS] ROOF REPLACEMENT. The process of removing the existing roof covering, repairing any damaged substrate and installing a new roof covering.

❖ This definition refers to the process of removing an existing roof system and replacing it with a new roofing system.

SEISMIC LOADING. The forces prescribed herein, related to the response of the structure to earthquake motions, to be used in the analysis and design of the structure and its components.

❖ This definition refers to the forces to be used in the seismic design of structures. This would apply to both structural and nonstructural components.

[BS] SUBSTANTIAL DAMAGE. For the purpose of determining compliance with the flood provisions of this code, damage of any origin sustained by a structure whereby the cost of restoring the structure to its before-damaged condition would equal or exceed 50 percent of the market value of the structure before the damage occurred.

❖ This term is used in the definition of "Substantial improvement." Substantial damage is a special case of substantial improvement, and if the cost of restoring damage equals or exceeds 50 percent of the market value of the structure, then compliance of the existing building is required. It is notable that a substantial damage determination is to be made regardless of what causes the damage. Buildings have sustained substantial damage due to flood, fire, wind, earthquake, deterioration and other causes.

[BS] SUBSTANTIAL IMPROVEMENT. For the purpose of determining compliance with the flood provisions of this code, any *repair*, *alteration*, *addition*, or improvement of a building or structure, the cost of which equals or exceeds 50 percent of the market value of the structure, before the improvement or *repair* is started. If the structure has sustained *substantial damage*, any repairs are considered *substantial improvement* regardless of the actual *repair* work performed. The term does not, however, include either:

1. Any project for improvement of a building required to correct existing health, sanitary, or safety code violations identified by the *code official* and that is the minimum necessary to ensure safe living conditions; or

2. Any *alteration* of a historic structure, provided that the *alteration* will not preclude the structure's continued designation as a historic structure.

❖ One of the long-range objectives of the National Flood Insurance Program (NFIP) is to reduce the exposure of older buildings that were built in flood hazard areas before local jurisdictions adopted flood hazard area maps and regulations. Section 105.3, Item 5 directs the applicant to state the valuation of the proposed work as part of the information submitted to obtain a permit. To make a determination as to whether a proposed repair, reconstruction, rehabilitation, addition or improvement constitutes substantial improvement or damage, the cost of the proposed work is to be compared to the market value of the building or structure before the work is started. In order to determine market value, the code official may require the applicant to provide such information, as allowed under Section 105.3. For additional guidance, refer to FEMA 213 and FEMA 311.

[BS] SUBSTANTIAL STRUCTURAL DAMAGE. A condition where one or both of the following apply:

1. In any story, the vertical elements of the lateral force-resisting system have suffered damage such that the lateral load-carrying capacity of the structure in any horizontal direction has been reduced by more than 33 percent from its predamage condition.

2. The capacity of any vertical gravity load-carrying component, or any group of such components, that supports more than 30 percent of the total area of the structure's floor(s) and roof(s) has been reduced more than 20 percent from its predamage condition and the remaining capacity of such affected elements, with respect to all dead and live loads, is less than 75 percent of that required by this code for new buildings of similar structure, purpose and location.

❖ This definition gives the specific parameters for evaluating when a building has sustained substantial structural damage. There are two separate criteria provided in the definition, either one of which will qualify a structure as being substantially damaged.

TECHNICALLY INFEASIBLE. An *alteration* of a facility that has little likelihood of being accomplished because the existing structural conditions require the removal or *alteration* of a load-bearing member that is an essential part of the structural frame, or because other existing physical or site constraints prohibit modification or addition of elements, spaces or features which are in full and strict compliance with the minimum requirements for new construction and which are necessary to provide accessibility.

❖ This term is defined in order to provide a basis for the application of accessibility provisions to existing buildings. Bringing any altered existing site or building into full compliance with all accessibility requirements applicable to new construction may require extraordinary effort because of existing physical characteristics. The code utilizes the concept of technical infeasibility to provide a basis for exceptions from strict compliance with the provisions for new construction in an existing building.

UNSAFE. Buildings, structures or equipment that are unsanitary, or that are deficient due to inadequate means of egress facilities, inadequate light and ventilation, or that constitute a fire hazard, or in which the structure or individual structural members meet the definition of "*Dangerous*," or that are otherwise *dangerous* to human life or the public welfare, or that involve illegal or improper occupancy or inadequate maintenance shall be deemed unsafe. A vacant structure that is not secured against entry shall be deemed unsafe.

❖ The term "unsafe" is defined in order to provide a basis for the code official to cause a condition that is hazardous to the health and welfare of individuals to be corrected, either by repair or demolition (see Section 115). Some unsafe buildings may be the result of unsafe or illegal occupancies. For example, prima facie evidence of an unsafe structure is an unsecured (open doors or windows) vacant structure. Because of the attractive nuisance they represent, all unsafe buildings must be either demolished or made safe and secure as deemed appropriate by the code official.

WORK AREA. That portion or portions of a building consisting of all reconfigured spaces as indicated on the construction documents. Work area excludes other portions of the building where incidental work entailed by the intended work must be performed and portions of the building where work not initially intended by the owner is specifically required by this code.

❖ This section specifically defines the area of all reconfigured spaces where work is expected to occur within the scope of a project. These areas are to be shown clearly on the construction documents. "Incidental" work areas are not required to be shown as work areas. Note that care needs to be taken in designating the work area. For instance, when a voluntary seismic upgrade (Level 2 alteration) is being made throughout a building, it would be reasonable to simply designate only the actual area of seismic upgrade to avoid placing the alteration in a Level 3 inappropriately. Since the work is voluntary, it seems inappropriate for such an upgrade to trigger Level 3 alteration requirements such as those for automatic sprinkler systems. In this example, only the actual floor area occupied by the columns, beams, walls, etc., that are being modified by the seismic upgrade would be included in the "work area." It would be inappropriate to include the floor area of an entire room simply because a wall or other structural element within that room was being altered. The key is that the definition states that the work area consist of "reconfigured spaces." Installing sprinklers or upgrading a structural element typically does not reconfigure a space.

Chapter 3:
Provisions for All Compliance Methods

General Comments

This chapter contains the provisions that explain how the code is intended to be applied. In addition, this chapter provides the procedures for seismic design and evaluation that apply throughout the code. These provisions were originally located in Chapter 1, but due to the technical nature of the seismic provisions and the fact that administration provisions are often heavily amended, a standalone chapter was necessary. Finally, this chapter provides some general sections regarding the applicability of other codes, building materials and occupancy classification.

In terms of the application of the code, this chapter explains the three main components of the code, which include the following:

- Prescriptive method.
- Work area method.
- Performance method.

It is intended that one method of compliance is chosen and applied in whole. The first of these methods is the prescriptive method, which is covered in Chapter 4. This chapter was originally excerpted from a portion of Chapter 34 of the *International Building Code*® (IBC®). Chapter 34 has been deleted from the IBC and Section 101.4.7 of that code refers users to the *International Existing Building Code*® (IEBC®) for existing building issues. Chapter 4 addresses additions, alterations, repairs, change of occupancy and accessibility in existing buildings. The requirements are fairly general, with a strong emphasis on structural evaluation and requirements for additions, repairs and alterations. Section 410, which addresses accessibility, provides a baseline of accessible improvement based on the type of work occurring in a building.

The second, core method introduced by the code is the work area method, which is addressed in Chapters 5 through 13. This concept was intended to provide more flexibility to encourage the reuse and continued use of existing buildings. More specifically, the provisions allow different levels of compliance based on the level of work occurring. Chapter 5 first classifies the type and level of work and then, based on that classification, specific provisions are applied. The various types and levels of work include the following:

- Repairs (Chapter 6)
- Alteration Level 1 (Chapter 7)
- Alteration Level 2 (Chapter 8)
- Alteration Level 3 (Chapter 9)
- Change of occupancy (Chapter 10)
- Additions (Chapter 11)
- Historic buildings (Chapter 12)
- Relocated or moved buildings (Chapter 13)

The final method provided in this code is the performance compliance method found in Chapter 14. Chapter 14 utilizes a scoring method to determine the overall safety level of a building. The main focus is on fire and life safety provisions, but base structural and accessibility requirements are also addressed. Specifically, Section 1401.2.5 addresses accessibility and Section 1401.4.1 addresses structural requirements. The accessibility provisions simply refer to the prescriptive method or work area method. The structural provisions are more basic than the prescriptive method and work area method. It is unclear whether this was intentionally different. The prescriptive and work area methods are more comprehensive and consistent with one another in terms of the structural requirements. The objective of this section is to provide an alternative compliance option that enables improvements to be made that will raise the score to a minimum level without strict compliance with the provisions of the IBC.

Chapter 3 also provides options related to seismic evaluation and design, which are intended to provide flexibility when addressing seismic design. These evaluation tools work with both the prescriptive and work area methods. The provisions help to determine which procedures and methods are to be applied when addressing seismic design in existing buildings. In some cases, the code requires compliance with the seismic design provisions of the IBC (Section 301.1.4.1), while in other sections, reduced seismic design provisions are permitted (Section 301.1.4.2). In both sections, specifics are provided regarding applicable procedures and methods. More detailed discussion about these methods is provided in the commentary to Sections 301.1.4, 301.1.4.1 and 301.1.4.2. In large part, these provisions are intended to facilitate an overall increase in seismic performance in existing buildings. Typically, buildings with higher occupant loads and those with increased importance due to their importance to the community will have more restrictive requirements.

Section 302 addresses generally applicable provisions for the overall code. Section 302.1 explains the applicability of this section to all aspects of the code. Section 302.2 notes that applicable requirements for existing situations are addressed in various *International Codes*® (I-Codes®) and the *National Electrical Code*® (NFPA 70). More specifically, codes such as the *International Plumbing Code*® (IPC®) have requirements for existing installations and how alterations or additions

PROVISIONS FOR ALL COMPLIANCE METHODS

are addressed. Other important examples are the retroactive provisions of Chapter 11 of the *International Fire Code*® (IFC®). Those provisions are applicable at all times regardless of whether a repair or alteration is being undertaken. These provisions remain applicable when applying this code. Additionally, this section states that where a conflict occurs, this code takes precedence.

Purpose

The purposes of this chapter are to define the three compliance options that are available to the users of the code, lay out the methods to be used for seismic design and evaluation throughout the code, and provide generally applicable minimum requirements for all three methods.

SECTION 301
ADMINISTRATION

301.1 General. The *repair, alteration, change of occupancy, addition* or relocation of all *existing buildings* shall comply with one of the methods listed in Sections 301.1.1 through 301.1.3 as selected by the applicant. Sections 301.1.1 through 301.1.3 shall not be applied in combination with each other. Where this code requires consideration of the seismic force-resisting system of an *existing building* subject to *repair, alteration, change of occupancy, addition* or relocation of *existing buildings*, the seismic evaluation and design shall be based on Section 301.1.4 regardless of which compliance method is used.

> **Exception:** Subject to the approval of the *code official*, *alterations* complying with the laws in existence at the time the building or the affected portion of the building was built shall be considered in compliance with the provisions of this code unless the building is undergoing more than a limited structural *alteration* as defined in Section 907.4.4. New structural members added as part of the *alteration* shall comply with the *International Building Code*. *Alterations* of *existing buildings* in *flood hazard areas* shall comply with Section 701.3.

❖ This section explains the options available to a designer or owner when dealing with construction related to existing buildings: prescriptive compliance method (Section 301.1.1), work area compliance method (Section 301.1.2) and performance compliance method (Section 301.1.3). This section also provides procedures for evaluation and design of seismic force-resisting systems of existing buildings where consideration of seismic forces is required by the structural provisions in the prescriptive and work area methods.

There is one alternative to using these three compliance methods that allows for compliance with the laws in existence at the time the structure was originally built, unless the building has sustained substantial structural damage or is undergoing more than a limited structural alteration. Repairs and alterations in flood hazard areas have additional requirements to the laws in existence at the time the structure was originally built.

301.1.1 Prescriptive compliance method. *Repairs, alterations, additions* and *changes of occupancy* complying with Chapter 4 of this code in buildings complying with the *International Fire Code* shall be considered in compliance with the provisions of this code.

❖ This section allows compliance in accordance with Chapter 4 of the code. These provisions are intended to prescribe specific minimum requirements for construction related to existing buildings, including additions, alterations, repairs, fire escapes, glass replacement, change of occupancy, historic buildings, moved structures and accessibility.

301.1.2 Work area compliance method. *Repairs, alterations, additions*, changes in occupancy and relocated buildings complying with the applicable requirements of Chapters 5 through 13 of this code shall be considered in compliance with the provisions of this code.

❖ This section allows compliance in accordance with Chapters 5 through 13 of the code. These chapters contain provisions based on a proportional approach to compliance where upgrades are triggered by the type and extent of the work.

301.1.3 Performance compliance method. *Repairs, alterations, additions*, changes in occupancy and relocated buildings complying with Chapter 14 of this code shall be considered in compliance with the provisions of this code.

❖ This section allows compliance in accordance with Chapter 14 of the code. This chapter provides a scoring method for evaluating a building based on fire safety, means of egress and general safety.

[BS] 301.1.4 Seismic evaluation and design procedures. The seismic evaluation and design shall be based on the procedures specified in the *International Building Code* or ASCE 41. The procedures contained in Appendix A of this code shall be permitted to be used as specified in Section 301.1.4.2.

❖ This section lists the documents that contain the provisions to be used for the seismic evaluation of an existing building as well as the design of any needed repairs. Since the scope of these documents varies considerably, brief descriptions are given below.

International Building Code® (IBC®)

The IBC is a comprehensive model building code with seismic provisions that are based, for the most part, on ASCE 7 as well as the National Earthquake Hazards Reduction Program (NEHRP) *Recommended Provisions for Seismic Regulations for New Buildings and Other Structures*. The requirements are intended to minimize the hazard to life for all buildings, increase the expected performance of higher occupancy buildings as compared to ordinary buildings and improve the capability of essential facilities to function during and after an earthquake. In addition to minimum seismic loading criteria, the earthquake design provisions include requirements for special inspection and testing as well as material-specific design requirements. Achieving the intended performance depends on a number of factors, including, for example, the structural framing type, configuration and construction materials.

The significant earthquake load concepts include the following:

1. The ground motions are based on a risk-targeted maximum considered earthquake (MCE_R) from ground motion response acceleration maps [IBC Figures 1613.3.1(1) through 1613.3.1(8)], which provide spectral response accelerations at short periods (S_S) and at a one-second period (S_1). These levels of ground motion are also used in ASCE 41.

2. Design for the effect of MCE_R: Considering the margin of safety inherent in seismic design practice, this achieves collapse prevention under MCE_R level ground motions. It is also intended that damage from the "design earthquake" ground motion would be repairable. For essential facilities (Risk Category IV) it is intended that damage from the "design earthquake" ground motion be relatively minor and allow continued occupancy and function of the facility. For higher ground motions, the intent is that there be a low probability of structural collapse.

3. Risk category and importance factors: The IBC assigns buildings to one of the four risk categories summarized in Commentary Figure 301.1.4(1). The intent is to provide increasingly higher performance as the risk category increases from I through IV. This is achieved in part by applying an importance factor in determining the design load. The importance factor specified in the load provisions of ASCE 7 directly impacts the calculation of seismic (as well as wind and snow) loads. The magnitude of the design load varies in proportion to the importance factor and a higher value is assigned to buildings with an occupancy that warrants a higher level of performance.

4. Nonlinear seismic behavior is accounted for through use of equivalent lateral forces that are reduced by a response modification factor (R). This approximates the internal forces under the design earthquake. The corresponding building displacements, however, must be increased by the deflection amplification factor (C_d) in meeting the drift limits. These factors are based on the type of seismic force-resisting system provided and are located in the referenced ASCE 7 load standard.

5. Detailing and limitations on the seismic force-resisting system are a function of a structure's seismic design category classification, which considers seismicity at the site, type of soil present at the site and the nature of the building occupancy. Since several code requirements and the *Guidelines for Seismic Retrofit of Existing Buildings* (GSREB) use the seismic design category as a threshold, but neither document contains the criteria for determining a building's seismic design category, Commentary Figure 301.1.4(2) depicts the steps used to determine a structure's seismic design category using the IBC seismic criteria.

Two levels of IBC seismic forces are used as the basis for the code requirements for seismic analysis and design. These two levels are either the full seismic force required by the IBC (see Section 301.1.4.1) or the reduced seismic force level, which is 75 percent of the full seismic forces required by the IBC (see Section 301.1.4.2) [see Commentary Figures 301.1.4(1) and 301.1.4(2)].

RISK CATEGORIES	NATURE OF OCCUPANCY	SEISMIC IMPORTANCE FACTOR FROM ASCE 7
I	Buildings and other structures that represent a low hazard to human life in the event of failure	1.0
II	Buildings and other structures except those listed in Categories I, III and IV	1.0
III	Buildings and other structures that represent a substantial hazard to human life in the event of failure	1.25
IV	Buildings and other structures designated as essential facilities	1.5

Figure 301.1.4(1)
RISK CATEGORIES AND IMPORTANCE FACTORS

PROVISIONS FOR ALL COMPLIANCE METHODS

Seismic Rehabilitation of Existing Buildings, ASCE 41

ASCE 41-13, *Seismic Evaluation and Retrofit of Existing Buildings*, is a combination of previous standards ASCE 41-06 and ASCE 31-03.

- ASCE 41-06—*Seismic Rehabilitation of Existing Buildings*.
- ASCE 31-03—*Seismic Evaluation of Existing Buildings*.

ASCE 41 was initially an updated version of FEMA 356, *Prestandard and Commentary for the Seismic Rehabilitation of Buildings*. FEMA 356 was essentially an updated version of FEMA 273, *NEHRP Guidelines for the Seismic Rehabilitation of Buildings*. As noted ASCE 41 now includes what was formerly ASCE 31. ASCE 31 was a consensus standard that was developed as a replacement for FEMA 310. It is an evaluation tool that provides a standardized process for identifying potential seismic deficiencies in existing buildings and includes an integrated commentary. It takes a three-tier approach to evaluation, starting with a screening phase (Tier 1) and proceeding to a detailed evaluation phase (Tier 3), if required.

The significant concepts in this standard include the following:

1. Discrete rehabilitation objectives are established based on target building performance levels at various levels of ground motion (earthquake hazard). These are summarized in Commentary Figure 301.1.4(3), which is reproduced from the ASCE 41 commentary.

---

1. Determine the mapped MCE spectral response acceleration at short periods, S_s, and at a 1-second period, S_1, for the site from Figures 1613.5(1) through 1613.5(4).

2. Determine the (soil) site class in accordance with Table 1613.5.2.

3. Determine the site coefficients, F_a and F_v, from Tables 1613.5.3(1) and 1613.5.3(2), respectively.

4. Determine the design spectral response acceleration at short periods, S_{DS}, and at a 1-second period, S_{D1}, as follows:

$$S_{DS} = (2/3)(F_a)(S_s)$$
$$S_{D1} = (2/3)(F_v)(S_1)$$

5. Determine the seismic design category as prescribed by Tables 1613.5.6(1) and 1616.5.6(2). The highest (most restrictive) of the seismic design categories from the two tables is the category assigned to the building.

Figure 301.1.4(2)
DETERMINATION OF SEISMIC DESIGN CATEGORY USING THE *INTERNATIONAL BUILDING CODE*

Note that this entire array of rehabilitation objectives is only available under a voluntary upgrade to the seismic force-resisting system in accordance with Section 807.6. Otherwise, the required performance level is established by the code [see Commentary Figure 301.1.4(3)].

2. Two specific levels of ground motion based on USGS ground motion maps are defined in ASCE 41. Basic Safety Earthquake-2 (BSE-2) is the ground motion based on the MCE and the mapped spectral response accelerations (S_S and S_1). Basic Safety Earthquake-1 (BSE-1) is MCE_R. BSE-1 and BSE-2 are identified in the left-hand (earthquake hazard level) column of Commentary Figure 301.1.4(4). ASCE 41 also provides procedures to establish other ground-

		\multicolumn{4}{c}{TARGET BUILDING PERFORMANCE LEVELS}
		Operational Performance Level (1-A)	Immediate Occupancy Performance Level (1-B)	Life Safety Performance Level (3-C)	Collapse Prevention Performance Level (5-E)
Earthquake Hazard Level	50%/50 year	a	b	c	d
	20%/50 year	e	f	g	h
	BSE-1 (approx. 10%/50 year)	i	j	k	l
	BSE-2 (approx. 2%/50 year)	m	n	o	p

Notes:
1. Each cell in the above matrix represents a discrete rehabilitation objective.
2. The rehabilitation objectives in the matrix above may be used to represent the three specific rehabilitation objectives defined in Sections 1.4.1, 1.4.2 and 1.4.3 as follows:

k + p = Basic safety objective (BSO)
k + p + any of a, e, i, b, f, j or n = enhanced objectives
o alone or n alone or m alone = enhanced objectives
k alone or p alone = limited objectives
c, g, d, h, l = limited objectives

Figure 301.1.4(3)
FEMA 356 REHABILITATION OBJECTIVES

motion levels that may be of interest in rehabilitations.

3. Section 1.4.1 of ASCE 41 establishes a rehabilitation objective that is referred to as the basic safety objective (BSO). The BSO requires the life safety performance level for BSE-1 and collapse prevention performance level for BSE-2. This dual objective corresponds to entries k and p as illustrated in Commentary Figure 301.1.4(4). The BSO approximates the risk to life safety that has traditionally been accepted in earthquake design and is comparable to the intended performance for new buildings under the IBC [see Commentary Figure 301.1.4(4)].

4. Systematic versus simplified rehabilitation methodologies: Commentary Figure 301.1.4(5), reprinted from the ASCE 41 commentary, provides an overview of the rehabilitation process under ASCE 41 [see Commentary Figure 301.1.4(5)].

5. The level of seismicity for a building is defined as low, moderate or high, based on the design short period spectral response acceleration (S_{DS}) and design spectral response acceleration at a one-second period (S_{D1}). These values are identical to the IBC design level ground motions.

6. The level of performance is established as life safety (LS), damage control, limited safety, collapse prevention or immediate occupancy (IO). Note that in complying with code requirements, Table 301.1.4.1 establishes the performance level that applies based on a building's occupancy classification.

7. ASCE 41 accounts for nonlinear response to earthquake ground motions by applying pseudo static lateral forces representing the forces required to impose the expected actual deformations of the structure in its yielded state under the design ground motion.

Appendix A—*Guidelines for the Seismic Retrofit of Existing Buildings* (GSREB)

Seismic retrofit guidelines have been developed and utilized in the western United States for many years. The 1997 edition of the *Uniform Code for Building Conservation* (UCBC) included three appendix chapters (see A1, A2 and A3 below) dealing with seismic strengthening of specific building types. In 2000, these three chapters were combined with two new chapters (see A4 and A5 below) and published as a standalone document, GSREB. The code has incorporated the GSREB in Appendix A. The applicability of each appendix chapter is stated in Section 301.1.4.2, Items 2.1 through 2.5. These chapters provide mitigation for the following areas of concern in existing buildings:

A1 Unreinforced Masonry Bearing Wall Buildings: This chapter applies to buildings with one or more unreinforced masonry-bearing walls.

A2 Reinforced Concrete and Reinforced Masonry Wall Buildings with Flexible Diaphragms: This chapter has requirements for wall anchorage systems for reinforced concrete or masonry walls that are laterally supported by flexible diaphragms.

A3 Strengthening of Cripple Walls and Sill Plate Anchorage of Light, Wood-Frame Residential Buildings: This chapter has requirements for perimeter foundations, sill plate connections and unbraced cripple walls.

A4 Wood-Frame Residential Buildings with Soft, Weak or Open-Front Walls: This chapter has

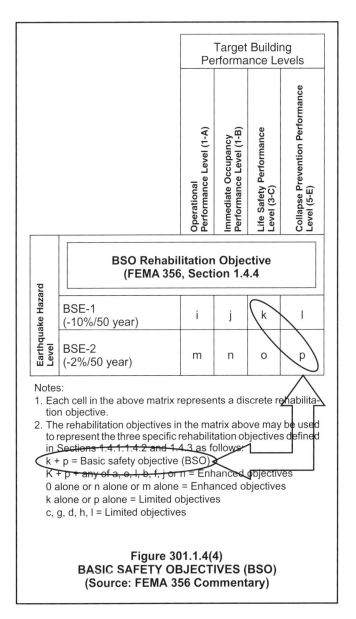

Figure 301.1.4(4)
BASIC SAFETY OBJECTIVES (BSO)
(Source: FEMA 356 Commentary)

PROVISIONS FOR ALL COMPLIANCE METHODS

requirements for wood-frame, multiple-unit residential buildings with soft, weak or open-front walls that are classified as Seismic Design Category C, D or E.

A5 Concrete Buildings: This chapter has requirements for concrete buildings. It is intended for buildings classified as Seismic Design Category B, C, D or E.

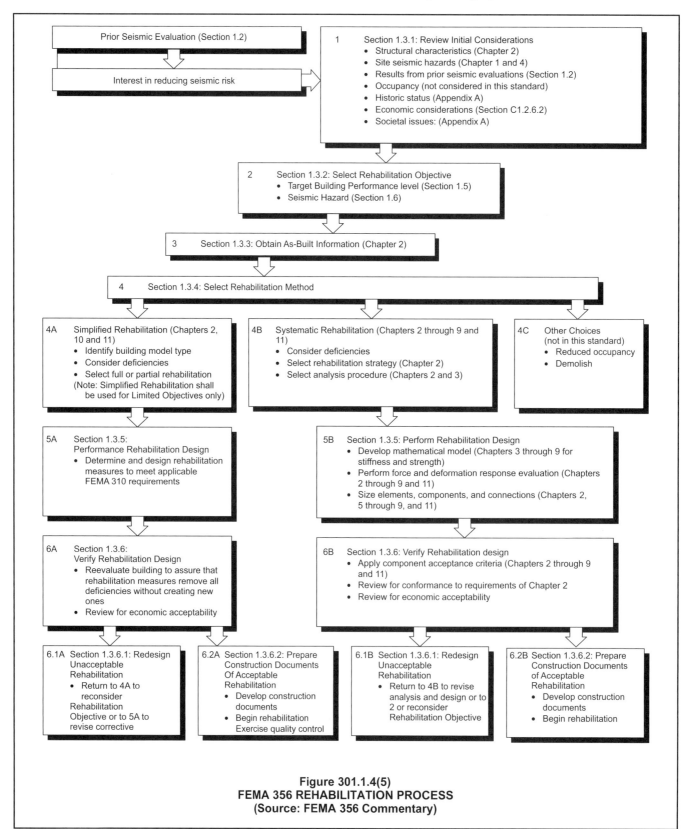

Figure 301.1.4(5)
FEMA 356 REHABILITATION PROCESS
(Source: FEMA 356 Commentary)

As noted previously, the scope and other thresholds include references to seismic criteria that correspond to the IBC as well (e.g., seismic design category). The GSREB is also based on IBC forces that require the use of a response modification coefficient (R) (see commentary, Section 301.1.4.1).

[BS] 301.1.4.1 Compliance with International Building Code-level seismic forces. Where compliance with the seismic design provisions of the *International Building Code* is required, the criteria shall be in accordance with one of the following:

1. One-hundred percent of the values in the *International Building Code*. Where the existing seismic force-resisting system is a type that can be designated as "Ordinary," values of R, Ω_0 and C_d used for analysis in accordance with Chapter 16 of the *International Building Code* shall be those specified for structural systems classified as "Ordinary" in accordance with Table 12.2-1 of ASCE 7, unless it can be demonstrated that the structural system will provide performance equivalent to that of a "Detailed," "Intermediate" or "Special" system.

2. ASCE 41, using a Tier 3 procedure and the two-level performance objective in Table 301.1.4.1 for the applicable risk category.

❖ Where the code requires the use of IBC-level seismic forces, it is intended to provide a level of earthquake performance comparable to new construction. The level of performance is typically a function of the building's occupancy classification (see commentary, Section 301.1.4 and Table 301.1.4.1).

One difficulty in applying the seismic requirements intended for new buildings to existing structures is the use of the response modification coefficient (R) in calculating the design seismic force. Under the IBC, the value of R that is obtained from the ASCE 7 standard is directly linked to the level of detailing required for any seismic force-resisting system. Systems are characterized as "ordinary," "intermediate" or "special" based on the extent of ductile detailing that is provided. In areas of moderate or higher seismicity (as reflected by a structure's seismic design category), the use of most ordinary systems (those with limited ductility) is typically restricted or prohibited. In an existing building, the system detailing is in place and the problem is in selecting an R-value that is consistent with the construction of that system. For this reason, Item 1 restricts R-values to be no greater than those listed for an ordinary system, unless there is clear evidence that a higher level of detailing has been provided.

By contrast, ASCE 41 avoids the previously discussed problem when assuming an R-value by instead considering the ductility of individual components. In Item 2, the code equates the use of full IBC seismic forces to the performance levels shown in Table 301.1.4.1 where ASCE 41 is used. The dual requirements listed in Table 301.1.4.1 are for BSE-1 and BSE-2 ground motions, which are discussed in the commentary to Section 301.1.4.

TABLE 301.1.4.1. See below.

❖ The performance levels listed in this table are comparable to those achieved under the IBC earthquake requirements. Risk Category III criteria are halfway between Risk Category II and Risk Category IV. This determination of Risk Category III is based on the definitions of "damage control" and "limited safety" within ASCE 41. This is analogous to the use of the seismic importance factor for new construction under the IBC.

[BS] 301.1.4.2 Compliance with reduced International Building Code-level seismic forces. Where seismic evaluation and design is permitted to meet reduced *International Building Code* seismic force levels, the criteria used shall be in accordance with one of the following:

1. The *International Building Code* using 75 percent of the prescribed forces. Values of R, Ω_0 and C_d used for analysis shall be as specified in Section 301.1.4.1 of this code.

2. Structures or portions of structures that comply with the requirements of the applicable chapter in Appendix A as specified in Items 2.1 through 2.5 and subject to the limitations of the respective Appendix A chapters shall be deemed to comply with this section.

    2.1. The seismic evaluation and design of unreinforced masonry bearing wall buildings in Risk Category I or II are permitted to be based on the procedures specified in Appendix Chapter A1.

    2.2. Seismic evaluation and design of the wall anchorage system in reinforced concrete and reinforced masonry wall buildings with flexible diaphragms in Risk Category I or II are permitted to be based on the procedures specified in Chapter A2.

[BS] TABLE 301.1.4.1
PERFORMANCE OBJECTIVES FOR USE IN ASCE 41 FOR COMPLIANCE WITH INTERNATIONAL BUILDING CODE-LEVEL SEISMIC FORCES

RISK CATEGORY (Based on IBC Table 1604.5)	STRUCTURAL PERFORMANCE LEVEL FOR USE WITH BSE-1N EARTHQUAKE HAZARD LEVEL	STRUCTURAL PERFORMANCE LEVEL FOR USE WITH BSE-2N EARTHQUAKE HAZARD LEVEL
I	Life Safety (S-3)	Collapse Prevention (S-5)
II	Life Safety (S-3)	Collapse Prevention (S-5)
III	Damage Control (S-2)	Limited Safety (S-4)
IV	Immediate Occupancy (S-1)	Life Safety (S-3)

PROVISIONS FOR ALL COMPLIANCE METHODS

2.3. Seismic evaluation and design of cripple walls and sill plate anchorage in residential buildings of light-frame wood construction in Risk Category I or II are permitted to be based on the procedures specified in Chapter A3.

2.4. Seismic evaluation and design of soft, weak, or open-front wall conditions in multiunit residential buildings of wood construction in Risk Category I or II are permitted to be based on the procedures specified in Chapter A4.

2.5. Seismic evaluation and design of concrete buildings assigned to Risk Category I, II or III are permitted to be based on the procedures specified in Chapter A5.

3. ASCE 41, using the performance objective in Table 301.1.4.2 for the applicable risk category.

❖ The lateral force-resisting systems in most older buildings are difficult, if not impossible, to upgrade to the same level of performance that is required of new construction. Where the code permits the use of this reduced seismic force level, it provides a means of achieving some improvement in earthquake performance of older buildings without making such efforts cost prohibitive. It also permits a building that is relatively close to current earthquake standards to comply outright. Item 1 establishes this level as 75 percent of the earthquake loading required in the design of new structures. The issues with R-value determination discussed in the commentary to Section 301.1.4.1 occur here as well.

The same approach is used in Appendix A (GSREB) and it is similar to the approach that has been used under FEMA 178. Therefore, Item 2 clarifies that Appendix A is a permitted alternative (see commentary, Section 301.1.4). Item 3 equates these reduced seismic forces to the level of performance required by Table 301.1.4.2 for only the BSE-1 under 41.

TABLE 301.1.4.2. See below.

❖ Item 3 of Section 301.1.4.2 references the values in Table 301.1.4.2 to use for compliance with reduced IBC-level seismic forces. The performance levels listed in this table are to allow for compliance by designing for seismic forces that are less than what is required by the IBC for those instances where the code allows this reduced level of compliance, such as for repairs. These performance levels are consistent with Item 1 of Section 301.1.4.2, which allows using 75 percent of the IBC-prescribed seismic forces. Note a addresses interpolation for Risk Category III. The ASCE 41 Tier 1 checklists are sufficiently conservative for the initial screening phase so that the less restrictive life safety performance level will not significantly compromise the overall seismic performance goals for Risk Category III buildings.

SECTION 302
GENERAL PROVISIONS

302.1 Applicability. The provisions of Section 302 apply to all alterations, repairs, additions, relocations of structures and changes of occupancy regardless of compliance method.

❖ This code contains three possible compliance paths: Prescriptive (Chapter 4), Work area (Chapters 5 through 13) and Performance (Chapter 14). For any of these paths, the requirements in Section 302 apply.

302.2 Additional codes. *Alterations, repairs, additions* and *changes of occupancy* to, or relocation of, *existing buildings* and structures shall comply with the provisions for *alterations, repairs, additions* and *changes of occupancy* or relocation, respectively, in this code and the *International Energy Conservation Code, International Fire Code, International Fuel Gas Code, International Mechanical Code, International Plumbing Code, International Property Maintenance Code, International Private Sewage Disposal Code, International Residential Code* and NFPA 70. Where provisions of the other codes conflict with provisions of this code, the provisions of this code shall take precedence.

❖ This section clarifies the relationship between this code and the IFC, *International Fuel Gas Code®* (IFGC®), *International Mechanical Code®* (IMC®), IPC, *International Property Maintenance Code®* (IPMC®), *International Private Sewage Disposal Code®* (IPSDC®), *International Residential Code®* (IRC®), *International Energy Conservation Code®* (IECC®) and NFPA 70. When alterations and repairs are made to existing mechanical and plumbing systems, the provisions of the I-Codes and NFPA 70 for alterations and repairs must be followed. Those codes indicate the extent to which existing systems must comply with the stated requirements. Where portions of existing building systems, such as plumbing, mechanical and electrical systems, are not being

[BS] TABLE 301.1.4.2
PERFORMANCE OBJECTIVES FOR USE IN ASCE 41 FOR COMPLIANCE WITH REDUCED INTERNATIONAL BUILDING CODE-LEVEL SEISMIC FORCES

RISK CATEGORY (Based on IBC Table 1604.5)	STRUCTURAL PERFORMANCE LEVEL FOR USE WITH BSE-1E EARTHQUAKE HAZARD LEVEL
I	Life Safety (S-3)
II	Life Safety (S-3)
III	Damage Control (S-2). See Note a
IV	Immediate Occupancy (S-1)

a. Tier 1 evaluation at the Damage Control performance level shall use the Tier 1 Life Safety checklists and Tier 1 Quick Check provisions midway between those specified for Life Safety and Immediate Occupancy performance.

altered or repaired, those systems may continue to exist without being upgraded as long as they are not hazardous or unsafe to the building occupants.

Another important element of this section is that this code will take precedence if a conflict occurs between one of the listed codes and this code. This is only as far as it concerns requirements for alteration, repairs, additions and change of occupancy. However, this would not address a situation where another code, such as the IFC, retroactively required changes to a building regardless if any repairs, alterations, additions or changes of occupancy were occurring.

302.3 Existing materials. Materials already in use in a building in compliance with requirements or approvals in effect at the time of their erection or installation shall be permitted to remain in use unless determined by the building official to be unsafe.

❖ This section is the same as Section 401.2.1. It has been repeated here as well to be generally applicable to all methods. If a material or system had been approved before the code took effect, it can continue to be used as long as it can be shown that the material or system is not detrimental to the health or safety of the building occupants or the public. Specifically, a material or system cannot be "unsafe" as defined in Chapter 2 and as addressed by Section 115. The code is not intended to be retroactive.

302.4 New and replacement materials. Except as otherwise required or permitted by this code, materials permitted by the applicable code for new construction shall be used. Like materials shall be permitted for *repairs* and *alterations*, provided no unsafe condition is created. Hazardous materials shall not be used where the code for new construction would not permit their use in buildings of similar occupancy, purpose and location.

❖ Similar to Section 302.3, this section is the same as Section 401.2.2. It has been repeated here to be generally applicable to all methods. There are two options for materials used in repairs to an existing building. Generally, the materials used for repairs should be those that are presently required or permitted for new construction in accordance with the I-Codes. It is also acceptable to use materials consistent with those that are already present, except where those materials pose a hazard. This allowance follows the general concept that any repair should not make a building more hazardous than it was prior to the repair. It is generally possible to repair a structure, its components and its systems with materials consistent with those materials that were used previously. However, where materials that are now deemed hazardous are involved in the repair work, they may no longer be used. For example, the code identifies asbestos and lead-based paint as two common hazardous materials that cannot be used in the repair process. Certain materials previously considered acceptable for building construction are now known to be threats to the health of occupants.

302.5 Occupancy and use. When determining the appropriate application of the referenced sections of this code, the occupancy and use of a building shall be determined in accordance with Chapter 3 of the *International Building Code*.

❖ This section provides a link to the occupancy classifications in the IBC. Any time a provision is based on occupancy classification, that classification is determined through IBC requirements, not from codes under which the building was originally built. Occupancy classifications have changed over the years and varied as to how they were named in previous codes.

In the early years of the last century, the essence of regulatory safeguards from fire was to provide a reasonable level of protection to property. The idea was that if property was adequately protected from fire, then the building occupants would also be protected.

From this outlook on fire safety, the concept of equivalent risk has evolved in the code. This concept maintains that, in part, an acceptable level of risk against the damages of fire, respective to a particular occupancy type (group), can be achieved by limiting the height and area of buildings containing such occupancies according to the building's construction type (i.e., its relative fire endurance).

The concept of equivalent risk involves three interdependent considerations: 1. The level of fire hazard associated with the specific occupancy of the facility; 2. The reduction of fire hazard by limiting the floor area(s) and the height of the building based on the fuel load (combustible contents and burnable building components); and 3. The level of overall fire resistance provided by the type of construction used for the building.

The interdependence of these fire safety considerations can be seen by first looking at IBC Tables 601 and 602, which show the fire-resistance ratings of the principal structural elements composing a building in relation to the five classifications for types of construction. Type I construction is the classification that generally requires the highest fire-resistance ratings for structural elements, whereas Type V construction, which is designated as a combustible type of construction, generally requires the least amount of fire-resistance-rated structural elements. If one then looks at IBC Tables 504.3, 504.4 and 504.5, the relationship among group classification, allowable heights and areas, and types of construction becomes apparent. Respective to each group classification, the greater the fire-resistance rating of structural elements, as represented by the type of construction, the greater the floor area and height allowances. The greater the potential fire hazards indicated as a function of the group, the lesser the height and area allowances for a particular construction type.

As a result of extensive research and advancements in fire technology, today's building codes are more comprehensive and complex regulatory instruments than they were in the earlier years of code development. While the principle of equivalent risk remains an important component in building codes, perspectives have changed and life safety is now the paramount fire issue. Even so, occupancy classification still plays a key part in organizing and prescribing the appropriate protection measures. As such, threshold requirements for fire protection and means of egress systems are based on occupancy classification (see Chapters 9 and 10 of the IBC).

Bibliography

The following resource materials were used in the preparation of the commentary for this chapter of the code.

FEMA 178, *NEHRP Handbook for the Seismic Evaluation of Buildings*. Washington, DC: Federal Emergency Management Agency, 2001.

FEMA 273, *NEHRP Guidelines for the Seismic Rehabilitation of Existing Buildings*. Washington, DC: Federal Emergency Management Agency, 2001.

FEMA 310, *Handbook for Seismic Evaluation of Buildings—A Prestandard*. Washington, DC: Federal Emergency Management Agency, 2001.

FEMA 356-00, *Prestandard and Commentary for the Seismic Rehabilitation of Buildings Guidelines for the Seismic Retrofit of Existing Buildings*. Washington, DC: Federal Emergency Management Agency, 2000.

Chapter 4:
Prescriptive Compliance Method

General Comments

This method of compliance is more simplistic and more administrative in nature than the work area method. It is derived from what was Chapter 34 of the *International Building Code®* (IBC®). It provides the basic requirements for additions (Section 402), alterations (Section 403), repairs (Section 404), change of occupancy (Section 407) and historic structures (Section 408). In addition, this chapter provides requirements for the use of fire escapes (Section 405) and glass replacement and window replacement (Section 406). Accessibility for existing buildings and facilities is addressed in Section 410.

Purpose

This chapter provides one of the three options of compliance available in the code for buildings and structures undergoing repair, alteration, addition or change of occupancy. This chapter originates from Chapter 34 of the IBC, Sections 3401 through 3409. Chapter 34 has been deleted from the 2015 IBC. The *International Existing Building Code®* (IEBC®) is now used for existing buildings. Section 101.4.7 of the IBC directs code users to the IEBC for existing buildings. There are also provisions from the other *International Codes®* (I-Codes®) dealing with system installations (electrical, energy, fuel gas, mechanical and plumbing), which have been duplicated in the code, as well. As a duplication of provisions, the ongoing code development maintenance will be accomplished by the code committee responsible for the code from which the provisions are being extracted and duplicated.

SECTION 401
GENERAL

401.1 Scope. The provisions of this chapter shall control the *alteration, repair, addition* and *change of occupancy* or relocation of *existing buildings* and structures, including *historic buildings* and structures as referenced in Section 301.1.1.

Exception: Existing bleachers, grandstands and folding and telescopic seating shall comply with ICC 300.

❖ This section states the scope of this chapter and references alternative methods of code compliance for alteration, repair, addition and change of occupancy of existing structures. This section also defines the responsibilities for maintenance, repairs, compliance with other codes and periodic testing. The exception sends the code user to ICC 300, *Standard for Bleachers, Folding and Telescopic Seating, and Grandstands* for alterations, repairs and additions to bleachers, grandstands, and folding and telescopic seating.

401.1.1 Compliance with other methods. *Alterations, repairs, additions* and *changes of occupancy* to or relocation of, *existing buildings* and structures shall comply with the provisions of this chapter or with one of the methods provided in Section 301.1.

❖ This section references Section 301.1 for the options available to deal with alterations, repairs, additions and changes of occupancy to existing structures. The following briefly describes the options available, other than compliance with this chapter:

1. Subject to the approval of the code official, repairs and alterations can comply with the requirements of the code at the time the building was built.

2. Repairs, alterations, additions and changes of occupancy can comply with the proportional approach where upgrades are triggered by the type and extent of the work. These work area method requirements are found in Chapters 5 through 13.

3. Repairs, alterations, additions and changes of occupancy can be designed in accordance with the compliance alternatives that are found in Chapter 14.

Please note that these options are separate and distinct and must not be combined in any way.

401.2 Building materials and systems. Building materials and systems shall comply with the requirements of this section.

❖ This section merely introduces the fact that materials and systems are required to comply with Sections 401.2.1 through 401.2.3. These sections deal with the use of existing materials, new and replacement materials and seismic force-resisting systems.

401.2.1 Existing materials. Materials already in use in a building in compliance with requirements or approvals in effect at the time of their erection or installation shall be permitted to remain in use unless determined by the building official to be unsafe per Section 115.

❖ If a material or system had been approved before the code took effect, it can continue to be used as long as it can be shown that the material or system is not detrimental to the health or safety of the building occupants or the public. Specifically, a material or system cannot be "unsafe" as defined in Chapter 2 and as addressed by Section 115. The code is not intended to be retroactive.

401.2.2 New and replacement materials. Except as otherwise required or permitted by this code, materials permitted by the applicable code for new construction shall be used. Like materials shall be permitted for *repairs* and *alterations*, provided no hazard to life, health or property is created. Hazardous materials shall not be used where the code for new construction would not permit their use in buildings of similar occupancy, purpose and location.

❖ There are two options for materials used in repairs to an existing building. Generally, the materials used for repairs should be those presently required or permitted for new construction in accordance with the I-Codes. It is also acceptable to use materials consistent with those that are already present, except where those materials pose a hazard. This allowance follows the general concept that any repair should not make a building more hazardous than it was prior to the repair. It is generally possible to repair a structure, its components and its systems with materials consistent with those materials that were used previously. However, where materials that are now deemed hazardous are involved in the repair work, they may no longer be used. For example, the code identifies asbestos and lead-based paint as two common hazardous materials that cannot be used in the repair process. Certain materials previously considered acceptable for building construction are now known to be a threat to the health of the occupants.

401.2.3 Existing seismic force-resisting systems. Where the existing seismic force-resisting system is a type that can be designated ordinary, values of R, Ω_0 and C_d for the existing seismic force-resisting system shall be those specified by the *International Building Code* for an ordinary system unless it is demonstrated that the existing system will provide performance equivalent to that of a detailed, intermediate or special system.

❖ This section provides guidance to engineers on selecting system-related design coefficients for existing seismic force systems. Previously, this guidance was located in each individual section related to alterations, additions, repairs and change of occupancy, but has been relocated to a more general section to ensure consistent application.

The intent is that existing systems should be considered "ordinary" by default. For seismic systems that may provide performance equivalent to that of a detailed, intermediate or special system, the code requires a demonstration of equivalence.

This only applies to systems for which there is a choice of "ordinary," "detailed," "intermediate" or "special" for the permitted seismic force systems. For example, light-frame shear walls are not categorized as "ordinary," "detailed," "intermediate" or "special." Those systems are acceptable, and the system coefficients specified for those systems in ASCE 7 are appropriate.

401.3 Dangerous conditions. The building official shall have the authority to require the elimination of conditions deemed *dangerous*.

❖ This section enables the building official to address dangerous conditions for any type of situation that may arise during renovations to existing buildings. "Dangerous" is specially defined in Chapter 2 and is related to structural stability. "Unsafe" is more general and is also specially defined in Chapter 2. Provisions for unsafe buildings are located in Section 115.

SECTION 402
ADDITIONS

402.1 General. *Additions* to any building or structure shall comply with the requirements of the *International Building Code* for new construction. Alterations to the *existing building* or structure shall be made to ensure that the *existing building* or structure together with the *addition* are no less conforming to the provisions of the *International Building Code* than the *existing building* or structure was prior to the *addition*. An *existing building* together with its *additions* shall comply with the height and area provisions of Chapter 5 of the *International Building Code*.

❖ The purpose of this section is to establish the guidelines for additions to existing buildings and structures.

An addition is an increase in the area or height of an existing building. When a new building is erected immediately adjacent to an existing building, and they are separated by a fire wall, it is considered a separate building, not an addition to the existing structure. The new building must be designed to comply with the technical provisions of Chapters 1 through 33 of the IBC; not with the provisions of this chapter. The existing building must be evaluated considering the elimination of the adjacent open space now occupied by the new building. An existing structure that is of a type of construction that does not comply with the height and area limitations of IBC Sections 504 and 506 may not be added to unless the type of construction is upgraded, or the allowable building height and area are increased through the installation of an automatic sprinkler system throughout both the existing building and the addition. A building with a proposed addition is to be evaluated based on the type of construction of the existing building or the addition, whichever is the lower type. When reviewing for compliance with IBC Sections 504 and 506, IBC Section

602.1.1 is also applicable. Commentary Figure 402.1 shows an example of this situation. If a sprinkler system is not planned for both the existing building and the addition, the proposed addition does not meet the requirements of IBC Table 506.2. The area limitation for the existing construction Type VB, Group F-1 is 8,500 square feet (790 m^2) and the total proposed area is 11,000 square feet (1022 m^2).

The area evaluation is based on Type VB construction, even though the proposed addition is Type VA construction, because the allowable area in IBC Table 506.2 for Type VB construction is less than the allowable area for Type VA construction. One option to satisfy the requirements is to upgrade the type of construction of the existing building to Type VA, which has an allowable area of 14,000 square feet (1301 m^2). Another solution is to install an automatic sprinkler system that complies with NFPA 13 throughout the building. The revised allowable area of the building would be 34,000 square feet (3159 m^2) because of the installation of the automatic sprinkler system. This would then exceed the proposed building area of 11,000 square feet (1022 m^2).

[BS] 402.2 Flood hazard areas. For buildings and structures in *flood hazard* areas established in Section 1612.3 of the *International Building Code*, or Section R322 of the *International Residential Code*, as applicable, any *addition* that constitutes *substantial improvement* of the existing structure shall comply with the flood design requirements for new construction, and all aspects of the existing structure shall be brought into compliance with the requirements for new construction for flood design.

For buildings and structures in *flood hazard areas* established in Section 1612.3 of the *International Building Code*, or Section R322 of the *International Residential Code*, as applicable, any *additions* that do not constitute *substantial improvement* of the existing structure are not required to comply with the flood design requirements for new construction.

❖ Reduction in exposure to flood hazards, including the exposure of older buildings, is one of the purposes for regulating flood plain development. Buildings or structures located in flood hazard areas are to be brought into compliance with the flood-resistance provisions of IBC Section 1612 or *International Residential Code*® (IRC®) Section R322, as applicable, where the value of improvements, including additions, exceeds a certain value. See Chapter 2 for the definition of "Substantial improvement."

Section 105.3 requires the applicant to state the valuation of proposed work, which is to include the total value of work, including materials and labor. If the proposed work will be performed on buildings in flood hazard areas, a determination must be made as to whether the proposed work constitutes a substantial improvement. If applicable, the value of work must include estimates of the value of the property owner's labor and the value of donated labor and materials.

To make a determination about whether a proposed addition constitutes a substantial improvement, the cost of the proposed work is to be compared to the market value of the building or structure before the work is started. In order to determine market value, the code official may require the applicant to provide an appraisal or use other methods acceptable to the Federal Emergency Management Agency (FEMA). For additional guidance see FEMA P-758, *Substantial Improvement/Substantial Damage Desk Reference*.

[BS] 402.3 Existing structural elements carrying gravity load. Any existing gravity load-carrying structural element for which an *addition* and its related alterations cause an increase in design gravity load of more than 5 percent shall be strengthened, supplemented, replaced or otherwise altered as needed to carry the increased gravity load required by the

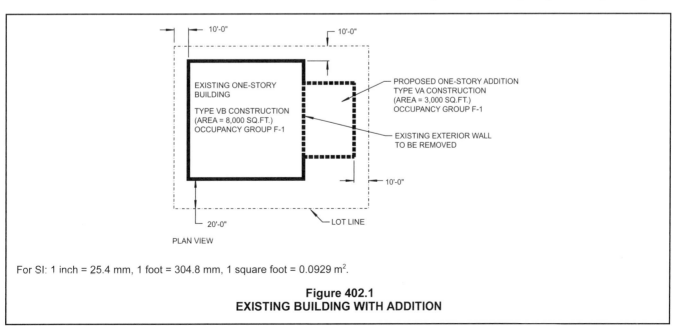

Figure 402.1
EXISTING BUILDING WITH ADDITION

International Building Code for new structures. Any existing gravity load-carrying structural element whose gravity load-carrying capacity is decreased shall be considered an altered element subject to the requirements of Section 403.3. Any existing element that will form part of the lateral load path for any part of the *addition* shall be considered an existing lateral load-carrying structural element subject to the requirements of Section 402.4.

❖ Wherever an addition to an existing building is made, the affected members of the original structure must be assessed to determine their ability to resist any increased forces. Because an addition typically adds new loads to the existing building, the architect or engineer responsible for the design of the project must analyze the existing gravity load-carrying elements to determine whether there are any structural components having loads increased by more than 5 percent, thus requiring reinforcement. Where an existing building was initially designed to support a future expansion, the code official should seek verification that the proposed addition will result in the affected building complying with current code requirements. Any new construction must comply with current code requirements.

[BS] 402.3.1 Design live load. Where the *addition* does not result in increased design live load, existing gravity load-carrying structural elements shall be permitted to be evaluated and designed for live loads approved prior to the *addition*. If the approved live load is less than that required by Section 1607 of the *International Building Code*, the area designed for the nonconforming live load shall be posted with placards of approved design indicating the approved live load. Where the *addition* does result in increased design live load, the live load required by Section 1607 of the *International Building Code* shall be used.

❖ It is not uncommon for the design live load requirements to change from the time a building is originally designed to when an addition is proposed. The live loads used in the original design may have been adequate for the building's initial use and may have been in compliance with all code requirements that were in effect at that time. Many years and many code changes can alter the status of the structural design criteria and code requirements, which does not mean the existing structural system is inadequate and cannot be used. It just means that when the live loads that were used for the design of the existing building are lower than those required by current standards, the design live loads used for the original design must be posted.

[BS] 402.4 Existing structural elements carrying lateral load. Where the *addition* is structurally independent of the existing structure, existing lateral load-carrying structural elements shall be permitted to remain unaltered. Where the *addition* is not structurally independent of the existing structure, the existing structure and its *addition* acting together as a single structure shall be shown to meet the requirements of Sections 1609 and 1613 of the *International Building Code*. For purposes of this section, compliance with ASCE 41, using a Tier 3 procedure and the two-level performance objective in Table 301.1.4.1 for the applicable risk category, shall be deemed to meet the requirements of Section 1613.

Exception: Any existing lateral load-carrying structural element whose demand-capacity ratio with the *addition* considered is no more than 10 percent greater than its demand-capacity ratio with the *addition* ignored shall be permitted to remain unaltered. For purposes of calculating demand-capacity ratios, the demand shall consider applicable load combinations with design lateral loads or forces in accordance with Sections 1609 and 1613 of the *International Building Code*. For purposes of this exception, comparisons of demand-capacity ratios and calculation of design lateral loads, forces and capacities shall account for the cumulative effects of *additions* and *alterations* since original construction.

❖ The requirements of this section do not affect an existing structure where the addition is structurally independent; in that case, the addition must comply with the seismic requirements for new structures. An existing building or structure where the addition will not be structurally independent must be carefully evaluated for its ability to withstand earthquake and wind loading. The seismic resistance of an existing building or structure cannot be lessened or pose any undue increase in the fire and life safety hazards of the building or structure. Because the addition and the existing structure resist lateral loads as a single structure, designing the entire structure to resist the total lateral seismic and wind load is necessary. This section specifically allows the use of ASCE 41 using a Tier 3 procedure and the two-level performance objectives of Section 301.1.4.1. The application of this procedure is also based on the risk category of the structure. This allowance provides more flexibility for existing buildings. ASCE 41 is specifically focused on the evaluation and rehabilitation of existing structures. See the commentary to Section 301.1.4 for more discussion on the origins and application of the standard.

The exception permits an addition that is not structurally independent without alteration to the seismic force-resisting system of the existing structure, provided the demand-capacity ratio is increased no more than 10 percent in any structural element. The demand is determined using the load combinations, including wind and seismic effects. By considering the demand-capacity ratio, any decrease in lateral resistance is accounted for as well.

402.5 Smoke alarms in existing portions of a building. Where an *addition* is made to a building or structure of a Group R or I-1 occupancy, the *existing building* shall be provided with smoke alarms in accordance with Section 1103.8 of the *International Fire Code*.

❖ This section provides a path to the *International Fire Code*® (IFC®) for the smoke alarm requirements in existing buildings. This section links the code directly to IFC Section 1103.8. The requirements in the IFC are retroactive and apply regardless of whether an alteration, addition or change of occupancy is occurring. Section 403.10 has a similar requirement, but for alterations.

SECTION 403
ALTERATIONS

403.1 General. Except as provided by Section 401.2 or this section, *alterations* to any building or structure shall comply with the requirements of the *International Building Code* for new construction. *Alterations* shall be such that the *existing building* or structure is no less conforming to the provisions of the *International Building Code* than the *existing building* or structure was prior to the *alteration*.

Exceptions:

1. An existing stairway shall not be required to comply with the requirements of Section 1011 of the *International Building Code* where the existing space and construction does not allow a reduction in pitch or slope.

2. Handrails otherwise required to comply with Section 1011.11 of the *International Building Code* shall not be required to comply with the requirements of Section 1014.6 of the *International Building Code* regarding full extension of the handrails where such extensions would be hazardous due to plan configuration.

❖ Alterations include renovations, which imply that something is changed in the structure. For example, the removal, rearrangement or replacement of partition walls in an office building is an alteration because, in part, of the possible impact on the means of egress, fire resistance or other life safety features of the building. Conversely, the replacement of damaged trim pieces on a door frame is considered a repair, not an alteration. Alterations are to conform to the requirements for a new structure. For example, consider the corridor of an office building that is to be extended 18 feet (5486 mm), as shown in Commentary Figure 403.1. The existing office building has no sprinkler system (and none is proposed), so the corridor extension walls and doors are to be fire-resistance rated in accordance with IBC Section 1020. This is applicable even if, for some reason, the walls and doors of the existing corridor system are not fire-resistance-rated construction. This section also indicates that unaltered portions of a structure are not required to comply with code provisions for new construction. However, also see Section 410, which requires areas outside of the alteration area to be revised in order to provide an accessible route.

The exceptions address issues that arise with stairways. While not specifically addressed, ramps may have similar concerns. According to Exception 1, if an existing stairway was built with a steeper rise/run ratio than permitted in the current code, the stairway can be replaced with the existing configuration. Enlarging the opening to achieve the current rise/run ratio and headroom would be considered technically infeasible. The principle is that not allowing for this option could result in stairways that were not maintained because they could not be brought up to current codes.

Exception 2 deals with an allowance to reduce

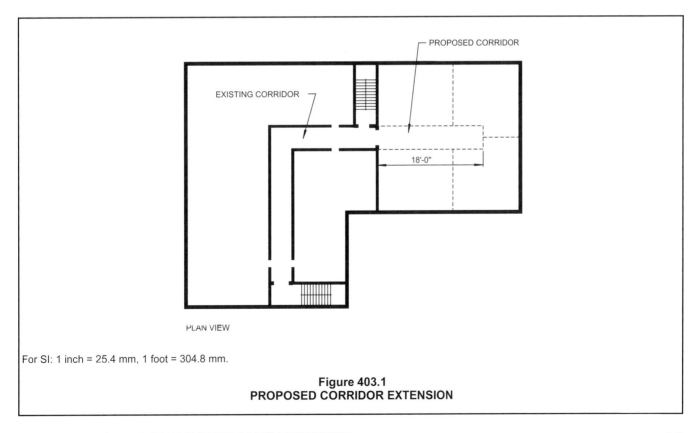

For SI: 1 inch = 25.4 mm, 1 foot = 304.8 mm.

**Figure 403.1
PROPOSED CORRIDOR EXTENSION**

handrail extensions when providing those extensions could become a protruding object for circulation or an obstruction for the means of egress. IBC Section 1014.6 requires the handrail extensions to continue in the direction of the stairway run. An example of an exempted configuration would be stairways having landings that are part of a corridor.

[BS] 403.2 Flood hazard areas. For buildings and structures in *flood hazard areas* established in Section 1612.3 of the *International Building Code*, or Section R322 of the *International Residential Code*, as applicable, any *alteration* that constitutes *substantial improvement* of the existing structure shall comply with the flood design requirements for new construction, and all aspects of the existing structure shall be brought into compliance with the requirements for new construction for flood design.

For buildings and structures in *flood hazard areas* established in Section 1612.3 of the *International Building Code*, or Section R322 of the *International Residential Code*, as applicable, any alterations that do not constitute *substantial improvement* of the existing structure are not required to comply with the flood design requirements for new construction.

❖ Reduction in exposure to flood hazards, including exposure of older buildings, is one of the purposes for regulating flood plain development. Buildings or structures that are located in flood hazard areas are to be brought into compliance with the flood-resistance provisions of IBC Section 1612 or IRC Section R322, as applicable, where the value of improvements, including alterations, exceeds a certain value. See Chapter 2 for the definition of "Substantial improvement."

Section 105.3 requires the applicant to state the valuation of the proposed work, which is to include the total value of the work, including materials and labor. If the proposed work will be performed on buildings in flood hazard areas, a determination must be made as to whether the proposed work constitutes a substantial improvement. If applicable, the value of the work must include estimates of the value of the property owner's labor and the value of donated labor and materials.

To make a determination about whether a proposed alteration constitutes a substantial improvement, the cost of the proposed work is to be compared to the market value of the building or structure before the work is started. In order to determine market value, the code official may require the applicant to provide an appraisal or use other methods acceptable to FEMA. For additional guidance, see FEMA P-758.

[BS] 403.3 Existing structural elements carrying gravity load. Any existing gravity load-carrying structural element for which an *alteration* causes an increase in design gravity load of more than 5 percent shall be strengthened, supplemented, replaced or otherwise altered as needed to carry the increased gravity load required by the *International Building Code* for new structures. Any existing gravity load-carrying structural element whose gravity load-carrying capacity is decreased as part of the *alteration* shall be shown to have the capacity to resist the applicable design gravity loads required by the *International Building Code* for new structures.

❖ Wherever an alteration to an existing building is made, the affected members must be assessed to determine their ability to resist the increased forces. Structural element forces may be increased by up to 5 percent, without the need to strengthen or replace that element.

[BS] 403.3.1 Design live load. Where the *alteration* does not result in increased design live load, existing gravity load-carrying structural elements shall be permitted to be evaluated and designed for live loads approved prior to the *alteration*. If the approved live load is less than that required by Section 1607 of the *International Building Code*, the area designed for the nonconforming live load shall be posted with placards of approved design indicating the approved live load. Where the *alteration* does result in increased design live load, the live load required by Section 1607 of the *International Building Code* shall be used.

❖ Where an existing building is undergoing alterations requiring changes to the structural loading, the applicable minimum design loads are those required by the code at the time of the original construction. It is not uncommon for the design live load requirements to change from the time a building is originally designed to when it undergoes a renovation. The live loads used in the original design may have been adequate for the building's initial use and may have been in compliance with all the code requirements that were in effect at that time. Many years and many code changes can impact the structural design parameters and code requirements, which does not mean the existing structural system is inadequate and cannot be used when the building is renovated. It just means that when the live loads used for the design of the existing building are lower than those required by current standards, the design live loads used for the original design must be posted. If the design live loads of the alteration are greater than those used for the existing building, then IBC Section 1607 would apply just as for new construction.

[BS] 403.4 Existing structural elements carrying lateral load. Except as permitted by Section 403.5, where the *alteration* increases design lateral loads in accordance with Section 1609 or 1613 of the *International Building Code*, or where the *alteration* results in a prohibited structural irregularity as defined in ASCE 7, or where the *alteration* decreases the capacity of any existing lateral load-carrying structural element, the structure of the altered building or structure shall be shown to meet the requirements of Sections 1609 and 1613 of the *International Building Code*. For purposes of this section, compliance with ASCE 41, using a Tier 3 procedure and the two-level performance objective in Table 301.1.4.1

for the applicable risk category, shall be deemed to meet the requirements of Section 1613 of the *International Building Code*.

Exception: Any existing lateral load-carrying structural element whose demand-capacity ratio with the *alteration* considered is no more than 10 percent greater than its demand-capacity ratio with the *alteration* ignored shall be permitted to remain unaltered. For purposes of calculating demand-capacity ratios, the demand shall consider applicable load combinations with design lateral loads or forces in accordance with Sections 1609 and 1613 of the *International Building Code*. For purposes of this exception, comparisons of demand-capacity ratios and calculation of design lateral loads, forces and capacities shall account for the cumulative effects of *additions* and *alterations* since original construction.

❖ An existing building or structure that is altered must be carefully evaluated for its ability to withstand earthquake and wind loads. The lateral resistance of an existing building or structure cannot be lessened or pose any undue increase in the fire and life safety hazards of the building or structure. This section of the code gives guidance with respect to the impact of an alteration on the lateral resistance of the structure. This section specifically allows the use of ASCE 41 using a Tier 3 procedure and the two-level performance objectives of Section 301.1.4.1. The application of this procedure is also based on the risk category of the structure. This allowance provides more flexibility for existing buildings. ASCE 41 is specifically focused on the evaluation and rehabilitation of existing structures. See the commentary to Section 301.1.4 for more discussion on the origins and application of the standard.

Alterations that affect existing structural elements to a lesser extent are permitted without requiring the existing structure to comply with the provisions for new structures, as long as the alteration itself complies. The exception allows alterations that increase the demand-capacity ratio on existing lateral load-carrying elements by no more than 10 percent or decrease the lateral resistance of existing structural elements by no more than 10 percent.

[BS] 403.4.1 Seismic Design Category F. Where the portion of the building undergoing the intended alteration exceeds 50 percent of the aggregate area of the building, and where the building is assigned to Seismic Design Category F, the structure of the altered building shall be shown to meet the earthquake design provisions of the *International Building Code*. For purposes of this section, the earthquake loads need not be taken greater than 75 percent of those prescribed in Section 1613 of the *International Building Code* for new buildings of similar occupancy, purpose and location. New structural members and connections required by this section shall comply with the detailing provisions of this code for new buildings of similar structure, purpose and location.

❖ This section is focused specifically on those buildings assigned to Seismic Design Category F, which is the most severe. Buildings in such areas are at a high risk for damage from a seismic event. Therefore, when a building is undergoing more than a 50-percent alteration, it was felt reasonable to also evaluate the lateral force-resisting system regardless of the extent of the structural alterations. This is intended to be consistent with Level 3 alterations in the work area method. The intention is that the alteration area is focused on reconfigured space. This section requires such buildings to be evaluated and analyzed to determine that the building's lateral load-resisting system meets the reduced IBC-level seismic forces. Section 907.4.3 addresses this topic with the work area method.

[BS] 403.5 Bracing for unreinforced masonry parapets upon reroofing. Where the intended alteration requires a permit for reroofing and involves removal of roofing materials from more than 25 percent of the roof area of a building assigned to Seismic Design Category D, E or F that has parapets constructed of unreinforced masonry, the work shall include installation of parapet bracing to resist out-of-plane seismic forces, unless an evaluation demonstrates compliance of such items. For purposes of this section, design seismic forces need not be taken greater than 75 percent of those that would be required for the design of similar nonstructural components in new buildings of similar purpose and location.

❖ The failure of parapets in unreinforced masonry-bearing (URM) wall buildings has been a recurring problem in areas that experience significant earthquakes. Because this poses a very real risk, the code requires these elements to be braced where the seismic hazard is deemed to be relatively high, as reflected in a building's seismic design category. Since the code does not provide the requirements for establishing the seismic design category, IBC Section 1613.3 must be used for this purpose. The code allows this parapet bracing to be designed using 75 percent of the seismic forces of other nonstructural elements of a building for similar purpose and location. It should be noted that this same requirement is found in Section 707.3.1, which allows compliance with Section 301.1.4.2.

An example of the prescribed parapet bracing is illustrated schematically in Commentary Figure 707.3.1. The approach shown—a steel brace anchored to the parapet wall and the roof structure—is a commonly used method of providing lateral support for a parapet. Other methods of strengthening a parapet can be used to mitigate the hazard, provided their use is approved as alternative methods of design and construction under Section 104.11.

Note that only the parapet bracing for seismic loads is required. Some buildings, especially those east of the Rocky Mountains, were not designed for seismic loads. They were only designed to resist wind loads. At the present time, the code does not require a review of the building for seismic loads unless the structure is altered or sustains substantial structural damage.

[BS] 403.6 Wall anchorage for unreinforced masonry walls in major alterations. Where the portion of the building undergoing the intended alteration exceeds 50 percent of the aggregate area of the building, the building is assigned to Seismic Design Category C, D, E or F, and the building's structural system includes unreinforced masonry walls, the alteration work shall include installation of wall anchors at the roof line to resist seismic forces, unless an evaluation demonstrates compliance of existing wall anchorage. For purposes of this section, design seismic forces need not be taken greater than 75 percent of those that would be required for the design of new buildings of similar structure, purpose and location.

❖ This section deals with unreinforced masonry walls in buildings where the alteration affects more than 50 percent of the building. This is intended to be consistent with Level 3 alterations in the work area method. The intention is that the alteration area is focused on reconfigured space.

The wall anchorage system includes the elements within the diaphragm required to develop wall anchorage forces, including: wall anchors, struts, subdiaphragms, cross ties and continuity ties. Note that if a structural evaluation indicates that the existing wall anchorage provided can already resist and transfer these forces, additional wall anchorage is not required. Section 907.4.5 addresses this topic with the work area method.

[BS] 403.7 Bracing for unreinforced masonry parapets in major alterations. Where the portion of the building undergoing the intended alteration exceeds 50 percent of the aggregate area of the building, and where the building is assigned to Seismic Design Category C, D, E or F, parapets constructed of unreinforced masonry shall have bracing installed as needed to resist out-of-plane seismic forces, unless an evaluation demonstrates compliance of such items. For purposes of this section, design seismic forces need not be taken greater than 75 percent of those that would be required for the design of similar nonstructural components in new buildings of similar purpose and location.

❖ The provisions found in this section are essentially the same as those found in Section 403.5. The reason for a duplication of the provision is related to the application of the provisions. Section 403.5 only applies if reroofing occurs and is only applicable to Seismic Design Categories D, E and F. This section also applies to Seismic Design Category C. These provisions will apply in buildings where the alteration affects more than 50 percent of the building. This is intended to be consistent with Level 3 alterations in the work area method. The intention is that the alteration area is focused on reconfigured space. It is felt that this bracing is a critical life safety measure that should be taken during larger alterations. Section 907.4.6 addresses this topic in the work area method.

[BS] 403.8 Roof diaphragms resisting wind loads in high-wind regions. Where the intended alteration requires a permit for reroofing and involves removal of roofing materials from more than 50 percent of the roof diaphragm of a building or section of a building located where the ultimate design wind speed is greater than 115 mph (51 m/s) in accordance with Figure 1609.3(1) of the *International Building Code* or in a special wind region as defined in Section 1609 of the *International Building Code*, roof diaphragms, connections of the roof diaphragm to roof framing members, and roof-to-wall connections shall be evaluated for the wind loads specified in Section 1609 of the *International Building Code*, including wind uplift. If the diaphragms and connections in their current condition are not capable of resisting at least 75 percent of those wind loads, they shall be replaced or strengthened in accordance with the loads specified in Section 1609 of the *International Building Code*.

❖ The removal of roofing provides an opportunity to inspect a portion of the structure that is otherwise concealed. In reroofing operations where more than 50 percent of the roof covering is removed and where the building is located where the ultimate design wind speed, V_{ult} [determined in accordance with IBC Table 1609.3(1)], is greater than 115 mph or in a special wind region, a roof diaphragm that is a part of the main windforce-resisting system is required to be evaluated for adequate strength to resist and transfer the wind loads in the IBC, including uplift loads. If the roof diaphragm is not able to resist 75 percent of the wind loads specified in the IBC, such connections are required to be strengthened or replaced. The 75-percent reduction is consistent with other structural allowances for existing conditions. Note that this section spells out which aspects of the roof need to be evaluated and possibly replaced and strengthened. This requirement is the same as that found in Section 707.3.2 in the work area method. These include the following:

- Roof diaphragms.
- Connections of the roof diaphragm to roof framing members.
- Roof-to-wall connections.

[BS] 403.9 Voluntary seismic improvements. *Alterations* to existing structural elements or *additions* of new structural elements that are not otherwise required by this chapter and are initiated for the purpose of improving the performance of the seismic force-resisting system of an existing structure or the performance of seismic bracing or anchorage of existing nonstructural elements shall be permitted, provided that an engineering analysis is submitted demonstrating the following:

1. The altered structure and the altered nonstructural elements are no less conforming to the provisions of the *International Building Code* with respect to earthquake design than they were prior to the *alteration*.

2. New structural elements are detailed as required for new construction.

3. New or relocated nonstructural elements are detailed and connected to existing or new structural elements as required for new construction.

4. The *alterations* do not create a structural irregularity as defined in ASCE 7 or make an existing structural irregularity more severe.

❖ This provision addresses the issue of upgrading existing structures voluntarily for improved seismic performance. It does not apply to situations where other code sections trigger full compliance with the code. Otherwise, it allows an owner to initiate an improvement to the seismic force-resisting system to the extent that it is viable to do so and provided the required engineering analysis is furnished. The intent is to encourage building owners to initiate upgrades to seismic systems that are considered prudent without making them cost prohibitive.

403.10 Smoke alarms. Individual sleeping units and individual dwelling units in Group R and I-1 occupancies shall be provided with smoke alarms in accordance with Section 1103.8 of the *International Fire Code*.

❖ This section is consistent with Section 402.5 but applies to alterations. As with Section 402.5, a reference to IFC Section 1103.8 is made for the requirements for smoke alarms in existing individual sleeping units and individual dwelling units in Group R and I-1 occupancies.

In the past, the code limited the level of building fire protection by isolating such requirements to work areas only. The structure of the IFC is that the retroactive requirements apply to all buildings, whether or not those buildings are undergoing an alteration.

403.11 Refuge areas. Where alterations affect the configuration of an area utilized as a refuge area, the capacity of the refuge area shall not be reduced below that required in Sections 403.11.1 through 403.11.3.

❖ This section was written with the concern that during alterations the capacity and arrangement of refuge areas, required for occupancies that limit movement of occupants or where occupants are incapable of self-preservation, may be reduced. Refuge areas are part of the defend-in-place strategy and are designed with a particular capacity in mind in each smoke compartment. The refuge areas created by horizontal exits also must be maintained as they use the same concept of moving occupants horizontally.

403.11.1 Smoke compartments. In Group I-2 and I-3 occupancies, the required capacity of the refuge areas for smoke compartments in accordance with Sections 407.5.1 and 408.6.2 of the *International Building Code* shall be maintained.

❖ The IBC requires that Group I-2 and I-3 occupancies provide the largest occupant load of the adjoining compartment. In Group I-2 occupancies, the criteria used to determine the capacity are 30 net square feet for each care recipient confined to a bed or stretcher and not less than 6 square feet for each ambulatory care recipient not confined to bed or stretcher and for other occupants. For Group I-3 occupancies, the IBC requires that 6 square feet per occupant shall be provided on both sides of the smoke barrier forming the smoke compartment.

403.11.2 Ambulatory care. In ambulatory care facilities required to be separated by Section 422.2 of the *International Building Code*, the required capacity of the refuge areas for smoke compartments in accordance with Section 422.4 of the *International Building Code* shall be maintained.

❖ Ambulatory care facilities are similar to Group I-2 occupancies and contain occupants that are incapable of self-preservation because of the procedures and surgeries done at such facilities. Therefore, the criteria are similar to that of Group I-2 occupancies and require 30 square feet for each nonambulatory care recipient. This area must be provided in the corridors, care recipient rooms, treatment rooms, lounge or dining areas and other low-hazard areas in the smoke compartment. Note that only ambulatory care facilities greater than 10,000 square feet require smoke compartments.

403.11.3 Horizontal exits. The required capacity of the refuge area for horizontal exits in accordance with Section 1026.4 of the *International Building Code* shall be maintained.

❖ This section is more general and addresses any occupancies using horizontal exits. Horizontal exits allow a similar concept to smoke compartments where occupants simply exit through doors into an adjoining area of the building separated by a fire barrier or fire wall. Section 1026.4 requires 3 square feet for each occupant. There are several specific exceptions for Group I-2 and I-3 occupancies that correlate with the criteria noted in Section 403.11.1

SECTION 404
REPAIRS

404.1 General. Buildings and structures, and parts thereof, shall be repaired in compliance with Sections 401.2 and 404. Work on nondamaged components that is necessary for the required *repair* of damaged components shall be considered part of the *repair* and shall not be subject to the requirements for *alterations* in this chapter. Routine maintenance required by Section 401.2, ordinary repairs exempt from permit in accordance with Section 105.2, and abatement of wear due to normal service conditions shall not be subject to the requirements for repairs in this section.

❖ The scope of this section differentiates repairs from alterations and clarifies that routine maintenance and any repairs not requiring permits would not have to meet these requirements.

In general, the section provides a logical method for evaluating damage and identifying cases where upgrades are warranted.

The provisions would require structural improvements meeting "the code for new structures," in certain cases, in the event of damage due to fire, structural overload, settlement, natural hazard or any other cause, no matter how extensive or disproportionate the damage. This section identifies conditions of damage that should warrant improvements to the structural system for purposes of increasing safety.

[BS] 404.2 Substantial structural damage to vertical elements of the lateral force-resisting system. A building that has sustained *substantial structural damage* to the vertical elements of its lateral force-resisting system shall be evaluated and repaired in accordance with the applicable provisions of Sections 404.2.1 through 404.2.3.

Exceptions:

1. Buildings assigned to Seismic Design Category A, B or C whose substantial structural damage was not caused by earthquake need not be evaluated or rehabilitated for load combinations that include earthquake effects.

2. One- and two-family dwellings need not be evaluated or rehabilitated for load combinations that include earthquake effects.

❖ This section provides requirements that apply where the damage threshold regarding the extent of the damage to the vertical elements of the lateral-force-resisting system in any story is exceeded. Substantial structural damage to the lateral system triggers the evaluation of the entire building for wind and seismic loads (see Section 404.2.1). The emphasis is placed on vertical elements, such as walls and columns, rather than horizontal elements, because the vertical elements of the lateral force-resisting system primarily determine the structure's response, particularly to earthquakes.

There are two exceptions that exempt certain combinations of buildings: seismic risk and damage from triggered seismic upgrades. Basic repair—that is, restoring the predamaged condition—is still required even for the exceptions outlined.

Exception 1 is for buildings in areas of low or moderate seismicity (Seismic Design Category A, B or C), where the damage was not caused by an earthquake and therefore would not be required to be evaluated or rehabilitated for load combinations that include earthquake effects. Where earthquakes are rare, it serves no significant public purpose to trigger seismic upgrades following damage caused by fire, collision, wind, etc.

Exception 2 is for one- and two-family dwellings, where the public risk is especially low even though the damage may be associated with earthquake effects. Similar exceptions are found in Section 606.2.2.

[BS] 404.2.1 Evaluation. The building shall be evaluated by a *registered design professional*, and the evaluation findings shall be submitted to the *building official*. The evaluation shall establish whether the damaged building, if repaired to its predamage state, would comply with the provisions of the *International Building Code* for wind and earthquake loads.

Wind loads for this evaluation shall be those prescribed in Section 1609 of the *International Building Code*. Earthquake loads for this evaluation, if required, shall be permitted to be 75 percent of those prescribed in Section 1613 of the *International Building Code*. Alternatively, compliance with ASCE 41, using the performance objective in Table 301.1.4.2 for the applicable risk category, shall be deemed to meet the earthquake evaluation requirement.

❖ The extent of repairs is based on an evaluation prepared by a registered design professional. Generally, the code's approach is that complete structural upgrades should be relatively rare. In this section, a structural upgrade is only triggered by substantial structural damage to the lateral system, and only when the evaluation shows that the predamaged building was substandard.

This provision refers to IBC Sections 1609 and 1613 for wind and earthquake loads, respectively. It permits reduced earthquake loading for both the evaluation and any required seismic rehabilitation to recognize that existing buildings cannot be expected to perform as well as newer buildings. This concept of using reduced earthquake loading is consistent with FEMA standards, as well as the legacy building codes. This section specifically allows the use of ASCE 41 using the performance objectives in Table 301.1.4.2. The application of this procedure is also based on the risk category of the structure. This allowance provides more flexibility for existing buildings. ASCE 41 is specifically focused on the evaluation and rehabilitation of existing structures. See the commentary to Section 301.1.4 for more discussion on the origins and application of the standard

[BS] 404.2.2 Extent of repair for compliant buildings. If the evaluation establishes compliance of the predamage building in accordance with Section 404.2.1, then repairs shall be permitted that restore the building to its predamage state.

❖ Where the evaluation establishes that the predamaged building meets the structural provisions of the code as provided for in Section 404.2.1, then the repairs may be limited to a restoration of the structural components based on the condition prior to the damage. This is similar to Section 606.2.2.2 for repairs under the work area method.

[BS] 404.2.3 Extent of repair for noncompliant buildings. If the evaluation does not establish compliance of the predamage building in accordance with Section 404.2.1, then the building shall be rehabilitated to comply with applicable provisions of the *International Building Code* for load combinations that include wind or seismic loads. The wind loads for the repair shall be as required by the building code in effect at the time of original construction, unless the damage was caused by wind, in which case the wind loads shall be as required by the *International Building Code*. Earthquake loads for this rehabilitation design shall be those required for the design of the predamage building, but not less than 75 percent of those prescribed in Section 1613 of the *International Building Code*. New structural members and connections required by this rehabilitation design shall comply with the detailing provisions of the *International Building Code* for new buildings of similar structure, purpose and location.

Alternatively, compliance with ASCE 41, using the performance objective in Table 301.1.4.2 for the applicable risk category, shall be deemed to meet the earthquake rehabilitation requirement.

❖ If the evaluation of the building, in accordance with Section 404.2.1, shows that, in the hypothetically repaired condition, the building would not comply with the established requirements, then the building must be rehabilitated as described in this section. The general requirement is to comply with the IBC load combinations. These load combinations establish the required strength of structural members.

The effects of wind and seismic loads warrant special consideration. In determining the level of compliance for repairs to buildings that have sustained substantial structural damage, it is important to determine if wind forces have caused that damage. If so, it is considered prudent to require repairs to wind-damaged buildings to comply with the wind load provisions from the current code. Seismic forces for the rehabilitation can be those required by the building code in effect at the time of the building's construction, but this may not be less than the reduced seismic force level, as described in Section 404.2.1. Section 404.2.1 and this section also allow the application of ASCE 41 (see commentary, Section 404.2.1).

[BS] 404.3 Substantial structural damage to gravity load-carrying components. Gravity load-carrying components that have sustained *substantial structural damage* shall be rehabilitated to comply with the applicable provisions of the *International Building Code* for dead and live loads. Snow loads shall be considered if the substantial structural damage was caused by or related to snow load effects. Existing gravity load-carrying structural elements shall be permitted to be designed for live loads approved prior to the damage. If the approved live load is less than that required by Section 1607 of the *International Building Code*, the area designed for the nonconforming live load shall be posted with placards of *approved* design indicating the *approved* live load. Nondamaged gravity load-carrying components that receive dead, live or snow loads from rehabilitated components shall also be rehabilitated or shown to have the capacity to carry the design loads of the *rehabilitation* design. New structural members and connections required by this rehabilitation design shall comply with the detailing provisions of the *International Building Code* for new buildings of similar structure, purpose and location.

❖ Substantial structural damage to gravity load-carrying elements, such as columns or bearing walls, must be repaired so that these members are adequate to resist the dead and live loads in accordance with current code requirements, as must other elements of the load path. Snow loads must also be included where the substantial damage is associated with the effects of snow load.

[BS] 404.3.1 Lateral force-resisting elements. Regardless of the level of damage to vertical elements of the lateral force-resisting system, if *substantial structural damage* to gravity load-carrying components was caused primarily by wind or earthquake effects, then the building shall be evaluated in accordance with Section 404.2.1 and, if noncompliant, rehabilitated in accordance with Section 404.2.3.

Exceptions:

1. One- and two-family dwellings need not be evaluated or rehabilitated for load combinations that include earthquake effects.

2. Buildings assigned to Seismic Design Category A, B or C whose substantial structural damage was not caused by earthquake need not be evaluated or rehabilitated for load combinations that include earthquake effects.

❖ In determining the extent of repairs to these gravity load-carrying elements that are not part of the lateral force-resisting system, it is important to determine if wind or earthquakes have caused the structural damage. If the substantial structural damage is caused by wind or earthquake, the lateral system must be checked even if no damage is apparent. Where this is the case, then the structure must be evaluated in accordance with Section 404.2.1.

There are two exceptions that acknowledge that applying this requirement in some cases is excessive and may have the effect of discouraging or delaying certain repairs by imposing the additional costs of seismic upgrade. Exception 1 exempts one- and two-family dwellings, where the risk to the public of poor earthquake performance is especially low. Exception 2 exempts buildings in areas of low or moderate seismicity, where the damage was caused by something other than an earthquake. Where earthquakes are rare, it serves no significant public purpose to trigger seismic upgrades following damage caused by fire, collision, wind and other events. Basic repair—that is, restoring to the predamaged condition—is still required.

[BS] 404.4 Less than substantial structural damage. For damage less than *substantial structural damage*, repairs shall be allowed that restore the building to its predamage state. New structural members and connections used for this *repair* shall comply with the detailing provisions of the *International Building Code* for new buildings of similar structure, purpose and location.

❖ For damage that is not deemed to be substantial structural damage, repairs are allowed that restore the building to its predamaged state. This is consistent with Section 606.2.1 for the work area method. New structural members and connections used for these repairs are still required to comply with the detailing provisions for new buildings of similar materials, purpose and location.

[BS] 404.5 Flood hazard areas. For buildings and structures in *flood hazard areas* established in Section 1612.3 of the *International Building Code*, or Section R322 of the *International Residential Code*, as applicable, any repair that constitutes *substantial improvement* or repair of *substantial damage* of the existing structure shall comply with the flood design requirements for new construction, and all aspects of the existing structure shall be brought into compliance with the requirements for new construction for flood design.

For buildings and structures in flood hazard areas established in Section 1612.3 of the *International Building Code*, or Section R322 of the *International Residential Code*, as applicable, any repairs that do not constitute *substantial improvement* or repair of *substantial damage* of the existing structure are not required to comply with the flood design requirements for new construction.

❖ Reduction in exposure to flood hazards, including exposure of older buildings, is one of the purposes for regulating flood plain development. Damaged buildings or structures that are located in flood hazard areas are to be brought into compliance with the flood-resistant provisions of IBC Section 1612 or IRC Section R322, as applicable, where the cost of repairs necessary to restore damage to its predamaged condition (even if less work is actually proposed) exceeds a certain value. See Section 202 for the definitions of "Substantial improvement" and "Substantial damage."

Section 105.3 requires the applicant to state the valuation of proposed work, which is to include the total value of work, including materials and labor. If the proposed work will be performed on buildings in flood hazard areas, the code official must determine whether the proposed work constitutes substantial improvement or repair of substantial damage. If applicable, the value of work must include estimates of the value of the property owner's labor and the value of donated labor and materials.

To make a determination about whether proposed repairs constitute repair of substantial damage, the cost of the proposed work is to be compared to the market value of the building or structure before the work is started. In order to determine market value, the code official may require the applicant to provide an appraisal or use other methods acceptable to FEMA. For additional guidance, see FEMA P-758, which advises that communities with many buildings in flood hazard areas become familiar with software developed to help make substantial damage determinations, especially in the post-disaster period. FEMA P-784, *Substantial Damage Estimator*, includes the software, user's manual and workbook.

SECTION 405
FIRE ESCAPES

405.1 Where permitted. Fire escapes shall be permitted only as provided for in Sections 405.1.1 through 405.1.4.

❖ Sections 405.1.1 through 405.1.4 address the current requirements for the use of exterior fire escapes. Their use and features, as defined in this section, should not be confused with the requirements for exterior exit stairways as defined in IBC Section 1027.

The use of exterior fire escapes as a means of egress was popular in building designs of the past. They were used for economy of construction and to "save" usable space in the buildings they served. Being located outside the building, they were also considered safer than the unenclosed interior exit stairways of the time. Over the years, exterior fire escapes lost their appeal for many reasons:

1. Fire escapes were never an integral part of the building's design; they were a necessary appendage hung from or attached to the building wall.

2. Fire escapes were most commonly constructed of cast-iron or steel, both of which required a high degree of maintenance to protect the corrodible materials from the effects of weather.

3. Because fire escapes are "open structures," they are subject to icing during the winter months. This makes them dangerous to use and, in extreme cases, they can become completely unusable.

4. People with a fear of heights can find exterior fire escapes difficult and sometimes impossible to use.

5. Fire escapes were an unpleasant sight and an unwanted element in the architectural design.

The use of exterior fire escapes is all but obsolete except for existing buildings with a clear deficiency in the means of egress that cannot be reasonably rectified in other ways.

This section indicates that fire escapes are permitted only on existing buildings. New fire escapes may be installed on existing buildings only where exterior stairs cannot be used. Section 405.2 addresses the location of fire escapes and Section 405.3 addresses the construction of fire escapes.

See also the requirements for fire escapes in IFC Section 1104.16. Note that these provisions are slightly different than those of Section 405.

405.1.1 New buildings. Fire escapes shall not constitute any part of the required means of egress in new buildings.

❖ Because of the inherent hazards and lack of dependability, open exterior fire escapes are not permitted as part of the required means of egress in new construction.

405.1.2 Existing fire escapes. Existing fire escapes shall continue to be accepted as a component in the means of egress in *existing buildings* only.

❖ Exterior fire escapes in existing buildings are acceptable as a component of the required means of egress because, in most cases, it would be physically impractical and economically prohibitive to retrofit older buildings with interior or exterior exit stairways that comply with all the code requirements for new construction.

405.1.3 New fire escapes. New fire escapes for *existing buildings* shall be permitted only where exterior stairways cannot be utilized due to lot lines limiting stairway size or due to the sidewalks, alleys or roads at grade level. New fire escapes shall not incorporate ladders or access by windows.

❖ Continuing corrosion of existing exterior fire escapes, as well as other factors that may affect the safety of the structures, often makes the replacement of these fire escapes necessary. The code official may determine that the means of egress in an existing building is unsafe, requiring some corrective action, which may be the installation of a new exterior fire escape. This section permits new exterior fire escapes for existing buildings where exterior stairways conforming to the requirements of IBC Section 1027 cannot be used because of lot lines that limit stairway size or encroachments on sidewalks, alleys or roads at grade level.

Because it was the accepted practice of the time, access to exterior fire escapes was quite often through windows located in corridors or individual dwelling units or offices. This practice is not permitted when replacement exterior fire escapes are used. The building must be altered to provide proper exit access to the fire escape.

The code does not permit the use of ladders as a component in the means of egress except as allowed in the IBC for control rooms or elevated facility observation rooms for Group I-3 occupancies; therefore, ladders can neither be used as a component part of an exterior fire escape, nor to access an exterior fire escape.

405.1.4 Limitations. Fire escapes shall comply with this section and shall not constitute more than 50 percent of the required number of exits nor more than 50 percent of the required exit capacity.

❖ For reasons of overall life safety, exterior fire escapes may be used as a component of the means of egress only when the total number used does not exceed one-half the required number of exits and more than one-half the total required exit capacity.

405.2 Location. Where located on the front of the building and where projecting beyond the building line, the lowest landing shall be not less than 7 feet (2134 mm) or more than 12 feet (3658 mm) above grade, and shall be equipped with a counterbalanced stairway to the street. In alleyways and thoroughfares less than 30 feet (9144 mm) wide, the clearance under the lowest landing shall be not less than 12 feet (3658 mm).

❖ In the past, exterior fire escapes were allowed to project beyond the property line and extend over sidewalks, alleyways and roads. Where such conditions are still accepted by the code (see Sections 405.1.2 and 405.1.3), the clearance to the lowest landing from grade level must be at least 7 feet (2134 mm), and not more than 12 feet (3658 mm) for exterior fire escapes located in front of the building. In alleyways and thoroughfares less than 30 feet (9144 mm) wide, the clearance under the lowest landing must be at least 12 feet (3658 mm). To facilitate these clearance requirements, exterior fire escapes are usually equipped with a counterbalanced stairway that retracts to a horizontal position when the fire escape is not in use.

405.3 Construction. The fire escape shall be designed to support a live load of 100 pounds per square foot (4788 Pa) and shall be constructed of steel or other approved *noncombustible materials*. Fire escapes constructed of wood not less than nominal 2 inches (51 mm) thick are permitted on buildings of Type V construction. Walkways and railings located over or supported by combustible roofs in buildings of Type III and IV construction are permitted to be of wood not less than nominal 2 inches (51 mm) thick.

❖ Traditionally and typically, exterior fire escapes are constructed of cast iron or steel, although the code does permit exterior fire escapes to be constructed of other noncombustible materials. The use of nominal 2-inch (51 mm) wood is permitted in Type V construction. Nominal 2-inch (51 mm) wood may also be used in Type III and IV construction when the exterior fire escape is supported by wood construction permitted by IBC Table 601.

405.4 Dimensions. Stairways shall be at least 22 inches (559 mm) wide with risers not more than, and treads not less than, 8 inches (203 mm) and landings at the foot of stairways not less than 40 inches (1016 mm) wide by 36 inches (914 mm) long, located not more than 8 inches (203 mm) below the door.

❖ The IBC's normal dimensions for interior and exterior exit stairways used in the means of egress do not apply to exterior fire escapes. The dimensional standard for exterior fire escapes has been in place for many years; thus, many exterior fire escapes in use were designed to comply with this standard. Many of these fire escapes are installed in places where it would be difficult, and often impossible, to retrofit a new fire escape if a new set of dimensional standards were employed. For this reason, it is judged appropriate to maintain the current standard. The minimum tread, riser, stair-width and landing-size dimensions for fire escape construction remain unchanged from the past.

405.5 Opening protectives. Doors and windows along the fire escape shall be protected with $^3/_4$-hour opening protectives.

❖ Safe exit access and exits are required for occupants using an exterior fire escape as a means of egress. Door and window openings in exterior walls adjacent to and along the path of travel of the exterior fire escape must be protected from the interior of the building with not less than a $^3/_4$-hour protective. This is similar to the requirements for protection of openings on interior exit stairways and ramps in IBC Section 1023.7 and exterior exit stairways and ramps in IBC Section 1027.6.

Where more-restrictive opening protectives are required by other sections elsewhere in the code, the more-restrictive protection must be used. The $^3/_4$-hour requirement for the exterior fire escape is the minimum when no other considerations are present (see IBC Tables 601, 602, 705.8 and 716.5).

SECTION 406
GLASS REPLACEMENT AND REPLACEMENT WINDOWS

406.1 Replacement glass. The installation or replacement of glass shall be as required for new installations.

❖ The technical provisions for glass and glazing are detailed in IBC Chapter 24. All technical requirements of the code for glass and glazing that apply to new construction also apply to all additions or alterations to existing buildings. There are no exceptions to this requirement. When glazing in an existing building is replaced or relocated within the same building, it must comply with current standards.

406.2 Replacement window opening control devices. In Group R-2 or R-3 buildings containing dwelling units, window opening control devices complying with ASTM F2090 shall be installed where an existing window is replaced and where all of the following apply to the replacement window:

1. The window is operable;
2. The window replacement includes replacement of the sash and the frame;
3. The top of the sill of the window opening is at a height less than 36 inches (915 mm) above the finished floor;
4. The window will permit openings that will allow passage of a 4-inch-diameter (102 mm) sphere when the window is in its largest opened position; and
5. The vertical distance from the top of the sill of the window opening to the finished grade or other surface below, on the exterior of the building, is greater than 72 inches (1829 mm).

The window opening control device, after operation to release the control device allowing the window to fully open, shall not reduce the minimum net clear opening area of the window unit to less than the area required by Section 1030.2 of the *International Building Code*.

Exceptions:

1. Operable windows where the top of the sill of the window opening is located more than 75 feet (22 860 mm) above the finished grade or other surface below, on the exterior of the room, space or building, and that are provided with window fall prevention devices that comply with ASTM F2006.
2. Operable windows with openings that are provided with window fall prevention devices that comply with ASTM F2090.

❖ This section requires that when windows are replaced, window opening control devices be pro-

vided to protect children from falls. There are several conditions that must be met before this section becomes applicable. The first condition is whether or not the window is even operable. If the window is not operable, the concern for safety from falling is not present. The second condition relates to whether the entire window is being replaced, including the sash and frame. If it is simply a repair of a pane of glass, the requirement would not be applicable. Item 3 addresses the specific location of the opening on the inside of the building [see Commentary Figure 702.4 (1)]. Item 4 focuses these provisions on not only operable windows but also windows with openings through which a child could actually fit. This criterion of a 4-inch-diameter sphere is used with guards in IBC Chapter 10 to avoid children fitting through. Finally, Item 5 addresses whether the window is high enough from the ground to be of concern for injury from falling [see Commentary Figure 702.4(2)]. This section applies to the following occupancies:

- Group R-2 occupancies containing dwelling units.
- Group R-3 occupancies containing dwelling units.

Note that the work area method also addresses one- and two-family dwellings. The control device cannot reduce to less than that required in IBC Section 1030.2. Section 702.4 also addresses this issue for the work area method. Note that the figures are located in Section 702 and simply referenced in this section.

406.3 Replacement window emergency escape and rescue openings. Where windows are required to provide *emergency escape* and *rescue openings* in Group R-2 and R-3 occupancies, replacement windows shall be exempt from the requirements of Sections 1030.2, 1030.3 and 1030.5 of the *International Building Code* provided the replacement window meets the following conditions:

1. The replacement window is the manufacturer's largest standard size window that will fit within the existing frame or existing rough opening. The replacement window shall be permitted to be of the same operating style as the existing window or a style that provides for an equal or greater window opening area than the existing window.
2. The replacement of the window is not part of a change of occupancy.

❖ This section focuses on how to address the replacement of windows intended to be emergency escape and rescue openings. More specifically, Group R-2 and R-3 occupancies in new construction require specifically sized windows for escape and rescue for certain windows in these structures. In existing buildings, meeting these requirements can be difficult. This section provides the necessary flexibility to these existing buildings to provide the largest window that can fit into the current opening without requiring full compliance with the IBC. There are three criteria that must be met. The first is that the window is the largest size the manufacturer can provide for that opening (the existing frame or rough opening). The second aspect is focused on the operating style of the existing window. It is permitted to remain the same but needs to provide equal or greater opening area. Finally, this allowance is not permitted if these windows are associated with a change of occupancy.

This section also provides permission for the use of window opening control devices as required in Section 406.2. The focus of Section 406.2 is on fall safety, whereas this section is focused on emergency escape. This provision avoids conflict between the two requirements.

Section 702.5 also addresses this issue for the work area method. Note that the figures are located in Section 702 and simply referenced in this section.

SECTION 407
CHANGE OF OCCUPANCY

407.1 Conformance. No change shall be made in the use or occupancy of any building unless such building is made to comply with the requirements of the *International Building Code* for the use or occupancy. Changes in use or occupancy in a building or portion thereof shall be such that the existing building is no less complying with the provisions of this code than the existing building or structure was prior to the change. Subject to the approval of the building official, the use or occupancy of *existing buildings* shall be permitted to be changed and the building is allowed to be occupied for purposes in other groups without conforming to all of the requirements of this code for those groups, provided the new or proposed use is less hazardous, based on life and fire risk, than the existing use.

> **Exception:** The building need not be made to comply with the seismic requirements for a new structure unless required by Section 407.4.

❖ A change of occupancy in an existing structure may change the level of inherent hazards that the code was initially intended to address.

Regardless of whether the change is to an occupancy considered to be more or less hazardous, this section applies the provisions of the code for new construction to an existing structure having a new occupancy. This is done so that the applicable code requirements adequately address the specific hazards of the new occupancy. For example, a change from an existing mercantile occupancy to a business occupancy renders all Group B provisions applicable to all portions of the structure where the occupancy has changed.

This section is one of the most frequently used provisions in the code for application to existing structures, since the occupancy in a building or structure is subject to change during the life of the building.

Note that this section does provide some flexibility to those buildings where the change of occupancy is to a less hazardous use. This requires some technical judgement as to what is considered less hazardous. Chapter 10, dealing with change of occupancy in the work area method, more specifically ranks these risks when addressing change of occupancy classification requirements.

The exception clarifies that seismic upgrades are only necessary as required by Section 407.4. Section 407.4 also focuses on those changes of occupancy that are to a higher risk category.

407.1.1 Change in the character of use. A change in occupancy with no change of occupancy classification shall not be made to any structure that will subject the structure to any special provisions of the applicable *International Codes*, without approval of the *building official*. Compliance shall be only as necessary to meet the specific provisions and is not intended to require the entire building be brought into compliance.

❖ This section focuses on buildings where the occupancy classification has not changed but its use has changed such that special use requirements apply. For example, the building is classified as a Group B office building and now is being converted to an ambulatory care facility. Ambulatory care facilities are Group B occupancies; however, they also have special requirements that must be met to address the risks found in such facilities. This section provides the authority to the code official to address these situations.

407.2 Certificate of occupancy. A certificate of occupancy shall be issued where it has been determined that the requirements for the new occupancy classification have been met.

❖ An existing building that has been classified into a new occupancy group must receive a certificate of occupancy before tenancy. The code requirements for one occupancy group are not always the same as those for the new occupancy group. The new occupancy must be inspected to verify that all the applicable code requirements have been met.

407.3 Stairways. An existing stairway shall not be required to comply with the requirements of Section 1011 of the *International Building Code* where the existing space and construction does not allow a reduction in pitch or slope.

❖ A stairway in an existing building does not have to be modified to comply with the dimensional provisions for width, tread and riser dimension, or landing size for new stairway construction. This provision is for the design of a particular stairway and is aimed at addressing that existing stairway only. This section, which is essentially an exception, is not intended to eliminate any of the other technical provisions that might apply to an occupancy group, such as means of egress, exit capacity, etc.

For example, when a building is altered or there is a change of tenant that results in a change in occupancy, the existing stairway is permitted to remain as it was prior to the alteration or change of occupancy. The means of egress capacity, however, must comply with the requirements for the new occupancy. The capacity may not be lessened to match the existing stairway if it is not sufficient to provide the capacity required by the new occupancy. An additional stairway or another approved means of egress must be provided to comply with the means of egress requirements for the new occupancy as indicated in Chapter 10.

[BS] 407.4 Structural. When a *change of occupancy* results in a structure being reclassified to a higher risk category, the structure shall conform to the seismic requirements for a new structure of the higher risk category. For purposes of this section, compliance with ASCE 41, using a Tier 3 procedure and the two-level performance objective in Table 301.1.4.1 for the applicable risk category, shall be deemed to meet the requirements of Section 1613 of the *International Building Code*.

Exceptions:

1. Specific seismic detailing requirements of Section 1613 of the *International Building Code* for a new structure shall not be required to be met where the seismic performance is shown to be equivalent to that of a new structure. A demonstration of equivalence shall consider the regularity, overstrength, redundancy and ductility of the structure.

2. When a change of use results in a structure being reclassified from Risk Category I or II to Risk Category III and the structure is located where the seismic coefficient, S_{DS}, is less than 0.33, compliance with the seismic requirements of Section 1613 of the *International Building Code* is not required.

❖ An existing building that undergoes a change of use and occupancy that places the building in a higher risk category must be evaluated for its seismic resistance. For this provision, the code is referring to risk categories listed in IBC Table 1604.5, which must be used to determine whether an existing building should be reclassified to a higher risk category. It is important that an existing building that previously contained ordinary uses and occupancies must be able to meet current code requirements relative to seismic loading if it will be used as an "essential facility." This section specifically allows the use of ASCE 41 using a Tier 3 procedure and the two-level performance objectives of Section 301.1.4.1. The application of this procedure is also based on the risk category of the structure. This allowance provides more flexibility for existing buildings. ASCE 41 is specifically focused on the evaluation and rehabilitation of existing structures. See the commentary to Section 301.1.4 for more discussion on the origins and application of the standard.

Without Exception 1, it would be impossible in many instances to make an existing structure comply with the seismic requirements for a new structure.

Exception 1 provides guidance for the code official and designer of areas that need to be investigated when compliance with the seismic requirements set forth in the code cannot be accomplished in a traditional manner.

Exception 2 has its origins in ASCE 7. The purpose is to permit a change from Risk Category I or II to Risk Category III for structures subjected to low earthquake accelerations without meeting the requirements for a new structure.

SECTION 408
HISTORIC BUILDINGS

408.1 Historic buildings. The provisions of this code that require improvements relative to a building's existing condition or, in the case of repairs, that require improvements relative to a building's predamage condition, shall not be mandatory for historic buildings unless specifically required by this section.

❖ This section provides an exception from code requirements when the building in question has historic value. The most important criterion for the application of this section is that the building must be essentially accredited as being of historic significance by a qualified party or agency (see the definition of "Historic building" in Section 202). Usually, this is done by a state or local authority after careful review of the historical value of the building. Most, if not all, states have such authorities, as do many local jurisdictions. The agencies with such authority can be located at the state or local government level or through the local chapter of the American Institute of Architects (AIA). This section also clarifies that the code generally does not apply unless specifically required by this section. There are two subsections. The first deals with addressing distinct hazards to life. The second deals with flood hazard areas and necessary upgrades as applicable. Note, as will be discussed, most historic buildings need not comply with the flood hazard requirements.

408.2 Life safety hazards. The provisions of this code shall apply to historic buildings judged by the building official to constitute a distinct life safety hazard.

❖ This section provides the authority to address distinct hazards to life safety. This can address issues such as structural instability or blocked or severely limited exits. Once a distinct hazard is established, the provisions of the code will apply. The extent of application is at the discretion of the code official.

[BS] 408.3 Flood hazard areas. Within flood *hazard areas* established in accordance with Section 1612.3 of the *International Building Code*, or Section R322 of the *International Residential Code*, as applicable, where the work proposed constitutes *substantial improvement*, the building shall be brought into compliance with Section 1612 of the *Interna-*

tional Building Code, or Section R322 of the *International Residential Code*, as applicable:

> **Exception:** *Historic buildings* need not be brought into compliance that are:
>
> 1. Listed or preliminarily determined to be eligible for listing in the National Register of Historic Places;
>
> 2. Determined by the Secretary of the U.S. Department of Interior as contributing to the historical significance of a registered historic district or a district preliminarily determined to qualify as an historic district; or
>
> 3. Designated as historic under a state or local historic preservation program that is approved by the Department of Interior.

❖ With respect to provisions applicable to buildings in flood hazard areas, the discretion given in Section 408.1 allowing the code official to waive the requirements for historic structures extends only to historic structures that meet the specific limitations in the exception to this section. Location in a historic district does not qualify a structure as historic. Care must be taken to ensure that work on such historic structures will not cause them to lose their continued listing or designation. If improvements or repairs cause a structure to lose its listing or designation as a historic building, then the code official must enforce the substantial improvement requirements. Owners of historic structures should consider measures to reduce flood damage to the extent practical. For additional guidance, see FEMA P-467-2, *Floodplain Management Bulletin on Historic Structures,* and FEMA P-758.

SECTION 409
MOVED STRUCTURES

409.1 Conformance. Structures moved into or within the jurisdiction shall comply with the provisions of the *International Building Code* for new structures.

❖ Moved structures are required to comply with the provisions applicable to new construction, which is generally intended to mean the IBC. The moved structure may comply with the alternative provisions of Chapter 14 instead of the code requirements for new structures, which may be particularly useful if the moved structure is older than the effective date of the adoption of the building codes in the jurisdiction. The fire separation distance of the moved structure must comply with requirements for new structures even if the compliance alternative provisions of Chapter 14 are used to meet the code requirements.

SECTION 410
ACCESSIBILITY FOR EXISTING BUILDINGS

410.1 Scope. The provisions of Sections 410.1 through 410.9 apply to maintenance, *change of occupancy*, *additions* and *alterations* to *existing buildings*, including those identified as *historic buildings*.

❖ The purpose of Section 410 is to establish minimum criteria for accessibility when dealing with existing buildings and facilities. The history and efforts involved are similar to those discussed in the commentary for IBC Chapter 11. Access to buildings and structures for people with physical disabilities has been a subject that the building codes have regulated since the early 1970s. They have consistently relied on a consensus national standard, ICC A117.1, *Accessible and Useable Buildings and Facilities*. Accessibility is not a new subject to the construction regulatory community. There has been a great deal of emphasis and awareness placed on the subject of accessibility through the passage of two federal laws. The Americans with Disabilities Act (ADA) and the Fair Housing Amendment Act (FHA) are federal regulations that affect building construction as it relates to accessibility.

The International Code Council® (ICC®) recognizes the value of consistency between federal laws and the codes. Efforts for coordination with federal accessibility requirements are ongoing. Representatives from interested accessibility groups, the Department of Housing and Urban Development (HUD) and the Architectural and Transportation Barriers Compliance Board (ATBCB, commonly referred to as the Access Board), have been attending and participating in the code change process for the IBC and ICC A117.1. In addition, the ICC has participated in the public comment process on the development of federal regulations for accessibility. Appendix B includes information found in the *Americans with Disabilities Act Accessibility Guidelines* (ADAAG) that cannot be enforced through the typical code enforcement process, but would provide beneficial information for the designer/owner for full compliance purposes. The ICC has worked toward, and will continue to strive for, accessibility regulations that reflect the highest possible degree of consistency with federal regulations and, more importantly, reasonable and appropriate provisions to meet the needs of people with disabilities.

Buildings undergoing alterations are generally expected to fully comply with the accessibility provisions. However, exceptions are then provided to indicate the conditions under which less than full accessibility is permitted.

For example, if a door and frame are removed and replaced, the door must meet the requirements for width, height, maneuvering clearances and hardware. If just the doorknob is being removed, it must be replaced with lever hardware.

If the area undergoing alteration does not contain a primary function, there are no additional requirements; however, if the area contains a primary function, there is an additional criterion that may require work not in the original scope to achieve accessibility. This additional criterion is to provide an accessible route to the altered area, including any toilets and drinking fountains that serve it. Requirements for an accessible route might specify that the door previously discussed be removed and replaced because it did not have adequate width or maneuvering clearances.

The principle behind this approach to upgrading existing buildings is that they will become more accessible over time. A valid time to work toward that goal is when a structure is being altered. Special considerations are offered because of the difficulty involved in dealing with existing facilities that may not have been built with accessibility for physically disabled persons in mind. For example, when a historically registered home is being made into a museum, if changing the front door to allow for wheelchair access would alter the historical significance, alternatives are offered in Section 410.9. Other examples include the alternatives offered in Section 410.7 if it is technically infeasible to provide full accessibility in an existing building. Please note that the term "technically infeasible" refers to either movement of a major structural element or other physical constraints. For example, a ramp to provide entrance or exit from a particular door may not be possible because of property lines or setback constraints.

410.2 Maintenance of facilities. A *facility* that is constructed or altered to be *accessible* shall be maintained *accessible* during occupancy.

❖ Continued compliance with the accessibility requirements of the code is dependent on maintenance of such facilities throughout the life of the building. For example, drinking fountains that are required to be accessible are of little value if they malfunction through deterioration or failure of any of the working parts. In other cases, inoperable elevators, locked accessible doors and obstructed accessible routes must be maintained such that they are readily usable by individuals with disabilities.

410.3 Extent of application. An *alteration* of an existing *facility* shall not impose a requirement for greater accessibility than that which would be required for new construction. *Alterations* shall not reduce or have the effect of reducing accessibility of a *facility* or portion of a *facility*.

❖ The purpose of this section is to clarify where alterations and scoping for alteration requirements apply. The requirements in Sections 410.6 and 410.7 do not impose a higher level of accessibility than the level required in new construction. At the same time, alterations cannot result in a lesser degree of accessibility than existed before the alterations were undertaken.

410.4 Change of occupancy. *Existing buildings* that undergo a change of group or occupancy shall comply with this section.

Exception: Type B dwelling or sleeping units required by Section 1107 of the *International Building Code* are not required to be provided in *existing buildings* and facilities undergoing a *change of occupancy* in conjunction with *alterations* where the *work area* is 50 percent or less of the aggregate area of the building.

❖ When an entire building undergoes a change of occupancy, the building must comply with the provisions in Section 410.4.2. If a portion of a building undergoes a change of occupancy, such as when there is a tenant change or a partial renovation where a space changes function, then the level of accessibility is addressed in Section 410.4.1.

The exception notes where Type B dwelling and sleeping units would not need to be included where there is a change of occupancy to a Group I or R building (see Section 1107 of the IBC). Where the change of occupancy does not involve extensive alterations (exceeding 50 percent of the square footage of the building), then Type B accessibility requirements would not need to be provided in the new or altered dwelling or sleeping units. This exception is similar to what is also found in the code for alterations where less than a Level 3 alteration occurs. A Level 3 alteration is greater than 50 percent of the square footage of the building. This, therefore, decreases the burden on many smaller projects. Type B dwelling units should be provided in larger alterations because, where major alterations are being performed, there is a prime opportunity to have those buildings move toward being able to serve a wider range of the population. With the population of the United States aging, there will be a steadily increasing demand for units that include accessibility features. See the commentary to Section 410.4.2, Section 410.6, Exception 4, and Section 410.8.8 for additional information.

410.4.1 Partial change in occupancy. Where a portion of the building is changed to a new occupancy classification, any *alterations* shall comply with Sections 410.6, 410.7 and 410.8.

❖ When a building undergoes a partial change of occupancy, such as where there is a tenant change or a change in function of a specific area, then the level of accessibility provided must achieve the same level as if that space were undergoing an alteration. Basically, the intent is that any spaces or elements being altered will meet new accessibility provisions unless technically infeasible (see Section 410.6). If the area changing occupancy is a primary function of the space, an evaluation of the accessible route, as well as bathrooms and drinking fountains serving this space, must be made. If these elements are not accessible, improvements must be made. However, there is a limit to the cost of the additional improvements to a maximum of 20 percent of the cost of the alteration (see Section 410.7). The reference to Section 410.8 provides for additional allowances because the designer/owner is still dealing with existing building constraints. For example, the accessible route could be provided by a platform lift (see Section 410.8.3), while in new construction, this option is limited (see IBC Sections 1009.5 and 1109.8).

410.4.2 Complete change of occupancy. Where an entire building undergoes a *change of occupancy*, it shall comply with Section 410.4.1 and shall have all of the following accessible features:

1. At least one accessible building entrance.
2. At least one accessible route from an accessible building entrance to *primary function* areas.
3. Signage complying with Section 1111 of the *International Building Code*.
4. Accessible parking, where parking is being provided.
5. At least one accessible passenger loading zone, when loading zones are provided.
6. At least one accessible route connecting accessible parking and accessible passenger loading zones to an accessible entrance.

Where it is *technically infeasible* to comply with the new construction standards for any of these requirements for a change of group or occupancy, the above items shall conform to the requirements to the maximum extent technically feasible.

Exception: The accessible features listed in Items 1 through 6 are not required for an accessible route to Type B units.

❖ For a project that involves a complete change of occupancy with alterations, full compliance with accessibility requirements is expected and reasonable for areas being altered, except where technical infeasibility can be demonstrated. If full compliance is technically infeasible, the element must be made accessible to the fullest extent that is feasible. This is consistent with the general approach that has always been taken relative to other matters regulated by the code.

In addition to accessibility requirements in the space, an accessible route is required to that space. Six items that make up that accessible route are listed. This section establishes that when an existing building undergoes a complete change of occupancy, compliance with those requirements, at minimum, is expected. That way, a person with mobility impairments would be able to arrive at the building (see Items 4 and 5), get to the accessible entrance (see Items 1, 3 and 6) and have at least one accessible route throughout the building to all the primary function areas (see Item 2). Changes between levels could be via a ramp (see Section 410.8.5), an elevator (see Section 410.8.2) or a platform lift (see Section 410.8.3). If the altered area would not be required to be served by an accessible route in new construction, an accessible route would not be required for a change of occupancy (see Section 410.3).

If the area undergoing a change of occupancy is being altered and contains a primary function area, the accessible route provisions in Section 410.7 are also applicable. This would not only require an accessible route to the change of occupancy area, but would also include possible upgrades to toilet rooms and drinking fountains that serve the area. These elements could use the 20-percent cap on cost offered in Section 410.7, Exception 1. If full compliance is technically infeasible, the element must be made accessible to the fullest extent that is feasible. See the commentary to Section 410.7 for additional information on the exceptions.

Typically, a building undergoing a complete change of occupancy is being, at least partially, gutted and undergoing alterations because of changes in function, possibly caused by increased occupant load, means of egress requirements, sprinkler requirements, or mechanical and plumbing changes. The intent is to create a balance between the change of occupancy meeting all new construction requirements for accessibility and the fact that the designer/owner is dealing with some existing building conditions.

This is not based on any specific provisions of the ADAAG, but parallels the intent of the requirements for the removal of barriers.

If the change of occupancy results in a Type B dwelling unit being provided, the exception permits the building to be exempt from providing the additional accessible route requirements listed in this section. This allowance addresses concerns of site impracticality for existing structures. This also reinforces the intent that the inclusion of Type B units is not meant to require elevators when alterations are performed on upper floors in nonelevator buildings (see the exceptions to Section 1107.7). These areas would have been exempted if built new under the FHA and IBC, and should continue to be exempted.

410.5 Additions. Provisions for new construction shall apply to *additions*. An *addition* that affects the accessibility to, or contains an area of, a *primary function* shall comply with the requirements in Section 410.7.

❖ Additions must comply with new construction. An addition, however, is also an alteration to an existing building; therefore, accessible route provisions for existing buildings are applicable (see commentary, Section 410.6). For example, a new dining area is added in a restaurant. All accessible elements within the parameter of the addition must be constructed to be accessible. If the route to or from the addition, or the bathrooms or drinking fountains that serve the addition, are in the existing building, the routes must be evaluated for accessibility. Section 410.7 specifies that the accessible route would include the route to these elements, the toilet rooms themselves, the fixtures in the toilet room and the drinking fountains.

410.6 Alterations. A *facility* that is altered shall comply with the applicable provisions in Chapter 11 of the *International Building Code*, unless *technically infeasible*. Where compliance with this section is *technically infeasible*, the *alteration* shall provide access to the maximum extent technically feasible.

Exceptions:

1. The altered element or space is not required to be on an accessible route, unless required by Section 410.7.

2. Accessible means of egress required by Chapter 10 of the *International Building Code* are not required to be provided in existing facilities.

3. The alteration to Type A individually owned dwelling units within a Group R-2 occupancy shall be permitted to meet the provision for a Type B dwelling unit.

4. Type B dwelling or sleeping units required by Section 1107 of the *International Building Code* are not required to be provided in *existing buildings* and facilities undergoing a *change of occupancy* in conjunction with *alterations* where the *work area* is 50 percent or less of the aggregate area of the building.

❖ The code approaches the application of accessibility provisions to a facility that is altered by broadly requiring full conformance to new construction, meaning full accessibility is expected. Exceptions are then provided to indicate the conditions under which less than full accessibility is permitted.

The circumstance under which full compliance with accessibility provisions is not required is when it is deemed to be technically infeasible (see the commentary for the definition of "Technically infeasible" in Section 202). This is considered reasonable since, if not provided for, plans for alterations may be otherwise abandoned by the building owner. The opportunity to upgrade and increase the current level of accessibility in an existing building would then be lost. This concern is also embodied in the requirement that an altered element or space is expected to be made accessible to the extent to which it is technically feasible to do so. In this manner, the code accomplishes the greatest degree of accessibility while recognizing the justifiable difficulties that may be involved in providing full accessibility. Alterations are not required to exceed new construction requirements (see Section 410.3).

In accordance with Exception 1, if the area undergoing alteration does not contain a primary function (see Section 410.7), there are no additional requirements past the original scope of the project. However, if the area contains a primary function, there are additional criteria to achieve accessibility that may require work not in the original scope of the project. The additional criteria are to provide an accessible route to the altered area, as well as improvements to any toilets and drinking fountains that serve the altered area.

Exception 2 indicates that accessible means of egress are not required as a result of undertaking alterations to existing buildings. Strict compliance with IBC Section 1009 is often technically infeasible. The requirement for a 48-inch (1219 mm) clear width between handrails would require many stairways to be widened. This often would entail movement of major structural elements in order to accomplish this alteration.

Exception 3 addresses the specific circumstances where an existing Type A dwelling unit is being altered. While Section 410.2 says that a level of accessibility must be maintained, in the situations where a Type A dwelling unit is "individually owned," such as a condominium, then it only needs to meet the technical requirements for a Type B dwelling unit (see ICC A117.1, Section 1004) when it is altered or remodeled by the owner. Type B units require a lesser level of accessibility than Type A units. For example, if an owner wanted to alter the bathroom in his or her unit, he or she would only be required to meet the lesser accessibility requirements of a Type B unit with the new construction. This exception would not be applicable to units that are for rent.

Exception 4 allows some partial building alterations in Group I and R structures to occur without providing Type B dwelling units and sleeping units (see Section 1107). Where the alteration does not exceed 50 percent of the square footage of the building, then Type B accessibility requirements would not need to be provided in the dwelling or sleeping units being altered (see Sections 410.8.6 and 410.8.7 for alterations where Accessible units and Type A units are required). This exception is similar to what is also found in the code for change of occupancy where less than a Level 3 alteration is occurring. A Level 3 alteration is greater than 50 percent of the square footage of the building. This decreases the burden on many smaller projects.

Type B dwelling units must be provided in larger alterations. Where major alterations are being performed, there is a prime opportunity to have those buildings move toward being able to serve a wider range of the population. With the population of the United States aging, there will be a steadily increasing demand for units that include accessibility features. In addition, while the Fair Housing Act (FHA) is only applicable to new construction, this law was enacted in 1991. Some buildings constructed after that time may not be in compliance with the FHA. When a major alteration/renovation is occurring, there is an opportunity to bring those buildings into compliance. See the commentary to Sections 410.4.2 (the exception), 410.8.6, 410.8.7 and 410.8.8 for additional information.

410.7 Alterations affecting an area containing a primary function. Where an *alteration* affects the accessibility to, or contains an area of *primary function*, the route to the *primary function* area shall be *accessible*. The accessible route to the *primary function* area shall include toilet facilities and drinking fountains serving the area of *primary function*.

Exceptions:

1. The costs of providing the *accessible* route are not required to exceed 20 percent of the costs of the *alterations* affecting the area of *primary function*.

2. This provision does not apply to *alterations* limited solely to windows, hardware, operating controls, electrical outlets and signs.

3. This provision does not apply to *alterations* limited solely to mechanical systems, electrical systems, installation or *alteration* of fire protection systems and abatement of hazardous materials.

4. This provision does not apply to *alterations* undertaken for the primary purpose of increasing the accessibility of a *facility*.

5. This provision does not apply to altered areas limited to Type B dwelling and sleeping units.

❖ An area containing a primary function is one in which a major activity for which the building or facility is intended is carried out (see the definition of "Primary function" in Section 202). For example, the lobby of a hotel in which the registration and check-out desk is located would be a primary function area. Other examples would be the dining area of a restaurant, the meeting rooms or exhibition halls in a conference center, virtually all office and work areas in a business building, and retail display areas in a mercantile occupancy. The key concept is that a primary function area is one that contains a major activity of the facility. Areas that contain activities not related to the main purpose of the facility would not be considered a primary function area. For example, a mechanical equipment room, storage closet, toilet facilities, corridors, lounges and locker rooms would not be considered primary function areas. With this background, it is clear that areas containing a primary function are clearly more critical in terms of the purpose for which people enter and use the facility; therefore, this section reflects that when such areas are altered or added, it is important to require that an accessible route to the primary function area be provided. When an accessible route to a primary function area is required by this section, an accessible route to such facilities, including any restrooms and drinking fountains serving the primary function area, must also be made accessible, even though such facilities and areas may not by themselves be considered primary function areas. This is not just the route to the bathrooms and drinking fountains; the bathrooms and drinking fountains themselves must also be improved when not accessible.

There are conditions under which it may not be reasonable to strictly enforce this requirement for an accessible route to an altered or added primary function area. Exception 1 approaches this by utilizing the cost of the alterations or addition as a basis for determining if providing a complete accessible route is reasonable. The requirement for a complete accessible route does not apply when the cost of providing it exceeds 20 percent of the cost of the alterations or addition to the primary function area. These costs are intended to be based on the actual costs of the planned alterations or addition to the primary function area before consideration of the cost of providing an accessible route. For example, if the planned alterations will cost $100,000, not including the cost of an accessible route to a primary function area, this exception would apply if the additional cost of providing the accessible route would exceed $20,000.

It is not the intent to exempt all requirements for accessibility when the total cost for providing the accessible route exceeds the 20-percent threshold. Improvements to the accessible route are required to the extent that costs do not exceed 20 percent of the cost to the planned alteration or addition. It is not required that the full 20 percent be spent. If the accessible route (including accessible bathrooms and drinking fountains) is already provided, no additional expenditure is required. Note that there is not a priority list given for where money should be spent on improving the accessible route. The logical progression is access to the site, accessible exterior routes to accessible entrances, access throughout the facility, access to services in the facility, toilet and bathing rooms and, finally, drinking fountains. Evaluation on how and where the money available should best be spent must be made on a case-by-case basis. For example, if an accessible route is not available to an upper level, and the cost of an elevator is more than 20 percent of the cost of the renovation, then other alternatives could be investigated, such as a platform lift or limited access elevator, or adding the elevator pit and shaft at this time, with elevator equipment added later. If all such items are in excess of the 20-percent limit, perhaps the money available could be spent toward making the toilet rooms accessible. The idea is that existing buildings would become fully accessible over time.

Exceptions 2 and 3 identify certain alterations that are not intended to trigger the requirement for providing an accessible route to a primary function area. Alterations limited to such elements as windows, hardware, operating controls, electrical outlets, signage, mechanical, electrical and fire protection systems, including alterations for the purpose of abating a hazardous materials circumstance, do not affect the usability of a primary function area in the same manner as alterations that affect the floor plan or the configuration, location or size of rooms or spaces. It is therefore considered unreasonable to require the installation of an accessible route when the scope of alterations is limited to that reflected in these exceptions.

Note that the costs for these items are not "backed out" of the total cost for the alteration before applying Exception 1. Exceptions 2 and 3 are alterations limited to the specific items referenced.

Exception 4 is intended to avoid penalizing a building owner who is undertaking alterations or additions for the purpose of increasing accessibility. It is appropriate to encourage owners to make such alterations without requiring them to do more work simply because they chose to increase the accessibility of the space. This could otherwise have the opposite effect of discouraging such alterations to avoid the expense of undertaking more work and expense than was originally planned. For example, federal law (ADA) requires that owners of existing buildings remove certain existing barriers to accessibility. Removal of such barriers may require a permit from the code official. It would be unreasonable to have such activity trigger the mandatory requirement for further alterations to accomplish accessibility beyond the originally planned work. In principle, the code takes the view that some extent of greater accessibility is positive progress and should be encouraged, not penalized.

Where the alterations result in Type B dwelling units being provided where none previously existed (Section 410.6, Exception 4), Exception 5 would obviate the need for any additional money to be spent toward providing an accessible route to those Type B dwelling or sleeping units. This is intended to encourage the creation of such units without penalizing building owners. Similar to Section 410.4.2, this exception is intended to address the concerns of site impracticality for providing accessible routes to and into existing buildings providing Type B dwelling units.

This also reinforces the intent that the inclusion of Type B units is not meant to require elevators when alterations are performed on upper floors in nonelevator buildings (see exceptions to Section 1107.7). These areas would have been exempted if built new under the FHA and IBC, and should continue to be exempted.

410.8 Scoping for alterations. The provisions of Sections 410.8.1 through 410.8.14 shall apply to *alterations* to *existing buildings* and *facilities*.

❖ The specific provisions of this section are intended to reflect conditions under which less than full accessibility, as would be required in new construction, is permitted in altered areas. As previously discussed, Section 410.6 requires alterations to comply with the full range of accessibility-related provisions of the code for new construction. This section reflects a reasonable set of conditions under which a different level of accessibility can be provided. Sections 410.8.1 through 410.8.14 are part of the code's coordination effort with ICC A117.1 and the recommendations for the ADAAG Review Federal Advisory Committee.

410.8.1 Entrances. *Accessible* entrances shall be provided in accordance with Section 1105.

> **Exception:** Where an *alteration* includes alterations to an entrance, and the *facility* has an *accessible* entrance, the altered entrance is not required to be *accessible*, unless required by Section 410.7. Signs complying with Section 1111 of the *International Building Code* shall be provided.

❖ This provision is contained here to point to the accessibility provisions of IBC Chapter 11 for entrances. A facility is not accessible if the entrances into it are inaccessible. IBC Section 1105 establishes reasonable criteria for providing accessible entrances. A facility is not required to have all of its entrances accessible in order to provide reasonable accommodation to disabled persons. If a facility has multiple public entrances, it is not considered unreasonable to require at least 60 percent of the entrances to be accessible. In addition to the 60-percent accessible public entrances requirement, entrances that have a specific function or provide access to only certain portions of the facility must be addressed.

If the building already has the accessible entrances required by IBC Section 1105, an entrance that is being altered is not required to be made accessible; however, if the entrance was required to be made accessible as part of the route to the altered primary function area, the exception would not apply. If entrances are not accessible, appropriate signage marking such entrances and directing persons with disabilities to the nearest accessible entrance is required.

410.8.2 Elevators. Altered elements of existing elevators shall comply with ASME A17.1 and ICC A117.1. Such elements shall also be altered in elevators programmed to respond to the same hall call control as the altered elevator.

❖ Requirements for new construction state that all elevators on an accessible route must be fully accessible in accordance with ICC A117.1. If a passenger elevator is altered, the altered element must be accessible in accordance with the requirements for existing elevators in Section 407 of ICC A117.1. If the altered elevator is part of a bank of elevators, the same element must be made accessible in every elevator that is part of that bank. The purpose of this requirement is to have consistency among elevators in a bank so that disabled people are not required to wait for a specific elevator when the general population can take the first available elevator. ICC A117.1 also provides accessibility requirements for limited use/limited access (LULA) elevators (ICC A117.1, Section 408) and private residence elevators (ICC A117.1, Section 409). While these types of elevators can serve as part of an accessible route, their use is limited by ASME A17.1.

410.8.3 Platform lifts. Platform (wheelchair) lifts complying with ICC A117.1 and installed in accordance with ASME A18.1 shall be permitted as a component of an accessible route.

❖ This section provides for the use of platform (wheelchair) lifts in existing buildings. In order to create an accessible route where there are changes in floor levels, the provisions for new construction would most often require the installation of an elevator or ramp. Platform lifts are allowed in new construction for limited conditions (see IBC Section 1109.8). If the space in an existing building precludes the installation of an elevator or ramp, a platform lift may be the only practical solution. Given the choice between no accessibility or accessibility by a platform lift, accessibility is preferable.

Previously, platform lift requirements were addressed in the elevator standard, ASME A17.1, but they are now addressed in their own standard, ASME A18.1. One of the many changes was the removal of the requirement for key operation, which previously discouraged independent utilization of platform lifts. Note that, in accordance with IBC Section 1009.5, platform lifts are also permitted for an accessible means of egress; however, accessible means of egress are not required in existing buildings undergoing alterations in accordance with IBC Section 410.6, Exception 2.

A chair lift is not acceptable as part of an accessible route in either new or existing construction. For examples of the difference between a chair lift and a platform lift, see the commentary to Section 1109.8 of the IBC.

410.8.4 Stairways and escalators in existing buildings. In *alterations*, *change of occupancy* or *additions* where an escalator or stairway is added where none existed previously and major structural modifications are necessary for installation, an accessible route shall be provided between the levels served by the escalator or stairways in accordance with Section 1104.4 of the *International Building Code*.

❖ If a stairway or escalator is added as part of an alteration in a location where one did not previously exist, the alteration must also include an accessible route between the same two levels. Where major structural modifications are required to install the stairway, there is no longer the argument that it is technically infeasible to add the accessible route to the other level. If an accessible route is already available between the two levels, or if the stairway or escalator is replacing an existing stairway or escalator, this requirement is not applicable. In conjunction with Section 410.3, if the requirement for the accessible route would be in excess of what is required for new construction, such as an accessible route to an area that was exempted by IBC Section 1103.2, 1104.4, 1107 or 1108, this requirement is not applicable. The intent is that if a route is provided between accessible levels for a nondisabled person to use, it is reasonable to also expect an accessible route.

410.8.5 Ramps. Where slopes steeper than allowed by Section 1012.2 of the *International Building Code* are necessitated by space limitations, the slope of ramps in or providing access to existing facilities shall comply with Table 410.8.5.

❖ This section recognizes the circumstances where, due to existing site or configuration constraints, a ramp with a slope of one unit vertical in 12 units horizontal (1:12) may not be feasible. A steeper slope is allowed where the elevation change does not exceed 6 inches (152 mm). Further ramp requirements, such as width, landings, etc., are set forth in IBC Section 1012.

TABLE 410.8.5
RAMPS

SLOPE	MAXIMUM RISE
Steeper than 1:10 but not steeper than 1:8	3 inches
Steeper than 1:12 but not steeper than 1:10	6 inches

For SI: 1 inch = 25.4 mm.

❖ In existing buildings, ramps that rise 3 inches (76 mm) or less may have a slope as steep as one unit vertical in eight units horizontal (1:8). In existing buildings, ramps that rise 6 inches (152 mm) or less may have a slope as steep as one unit vertical in 10 units horizontal (1:10). If it is possible to provide a lesser slope, it is desirable to do so. These steeper slopes should only be utilized when the one unit vertical in 12 units horizontal (1:12) slope is not possible.

410.8.6 Accessible dwelling or sleeping units. Where Group I-1, I-2, I-3, R-1, R-2 or R-4 dwelling or sleeping units are being altered or added, the requirements of Section 1107 of the *International Building Code* for Accessible units apply only to the quantity of spaces being altered or added.

❖ This section sets forth the rate for providing Accessible dwelling or sleeping units in Groups I-1, I-2, I-3, R-1, R-2 and R-4 where such facilities are altered or where units are added. Assuming that Accessible units are not already provided, the number of Accessible units to be incorporated into each alteration is based on the number being altered or added. For example, if a nursing home was being altered a portion at a time, 50 percent of the units being altered each time would be required to be fully wheelchair accessible. It is not the intent that all units being altered are required to be Accessible units until 50 percent of the units in the entire facility are Accessible. The total number of Accessible units in the facility is not required to exceed that required for new construction. It is unreasonable to require a greater level of accessibility in an existing building than is required in new construction.

410.8.7 Type A dwelling or sleeping units. Where more than 20 Group R-2 dwelling or sleeping units are being altered or added, the requirements of Section 1107 of the *International Building Code* for Type A units apply only to the quantity of the spaces being altered or added.

❖ Type A units are required in new construction where 20 or more apartments (including condominium style) are constructed on a site. Group R-2 requirements for

Type A units would also include convents and monasteries and could include some townhouse-style units (see the commentary to IBC Section 1107.6.2.2.1 for additional information).

Note that this requirement is applicable when 20 or more units are included in the construction project; not when an individual owner performs alterations to his or her condo when there are 20 or more units in the building. This section sets forth the rate for providing Type A dwelling or sleeping units in Group R-2 facilities in an addition where there is a change to the function of a space (i.e., a change of occupancy, such as creating apartments in an old warehouse, increasing the number of apartments by changing a storage area to apartment units or reconfiguring larger units into smaller units) or an alteration. Assuming that Type A units are not already provided, the number of Type A units required is based on the number being added or altered. For example, if a story was being added to an apartment building or a story in an apartment building was being altered, the number of Type A units required would be based on the number of units in the new story or the area being altered, not the number of units in the entire building. If Type A units are provided, the total number of Type A units in the facility is not required to exceed that required for new construction, as indicated in Section 410.3. It is unreasonable to require a greater level of accessibility in an existing building than is required in new construction.

410.8.8 Type B dwelling or sleeping units. Where four or more Group I-1, I-2, R-1, R-2, R-3 or R-4 dwelling or sleeping units are being added, the requirements of Section 1107 of the *International Building Code* for Type B units apply only to the quantity of the spaces being added. Where Group I-1, I-2, R-1, R-2, R-3 or R-4 dwelling or sleeping units are being altered and where the work area is greater than 50 percent of the aggregate area of the building, the requirements of Section 1107 of the *International Building Code* for Type B units apply only to the quantity of the spaces being altered.

❖ In new construction, Type B units are required where four or more dwelling units or sleeping units are constructed together and those units are "intended to be occupied as a residence" (see commentary, IBC Section 1107). The IRC references IBC Group R-3 for accessibility requirements for IRC units (see the definition of "Townhouse" and IRC Section R320.1).

This section sets forth the rate for providing Type B dwelling or sleeping units in Group I-1, I-2, R-1, R-2, R-3 or R-4 facilities. Type B units are required where more than four units are added in an addition or where major alterations are occurring that affect more the 50 percent of the area of the building. The Type B units in the facility are not required to exceed that required for new construction, as indicated in Section 410.3 and IBC Section 1107.7. It is unreasonable to require a greater level of accessibility in an existing building than is required in new construction. In consideration of existing site constraints, Sections 410.4.2 and 410.7 also include exceptions for the additional accessible route requirements in structures where Type B units are added as part of an alteration or change of occupancy. Note that this exception is not applicable to additions where Type B units are required.

410.8.9 Jury boxes and witness stands. In *alterations*, accessible wheelchair spaces are not required to be located within the defined area of raised jury boxes or witness stands and shall be permitted to be located outside these spaces where the ramp or lift access restricts or projects into the means of egress.

❖ This exception for jury boxes and witness stands is consistent with Sections 231 and 808 of the 2010 ADA *Standards for Accessible Design*. The intent is that if ramp access to a jury box or witness stand would have the ramp limiting or blocking the means of egress for the general population in the space, alternative locations for potential jurors or witnesses are viable. The same alternative is available if a platform lift would effectively be an obstruction for the general means of egress from the courtroom.

410.8.10 Toilet rooms. Where it is *technically infeasible* to alter existing toilet and bathing rooms to be *accessible*, an *accessible* family or assisted-use toilet or bathing room constructed in accordance with Section 1109.2.1 of the *International Building Code* is permitted. The family or assisted-use toilet or bathing room shall be located on the same floor and in the same area as the existing toilet or bathing rooms. At the inaccessible toilet and bathing rooms, provide directional signs indicating the location of the nearest family or assisted-use toilet room or bathing room. These directional signs shall include the International Symbol of Accessibility and sign characters shall meet the visual character requirements in accordance with ICC A117.1.

❖ This section deals with circumstances in which it is technically infeasible to alter existing toilet facilities to be accessible. Where new bathrooms are created in existing buildings or existing bathrooms are part of an alteration, both the men's and women's facilities would be required to be accessible. An alternative solution when it is technically infeasible to alter the existing toilet rooms would be the creation of a single unisex toilet or bathing room containing accessible facilities. The requirements for family or assisted-use toilet facilities in IBC Sections 1109.2.1 through 1109.2.1.7 provide guidance on what is required in such toilet rooms. If this alternative is selected, the room must be located on the same floor and in the same area as the existing toilet or bathroom. This is the best alternative to providing fully complying separate men's and women's facilities. One might argue that it is technically infeasible to accomplish either of these alternatives, since the alternative to altering existing facilities involves creation of an additional toilet or bathroom that was not otherwise contemplated. This would not be a persuasive argument, since there is likely to be space available somewhere in the facility to commit for use as a toilet or bathroom. In any case, the intent is that some form of an accessible toilet room or bathing facility is necessary and must be provided.

It should be noted that this alternative is not offered as a choice between making the existing separate-sex toilet rooms accessible or providing a family or assisted-use accessible toilet room. The existing separate-sex toilet rooms must be altered when it is technically feasible. Consideration of this alternative is only available when altering the existing toilet rooms is technically infeasible (see the definition of "Technically infeasible" in Section 202).

Signage must be provided at the inaccessible toilet rooms in accordance with IBC Sections 1110.1 and 1110.2 to notify disabled persons where a facility is not accessible and direct them to the nearest accessible facilities.

410.8.11 Dressing, fitting and locker rooms. Where it is *technically infeasible* to provide accessible dressing, fitting or locker rooms at the same location as similar types of rooms, one accessible room on the same level shall be provided. Where separate-sex facilities are provided, accessible rooms for each sex shall be provided. Separate-sex facilities are not required where only unisex rooms are provided.

❖ This section takes a similar approach for dressing rooms as provided for in Section 410.8.10 for toilet rooms and bathing facilities. If it is technically infeasible to alter existing dressing rooms to be accessible, then space elsewhere on the level must be committed to providing no less than one accessible dressing room. In this case, if the existing dressing rooms provide separate rooms for each sex, then no less than one accessible dressing room for each sex must be provided.

410.8.12 Fuel dispensers. Operable parts of replacement fuel dispensers shall be permitted to be 54 inches (1370 mm) maximum, measuring from the surface of the vehicular way where fuel dispensers are installed on existing curbs.

❖ The requirements for new fuel dispensers (i.e., gas pumps) can be found in Section 1109.14 of the IBC. Basically, the idea is that the controls must be within the reach ranges for someone standing on the parking lot surface. However, many existing facilities have gas pumps located on raised islands as a feature for protection of the pumps from accidental contact. This section would allow the new gas pump with the reach range of 15 inches (380 mm) to 48 inches (1220 mm) to be located on top of a 6-inch (150 mm) curb and still meet the maximum reach of 54 inches (1370 mm).

410.8.13 Thresholds. The maximum height of thresholds at doorways shall be $^3/_4$ inch (19.1 mm). Such thresholds shall have beveled edges on each side.

❖ Thresholds at doorways may be $^3/_4$ inch (19.1 mm) maximum in existing buildings. In new construction, a typical threshold is $^1/_2$ inch (12.7 mm) maximum in accordance with the IBC. This section recognizes that such things as differences in floor materials may create changes in elevation greater than that allowed in new construction. Edges of thresholds must be beveled to allow for the passage of a wheelchair.

410.8.14 Amusement rides. Where the structural or operational characteristics of an amusement ride are altered to the extent that the amusement ride's performance differs from that specified by the manufacturer or the original design, the amusement ride shall comply with requirements for new construction in Section 1110.4.8 of the *International Building Code*.

❖ To the extent that amusement rides are subject to the accessibility requirements of IBC Section 1110.4.8, they should be accessible and usable by individuals with disabilities. These scoping provisions are flexible, permitting latitude in terms of the method of access (e.g., transfer seat, roll-on seat or transfer device to lift the rider). Mobile and portable rides are exempted. Rides without seats, designed for children who are assisted onto the ride or controlled by the user are also exempted under IBC Section 1110.4.8.2 from providing wheelchair transfer spaces. Technical criteria can be found in the 2009 ICC A117.1, Section 1102 and includes accessible routes, load and unload areas, wheelchair spaces on rides, seats for transfer, and transfer devices. Recognizing the technical issues associated with existing amusement rides, compliance with accessibility requirements is required when the amusement ride is extensively altered.

410.9 Historic buildings. These provisions shall apply to *facilities* designated as historic structures that undergo *alterations* or a *change of occupancy*, unless *technically infeasible*. Where compliance with the requirements for accessible routes, entrances or toilet rooms would threaten or destroy the historic significance of the *facility*, as determined by the applicable governing authority, the alternative requirements of Sections 410.9.1 through 410.9.4 for that element shall be permitted.

Exception: Type B dwelling or sleeping units required by Section 1107 of the *International Building Code* are not required to be provided in historical buildings.

❖ The regulations of individual adopting jurisdictions may provide additional guidance or limitations on the use of this section. Basic ADA requirements permit exceptions and alternatives for historic buildings. These minimum requirements provide reasonable accommodations for building users.

For this section to be applicable, the building must be registered as historic. Historic buildings are treated much the same as provided for in Sections 410.4 and 410.6, in that a historic building that is altered or has undergone a change of occupancy is expected to comply with accessibility requirements, unless technical infeasibility can be demonstrated; however, this section also goes on to acknowledge that the historic character of a building may be adversely affected by strict compliance with accessibility provisions. For example, compliance with door width requirements may necessitate the removal of an existing set of doors that is critical to the historic character of the building. To assist the code official in determining if the required provisions are detrimental to the historic significance, recommended guidelines have been incorporated into Appendix B, Section B101.

This section is intended to exempt such conditions in order to maintain the historic character of the building. Because limited extent of accessibility is desired in all facilities, Sections 410.9.1 through 410.9.4 allow for alternatives. In addition, there is an exception that would exempt the Type B dwelling or sleeping unit requirements for historical buildings, including when the alteration exceeds 50 percent of the building. This exception is important in that owners will not need to demonstrate that the addition of such units will threaten or destroy the historical character of the building.

In an effort to coordinate with the 2010 ADA *Standards for Accessible Design* requirements, items that cannot be enforced through the typical code enforcement process, but are associated with historical buildings, have been included in Appendix B Section B101.5.

410.9.1 Site arrival points. At least one accessible route from a site arrival point to an accessible entrance shall be provided.

❖ Full compliance would require an accessible route from all site arrival points. If this requirement would adversely affect the historical significance of the building, the alternative available is to provide an accessible route from one site arrival point to an accessible entrance.

410.9.2 Multilevel buildings and facilities. An accessible route from an accessible entrance to public spaces on the level of the accessible entrance shall be provided.

❖ It is not required in building alteration that accessibility to spaces above or below the level of accessible entrance be provided. Full compliance for new construction might require an accessible route to levels above or below, as well as throughout, the entrance level. If this requirement would adversely affect the historical significance of the building, the alternative is to provide an accessible route from the accessible entrance to all spaces open to the public on the entrance level. If elevators are provided, but are not accessible, signage in accordance with IBC Section 1110.2 is required.

410.9.3 Entrances. At least one main entrance shall be accessible.

Exceptions:

1. If a main entrance cannot be made accessible, an accessible nonpublic entrance that is unlocked while the building is occupied shall be provided; or

2. If a main entrance cannot be made accessible, a locked accessible entrance with a notification system or remote monitoring shall be provided.

Signs complying with Section 1111 of the *International Building Code* shall be provided at the primary entrance and the accessible entrance.

❖ Although uniform building access is the rule for new construction, older buildings may have main entrances that are both inaccessible and a significant element of the building's character. In these cases, providing access by an alternative route is deemed to meet the primary intent of the accessibility regulations.

Full compliance would require 60 percent of the entrances to be accessible. If this requirement would adversely affect the historical significance of the building, only one main entrance is required to be made accessible. If a main entrance cannot be made accessible, then an employee or service entrance may serve as the accessible entrance, provided that it remains unlocked when the building is open. Alternatively, a locked entrance, where monitoring or a notification system is available, could be provided. Signage must be provided at inaccessible entrances in accordance with IBC Sections 1111.1 and 1111.2.

410.9.4 Toilet and bathing facilities. Where toilet rooms are provided, at least one accessible family or assisted-use toilet room complying with Section 1109.2.1 of the *International Building Code* shall be provided.

❖ Full compliance would require an accessible toilet/bathing facility at each location where toilet/bathing facilities are provided. If altering the existing facilities to be accessible would adversely affect the historical significance of the building, only one unisex bathroom that complies with the accessible family or assisted-use toilet/bathing requirements in IBC Section 1109.2.1 is required. Signage must be provided at inaccessible toilet rooms in accordance with IBC Section 1110.2.

Bibliography

The following resource materials were used in the preparation of the commentary for this chapter of the code.

24 CFR, *Fair Housing Accessibility Guidelines* (FHAG). Washington, DC: Department of Housing and Urban Development, 1991.

36 CFR, Parts 1190 and 1191 Final Rule, *The Americans with Disabilities Act (ADA) Accessibility Guidelines; Architectural Barriers Act (ABA) Accessibility Guidelines*. Washington, DC: Architectural and Transportation Barriers Compliance Board, July 23, 2004.

42 USC 3601-88, *Fair Housing Amendments Act* (FHAA). Washington, DC: United States Code, 1988.

2010 *ADA Standards for Accessible Design*. Washington, DC: Department of Justice, September 15, 2010.

ASCE 7-10, *Minimum Design Loads for Buildings and Other Structures.* American Society of Civil Engineers, 2011.

DOJ 28 CFR, Part 36-91, *Americans with Disabilities Act* (ADA). Washington, DC: Department of Justice, 1991.

FEMA 311-07, *Residential Substantial Damage Estimators In Your Community.* Washington, DC: Federal Emergency Management Agency, 2007.

FEMA P-758, *Substantial Improvement/Substantial Damage Desk Reference*. Washington, DC: Federal Emergency Management Agency, 2009.

NFPA 13-07, *Installation of Sprinkler Systems*. Quincy, MA: National Fire Protection Association, 2007.

Chapter 5:
Classification of Work

General Comments

This chapter provides an overview of the process for the repair, alteration and restoration of existing buildings. A brief description is provided that identifies the differences between the three levels of alterations. In addition, the topics of additions, historic buildings and relocated buildings are mentioned.

Purpose

This chapter enables the contractor, design professional or code official to easily identify the classification of and the associated chapter in the code for building alterations, additions and repairs.

SECTION 501
GENERAL

501.1 Scope. The provisions of this chapter shall be used in conjunction with Chapters 6 through 13 and shall apply to the *alteration*, *repair*, *addition* and *change of occupancy* of existing structures, including historic and moved structures, as referenced in Section 301.1.2. The work performed on an *existing building* shall be classified in accordance with this chapter.

❖ This section establishes when the regulations contained in the code must be followed, whether all or in part. Something must happen (modification to an existing building or allowing an existing building or structure to become unsafe) for the code to be applicable. The code is not a maintenance document requiring periodic inspections that will, in turn, result in an enforcement action. Periodic inspections are addressed by the *International Fire Code*® (IFC®).

501.1.1 Compliance with other alternatives. *Alterations*, *repairs*, *additions* and *changes of occupancy* to existing structures shall comply with the provisions of Chapters 6 through 13 or with one of the alternatives provided in Section 301.1.

❖ This section offers the code user compliance options. In addition to being able to use the provisions listed in Chapters 6 through 13, there are three other compliance alternatives listed under Section 301.1.

First, Section 301.1.1 offers a prescriptive compliance method for repairs, alterations, additions and changes of occupancy as long as they comply with Chapter 3 of the code and the IFC. Second, Section 301.1.2 describes the work area compliance method that requires compliance with the applicable provisions of Chapters 6 through 13. Third, Section 301.1.3 describes the performance compliance method that requires compliance with Chapter 14.

501.2 Work area. The *work area*, as defined in Chapter 2, shall be identified on the construction documents.

❖ As defined in Chapter 2, a "Work area" is the area of all reconfigured spaces where work is occurring within the scope of a project. These areas are to be shown clearly on the construction documents. Work areas exclude other portions of the building where incidental work is ongoing.

SECTION 502
REPAIRS

502.1 Scope. *Repairs,* as defined in Chapter 2, include the patching or restoration or replacement of damaged materials, elements, *equipment or fixtures* for the purpose of maintaining such components in good or sound condition with respect to existing loads or performance requirements.

❖ This section describes repairs to existing structures, including the restoration or replacement of damaged materials to good or sound condition, as they apply to existing loading and the performance requirements of any part of the building with any of the materials and methods listed in Section 502.1.

502.2 Application. *Repairs* shall comply with the provisions of Chapter 6.

❖ Chapter 6 provides the guidelines for repairs to existing structures. It covers topics ranging from building elements and materials to fire protection and accessibility. The main focus of Chapter 6 is covered in the provisions of Section 606.

502.3 Related work. Work on nondamaged components that is necessary for the required *repair* of damaged components shall be considered part of the *repair* and shall not be subject to the provisions of Chapter 7, 8, 9, 10 or 11.

❖ This section makes it clear that portions of a building, while they may be intact, that have to be removed to repair or replace another part of the structure can be considered part of the repair even though they were not damaged and unserviceable. For example, a building that was not properly flashed around penetrations to the exterior envelope develops water intrusion. The water has resulted in the deterioration of the wall structure, which will have to be removed and replaced as necessary. Even though the siding covering the damaged portion of the structure is serviceable, it is considered part of the related work as it has to be removed and then replaced in order to fix the damaged portions.

SECTION 503
ALTERATION—LEVEL 1

503.1 Scope. Level 1 alterations include the removal and replacement or the covering of existing materials, elements, equipment, or fixtures using new materials, elements, equipment, or fixtures that serve the same purpose.

❖ Level 1 alterations represent the most basic or foremost level of building alterations. This includes the removal and replacement or the covering of existing materials, elements, equipment or fixtures. An example would be the addition of a new roof to an existing building. Another example would be the removal of aluminum siding exterior finish to be replaced with vinyl siding.

503.2 Application. Level 1 *alterations* shall comply with the provisions of Chapter 7.

❖ Chapter 7 describes, in detail, the requirements for Level 1 alterations. It is important to note that historic buildings must also comply with this chapter unless there is a modification noted in Chapter 12. Note that Level 1 alterations do not involve the reconfiguration of space.

SECTION 504
ALTERATION—LEVEL 2

504.1 Scope. Level 2 *alterations* include the reconfiguration of space, the addition or elimination of any door or window, the reconfiguration or extension of any system, or the installation of any additional equipment.

❖ Chapter 8 describes, in detail, the requirements for Level 2 alterations. The exception to Section 801.1 allows buildings undergoing alterations that are exclusively the result of compliance with the accessibility requirements of Section 705.2 to comply with Chapter 7.

504.2 Application. Level 2 *alterations* shall comply with the provisions of Chapter 7 for Level 1 *alterations* as well as the provisions of Chapter 8.

❖ In addition to the provisions listed in Chapter 8, Level 2 alterations are also required to meet all of the provisions of Chapter 7, Level 1 alterations. This requirement effectively compounds the requirements for someone planning to alter an existing structure. For example, if during the process of replacing the aluminum siding on a building with vinyl siding, the building owner decides to eliminate one of four windows from a room, then this project would be classified as a Level 2 alteration and would, therefore, be required to meet the provisions of Chapters 7 and 8.

SECTION 505
ALTERATION—LEVEL 3

505.1 Scope. Level 3 *alterations* apply where the work area exceeds 50 percent of the *building area.*

❖ Any time the work area, as defined in Section 202, exceeds one-half of the aggregate building area, it is considered to be a Level 3 alteration and, therefore, has to meet the requirements of Chapter 9. In the code, a work area encompasses all portions of the existing building that are proposed to be reconfigured.

505.2 Application. Level 3 *alterations* shall comply with the provisions of Chapters 7 and 8 for Level 1 and 2 *alterations*, respectively, as well as the provisions of Chapter 9.

❖ Any project that qualifies as a Level 3 alteration project must meet all of the requirements of Chapters 7, 8 and 9.

SECTION 506
CHANGE OF OCCUPANCY

506.1 Scope. *Change of occupancy* provisions apply where the activity is classified as a *change of occupancy* as defined in Chapter 2.

❖ A change of occupancy in an existing structure may change the level of inherent hazards that the code was initially intended to address.
  This is done so that the applicable code requirements adequately address the specific hazards of the new occupancy. For example, a change from an existing mercantile occupancy to a business occupancy renders all Group B provisions applicable to all portions of the structure where the occupancy has changed.

506.2 Application. *Changes of occupancy* shall comply with the provisions of Chapter 10.

❖ This section indicates compliance with Chapter 10 for changes of occupancy. Chapter 10 contains provisions frequently used for existing structures since the occupancy in a building or structure often changes during the life of the building.

SECTION 507
ADDITIONS

507.1 Scope. Provisions for *additions* shall apply where work is classified as an *addition* as defined in Chapter 2.

❖ Any project that would increase the floor area, the number of stories in a building or the height of a structure would qualify as an addition.

507.2 Application. *Additions* to *existing buildings* shall comply with the provisions of Chapter 11.

❖ Additions to existing structures are specifically covered in Chapter 11 of the code.

SECTION 508
HISTORIC BUILDINGS

508.1 Scope. *Historic building* provisions shall apply to buildings classified as historic as defined in Chapter 2.

❖ The most important criterion for the application of this section is that the building must be certified as being of historic significance by a qualified party or agency. Usually this is done by a state or local authority after careful review of the historical value of the building. Most, if not all, states have such authorities, as do many local jurisdictions. The agencies with such authority can be located at the state or local government level or through the local chapter of the American Institute of Architects (AIA). Other considerations for classification as a historical building include the structural condition of the building (i.e., whether it is structurally sound), its proposed use, its impact on life safety and how the intent of the code, if not the letter, will be achieved.

508.2 Application. Except as specifically provided for in Chapter 12, *historic buildings* shall comply with applicable provisions of this code for the type of work being performed.

❖ Chapter 12 covers the various aspects of existing historic structures and includes specific sections on repairs, fire safety, alterations, change of occupancy and structure. In the absence of any specific requirements or provisions in Chapter 12, the remainder of the code is applicable to work proposed for historic buildings.

SECTION 509
RELOCATED BUILDINGS

509.1 Scope. Relocated building provisions shall apply to relocated or moved buildings.

❖ Any structure that is relocated or moved to a different lot or a new location on the same lot falls within the scope of Section 509.

Structures that are relocated or moved are required to comply with the provisions applicable for new construction. The moved structure may comply with the alternative provisions of Chapter 14 instead of the code requirements for new structures, which may be particularly useful if the moved structure is older than the effective date of the adoption of the building codes in the jurisdiction. The fire separation distance of the moved structure must also comply with the requirements for new structures even if the compliance alternative provisions in Chapter 14 are used to meet the code requirements.

509.2 Application. Relocated buildings shall comply with the provisions of Chapter 13.

❖ The requirements for relocated or moved buildings are found in Chapter 13 of the code.

Chapter 6: Repairs

General Comments

Chapter 6 governs the repair of existing buildings. The provisions define conditions under which repairs may be made using materials and methods like those of the original construction or the extent to which repairs must comply with requirements for new buildings.

It should be noted that buildings such as those completely destroyed by fire would typically no longer be considered repairs and would instead need to be constructed as a new building in accordance with the *International Building Code*® (IBC®). There are, however, many situations that would still allow the damage to be dealt with as a repair. An example may be a small kitchen fire where the cabinets may need replacing and the walls in that area may also need replacing.

Purpose

Repairs to an existing structure must be made with the proper materials in a manner that will safeguard the public and ensure the building does not become a hazard to life, health or property.

SECTION 601 GENERAL

601.1 Scope. Repairs as described in Section 502 shall comply with the requirements of this chapter. Repairs to *historic buildings* need only comply with Chapter 12.

❖ Repairs are described in Section 502 as the patching, restoration or replacement of damaged materials, elements, equipment or fixtures for the purpose of maintaining such materials and elements in good or sound condition with respect to existing loads or performance requirements. The scoping provisions of this section refer the user to Section 502 to be certain the work classification is "repair" and Chapter 6 is the appropriate chapter to be used. There are additional provisions specific to repairs in historic buildings and the code user is referred to Chapter 12 for the possible applicability of those provisions in addition to the requirements of Chapter 6. It is the intent of the code to allow original materials and construction methods for repairs to historical buildings in order to limit any negative impact on the structure's historical significance.

601.2 Conformance. The work shall not make the building less conforming than it was before the *repair* was undertaken.

❖ The general limitation on repairs is that the level of safety, health and public welfare of the existing building must not be reduced by any work being performed. This requirement can be broadly interpreted, as its applications vary on a case-by-case situation, but the level of safety provided by the structure and systems, such as plumbing and mechanical, is not to be decreased in the course of making repairs.

[BS] 601.3 Flood hazard areas. In flood hazard areas, repairs that constitute *substantial improvement* shall require that the building comply with Section 1612 of the *International Building Code*, or Section R322 of the *International Residential Code*, as applicable.

❖ If located in designated flood hazard areas, buildings and structures that are damaged by any cause are to be examined to determine if the damage constitutes substantial damage, in which the cost of repairing/restoring the building or structure to its predamaged condition equals or exceeds 50 percent of its market value before the damage occurred. All substantial improvements and repairs of buildings and structures that are substantially damaged are to meet the flood-resistant provisions of the IBC or *International Residential Code*® (IRC®).

SECTION 602 BUILDING ELEMENTS AND MATERIALS

602.1 Existing building materials. Materials already in use in a building in compliance with requirements or approvals in effect at the time of their erection or installation shall be permitted to remain in use unless determined by the *code official* to render the building or structure unsafe or *dangerous* as defined in Chapter 2.

❖ A material or system approved before the code took effect can continue to be used as long as it can be shown that the material or system is not detrimental to the health or safety of the building occupants or the public. In other words, the code is not retroactive.

602.2 New and replacement materials. Except as otherwise required or permitted by this code, materials permitted by the applicable code for new construction shall be used. Like materials shall be permitted for *repairs* and *alterations*, provided no *dangerous* or *unsafe* condition, as defined in Chapter 2, is created. Hazardous materials, such as asbestos and lead-based paint, shall not be used where the code for new construction would not permit their use in buildings of similar occupancy, purpose and location.

❖ There are two options for materials used in repairs to an existing building. Generally, the materials used for repairs should be those that are presently required or permitted for new construction under the *International Codes*® (I-Codes®). It is also acceptable to use materials consistent with those that are already present, except where those materials pose a hazard. This allowance follows the general concept that any repair should not make a building more hazardous than it was prior to the repair. It is generally possible to repair a structure, its components and its systems with materials consistent with those materials that were used previously. However, where materials that are now deemed hazardous were used in areas subject to repair, they may no longer be used. For example, the code identifies asbestos and lead-based paint as two common hazardous materials that cannot be used in the repair process. Certain materials previously considered acceptable for building construction are now a threat to the health of the occupants.

602.3 Glazing in hazardous locations. Replacement glazing in hazardous locations shall comply with the safety glazing requirements of the *International Building Code* or *International Residential Code* as applicable.

Exception: Glass block walls, louvered windows, and jalousies repaired with like materials.

❖ When glazing in an existing building is replaced within the same building, it must comply with the current requirements and standards of the IBC or the IRC, as applicable. This includes installing new glass in an existing window, door or other type of opening, even where the glass replaced did not comply with the standards of the code.

Glass block walls are described in Chapter 21 of the IBC, which eliminates the Consumer Product Safety Commission (CPSC) test requirement. Glass block walls are not required to meet the test requirements of CPSC 16 CFR, Part 1201 for safety glazing; however, there are still safety requirements placed on the installation of the glass block.

Louvered and jalousie windows are exempt from safety glazing requirements in all applications, including those where a flat plane of glass is otherwise required to be safety glass. This exemption is based on records that show the injuries associated with this use of glass are primarily from persons impacting the glass edge with no cutting or piercing injuries resulting from glass breakage. Safety glass would not have an effect on the type of injury. There are also practical production reasons associated with fabricating safety glazing for the relatively long, thin slats.

SECTION 603
FIRE PROTECTION

603.1 General. Repairs shall be done in a manner that maintains the level of fire protection provided.

❖ Any level of fire protection that currently exists in a building must not be adversely affected as a result of any repair. For example, repairing the existing ceiling and sprinkler heads or repairing the fire alarm system equipment must ultimately provide the same level of coverage and protection that existed prior to the repairs being undertaken.

SECTION 604
MEANS OF EGRESS

604.1 General. Repairs shall be done in a manner that maintains the level of protection provided for the means of egress.

❖ Any level of protection provided by the means of egress that currently exists in a building must not be adversely affected as a result of any repair. For example, repairing the walls and doors of a corridor must ultimately provide the same level of protection that existed prior to the repairs being undertaken.

SECTION 605
ACCESSIBILITY

605.1 General. Repairs shall be done in a manner that maintains the level of accessibility provided.

❖ The level of accessibility that currently exists in a building must not be adversely affected as a result of any repair. Continued compliance with the accessibility requirements of the code is dependent on the maintenance of such facilities throughout the life of the building. For example, drinking fountains that are required to be accessible are of little value if they malfunction through the deterioration or failure of any of the working parts. In other cases, inoperable elevators, locked accessible doors and obstructed accessible routes must be maintained such that they are readily usable by individuals with disabilities.

SECTION 606
STRUCTURAL

[BS] 606.1 General. Structural repairs shall be in compliance with this section and Section 601.2. Regardless of the extent of structural or nonstructural damage, *dangerous* conditions shall be eliminated. Regardless of the scope of *repair*, new structural members and connections used for *repair* or *rehabilitation* shall comply with the detailing provisions of the *International Building Code* for new buildings of similar structure, purpose and location.

❖ This section gives the requirements that pertain to structural materials and elements in need of repair;

Section 606.2.1 addresses repairs for less than substantial structural damage; Section 606.2.2 addresses repairs for substantial structural damage to vertical elements of the lateral force-resisting system; and Section 606.2.3 addresses repairs for substantial structural damage to gravity load-carrying components. This section also requires dangerous conditions to be eliminated. See the definition of "Dangerous" in Section 202 for a list of the conditions that must always be corrected regardless of the extent of damage. Finally, regardless of the scope of work, new connections and new structural members must be in compliance with the IBC.

[BS] 606.2 Repairs to damaged buildings. Repairs to damaged buildings shall comply with this section.

❖ Buildings can suffer damage from numerous sources. Natural disasters, such as earthquakes, floods, hurricanes and tornadoes, can cause extensive damage over widespread areas, depending on the severity of the event. Water intrusion due to a failure in the building envelope, termite infestations or exposure to corrosive chemicals can all lead to the deterioration of structural members over time. For the most part, this section does not differentiate between the possible causes of the damage (the exceptions are Sections 606.2.2.3 and 606.2.3.1). Needless to say, determining the root cause of any damage would be advisable so that an owner can ascertain the risk of a recurrence and, if necessary, develop a plan to address that risk.

The primary concern in determining how repairs are to be accomplished is establishing the extent of the damage that has been sustained to see if it exceeds either of the thresholds contained in the definition of "Substantial structural damage." Where it does not, Section 606.2.4 applies and the repairs can typically be limited to restoring the building to its predamaged state. For buildings that have suffered substantial structural damage, the approach to repairs is dependent on whether that damage is to elements of the lateral system (see Section 606.2.2) or only to elements of the gravity system (see Section 606.2.3). This parallels the classes of substantial structural damage defined in Section 202. The definition of "Substantial structural damage" would itself necessitate some preliminary level of structural evaluation.

There are two repair requirements that are not related to the extent of the damage. Dangerous conditions must always be eliminated (see Section 606.2.1) and any new structural members and connections must meet the code requirements for new construction.

[BS] 606.2.1 Repairs for less than substantial structural damage. For damage less than *substantial structural damage*, the damaged elements shall be permitted to be restored to their predamage condition.

❖ For damage less than substantial structural damage, repairs are allowed that restore the building to its pre-damaged state using materials and strengths that existed prior to the damage. Again, as required in Section 606.1, new structural members and connections used for this repair must comply with the detailing provisions of the IBC for new buildings of similar structure, purpose and location.

[BS] 606.2.2 Substantial structural damage to vertical elements of the lateral force-resisting system. A building that has sustained *substantial structural damage* to the vertical elements of its lateral force-resisting system shall be evaluated in accordance with Section 606.2.2.1, and either repaired in accordance with Section 606.2.2.2 or repaired and rehabilitated in accordance with Section 606.2.2.3, depending on the results of the evaluation.

Exceptions:

1. Buildings assigned to Seismic Design Category A, B, or C whose substantial structural damage was not caused by earthquake need not be evaluated or rehabilitated for load combinations that include earthquake effects.

2. One- and two-family dwellings need not be evaluated or rehabilitated for load combinations that include earthquake effects.

❖ This section provides requirements that apply where the damage threshold based on the extent of damage to vertical elements of the lateral force-resisting system in any story is exceeded. Substantial structural damage to the lateral system triggers the evaluation of the whole building for wind and seismic loads (see Section 606.2.2.1). The emphasis is placed on vertical elements, such as walls and columns, rather than horizontal elements, because the vertical elements of the lateral force-resisting system determine the structure's response, particularly to earthquakes. Based on the results of this evaluation, the lateral force-resisting system is required to either be repaired (see Section 606.2.2.2), or repaired and rehabilitated (see Section 606.2.2.3).

There are two exceptions that exempt certain combinations of buildings: seismic risk and damage from triggered seismic upgrades. Basic repair—that is, restoring the predamaged condition—is still required even for the exceptions outlined.

Exception 1 is for buildings in areas of low or moderate seismicity (Seismic Design Category A, B or C), where the damage was not caused by an earthquake and therefore would not be required to be evaluated or rehabilitated for load combinations that include earthquake effects. Where earthquakes are rare, it serves no significant public purpose to trigger seismic upgrades following damage caused by fire, collision, wind, etc.

Exception 2 is for one- and two-family dwellings, where the public risk is especially low even though the damage may be associated with earthquake effects.

Similar exceptions are found in Section 404.2.

[BS] 606.2.2.1 Evaluation. The building shall be evaluated by a registered design professional, and the evaluation findings shall be submitted to the *code official*. The evaluation shall establish whether the damaged building, if repaired to its predamage state, would comply with the provisions of the *International Building Code* for load combinations that include wind or earthquake effects, except that the seismic forces shall be the reduced *International Building Code*-level seismic forces.

❖ This section contains the requirements for the building evaluation mandated by Section 606.2.2. For the purpose of establishing the structure's adequacy, the evaluation is to utilize the strength and stiffness of the original, predamaged structure.

The structural evaluation must assess the entire structure for compliance with IBC wind and seismic load combinations. This means that the evaluation must demonstrate compliance for all loads that are applicable to new construction. The only exception is that the approaches permitted for the reduced seismic forces in Section 301.1.4.2 may be used for the evaluation of seismic effects on the structure. Since no other exceptions are stated, the implication is that new construction requirements for material strength and detailing must be satisfied for loading considerations other than earthquakes.

If this evaluation indicates that the building's lateral force-resisting system complies with the IBC using the permitted reduced seismic design criteria (see Section 301.1.4.2), then it need only be repaired to the predamaged condition. Otherwise, the lateral force-resisting system must be rehabilitated in accordance with Section 606.2.2.3.

[BS] 606.2.2.2 Extent of repair for compliant buildings. If the evaluation establishes that the building in its predamage condition complies with the provisions of Section 606.2.2.1, then the damaged elements shall be permitted to be restored to their predamage condition.

❖ Where the evaluation provided for in Section 606.2.2.1 establishes that the predamaged building meets the structural provisions of the IBC, the repairs may be limited to a restoration of the structural components.

[BS] 606.2.2.3 Extent of repair for noncompliant buildings. If the evaluation does not establish that the building in its predamage condition complies with the provisions of Section 606.2.2.1, then the building shall be rehabilitated to comply with the provisions of this section. The wind loads for the *repair* and *rehabilitation* shall be those required by the building code in effect at the time of original construction, unless the damage was caused by wind, in which case the wind loads shall be in accordance with the *International Building Code*. The seismic loads for this *rehabilitation* design shall be those required by the building code in effect at the time of original construction, but not less than the reduced *International Building Code*-level seismic forces.

❖ If the evaluation of the building, when hypothetically repaired to its predamaged condition, does not comply with the requirements established in Section 606.2.2.1, then the building must be rehabilitated as described in this section. The general requirement is to comply with the IBC load combinations, which are used to establish the required strength of structural members. The effects of wind and seismic loads warrant special consideration.

In determining the level of compliance for repairs to buildings that have sustained substantial structural damage, it is important to determine if wind forces have caused that damage. If so, it is considered prudent to require the repairs of wind damage to use wind loading that is the higher of the building code in effect at the time of original construction or the IBC. If not, the wind loading in effect at the time of the original construction is the basis for design. Seismic forces for the rehabilitation can be those required by the building code in effect at the time of the building's construction, but this may not be less than the reduced IBC seismic force level.

[BS] 606.2.3 Substantial structural damage to gravity load-carrying components. Gravity load-carrying components that have sustained *substantial structural damage* shall be rehabilitated to comply with the applicable provisions for dead and live loads in the *International Building Code*. Snow loads shall be considered if the *substantial structural damage* was caused by or related to snow load effects. Undamaged gravity load-carrying components that receive dead, live or snow loads from rehabilitated components shall also be rehabilitated if required to comply with the design loads of the *rehabilitation* design.

❖ Substantial structural damage to gravity load-carrying elements, such as columns or bearing walls, must be repaired so that these members are adequate to resist the dead and live loads in accordance with current code requirements, as must other elements of the load path. Snow loads must also be considered if the substantial damage was in any way caused by snow load effects.

[BS] 606.2.3.1 Lateral force-resisting elements. Regardless of the level of damage to gravity elements of the lateral force-resisting system, if substantial structural damage to gravity load-carrying components was caused primarily by wind or seismic effects, then the building shall be evaluated in accordance with Section 606.2.2.1 and, if noncompliant, rehabilitated in accordance with Section 606.2.2.3.

Exceptions:

1. Buildings assigned to Seismic Design Category A, B, or C whose substantial structural damage was not caused by earthquake need not be evaluated or rehabilitated for load combinations that include earthquake effects.

2. One- and two-family dwellings need not be evaluated or rehabilitated for load combinations that include earthquake effects.

❖ In determining the extent of repairs to these gravity load-carrying elements that are not part of the lateral force-resisting system, it is important to determine if wind or earthquakes have caused the structural damage. If the substantial structural damage is due to wind or earthquakes, then the lateral system is suspect and it must be checked even if no damage is apparent. Where this is the case, then the structure must be evaluated in accordance with Section 606.2.2.1.

There are two exceptions that acknowledge that applying this requirement in some cases is excessive and may have the effect of discouraging or delaying certain repairs by imposing the additional costs of a seismic upgrade. Exception 1 exempts buildings in areas of low or moderate seismicity, where the damage was caused by something other than an earthquake. Where earthquakes are rare, it serves no significant public purpose to trigger seismic upgrades following damage caused by fire, collision, wind and other events. Exception 2 exempts one- and two-family dwellings, where the risk to the public of poor earthquake performance is especially low. Basic repair—that is, restoring to the predamaged condition—is still required. Similar exceptions are found in Section 404.3.1.

[BS] 606.2.4 Flood hazard areas. In *flood hazard* areas, buildings that have sustained *substantial damage* shall be brought into compliance with Section 1612 of the *International Building Code*, or Section R322 of the *International Residential Code*, as applicable.

❖ The definition of "Flood hazard area" provided in Section 202 is taken from the IBC and establishes where this provision is applicable. If located in designated flood hazard areas, buildings that are damaged by any cause must be examined to determine if the damage constitutes substantial damage (see the definition of "Substantial damage" in Section 202). Buildings are considered to have sustained substantial damage when the cost of repairing the building to its predamaged condition is 50 percent or more of its market value before the damage occurred. Buildings determined to be substantially damaged must meet the flood-resistant provisions of the IBC (see FEMA-213, *Answers to Questions About Substantially Damaged Buildings*) or IRC Section R322, as applicable.

Section 1612 of the IBC and Section R322 of the IRC address requirements for buildings in designated flood hazard areas. The design and construction is required to be in accordance with ASCE 24, which in turn references the flood loading given in ASCE 7. Through use of these IBC and IRC provisions, communities meet a significant portion of the flood plain management regulation requirements necessary to participate in the National Flood Insurance Program (NFIP).

SECTION 607
ELECTRICAL

607.1 Material. Existing electrical wiring and equipment undergoing *repair* shall be allowed to be repaired or replaced with like material.

❖ In essence, this section states that existing wiring systems can be maintained in the same manner in which they were installed. Repairs are to be made with materials and components that do not in any way make the existing wiring system less safe. Materials and components can be replaced with items of equal quality and integrity or with items of superior quality and integrity. The intent is to allow necessary repairs without subjecting the system to new construction requirements. A wiring system material or component that is obsolete or no longer recognized by current codes is permitted to be used for the purpose of making repairs provided that it is consistent with the existing materials and components, and the system is not made any less safe.

607.1.1 Receptacles. Replacement of electrical receptacles shall comply with the applicable requirements of Section 406.4(D) of NFPA 70.

❖ This section ties the replacement of receptacle devices to the provisions of NFPA 70, which do not always allow replacement with similar devices. For example, where a grounding means exists, ungrounded-type (two conductor) receptacles must be replaced only with grounding-type (three conductor) receptacles. Similarly, receptacles in locations where ground-fault circuit-interrupter (GFCI) protection is required must be replaced only with GFCI-type receptacles or the branch circuit must provide such GFCI protection. See Sections 406.4(D)(1) through (D)(6) of NFPA 70 for more replacement provisions.

607.1.2 Plug fuses. Plug fuses of the Edison-base type shall be used for replacements only where there is no evidence of over fusing or tampering per applicable requirements of Section 240.51(B) of NFPA 70.

❖ This section does not allow Edison-base-type (screw base) fuses to replace existing fuses, except where there is no reason to believe that the wrong size fuses have been or are being used or where there is no evidence of attempts to defeat the protection afforded by the fuses. Edison-base fuses are plug-style fuses with the same screw thread base as the common incandescent lamp. Such fuses have ampere ratings of 30 amps and less, and are interchangeable, meaning that occupants are not prevented from inserting fuses that are rated higher than the capacity of the wiring they are intended to protect; therefore, fire hazards are likely to be created.

Evidence of oversized fuses being used or tampered with is the code's justification for prohibiting the installation of any new Edison-base plug fuses in existing fuseholders. Obviously, the hazardous condition of overfusing or tampering must not be allowed to continue to exist. In such cases, replacement fuses must be Type S fuses, which are designed to thwart overfusing and attempts at tampering or bypassing.

607.1.3 Nongrounding-type receptacles. For replacement of nongrounding-type receptacles with grounding-type receptacles and for branch circuits that do not have an equipment grounding conductor in the branch circuitry, the grounding conductor of a grounding-type receptacle outlet shall be permitted to be grounded to any accessible point on the grounding electrode system or to any accessible point on the grounding electrode conductor in accordance with Section 250.130(C) of NFPA 70.

❖ This section is a recognition of the provisions of Section 250.130(C) of NFPA 70. Section 607.1.1 addresses circumstances where existing receptacles cannot be replaced with like receptacles and this section describes one of the options specified in Section 406.4(D) of NFPA 70.

607.1.4 Group I-2 receptacles. Non-"hospital grade" receptacles in patient bed locations of Group I-2 shall be replaced with "hospital grade" receptacles, as required by NFPA 99 and Article 517 of NFPA 70.

❖ This section reflects the intent of Section 517.18(B) of NFPA 70 and it's an exception.

607.1.5 Grounding of appliances. Frames of electric ranges, wall-mounted ovens, counter-mounted cooking units, clothes dryers and outlet or junction boxes that are part of the existing branch circuit for these appliances shall be permitted to be grounded to the grounded circuit conductor in accordance with Section 250.140 of NFPA 70.

❖ This section is a recognition of a provision of Section 250.140 of NFPA 70 that parallels the intent of the code to allow repairs to be consistent with the original installation. Although allowed in the past, grounding of appliances to the grounded circuit conductor is considered to be unnecessarily risky and, therefore, is now allowed only for existing wiring installations under specified conditions that serve to limit the risk to an acceptable level.

SECTION 608
MECHANICAL

608.1 General. Existing mechanical systems undergoing *repair* shall not make the building less conforming than it was before the *repair* was undertaken.

❖ This section is essentially a referencing section to direct the user's attention to the possibility of other sections that might be relevant to mechanical systems. Repair work must not alter the nature of appliances and equipment in a way that would invalidate the listing or conditions of approval.

608.2 Mechanical draft systems for manually fired appliances and fireplaces. A mechanical draft system shall be permitted to be used with manually fired appliances and fireplaces where such a system complies with all of the following requirements:

1. The mechanical draft device shall be listed and installed in accordance with the manufacturer's installation instructions.

2. A device shall be installed that produces visible and audible warning upon failure of the mechanical draft device or loss of electrical power at any time that the mechanical draft device is turned on. This device shall be equipped with a battery backup if it receives power from the building wiring.

3. A smoke detector shall be installed in the room with the appliance or fireplace. This device shall be equipped with a battery backup if it receives power from the building wiring.

❖ This section contains a remedy for existing chimneys that do not produce sufficient draft. Some chimneys fail to produce sufficient draft intermittently because of wind speed and direction or outdoor temperatures. Gas-fired and oil-fired appliances can be interlocked to the exhauster to immediately shut off the flow of fuel if there is a power failure or malfunction of the exhauster. Obviously, the same interlock cannot be used in a wood-burning fireplace or stove. This section allows the use of mechanical draft systems with solid fuel-burning appliances and fireplaces if certain requirements are met.

The first requirement is that the draft device must be listed for this application and installed in compliance with the manufacturer's installation instructions. This will ensure that the draft system is installed in the same way it was tested in the laboratory of the listing agency.

The second requirement is that a visible and audible alarm be installed to warn occupants upon failure of the mechanical draft device or loss of electrical power. If the exhauster fails to operate, the solid fuel will continue to burn and produce smoke and other products of combustion. The occupants of the building must be warned that the potentially deadly products of combustion might be spilling into the living space. This is especially important in residential occupancies where the occupants may have gone to sleep with the fireplace or appliance still burning.

The third requirement is that a smoke detector must be installed in the room with the appliance to provide further warning should the exhauster fail to operate properly. This section is not intended to allow mechanical draft systems (chimney exhausters) as a substitute for proper chimney design and construction. It is intended to apply to existing chimneys that fail to produce the required draft.

SECTION 609
PLUMBING

609.1 Materials. Plumbing materials and supplies shall not be used for repairs that are prohibited in the *International Plumbing Code*.

❖ This section essentially refers the user to the *International Plumbing Code*® (IPC®) in order to determine prohibited materials for use in the repair of plumbing systems. For example, Chapters 7 and 8 of the IPC contain prohibited plumbing system joint and connection methods that would also be applicable to repair work.

609.2 Water closet replacement. The maximum water consumption flow rates and quantities for all replaced water closets shall be 1.6 gallons (6 L) per flushing cycle.

Exception: Blowout-design water closets [3.5 gallons (13 L) per flushing cycle].

❖ Federal legislation mandates the design and use of water closets to have flow rates restricted to 1.6 gallons (6.1 L) per flushing cycle, except where blowout-type fixtures are utilized. Blowout-designed water closets are exempt from the 1.6-gallon (6.1 L) requirement because such fixtures depend on high-volume and high-velocity water flow to evacuate the contents of the bowl. Therefore, such fixture designs may not be able to function at lower consumption rates.

Bibliography

The following resource materials were used in the preparation of the commentary for this chapter of the code.

ATC 20, *Procedures for Postearthquake Safety Evaluation of Buildings*. Redwood City, CA: Applied Technology Council, 1989.

ATC 20-2, *Addendum to the ATC 20 Postearthquake Building Safety Evaluation Procedures*. Redwood City, CA: Applied Technology Council, 1995.

ATC 40, *Seismic Evaluation and Retrofit of Concrete Buildings*. Redwood City, CA: Prepared by the Applied Technology Council for the California Seismic Safety Commission, 1996.

Chang, David and Hamid Naderi. "International Existing Building Code." *Structural Engineer*. October 2003, pp 22-25.

Chen, Wai-Fah and Charles Scawthorn. *Earthquake Engineering Handbook*. Boca Raton, FL: CRC Press LLC, 2003.

FEMA 178, *NEHRP Handbook for the Seismic Evaluation of Existing Buildings*. Washington, DC: Federal Emergency Management Agency, 1992.

FEMA 213, *Answers to Questions About Substantially Damaged Buildings*. Washington, DC: Federal Emergency Management Agency, 1991.

FEMA 259, *Engineering Principles and Practices for Retrofitting Flood-prone Residential Buildings*. Washington, DC: Federal Emergency Management Agency, 1995.

FEMA 273, *NEHRP Guidelines for the Seismic Rehabilitation of Buildings*. Washington, DC: Federal Emergency Management Agency, 1997.

FEMA 274, *NEHRP Commentary on the Guidelines for the Seismic Rehabilitation of Buildings*. Washington, DC: Federal Emergency Management Agency, 1997.

FEMA 310, *Handbook for the Seismic Evaluation of Buildings—A Prestandard*. Washington, DC: Federal Emergency Management Agency, 1998.

FEMA 311, *Guidance on Estimating Substantial Damage Using the NFIP Residential Substantial Damage Estimator*. Washington, DC: Federal Emergency Management Agency, 1998.

FEMA 312, *Homeowner's Guide to Retrofitting—Six Ways to Protect Your House From Flooding*. Washington, DC: Federal Emergency Management Agency, 1998.

FEMA 356, *Prestandard and Commentary for the Seismic Rehabilitation of Buildings*. Washington, DC: Federal Emergency Management Agency, 2000.

FEMA 450-1, *NEHRP Recommended Provisions for Seismic Regulations for New Buildings and Other Structures, Part 1*. Washington, DC: Prepared by the Building Seismic Safety Council for the Federal Emergency Management Agency, 2004.

FEMA 450-2, *NEHRP Recommended Provisions for Seismic Regulations for New Buildings and Other Structures Commentary, Part 2 Commentary*. Washington, DC: Prepared by the Building Seismic Safety Council for the Federal Emergency Management Agency, 2004.

Ghosh, S.K. and David A. Fanella. *Seismic and Wind Design of Concrete Buildings (2000 IBC, ASCE 7-98, ACI 318-99)*. Washington, DC: International Code Council, 2003.

Leyendecker, E.V., A.D. Frankel, K.S. Rustales. *Seismic Design Parameters*, U.S. Geological Survey Open-File Report 01-437. Washington, DC: United States Geological Survey, 2001.

Naeim, Farzad. *The Seismic Design Handbook*. Norwell, MA: Kluwer Academic Publishers, 2001.

Recommended Lateral Force Requirements and Commentary (SEAOC Blue Book). Sacramento, CA: Structural Engineers Association of California, 1999.

Reducing Flood Losses Through the International Code Series: Meeting the Requirements of the National Flood Insurance Program. Washington, DC: International Code Council, May 2000.

Taly, Narendra. *Loads and Loads Paths in Buildings Principles of Structural Design*. Washington, DC: International Code Council, 2003.

Williams, Alan. *Seismic and Wind Forces Structural Design Examples*. Washington, DC: International Code Council, 2003.

Chapter 7:
Alterations—Level 1

General Comments

This chapter provides the technical requirements for existing buildings that undergo Level 1 alterations as described in Section 503, including the replacement or covering of existing materials, elements, equipment or fixtures using new materials for the same purpose.

This chapter, similar to other chapters of the code, covers all building-related subjects, such as structural, mechanical, plumbing, electrical and accessibility, as well as the fire and life safety issues when the alterations are classified as Level 1. Sections 701 and 702 are related to scoping building elements and materials. It should be noted that, in the interest of being brief and avoiding the presentation of materials that are repetitive in nature in various chapters, Section 801.2 requires that Level 2 alterations comply not only with Chapter 8, but also with Chapter 7. Similarly, Section 901.2 requires that Level 3 alterations comply with Chapters 7 and 8, as well as Chapter 9. As such, Chapter 7 is applicable to all levels of alteration work. Section 702, Building Elements and Materials, covers, in detail, elements such as interior finishes and carpeting. Section 702 also refers to the *International Building Code®* (IBC®), *International Energy Conservation Code®* (IECC®), *International Mechanical Code®* (IMC®) and *International Plumbing Code®* (IPC®) for new materials. The remainder of the chapter is related to fire protection, means of egress, accessibility, structural and energy conservation.

Level 1 alterations are considered the least drastic. Known in past codes simply as "alterations," the classification was broken into three levels based on the fact that minor alterations, not including space reconfiguration and extensive alterations that might include relocation of walls and partitions in the entire building, should be treated differently and with different threshold levels for requiring upgrades or improvements to the building or spaces in the building.

Purpose

The purpose of this chapter is to provide detailed requirements and provisions to identify mandated improvements in existing building elements, building spaces and the building's structural system. This chapter is distinguished from Chapters 8 and 9 by only involving the replacement of building components with new components. Level 2 alterations involve more space reconfiguration, while Level 3 alterations involve space reconfiguration exceeding 50 percent of the building area.

SECTION 701
GENERAL

701.1 Scope. Level 1 *alterations* as described in Section 503 shall comply with the requirements of this chapter. Level 1 *alterations* to *historic buildings* shall comply with this chapter, except as modified in Chapter 12.

❖ Level 1 alterations are described in Section 503 as the type of alterations that include the removal and replacement or the covering of existing materials and elements. The scoping provisions of this section refer the user to Section 503 to be certain that the work classification is Level 1 alterations and that Chapter 7 is the appropriate chapter to be used. Historic buildings have unique allowances and requirements in Chapter 12. The provisions of this chapter apply but the requirements of this chapter may be superseded by allowances or requirements in Chapter 12 that are specific to historic buildings.

701.2 Conformance. An *existing building* or portion thereof shall not be altered such that the building becomes less safe than its existing condition.

Exception: Where the current level of safety or sanitation is proposed to be reduced, the portion altered shall conform to the requirements of the *International Building Code*.

❖ The current level of safety or level of compliance with regulatory provisions in a building is not, in general, allowed to be reduced, regardless of the type of work taking place in the building. For example, a Class B interior wall finish in an exit passageway will not be allowed to be replaced with a finish material that has a Class C rating. The exception describes the only situation where the reduction of the level of safety or the level of compliance is allowed. This is the condition where the existing level of compliance is above the level required by the IBC and the reduced level of

ALTERATIONS—LEVEL 1

compliance still meets or exceeds the IBC requirements. Some other examples where the reduction of level of compliance is not allowed might be to meet water supply requirements (see Commentary Figure 701.2), and to meet electrical loads or reduction in existing light and ventilation to below the levels required in the IBC, or below the existing levels, whichever is lower.

[BS] 701.3 Flood hazard areas. In *flood hazard areas*, *alterations* that constitute *substantial improvement* shall require that the building comply with Section 1612 of the *International Building Code*, or Section R322 of the *International Residential Code*, as applicable.

❖ When alterations to existing buildings that are located in flood hazard areas are proposed, determinations are to be made as to whether the proposed work is a substantial improvement. If the proposed alterations are determined to be substantial improvements, then the existing building is to be brought into compliance with the flood-resistant provisions of the IBC, which requires the design and construction of the building to comply with ASCE 24.

SECTION 702
BUILDING ELEMENTS AND MATERIALS

702.1 Interior finishes. All newly installed interior wall and ceiling finishes shall comply with Chapter 8 of the *International Building Code*.

❖ Newly installed interior finish materials are required to comply with the flame spread and smoke-developed index requirements of the IBC. The level of developed smoke, surface burning, flame spread and toxic byproducts of combustion are critical elements to be considered in fire situations. For this reason, the code explicitly outlines the criteria for interior finishes. These are requirements that have potentially dramatic effects on building and occupant safety in fire situations.

702.2 Interior floor finish. New interior floor finish, including new carpeting used as an interior floor finish material, shall comply with Section 804 of the *International Building Code*.

❖ The same description and reasoning provided in Section 702.1 apply here to materials newly installed as floor finish materials, including, but not limited to, carpeting.

702.3 Interior trim. All newly installed interior trim materials shall comply with Section 806 of the *International Building Code*.

❖ The same description and reasoning provided in Section 702.1 apply here to materials newly installed as interior trim materials.

702.4 Window opening control devices. In Group R-2 or R-3 buildings containing dwelling units and one- and two-family dwellings and townhouses regulated by the *International Residential Code*, window opening control devices complying with ASTM F2090 shall be installed where an existing window is replaced and where all of the following apply to the replacement window:

1. The window is operable;

2. The window replacement includes replacement of the sash and the frame;

3. One of the following applies:

    3.1. In Group R-2 or R-3 buildings containing dwelling units, the top of the sill of the window opening is at a height less than 36 inches (915 mm) above the finished floor; or

    3.2. In one- and two-family dwellings and townhouses regulated by the *International Residential Code*, the top sill of the window opening is at a height less than 24 inches (610 mm) above the finished floor;

4. The window will permit openings that will allow passage of a 4-inch-diameter (102 mm) sphere when the window is in its largest opened position; and

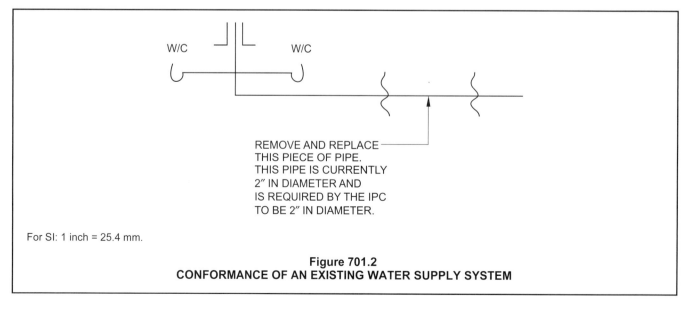

For SI: 1 inch = 25.4 mm.

Figure 701.2
CONFORMANCE OF AN EXISTING WATER SUPPLY SYSTEM

5. The vertical distance from the top of the sill of the window opening to the finished grade or other surface below, on the exterior of the building, is greater than 72 inches (1829 mm).

The window opening control device, after operation to release the control device allowing the window to fully open, shall not reduce the minimum net clear opening area of the window unit to less than the area required by the *International Building Code*.

Exceptions:

1. Operable windows where the top of the sill of the window opening is located more than 75 feet (22 860 mm) above the finished grade or other surface below, on the exterior of the room, space or building, and that are provided with window fall prevention devices that comply with ASTM F2006.

2. Operable windows with openings that are provided with window fall prevention devices that comply with ASTM F2090.

❖ This section requires that, when windows are replaced, window opening control devices be provided to protect children from falls. There are several conditions that must be met before this section becomes applicable. The first condition is that the window must be operable. If the window is not operable then the concern for safety from falling is not present. The second condition relates to whether the entire window including the sash and frame is being replaced. If it is simply a repair of a pane of glass then the requirement would not be applicable. Item 3 addresses the specific location of the opening on the inside of the building [see Commentary Figure 702.4(1)]. Item 4 focuses these provisions on not only operable windows but also windows through which a child could actually fit. If a 4-inch-diameter sphere can pass through a window opening, so can a child; thus, guards must be installed. Finally, Item 5 addresses whether the window is high enough from the ground to be of concern for injury from falling [see Commentary Figure 702.4(2)]. This section applies to the following occupancies.

- Group R-2 Occupancies containing dwelling units.
- Group R-3 Occupancies containing dwelling units.
- One- and two-family dwellings and townhouses.

The control device can not reduce to less than that required in IBC Section 1030.2. Section 406.2 also addresses this issue for the prescriptive method.

702.5 Emergency escape and rescue openings. Where windows are required to provide emergency escape and rescue openings in Group R-2 and R-3 occupancies and one- and two-family dwellings and townhouses regulated by the *International Residential Code*, replacement windows shall be exempt from the requirements of Sections 1030.2, 1030.3 and 1030.5 of the *International Building Code* and Sections R310.21 and R310.2.3 of the *International Residential Code* accordingly, provided the replacement window is the manufacturer's largest standard size window that will fit within the existing frame or existing rough opening. The replacement window shall be permitted to be of the same operating style as the existing window or a style that provides for an equal or greater window opening area than the existing window.

Window opening control devices complying with ASTM F2090 shall be permitted for use on windows required to provide *emergency escape* and *rescue openings*.

❖ This section focuses on how to address the replacement of windows intended to be emergency escape and rescue openings. More specifically, Groups R-2

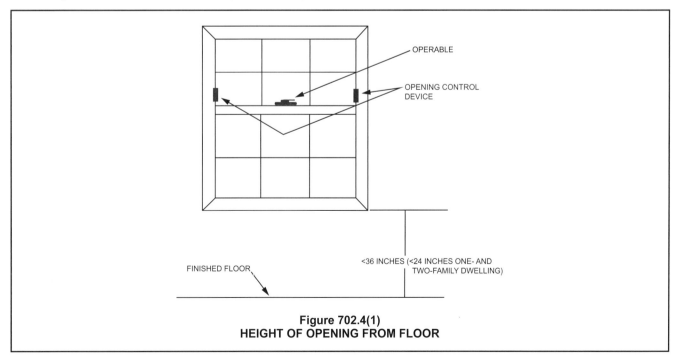

Figure 702.4(1)
HEIGHT OF OPENING FROM FLOOR

and R-3 and one- and two-family dwellings in new construction require certain windows to be specifically sized for escape and rescue in these structures. Meeting these requirements in existing buildings can be difficult. This section allows existing buildings to provide the largest window that can fit into the current opening without requiring full compliance with the IBC or *International Residential Code*® (IRC®), as applicable. There are two criteria that must be met. The first is that the window is the largest size that the manufacturer can provide for that opening (the existing frame or rough opening). The second permits the operating style of the original window to be replicated, but the replacement window must provide equal or greater opening area.

This section also permits the use of window opening control devices as required in Section 702.4. The focus of Section 702.4 is on fall safety, whereas this section is focused on emergency escape. This provision avoids conflict between the two requirements.

702.6 Materials and methods. All new work shall comply with the materials and methods requirements in the *International Building Code*, *International Energy Conservation Code*, *International Mechanical Code*, and *International Plumbing Code*, as applicable, that specify material standards, detail of installation and connection, joints, penetrations, and continuity of any element, component, or system in the building.

❖ Materials and methods requirements refer to the requirements of various codes, such as the IBC, IMC and IPC, that specify material standards, details of installation and connection, joints, penetrations and continuity of any element, component or system in the building. This description, which is not presented in the form of a definition, is so broad that there are numerous sections in the *International Codes*® (I-Codes®) that are considered to be related to materials and methods. One way of dealing with this section might have been to list every section in every I-Code that deals with materials and methods. As this would have been a long list of sections, the committee and the membership chose to address this issue in general terms rather than listing such sections from the I-Codes, except for the materials and methods sections of the *International Fuel Gas Code*® (IFGC®) that have been specifically listed in Section 702.1 of the code.

All new work in Level 1 alterations must comply with the materials and methods requirements of the I-Codes.

For example, the sheetrock from one side of an existing corridor is to be removed and replaced with new sheetrock and finish paneling. Regardless of how the existing sheetrock was attached and what the characteristics of the finish material were, the new sheetrock must comply with material referenced standards in IBC Section 2506 and its installation must be in accordance with IBC Table 2508.1. Accordingly, the interior finish paneling used must be of a class in compliance with IBC Table 803.11.

[FG] 702.6.1 International Fuel Gas Code. The following sections of the *International Fuel Gas Code* shall constitute the fuel gas materials and methods requirements for Level 1 alterations.

1. All of Chapter 3, entitled "General Regulations," except Sections 303.7 and 306.

2. All of Chapter 4, entitled "Gas Piping Installations," except Sections 401.8 and 402.3.

    2.1. Sections 401.8 and 402.3 shall apply when the work being performed increases the load on the system such that the existing pipe does not

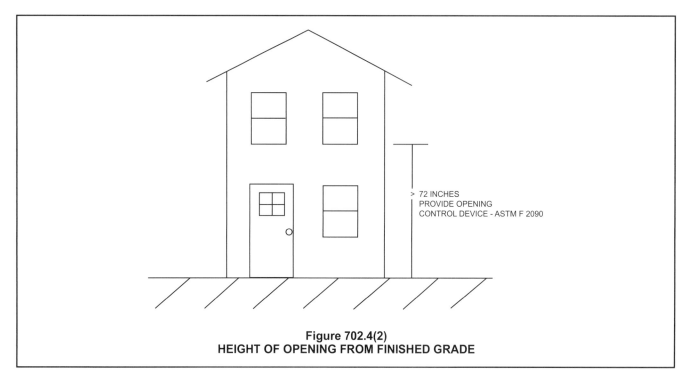

Figure 702.4(2)
HEIGHT OF OPENING FROM FINISHED GRADE

meet the size required by code. Existing systems that are modified shall not require resizing as long as the load on the system is not increased and the system length is not increased even if the altered system does not meet code minimums.

3. All of Chapter 5, entitled "Chimneys and Vents."
4. All of Chapter 6, entitled "Specific Appliances."

❖ Any alteration of fuel gas equipment or piping that falls under the category of Level 1 alterations must comply with the materials and methods requirements of the IFGC. This section identifies those sections of the IFGC that are considered to be related to materials and methods. These, with the exception of four specific sections, include all of Chapters 3, 4, 5 and 6 of the IFGC.

SECTION 703
FIRE PROTECTION

703.1 General. *Alterations* shall be done in a manner that maintains the level of fire protection provided.

❖ Any level of fire protection that currently exists in a building must not be adversely affected or lessened as a result of any alteration. For example, removing and replacing the existing ceiling and rearranging some fire sprinkler heads should ultimately provide the same level of sprinkler coverage and protection that existed prior to the alterations being undertaken. Section 701.2, however, allows a reduction in safety and sanitation if it complies with the IBC. Section 803.6 allows a reduction in fire-resistance ratings in a building where an automatic sprinkler system is installed. This is based on an analysis and the documentation that must be reviewed and approved by the code official. In this context, sprinklers are addressed in Chapter 8, as the installation of new systems is considered a Level 2 alteration.

SECTION 704
MEANS OF EGRESS

704.1 General. *Alterations* shall be done in a manner that maintains the level of protection provided for the means of egress.

❖ Any level of safety that currently exists in a building's means of egress system must not be adversely affected or lessened as a result of any alteration. This means that any new building element that is replacing the old one or new elements that are covering existing elements must be at least as safe as the old one. Chapter 7 must be checked for those cases where the building elements or components being installed as part of the alteration might need to comply with the IBC or the IRC. For example, replacement interior finishes are not allowed to simply duplicate the old one; they must meet the IBC requirements based on Section 702.1.

SECTION 705
ACCESSIBILITY

705.1 General. A *facility* that is altered shall comply with the applicable provisions in Sections 705.1.1 through 705.1.14, and Chapter 11 of the *International Building Code* unless it is *technically infeasible*. Where compliance with this section is *technically infeasible*, the alteration shall provide access to the maximum extent that is technically feasible.

A *facility* that is constructed or altered to be accessible shall be maintained accessible during occupancy.

Exceptions:

1. The altered element or space is not required to be on an accessible route unless required by Section 705.2.
2. Accessible means of egress required by Chapter 10 of the *International Building Code* are not required to be provided in existing *facilities*.
3. Type B dwelling or sleeping units required by Section 1107 of the *International Building Code* are not required to be provided in existing *facilities* undergoing less than a Level 3 *alteration*.
4. The alteration to Type A individually owned dwelling units within a Group R-2 occupancy shall meet the provisions for Type B dwelling units.

❖ The purpose of Section 705.1 is to establish the minimum criteria for accessibility when dealing with existing buildings and facilities that are being renovated or altered. The history and efforts involved are similar to that discussed in the commentary to Chapter 11 of the IBC. Access to buildings and structures for people with physical disabilities has been a subject regulated by the building codes since the early 1970s. The codes have consistently relied on a consensus national standard, ICC A117.1, as the technical basis for accessibility. There has been a great deal of emphasis and awareness placed on the subject of accessibility through the passage of two federal laws. The Americans with Disabilities Act (ADA) and the Fair Housing Amendment Act (FHAA) are federal regulations that affect building construction as it relates to accessibility.

The International Code Council® (ICC®) recognizes the value of consistency between federal laws and the codes. Efforts for coordination with the federal accessibility requirements are ongoing. Representatives from interested accessibility groups, the Department of Housing and Urban Development (HUD) and the Architectural and Transportation Barriers Compliance Board (ATBCB, commonly referred to as the Access Board), have been attending and participating in the code change process for the IBC and ICC A117.1. In addition, the ICC has participated in the public comment process on the development of federal regulations for accessibility. Appendix B includes information found in the *Americans with Disabilities Act Accessibility Guidelines* (ADAAG) that cannot be enforced through the typical code enforcement process, but would provide beneficial information for the designer/owner for full compliance purposes. The ICC has worked toward, and will continue to strive for, accessibility regulations that reflect the highest possible degree of consistency with federal regulations and, more importantly, reasonable and appropriate provisions to meet the needs of people with disabilities.

The code approaches the application of accessibility provisions to a facility that is altered by broadly requiring full conformance to new construction, meaning full accessibility is expected (see Section 705.1.13). Exceptions are then provided to indicate the conditions under which less than full accessibility is permitted.

For example, if a door and frame are removed and replaced, the door must meet the requirements for width, height, maneuvering clearances and hardware. If just the doorknob is being removed, it must be replaced with lever hardware.

The circumstance under which full compliance with accessibility provisions is not required is when it is deemed to be technically infeasible (see the commentary to the definition of "Technically infeasible" in Section 202). This is considered reasonable since, if not provided for, plans for alterations may be otherwise abandoned by the building owner. The opportunity to upgrade and increase the current level of accessibility in an existing building would then be lost. This concern is also embodied in the requirement that an altered element or space is expected to be made accessible to the extent to which it is technically feasible to do so. In this manner, the code accomplishes the greatest degree of accessibility while at the same time recognizing the justifiable difficulties that may be involved in providing full accessibility in existing buildings.

In accordance with Exception 1, if the area undergoing alteration does not contain a primary function (see Section 705.2), there are no additional requirements past the original scope of the project. However, if the area contains a primary function, there are additional criteria to achieve accessibility that may require work not in the original scope of the project. These additional criteria are to provide an accessible route to the altered area, as well as improvements to any toilets and drinking fountains that serve the altered area. Requirements for an accessible route might specify, for instance, that the door discussed in the earlier example be removed and replaced because it does not have adequate width or maneuvering clearances.

Exception 2 indicates that accessible means of egress are not required as a result of undertaking alterations to existing buildings. Strict compliance with IBC Section 1009 is often technically infeasible. The requirement for a 48-inch (1219 mm) clear width between handrails would require many stairways to be widened. This often would entail the movement of major structural elements in order to accomplish this alteration. Note that this is not an exception for the accessible entrance requirements.

Exception 3 exempts existing buildings with a Level 1 or 2 alteration from having to provide Type B dwelling and sleeping units. Since this section is referenced from Chapter 10, this would include existing buildings undergoing a change of occupancy. Note that Chapter 10 has specific requirements for Type B dwelling and sleeping units that are both undergoing a change of occupancy classification and have a Level 3 alteration associated with the change of occupancy. Additions that contain four or more dwelling and sleeping units would be required to provide Type B units in accordance with IBC Section 1107. It should be noted that Accessible and Type A dwelling and sleeping units are required in existing residential and institutional buildings undergoing Level 1 or greater alterations (see Section 705.1.8 and the definitions in IBC Section 201 for "Accessible unit" and "Dwelling or sleeping unit, Type A").

The Type B dwelling units must be provided in Level 3 alterations. Where Level 3 alterations are being performed, there is a prime opportunity to have those buildings move toward being able to serve a wider range of the population. With the population of the United States aging, there will be a steadily increasing demand for units that include accessibility features. In addition, while the FHA is only applicable to new construction, this law was enacted in 1991. Some buildings constructed after that time may not be in compliance with the FHA. When a major alteration/renovation occurs, there is an opportunity to bring those buildings into compliance. See the commentary to Sections 705.1.7, 705.1.8, 1105.2, 1105.3 and 1105.4, and the exceptions to Sections 1012.8 and 1012.8.2 for additional information.

Exception 4 addresses the specific circumstances when an existing Type A dwelling unit is being altered. While Section 410.2 says that a level of accessibility must be maintained, in the situations where a Type A dwelling unit is "individually owned," such as a condominium, then it only needs to meet the technical requirements for a Type B dwelling unit (see ICC A117.1, Section 1004) when it is altered or remodeled by the owner. Type B units require a lesser level of accessibility than Type A units.

For example, if an owner wanted to alter the bathroom in his or her unit, he or she would only be required to meet the lesser accessibility requirements of a Type B unit with the new construction. This exception would not be applicable to units that are for rent.

The specific provisions of the following subsections are intended to reflect conditions under which less than full accessibility, as would be required in new construction, is permitted in altered areas. As previously discussed, Section 705.1 requires alterations to comply with the full range of accessibility-related provisions of the code for new construction. The exceptions and subsections reflect a reasonable set of conditions under which a different level of accessibility can be provided. Sections 705.1.1 through 705.1.14 are part of the IBC coordination effort with the ICC A117.1 accessibility standard and the recommendations for the ADAAG Review Federal Advisory Committee.

705.1.1 Entrances. Where an *alteration* includes alterations to an entrance, and the *facility* has an accessible entrance on an accessible route, the altered entrance is not required to be accessible unless required by Section 705.2. Signs complying with Section 1111 of the *International Building Code* shall be provided.

❖ This provision is contained here to point to the accessibility provisions of IBC Chapter 11 for entrances. A facility is not accessible if the entrances into it are inaccessible. IBC Section 1105 establishes reasonable criteria for providing accessible entrances. A facility is not required to have all of its entrances accessible in order to provide reasonable accommodation to disabled persons. If a facility has multiple public entrances, as a minimum, it is not considered unreasonable to require at least 60 percent of the entrances to be accessible. In addition to the 60-percent accessible public entrances, entrances that have a specific function or provide access to only certain portions of the facility must be addressed.

If the building already has the accessible entrances required by IBC Section 1105, an entrance that is being altered is not required to be made accessible. An exception to this would be if the entrance was required to be made accessible as part of the route to the altered primary function area. If not all entrances are accessible, provide appropriate signage to notify persons with disabilities when an entrance is or is not accessible and, if not accessible, direct them to the nearest accessible entrance.

705.1.2 Elevators. Altered elements of existing elevators shall comply with ASME A17.1/CSA B44 and ICC A117.1. Such elements shall also be altered in elevators programmed to respond to the same hall call control as the altered elevator.

❖ Requirements for new construction state that all elevators on an accessible route must be fully accessible in accordance with ICC A117.1. If a passenger elevator is altered, the altered element must be accessible in accordance with the requirements for existing elevators in Section 407 of ICC A117.1. If the altered elevator is part of a bank of elevators, the same element must be made accessible in every elevator that is part of that bank. The purpose of this requirement is to have consistency among elevators in a bank so that disabled people are not required to wait for a specific elevator, whereas the general population can take the first available elevator. The ICC A117.1 also provides accessibility requirements for limited use/limited access (LULA) elevators (ICC A117.1, Section 408) and private residence elevators (ICC A117.1, Section 409). While these types of elevators can serve as part of an accessible route, their use is limited by ASME A17.1.

705.1.3 Platform lifts. Platform (wheelchair) lifts complying with ICC A117.1 and installed in accordance with ASME A18.1 shall be permitted as a component of an accessible route.

❖ This section provides for the use of platform (wheelchair) lifts in existing buildings. In order to create an accessible route where there are changes in floor levels, the provisions for new construction would most often require the installation of an elevator or ramp. Platform lifts are allowed in new construction for limited conditions (see IBC Section 1109.8). If the space in an existing building precludes the installation of an elevator or ramp, a platform lift may be the only practical solution. Given the choice between no accessibility or accessibility by a platform lift, accessibility is preferred. Previously, platform lift requirements were addressed in the elevator standard, ASME A17.1, but they are now addressed in their own standard, ASME A18.1. One of the many changes was the removal of the requirement for key operation, which previously discouraged independent utilization of platform lifts. Note that in accordance with IBC Section 1009.5, platform lifts are also permitted for an accessible means of egress in some locations. However, accessible means of egress are not required in existing buildings in accordance with Section 705.1, Exception 2.

A chair lift is not acceptable as part of an accessible route in either new or existing construction. For examples of the difference between a chair lift and a platform lift, see the commentary to IBC Section 1109.8.

705.1.4 Ramps. Where steeper slopes than allowed by Section 1012.2 of the *International Building Code* are necessitated by space limitations, the slope of ramps in or providing access to existing facilities shall comply with Table 705.1.4.

❖ This section recognizes the circumstances where, because of existing site or configuration constraints, a ramp with a slope of one unit vertical in 12 units horizontal (1:12) may not be feasible. A steeper slope is allowed where the elevation change does not exceed 6 inches (152 mm). The remainder of ramp requirements, such as width, landings, etc., is set forth in IBC Section 1010.

**TABLE 705.1.4
RAMPS**

SLOPE	MAXIMUM RISE
Steeper than 1:10 but not steeper than 1:8	3 inches
Steeper than 1:12 but not steeper than 1:10	6 inches

For SI: 1 inch = 25.4 mm.

❖ In existing buildings, ramps that rise 3 inches (76 mm) or less may have a slope as steep as one unit vertical in eight units horizontal (1:8). In existing buildings, ramps that rise 6 inches (152 mm) or less may have a slope as steep as one unit vertical in 10 units horizontal (1:10). If it is possible to provide a lesser slope, it is desirable to do so. These steeper slopes should only be utilized when the one unit vertical in 12 units horizontal (2 percent) (1:12) slope is not possible.

705.1.5 Dining areas. An accessible route to raised or sunken dining areas or to outdoor seating areas is not required provided that the same services and decor are provided in an accessible space usable by any occupant and not restricted to use by people with a disability.

❖ This section is intended to provide equal access to dining services for a disabled individual without segregation in existing buildings undergoing alterations. If equivalent dining services are available on an accessible level, an accessible route to other levels, or to outside dining, is not required as part of the alteration to an existing dining area. For example, where a snack bar is located on one level, while full dining is provided on two other levels, an accessible route would be required to the snack bar level and one of the full dining levels.

705.1.6 Jury boxes and witness stands. In *alterations*, accessible wheelchair spaces are not required to be located within the defined area of raised jury boxes or witness stands and shall be permitted to be located outside these spaces where ramp or lift access poses a hazard by restricting or projecting into a required means of egress.

❖ This exception for jury boxes and witness stands is consistent with Sections 231 and 808 of the ADA *Standards for Accessible Design*. The intent is that if ramp access to a jury box or witness stand would have the ramp limiting or blocking the means of egress for the general population in the space, alternative locations for potential jurors or witnesses are viable. The same alternative is available if a platform lift would effectively be an obstruction for the general means of egress from the courtroom.

705.1.7 Accessible dwelling or sleeping units. Where Group I-1, I-2, I-3, R-1, R-2 or R-4 dwelling or sleeping units are being altered, the requirements of Section 1107 of the *International Building Code* for Accessible units apply only to the quantity of the spaces being altered.

❖ This section sets forth the rate for providing Accessible dwelling or sleeping units in Groups I-1, I-2, I-3, R-1, R-2 and R-4 when such facilities are altered (also see Section 1105.2, and the definitions of "Dwelling unit," "Sleeping unit" and "Accessible unit" in IBC Chapter 2). Assuming that the required number of Accessible units is not already provided, the number of Accessible units to be incorporated into each alteration is based on the number being altered. For example, if a nursing home was being altered a portion at a time, 50 percent of the units being altered each time would be required to be wheelchair accessible. It is not the intent that all units being altered are required to be Accessible units until 50 percent of the units in the entire facility are Accessible units. The total number of Accessible units in the facility is not required to exceed that required for new construction, as indicated in Section 705.1.7. It is unreasonable to require a greater level of accessibility in an existing building than is required in new construction. The technical criteria for Accessible units is found in ICC A117.1, Section 1002.

705.1.8 Type A dwelling or sleeping units. Where more than 20 Group R-2 dwelling or sleeping units are being altered, the requirements of Section 1107 of the *International Building Code* for Type A units and Chapter 9 of the *International Building Code* for visible alarms apply only to the quantity of the spaces being altered.

❖ Type A units are required in new construction when 20 or more apartments (including condominium style) are constructed on a site. Group R-2 requirements for Type A units would also include convents and monasteries and could include some townhouse-style units (see the commentary to IBC Section 1107.6.2.2.1 for additional information).

Note that this requirement is applicable when 20 or more units are included in the construction project; not when an individual owner performs alterations to his or her condo when there are 20 or more units in the building (also see Section 1105.3, and the definitions of "Dwelling unit," "Sleeping unit" and "Type A unit" in IBC Chapter 2). Assuming that Type A units are not already provided, the number of Type A units required is based on the number being altered. For example, if a story in an apartment building was being altered, the number of Type A units required would be based on the number of units in the area being altered, not the number of units in the entire building. If Type A units are provided, the total number of Type A units in the facility is not required to exceed that required for new construction, as indicated in Section 705.1.13. It is unreasonable to require a greater level of accessibility in an existing building than is required in new construction.

705.1.9 Toilet rooms. Where it is technically infeasible to alter existing toilet and bathing rooms to be accessible, an accessible family or assisted-use toilet or bathing room constructed in accordance with Section 1109.2.1 of the *International Building Code* is permitted. The family or assisted-use toilet or bathing room shall be located on the same floor and in the same area as the existing toilet or bathing rooms. At the inaccessible toilet and bathing rooms, directional signs indicating the location of the nearest family or assisted-use toilet room or bathing room shall be provided. These directional signs shall include the International Symbol of Accessibility and sign characters shall meet the visual character requirements in accordance with ICC A117.1.

❖ This section deals with circumstances in which it is technically infeasible to alter existing toilet facilities to be accessible. Where new bathrooms are created in existing buildings or existing bathrooms are part of an alteration, both the men's and women's facilities would be required to be accessible. An alternative solution when it is technically infeasible to alter the existing toilet rooms would be the creation of a single unisex toilet or bathing room containing accessible facilities. The requirements for family or assisted-use toilet facilities in IBC Sections 1109.2.1 through 1109.2.1.7 provide guidance on what is required in this toilet room. If this alternative is selected, the room must be located on the same floor and in the same area as the existing toilet or bathroom. This is the best alternative to providing fully complying separate men's and women's facilities. One might argue that it is technically infeasible to accomplish either of these alternatives, since the alternative to altering existing facilities involves creation of an additional toilet or bathroom that was not otherwise contemplated. This would not be a persuasive argument, since there is likely to be space available somewhere in the facility to commit for use as a toilet or bathroom. In any case, the intent is that some form of an accessible toilet room or bathing facility is necessary and must be provided.

It should be noted that this alternative is not offered as a choice between making the existing separate-sex toilet rooms accessible or providing a family or assisted-use accessible toilet room. The existing separate-sex toilet rooms must be altered when it is technically feasible. Consideration of the alternative is only available when altering the existing toilet rooms is technically infeasible (see the definition for "Technically infeasible" in Section 202).

Signage must be provided at the inaccessible toilet rooms in accordance with IBC Sections 1110.1 and 1110.2 to notify disabled persons that a facility is not accessible and direct them to the nearest accessible facilities.

705.1.10 Dressing, fitting and locker rooms. Where it is *technically infeasible* to provide accessible dressing, fitting, or locker rooms at the same location as similar types of rooms, one accessible room on the same level shall be provided. Where separate sex facilities are provided, accessible rooms for each sex shall be provided. Separate sex facilities are not required where only unisex rooms are provided.

❖ This section takes a similar approach for dressing rooms to that provided in Section 705.1.9 for toilet and bathing facilities. If it is technically infeasible to alter existing dressing rooms to be accessible, then space elsewhere on the level must be committed to providing not less than one accessible dressing room. In this case, if the existing dressing rooms provide separate rooms for each sex, then not less than one accessible dressing room for each sex must be provided.

705.1.11 Fuel dispensers. Operable parts of replacement fuel dispensers shall be permitted to be 54 inches (1370 mm) maximum measured from the surface of the vehicular way where fuel dispensers are installed on existing curbs.

❖ The requirements for new fuel dispensers (i.e., gas pumps) can be found in IBC Section 1109.14. Basically, the idea is that the controls must be in the reach ranges for someone standing on the parking lot surface. However, many existing facilities have gas pumps located on raised islands as a feature for protection of the pumps from accidental contact. This section would allow a new gas pump with a reach range of 15 inches (380 mm) to 48 inches (1220 mm) to be located on top of a 6-inch (150 mm) curb and still meet the maximum reach of 54 inches (1370 mm).

705.1.12 Thresholds. The maximum height of thresholds at doorways shall be $^3/_4$ inch (19.1 mm). Such thresholds shall have beveled edges on each side.

❖ Thresholds at doorways may be $^3/_4$ inch (19.1 mm) maximum in existing buildings. In new construction, a typical threshold is $^1/_2$ inch (12.7 mm) maximum in accordance with IBC Section 1008.1.7. This section recognizes that such things as differences in floor materials may create changes in elevation greater than that allowed in new construction. Edges of thresholds greater than $^1/_4$ inch (6.4 mm) must be beveled to allow for the passage of a wheelchair.

705.1.13 Extent of application. An *alteration* of an existing element, space, or area of a *facility* shall not impose a requirement for greater accessibility than that which would be required for new construction. *Alterations* shall not reduce or have the effect of reducing accessibility of a *facility* or portion of a *facility*.

❖ The purpose of this section is to clarify to which level the requirements of Sections 705.1 and 705.2 apply. The requirements are not intended to impose a higher level of accessibility than that required in new construction. At the same time, alterations cannot result in a lesser degree of accessibility than existed before alterations were undertaken.

705.1.14 Amusement rides. Where the structural or operational characteristics of an amusement ride are altered to the extent that the amusement ride's performance differs from that specified by the manufacturer or the original design, the amusement ride shall comply with requirements for new construction in accordance with Section 1110.4.8 of the *International Building Code*.

❖ To the extent that amusement rides are subject to the accessibility requirements of IBC Section 1110.4.8, they should be accessible and usable by individuals with disabilities. These scoping provisions are flexible, permitting latitude in terms of the method of access (e.g., transfer seat, roll-on seat or transfer device to lift the rider). Mobile and portable rides are exempted. Rides without seats, those designed for children who are assisted onto the ride and those rides controlled by the user are also exempted under IBC Section 1110.4.8.2 from providing wheelchair transfer spaces. Technical criteria can be found in the 2009 edition of the ICC A117.1, Section 1102 and includes accessible routes, load and unload areas, wheelchair spaces on rides, seats for transfer and transfer devices. Recognizing the technical issues associated with existing amusement rides, compliance with accessibility requirements is required when the amusement ride is extensively altered.

705.2 Alterations affecting an area containing a primary function. Where an *alteration* affects the accessibility to a, or contains an area of, *primary function*, the route to the primary function area shall be accessible. The accessible route to the *primary function* area shall include toilet facilities and drinking fountains serving the area of *primary function*.

Exceptions:

1. The costs of providing the accessible route are not required to exceed 20 percent of the costs of the alterations affecting the area of *primary function*.

2. This provision does not apply to *alterations* limited solely to windows, hardware, operating controls, electrical outlets and signs.

3. This provision does not apply to *alterations* limited solely to mechanical systems, electrical systems, installation or *alteration* of fire protection systems and abatement of hazardous materials.

4. This provision does not apply to *alterations* undertaken for the primary purpose of increasing the accessibility of a *facility*.

5. This provision does not apply to altered areas limited to Type B dwelling and sleeping units.

❖ An area containing a primary function is one in which a major activity for which the building or facility is intended is carried out (see the definition of "Primary function" in Section 202). For example, the lobby of a hotel in which the registration and check-out desk is located would be a primary function area. Other examples would be the dining area of a restaurant, the meeting rooms or exhibition halls in a conference center, virtually all office and work areas in a business building, and retail display areas in a mercantile occupancy. The key concept is that a primary function area is one that contains a major activity of the facility. Areas that contain activities not related to the main purpose of the facility would not be considered a primary function area. For example, a mechanical equipment room, storage closet, toilet facilities, corridors, lounges and locker rooms would not be considered primary function areas. With this background, it is clear that areas containing a primary function are more critical in terms of the purpose for which people enter and use the facility; therefore, this section reflects that when such areas are altered or added, it is important to require that an accessible route to the primary function area be provided. Where an accessible route to a primary function area is required by this section, an accessible route to facilities such as any restrooms and drinking fountains serving the primary function area must also be made accessible, even though such facilities and areas may not by themselves be considered primary function areas. This is not just the route to the bathroom and drinking fountains; the bathrooms and drinking fountains themselves must also be improved when not accessible.

There are conditions under which it may not be reasonable to strictly enforce this requirement. Exception 1 approaches this by utilizing the cost of the alterations or addition as a basis for determining if providing a complete accessible route is reasonable. The requirement does not apply when the cost of providing an accessible route exceeds 20 percent of the cost of the alterations or addition to the primary function area. These costs are intended to be based on the actual costs of the planned alterations or addition to the primary function area before consideration of the cost of providing an accessible route. For example, if the planned alterations will cost $100,000, not including the cost of an accessible route to a primary function area, this exception would apply if the additional cost of providing the accessible route would exceed $20,000.

It is not the intent to exempt all requirements for accessibility when the total cost for providing the accessible route exceeds the 20-percent threshold. Improvements to the accessible route are required to the extent that costs do not exceed 20 percent of the cost to the planned alteration or addition. It is not required that the full 20 percent be spent. If the accessible route (including accessible bathrooms and drinking fountains) is already provided, no additional expenditure is required. Note that there is not a priority list given for where money should be spent on improving the accessible route. The logical progression is access to the site, accessible exterior routes to accessible entrances, access throughout the facility, access to services in the facility, toilet and bathing rooms and, finally, drinking fountains. Evaluation of how and where the money available should best be spent must be made on a case-by-case basis. For example, if an accessible route is not available to an upper level, and the cost of an elevator is more than 20 percent of the cost of the renovation, then other alternatives could be investigated, such as a platform lift or limited access elevator, or adding the elevator pit and shaft at this time, with elevator equipment added later. If all such items are in excess of the 20-percent limit, perhaps the money available could be spent toward making the toilet rooms accessible. The idea is that existing buildings would become fully accessible over time.

Exceptions 2 and 3 identify certain alterations that are not intended to trigger the requirement for providing an accessible route to a primary function area. Alterations limited to elements such as windows, hardware, operating controls, electrical outlets, signage, mechanical, electrical and fire protection systems, including alterations for the purpose of abating a hazardous materials circumstance, do not affect the usability of a primary function area in the same manner as alterations that affect the floor plan or the configuration, location or size of rooms or spaces. It is therefore considered unreasonable to require the installation of an accessible route when the scope of alterations is limited to that reflected in these exceptions.

Note that the costs for these items are not "backed out" of the total cost for the alteration before applying Exception 1. Exceptions 2 and 3 are alterations limited to the specific items referenced.

Exception 4 is intended to avoid penalizing a building owner who is undertaking alterations or additions for the purpose of increasing accessibility. It is appropriate to encourage owners to make such alterations without requiring them to do more work simply because they chose to increase the accessibility of the space. This could otherwise have the opposite effect of discouraging such alterations to avoid the expense of undertaking more work than was originally planned. For example, federal law (ADA) requires that owners of existing buildings remove certain existing barriers to accessibility. Removal of such barriers may require a permit from the code official. It would be unreasonable to have such activity trigger the mandatory requirement for further alterations to accomplish accessibility beyond the originally planned work. In principle, the code takes the view that some extent of greater accessibility is positive progress and should be encouraged, not penalized.

Where the alterations result in Type B dwelling units being provided where none previously existed (Section 410.6, Exception 4), Exception 5 would eliminate the need for any additional money to be spent toward providing an accessible route to those units. This is intended to encourage the creation of such units without penalizing building owners. Similar to Section 410.4.2, this exception is intended to address the concerns of site impracticality for providing accessible routes to and into existing buildings providing Type B dwelling units.

This also reinforces the intent that the inclusion of Type B units is not meant to require elevators when alterations are performed on upper floors in buildings not containing elevators (see exceptions to Section 1107.7). These areas would have been exempted if built new under the FHA and IBC, and should continue to be exempted.

SECTION 706
REROOFING

[BS] 706.1 General. Materials and methods of application used for recovering or replacing an existing roof covering shall comply with the requirements of Chapter 15 of the *International Building Code*.

> **Exception:** Reroofing shall not be required to meet the minimum design slope requirement of one-quarter unit vertical in 12 units horizontal (2-percent slope) in Section 1507 of the *International Building Code* for roofs that provide positive roof drainage.

❖ This section simply states that when a roof is replaced or recovered it must comply with IBC Chapter 15 for the materials and methods used, with an exception for low-sloped roofs. This section does not mandate that the entire roof be replaced but simply that the portion being replaced comply with IBC Chapter 15.

For low-sloped roofs, the exception indicates that reroofing (i.e., recovering or replacement) is not required to meet the $1/_4$:12 minimum slope requirement of IBC Section 1507, provided that the roof has positive drainage. The term "Positive roof drainage" is defined in IBC Chapter 2 as "the drainage condition in which consideration has been made for all loading deflections of the roof deck, and additional roof slope has been provided to ensure drainage of the roof area within 48 hours of precipitation."

[BS] 706.2 Structural and construction loads. Structural roof components shall be capable of supporting the roof-covering system and the material and equipment loads that will be encountered during installation of the system.

❖ The structural integrity of the roof must be maintained during reroofing operations, which can significantly contribute to the loading of the roof due to workers and material being present. The roof support system must be able to structurally support all additional layers of roof covering material.

[BS] 706.3 Recovering versus replacement. New roof coverings shall not be installed without first removing all existing layers of roof coverings down to the roof deck where any of the following conditions occur:

1. Where the existing roof or roof covering is water soaked or has deteriorated to the point that the existing roof or roof covering is not adequate as a base for additional roofing.

2. Where the existing roof covering is wood shake, slate, clay, cement or asbestos-cement tile.

3. Where the existing roof has two or more applications of any type of roof covering.

   Exceptions:

   1. Complete and separate roofing systems, such as standing-seam metal roof systems, that are designed to transmit the roof loads directly to the building's structural system and that do not rely on existing roofs and roof coverings for support, shall not require the removal of existing roof coverings.

   2. Metal panel, metal shingle and concrete and clay tile roof coverings shall be permitted to be installed over existing wood shake roofs when applied in accordance with Section 706.4.

   3. The application of a new protective coating over an existing spray polyurethane foam roofing system shall be permitted without tear-off of existing roof coverings.

   4. Where the existing roof assembly includes an ice barrier membrane that is adhered to the roof deck, the existing ice barrier membrane

shall be permitted to remain in place and covered with an additional layer of ice barrier membrane in accordance with Section 1507 of the *International Building Code*.

❖ This section determines where all layers of previously installed roof covering systems must be removed prior to the installation of the new roof covering system. Note that roof coverings need to be removed down to the roof decking or sheathing.

Where the existing roof or roof covering is water soaked, it must be allowed to dry completely so as not to trap moisture beneath the new layer of covering. This could cause a rapid deterioration of the new covering material, as well as the existing sheathing. The existing covering is required to be removed if it cannot adequately dry out or if its physical properties have been permanently altered.

Wood shake, slate, clay, cement or asbestos-cement tile types of existing roof coverings historically do not make an adequate base for new roof coverings and could prevent the new covering from making a weather-tight seal. They could also allow penetration of water, snow, etc. These types of existing coverings must always be removed.

When the existing roof has two or more layers of any type of covering system, all layers need to be removed to enable the inspector and contractor to verify that the existing sheathing is not water damaged and still capable of providing an adequate nailing base.

Exception 1 states that new roofing systems designed to transmit all roof loads directly to the structural supports of the building do not necessitate removal of the existing roofing system. Exception 2 allows certain roof coverings, including metal, concrete panel and clay, to be placed over wood shingle and shake roofs only if any concealed combustible spaces are properly addressed in accordance with IBC Section 1510.4 (see commentary, IBC Section 1510.4). Exception 3 refers to the practice of "recoating," in which a new protective coating is placed over an existing spray polyurethane foam roofing system. Recoating can add many years to the effective life of a spray polyurethane foam roofing system without the associated downside of adding significant weight associated with other reroofing-type systems.

When removing roof coverings, it can be difficult, if not impossible, to remove an existing layer of adhered ice barrier membrane without damaging or replacing the roof deck. Exception 4 accounts for this condition by allowing an existing adhered ice barrier membrane to remain in place and be covered with a new ice barrier membrane as required in IBC Section 1507, followed by the installation of the new primary roof-covering material.

[BS] 706.4 Roof recovering. Where the application of a new roof covering over wood shingle or shake roofs creates a combustible concealed space, the entire existing surface shall be covered with gypsum board, mineral fiber, glass fiber or other approved materials securely fastened in place.

❖ "Roof recovering" is defined as the process of installing an additional roof covering over a prepared existing roof covering without removing the existing roof covering. Where recovering over wood shingles or shakes creates a combustible concealed space, the code requires that the entire surface of the wood shakes and shingles be covered with a material that will reduce the possibility of such materials adding fuel to a fire in such a space.

[BS] 706.5 Reinstallation of materials. Existing slate, clay or cement tile shall be permitted for reinstallation, except that damaged, cracked or broken slate or tile shall not be reinstalled. Existing vent flashing, metal edgings, drain outlets, collars and metal counterflashings shall not be reinstalled where rusted, damaged or deteriorated. Aggregate surfacing materials shall not be reinstalled.

❖ This section places conditions on the reuse of roof-covering materials. Historically, various types of materials have been removable without substantially damaging the material. Materials such as wood shingles and shakes, roll roofing and asphalt shingles are usually torn or cracked and cannot be reused. Fastener holes also violate the integrity of the material. Before reuse is allowed, materials such as slate, clay or cement tile should be examined thoroughly for cracks and deterioration.

[BS] 706.6 Flashings. Flashings shall be reconstructed in accordance with approved manufacturer's installation instructions. Metal flashing to which bituminous materials are to be adhered shall be primed prior to installation.

❖ Flashings to be reused or reconstructed must be in accordance with the manufacturer's installation instructions. Metal flashings that are to be reused for bituminous materials must be primed in accordance with the manufacturer's instructions.

SECTION 707
STRUCTURAL

[BS] 707.1 General. Where *alteration* work includes replacement of equipment that is supported by the building or where a reroofing permit is required, the provisions of this section shall apply.

❖ These requirements apply to alterations involving reroofing or the replacement of equipment that is supported by the building.

[BS] 707.2 Addition or replacement of roofing or replacement of equipment. Where addition or replacement of roofing or replacement of equipment results in additional dead loads, structural components supporting such reroofing or equipment shall comply with the gravity load requirements of the *International Building Code*.

Exceptions:

1. Structural elements where the additional dead load from the roofing or equipment does not increase the force in the element by more than 5 percent.

2. Buildings constructed in accordance with the *International Residential Code* or the conventional light-frame construction methods of the *International Building Code* and where the dead load from the roofing or equipment is not increased by more than 5 percent.

3. Addition of a second layer of roof covering weighing 3 pounds per square foot (0.1437 kN/m2) or less over an existing, single layer of roof covering.

❖ Where reroofing, the addition of a new layer of roofing or replacement of equipment results in a net increase in the supported dead load; the affected structural components must be checked to verify that the gravity load requirements of the IBC for a new structure are satisfied. The reference to only gravity loads is an indication that an analysis for the effects of lateral loads, such as seismic, would not be a necessity for this level of alteration.

There are three exceptions to satisfy full compliance with the IBC. Exception 1 allows an increase in the force in a structural element of up to 5 percent due to the additional dead load from the roofing or equipment. Exception 2 allows additional dead loads from roof coverings or replacement equipment to be increased by 5 percent beyond the original dead load of the roof coverings and equipment without having to do any further analysis, such as computation of structural member stress increases. Exceptions 1 and 2 do not specifically address the cumulative effects from successive alterations; for instance, it would be prudent to limit cumulative increases under these exceptions to a total of 5 percent unless it is clearly documented that the structure has additional capacity.

Exception 3 allows the addition of a second layer of roof covering over an existing single layer of roof covering as long as the new roof covering does not weigh more than 3 pounds per square foot (0.1437 kN/m^2). Under this exception, even though the new layer of roof covering might be adding more than 5 percent of the original dead load, there is still no analysis required. This exception recognizes a common practice in many jurisdictions across the country where the addition of a new lightweight layer of roof covering over an existing single layer has been allowed without known life safety issues. Note that Section 706 may have other limitations on reroofing.

[BS] 707.3 Additional requirements for reroof permits. The requirements of this section shall apply to *alteration* work requiring reroof permits.

❖ This section refers the code user to Sections 707.3.1 and 707.3.2 for additional requirements for reroofing projects where a permit has been issued.

[BS] 707.3.1 Bracing for unreinforced masonry bearing wall parapets. Where a permit is issued for reroofing for more than 25 percent of the roof area of a building assigned to Seismic Design Category D, E or F that has parapets constructed of unreinforced masonry, the work shall include installation of parapet bracing to resist the reduced *International Building Code* level seismic forces as specified in Section 301.1.4.2 of this code, unless an evaluation demonstrates compliance of such items.

❖ The failure of parapets in unreinforced masonry-bearing (URM) wall buildings has been a recurring problem in areas that experience significant earthquakes. Because this poses a very real risk, the code requires these elements to be braced where the seismic hazard is deemed to be relatively high as reflected in a building's seismic design category. Since the code does not provide the requirements for establishing the seismic design category, IBC Section 1613.3 must be used for this purpose. For an explanation of this process, see the commentary to Section 301.1.4. The code allows this parapet bracing to be designed using the reduced seismic forces of Section 301.1.4.2 and the design procedures listed in Section 301.1.4. See ASCE 41 or Section A113.6 of *Guidelines for Seismic Evaluation of Existing Buildings* (GSREB) in Appendix A of the code for specific requirements for parapets.

An example of the prescribed parapet bracing is illustrated schematically in Commentary Figure 707.3.1. The approach shown—a steel brace anchored to the parapet wall and the roof structure—is commonly used to provide lateral support for a parapet. Other methods of strengthening a parapet can be used to mitigate the hazard, provided their uses are approved as alternative methods of design and construction under Section 104.11.

Note that only the parapet bracing for seismic loads is required. Some buildings, especially those east of the Rocky Mountains, were not designed for seismic loads. They were only designed to resist wind loads. At the present time, the code does not require a review of the building for seismic loads.

ALTERATIONS—LEVEL 1

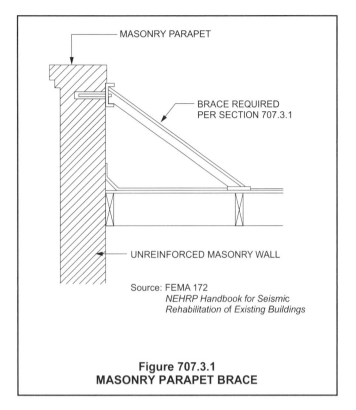

Figure 707.3.1
MASONRY PARAPET BRACE

[BS] 707.3.2 Roof diaphragms resisting wind loads in high-wind regions. Where roofing materials are removed from more than 50 percent of the roof diaphragm or section of a building located where the ultimate design wind speed, V_{ult}, determined in accordance with Figure 1609.3(1) of the *International Building Code*, is greater than 115 mph (51 m/s) or in a special wind region, as defined in Section 1609 of the *International Building Code*, roof diaphragms, connections of the roof diaphragm to roof framing members, and roof-to-wall connections shall be evaluated for the wind loads specified in the *International Building Code*, including wind uplift. If the diaphragms and connections in their current condition are not capable of resisting at least 75 percent of those wind loads, they shall be replaced or strengthened in accordance with the loads specified in the *International Building Code*.

❖ The removal of roofing provides an opportunity to inspect a portion of the structure that is otherwise concealed. Such inspections must take place where more than 50 percent of the roof covering is removed from a building located in a special wind region or where the ultimate design wind speed, V_{ult}, determined in accordance with IBC Table 1609.3(1), is greater than 115 mph.

A roof diaphragm that is a part of the main wind-force-resisting system is required to have adequate strength to resist and transfer the wind loads referenced in the IBC, including uplift loads. If the roof diaphragm is not able to resist 75 percent of the wind loads specified in the IBC, such connections are required to be strengthened or replaced. The 75-percent reduction is consistent with other structural allowances for existing conditions. Note that this section spells out which aspects of the roof need to be evaluated and possibly replaced and strengthened. These include the following:

- Roof diaphragms.
- Connections of the roof diaphragm to roof framing members.
- Roof-to-wall connections.

This requirement is also found in the prescriptive method in Section 403.8.

SECTION 708
ENERGY CONSERVATION

708.1 Minimum requirements. Level 1 *alterations* to *existing buildings* or structures are permitted without requiring the entire building or structure to comply with the energy requirements of the *International Energy Conservation Code* or *International Residential Code*. The *alterations* shall conform to the energy requirements of the *International Energy Conservation Code* or *International Residential Code* as they relate to new construction only.

❖ A building that undergoes Level 1 alterations is required to meet a certain level of energy compliance. The level of compliance depends on the extent of the alterations taking place. As Level 1 alterations are those alterations where certain building elements are removed and replaced with new elements or where old elements are covered with new elements, then any such new element is required to meet the applicable energy provisions of the IECC or IRC, in the case of buildings that fall under the scoping of the IRC. Elements in the building that are not being affected do not need to be evaluated and do not need to comply with the energy provisions. Essentially, the entire building is not required to meet the energy provisions; only a degree of possible improvement in the energy performance of the building is intended to be achieved by making the new elements meet the IECC or IRC requirements. The language in this section is intended to be consistent with the requirements in the IECC and IRC for alterations.

Bibliography

The following resource materials were used in the preparation of the commentary for this chapter of the code.

24 CFR, *Fair Housing Accessibility Guidelines* (FHAG). Washington, DC: Department of Housing and Urban Development, 1991.

36 CFR, Parts 1190 and 1191 Final Rule, *The Americans with Disabilities Act (ADA) Accessibility Guidelines; Architectural Barriers Act (ABA) Accessibility Guidelines*. Washington, DC: Architectural and Transportation Barriers Compliance Board, July 23, 2004.

42 USC 3601-88, *Fair Housing Amendments Act* (FHA). Washington, DC: United States Code, 1988.

2010 *ADA Standard for Accessible Design*. Washington, DC: Department of Justice, September 15, 2010.

ASCE 24-13, *Flood-resistance Design and Construction Standard*. Reston, VA: American Society of Civil Engineers, 2005.

DOJ 28 CFR, Part 36-91, *Americans with Disabilities Act* (ADA). Washington, DC: Department of Justice, 1991.

Guidelines for Seismic Retrofit of Existing Buildings. Birmingham, AL: Southern Building Code Congress International, 2000.

Chapter 8:
Alterations—Level 2

General Comments

This chapter provides the technical requirements for those existing buildings that undergo Level 2 alterations. Reference is made to Level 2 alterations as described in Section 504, which includes the reconfiguration of space. Installation of additional equipment that did not previously exist, or the addition or elimination of doors and windows, are also considered Level 2 alterations and, hence, are covered by this chapter.

This chapter, similar to other chapters of the code, covers all building-related subjects, such as structural, mechanical, plumbing, electrical, accessibility and fire and life safety issues in Level 2 alterations. Sections 801, 802 and 803 are related to scoping, special use and occupancy, and building elements and materials. In the interest of being brief and avoiding the presentation of materials that are repetitive in nature in various chapters, Section 801.2 requires that Level 2 alterations comply not only with Chapter 8, but also with Chapter 7. As such, any alteration work that is classified as Level 2 is assumed to include everything related to Level 1 alterations plus the reconfiguration of space. Section 803, Building Elements and Materials, covers elements such as existing vertical openings and when such openings might be required to be enclosed, stairway enclosures, interior finishes and guards. The remainder of the chapter is related to fire protection, means of egress, accessibility, structural, electrical, mechanical and plumbing provisions. While the means of egress, Section 805, covers subjects such as the minimum number of exits, fire escape requirements and construction details, corridor openings and exit signs, the structural part of this chapter, Section 807, addresses subjects such as new and existing structural members, gravity loads, lateral loads and snow drift loads.

The level of improvements required as a result of Level 2 alterations is extensive when compared to Level 1 alterations because the extent of alterations is more drastic and space reconfiguration is involved. The level typically known as "alterations" in the *International Building Code*® (IBC®) and previous legacy building codes was divided into Levels 1, 2 and 3 because minor alterations that do not include space reconfiguration and extensive alterations such as the relocation of walls and partitions should be treated differently and with different threshold levels for requiring upgrades or improvements to the building or spaces in the building.

Purpose

The purpose of this chapter is to detail required improvements in existing building elements, building spaces and the building structural system. This chapter is distinguished from Chapters 7 and 9 by involving space reconfiguration that could be up to and including 50 percent of the area of the building. In contrast, Level 1 alterations do not involve space reconfiguration and Level 3 alterations involve extensive space reconfiguration exceeding 50 percent of the building area. Depending on the nature of alteration work, its location in the building and whether it encompasses one or more tenants, improvements and upgrades could be required for the open floor penetrations, sprinkler system or the installation of additional means of egress, such as stairs or fire escapes. At times, and under certain situations, this chapter is also intended to improve the safety of certain building features beyond the work area and in other parts of the building where no alteration work might be taking place.

SECTION 801
GENERAL

801.1 Scope. Level 2 *alterations* as described in Section 504 shall comply with the requirements of this chapter.

> **Exception:** Buildings in which the reconfiguration is exclusively the result of compliance with the accessibility requirements of Section 705.2 shall be permitted to comply with Chapter 7.

❖ Any alteration work that results in the reconfiguration of space or otherwise falls under the classification of Level 2 alterations, as described in Section 504, must comply with Chapter 8 (see Commentary Figure 801.1). The exception is intended to encourage existing buildings to improve accessibility for the disabled by allowing any such alterations (which are solely for compliance with the accessibility provisions of Section 705) to only comply with Chapter 7 provisions, as if no space reconfiguration has taken place. This is an advantage for improving accessibility in buildings because Chapter 7 is related to alterations that do not encompass any space reconfiguration and, as such, has a lower level of compliance requirement and higher thresholds for triggering other improvements.

Example 801.1:

An existing office building does not comply with current accessibility requirements. The owner wishes to make the building completely accessible to the disabled by installing ramps, enlarging toilet rooms and widening doorways.

Q: What, if any, additional code requirements are triggered?

A: All affected areas can be rebuilt with similar materials, except glass at human impact areas, which must comply with safety glass requirements (see Section 602).

801.2 Alteration Level 1 compliance. In addition to the requirements of this chapter, all work shall comply with the requirements of Chapter 7.

❖ Chapters 8 and 9 have cascading effects by requiring that each level of alteration must comply with the chapter for the lower level of alteration. Accordingly, an alteration project that is classified as Level 2 must comply with Chapters 7 and 8. Similarly, an alteration project classified as Level 3 must comply with Chapters 7, 8 and 9. This is to eliminate the repetition of various requirements from Chapter 7 in Chapters 8 and 9 and from Chapter 8 in Chapter 9, as a Level 2 alteration, in general, includes various aspects of a Level 1 alteration; and a Level 3 alteration, in general, includes various aspects of Level 1 and 2 alterations. The code drafting committee decided to make reference to these chapters rather than repeat all such provisions again.

801.3 Compliance. All new construction elements, components, systems, and spaces shall comply with the requirements of the *International Building Code*.

Exceptions:

1. Windows may be added without requiring compliance with the light and ventilation requirements of the *International Building Code*.

2. Newly installed electrical equipment shall comply with the requirements of Section 808.

3. The length of dead-end corridors in newly constructed spaces shall only be required to comply with the provisions of Section 805.6.

4. The minimum ceiling height of the newly created habitable and occupiable spaces and corridors shall be 7 feet (2134 mm).

❖ New elements, components and spaces created as a result of Level 2 alteration work must comply with the IBC or other applicable code, as Section 101.4 of the IBC makes reference to other *International Codes*® (I-Codes®), such as the *International Plumbing Code*® (IPC®), *International Mechanical Code*® (IMC®) and *International Fire Code*® (IFC®). Exception 1 allows new windows to be added without complying with IBC light and ventilation requirements, so as not to discourage a building owner from making improvements in the current amount of natural light and ventilation. Exceptions 2 and 3 refer the user to Sections 805.6 and 808, which detail provisions for dead-end corridors and electrical equipment. Exception 4 allows the minimum ceiling height of newly created spaces to be lowered

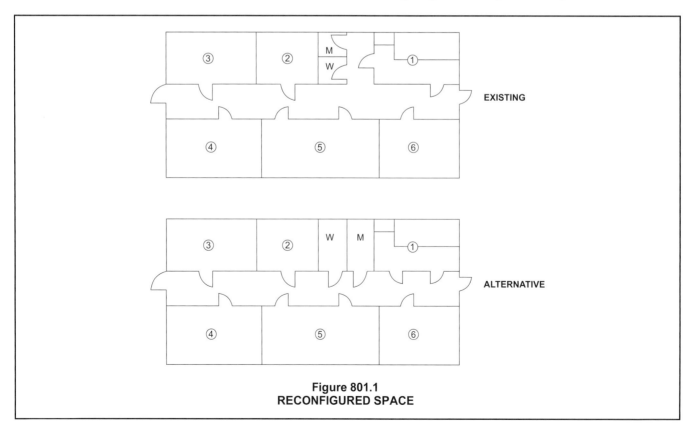

**Figure 801.1
RECONFIGURED SPACE**

from the IBC requirement of 7 feet, 6 inches (2289 mm) to 7 feet (2134 mm), recognizing that even though the space is newly created, it is in an existing building that might have framing members prohibiting higher ceilings.

SECTION 802
SPECIAL USE AND OCCUPANCY

802.1 General. *Alteration* of buildings classified as special use and occupancy as described in the *International Building Code* shall comply with the requirements of Section 801.1 and the scoping provisions of Chapter 1 where applicable.

❖ Special use and occupancy buildings described in Chapter 4 of the IBC, such as covered mall buildings and high-rise buildings, are treated the same as any other building when the work taking place is a Level 2 alteration. This section specifically and clearly provides this information because at higher levels of alteration work, or where there is a change of occupancy, some special use and occupancy buildings are required to comply with the IBC, either completely or in particular alteration areas (see Sections 802 and 902). By reference to Section 801.1 and Chapter 1, this section makes indirect reference to Section 101.2.

SECTION 803
BUILDING ELEMENTS AND MATERIALS

803.1 Scope. The requirements of this section are limited to work areas in which Level 2 alterations are being performed and shall apply beyond the work area where specified.

❖ This section provides the scoping provisions for vertical openings, smoke barriers, interior finish and guards by requiring that the work areas must comply with this section, and that at times there are supplementary requirements that apply to spaces beyond the work areas. Such supplementary requirements are found in Sections 803.2.2, 803.2.3 and 803.4.1 where the extent of compliance beyond the work area is identified.

803.2 Vertical openings. Existing vertical openings shall comply with the provisions of Sections 803.2.1, 803.2.2 and 803.2.3.

❖ The purpose of this section is to provide requirements for vertical openings in Level 2 alterations. Sections 803.2.1 and 803.2.2 provide these requirements.

803.2.1 Existing vertical openings. All existing interior vertical openings connecting two or more floors shall be enclosed with approved assemblies having a fire-resistance rating of not less than 1 hour with approved opening protectives.

Exceptions:

1. Where vertical opening enclosure is not required by the *International Building Code* or the *International Fire Code*.

2. Interior vertical openings other than stairways may be blocked at the floor and ceiling of the *work area* by installation of not less than 2 inches (51 mm) of solid wood or equivalent construction.

3. The enclosure shall not be required where:
    3.1. Connecting the main floor and mezzanines; or
    3.2. All of the following conditions are met:
        3.2.1. The communicating area has a low hazard occupancy or has a moderate hazard occupancy that is protected throughout by an automatic sprinkler system.
        3.2.2. The lowest or next to the lowest level is a street floor.
        3.2.3. The entire area is open and unobstructed in a manner such that it may be assumed that a fire in any part of the interconnected spaces will be readily obvious to all of the occupants.
        3.2.4. Exit capacity is sufficient to provide egress simultaneously for all occupants of all levels by considering all areas to be a single floor area for the determination of required exit capacity.
        3.2.5. Each floor level, considered separately, has at least one-half of its individual required exit capacity provided by an exit or exits leading directly out of that level without having to traverse another communicating floor level or be exposed to the smoke or fire spreading from another communicating floor level.

4. In Group A occupancies, a minimum 30-minute enclosure shall be provided to protect all vertical openings not exceeding three stories.

5. In Group B occupancies, a minimum 30-minute enclosure shall be provided to protect all vertical openings not exceeding three stories. This enclosure, or the enclosure specified in Section 803.2.1, shall not be required in the following locations:
    5.1. Buildings not exceeding 3,000 square feet (279 m^2) per floor.
    5.2. Buildings protected throughout by an approved automatic fire sprinkler system.

6. In Group E occupancies, the enclosure shall not be required for vertical openings not exceeding three stories when the building is protected throughout by an approved automatic fire sprinkler system.

7. In Group F occupancies, the enclosure shall not be required in the following locations:

   7.1. Vertical openings not exceeding three stories.

   7.2. Special purpose occupancies where necessary for manufacturing operations and direct access is provided to at least one protected stairway.

   7.3. Buildings protected throughout by an approved automatic sprinkler system.

8. In Group H occupancies, the enclosure shall not be required for vertical openings not exceeding three stories where necessary for manufacturing operations and every floor level has direct access to at least two remote enclosed stairways or other approved exits.

9. In Group M occupancies, a minimum 30-minute enclosure shall be provided to protect all vertical openings not exceeding three stories. This enclosure, or the enclosure specified in Section 803.2.1, shall not be required in the following locations:

   9.1. Openings connecting only two floor levels.

   9.2. Occupancies protected throughout by an approved automatic sprinkler system.

10. In Group R-1 occupancies, the enclosure shall not be required for vertical openings not exceeding three stories in the following locations:

    10.1. Buildings protected throughout by an approved automatic sprinkler system.

    10.2. Buildings with less than 25 dwelling units or sleeping units where every sleeping room above the second floor is provided with direct access to a fire escape or other approved second exit by means of an approved exterior door or window having a sill height of not greater than 44 inches (1118 mm) and where:

        10.2.1. Any exit access corridor exceeding 8 feet (2438 mm) in length that serves two means of egress, one of which is an unprotected vertical opening, shall have at least one of the means of egress separated from the vertical opening by a 1-hour fire barrier; and

        10.2.2. The building is protected throughout by an automatic fire alarm system, installed and supervised in accordance with the *International Building Code*.

11. In Group R-2 occupancies, a minimum 30-minute enclosure shall be provided to protect all vertical openings not exceeding three stories. This enclosure, or the enclosure specified in Section 803.2.1, shall not be required in the following locations:

    11.1. Vertical openings not exceeding two stories with not more than four dwelling units per floor.

    11.2. Buildings protected throughout by an approved automatic sprinkler system.

    11.3. Buildings with not more than four dwelling units per floor where every sleeping room above the second floor is provided with direct access to a fire escape or other approved second exit by means of an approved exterior door or window having a sill height of not greater than 44 inches (1118 mm) and the building is protected throughout by an automatic fire alarm system complying with Section 804.4.

12. One- and two-family dwellings.

13. Group S occupancies where connecting not more than two floor levels or where connecting not more than three floor levels and the structure is equipped throughout with an approved automatic sprinkler system.

14. Group S occupancies where vertical opening protection is not required for open parking garages and ramps.

❖ All existing vertical openings connecting two or more floors must be enclosed with assemblies of 1-hour fire-resistance-rated construction and approved protected openings. Even without its 14 exceptions, this provision is more relaxed than the IBC shaft enclosure requirements that require 2-hour fire-resistance-rated enclosures for buildings four stories or more in height. Additionally, the user should remember the scoping provisions of Section 803.1, which indicates that the enclosure requirements triggered under Level 2 alterations apply only to work areas. As such, if the alteration is on the first floor of a multiple-story building, and the vertical opening is in the work area, then only the portion of the vertical opening in the first floor, not the entire opening from the bottom to top floor, is required to be enclosed.

Exception 1 is provided so that the provisions of the code in relation to vertical openings are not more restrictive than those of the IBC and IFC. The IFC regulates vertical openings in existing buildings in Section 1103.4 of that code. This section allows vertical openings without enclosures where they connect only two stories in other than hazardous or institutional occupancies.

Exception 2 allows vertical openings other than stairways to be blocked by solid wood of 2 inches (51 mm) in thickness or equivalent construction. The 2 inches (51 mm) used here is the nominal thickness of the solid wood members.

Exception 3.1 is a repetition of IBC Section 712.1.11. Exception 3.2 provides five conditions that, if all are met simultaneously, do not require the enclosure of vertical openings. The combination of these five conditions is equivalent to a situation where the occupancy is a low-hazard occupancy, fire sprinklers are present and there is sufficient means-of-egress capacity.

Exceptions 4 through 11 and 13 allow either a 30-minute protection for vertical openings in lieu of 1-hour protection or allow the vertical opening to remain open for buildings three stories or less in height with certain features such as sprinklers or multiple remote means of egress.

Exceptions 12 and 14 allow existing vertical openings to remain open in single-family, duplex buildings and open parking garages. The IBC and IFC exception is listed under Exception 1. As can be seen, the exceptions within the IBC and the code allow for numerous methods of dealing with vertical openings so that the building owners and designers have many options to provide a minimum level of protection intended by the codes.

803.2.2 Supplemental shaft and floor opening enclosure requirements. Where the *work area* on any floor exceeds 50 percent of that floor area, the enclosure requirements of Section 803.2 shall apply to vertical openings other than stairways throughout the floor.

Exception: Vertical openings located in tenant spaces that are entirely outside the *work area*.

❖ Unenclosed vertical openings and shafts, other than stairways that would typically be required to be enclosed by the IBC, are required to be enclosed throughout a floor where the work area on that floor exceeds 50 percent of the floor area. Stairways are not included here because they are specifically addressed in Section 803.2.3. This is the first of the supplemental requirements for vertical enclosures to which the code user is referred in Section 803.1. The proportionality philosophy of the code is applied here, as it is believed that a work area extending to more than 50 percent of the floor area of a story is large enough to trigger additional vertical openings, other than stairways, to be enclosed. As the undertaking of such work in tenant spaces, which are completely outside of the work area and are not within the scope of the originally proposed alteration, is disruptive and an unacceptable practice, the exception eliminates this requirement in such tenant spaces.

Example 803.2.2:

An existing four-story retail building is being altered on the third floor. The alterations involve the reconfiguration of space in about 70 percent of the third floor. There are two enclosed stairways, an open escalator and an elevator without protected enclosures. The building is not sprinklered.

Q: Will the escalator and the elevator be required to be protected or enclosed in accordance with the IBC?

A: Since the work area on the third floor exceeds 50 percent of the area of the third floor, both the elevator and the escalator must be protected even though the elevator is outside of the work area. The protection is needed on the third floor only (see Section 803.1 and Commentary Figure 803.2.2).

803.2.3 Supplemental stairway enclosure requirements. Where the *work area* on any floor exceeds 50 percent of that floor area, stairways that are part of the means of egress serving the *work area* shall, at a minimum, be enclosed with smoke-tight construction on the highest *work area* floor and all floors below.

Exception: Where stairway enclosure is not required by the *International Building Code* or the *International Fire Code*.

❖ Stairways serving as part of the means-of-egress system must be enclosed, where possible, to increase the degree of protection for building occupants. The triggering mechanism in this section is the percentage of the work area on any given floor. If the work area exceeds 50 percent of the floor area on any floor, then stairways that are part of the means of egress and serve that work area must be enclosed, even though the stairways may not be located in the work area. This provision could affect several stairways, all of which might be located outside of the proposed work area. For example, a multiple-story building contains three required stairways, all of which are unenclosed. The work area on an upper floor that exceeds 50 percent of that story area triggers the requirement for the enclosure of all three stairways if they each serve as one of the exits for that work area. The enclosure need only be smoke tight and need not provide any fire-resistance rating, and the openings into such a smoke-tight enclosure need not be protected with fire-resistant assemblies, only smoke-protected assemblies. The extent of the enclosure is specified to be from the floor at which the work areas have triggered this requirement and all floors below it. The exception is a reminder that stairways not required to be enclosed by the IBC or IFC are not affected by this section.

803.3 Smoke compartments. In Group I-2 occupancies where the work area is on a story used for sleeping rooms for more than 30 patients, the story shall be divided into not less than two compartments by smoke barrier walls in accordance with Section 407.5 of the *International Building Code* as required for new construction.

❖ The smoke compartmentation and related smoke barrier provisions in alterations classified as Level 2 apply only to Group I-2 occupancies, such as hospitals, nursing homes, mental hospitals and detoxification facilities. In such facilities, this section requires that a story used for sleeping rooms for more than 30 patients be divided into at least two compartments by the use of smoke barriers. The section then references Section 407.5 for smoke compartment size and other related limitations. More specifically, no such compartment can be larger than 22,500 square feet (2093 m^2) for Group I-2 Condition 1 and 40,000 square feet for Group I-2 Condition 2. Realizing that these compartments could be very long and narrow spaces, the travel distance from any point to reach a door in the smoke barrier is limited to 200 feet (60 960 mm). The trigger used to determine when a second compartment is required differs from the IBC. The IBC requires compartmentation in all sleeping areas regardless of occupant load, and in nonsleeping areas with an occupant load of 50 or more (see commentary, IBC Section 407.5).

803.4 Interior finish. The interior finish of walls and ceilings in exits and corridors in any *work area* shall comply with the requirements of the *International Building Code*.

Exception: Existing interior finish materials that do not comply with the interior finish requirements of the *International Building Code* shall be permitted to be treated with an approved fire-retardant coating in accordance with the manufacturer's instructions to achieve the required rating.

❖ The interior finish of corridor and exit walls and ceilings in work areas must meet the flame spread and smoke-development limitations and requirements of the IBC. "Corridors" and "Exits" are defined in IBC Section 202, and flame-spread and smoke-developed classifications are determined from IBC Table 803.11 and based on ASTM E84. To be able to maintain some interior finish materials that do not comply with the flame-spread and smoke-development requirements of the code, the exception provides the option of available technology for the use of fire-retardant treatment of such surfaces as long as the code official approves the coating and it is applied in accordance with the manufacturer's instructions to achieve the required rating.

803.4.1 Supplemental interior finish requirements. Where the *work area* on any floor exceeds 50 percent of the floor area, Section 803.4 shall also apply to the interior finish in

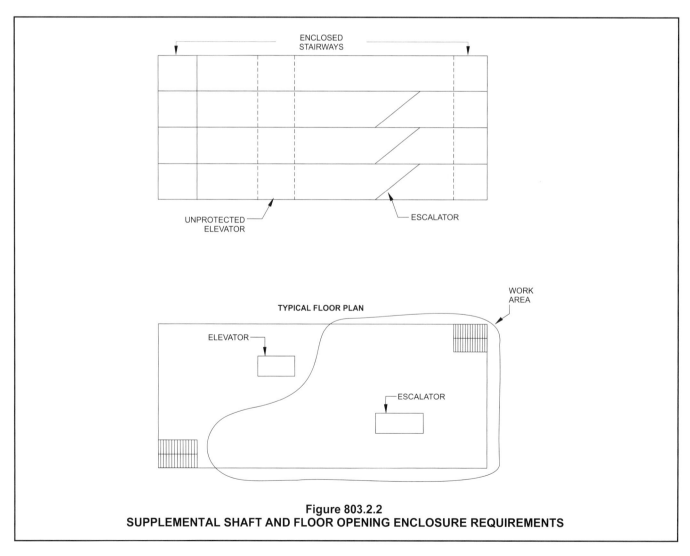

Figure 803.2.2
SUPPLEMENTAL SHAFT AND FLOOR OPENING ENCLOSURE REQUIREMENTS

exits and corridors serving the *work area* throughout the floor.

Exception: Interior finish within tenant spaces that are entirely outside the *work area*.

❖ The interior finish requirements discussed in Section 803.4 are also applicable in corridors and exits that are outside of the work area but serve that same work area, if the work area on any floor exceeds 50 percent of the floor area at that level. Similar to other supplemental requirements, this extended area of code application is not required in tenant spaces that are entirely outside of the work area.

803.5 Guards. The requirements of Sections 803.5.1 and 803.5.2 shall apply in all *work areas*.

❖ This is the charging paragraph for guard requirements. The detailed requirements are given in Sections 803.5.1 and 803.5.2.

803.5.1 Minimum requirement. Every portion of a floor, such as a balcony or a loading dock, that is more than 30 inches (762 mm) above the floor or grade below and is not provided with guards, or those in which the existing guards are judged to be in danger of collapsing, shall be provided with guards.

❖ To reduce the potential of an accidental fall, the IBC requires guards where open sides of walking surfaces, mezzanines, stairways, ramps, landings and other similar areas are located more than 30 inches (702 mm) above the floor or grade below. The code, in this section, uses the same criteria to require new guards where none existed before or to require that dangerous guards be replaced. This requirement is applicable only in the work areas and does not contain supplementary provisions. Replacement of existing guards is required if they are judged to be in danger of collapsing. This sentence makes it clear that the replacement of guards is triggered if anyone, including the owner, designer or code official, judges the guard to be in danger of collapsing. It needs to be stressed, though, that guards that could potentially collapse or could be considered unsafe, as defined in Section 202, are required to be replaced on the basis of Section 115.1, whether such guards are in the work area or not. Section 116 of the IBC and Section 110 of the IFC have the same effect and could require the replacement of guards anywhere in the building, even if no alterations are planned.

Where this section triggers the installation of new guards or the replacement of existing guards, the new guards must comply with the IBC for height, spacing between balusters and structural strength (see IBC Sections 1016 and 1607.8).

Example 803.5.1:

An existing restaurant has an elevated platform for performances, an elevated dining area and an outside dining deck, all of which require guards in accordance with the IBC. The performance area has no guards and the elevated dining area has existing guards that are in acceptable physical condition, but the spacing between the rails is $5^1/_2$ inches (140 mm). The outside dining deck has guards that are deteriorated and might collapse under applied loads. The owner plans to make some rearrangements of the kitchen, toilets and the elevated dining deck, which are all in the same area.

Q: What requirements related to guards are applicable?

A: 1. The guards for the elevated dining area do not require any upgrading, as they are in acceptable physical shape.

A: 2. The elevated performance area and the outside dining deck are not within the work area and, as such, are not covered by this section of the code. Both of these areas, however, could most likely be considered unsafe under the provisions of Section 115.1 and, accordingly, guards in full compliance with the IBC must be installed in both locations.

803.5.2 Design. Where there are no guards or where existing guards must be replaced, the guards shall be designed and installed in accordance with the *International Building Code*.

❖ The strength, load-carrying capacity and maximum opening requirements for guards are given by the IBC. This is consistent with the requirements of Section 702.6 for new building materials and equipment.

803.6 Fire-resistance ratings. Where approved by the code official, buildings where an automatic sprinkler system installed in accordance with Section 903.3.1.1 or 903.3.1.2 of the *International Building Code* has been added, and the building is now sprinklered throughout, the required fire-resistance ratings of building elements and materials shall be permitted to meet the requirements of the current building code. The building is required to meet the other applicable requirements of the *International Building Code*.

Plans, investigation and evaluation reports, and other data shall be submitted indicating which building elements and materials the applicant is requesting the code official to review and approve for determination of applying the current building code fire-resistance ratings. Any special construction features, including fire-resistance-rated assemblies and smoke-resistive assemblies, conditions of occupancy, means-of-egress conditions, fire code deficiencies, approved modifications or approved alternative materials, design and methods of construction, and equipment applying to the building that impact required fire-resistance ratings shall be identified in the evaluation reports submitted.

❖ This section provides allowances in fire-resistance-rating requirements where a building is equipped throughout with an automatic sprinkler system. This installation could be related to an alteration or compliance with retroactive code requirements. These reductions are allowed to be equivalent to the IBC. A concern with such allowances is how to ensure that such reductions in fire resistance do not result in an overall reduction of the existing levels of protection.

This section provides a process to review the required levels of protection with a sprinkler system as opposed to without. The process is that once an existing building is sprinklered throughout and meets other applicable requirements of the IBC, plans, investigation and evaluation reports, and other data can be submitted seeking approval of the code official for the assignment of the new fire-resistance ratings, which might result in a reduction, or potentially an increase.

The provisions require that the evaluation reports identify the following: any special construction features, conditions of occupancy, approved modifications or alternative materials, design and methods of construction, and equipment applying to the building that impact required fire-resistance ratings. This is to ensure special conditions are identified that may prevent a reduction in fire-resistance ratings.

SECTION 804
FIRE PROTECTION

804.1 Scope. The requirements of this section shall be limited to work areas in which Level 2 *alterations* are being performed, and where specified they shall apply throughout the floor on which the *work areas* are located or otherwise beyond the *work area*.

❖ The scoping provisions of fire protection are the same as all other subjects in Chapter 8. The requirements apply in work areas. At times, there are supplemental requirements that apply to areas outside of the work area.

804.1.1 Corridor ratings. Where an approved automatic sprinkler system is installed throughout the story, the required fire-resistance rating for any corridor located on the story shall be permitted to be reduced in accordance with the *International Building Code*. In order to be considered for a corridor rating reduction, such system shall provide coverage for the stairway landings serving the floor and the intermediate landings immediately below.

❖ Corridor reductions on a fully sprinklered floor provide a significant incentive to sprinklering the floor and a reasonable level of safety for the occupants of the floor. Once water is brought to the floor, trade-offs for rated corridor doors and dampers, plus the increase in multiple leasing design options, support an owner's decision to sprinkler. To be considered a fully sprinklered floor, the sprinkler system also needs to include the exit stairway(s). Once the sprinkler risers are provided in an existing building, future alterations may also easily be sprinklered by utilizing the riser. This trade-off is permitted for automatic sprinkler systems that are required as a result of the alteration, and in voluntary installations.

804.2 Automatic sprinkler systems. Automatic sprinkler systems shall be provided in accordance with the requirements of Sections 804.2.1 through 804.2.5. Installation requirements shall be in accordance with the *International Building Code*.

❖ The sprinkler requirements are grouped into four sections: high-rise buildings; Groups A, E, F-1, H, I, M, R-1, R-2, R-4, S-1 and S-2; windowless stories; and buildings and areas listed in IBC Table 903.2.11.6.

804.2.1 High-rise buildings. In high-rise buildings, work areas that have exits or corridors shared by more than one tenant or that have exits or corridors serving an occupant load greater than 30 shall be provided with automatic sprinkler protection in the entire *work area* where the *work area* is located on a floor that has a sufficient sprinkler water supply system from an existing standpipe or a sprinkler riser serving that floor.

❖ The requirement in high-rise buildings is dependent on the availability of a sufficient sprinkler water supply from an existing standpipe, or a sprinkler riser serving the floor where the alteration work area is located. These criteria are included because of the complexity and extensive cost involved in completely constructing new sprinkler water mains, risers, and associated piping and equipment in high-rise buildings. Where there are existing standpipe or sprinkler risers with a sufficient water supply serving a particular floor, work areas on that floor must be sprinklered if such work areas have exits or corridors shared by more than one tenant, or exits or corridors serving an occupant load greater than 30.

804.2.1.1 Supplemental automatic sprinkler system requirements. Where the *work area* on any floor exceeds 50 percent of that floor area, Section 804.2.1 shall apply to the entire floor on which the *work area* is located.

> **Exception:** Occupied tenant spaces that are entirely outside the work area.

❖ If the work area on a floor exceeds 50 percent of that floor area, sprinklers as previously discussed must be installed throughout the floor except in occupied tenant areas that are entirely outside the work area. Note that the 50-percent rule applies to the floor area only and, therefore, this does not become a Level 3 alteration until the work area exceeds 50 percent of the aggregate area of all floors.

804.2.2 Groups A, B, E, F-1, H, I, M, R-1, R-2, R-4, S-1 and S-2. In buildings with occupancies in Groups A, B, E, F-1, H, I, M, R-1, R-2, R-4, S-1 and S-2, work areas that have exits or corridors shared by more than one tenant or that have exits or corridors serving an occupant load greater than 30 shall be provided with automatic sprinkler protection where all of the following conditions occur:

1. The *work area* is required to be provided with automatic sprinkler protection in accordance with the *Inter-*

national Building Code as applicable to new construction; and

2. The *work area* exceeds 50 percent of the floor area.

    Exception: If the building does not have sufficient municipal water supply for design of a fire sprinkler system available to the floor without installation of a new fire pump, work areas shall be protected by an automatic smoke detection system throughout all occupiable spaces other than sleeping units or individual dwelling units that activates the occupant notification system in accordance with Sections 907.4, 907.5 and 907.6 of the *International Building Code*.

❖ While, in general, automatic sprinkler protection for various occupancies in the IBC is a function of the "fire area" size, in the code it is a function of four conditions in select occupancies. This section establishes the criteria for all occupancies, except for Group F-2 and U occupancies, because these groups are not required to be sprinklered in the IBC unless they are in special buildings, such as covered malls, high rises and the so-called windowless stories. The foremost consideration regarding potential sprinkler systems is whether such a work area would be required to be sprinklered under the IBC provisions for new construction. If the IBC does not require a sprinkler system for the space or building under consideration, neither will the code. Once this is established, the other conditions to address include: whether the work area has exits or corridors shared by more than one tenant, or has exits or corridors that serve an occupant load greater than 30; and whether the work area exceeds 50 percent of the floor area. The 30-occupant-load threshold in this section is taken from IBC Table 1020.1, which uses this threshold in most occupancies to require either sprinklers or rated corridors.

One exception to these requirements states that if the building does not have a sufficient municipal water supply for a sprinkler system at the floor where the work area is located, then sprinklers are not required; however, that same exception does require an automatic smoke detection system throughout the work area. The smoke detection coverage is required throughout all occupiable spaces other than areas already required to install smoke alarms.

804.2.2.1 Mixed uses. In work areas containing mixed uses, one or more of which requires automatic sprinkler protection in accordance with Section 804.2.2, such protection shall not be required throughout the *work area* provided that the uses requiring such protection are separated from those not requiring protection by fire-resistance-rated construction having a minimum 2-hour rating for Group H and a minimum 1-hour rating for all other occupancy groups.

❖ Designers and owners who do not wish to sprinkler entire work areas simply because one of a multiple number of mixed uses requires automatic sprinklers are provided an option to separate the various occupancies. This concept is very similar to the concept in IBC Section 508, where two options of nonseparated uses and separated uses are provided. Here, the code user is given the option of separating the occupancies by a 1-hour fire-resistance rating (2-hour rating in Group H occupancies) and sprinkler only the occupancy that has been triggered to provide sprinklers, or provide the automatic sprinkler in the entire work area and multiple occupancies in such work areas, thus eliminating the need for separation of the various mixed occupancies.

Example 804.2.2.1:

Q: Which occupancies or tenants must be provided with a sprinkler system?

A: First, the two conditions of Section 804.2.2 must be verified. Condition 2 is present. For Condition 1, IBC Section 903 requires sprinklers for tenants 2, 5, 7 and 10. Tenants 5 and 7, however, are outside of the work area and, as such, are not going to be required to provide sprinklers (see Section 804.1). Tenants 2 and 10 are in the work area. Accordingly, all of the tenants in the work area (tenants 1, 2, 3, 8, 9, 10 and 11) will be required to be sprinklered, even though tenants 1, 3, 8, 9 and 11 are not individually required to be sprinklered by the IBC. If the owner wishes to only sprinkler tenants 2 and 10, then full-height, 1-hour-rated partitions must be provided to separate tenants 2 and 10 from the adjacent tenants 1, 3, 9 and 11 (see Commentary Figure 804.2.2.1).

804.2.3 Windowless stories. Work located in a windowless story, as determined in accordance with the *International Building Code*, shall be sprinklered where the work area is required to be sprinklered under the provisions of the *International Building Code* for newly constructed buildings and the building has a sufficient municipal water supply without installation of a new fire pump.

❖ In the context of the code, a windowless story is a reference to IBC Section 903.2.11.1. In this context, work areas in any story or basement larger than 1,500 square feet (139 m^2) that do not have access to openings, as described in Items 1 and 2 of IBC Section 903.2.11.1, are required to be sprinklered. The only exception is if there is not a sufficient municipal water supply available to the building without the need for a new fire pump.

804.2.4 Other required automatic sprinkler systems. In buildings and areas listed in Table 903.2.11.6 of the *International Building Code*, *work areas* that have exits or corridors shared by more than one tenant or that have exits or corridors serving an occupant load greater than 30 shall be provided with an automatic sprinkler system under the following conditions:

1. The *work area* is required to be provided with an automatic sprinkler system in accordance with the *International Building Code* applicable to new construction; and

2. The building has sufficient municipal water supply for design of an automatic sprinkler system available to the floor without installation of a new fire pump.

❖ Special buildings and occupancies prompting separate consideration of additional suppression system requirements include: covered malls, atriums, stages and others listed in IBC Table 903.2.11.6; and dry cleaning plants, spray booths and others listed in IFC Table 903.2.11.6. The first condition to be considered for an area to require a sprinkler system is whether such a work area would be required to be sprinklered under the provisions of the IBC for new construction. If the IBC does not require a sprinkler system for the space or building under consideration, neither will the code. Once this is established, other conditions are evaluated, such as whether the work area has exits or corridors shared by more than one tenant or serving an occupant load greater than 30; and whether the building, without consideration for a new fire pump, has a sufficient municipal water supply for a sprinkler system at the floor where the work area is located. The 30-occupant-load threshold in this section is taken from IBC Table 1020.1, which uses this threshold in most occupancies to require either sprinklers or rated corridors.

804.2.5 Supervision. Fire sprinkler systems required by this section shall be supervised by one of the following methods:

1. Approved central station system in accordance with NFPA 72;
2. Approved proprietary system in accordance with NFPA 72;
3. Approved remote station system of the jurisdiction in accordance with NFPA 72; or
4. When approved by the *code official*, approved local alarm service that will cause the sounding of an alarm in accordance with NFPA 72.

Exception: Supervision is not required for the following:

1. Underground gate valve with roadway boxes.
2. Halogenated extinguishing systems.
3. Carbon dioxide extinguishing systems.
4. Dry- and wet-chemical extinguishing systems.

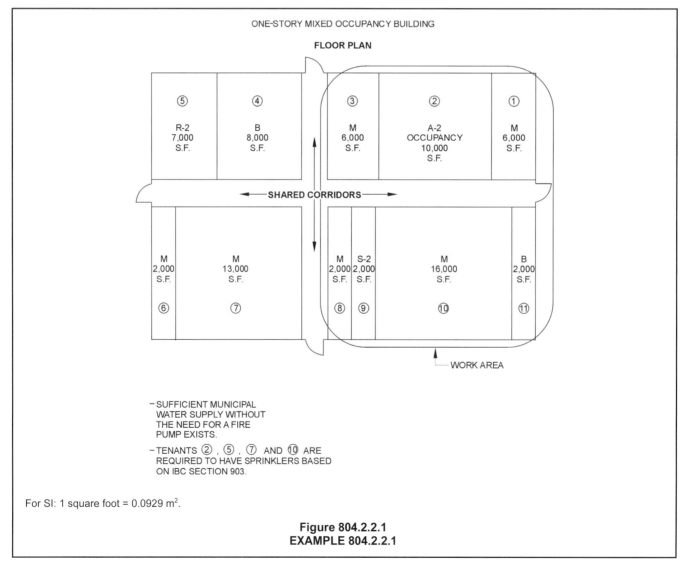

For SI: 1 square foot = 0.0929 m².

**Figure 804.2.2.1
EXAMPLE 804.2.2.1**

5. Automatic sprinkler systems installed in accordance with NFPA 13R where a common supply main is used to supply both domestic and automatic sprinkler systems and a separate shutoff valve for the automatic sprinkler system is not provided.

❖ If a sprinkler system is required based on Section 804.2.1, 804.2.2, 804.2.3 or 804.2.4, then such a system must be supervised in accordance with NFPA 72, the same standard used for the supervision of sprinklers in new buildings constructed under the IBC. This section identifies the four acceptable methods described in NFPA 72: central station system, proprietary systems, remote station system of the jurisdiction or local alarm service that will cause the sounding of an alarm. Five exceptions are provided where supervision is not required; these five exceptions are similar to the exceptions found in IBC Sections 903.4 and 903.4.1.

804.3 Standpipes. Where the *work area* includes exits or corridors shared by more than one tenant and is located more than 50 feet (15 240 mm) above or below the lowest level of fire department access, a standpipe system shall be provided. Standpipes shall have an approved fire department connection with hose connections at each floor level above or below the lowest level of fire department access. Standpipe systems shall be installed in accordance with the *International Building Code*.

Exceptions:

1. No pump shall be required provided that the standpipes are capable of accepting delivery by fire department apparatus of a minimum of 250 gallons per minute (gpm) at 65 pounds per square inch (psi) (946 L/m at 448KPa) to the topmost floor in buildings equipped throughout with an automatic sprinkler system or a minimum of 500 gpm at 65 psi (1892 L/m at 448KPa) to the topmost floor in all other buildings. Where the standpipe terminates below the topmost floor, the standpipe shall be designed to meet (gpm/psi) (L/m/KPa) requirements of this exception for possible future extension of the standpipe.

2. The interconnection of multiple standpipe risers shall not be required.

❖ Once a work area encompasses exits or corridors that are shared by more than one tenant, the standpipe requirement is triggered only by work area height criteria, similar to building height criteria in IBC Section 905.3.1. The standpipe triggers in the IBC for new buildings contain additional criteria based on Group A occupancies (see IBC Section 905.3.2), covered and open mall buildings (see IBC Section 905.3.3), underground buildings (see IBC Section 905.3.5), and helistops and heliports (see IBC Section 905.3.6), none of which are used as the basis for standpipe triggers in the code. Where the work area that includes exits or corridors serving more than one tenant is located more than 50 feet (15 240 mm) above or below the lowest level of fire department vehicle access, a standpipe system is required. Such standpipes must have a fire department connection with a hose connection at each floor level up to the level where the work area is located in accordance with the IBC. Even though the class of the standpipe system is not specified in this section, because the trigger is based on work area height, the standpipe intended is a Class III standpipe. The exceptions in IBC Section 905.3.1 that allow a Class I standpipe in lieu of a Class III would also be allowed here. Exception 1 allows the installation of a standpipe system without a fire pump if sufficient water with sufficient pressure could be delivered by means of fire department apparatus pumps to the topmost floor of the building. Under this exception, the location of the work area that has triggered the standpipe requirement is irrelevant, and the topmost floor must be considered whether the work area is located at the topmost floor or not. Exception 2 recognizes the fact that over the life of a building, and as a result of various alteration work in different parts of the building, multiple independent standpipe risers might be present and these are allowed to remain independent and function separately as needed.

804.4 Fire alarm and detection. An approved fire alarm system shall be installed in accordance with Sections 804.4.1 through 804.4.3. Where automatic sprinkler protection is provided in accordance with Section 804.2 and is connected to the building fire alarm system, automatic heat detection shall not be required.

An approved automatic fire detection system shall be installed in accordance with the provisions of this code and NFPA 72. Devices, combinations of devices, appliances, and equipment shall be approved. The automatic fire detectors shall be smoke detectors, except that an approved alternative type of detector shall be installed in spaces such as boiler rooms, where products of combustion are present during normal operation in sufficient quantity to actuate a smoke detector.

❖ Fire alarm system requirements are solely based on the nature of occupancy. Fire detection system requirements are based on specific features to perform specific functions as described in the IBC, such as elevator recall (see IBC Section 3003.2) or smokeproof enclosure ventilation (see IBC Section 909.20). These requirements apply within the work area only. Such devices, where required, much the same as the IBC, must be installed in accordance with NFPA 72 and need not be anything other than smoke detectors, with certain minor exceptions.

804.4.1 Occupancy requirements. A fire alarm system shall be installed in accordance with Sections 804.4.1.1 through 804.4.1.7. Existing alarm-notification appliances shall be automatically activated throughout the building. Where the building is not equipped with a fire alarm system, alarm-notification appliances within the *work area* shall be provided and automatically activated.

Exceptions:

1. Occupancies with an existing, previously approved fire alarm system.
2. Where selective notification is permitted, alarm-notification appliances shall be automatically activated in the areas selected.

❖ As previously mentioned, fire alarm system requirements are based on occupancy classification as determined by Sections 804.4.1.1 through 804.4.1.7. If the automatic alarm system is required in the work area in a certain occupancy and alarm notification devices already exist in other parts of the building, then all such existing devices must be connected to the new alarm system and be automatically activated on the activation of the new alarm system. If there are no such existing alarms or notification devices, then the new alarm and its notification devices are only required in the work area. Unless the supplemental fire alarm system requirements of Section 804.4.2 are applicable, fire alarm systems are required in work areas only, in Groups E, I-1 residential care/assisted-living, I-2, I-3, R-1, R-2 and R-4 residential care/assisted-living facilities. The triggers are the same as the IFC retroactive triggers for the same occupancies.

804.4.1.1 Group E. A fire alarm system shall be installed in *work areas* of Group E occupancies as required by the *International Fire Code* for existing Group E occupancies.

❖ A building with a maximum of 1,000 square feet (93 m²) in area that contains a single classroom and is located 50 feet (15 240 mm) or more from another building, and any building with an occupant load less than 50, are not required to have an alarm system. Work areas in Group E buildings are required to be provided with a manual fire alarm system. It must be noted, however, that in this case, the requirements of the IFC are more stringent than the code. If a jurisdiction enforces the retroactive provisions of the IFC, the entire Group E occupancy (and not just the work area) is required to be provided with a manual fire alarm system based on the size and occupant load criteria just discussed.

804.4.1.2 Group I-1. A fire alarm system shall be installed in *work areas* of Group I-1 residential care/assisted living facilities as required by the *International Fire Code* for existing Group I-1 occupancies.

❖ A manual fire alarm system must be installed in work areas of all Group I-1 residential care/assisted-living facilities. IFC Section 1103.7.2 requires this same system retroactively in the entire Group I-1 occupancy, which, if enforced, is more comprehensive than the code's requirement.

804.4.1.3 Group I-2. A fire alarm system shall be installed throughout Group I-2 occupancies as required by the *International Fire Code*.

❖ A manual fire alarm system is required to be provided in the entire Group I-2 occupancy where there is a Level 2 alteration. This is consistent with IFC retroactive provisions that require the entire existing Group I-2 occupancies to be provided with an automatic fire alarm system (see IFC Sections 907.2.6.2 and 1105.9).

804.4.1.4 Group I-3. A fire alarm system shall be installed in *work areas* of Group I-3 occupancies as required by the *International Fire Code*.

❖ The work areas of Group I-3 must be provided with a manual and automatic fire alarm system. IFC Sections 907.2.6.3 and 1103.7.4 require this same system retroactively in the entire Group I-3 occupancy, which, if enforced, is more comprehensive than the code's requirement.

804.4.1.5 Group R-1. A fire alarm system shall be installed in Group R-1 occupancies as required by the *International Fire Code* for existing Group R-1 occupancies.

❖ The fire alarm requirement is again referred to the retroactive provisions of the IFC. The IFC deals with Group R-1, hotels and motels, and Group R-1, boarding and rooming houses, differently. Boarding and rooming houses are required to be provided with a fire alarm system in the work area, regardless of the size of the building, height or number of stories. For hotels and motels, the trigger is set for buildings that are more than three stories in height or with more than 20 guestrooms. There is also an exception that is related to buildings that are less than two stories in height (see Sections 1103.7.5 through 1103.7.5.2.1 of the IFC).

804.4.1.6 Group R-2. A fire alarm system shall be installed in *work areas* of Group R-2 apartment buildings as required by the *International Fire Code* for existing Group R-2 occupancies.

❖ The fire alarm system trigger for apartments is similar to that of hotels and motels, and is set at buildings of more than three stories in height or containing more than 16 dwelling units. IFC Section 1103.7.6 includes three exceptions that are also applicable under the code for Group R-2 apartment buildings.

804.4.1.7 Group R-4. A fire alarm system shall be installed in *work areas* of Group R-4 residential care/assisted living facilities as required by the *International Fire Code* for existing Group R-4 occupancies.

❖ Other than the three exceptions found under IFC Section 1103.7.7, work areas in Group R-4 residential care/assisted-living facilities must be provided with a fire alarm system.

804.4.2 Supplemental fire alarm system requirements. Where the *work area* on any floor exceeds 50 percent of that floor area, Section 804.4.1 shall apply throughout the floor.

> **Exception:** Alarm-initiating and notification appliances shall not be required to be installed in tenant spaces outside of the *work area*.

❖ The fire alarm requirements previously discussed for various occupancies apply to the work area only (except where the retroactive provisions of the IFC govern and are enforced by the jurisdiction, as previously discussed), unless the work area on any floor exceeds 50 percent of that floor area. In those cases, the fire alarm system must be provided beyond the work area throughout that floor. Where this provision applies and there are multiple tenants on the floor required to be equipped with a fire alarm system, the extension of the alarm system into tenant areas that are not part of the alteration work is very disruptive and, in most cases, acts as a major deterrent to undertaking certain levels of alteration. For this reason, the exception eliminates the requirement for the alarm-initiating and notification appliances to be installed in tenant spaces that are not part of the alteration work.

804.4.3 Smoke alarms. Individual sleeping units and individual dwelling units in any *work area* in Group R and I-1 occupancies shall be provided with smoke alarms in accordance with the *International Fire Code*.

> **Exception:** Interconnection of smoke alarms outside of the *work area* shall not be required.

❖ Smoke alarms in sleeping and dwelling units in the work area must be provided as required by the IFC. Sections 1103.8 and 907.2.11 of the IFC are the applicable sections for smoke alarms. According to these sections, smoke alarms are required in sleeping rooms, in every room in the path of the means of egress from sleeping areas and in each story and basement. There are several exceptions to these smoke alarm placement requirements for buildings that comply with previous codes or that have smoke alarms connected to an automatic fire alarm system. Interconnection of smoke alarms is still required, unless such smoke alarms are located outside of the work area or qualify under Exception 2 of IFC Section 1103.8.2 for Groups R-2 and R-4 only. Also note that exceptions apply to the power source requirements in existing buildings in IFC Section 1103.8.3.

SECTION 805
MEANS OF EGRESS

805.1 Scope. The requirements of this section shall be limited to work areas that include exits or corridors shared by more than one tenant within the *work area* in which Level 2 *alterations* are being performed, and where specified they shall apply throughout the floor on which the *work areas* are located or otherwise beyond the *work area*.

❖ This section is entirely devoted to the means-of-egress requirements in existing buildings undergoing Level 2 or 3 alterations. The requirements of this section apply to the work area only, unless there are specific sections addressing supplemental requirements; in which case, the requirements will apply throughout the floor under consideration or beyond. The provisions of this section are applicable only when the alteration work area includes exits or corridors shared by more than one tenant. As such, in multiple-tenant buildings where the reconfiguration of space takes place in one of the tenant spaces and does not include corridors or exits that affect others, the tenant space undergoing alterations need only comply with the means-of-egress requirements of Section 704. In essence, if there are any egress conditions that are considered unsafe, the conditions must be remedied. It is not necessary to confirm that the level of egress safety has not been reduced compared to the current conditions.

805.2 General. The means of egress shall comply with the requirements of this section.

> **Exceptions:**
>
> 1. Where the *work area* and the means of egress serving it complies with NFPA 101.
>
> 2. Means of egress conforming to the requirements of the building code under which the building was constructed shall be considered compliant means of egress if, in the opinion of the *code official*, they do not constitute a distinct hazard to life.

❖ The provisions of Section 805 are intended to address improvements in the means of egress that are crucial for the safe egress of occupants. With the exception of very old buildings, most existing buildings designed under the building codes of a jurisdiction provide a certain degree of safe egress. Accordingly, unless the code official finds that all or parts of the means-of-egress system constitute a distinct hazard, the means of egress does not need to comply with Section 805 as long as it complies with the requirements of the building code under which it was built, including NFPA 101. Where the code official finds a distinct hazard, the provisions of Section 805 could be applied or whatever other remedy determined by the code official must be followed.

805.3 Number of exits. The number of exits shall be in accordance with Sections 805.3.1 through 805.3.3.

❖ The number of exits in any building is one of the most important factors in the safe egress of occupants. The IBC and IFC make it unlawful to reduce the number of exits in a building.

805.3.1 Minimum number. Every story utilized for human occupancy on which there is a *work area* that includes exits or corridors shared by more than one tenant within the *work area* shall be provided with the minimum number of exits based on the occupancy and the occupant load in accordance with the *International Building Code*. In addition, the exits shall comply with Sections 805.3.1.1 and 805.3.1.2.

❖ Given that the number of exits is so important, this section requires that every story with a work area that falls within the limitations discussed in Section 805.1 be provided with a minimum number of exits as required for new construction (in accordance with the IBC). The code does, however, provide exceptions that allow single-exit construction, many of which are common to the IBC, as listed in Section 805.3.1.1. In addition, fire escapes can be used as a solution where more than one exit is required, as provided in Section 805.3.1.2.

805.3.1.1 Single-exit buildings. Only one exit is required from buildings and spaces of the following occupancies:

1. In Group A, B, E, F, M, U and S occupancies, a single exit is permitted in the story at the level of exit discharge when the occupant load of the story does not exceed 50 and the exit access travel distance does not exceed 75 feet (22 860 mm).

2. Group B, F-2, and S-2 occupancies not more than two stories in height that are not greater than 3,500 square feet per floor (326 m^2), when the exit access travel distance does not exceed 75 feet (22 860 mm). The minimum fire-resistance rating of the exit enclosure and of the opening protection shall be 1 hour.

3. Open parking structures where vehicles are mechanically parked.

4. In Group R-4 occupancies, the maximum occupant load excluding staff is 16.

5. Groups R-1 and R-2 not more than two stories in height, when there are not more than four dwelling units per floor and the exit access travel distance does not exceed 50 feet (15 240 mm). The minimum fire-resistance rating of the exit enclosure and of the opening protection shall be 1 hour.

6. In multilevel dwelling units in buildings of occupancy Group R-1 or R-2, an exit shall not be required from every level of the dwelling unit provided that one of the following conditions is met:

    6.1. The travel distance within the dwelling unit does not exceed 75 feet (22 860 mm); or

    6.2. The building is not more than three stories in height and all third-floor space is part of one or more dwelling units located in part on the second floor; and no habitable room within any such dwelling unit shall have a travel distance that exceeds 50 feet (15 240 mm) from the outside of the habitable room entrance door to the inside of the entrance door to the dwelling unit.

7. In Group R-2, H-4, H-5 and I occupancies and in rooming houses and child care centers, a single exit is permitted in a one-story building with a maximum occupant load of 10 and the exit access travel distance does not exceed 75 feet (22 860 mm).

8. In buildings of Group R-2 occupancy that are equipped throughout with an automatic fire sprinkler system, a single exit shall be permitted from a basement or story below grade if every dwelling unit on that floor is equipped with an approved window providing a clear opening of at least 5 square feet (0.47 m^2) in area, a minimum net clear opening of 24 inches (610 mm) in height and 20 inches (508 mm) in width, and a sill height of not more than 44 inches (1118 mm) above the finished floor.

9. In buildings of Group R-2 occupancy of any height with not more than four dwelling units per floor; with a smokeproof enclosure or outside stairway as an exit; and with such exit located within 20 feet (6096 mm) of travel to the entrance doors to all dwelling units served thereby.

10. In buildings of Group R-3 occupancy equipped throughout with an automatic fire sprinkler system, only one exit shall be required from basements or stories below grade.

❖ There are 10 specific conditions described where the building or a certain story could be provided with only one single exit. Several of these 10 conditions are either the same or similar to various provisions in the IBC. For example, Item 1 in this section is similar to the first row in IBC Table 1006.3.2(2). The only difference is that Item 1 of this section allows the condition at the level of exit discharge in multiple-story buildings while IBC Table 1006.3.2(2) allows this condition in a one-story building only. Item 3 of this section is the same as Exception 3 to IBC Section 1006.3.2, and Item 7 is the same as the third row in IBC Table 1006.3.2(2).

The characteristics considered for a building to be of single-exit construction include occupancy, maximum height of building above grade plane, maximum occupants or dwelling units per floor, and exit access travel distance per floor. The occupant load of each floor is determined in accordance with the provisions of IBC Section 1004.1. The exit access travel distance is measured along the natural and unobstructed path to the exit, as described in IBC Section 1017.1.

805.3.1.2 Fire escapes required. For other than Group I-2, where more than one exit is required, an existing or newly constructed fire escape complying with Section 805.3.1.2.1 shall be accepted as providing one of the required means of egress.

❖ The use of fire escapes as an element of the required means of egress has not been allowed by building codes in many years. The typical problems with fire escapes are the difficulty of access in conditions of panic, the minimal width and their location outside the building where they can be subjected to ice, snow and slippery conditions. In existing buildings, however, fire escapes offer a solution where the construction of a new exit stair may be impossible because of the proximity of the exterior walls to property lines; the infeasibility of penetrating or rearranging interior or exterior structural elements; the prohibitive cost of major restructuring of building elements; or the impact on other relevant occupancy issues. As such, this section provides for the use of existing or newly constructed fire escapes to be used as a required element of the means-of-egress system in cases where more than one exit is required and the fire escape meets the construction requirements of this section. The fire escapes can be counted only as one of the required means of egress and not both. Specific detail requirements and construction requirements are given in Sections 805.3.1.2.1 and 805.3.1.2.2.

Since this is located in the alterations section of the code, it is important to stress that this could not be used for a second exit for a vertical addition. Additions are addressed by Chapter 11 and are essentially considered a new building in terms of the application of the IBC.

Group I-2 occupancies are specifically restricted from using the fire escape allowance for an existing building. The restriction is somewhat intuitive based on the type of occupancy, but in order to be consistent with federal licensing requirements this specific restriction is necessary.

805.3.1.2.1 Fire escape access and details. Fire escapes shall comply with all of the following requirements:

1. Occupants shall have unobstructed access to the fire escape without having to pass through a room subject to locking.

2. Access to a new fire escape shall be through a door, except that windows shall be permitted to provide access from single dwelling units or sleeping units in Group R-1, R-2 and I-1 occupancies or to provide access from spaces having a maximum occupant load of 10 in other occupancy classifications.

    2.1. The window shall have a minimum net clear opening of 5.7 square feet (0.53 m^2) or 5 square feet (0.46 m^2) where located at grade.

    2.2. The minimum net clear opening height shall be 24 inches (610 mm) and net clear opening width shall be 20 inches (508 mm).

    2.3. The bottom of the clear opening shall not be greater than 44 inches (1118 mm) above the floor.

    2.4. The operation of the window shall comply with the operational constraints of the *International Building Code*.

3. Newly constructed fire escapes shall be permitted only where exterior stairways cannot be utilized because of lot lines limiting the stairway size or because of the sidewalks, alleys, or roads at grade level.

4. Openings within 10 feet (3048 mm) of fire escape stairways shall be protected by fire assemblies having minimum $^3/_4$-hour fire-resistance ratings.

    Exception: Opening protection shall not be required in buildings equipped throughout with an approved automatic sprinkler system.

5. In all buildings of Group E occupancy, up to and including the 12th grade, buildings of Group I occupancy, rooming houses and childcare centers, ladders of any type are prohibited on fire escapes used as a required means of egress.

❖ This section provides the access and detail conditions that must be followed in order to make a fire escape an acceptable second exit.

Newly constructed fire escapes are allowed only if the construction of exterior stairs is impossible because of property lines, sidewalks, alleys or roads. If allowed to be used by this section, then a fire escape must be readily accessible without any obstructions, and access to it must be through a door with the exception of a single dwelling unit, sleeping unit or spaces with a maximum occupant load of 10, in which case access through a window is acceptable. The requirements for the size and location of a window providing access to the fire escape mirrors the requirement for emergency escape and rescue openings contained in IBC Chapter 10. The height of the bottom of the clear opening is limited to 44 inches (1118 mm) or less such that it can be used effectively as an emergency escape.

If security grilles, decorations or similar devices are installed on escape windows, such items must be readily removable to permit occupant escape without the use of any tools, keys or a force greater than that required for the normal operation of the window.

Where bars, grilles or grates are placed over the emergency escape and rescue opening, it is important that they are easily removable. Thus, the requirements for ease of operation are the same as those for windows.

All openings within 10 feet (3048 mm) of fire escape stairways used for a required means of egress must be protected by fire assemblies having a minimum $^3/_4$-hour fire-resistance rating. This is consistent with protection of openings adjacent to exterior exit stairways required by IBC Section 1023.7. Note that the exterior exit stairway requirements in Section 1027.6 reference the interior exit stairway requirements for this purpose. Much the same as IBC Section 1023.7, the protection of openings within 10 feet (3048 mm) of fire escapes is not required for buildings equipped throughout with an approved automatic sprinkler system in accordance with NFPA 13 or 13R.

Example 805.3.1.2.1:

An existing two-story downtown building is used for shops on the first floor and the second floor is used for offices of several tenants. The owner wishes to rearrange, add interior partitions to and alter about 80 percent of the second floor. The second floor has only one stairway and, because of its current uses and the building location on the lot, is not able to add a second stairway. The area of the second floor is such that it would require two stairways according to the IBC.

Q: Would the alteration trigger the requirement of the second stairway?

A: This would require the construction of a second stairway; however, it would allow fire escapes to be utilized in this case (see Sections 805.3 and 805.3.1.2).

805.3.1.2.2 Construction. The fire escape shall be designed to support a live load of 100 pounds per square foot (4788 Pa) and shall be constructed of steel or other approved *noncombustible materials*. Fire escapes constructed of wood not less than nominal 2 inches (51 mm) thick are permitted on buildings of Type V construction. Walkways and railings located over or supported by combustible roofs in buildings of Types III and IV construction are permitted to be of wood not less than nominal 2 inches (51 mm) thick.

❖ Sections 805.3.1.2.2 and 805.3.1.2.3 provide the minimum structural and dimensional requirements for new fire escapes. Existing fire escapes are not restricted to compliance with Sections 805.3.1.2.2 and 805.3.1.2.3, but must be inspected and evaluated to be sure no dangerous conditions exist and that they will function as intended for a required means of egress.

805.3.1.2.3 Dimensions. Stairways shall be at least 22 inches (559 mm) wide with risers not more than, and treads not less than, 8 inches (203 mm). Landings at the foot of stairways shall be not less than 40 inches (1016 mm) wide by 36 inches (914 mm) long and located not more than 8 inches (203 mm) below the door.

❖ Sections 805.3.1.2.2 and 805.3.1.2.3 provide the minimum structural and dimensional requirements for new fire escapes. Existing fire escapes are not restricted to compliance with Sections 805.3.1.2.2 and 805.3.1.2.3, but must be inspected and evaluated to be sure no dangerous conditions exist and that they will function as intended for a required means of egress.

805.3.2 Mezzanines. Mezzanines in the *work area* and with an occupant load of more than 50 or in which the travel distance to an exit exceeds 75 feet (22 860 mm) shall have access to at least two independent means of egress.

Exception: Two independent means of egress are not required where the travel distance to an exit does not exceed 100 feet (30 480 mm) and the building is protected throughout with an automatic sprinkler system.

❖ The means-of-egress requirements for mezzanines are almost exactly the same as the provisions found in the IBC. This applies to those mezzanines that are in the work area. The triggers for requiring two independent means of egress from mezzanines are the occupant load being greater than 50 or the travel distance being greater than 75 feet (22 860 mm), with an exception for sprinklered buildings allowing travel distance up to 100 feet (30 480 mm). Similar provisions are found in IBC Sections 505.2.2 and 1006.2.1. The code is somewhat less restrictive than the IBC since the occupant load trigger is set at over 50 regardless of occupancy classification, whereas IBC Table 1006.2.1 has a lower level of threshold trigger for hazardous, industrial, residential and storage occupancies. Additionally, the code limitation of 75 feet (22 860 mm) [or 100 feet (30 480 mm) in sprinklered buildings] is the measurement of travel distance, whereas the similar provisions in IBC Section 1006.2.1 are the measurement of the common path of egress travel.

805.3.3 Main entrance—Group A. All buildings of Group A with an occupant load of 300 or more shall be provided with a main entrance capable of serving as the main exit with an egress capacity of at least one-half of the total occupant load. The remaining exits shall be capable of providing one-half of the total required exit capacity.

Exception: Where there is no well-defined main exit or where multiple main exits are provided, exits shall be permitted to be distributed around the perimeter of the building provided that the total width of egress is not less than 100 percent of the required width.

❖ Assembly occupancy egress, particularly the capacity of the main entrance/main exit, is so critical for assembly occupancies with a high occupant load that the code, in this section, provides the same exact requirements found in Section 1029.2 of the IBC for new construction. Accordingly, when Level 2 or 3 alteration work is taking place in assembly occupancies with an occupant load greater than 300, the main entrance/exit must provide sufficient width to accommodate one-half of the total occupant load.

Example 805.3.3:

An existing community hall shown in Commentary Figure 805.3.3 is to undergo some reconfiguration of its interior spaces. There are currently several 3-foot (914 mm) doors that provide the needed exit width and separation of exit requirements based on the IBC.

Q: Are there any additional requirements related to the exterior doors that must be complied with?

A: Yes, but only for a Level 3 alteration since this only affects a single tenant (Section 805.1).

A main entrance must be provided that is wide enough to accommodate one-half of the total occupant load (450/2 = 225). This is a width of 225 × 0.2 = 45 inches (1143 mm) (see Section 1005.1 of the IBC and Commentary Figure 805.3.3). Note that this is applicable only if considered a Level 3 alteration. Section 805 only applies where there is more than one tenant. Chapter 9 and specifically Section 901.2 note that the requirements of Section 805 apply whether or not they include more than one tenant.

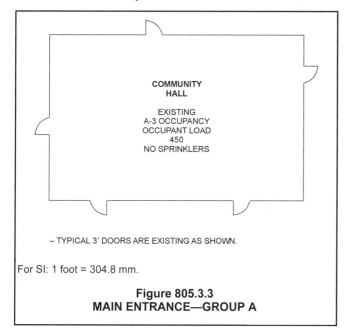

– TYPICAL 3′ DOORS ARE EXISTING AS SHOWN.

For SI: 1 foot = 304.8 mm.

**Figure 805.3.3
MAIN ENTRANCE—GROUP A**

805.4 Egress doorways. Egress doorways in any *work area* shall comply with Sections 805.4.1 through 805.4.5.

❖ This is the charging paragraph for the requirements for egress doorways in Level 2 or 3 alterations. The specific requirements are given in Sections 805.4.1 through 805.4.5, relating the number of egress doorways, door swing, door closing, panic hardware and emergency power in detention facilities.

805.4.1 Two egress doorways required. Work areas shall be provided with two egress doorways in accordance with the requirements of Sections 805.4.1.1 and 805.4.1.2.

❖ The triggers for requiring two egress doorways are similar in some respects to the IBC triggers. Section 805.4.1.1 relates these triggers for all occupancies, and Section 805.4.1.2 deals with Group I-2 occupancies.

805.4.1.1 Occupant load and travel distance. In any *work area*, all rooms and spaces having an occupant load greater than 50 or in which the travel distance to an exit exceeds 75 feet (22 860 mm) shall have a minimum of two egress doorways.

Exceptions:

1. Storage rooms having a maximum occupant load of 10.

2. Where the *work area* is served by a single exit in accordance with Section 805.3.1.1.

❖ The discussion in Section 805.3 demonstrated how several criteria for requiring only one means of egress are also found in the IBC for new construction. The same is true here: the occupant load being greater than 50 and travel distance exceeding 75 feet (22 860 mm) are the triggers used to require at least two egress doorways for all occupancies.

Exception 1 to Section 805.4.1.1 addresses storage rooms with small occupant loads that might be of such dimensions that the travel distance might exceed 75 feet (22 860 mm). Realizing that many storage rooms are generally not occupied regularly, and when occupied, such occupants are either employees or individuals who are familiar with the surroundings, this exception allows storage rooms of any size and any length of travel distance to have only one egress doorway as long as the occupant load does not exceed 10.

Example 805.4.1.1:

An existing grocery store as shown in Commentary Figure 805.4.1.1 is undergoing Level 2 or 3 alterations. The storage area has a travel distance of 90 feet (27 432 mm) within the storage room and an overall travel distance of 10 feet (3048 mm) to exit the building. There is currently one egress doorway from the storage room to the retail area.

Q: Is a second egress doorway required from the storage area?

A: No. The storage room has an occupant load of 10 (3,000/300 = 10) and falls under Exception 1.

- Storage room occupant load = 10 (see Table 1004.1.2 of the IBC).
- Maximum travel distance in storage room = 90 feet (27 432 mm).

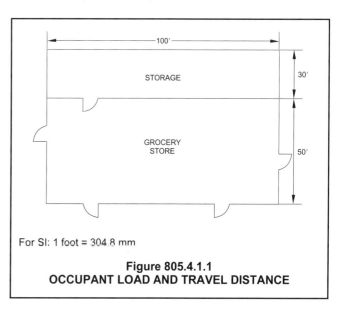

For SI: 1 foot = 304.8 mm

**Figure 805.4.1.1
OCCUPANT LOAD AND TRAVEL DISTANCE**

ALTERATIONS—LEVEL 2

- One egress doorway from the storage room is acceptable (see Section 805.4.1.1, Exception 1 and Commentary Figure 805.4.1.1).

Note that even if the travel distances were exceeded, this is a single-tenant building and unless the alteration is Level 3 this requirement would not apply. Section 805 only applies where there is more than one tenant. Chapter 9 and specifically Section 901.2 note that the requirements of Section 805 apply whether or not they include more than one tenant.

Exception 2 refers the user to Section 805.3.1.1 so that a building or space that qualifies under any of the 10 items listed would still be required to have one egress doorway.

805.4.1.2 Group I-2. In buildings of Group I-2 occupancy, any patient sleeping room or suite of patient rooms greater than 1,000 square feet (93 m^2) within the *work area* shall have a minimum of two egress doorways.

❖ Group I-2 occupancies have additional criteria for patient sleeping rooms or suites in this section. This additional criterion for Group I-2 patient sleeping areas is taken directly from a part of IBC Section 407.4.4.6.1, and recognizes the low patient-to-staff ratio of these facilities where staff is directly responsible for the safety of patients in the event of a fire.

805.4.2 Door swing. In the *work area* and in the egress path from any *work area* to the exit discharge, all egress doors serving an occupant load greater than 50 shall swing in the direction of exit travel.

❖ Door swings must be considered not only in the work area, but all the way from the work area along the path of egress to the exit discharge. Again, door swing is such a critical element of safe egress that existing doors must comply just as new doors do in accordance with the IBC. The threshold of "serving an occupant load greater than 50" of this section is taken directly from IBC Section 1010.1.2.1.

Example 805.4.2:

An existing three-story office building as shown in Commentary Figure 805.4.2 is undergoing Level 2 or 3 alterations on the third floor. Stairway 1 is the most obvious, the closest and the most likely path of egress for the occupants in the work area under consideration.

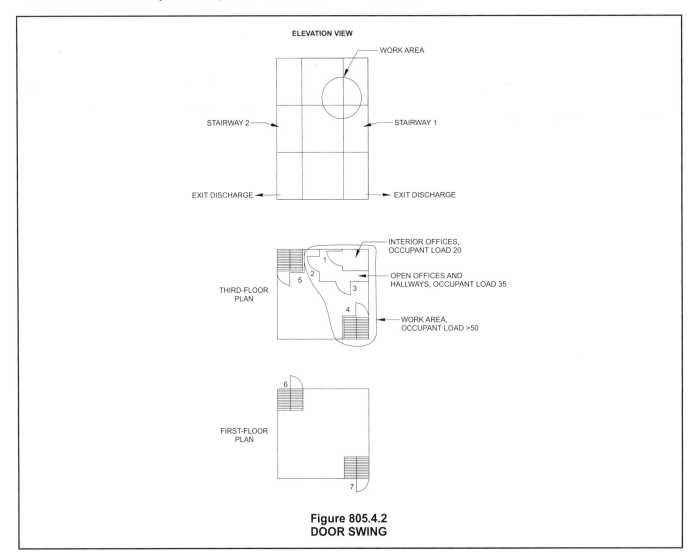

**Figure 805.4.2
DOOR SWING**

Q: Which doors are required to swing in the direction of egress travel?

A: Doors 2, 3, 4 and 7 in Commentary Figure 805.4.2 must swing in the direction of egress travel as they serve an occupant load greater than 50 and are either in the work area or along the egress path to the exit discharge (door 1 serves an occupant load less than 50, and doors 5 and 6 are not along the path of egress of the work area and, as such, are not required to comply with Section 805.4.2) (see Commentary Figure 805.4.2).

805.4.2.1 Supplemental requirements for door swing. Where the *work area* exceeds 50 percent of the floor area, door swing shall comply with Section 805.4.2 throughout the floor.

> **Exception:** Means of egress within or serving only a tenant space that is entirely outside the *work area*.

❖ Door swings on the entire floor, except doors that are in a tenant space located entirely outside the work area, must swing in the direction of egress travel if the work area on that floor exceeds 50 percent of the area of that floor and if the occupant load the doors serve is greater than 50. Accordingly, all doors along the means of egress to the exit discharge must also swing in the direction of egress.

Example 805.4.2.1:

An existing two-story building has three tenants on the third floor as shown in Commentary Figure 805.4.2.1. The owner plans some alterations of the second floor. The work area will be more than 50 percent of the second floor area.

Q: Which doors must be brought into compliance with the door swing in the direction of egress travel?

A: Doors 4, 5, 6, 7, 10, 11, 12 and 13 in Commentary Figure 805.4.2.1 must swing in the direction of egress travel. Doors 6, 7 and 10 serve an occupant load of more than 50 and are in the work area. Door 11

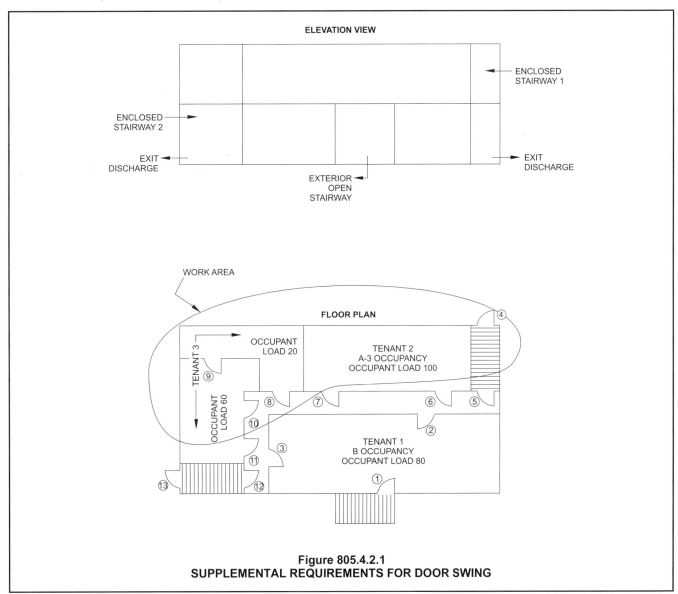

Figure 805.4.2.1
SUPPLEMENTAL REQUIREMENTS FOR DOOR SWING

serves an occupant load of more than 50 and is outside of the work area; however, because the work area is greater than 50 percent of the second-floor area, this door must also swing in the direction of egress. Doors 4, 5, 12 and 13 are on the egress path from the work area to exit discharge.

Doors 1, 2, 3, 8 and 9 are not required to change their swing to the direction of egress if they currently swing inward. Doors 1, 2 and 3 serve an occupant load greater than 50, but are in a tenant space completely outside of the work area and fall under the exception. Door 8 serves an occupant load less than 50, and door 9 is a convenience door, connecting the two parts of tenant 3 (see Commentary Figure 805.4.2.1).

805.4.3 Door closing. In any *work area*, all doors opening onto an exit passageway at grade or an exit stairway shall be self-closing or automatic-closing by listed closing devices.

Exceptions:

1. Where exit enclosure is not required by the *International Building Code*.

2. Means of egress within or serving only a tenant space that is entirely outside the *work area*.

❖ The same concepts discussed for door swing in the work area and beyond the work area when supplemental provisions are applicable are true for door closings. Doors opening into exit passageways and exit stairway enclosures are required to be self-closing or automatic closing by means of listed devices.

805.4.3.1 Supplemental requirements for door closing. Where the *work area* exceeds 50 percent of the floor area, doors shall comply with Section 805.4.3 throughout the exit stairway from the *work area* to, and including, the level of exit discharge.

❖ If the work area on the floor exceeds 50 percent of the area of that floor and if the occupant load it serves is greater than 50, all doors along the means of egress including the level of exit discharge must also be self-closing or automatic closing.

805.4.4 Panic hardware. In any *work area*, and in the egress path from any *work area* to the exit discharge, in buildings or portions thereof of Group A assembly occupancies with an occupant load greater than 100, all required exit doors equipped with latching devices shall be equipped with approved panic hardware.

❖ The panic hardware provisions are only applicable to Group A occupancies and are much the same as IBC Section 1010.1.10. The triggers for requiring panic hardware are the presence of latching devices on the door and the occupant load of the assembly occupancy being greater than 100. The panic hardware requirements are required in the work areas and the path of egress from the work area to the exit discharge. Panic hardware requirements could be applicable in other parts of the space or building beyond the work area and the path of egress when the supplemental provisions apply.

805.4.4.1 Supplemental requirements for panic hardware. Where the *work area* exceeds 50 percent of the floor area, panic hardware shall comply with Section 805.4.4 throughout the floor.

Exception: Means of egress within a tenant space that is entirely outside the *work area*.

❖ Similar to the supplemental requirements for door swing and door closing, work exceeding 50 percent of the floor area triggers a more restrictive requirement for panic hardware, as well. Except for means of egress in an unaffected tenant space, the 50-percent threshold triggers a requirement for panic hardware throughout the work area floor.

805.4.5 Emergency power source in Group I-3. Power-operated sliding doors or power-operated locks for swinging doors shall be operable by a manual release mechanism at the door. Emergency power shall be provided for the doors and locks in accordance with Section 2702 of the *International Building Code*.

Exceptions:

1. Emergency power is not required in facilities with 10 or fewer locks complying with the exception to Section 408.4.1 of the *International Building Code*.

2. Emergency power is not required where remote mechanical operating releases are provided.

❖ This threshold for emergency power for doors with remote power unlocking equipment is the same as given in IBC Section 408.4.2 for Group I-3 facilities.

805.5 Openings in corridor walls. Openings in corridor walls in any *work area* shall comply with Sections 805.5.1 through 805.5.4.

Exception: Openings in corridors where such corridors are not required to be rated in accordance with the *International Building Code*.

❖ This is the charging paragraph for the requirements for openings in corridor walls in Level 2 or 3 alterations. Sections 805.5.1 through 805.5.4 relate specific requirements for corridor doors, transoms and other corridor openings such as windows, and supplemental requirements. The protections prescribed in these four sections are required only if the corridor under consideration is required to be a fire-resistance-rated corridor in accordance with the IBC. The overall goal is to eliminate certain unsuitable doors and openings in rated corridors in a Level 2 or 3 work area, even if there was no intention by the scope of the alteration work to do anything to these elements.

805.5.1 Corridor doors. Corridor doors in the *work area* shall not be constructed of hollow core wood and shall not contain louvers. All dwelling unit or sleeping unit corridor doors in work areas in buildings of Groups R-1, R-2, and I-1 shall be at least $1^3/_8$-inch (35 mm) solid core wood or approved equivalent and shall not have any glass panels, other than approved wired glass or other approved glazing material in metal frames. All dwelling unit or sleeping unit corridor doors in *work areas* in buildings of Groups R-1, R-2, and I-1 shall be equipped with approved door closers. All replacement doors shall be $1^3/_4$-inch (44 mm) solid bonded wood core or approved equivalent, unless the existing frame will accommodate only a $1^3/_8$-inch (35 mm) door.

Exceptions:

1. Corridor doors within a dwelling unit or sleeping unit.

2. Existing doors meeting the requirements of *Guidelines on Fire Ratings of Archaic Materials and Assemblies* (IEBC Resource A) for a rating of 15 minutes or more shall be accepted as meeting the provisions of this requirement.

3. Existing doors in buildings protected throughout with an approved automatic sprinkler system shall be required only to resist smoke, be reasonably tight fitting, and shall not contain louvers.

4. In group homes with a maximum of 15 occupants and that are protected with an approved automatic detection system, closing devices may be omitted.

5. Door assemblies having a fire protection rating of at least 20 minutes.

❖ In cases where doors are being replaced, these provisions require that a certain level of fire or smoke protection be provided. As such, no hollow core wood doors and no louvers are allowed in a rated corridor in the work area. In addition to this prohibition, if the occupancy of the building undergoing alteration is Group B-1, R-1, or R-2 and the corridor under consideration serves the dwelling or sleeping units of these occupancies, all doors in the rated corridors must be equipped with approved closers and be at least $1^3/_8$-inch (35 mm) solid core wood or an approved equivalent. Further, these doors cannot contain any glass panels unless tested assemblies or approved wired glass is used. Approved wired glass is generally considered to be $^1/_4$-inch (6.4 mm) wired glass.

Once the doors that are to be replaced have been identified, then the replacement doors must be at least $1^3/_4$-inch (45 mm) solid-bonded wood core or an approved equivalent. One-and-three-eighths-inch (35 mm) solid wood doors are allowed for such replacement doors if the existing frame can only accommodate a $1^3/_8$-inch-thick (35 mm) door. Five exceptions have been provided for corridor door requirements that are self-explanatory. The replacement doors have a greater protection requirement compared to those doors that are not being replaced but happen to be in the work area because it is unreasonable to require every existing door that provides a certain level of protection to be replaced within the entire work area, even though they provide a certain level of protection with a $1^3/_8$-inch (35 mm) solid core door or equivalent.

805.5.2 Transoms. In all buildings of Group I-1, I-2, R-1 and R-2 occupancies, all transoms in corridor walls in work areas shall be either glazed with $^1/_4$-inch (6.4 mm) wired glass set in metal frames or other glazing assemblies having a fire protection rating as required for the door and permanently secured in the closed position or sealed with materials consistent with the corridor construction.

❖ Transoms in corridors are specifically regulated only in Group I-1, I-2, R-1 and R-2 occupancies. It should be noted again that the applicability here is only in corridors that would be required to be rated. The IBC requires that glass sidelites and transoms perform the same as fire windows and does not consider these elements part of the door assembly. Based on this approach, sidelites and transoms in occupancies other than Group I-1, I-2, R-1 and R-2 would be covered in Section 805.5.3, which addresses sash, grilles and windows. Transoms must meet one of the following requirements:

1. Use $^1/_4$-inch (6.4 mm) wired glass in metal frame.

2. A minimum of 20-minute fire-resistant-protected glazing fixed assembly.

3. Do away with the transom and seal it with wall construction similar to the existing corridor wall.

805.5.3 Other corridor openings. In any *work area*, any other sash, grille, or opening in a corridor and any window in a corridor not opening to the outside air shall be sealed with materials consistent with the corridor construction.

❖ Openings other than doors, such as sash, grille, louver and windows, that open into a fire-resistance-rated corridor and do not open to the outside air, must be covered and sealed the same as the existing corridor wall construction. This is required to protect the corridor from potential smoke contamination from adjacent rooms. The reference to such openings "not opening to the outside air" is consistent with IBC Section 708.5, which allows corridor exterior walls and their openings to comply with the fire-resistance rating based primarily on proximity to property lines, otherwise known as "fire separation distance." Two questions arising from this provision are why windows in a rated corridor are required to be sealed with a wall construction, and is there a way to allow such windows to remain in place and not be covered with wall construction? The code in this section intends to improve the safety and protection of corridors and does not intend to necessarily eliminate all windows opening into a corridor. As such, if it is desired for such windows to remain, then the window and its glazing must comply with the fire window and fire-protection-rated glazing requirements of IBC Section 716.5.8. It must be clarified that sidelites and transoms are not considered part of the door assembly and are regulated by this section.

805.5.3.1 Supplemental requirements for other corridor opening. Where the *work area* exceeds 50 percent of the floor area, Section 805.5.3 shall be applicable to all corridor windows, grills, sashes, and other openings on the floor.

> **Exception:** Means of egress within or serving only a tenant space that is entirely outside the *work area*.

❖ The supplemental provisions, where the work area exceeds 50 percent of the floor area, are the same as discussed in all previous sections. As in all other areas, the 50-percent threshold brings a more serious concern regarding the alteration and, thus, more restrictive requirements are applied throughout the floor, even in areas where no alteration is being performed. For other corridor openings, however, the tenant spaces that are unaffected do not require these alterations.

805.5.4 Supplemental requirements for corridor openings. Where the *work area* on any floor exceeds 50 percent of the floor area, the requirements of Sections 805.5.1 through 805.5.3 shall apply throughout the floor.

❖ The supplemental provisions, where the work area exceeds 50 percent of the floor area, are the same as discussed in all previous sections. As in all other areas, the 50-percent threshold brings a more serious concern regarding the alteration and, thus, more restrictive requirements are applied throughout the floor, even in areas where no alteration is being performed. This paragraph does not exclude tenant spaces that are unaffected by the alterations; therefore, the doors, openings and transoms must be modified according to the applicable section. This paragraph supersedes the exception given in Section 805.5.3.1.

805.6 Dead-end corridors. Dead-end corridors in any *work area* shall not exceed 35 feet (10 670 mm).

> **Exceptions:**
>
> 1. Where dead-end corridors of greater length are permitted by the *International Building Code*.
> 2. In other than Group A and H occupancies, the maximum length of an existing dead-end corridor shall be 50 feet (15 240 mm) in buildings equipped throughout with an automatic fire alarm system installed in accordance with the *International Building Code*.
> 3. In other than Group A and H occupancies, the maximum length of an existing dead-end corridor shall be 70 feet (21 356 mm) in buildings equipped throughout with an automatic sprinkler system installed in accordance with the *International Building Code*.
> 4. In other than Group A and H occupancies, the maximum length of an existing, newly constructed, or extended dead-end corridor shall not exceed 50 feet (15 240 mm) on floors equipped with an automatic sprinkler system installed in accordance with the *International Building Code*.

❖ The dead-end corridor limitations are only applicable where more than one exit or exit access doorway is required. The typical existing dead-end length allowed is 35 feet (10 670 mm) with exceptions that allow lengths up to 70 feet (21 336 mm). Existing dead ends in Group A and H occupancies are limited to 35 feet (10 670 mm), unless Exception 3 of IBC Section 1020.4 could apply. This is the exception that allows unlimited dead-end length as long as the length of the corridor is less than two and one-half times the least width of the dead-end corridor.

The installation of an automatic fire alarm system would increase the allowed length of an existing dead end to as much as 50 feet (15 240 mm), and the installation of an automatic sprinkler system throughout the building would increase the allowed length to 70 feet (21 336 mm). These two allowances are only applicable to existing corridors that happen to be longer than allowed by the IBC and are intended to encourage the installation of fire alarms and fire sprinklers throughout the building where such a corridor exists. Existing dead-end corridors must be brought into compliance with the 35-foot (10 670 mm) length limitation if these exceptions are not used.

805.7 Means-of-egress lighting. Means-of-egress lighting shall be in accordance with this section, as applicable.

❖ This is the charging paragraph for the requirements for means-of-egress lighting in Level 2 or 3 alterations. The specific requirements are given in Sections 805.7.1 and 805.7.2, relating the requirements for artificial lighting and supplemental requirements, respectively.

805.7.1 Artificial lighting required. Means of egress in all work areas shall be provided with artificial lighting in accordance with the requirements of the *International Building Code*.

❖ The means of egress in the work area and beyond the work area in the entire floor, if the work area exceeds 50 percent of the floor area, must be illuminated in accordance with the requirements of the IBC. This illumination must be provided by artificial lighting and must be a minimum of 1 footcandle (11 lux) at the floor level. There are some exceptions to this requirement that can be found in Sections 1006.1 and 1006.2 of the IBC, which would apply here.

805.7.2 Supplemental requirements for means-of-egress lighting. Where the *work area* on any floor exceeds 50 percent of that floor area, means of egress throughout the floor shall comply with Section 805.7.1.

> **Exception:** Means of egress within or serving only a tenant space that is entirely outside the *work area*.

❖ The supplemental provisions, where the work area exceeds 50 percent of the floor area, are the same here as discussed in all previous sections. As in all other areas, the 50-percent threshold brings a more serious concern regarding the alteration and, thus, more restrictive requirements are applied throughout the floor, even in areas where no alteration is being performed. However, the tenant spaces that are unaffected do not require these alterations.

805.8 Exit signs. Exit signs shall be in accordance with this section, as applicable.

❖ This is the charging paragraph for the requirements for exit signs in Level 2 or 3 alterations. The specific requirements are given in Sections 805.8.1 and 805.8.2.

805.8.1 Work areas. Means of egress in all work areas shall be provided with exit signs in accordance with the requirements of the *International Building Code*.

❖ Exit signs are a critical element of directing occupants to safety outside the building. Exit signs must be installed in all Level 2 or 3 alteration work areas and beyond, where supplemental rules apply, in accordance with the IBC as if this was new construction. The exit sign provisions are found in IBC Section 1013, and it is important to note that such requirements as illumination and power source apply to exit signs being installed in an alteration.

805.8.2 Supplemental requirements for exit signs. Where the *work area* on any floor exceeds 50 percent of that floor area, means of egress throughout the floor shall comply with Section 805.8.1.

Exception: Means of egress within a tenant space that is entirely outside the *work area*.

❖ The supplemental provisions, where the work area exceeds 50 percent of the floor area, are the same here as discussed in all previous sections. As in all other areas, the 50-percent threshold brings a more serious concern regarding the alteration and, thus, more restrictive requirements are applied throughout the floor, even in areas where no alteration is being performed. However, the tenant spaces that are unaffected do not require these alterations.

805.9 Handrails. The requirements of Sections 805.9.1 and 805.9.2 shall apply to handrails from the *work area* floor to, and including, the level of exit discharge.

❖ This is the charging paragraph for the requirements for handrails in Level 2 or 3 alterations. The specific requirements are given in Sections 805.9.1 and 805.9.2, relating the minimum requirements for handrails and design requirements, respectively.

805.9.1 Minimum requirement. Every required exit stairway that is part of the means of egress for any *work area* and that has three or more risers and is not provided with at least one handrail, or in which the existing handrails are judged to be in danger of collapsing, shall be provided with handrails for the full length of the stairway on at least one side. All exit stairways with a required egress width of more than 66 inches (1676 mm) shall have handrails on both sides.

❖ Complying handrails must be installed in all required exit stairways that serve a work area. If additional stairways are provided that are not required by the IBC as an element of means of egress, then the handrail provisions of this section are not applicable to such stairways. The degree of compliance required is limited to a handrail on one side of the stairway only until the occupant load served is so high that the required width of the stairway would be a minimum of 66 inches (1676 mm). At this level of occupant load, complying handrails must be provided on both sides of the stairway. Intermediate handrails are not required as an improvement criterion under the code. This section requires complying handrails for those stairs that have three or more risers. This trigger level is an attempt to be compatible with an exception in some of the legacy building codes that did not require a handrail for stairways containing less than four risers in certain occupancies. The handrail requirement is applicable to required exit stairways that serve the work area, starting at the work area all the way to the level of exit discharge, including the exit discharge itself.

805.9.2 Design. Handrails required in accordance with Section 805.9.1 shall be designed and installed in accordance with the provisions of the *International Building Code*.

❖ Once a handrail is required by this section, its structural strength, height, continuity and other design features must comply with the IBC.

Example 805.9.2:

An existing two-story office building has two enclosed stairways: one convenience stairway, a mezzanine stair and a stair outside the building as shown in Commentary Figure 805.9.2. Stairs either have no handrails or have handrails that are not in compliance with the IBC.

Q: Which stairs must be provided with new complying handrails to bring the current handrails up to the IBC requirements?

A: Stairs 1 and 2 must be provided with new complying handrails to bring their existing handrails up to the IBC requirements. These stairways serve the work area and are both part of the required means-of-egress pathway (see Commentary Figure 805.9.2).

805.10 Refuge areas. Where alterations affect the configuration of an area utilized as a refuge area, the capacity of the refuge area shall not be reduced below that required in Sections 805.10.1 and 805.10.2.

❖ This section was written to prevent reduced capacity and arrangement of refuge areas, required for occupancies that limit movement of occupants or where occupants are incapable of self-preservation, during

ALTERATIONS—LEVEL 2

alterations. Refuge areas are part of the defend-in-place strategy and are designed with a particular capacity in mind in each smoke compartment. Three different occupancy types in the I-Codes have such requirements, which are specifically and individually addressed in Sections 805.10.1.1 through 805.10.1.3. The horizontal exits are based on a strategy similar to defend-in-place but can be applied to any occupancy. The refuge areas created by horizontal exits also need to be maintained.

805.10.1 Capacity. The required capacity of refuge areas shall be in accordance with Sections 805.10.1.1 through 805.10.1.3.

❖ The three subsections address the necessary capacity for each of the occupancies individually since the requirements for refuge areas vary. Essentially, each references the appropriate section in the IBC for such requirements.

805.10.1.1 Group I-2. In Group I-2 occupancies, the required capacity of the refuge areas for smoke compartments in accordance with Section 407.5.1 of the *International Building Code* shall be maintained.

❖ The IBC requires that Group I-2 occupancies provide the largest occupant load of the adjoining compartments. The criteria used to determine the capacity are 30 net square feet for each care recipient confined to a bed or stretcher and not less than 6 square feet for each ambulatory care recipient not confined to bed or stretcher and for other occupants.

805.10.1.2 Group I-3. In Group I-3 occupancies, the required capacity of the refuge areas for smoke compartments in accordance with Section 408.6.2 of the *International Building Code* shall be maintained.

❖ For Group I-3 occupancies, the IBC requires that 6 square feet per occupant be provided on both sides of the smoke barrier forming the smoke compartment. This will include the total of adjoining smoke compartments. Occupants in Group I-3 are typically ambulatory but their movement is limited because of the nature of Group I-3 occupancies.

805.10.1.3 Ambulatory care. In ambulatory care facilities required to be separated by Section 422.2 of the *International Building Code*, the required capacity of the refuge areas for smoke compartments in accordance with Section 422.4 of the *International Building Code* shall be maintained.

❖ Ambulatory care facilities are similar to Group I-2 occupancies and contain occupants that are incapable of self-preservation because of the procedures and surgeries done at such facilities. Therefore, the criterion is similar to that of Group I-2 occupancies and requires 30 square feet for each nonambulatory care recipient. This area must be provided in the corridors, care recipient rooms, treatment rooms, lounge or dining areas and other low-hazard areas in the smoke compartment. Note that only larger ambulatory care facilities greater than 10,000 square feet require smoke compartments.

805.10.2 Horizontal exits. The required capacity of the refuge area for horizontal exits in accordance with Section 1026.4 of the *International Building Code* shall be maintained.

❖ This section is more general and addresses any occupancies using horizontal exits. Horizontal exits allow a similar concept to smoke compartments where occupants simply exit through doors into an adjoining area of the building separated by a fire barrier or fire wall. Section 1026.4 requires 3 square feet for each occupant. There are several specific exceptions for Group I-2 and I-3 occupancies that correlate with the criteria noted in Sections 805.10.1.1 and 805.10.1.2.

805.11 Guards. The requirements of Sections 805.11.1 and 805.11.2 shall apply to guards from the *work area* floor to,

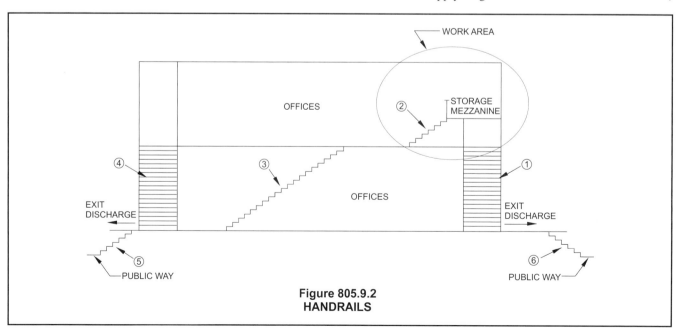

Figure 805.9.2
HANDRAILS

and including, the level of exit discharge but shall be confined to the egress path of any *work area*.

❖ This is the charging paragraph for the requirements for guards in Level 2 or 3 alterations. The specific requirements are given in Sections 805.11.1 and 805.11.2, relating the minimum location and design requirements for guards, respectively.

805.11.1 Minimum requirement. Every open portion of a stairway, landing, or balcony that is more than 30 inches (762 mm) above the floor or grade below and is not provided with guards, or those portions in which existing guards are judged to be in danger of collapsing, shall be provided with guards.

❖ The criteria and triggers for guards are much the same as those for handrails. IBC-compliant guards must be installed along the path of egress from the work area to and including the level of exit discharge, and anywhere at stairways, landings or balconies where an elevation difference of more than 30 inches (762 mm) exists and a complying guard is not present.

805.11.2 Design. Guards required in accordance with Section 805.11.1 shall be designed and installed in accordance with the *International Building Code*.

❖ Once a guard is required by this section, its structural strength, height, size of openings and other design features must comply with the IBC.

SECTION 806
ACCESSIBILITY

806.1 General. A building, *facility*, or element that is altered shall comply with this section and Section 705.

❖ A Level 2 or 3 alteration would include the reconfiguration of space, while a Level 1 alteration would not. Therefore, alterations covered under this section would have all accessibility requirements in Section 705, as well as addressing the possible need for an elevator where a stairway or escalator is being added.

806.2 Stairways and escalators in existing buildings. In *alterations* where an escalator or stairway is added where none existed previously, an accessible route shall be provided in accordance with Sections 1104.4 and 1104.5 of the *International Building Code*.

❖ If a stairway or escalator is added as part of an alteration in a location where one did not previously exist, the alteration must also include an accessible route between the same two levels. If an accessible route is already available between the two levels, this requirement is not applicable. If the stairway or escalator is replacing an existing stairway or escalator, this requirement is not applicable. In conjunction with Section 705.1.13, if the requirement for the accessible route would be in excess of what is required for new construction, such as an accessible route to an area that was exempted by IBC Section 1103.2, 1104, 1107 or 1108, this requirement is not applicable. The intent is that if a route is provided between accessible levels for a nondisabled person to use, it is reasonable to also expect an accessible route.

SECTION 807
STRUCTURAL

[BS] 807.1 General. Structural elements and systems within buildings undergoing Level 2 *alterations* shall comply with this section.

❖ This section directs the code user to the requirements contained in Sections 807.2 through 807.6. These structural requirements apply to alterations involving the addition of equipment or where a reconfiguration of building space results in an increased minimum live load required by IBC Section 1607. Note that the structural requirements for Level 1 alterations in Section 707 are also applicable according to Section 504.2.

[BS] 807.2 New structural elements. New structural elements in *alterations*, including connections and anchorage, shall comply with the *International Building Code*.

❖ Any new structural element that is added in the course of alteration work, including its connections to the existing structure, must comply with IBC requirements for new construction. There is no exception to this requirement.

[BS] 807.3 Minimum design loads. The minimum design loads on existing elements of a structure that do not support additional loads as a result of an *alteration* shall be the loads applicable at the time the building was constructed.

❖ Unless an alteration adds loads to an existing structural member, there is no need to reevaluate it against current code-loading criteria. This is in contrast to the treatment of existing members that have increased loading in accordance with Section 807.4. The applicable loads of the code at the time the building was built apply.

[BS] 807.4 Existing structural elements carrying gravity loads. *Alterations* shall not reduce the capacity of existing gravity load-carrying structural elements unless it is demonstrated that the elements have the capacity to carry the applicable design gravity loads required by the *International Building Code*. Existing structural elements supporting any additional gravity loads as a result of the *alterations*, including the effects of snow drift, shall comply with the *International Building Code*.

Exceptions:

1. Structural elements whose stress is not increased by more than 5 percent.
2. Buildings of Group R occupancy with not more than five dwelling or sleeping units used solely for residential purposes where the *existing building* and its *alteration* comply with the conventional light-frame construction methods of the *International Building*

Code or the provisions of the *International Residential Code*.

❖ In the course of reconfiguring building spaces, it is often necessary to relocate or add equipment. Further, sometimes a change in use (without actually changing the occupancy classification) necessitates use of a higher design live load. Consequently, existing structural members that are subjected to increased loads must comply with the requirements in this section for gravity loads. This section requires the evaluation of these additional loads, including the effects of snow drift, to determine if these loads decrease the capacity of the existing structural elements. Where the alterations result in a net increase in the gravity load that is supported by the existing structure, including the effects of snow drift, the affected structural components must be checked to verify they satisfy the requirements of the IBC for a new structure. There are two exceptions to full compliance with the IBC. One permits smaller Group R buildings and their alterations that comply with either the *International Residential Code*® (IRC®) or the conventional light-frame construction provisions of the IBC. Another exception allows additional gravity loads that do not increase stresses in affected structural elements by more than 5 percent. Allowing overstresses of up to 5 percent in existing structural members has been a long-standing rule of thumb used by structural engineers. This exception does not specifically address the cumulative effects from successive alterations, but it would be prudent to limit cumulative increases under this exception to a total of 5 percent unless it is clearly documented that a structure has additional capacity.

[BS] 807.5 Existing structural elements resisting lateral loads. Except as permitted by Section 807.6, where the alteration increases design lateral loads, or where the alteration results in prohibited structural irregularity as defined in ASCE 7, or where the alteration decreases the capacity of any existing lateral load-carrying structural element, the structure of the altered building or structure shall be shown to meet the wind and seismic provisions of the *International Building Code*. Reduced *International Building Code*-level seismic forces in accordance with Section 301.1.4.2 shall be permitted.

> **Exception:** Any existing lateral load-carrying structural element whose demand-capacity ratio with the alteration considered is not more than 10 percent greater than its demand-capacity ratio with the alteration ignored shall be permitted to remain unaltered. For purposes of calculating demand-capacity ratios, the demand shall consider applicable load combinations with design lateral loads or forces in accordance with *International Building Code* Sections 1609 and 1613. Reduced *International Building Code*-level seismic forces in accordance with Section 301.1.4.2 shall be permitted. For purposes of this exception, comparisons of demand-capacity ratios and calculation of design lateral loads, forces and capacities shall account for the cumulative effects of additions and alterations since original construction.

❖ This section gives guidance with respect to the impact of an alteration on the lateral resistance of the structure. Alterations that affect existing structural elements to a lesser extent are permitted without requiring the existing structure to comply with the provisions for new structures, as long as the alteration itself complies. An existing building or structure that is altered must be carefully evaluated for its ability to withstand earthquake and wind loads. The lateral resistance of an existing building or structure cannot be lessened or pose any undue increase in the fire and life safety hazards of the building or structure. The exception allows alterations that increase the demand-capacity ratio on existing lateral load-carrying elements by no more than 10 percent or decrease the lateral resistance of existing structural elements by no more than 10 percent.

[BS] 807.6 Voluntary lateral force-resisting system alterations. *Alterations* of existing structural elements and additions of new structural elements that are initiated for the purpose of increasing the lateral force-resisting strength or stiffness of an existing structure and that are not required by other sections of this code shall not be required to be designed for forces conforming to the *International Building Code*, provided that an engineering analysis is submitted to show that:

1. The capacity of existing structural elements required to resist forces is not reduced;

2. The lateral loading to existing structural elements is not increased either beyond its capacity or more than 10 percent;

3. New structural elements are detailed and connected to the existing structural elements as required by the *International Building Code*;

4. New or relocated nonstructural elements are detailed and connected to existing or new structural elements as required by the *International Building Code*; and

5. A *dangerous* condition as defined in this code is not created. Voluntary *alterations* to lateral force-resisting systems conducted in accordance with Appendix A and the referenced standards of this code shall be permitted.

❖ This section addresses the issue of upgrading a building's lateral force-resisting system voluntarily to improve resistance to wind and seismic forces. It does not apply in situations where other code sections require compliance with the IBC or other specific minimum wind or seismic design loads. This section allows an owner to initiate an improvement to the lateral force-resisting system to the extent it is viable to do so and provided the requisite engineering analysis is furnished. Since no minimum load requirement is established by the code, the building owner and the design professional have the latitude to establish performance goals and objectives. Thus, an owner can do something to mitigate the hazard of future earthquakes or wind storms in an existing building without being discouraged from doing so by incurring prohibitive costs.

For example, the retrofit of an existing building to improve performance in an earthquake to a level of resistance less than that required for a new building considers the overall benefit of hazard mitigation. It is an often-used approach to managing risk in areas that are susceptible to frequent earthquakes. This is the thrust of programs such as the Federal Emergency Management Agency's (FEMA) Project Impact. The *Guidelines for the Seismic Retrofit of Existing Buildings* (GSREB) in Appendix A and the other referenced standards for seismic evaluation and rehabilitation have been developed for this purpose, and this section of the code recognizes their use. A voluntary seismic upgrade of a building structure can, therefore, consider the full range of rehabilitation objectives available under, for instance, ASME 41 (see commentary, Section 301.1.4).

SECTION 808
ELECTRICAL

808.1 New installations. All newly installed electrical equipment and wiring relating to work done in any work area shall comply with all applicable requirements of NFPA 70 except as provided for in Section 808.3.

❖ In the course of alterations that involve the reconfiguration of space, there is sometimes a need to rewire or replace electrical equipment or fixtures. This section requires that all newly installed electrical equipment and wiring comply with NFPA 70 in the work area. This is reasonable since the space is being reconfigured and the alteration is substantial enough to justify upgraded electrical in these areas. Section 808.3 provides some exceptions to this for residential occupancies.

808.2 Existing installations. Existing wiring in all work areas in Group A-1, A-2, A-5, H and I occupancies shall be upgraded to meet the materials and methods requirements of Chapter 7.

❖ Faulty and damaged wiring or wiring that might not be suitable for certain applications contributes to many fires that might have electrical origins. To reduce the risk of such fires, this section requires upgrading electrical wiring in high occupant load, hazardous and institutional occupancies. The electrical wiring in the entire work area, in Group A-1, A-2, A-5, H and I occupancies, must be examined and, where such wiring does not comply with the materials and methods provisions of NFPA 70, it must be upgraded to comply. It should be noted that electrical wiring in partitions and other elements that are not being altered or relocated must also comply as long as they are within the work area. The wiring materials and methods are mostly covered in Chapter 3 of NFPA 70.

808.3 Residential occupancies. In Group R-2, R-3 and R-4 occupancies and buildings regulated by the *International Residential Code*, the requirements of Sections 808.3.1 through 808.3.7 shall be applicable only to work areas located within a dwelling unit.

❖ This is the charging paragraph for the requirements for electrical work in residential occupancies in Level 2 or 3 alterations. The specific requirements are given in Sections 808.3.1 through 808.3.7.

The work areas within a dwelling unit in Group R-2, R-3 and R-4 occupancies are required to meet a certain minimum number of receptacle outlets, ground-fault circuit interrupters (GFCI) and lighting outlets. This is true regardless of whether the listed occupancies are being regulated by the IBC or IRC. Kitchens, laundry areas, sleeping rooms, study rooms or other similar rooms are required to provide at least one or two duplex receptacle outlets.

Closets, storage areas, hallways, garages and basements are exempt from this requirement, even if they are within the work area. Even though receptacle outlets are needed and used in these areas, their use is not as frequent or as necessary compared to areas that are required to provide such receptacles under this section.

Where new receptacle outlets are installed, and if NFPA 70 requires GFCI for the location under consideration, then such GFCI must also be installed.

Lighting outlets must be provided in: detached garages that are served by electric power, attached garages, bathrooms, hallways, stairways, utility rooms, basements used for storage or mechanical equipment, exits and outdoor entrances that are within the work area.

Example 808.3:

Q: A homeowner wishes to remodel and reconfigure some interior spaces including a bathroom, laundry and the living room. There is currently no light fixture outside the front door. Is a light fixture required to be installed outside of the front door?

A. Yes. Even though the exterior wall at the front door is not being reconfigured, it is located at the wall, which is part of the alteration work area and, as such, is required to be provided with a light fixture (see Commentary Figure 808.3).

Electrical service equipment that is located in the work area and in the dwelling unit must be adjusted or relocated, or its surrounding obstructions must be removed or relocated such that the minimum clearances required by NFPA 70 are provided. Providing this clearance in compliance with NFPA 70 addresses the safety of repair personnel.

808.3.1 Enclosed areas. All enclosed areas, other than closets, kitchens, basements, garages, hallways, laundry areas, utility areas, storage areas and bathrooms shall have a minimum of two duplex receptacle outlets or one duplex receptacle outlet and one ceiling or wall-type lighting outlet.

❖ See the commentary to Section 808.3.

808.3.2 Kitchens. Kitchen areas shall have a minimum of two duplex receptacle outlets.

❖ See the commentary to Section 808.3.

808.3.3 Laundry areas. Laundry areas shall have a minimum of one duplex receptacle outlet located near the laundry equipment and installed on an independent circuit.

❖ See the commentary to Section 808.3.

808.3.4 Ground fault circuit interruption. Newly installed receptacle outlets shall be provided with ground fault circuit interruption as required by NFPA 70.

❖ See the commentary to Section 808.3.

808.3.5 Minimum lighting outlets. At least one lighting outlet shall be provided in every bathroom, hallway, stairway, attached garage, and detached garage with electric power, and to illuminate outdoor entrances and exits.

❖ See the commentary to Section 808.3.

808.3.6 Utility rooms and basements. At least one lighting outlet shall be provided in utility rooms and basements where such spaces are used for storage or contain equipment requiring service.

❖ See the commentary to Section 808.3.

808.3.7 Clearance for equipment. Clearance for electrical service equipment shall be provided in accordance with the NFPA 70.

❖ See the commentary to Section 808.3.

SECTION 809
MECHANICAL

809.1 Reconfigured or converted spaces. All reconfigured spaces intended for occupancy and all spaces converted to habitable or occupiable space in any *work area* shall be provided with natural or mechanical ventilation in accordance with the *International Mechanical Code*.

Exception: Existing mechanical ventilation systems shall comply with the requirements of Section 809.2.

❖ This section is consistent with the basic premise for new building elements and materials given in Chapter 7, Section 702—that replacing elements is required to comply with the applicable code for new construction. The amount of mechanical ventilation must be provided in accordance with the IMC as required for new construction. This could be that air-moving equipment would need to be altered, as well. However, the exception allows special minimum provisions for existing mechanical equipment that would not require increasing the capacities for existing air-moving equipment, as long as minimum ventilation rates are provided.

809.2 Altered existing systems. In mechanically ventilated spaces, existing mechanical ventilation systems that are altered, reconfigured, or extended shall provide not less than 5 cubic feet per minute (cfm) (0.0024 m^3/s) per person of outdoor air and not less than 15 cfm (0.0071 m3/s) of ventilation air per person; or not less than the amount of ventilation air determined by the Indoor Air Quality Procedure of ASHRAE 62.

❖ This section essentially gives alternative minimum requirements to the IBC for ventilation in work areas

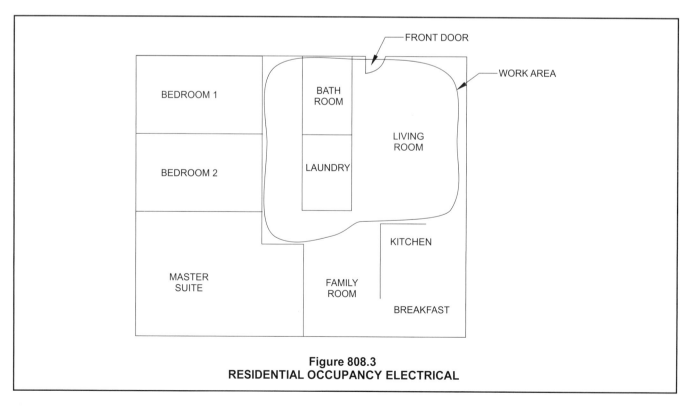

Figure 808.3
RESIDENTIAL OCCUPANCY ELECTRICAL

of Level 2 or 3 alterations. However, this does not really reduce the amount of engineering work that needs to be done in an altered space. The referenced ASHRAE 62 must be consulted for the overarching ventilation requirements, and then the minimums stated in this section must be provided. Therefore, even in a situation where the intent was not to change the air-moving equipment, but only the configuration of the ductwork or air channels, the minimum requirements could necessitate changes to air-moving equipment capabilities.

809.3 Local exhaust. All newly introduced devices, equipment, or operations that produce airborne particulate matter, odors, fumes, vapor, combustion products, gaseous contaminants, pathogenic and allergenic organisms, and microbial contaminants in such quantities as to affect adversely or impair health or cause discomfort to occupants shall be provided with local exhaust.

❖ This is a straightforward, common-sense requirement that will readily be observed, and probably already dealt with in the manufacturer's installation instructions for mechanical equipment. However, the code official should carefully review the particular details for all such equipment to ensure that appropriate measures are taken for local exhaust.

SECTION 810
PLUMBING

810.1 Minimum fixtures. Where the occupant load of the story is increased by more than 20 percent, plumbing fixtures for the story shall be provided in quantities specified in the *International Plumbing Code* based on the increased occupant load.

❖ Alterations involving the reconfiguration of spaces could, at times, result in an increased occupant load. The minimum number of plumbing fixtures is not affected and does not need to be reviewed or revised as long as the increased occupant load is 20 percent or less of the existing occupant load. This criterion is applied story by story and not to the building as a whole, because in most cases building occupants normally use the toilet facilities on the floor where they work or live and rarely travel to another floor to use the plumbing facilities. Once the 20-percent increased occupant load threshold is passed, then the entire occupant load of the story after alterations must be considered and the minimum plumbing fixtures, as required by the IPC, must be provided.

Example 810.1:

An existing mixed occupancy building as shown in Commentary Figure 810.1 is to undergo alterations to reconfigure certain areas and increase the occupant load as shown. There are four water closets currently provided in each restroom.

Q: What is the impact of the alterations on the required number of water closets?

A: Current occupant load is 650 (based on IBC Table 2902.1).

Offices: 100 occupant load, two water closets for male and two for female.

Restaurant: 300 occupant load, two water closets for male and two for female.

Retail: 250 occupant load, one water closet for male and one for female.

Total number of water closets in each restroom is five.

Case 1:

Office occupant load (100) + restaurant (375) + 292 = 767

Percent change of occupant load for the story = (767 - 650)/650 = 18 percent

The number of water closets does not need to be increased; there can remain four water closets in each male and female restroom.

Case 2:

Office occupant load (100) + restaurant (395) + 292 = 787

Percent change of occupant load for the story = (787 - 650)/650 = 21 percent

The number of water closets needs to be increased:

Offices: 100 occupant load, two water closets for male and two for female.

Restaurant: 395 occupant load, three water closets for male and three for female.

Retail: 292 occupant load, one water closet for male and one for female.

The number of water closets must be increased to six in each male and female restroom (see Commentary Figure 810.1).

SECTION 811
ENERGY CONSERVATION

811.1 Minimum requirements. Level 2 *alterations* to *existing buildings* or structures are permitted without requiring the entire building or structure to comply with the energy requirements of the *International Energy Conservation Code* or *International Residential Code*. The *alterations* shall conform to the energy requirements of the *International Energy Conservation Code* or *International Residential Code* as they relate to new construction only.

❖ A building that undergoes Level 2 alterations is required to meet a certain level of energy compliance. The level of compliance depends on the extent of the alterations taking place. As Level 2 alterations are those alterations where the reconfiguration of space takes place, new doors or windows are installed, or existing systems are extended, any such new ele-

ALTERATIONS—LEVEL 2

ment is required to meet the applicable energy provisions of the *International Energy Conservation Code®* (IECC®) or the IRC in the case of buildings that fall under the scoping of the IRC. Elements in the building that are not being affected do not need to be evaluated and do not need to comply with the energy provisions. Essentially, the entire building is not required to meet the energy provisions; only a degree of possible improvement in the energy performance of the building is intended to be achieved by making the new elements meet the IECC or IRC requirements. In certain cases where the reconfiguration of space might have resulted in the creation of new spaces, the newly created space should be evaluated as a whole for compliance with the energy provisions, even though some of the elements in the space might actually not have been altered. Likewise, in a case where an existing mechanical system is being extended to other areas or new ductwork is being installed to reconfigure and reroute the ducts to various spaces, it is only required to have the new elements meet the energy provisions and not the entire system. The language in this section is intended to be consistent with the requirements in the IECC and the IRC for alterations.

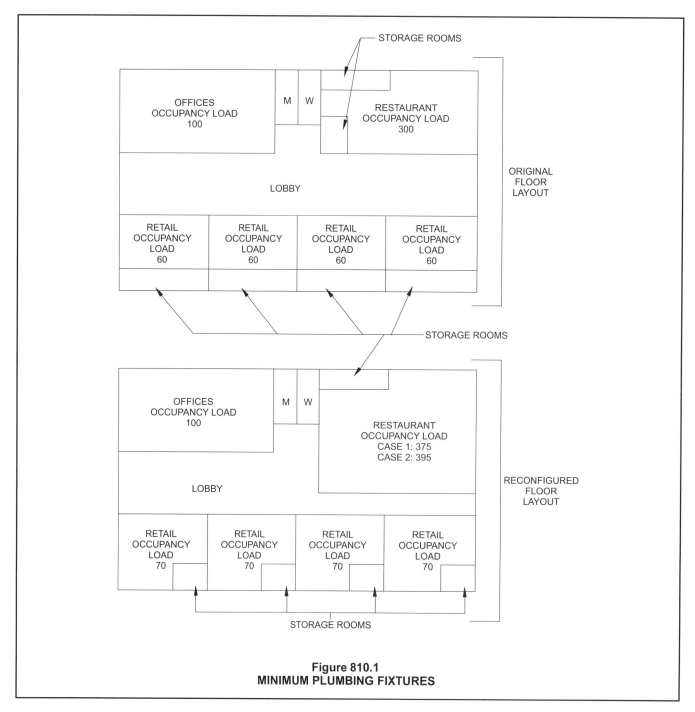

Figure 810.1
MINIMUM PLUMBING FIXTURES

Chapter 9:
Alterations—Level 3

General Comments

This chapter provides the technical requirements for existing buildings that undergo Level 3 alterations. Reference is made to Level 3 alterations as described in Section 505, which includes any alteration that involves more than 50 percent of the aggregate area of the building.

This chapter, similar to other chapters of the code, covers all building-related subjects, such as structural, means of egress, accessibility and energy conservation, as well as the fire and life safety issues when the alterations are classified as Level 3. Sections 901, 902 and 903 are related to scoping, special use and occupancy, and building elements and materials. In the interest of being brief and avoiding the presentation of materials that are repetitive in nature in various chapters, Section 901.2 requires that Level 3 alterations comply not only with Chapter 9, but also with Chapters 7 and 8. As such, any alteration work that is classified as Level 3 is assumed to include everything related to Level 1 and 2 alterations plus the provisions of this chapter. Section 903, Building Elements and Materials, covers, in detail, elements such as existing shafts and vertical openings, including when such openings might be required to be enclosed, fire partitions and interior finishes. The remainder of the chapter is related to fire protection, means of egress, accessibility, structural, electrical, mechanical and plumbing provisions. Section 905 requires additional means of egress lighting and exit signs beyond what is required. Keep in mind that Section 805 covers a wide range of means of egress subjects, such as the minimum number of exits, fire escape requirements and construction details, corridor openings and exit signs, which are also applicable here. The structural provisions for Level 3 alterations are contained in Section 907.

Known in past codes simply as "alterations," the classification was broken into three levels based on the facts that minor alterations, not including space reconfiguration and extensive alterations that might include relocation of walls and partitions in the entire building, should be treated differently and with different threshold levels for requiring upgrades or improvements to the building or spaces in the building.

Purpose

The purpose of this chapter is to provide detailed requirements and provisions to identify mandated improvements in existing building elements, building spaces and building structural system. This chapter is distinguished from Chapters 7 and 8 by involving alterations that cover 50 percent of the aggregate area of the building. Level 1 alterations do not involve space reconfiguration, while Level 2 alterations involve space reconfiguration that does not exceed 50 percent of the building area. Depending on the nature of alteration work, its location in the building and whether it encompasses one or more tenants, improvements and upgrades could be required for open-floor penetrations, sprinkler systems or the installation of additional means of egress, such as stairs or fire escapes. Level 3 alteration requirements no longer look at occupant load or whether exits are shared by more than one tenant when applying requirements such as sprinklers, as is done in Chapter 8 for Level 2 alterations. Additionally, this chapter is intended to improve the safety of certain building features beyond the work area and in other parts of the building where no alteration work might be taking place.

SECTION 901
GENERAL

901.1 Scope. Level 3 *alterations* as described in Section 505 shall comply with the requirements of this chapter.

❖ Any alteration work that results in reconfiguration of space or otherwise falls under the classification of Level 3 alterations, as described in Section 505, must comply with Chapter 9. The exception of Section 901.2 is intended to encourage improved accessibility in existing buildings by allowing any alterations that are solely for compliance with the accessibility provisions of Section 705 to only comply with Chapter 7 provisions as if no space reconfiguration has taken place. This is an advantage for improving the accessibility of buildings since Chapter 7 is related to alterations that do not encompass any space reconfiguration and, as such, has a lower level of compliance requirements and lower threshold for triggering other improvements.

901.2 Compliance. In addition to the provisions of this chapter, work shall comply with all of the requirements of Chapters 7 and 8. The requirements of Sections 803, 804 and 805 shall apply within all *work areas* whether or not they include exits and corridors shared by more than one tenant and regardless of the occupant load.

Exception: Buildings in which the reconfiguration of space affecting exits or shared egress access is exclusively the result of compliance with the accessibility requirements of Section 705.2 shall not be required to comply with this chapter.

❖ Chapters 8 and 9 have cascading effects in that each level of alteration must comply with the requirements for lower levels of alteration. Accordingly, an alteration project that is classified as Level 2 must comply with Chapter 7, as well as with Chapter 8. Similarly, an alteration project classified as Level 3 must comply with Chapters 7, 8 and 9. This is to eliminate the repetition of various requirements from Chapter 7 in Chapters 8 and 9, and from Chapter 8 in Chapter 9. The code drafting committee decided to make reference to these chapters rather than repeat all such provisions again.

Keep in mind that Section 801.3 is applicable in this chapter, as well. Section 801.3 requires that new elements, components and spaces created as a result of Level 2 alteration work must comply with the *International Building Code*® (IBC®) or other applicable codes, just as Section 101.4 of the IBC makes reference to other *International Codes*® (I-Codes®) such as the *International Plumbing Code*® (IPC®), *International Mechanical Code*® (IMC®) and *International Fire Code*® (IFC®). Please review the requirements and commentary in Section 801.3, including the exceptions, which also apply.

The exception allows for when a Level 1 alteration to a primary function area prompts improvement to the route, bathrooms and drinking fountains (in accordance with Section 705.2), and, thus, a reconfiguration of the space. Level 3 triggers other requirements, such as sprinklers, when a space is reconfigured. This exception allows the project to remain a Level 1 alteration.

SECTION 902
SPECIAL USE AND OCCUPANCY

902.1 High-rise buildings. Any building having occupied floors more than 75 feet (22 860 mm) above the lowest level of fire department vehicle access shall comply with the requirements of Sections 902.1.1 and 902.1.2.

❖ This is the charging paragraph for high-rise requirements. The specific requirements contained herein pertain to the recirculating or exhausting of air (see Section 902.1.1) or elevators (see Section 902.1.2). A high-rise building is defined as a building having an occupied floor located more than 75 feet (22 860 mm) above the lowest level of fire department vehicle access, which is the same as the IBC criteria for identifying a high-rise building. In high-rise buildings undergoing a Level 3 alteration, the subjects of recirculating air and elevators must specifically be addressed. All other provisions of Chapter 9 still apply to high-rise buildings, but these two are specific only to high rises. The notable difference between this section and IBC Section 403, is that IBC Section 403.1 has five exceptions. The five exceptions are buildings that do not fall under the high-rise provisions of IBC Section 403. The code does not contain any of these five exceptions. So whether the building under consideration is an airport traffic control tower (see IBC Section 403.1, Exception 1), an open parking garage (see IBC Section 403.1, Exception 2), a Group A-5 occupancy (see IBC Section 403.1, Exception 3), a low-hazard special industrial occupancy (see IBC Section 403.1, Exception 4) or a Group H-1, H-2 or H-3 occupancy (see IBC Section 403.1, Exception 5), the recirculating air and elevator recall requirements of Sections 902.1.1 and 902.1.2 are applicable. Even though this appears to be more restrictive than the IBC, it really is not, because in most buildings the five exceptions in IBC Section 403.1 have a special section in the IBC that they must comply with (see IBC Sections 406.5, 412 and 415). Also note that IFC Section 1103.3.2 requires the implementation of Phase 1 recall and Phase 2 emergency in-car operation for all existing elevators that travel a distance more than 25 feet, regardless of whether an alteration is being undertaken in the building.

902.1.1 Recirculating air or exhaust systems. When a floor is served by a recirculating air or exhaust system with a capacity greater than 15,000 cubic feet per minute (701 m3/s), that system shall be equipped with approved smoke and heat detection devices installed in accordance with the *International Mechanical Code.*

❖ This section applies to a high-rise building undergoing a Level 3 alteration, as described in Section 505, which has a floor served by a recirculating air or exhaust system with a capacity of greater than 15,000 cubic feet per minute (701 m^3/s). Such a system is required to be equipped with approved smoke and heat detection devices. This is clearly a condition where large mechanical devices have the potential of spreading products of combustion quickly through a particular floor and beyond. Over the years, it has been recognized that many injuries and deaths in high-rise fires have been the result of smoke inhalation due to smoke penetrating the floors above the floor of fire origin. The smoke and heat detection devices required to be installed in accordance with the IMC are intended to shut down the mechanical circulating air or exhaust system and to reduce the potential spread of smoke.

902.1.2 Elevators. Where there is an elevator or elevators for public use, at least one elevator serving the *work area* shall comply with this section. Existing elevators with a travel distance of 25 feet (7620 mm) or more above or below the main floor or other level of a building and intended to serve the needs of emergency personnel for fire-fighting or rescue purposes shall be provided with emergency operation in accordance with ASME A17.3. New elevators shall be provided with Phase I emergency recall operation and Phase II emergency in-car operation in accordance with ASME A17.1.

❖ The elevator recall provisions found in this section are taken from the IFC and are intended to accommodate the needs of emergency personnel for firefighting or rescue purposes. Elevators are often used by emergency responders to access a building's various floor levels when responding to fires and other emergencies; because of these needs, the elevators must be capable of providing certain functions, such as recall and emergency operation.

Existing elevators that travel 25 feet (7620 mm) or more above or below the main level should, at minimum, be equipped with emergency operation capabilities that comply with ASME A17.3. New elevator installations are held to more restrictive requirements and must have both emergency recall and emergency in-car operation to comply with ASME A17.1 for any amount of travel distance. The ASME A17 standards are safety codes for elevators and escalators: ASME A17.1 is for new elevator installations and A17.3 is for existing elevators.

902.2 Boiler and furnace equipment rooms. Boiler and furnace equipment rooms adjacent to or within Groups I-1, I-2, I-4, R-1, R-2 and R-4 occupancies shall be enclosed by 1-hour fire-resistance-rated construction.

Exceptions:

1. Steam boiler equipment operating at pressures of 15 pounds per square inch gauge (psig) (103.4 KPa) or less is not required to be enclosed.

2. Hot water boilers operating at pressures of 170 psig (1171 KPa) or less are not required to be enclosed.

3. Furnace and boiler equipment with 400,000 British thermal units (Btu) (4.22×10^8 J) per hour input rating or less is not required to be enclosed.

4. Furnace rooms protected with an automatic sprinkler system are not required to be enclosed.

❖ The boiler and furnace room separation requirements are based on the separation concept of IBC Section 509, but the trigger mechanism is different from that of the IBC. The IBC requires any boiler or furnace room with a single piece of equipment larger than 400,000 Btu per hour (117 200 watts) input to be separated from the main areas by a fire barrier, whereas this code additionally requires any such room with high-pressure-type equipment be separated from certain uses. High pressure is considered 15 pounds-per-square-inch gauge (psig) (103.4 kPa) for steam equipment and 170 psig (1171 kPa) for hot water equipment. The uses from which the separation is required are those that are residential in nature or occupancies containing occupants who are not capable of self-preservation such as hospitals and day care facilities. Furnace rooms protected with an automatic sprinkler system are not required to be separated.

SECTION 903
BUILDING ELEMENTS AND MATERIALS

903.1 Existing shafts and vertical openings. Existing stairways that are part of the means of egress shall be enclosed in accordance with Section 803.2.1 from the highest *work area* floor to, and including, the level of exit discharge and all floors below.

❖ This section addresses the enclosure of existing stairways only. Other shafts and floor openings are regulated exactly the same as under Level 2 alterations. The code user is referred to Section 803.2.1 for stairway enclosure methods and triggers. Such enclosures, though, must be provided from the highest work area floor all the way to the level of exit discharge and all the floors below the level of exit discharge. Note that these enclosure requirements would also include the level of exit discharge. This requirement is only applicable to stairways that are part of the means of egress and not other stairways used for convenience, attending equipment or those that have other specific purposes.

903.2 Fire partitions in Group R-3. Fire separation in Group R-3 occupancies shall be in accordance with Section 903.2.1.

❖ This is the charging paragraph for fire separations between dwelling units in Group R-3 occupancies, sending the code user to Section 903.2.1. It should be noted that the title of this section is somewhat misleading, referring to "fire partitions," which are defined and detailed specifically in the IBC. However, the regulatory text describes the required separation between the dwelling units, never actually requiring construction of fire partitions as described in the IBC.

903.2.1 Separation required. Where the *work area* is in any attached dwelling unit in Group R-3 or any multiple single-family dwelling (townhouse), walls separating the dwelling units that are not continuous from the foundation to the underside of the roof sheathing shall be constructed to provide a continuous fire separation using construction materials consistent with the existing wall or complying with the requirements for new structures. All work shall be performed on the side of the dwelling unit wall that is part of the *work area*.

Exception: Where *alterations* or *repairs* do not result in the removal of wall or ceiling finishes exposing the structure, walls are not required to be continuous through concealed floor spaces.

❖ The separation of dwelling units in duplex and townhouse buildings is found in Chapter 3 of the *International Residential Code®* (IRC®) for new construction. The code in this section addresses the vertical separation aspect only. If the alteration includes the removal of wall or ceiling finishes that expose the wall structure, then the separation wall must be evaluated for continuity from the foundation to roof deck. The existing wall construction is allowed to remain and be continued to the roof deck even if it does not provide the IRC's required 1-hour fire-resistance rating.

It is intended that this requirement not create a burden for the residents of adjacent dwellings if they are not involved in the alteration project; as such, it is allowed that the extension of such walls to the roof deck be placed on the side where an alteration project is taking place and not intrude in the neighboring units.

903.3 Interior finish. Interior finish in exits serving the *work area* shall comply with Section 803.4 between the highest floor on which there is a *work area* to the floor of exit discharge.

❖ The interior finish materials in Level 3 alterations are regulated the same as Level 2 alterations. The reader is referred to Section 803.4, where the code requires interior finishes in exits and corridors serving the work area to comply with the interior finish requirements of the IBC. The only additional requirement in Level 3 alterations is the extent of coverage. Exits serving the work area must comply with the interior finish requirements of Section 803.4, from the highest floor where there is any work area to the floor of exit discharge. Any corridors in the work area must also comply with Section 803.4, as Section 901.2 requires all Level 3 alterations to comply with Chapters 7, 8 and 9. Note that Chapter 8 of the IFC has interior finish requirements for existing buildings.

SECTION 904
FIRE PROTECTION

904.1 Automatic sprinkler systems. An automatic sprinkler system shall be provided in a work area where required by Section 804.2 or this section.

❖ Automatic sprinkler requirements are referenced by Section 804.2 where the requirements for various occupancies, mixed uses and windowless stories are provided. This section has specific requirements for high-rise buildings, and rubbish and linen chutes. When Section 804.2 requires an automatic sprinkler system for a Level 2 alteration, the system would also be required for a Level 3 alteration throughout the work area.

904.1.1 High-rise buildings. An automatic sprinkler system shall be provided in work areas where the high-rise building has a sufficient municipal water supply for the design and installation of an automatic sprinkler system at the site.

❖ Where there is sufficient municipal water to supply an automatic sprinkler system, then all of the work areas in the high-rise building must be sprinklered. "At the site" refers to the availability of water at the property line and not at the building itself. In certain high-rise, Level 3 alteration projects, this could have the effect of requiring the design and installation of a new fire line main and associated vaults, valves and other related components from the municipal water main to the building. Sprinklers could be required beyond the work areas on any floor where the work area exceeds 50 percent of that floor area. This extension of sprinklers is not required in tenant spaces that are entirely outside of the work area (see Commentary Figure 904.1.1).

Example:

Sprinklers required in:

Level 1: Tenants 1 and 3 (more than 50 percent of floor—tenant 2 is entirely outside the work area).

Level 2: Parts of tenants 1, 2 and 3 that are within the work area (work areas only—less than 50 percent of floor area).

Level 3: Tenant 2 (work area not more than 50 percent—tenant 1 is entirely outside of work area).

Level 4: Tenants 1, 2, 3 and 4 (work area is more than 50 percent of floor area—tenants 2 and 4 are not entirely outside the work area).

Level 5: Tenants 1, 2, 3 and 4 (work area is more than 50 percent of the area—tenants 2 and 4 are not entirely outside the work area).

904.1.2 Rubbish and linen chutes. Rubbish and linen chutes located in the *work area* shall be provided with automatic sprinkler system protection or an approved automatic fire-extinguishing system where protection of the rubbish and linen chute would be required under the provisions of the *International Building Code* for new construction.

❖ In any building undergoing a Level 3 alteration, the rubbish and linen chutes in the work areas must be provided with sprinklers in accordance with the IBC. The IBC requires an automatic sprinkler system at the top and in the terminal rooms of the rubbish and linen chutes; and, additional sprinkler heads are required at alternate floors and at the lowest intake. The IBC also requires protection at alternating floors where the chute extends more than one floor below the lowest intake. Accordingly, in a four-story building where there is a linen chute from the fourth floor to the first and there are alterations on all floors, the code sprinkler requirement for the chute will be exactly the same as the IBC. Chute sprinklers must be available for servicing (see Commentary Figure 904.1.2).

904.1.3 Upholstered furniture or mattresses. *Work areas* shall be provided with an automatic sprinkler system in accordance with the *International Building Code* where any of the following conditions exist:

1. A Group F-1 occupancy used for the manufacture of upholstered furniture or mattresses exceeds 2,500 square feet (232 m^2).

2. A Group M occupancy used for the display and sale of upholstered furniture or mattresses exceeds 5,000 square feet (464 m^2).

3. A Group S-1 occupancy used for the storage of upholstered furniture or mattresses exceeds 2,500 square feet (232 m^2).

❖ This section requires sprinklers for the protection of upholstered furniture and mattresses. The requirements mirror those required in the IFC and IBC in Sections 903.2.4 (Item 4), 903.2.7 (Item 4) and 903.2.9 (Item 5). The application of this section will only require an automatic sprinkler system in the work area versus throughout the building as required by the IFC and IBC. Note that this requirement is based on the size of the occupancy versus the size of the fire area. For example, consider a 9,000-square-foot Group M Occupancy, with a 4,500-square-foot work area. Although the work area is less than 5,000 square feet, the Group M occupancy is 9,000 square feet. Therefore, the 4,500-square-foot work area would need to be provided with an automatic sprinkler system.

904.2 Fire alarm and detection systems. Fire alarm and detection shall be provided in accordance with Section 907 of the *International Building Code* as required for new construction.

❖ This section essentially requires fire alarm systems as if the building were new. The requirements do not state "throughout the building," as it is not always the case that the fire alarm requirements would apply throughout the building. The reference to IBC Section 907 allows this to be dealt with as applicable. Most of the fire alarm requirements are based on occupancy rather than fire area, which is typically the basis for automatic sprinkler systems. Sections 904.2.1 and 904.2.2 are exceptions to this more general compliance requirement with the IBC, which focus on installation of such systems only in the work area.

904.2.1 Manual fire alarm systems. Where required by the *International Building Code*, a manual fire alarm system shall be provided throughout the *work area*. Alarm notification

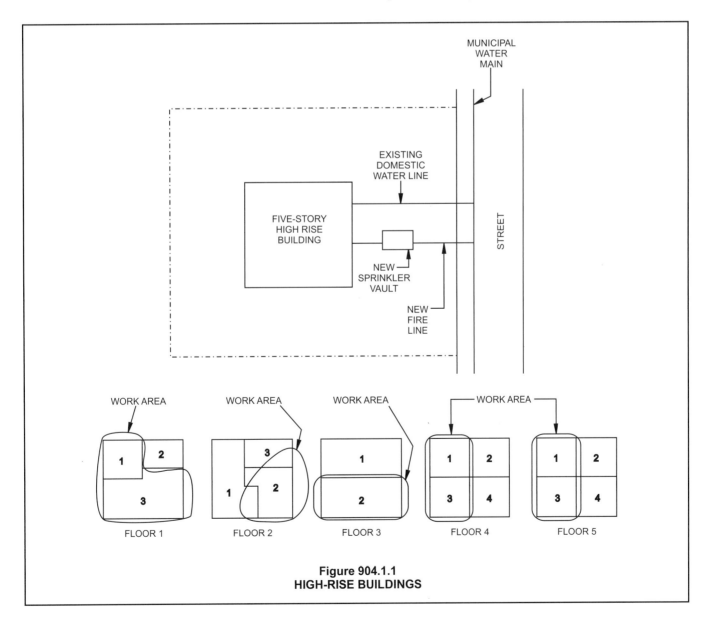

Figure 904.1.1
HIGH-RISE BUILDINGS

ALTERATIONS—LEVEL 3

appliances shall be provided on such floors and shall be automatically activated as required by the *International Building Code*.

Exceptions:

1. Alarm-initiating and notification appliances shall not be required to be installed in tenant spaces outside of the *work area*.
2. Visual alarm notification appliances are not required, except where an existing alarm system is upgraded or replaced or where a new fire alarm system is installed.

❖ Manual fire alarm systems must be provided throughout the work areas in any occupancy where the IBC requires a manual fire alarm system. Note that this section only requires such systems in the work area. IBC Section 907.2, for example, requires a manual fire alarm system in Group A, B, E and M occupancies that exceed certain thresholds. This is required in the work areas on any floor regardless of the size of the work area on a particular floor. There are two exceptions for this requirement. Exception 1 addresses the typical exception that has been seen in many sections of the code, which is an exemption for tenant spaces that are completely outside of the work area. Exception 2 is an exemption for visual alarm notification when not upgrading an existing system or installing a new system.

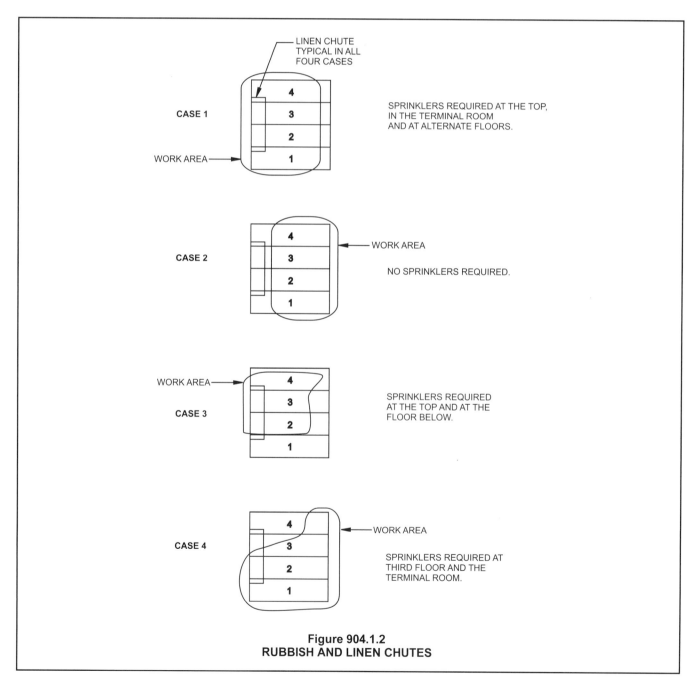

Figure 904.1.2
RUBBISH AND LINEN CHUTES

904.2.2 Automatic fire detection. Where required by the *International Building Code* for new buildings, automatic fire detection systems shall be provided throughout the *work area*.

❖ IBC Section 907.2 and its subsections require automatic fire detection systems for certain occupancies and in certain locations in a building, such as smoke detector requirements in mechanical equipment, electrical, transformer, telephone equipment and similar rooms. In some instances, sprinkler protection or other protective measures can substitute for such automatic fire detection systems. The provisions of this section allow existing buildings that are not equipped with an automatic sprinkler system, automatic fire detection system or other specific safety equipment to undergo Level 3 alterations as long as some improvement in the life safety systems of the building is achieved. For example, an existing Group I-2 nursing home undergoing a Level 3 alteration will be required to be equipped with an automatic fire detection system in the corridors and other spaces that are permitted to be open to such corridors, in accordance with IBC Section 907.2.6.2.

SECTION 905
MEANS OF EGRESS

905.1 General. The means of egress shall comply with the requirements of Section 805 except as specifically required in Sections 905.2 and 905.3.

❖ The means-of-egress requirements for Level 3 alterations are the same as the Level 2 alterations found in Section 805, with the exception that the means-of-egress lighting and exit signs must meet an additional criterion. Also, in accordance with Section 901.2, the means of egress requirements of Section 805 will apply regardless of whether the work area includes more than one tenant or a certain occupant load.

905.2 Means-of-egress lighting. Means of egress from the highest *work area* floor to the floor of exit discharge shall be provided with artificial lighting within the exit enclosure in accordance with the requirements of the *International Building Code*.

❖ The additional criteria for means-of-egress lighting requires that, in interior exit stairways, illumination be provided from the highest floor where there is any work area down to the level of exit discharge. The intensity of illumination, emergency power and performance of the system are detailed in the IBC requirements.

905.3 Exit signs. Means of egress from the highest *work area* floor to the floor of exit discharge shall be provided with exit signs in accordance with the requirements of the *International Building Code*.

❖ The additional criteria for exit signs require that the means of egress from the highest floor where there is any work area to the level of exit discharge be provided with exit signs. The location, height, illumination and other requirements of such exit signs must comply with the IBC.

SECTION 906
ACCESSIBILITY

906.1 General. A building, *facility* or element that is altered shall comply with this section and Sections 705 and 806.

❖ Alterations covered under this section must meet all accessibility requirements in Section 705, as well as two additional requirements covered in Section 806.

906.2 Type B dwelling or sleeping units. Where four or more Group I-1, I-2, R-1, R-2, R-3 or R-4 dwelling or sleeping units are being altered, the requirements of Section 1107 of the *International Building Code* for Type B units and Chapter 9 of the *International Building Code* for visible alarms apply only to the quantity of the spaces being altered.

Exception: Group I-1, I-2, R-2, R-3 and R-4 dwelling or sleeping units where the first certificate of occupancy was issued before March 15, 1991 are not required to provide Type B dwelling or sleeping units.

❖ This section sets forth the rate for providing Type B dwelling or sleeping units in Group I-1, I-2, R-2, R-3 or R-4 facilities that are undergoing a Level 3 alteration. Type B units are required when four or more units are added in the existing shell or extensively altered. The Type B units in the facility are not required to exceed that required for new construction, as indicated in Section 705.1.13. Section 705.2 also includes exceptions for the additional accessible route requirements in structures where Type B units are added as part of an alteration or change of occupancy. It is unreasonable to require a greater level of accessibility in an existing building than is required in new construction. Note that Section 1105.4 requires compliance with IBC Section 1107 when four or more dwelling or sleeping units are being added as part of an addition, but would not include the alteration of existing dwelling or sleeping units or a change of occupancy where dwelling and sleeping units were created in an existing building.

SECTION 907
STRUCTURAL

[BS] 907.1 General. Where buildings are undergoing Level 3 *alterations* including structural *alterations*, the provisions of this section shall apply.

❖ These structural requirements apply to alterations where the work area exceeds 50 percent of the aggregate area of the building. Note that the structural requirements for Level 1 and 2 alterations in Sections 706 and 807, respectively, are also applicable to Level 3 alterations according to Section 505.2.

[BS] 907.2 New structural elements. New structural elements shall comply with Section 807.2.

❖ Any new structural element that is added in the course of alteration work, including its connections to the existing structure, is required to comply with Section 807.2. Section 807.2 requires new structural elements to comply with the IBC. There is no exception to this requirement.

[BS] 907.3 Existing structural elements carrying gravity loads. Existing structural elements carrying gravity loads shall comply with Section 807.4.

❖ Existing gravity load-carrying elements are required to comply with Section 807.4. Section 807.4 requires the evaluation of these additional loads, including the effects of snow drift, to determine if these loads decrease the capacity of existing structural elements. See the commentary to Section 807.4 for further discussion of the specific requirements.

[BS] 907.4 Existing structural elements resisting lateral loads. All existing elements of the lateral force-resisting system shall comply with this section.

Exceptions:

1. Buildings of Group R occupancy with no more than five dwelling or sleeping units used solely for residential purposes that are altered based on the conventional light-frame construction methods of the *International Building Code* or in compliance with the provisions of the *International Residential Code*.

2. Where such *alterations* involve only the lowest story of a building and the *change of occupancy* provisions of Chapter 10 do not apply, only the lateral force-resisting components in and below that story need comply with this section.

❖ This section applies to all Level 3 alterations. This section also applies to Level 2 alterations where the demand-capacity ratio of an existing lateral load-resisting element, with a proposed alteration considered, exceeds 10 percent of its demand-capacity ratio with the alteration ignored. Where this occurs, it is necessary to have an engineering evaluation of the building. The loading criteria to be used in the evaluation is dependent on the extent of the alteration work that is undertaken. The code assumes that the structural members were properly designed for the applicable loads at the time of construction.

The seismic base shear depends on several variables, such as the mapped spectral accelerations at the site, the site soil coefficients, the importance factor based on the nature of the occupancy, as well as the structure's effective seismic weight and response modification factor. For work that is considered an alteration, it is primarily the structure's effective seismic weight that is of interest, as the other variables would likely remain fixed. The effective seismic weight (*W*) used to determine the base shear consists of the total weight of the structure, all equipment and other permanently attached items, as well as a percentage of the design live load for storage uses. It reflects the weight of all dead loads and contents that might reasonably be expected to be attached to the structure at the time the design earthquake occurs.

While equipment weight and some gravity loads are calculable to a 5-percent margin, seismic loads and response characteristics of existing materials are not realistically calculable with that precision and, as such, the trigger for possible seismic upgrades of Sections 807.6 and 907.4 is set at a rational 10-percent increase in seismic base shear or decrease in seismic base shear capacity. As it is possible that various alterations over the years might increase the seismic base shear or decrease the seismic base shear capacity in a cumulative manner beyond the 10-percent trigger level, the code makes it explicitly clear that the 10-percent threshold is calculated relative to the original building construction and not the latest alteration. Exceptions 1 and 2 are related to the evaluation requirement, one of which permits smaller Group R buildings that are altered to comply with either the IRC or the conventional light-frame construction provisions of the IBC. Exception 2 pertains to multiple-story buildings, where alterations are made to only the lowest story. This permits compliance with the IBC's wind and seismic criteria to be limited to the lateral force-resisting elements in that story and below. While the impact of wind and seismic loading contributed by the stories above must be considered, no evaluation of those stories is required. The code assumes the upper stories have been designed for wind and earthquakes according to the code in effect at the time of construction.

[BS] 907.4.1 Evaluation and analysis. An engineering evaluation and analysis that establishes the structural adequacy of the altered structure shall be prepared by a registered design professional and submitted to the *code official*.

❖ A building undergoing alterations that result in a seismic base shear increase greater than 10 percent and exceed the area threshold established in this section requires an engineering evaluation of the structure as described in this section. The evaluation must demonstrate that the altered structure complies with the IBC wind-loading criteria and with reduced IBC seismic forces in Section 301.1.4.2. The latter permits the use of the *Guidelines for Seismic Retrofit of Existing Buildings* (GSREB) in Appendix A of the code and ASCE 41 (see Commentary Section 301.1.4).

[BS] 907.4.2 Substantial structural alteration. Where more than 30 percent of the total floor and roof areas of the building or structure have been or are proposed to be involved in structural *alteration* within a 5-year period, the evaluation and analysis shall demonstrate that the lateral load-resisting system of the altered building or structure complies with the *International Building Code* for wind loading and with reduced *International Building Code*-level seismic forces in accordance with **Section** 301.1.4.2. The areas to be counted toward the 30 percent shall be those areas tributary to the vertical load-carrying components, such as joists, beams, columns, walls and other structural components that have been or will be removed, added or altered, as well as areas such as mezzanines, penthouses, roof structures and in-filled courts and shafts.

❖ The area threshold of 30 percent of the floor and roof areas would likely include all Level 3 alterations, since Section 505.1 establishes that these are in excess of 50 percent of the aggregate area of the building. However, the 30-percent threshold is focused on structural alterations versus alterations in general. The additional stipulation that all structural alterations occurring over a 5-year period must be considered prevents a building owner from breaking a large alteration project into a series of smaller alterations in hopes of complying with less-stringent requirements. The clarification of exactly which areas contribute to meeting the area threshold of 30 percent is important. For instance, mezzanines complying with IBC Section 505 do not contribute to the building area but will contribute to the 30-percent criteria.

[BS] 907.4.3 Seismic Design Category F. Where the building is assigned to Seismic Design Category F, the evaluation and analysis shall demonstrate that the lateral load-resisting system of the altered building or structure complies with reduced *International Building Code*-level seismic forces in accordance with **Section** 301.1.4.2 and with the wind provisions applicable to a limited structural alteration.

❖ This section is focused specifically on those buildings assigned to Seismic Design Category F. Seismic Design Category F is the most severe. Buildings in such areas are at high risk for damage from a seismic event. Therefore, when a building is undergoing over 50-percent alteration (Level 3), it was felt reasonable to also evaluate the lateral force-resisting system regardless of how extensive the structural alterations. This section requires such buildings to be evaluated and analyzed to determine that the building's lateral load-resisting system meets the reduced IBC-level seismic forces. Since the hazard is seismic, the building only needs to meet the limited structural alteration requirements for wind. Note that this section is consistent with Section 403.4.1 in the prescriptive method.

[BS] 907.4.4 Limited structural alteration. Where the work does not involve a substantial structural *alteration* and the building is not assigned to Seismic Design Category F, the existing elements of the lateral load-resisting system shall comply with Section 807.5.

❖ A building undergoing alterations not exceeding the threshold established in Section 907.4.2 is required to comply with Section 807.5. See the commentary to Section 807.5 for additional information. This does not apply to buildings located in Seismic Design Category F. Those buildings must meet the reduced seismic loads and IBC wind loads regardless of the level of structural alteration. See the commentary to Section 907.4.3.

[BS] 907.4.5 Wall anchors for concrete and masonry buildings. For any building assigned to Seismic Design Category D, E or F with a structural system consisting of concrete or reinforced masonry walls with a flexible roof diaphragm and any building assigned to Seismic Design Category C, D, E or F with a structural system consisting of unreinforced masonry walls with any type of roof diaphragm, the alteration work shall include installation of wall anchors at the roof line to resist the reduced *International Building Code*-level seismic forces in accordance with **Section** 301.1.4.2, unless an evaluation demonstrates compliance of existing wall anchorage.

❖ This section deals with both reinforced concrete or masonry walls and unreinforced masonry walls. The two conditions where wall anchors may be necessary for wall anchorage systems if the existing system does not meet the reduced seismic level forces are:

- Buildings in Seismic Design Category D, E or F having a structural system consisting of reinforced concrete or masonry walls and a flexible roof diaphragm.
- Buildings in Seismic Design Category C, D, E or F having a structural system consisting of unreinforced masonry.

The wall anchorage system includes the elements in the diaphragm required to develop wall anchorage forces, including: wall anchors, struts, subdiaphragms, cross ties and continuity ties. Note that if a structural evaluation indicates that the existing wall anchorage provided can already resist and transfer these forces, additional wall anchorage is not required. This section is consistent with Section 403.6 of the prescriptive method.

[BS] 907.4.6 Bracing for unreinforced masonry parapets. Parapets constructed of unreinforced masonry in buildings assigned to Seismic Design Category C, D, E or F shall have bracing installed as needed to resist the reduced *International Building Code*-level seismic forces in accordance with **Section** 301.1.4.2, unless an evaluation demonstrates compliance of such items.

❖ The provisions found in this section are essentially the same as those found in Section 707.3.1. The reason for the duplication is related to the application of the provisions. Section 707.3.1 only applies if reroofing is occurring and is only applicable to Seismic Design Categories D, E and F. This section also applies to Seismic Design Category C. These provisions apply any time a Level 3 alteration occurs. It is felt that this bracing is a critical life safety measure that should be taken during larger alterations. This section is consistent with Section 403.7 of the prescriptive method.

SECTION 908
ENERGY CONSERVATION

908.1 Minimum requirements. Level 3 *alterations* to *existing buildings* or structures are permitted without requiring the entire building or structure to comply with the energy requirements of the *International Energy Conservation Code* or *International Residential Code*. The *alterations* shall conform to the energy requirements of the *International Energy Conservation Code* or *International Residential Code* as they relate to new construction only.

❖ A building that undergoes Level 3 alterations is required to meet a certain level of energy compliance. The level of compliance depends on the extent of the alterations taking place. As Level 3 alterations are those alterations where reconfiguration of space takes place in more than 50 percent of the building aggregate area, then any new element within the work area, as defined in Chapter 2, is required to meet the applicable energy provisions of the *International Energy Conservation Code*® (IECC®), or the IRC in the case of buildings that fall under its scoping. Elements in the building that are not being affected need not be evaluated and need not comply with the energy provisions. The entire building is not required to meet the energy provisions, but improvement in the energy performance of the building is intended to be achieved by making the new elements meet the IECC or IRC requirements. In those cases where the reconfiguration of space might have resulted in the creation of new spaces, the newly created spaces should be evaluated as a whole for compliance with the energy provisions, even though some of the elements in the space might not have been altered. Likewise, in a case where an existing mechanical system is being extended to other areas or new duct work is being installed to reconfigure and reroute the ducts to various spaces, it is only required to have the new elements meet the energy provisions and not the entire system. Newly installed systems must meet the energy provisions of the IECC or the IRC for buildings under the scoping of the IRC. The language in this section is intended to be consistent with the requirements in the IECC and IRC for alterations.

Chapter 10:
Change of Occupancy

General Comments

This chapter deals with the special situations involved in an existing building when a change of occupancy occurs. A change of occupancy is not to be confused with a change of occupancy classification. The *International Building Code®* (IBC®) defines different occupancy classifications in Chapter 3, and special occupancy requirements in Chapter 4. Within specific occupancy classifications there can be many different types of actual activities that can take place. For instance, a Group A-3 occupancy classification deals with a wide variety of activities, including bowling alleys and courtrooms, indoor tennis courts and dance halls. When a facility changes use from a bowling alley to a dance hall, the occupancy classification remains A-3, but the different uses could lead to drastically different code requirements. Therefore, this chapter deals with the special circumstances that are associated with a change in the use of a building within the same occupancy classification, as well as a change of occupancy classification.

As a general rule, when a change of occupancy classification occurs, the requirements of Chapter 9 for Level 3 alterations apply, along with all of the provisions of Section 1012. However, there are exceptions to this rule as well, as provided in Section 1012.4, which deal with a change of occupancy where the change is clearly to a lesser hazard.

Purpose

The purpose of this chapter is to provide regulations for the circumstances when an existing building is subject to a change of occupancy or a change of occupancy classification. As previously discussed, a change of occupancy is defined in Chapter 2 of the code:

CHANGE OF OCCUPANCY. A change in the use of the building or a portion of a building. A change of occupancy shall include any change of occupancy classification, any change from one group to another group within an occupancy classification or any change in use within a group for a specific occupancy classification.

In either case, some scenarios exist that are not dealt with in Chapters 7, 8 and 9 but must be addressed nevertheless. Changes in occupancy or occupancy classification change the activities or activity levels within a building, which can have implications on the life safety hazards associated with this change in activity. A change from a dance hall to a nightclub is a change in occupancy classification that has clear implications with regard to risks to life. A change from a restaurant to a nightclub is not a change in occupancy classification, but is, nevertheless, a significant change in use and activity levels with different related life safety risks.

SECTION 1001
GENERAL

1001.1 Scope. The provisions of this chapter shall apply where a *change of occupancy* occurs, as defined in Section 202.

❖ By definition, a "Change of occupancy" is the result of a change of use of the building that involves a change in application of the requirements of the code. Chapter 10 would be applicable in the case of a structure where the use has changed but the occupancy classification has not. The scope of this chapter also includes a change of occupancy classification, in addition to a change in the occupancy group designation.

1001.2 Certificate of occupancy. A change of occupancy or a change of occupancy within a space where there is a different fire protection system threshold requirement in Chapter 9 of the *International Building Code* shall not be made to any structure without the approval of the *code official*. A certificate of occupancy shall be issued where it has been determined that the requirements for the change of occupancy have been met.

❖ A change of occupancy without a change in occupancy classification is a change in the level and/or type of use in a building that could have life safety implications. For instance, imagine that a community center is sold and converted to a dance hall. Both of these are considered by the IBC as being an Occupancy Classification A-3, but the requirements of the code to each facility could be considerably different. The dance hall might have a higher occupant load and, therefore, require different egress facilities. This section specifically states that if the change in occupancy would result in a different fire protection threshold in IBC Chapter 9, then it must be addressed. For instance, if the occupant load in the above example increases from 250 to 300, then an automatic sprinkler system would be required in accordance with IBC Section 903.2.1.3.

1001.2.1 Change of use. Any work undertaken in connection with a change in use that does not involve a *change of occupancy* classification or a change to another group within an occupancy classification shall conform to the applicable requirements for the work as classified in Chapter 5 and to the requirements of Sections 1002 through 1011.

Exception: As modified in Section 1205 for *historic buildings*.

❖ A change of occupancy that is only a change in use can be dealt with as a Level 1, 2 or 3 alteration, in accordance with Chapter 7, 8 or 9 depending on the extent of the remodel, and with the particular requirements contained in Sections 1007, Structural; 1008, Electrical; 1009, Mechanical; and 1010, Plumbing. The exception provides for special considerations for historic buildings, as given in Section 1205.

1001.2.2 Change of occupancy classification or group. Where the occupancy classification of a building changes, the provisions of Sections 1002 through 1012 shall apply. This includes a *change of occupancy* classification and a change to another group within an occupancy classification.

❖ When a use of a building changes and the occupancy classification changes, the provisions of Sections 1002 through 1012 will apply. For example, if an old downtown hardware store Occupancy Classification M is sold and remodeled into a condominium project Occupancy Classification R, then it is important to protect the new residents with all the required code provisions that may not have been in place when the property was utilized as a hardware store. Sections 1001.2 and 1012 require addressing IBC Chapter 9 requirements if the thresholds are exceeded through the change of occupancy classification or group.

1001.2.2.1 Partial change of occupancy. Where the occupancy classification or group of a portion of an *existing building* is changed, Section 1012 shall apply.

❖ Circumstances occur where only a portion of a building is changed to a different occupancy classification. This section refers to Section 1012, where Sections 1012.1.1 and 1012.1.2 address a partial change of occupancy for two options of separated and nonseparated conditions.

1001.3 Certificate of occupancy required. A certificate of occupancy shall be issued where a *change of occupancy* occurs that results in a different occupancy classification as determined by the *International Building Code*.

❖ A change of occupancy has historically been treated as severely as if the construction were new. In the code, the change of occupancy is treated as the most extensive alteration (Level 3); therefore, this is considered nearly as drastic as if the building were rebuilt. Ergo, the occupancy requires that a certificate of occupancy be issued again, as if the construction were new.

SECTION 1002
SPECIAL USE AND OCCUPANCY

1002.1 Compliance with the building code. Where the character or use of an *existing building* or part of an *existing building* is changed to one of the following special use or occupancy categories as defined in the *International Building Code*, the building shall comply with all of the applicable requirements of the *International Building Code*:

1. Covered and open mall buildings.
2. Atriums.
3. Motor vehicle-related occupancies.
4. Aircraft-related occupancies.
5. Motion picture projection rooms.
6. Stages and platforms.
7. Special amusement buildings.
8. Incidental use areas.
9. Hazardous materials.
10. Ambulatory care facilities.
11. Group I-2 occupancies.

❖ The IBC contains some special types of construction or occupancies. These are examples where the alteration to the building does not actually involve a change in occupancy classification, but involves a significant change to these special occupancies that are regulated by very specific provisions in Chapter 4 of the IBC. As such, the code cannot be applied to these specific provisions because such a change is outside the scope of the code. For instance, a large retail store could be converted to a multiple-tenant facility where the owner would like to use a covered mall concept. The utilization of the general provisions of the code would not adequately deal with the specific considerations for life safety envisioned as necessary in the special occupancy provisions of the IBC. Therefore, the designer must simply follow the provisions of the IBC as if the construction were new.

1002.2 Underground buildings. An underground building in which there is a change of use shall comply with the requirements of the *International Building Code* applicable to underground structures.

❖ Similar to the discussion for special occupancies given in Section 1002.1, underground buildings are dealt with in a specific manner in Chapter 4 of the IBC. The difference here is that, wherever there is a change of occupancy in an underground building, only the IBC requirements applicable to underground buildings apply versus full compliance with the IBC. The building is otherwise dealt with as a change of occupancy in accordance with the code.

SECTION 1003
BUILDING ELEMENTS AND MATERIALS

1003.1 General. Building elements and materials in portions of buildings undergoing a *change of occupancy* classification shall comply with Section 1012.

❖ Chapters 7, 8, 9 and 10 are all organized generally in the same manner: Special Use and Occupancy; Building Elements and Materials; Fire Protection; Means of Egress; Accessibility; and so on. Consistent with the stated requirements of Section 1001.2.1, a change of occupancy not involving a change of occupancy classification must be dealt with as a Level 1, 2 or 3 alteration; therefore, the requirements contained in those chapters for building elements and materials would apply, depending on the applicable requirements based on the classification of the work. Buildings or portions thereof that undergo a change of occupancy classification must further comply with the additional requirements in Section 1012.

SECTION 1004
FIRE PROTECTION

1004.1 General. Fire protection requirements of Section 1012 shall apply where a building or portions thereof undergo a *change of occupancy* classification or where there is a change of occupancy within a space where there is a different fire protection system threshold requirement in Chapter 9 of the *International Building Code*.

❖ Chapters 7, 8, 9 and 10 are all organized generally in the same manner: Special Use and Occupancy; Building Elements and Materials; Fire Protection; Means of Egress; Accessibility; and so on. Consistent with the stated requirements of Section 1001.2.1, a change of occupancy not involving a change of occupancy classification must be dealt with as a Level 1, 2 or 3 alteration, based on the classification of the work. This section also requires a review of the IBC Chapter 9 fire protection thresholds. If they are exceeded, a new fire protection system such as an automatic sprinkler system or fire alarm system may be required despite there not being a change of occupancy classification. Generally, buildings or portions thereof that undergo a change of occupancy classification must further comply with the additional requirements in Section 1012.

SECTION 1005
MEANS OF EGRESS

1005.1 General. Means of egress in portions of buildings undergoing a *change of occupancy* classification shall comply with Section 1012.

❖ Chapters 7, 8, 9 and 10 are all organized generally in the same manner: Special Use and Occupancy; Building Elements and Materials; Fire Protection; Means of Egress; Accessibility; and so on. Consistent with the stated requirements of Section 1001.2.1, a change of occupancy not involving a change of occupancy classification must be dealt with as a Level 1, 2 or 3 alteration; therefore, the requirements contained in those chapters for means of egress would apply, depending on the applicable requirements based on the classification of the work. Buildings or portions thereof that undergo a change of occupancy classification must further comply with the additional requirements in Section 1012.

SECTION 1006
ACCESSIBILITY

1006.1 General. Accessibility in portions of buildings undergoing a *change of occupancy* classification shall comply with Section 1012.8.

❖ When the use in a structure or a portion of a structure changes, the space must comply with new construction unless technically infeasible. For additional information, see Section 1012.5.

Chapters 7, 8, 9 and 10 are all organized in the same manner: Special Use and Occupancy; Building Elements and Materials; Fire Protection; Means of Egress; Accessibility; and so on. Consistent with the stated requirements of Section 1001.2.1, a change in occupancy not involving a change of occupancy classification must be dealt with as a Level 1, 2 or 3 alteration; therefore, the requirements contained in those chapters for accessibility would apply, depending on the applicable requirements based on the classification of the work. Buildings or portions thereof that undergo a change of occupancy classification must further comply with the additional requirements in Section 1012.

SECTION 1007
STRUCTURAL

[BS] 1007.1 Gravity loads. Buildings or portions thereof subject to a *change of occupancy* where such change in the nature of occupancy results in higher uniform or concentrated loads based on Table 1607.1 of the *International Building Code* shall comply with the gravity load provisions of the *International Building Code*.

Exception: Structural elements whose stress is not increased by more than 5 percent.

❖ These structural requirements apply where the change of occupancy results in an increased minimum live load required by IBC Section 1607. The exception allows increased gravity design loads that do not increase stresses in affected structural elements by more than 5 percent. Allowing overstresses of up to 5 percent in existing structural members has been a long-standing rule of thumb used by structural engineers.

[BS] 1007.2 Snow and wind loads. Buildings and structures subject to a *change of occupancy* where such change in the nature of occupancy results in higher wind or snow risk categories based on Table 1604.5 of the *International Building Code* shall be analyzed and shall comply with the applicable wind or snow load provisions of the *International Building Code*.

> **Exception:** Where the new occupancy with a higher risk category is less than or equal to 10 percent of the total building floor area. The cumulative effect of the area of occupancy changes shall be considered for the purposes of this exception.

❖ This section addresses a change of occupancy that results in higher wind or snow risk categories than required for the current occupancy. The risk categories listed in Table 1604.5 of the IBC are directly proportional to the risk categories for snow and wind that are listed in the ASCE 7 design standard, which is referenced in the IBC for determining wind and snow loads (also see IBC Sections 1609 for wind and 1608 for snow). These risk categories are a function of the building's occupancy. The intent is to achieve a higher level of performance for the facilities that have higher risk categories; thus, the risk categories generally increase with the importance of the facility. The facilities listed as Category IV are considered essential facilities and are assigned the highest risk category. The facilities are essential in that their continuous use is needed. For example, fire, rescue and police stations and emergency vehicle garages are needed to be in service during and immediately after a major windstorm or earthquake event.

Where this applies, the structure must be analyzed for snow and wind loads and shown to be adequate under the provisions of the IBC for new buildings. The impact this may have on the existing structure and the minimum design loads depends on the magnitude of the risk category for the new (proposed) occupancy category relative to those required for the current occupancy category. Other less apparent impacts will result due to the fact that the basic design loads prior to considering the risk category (e.g., wind pressure) required by the IBC will often be different than the predecessor model code used in the building's original design.

The exception allows a change of occupancy to an occupancy requiring a higher risk category without a structural analysis where this occupancy comprises only a relatively small portion of the entire building.

[BS] 1007.3 Seismic loads. *Existing buildings* with a *change of occupancy* shall comply with the seismic provisions of Sections 1007.3.1 and 1007.3.2.

❖ A change of occupancy must comply with the seismic provisions of Sections 1007.3.1 and 1007.3.2.

[BS] 1007.3.1 Compliance with International Building Code-level seismic forces. Where a building or portion thereof is subject to a *change of occupancy* that results in the building being assigned to a higher risk category based on Table 1604.5 of the *International Building Code*, the building shall comply with the requirements for *International Building Code*-level seismic forces as specified in Section 301.1.4.1 for the new risk category.

> **Exceptions:**
> 1. Where approved by the *code official*, specific detailing provisions required for a new structure are not required to be met where it can be shown that an equivalent level of performance and seismic safety is obtained for the applicable risk category based on the provision for reduced *International Building Code*-level seismic forces as specified in Section 301.1.4.2.
> 2. Where the area of the new occupancy with a higher hazard category is less than or equal to 10 percent of the total building floor area and the new occupancy is not classified as Risk Category IV. For the purposes of this exception, buildings occupied by two or more occupancies not included in the same risk category, shall be subject to the provisions of Section 1604.5.1 of the *International Building Code*. The cumulative effect of the area of occupancy changes shall be considered for the purposes of this exception.
> 3. Unreinforced masonry bearing wall buildings in Risk Category III when assigned to Seismic Design Category A or B shall be allowed to be strengthened to meet the requirements of Appendix Chapter A1 of this code [Guidelines for the Seismic Retrofit of Existing Buildings (GSREB)].

❖ The two types of occupancy changes that make it necessary to comply with IBC-level seismic forces in accordance with Section 301.1.4.1 are as follows:
  • The change of occupancy results in the new occupancy requiring a higher risk category (see IBC Table 1604.5). The explanation is the same as that given in the commentary to Section 1007.2.
  • The change of occupancy results in the new occupancy having a higher hazard category in accordance with Table 1012.4.

The code recognizes that the risk to safety is likely to be greater in the previously listed situations and conformance to the IBC seismic forces in accordance with Section 301.1.4.1 is mandated, unless compliance with any of the exceptions can be demonstrated. Section 301.1.4.1 contains two methods of compliance, either of which can be utilized. See the commentary to Section 301.1.4.1 for more information.

Exception 1 is based on a similar exception to Chapter 16 of the IBC and recognizes the difficulty of making an older structure comply with the seismic requirements for a new structure. It provides general guidance to the code official and designer on areas that need to be investigated when compliance with the seismic requirements for new buildings is not feasible. See Section 301.1.4.2 for the permitted evaluation and rehabilitation procedures.

Exception 2 contains requirements and exceptions that apply to buildings housing two or more occupancies that are classified in different risk categories. The general requirement for a change of occupancy resulting in a mixed-use building is that the entire structure must be designed in accordance with the requirements of the most critical risk (higher hazard classification) category, because it resists the earthquake forces as a unit. This is consistent with IBC Chapter 16. The actual exception to this rule is where a new occupancy having a risk category comprises no more than 10 percent of the total building area and it is not considered an essential facility. Through reference to IBC Section 1604.5.1, this exception further allows multiple occupancies that are structurally separated to be separately classified and designed independently using the risk categories applicable to each occupancy category.

Exception 3 refers to Chapter A1 of Appendix A, *Guidelines for Seismic Retrofit of Existing Buildings* (GSREB), which is applicable to unreinforced masonry-bearing (URM) wall buildings. These buildings would likely fall into the category of buildings already exempt under Exception 2. Nevertheless, this exception provides a direct link to the GSREB requirements for URM buildings that undergo a change of occupancy. Section A102.2 limitations make Chapter A1 applicable to Risk Categories I and II in IBC Table 1604.5.

[BS] 1007.3.2 Access to Risk Category IV. Where a *change of occupancy* is such that compliance with Section 1007.3.1 is required and the building is assigned to Risk Category IV, the operational access to the building shall not be through an adjacent structure, unless that structure conforms to the requirements for Risk Category IV structures. Where operational access is less than 10 feet (3048 mm) from either an interior lot line or from another structure, access protection from potential falling debris shall be provided by the owner of the Risk Category IV structure.

❖ This section is intended to ensure access for the occupant to essential facilities after an earthquake by prohibiting access through an adjacent structure that does not comply with the seismic criteria for essential facilities. These requirements are the result of lessons learned from previous earthquakes in which essential facilities have been rendered out of service because of the failure of an adjacent structure or a structure's exterior components. These requirements are based on ASCE 7, as well as FEMA 368.

SECTION 1008
ELECTRICAL

1008.1 Special occupancies. Where the occupancy of an *existing building* or part of an *existing building* is changed to one of the following special occupancies as described in NFPA 70, the electrical wiring and equipment of the building or portion thereof that contains the proposed occupancy shall comply with the applicable requirements of NFPA 70 whether or not a *change of occupancy* group is involved:

1. Hazardous locations.
2. Commercial garages, *repair*, and storage.
3. Aircraft hangars.
4. Gasoline dispensing and service stations.
5. Bulk storage plants.
6. Spray application, dipping, and coating processes.
7. Health care facilities.
8. Places of assembly.
9. Theaters, audience areas of motion picture and television studios, and similar locations.
10. Motion picture and television studios and similar locations.
11. Motion picture projectors.
12. Agricultural buildings.

❖ Similar to the discussion in Section 1002.1 of the code regarding the special occupancy provisions of IBC Chapter 4, the *National Electrical Code®*, NFPA 70, contains provisions for special types of constructions or occupancies. These are examples where the alteration to the building does not actually involve a change of occupancy classification, but involves a significant change to these special occupancies that are regulated by very specific provisions in NFPA 70. As such, the code cannot be applied to these specific provisions because such a change is outside the scope of the code. For instance, a large hospital could be renovated to change some of the spaces to a new emergency room or new testing facility. The utilization of the general provisions of the code would not adequately deal with the specific considerations for life safety envisioned as necessary in the special occupancy provisions of NFPA 70. Therefore, the designer must simply follow the provisions of NFPA 70 without modification.

1008.2 Unsafe conditions. Where the occupancy of an *existing building* or part of an *existing building* is changed, all unsafe conditions shall be corrected without requiring that all parts of the electrical system comply with NFPA 70.

❖ Buildings, structures or premises that constitute a fire hazard or are otherwise dangerous to human life, or that, in relation to existing use, constitute a hazard to safety, health or public welfare, by reason of inadequate maintenance, dilapidation, obsolescence, fire hazard, disaster damage or abandonment as specified in the code or any other ordinance, are unsafe conditions.

The intent of this section is to address unsafe electrical conditions before they can lead to injury or death. The code specifically requires only the unsafe wiring condition to be brought up to the current provisions of NFPA 70. Any additional existing wiring that is isolated from the damaged area is not required to be repaired or replaced.

Example: On a typical downtown street above the main street commercial occupancies there are some older existing apartments. Prior to the rental of one particular unit, an inspector notices a damaged receptacle with wiring exposed. According to Section 1008.2 of the code, this outlet must be properly repaired or replaced in accordance with NFPA 70. The remainder of the existing electrical system would not be required to be changed unless it was damaged.

1008.3 Service upgrade. Where the occupancy of an *existing building* or part of an *existing building* is changed, electrical service shall be upgraded to meet the requirements of NFPA 70 for the new occupancy.

❖ If an existing structure or a portion thereof undergoes a change of occupancy, this section requires the wiring supporting the changed area to meet NFPA 70.

1008.4 Number of electrical outlets. Where the occupancy of an *existing building* or part of an *existing building* is changed, the number of electrical outlets shall comply with NFPA 70 for the new occupancy.

❖ The number of outlets that are inherent to an area undergoing a change of occupancy must meet NFPA 70 requirements for the new occupancy.

Example: A room is undergoing a change from meeting room usage to a kitchen. The old rule of being able to go in any direction along the wall no more than 6 feet (1829 mm) and reach an outlet is no longer sufficient. There would have to be outlets added such that no point along the counter wall line is more than 24 inches (610 mm) from an outlet.

SECTION 1009
MECHANICAL

1009.1 Mechanical requirements. Where the occupancy of an *existing building* or part of an *existing building* is changed such that the new occupancy is subject to different kitchen exhaust requirements or to increased mechanical ventilation requirements in accordance with the *International Mechanical Code*, the new occupancy shall comply with the respective *International Mechanical Code* provisions.

❖ This section deals with two specific scenarios that could occur in a change of occupancy that does not qualify as a change of occupancy classification.
   The first scenario is a change in kitchen equipment that would require different exhaust requirements. A simple change in the size of a grill would require a different hood and levels of exhaust. This type of change meets the definition of a "Change of occupancy."
   The second scenario is a change in ventilation requirements. Many changes of occupancy could lead to changes in the occupant load of the building. An increase in the occupant load could mean an increase in ventilation requirements. Therefore, ventilation levels would need to be based on the requirements of the *International Mechanical Code®* (IMC®).

SECTION 1010
PLUMBING

1010.1 Increased demand. Where the occupancy of an *existing building* or part of an *existing building* is changed such that the new occupancy is subject to increased or different plumbing fixture requirements or to increased water supply requirements in accordance with the *International Plumbing Code*, the new occupancy shall comply with the intent of the respective *International Plumbing Code* provisions.

❖ Many changes of occupancy could lead to changes in the occupant load of the building. An increase in the occupant load could mean an increase in needed plumbing fixtures. Therefore, fixture levels would need to be based on the intent of the *International Plumbing Code®* (IPC®).

1010.2 Food-handling occupancies. If the new occupancy is a food-handling establishment, all existing sanitary waste lines above the food or drink preparation or storage areas shall be panned or otherwise protected to prevent leaking pipes or condensation on pipes from contaminating food or drink. New drainage lines shall not be installed above such areas and shall be protected in accordance with the *International Plumbing Code*.

❖ When a change of occupancy involves new food-handling facilities, the health safety issues associated with the integrity of the existing plumbing equipment and facilities must be addressed. Again, the code does not contain special provisions for such a specific issue. Therefore, the facilities must be modified in accordance with the intent of the IPC for new facilities.

1010.3 Interceptor required. If the new occupancy will produce grease or oil-laden wastes, interceptors shall be provided as required in the *International Plumbing Code*.

❖ When a change of occupancy involves new facilities that produce grease or oil-laden wastes, the issues associated with the integrity of the existing plumbing equipment and facilities must be addressed. Again, the code does not contain special provisions for such a specific issue. Therefore, the facilities must be modified in accordance with the intent of the IPC for new facilities.

1010.4 Chemical wastes. If the new occupancy will produce chemical wastes, the following shall apply:

1. If the existing piping is not compatible with the chemical waste, the waste shall be neutralized prior to entering the drainage system, or the piping shall be changed to a compatible material.

2. No chemical waste shall discharge to a public sewer system without the approval of the sewage authority.

❖ When a change of occupancy involves new facilities that produce chemical wastes, issues associated with the integrity of existing plumbing equipment and facilities must be addressed. In this case, the performance requirements essentially address the intent of the IPC for health safety, life safety and environmental safety by requiring that the piping be made to handle the particular chemicals or the chemicals be treated so as to not harm the piping, and that the impact on the public sewer system be addressed.

1010.5 Group I-2. If the occupancy group is changed to Group I-2, the plumbing system shall comply with the applicable requirements of the *International Plumbing Code*.

❖ Similar to the provisions given in Section 1002.1 for special occupancies in the IBC, the utilization of the general provisions of the code would not adequately deal with the specific considerations for life safety and health safety envisioned as necessary in the specific provisions of the IPC for the plumbing facilities in a Group I-2 occupancy. Therefore, the designer must simply follow the provisions of the IPC as if the construction were new.

SECTION 1011
OTHER REQUIREMENTS

1011.1 Light and ventilation. Light and ventilation shall comply with the requirements of the *International Building Code* for the new occupancy.

❖ Many changes of occupancy could lead to changes in the occupant load of the building. An increase in the occupant load could mean an increase in needed lighting or ventilation. Therefore, the requirements of the IBC applicable to new construction are applied to light and ventilation.

SECTION 1012
CHANGE OF OCCUPANCY CLASSIFICATION

1012.1 General. The provisions of this section shall apply to buildings or portions thereof undergoing a change of occupancy classification. This includes a change of occupancy classification within a group as well as a change of occupancy classification from one group to a different group or where there is a change of occupancy within a space where there is a different fire protection system threshold requirement in Chapter 9 of the *International Building Code*. Such buildings shall also comply with Sections 1002 through 1011. The application of requirements for the change of occupancy shall be as set forth in Sections 1012.1.1 through 1012.1.4. A *change of occupancy*, as defined in Section 202, without a corresponding change of occupancy classification shall comply with Section 1001.2.

❖ Buildings that undergo a change of occupancy classification must meet the requirements of Section 1012. This would include a Group A-2 occupancy changing to a Group A-3 occupancy, as well as a Group B occupancy that changes to a Group M occupancy classification group.

1012.1.1 Compliance with Chapter 9. The requirements of Chapter 9 shall be applicable throughout the building for the new occupancy classification based on the separation conditions set forth in Sections 1012.1.1.1 and 1012.1.1.2.

❖ This section contains provisions that allow a change of occupancy classification within an existing building as defined in IBC Section 302.1, based on whether the proposed change of occupancy is separated from adjacent spaces with fire barriers as required in the IBC. Conditions to allow that change of occupancy classification require a review of the existing building, or area of the proposed change of occupancy, to be conducted to determine compliance with the requirements of Chapter 9. This essentially triggers a review against the requirements of Chapters 7, 8 and 9, and Sections 1002 through 1012. All deficiencies disclosed by the review are required to be brought into compliance with the provisions of the code. Design and detail of the corrective construction must be incorporated into the scope of work and identified in the construction documents. The basic philosophy of the code continues to be that as the level of alteration activity increases in scope or area of the building, the level of compliance with new construction code requirements also increases.

1012.1.1.1 Change of occupancy classification without separation. Where a portion of an *existing building* is changed to a new occupancy classification or where there is a change of occupancy within a space where there is a different fire protection system threshold requirement in Chapter 9 of the *International Building Code*, and that portion is not separated from the remainder of the building with fire barriers having a fire-resistance rating as required in the *International Building Code* for the separate occupancy, the entire building shall comply with all of the requirements of Chapter 9 applied throughout the building for the most restrictive occupancy classification in the building and with the requirements of this chapter.

❖ Requirements of this subsection and Section 1012.1.1.2 provide alternatives for the evaluation process where the change of occupancy occurs in only a portion of an existing building. These two sections incorporate two of the philosophies for mixed-use occupancies found in IBC Section 508.

If the portion of the building that is undergoing the change of occupancy classification is not separated from the remainder of the building by fire barriers, this section allows the two or more uses to remain unseparated from each other. However, the entire building must then comply with the requirements of Chapter 9 and those of this chapter for the most restrictive occupancy group. This also applies to exceeding IBC Chapter 9 thresholds. This philosophy is consistent with the approach made for mixed uses, nonseparated, as given in IBC Section 508.3. Note that this could require construction activity in existing occupied spaces or portions of the building not otherwise impacted by the change of occupancy classification.

1012.1.1.2 Change of occupancy classification with separation. Where a portion of an *existing building* is changed to a new occupancy classification or where there is a change of occupancy within a space where there is a different fire protection system threshold requirement in Chapter 9 of the *International Building Code,* and that portion is separated from the remainder of the building with fire barriers having a fire-resistance rating as required in the *International Building Code* for the separate occupancy, that portion shall comply with all of the requirements of Chapter 9 for the new occupancy classification and with the requirements of this chapter.

❖ When the change of occupancy classification is wholly contained in a portion of the building that is completely separated from the other uses occupying the building by fire barriers as required by IBC Table 508.4, compliance with the requirements of Chapter 9 and Sections 1002 through 1012 is only required for that portion of the building occupied by the new occupancy classification. This also applies to exceeding IBC Chapter 9 thresholds. The remaining portion(s) of the building that are legally occupied are permitted to continue occupancy without change, until activities specifically covered in the code occur.

1012.1.2 Fire protection and interior finish. The provisions of Sections 1012.2 and 1012.3 for fire protection and interior finish, respectively, shall apply to all buildings undergoing a change of occupancy classification.

❖ When a building undergoes a change of occupancy classification, it is important that the interior finishes and fire protection systems meet the requirements for the new occupancy classification. This is especially the case when the provisions for the new occupancy classification are more restrictive than the existing occupancy classification. This section guarantees the appropriate level of protection for the building's occupants.

1012.1.3 Change of occupancy classification based on hazard category. The relative degree of hazard between different occupancy classifications shall be determined in accordance with the categories specified in Tables 1012.4, 1012.5 and 1012.6. Such a determination shall be the basis for the application of Sections 1012.4 through 1012.7.

❖ All of the classified occupancy groups were evaluated in three areas in which the degree of inherent hazards of an occupancy group can differ. These evaluations were reviewed, and like results were grouped and assigned a relative rating number, 1 through 4 or 5, depending on the element being evaluated. The lower the number, the higher the relative hazard of the occupancy classification. Aspects of a building that were considered were: fire and life safety, including the ability to exit; the height and area of the building; and the need to protect surrounding buildings from fire conflagration. Occupancies with equivalent relative hazard numbers were then grouped together to form Tables 1012.4, 1012.5 and 1012.6 of Sections 1012.4 through 1012.7. These tables are used to determine whether a change of occupancy classification is to a higher, lesser or equivalent occupancy classification, in relation to the relative hazard of the use.

1012.1.4 Accessibility. All buildings undergoing a change of occupancy classification shall comply with Section 1012.8.

❖ See the commentary to Section 1012.8.

1012.2 Fire protection systems. Fire protection systems shall be provided in accordance with Sections 1012.2.1 and 1012.2.2.

❖ Fire protection systems must comply with the fire protection requirements of Sections 1012.2.1 and 1012.2.2.

1012.2.1 Fire sprinkler system. Where a change in occupancy classification occurs or where there is a change of occupancy within a space where there is a different fire protection system threshold requirement in Chapter 9 of the *International Building Code* that requires an automatic fire sprinkler system to be provided based on the new occupancy in accordance with Chapter 9 of the *International Building Code,* such system shall be provided throughout the area where the *change of occupancy* occurs.

❖ Depending on the new occupancy classification, the need for an automatic sprinkler system may depend on not only the occupancy, but also the occupant load, fuel load, height and area of the building, as well as fire-fighting capabilities. IBC Section 903.2 addresses all occupancy conditions requiring automatic sprinkler systems. IBC Section 903.3 contains the installation requirements for all sprinkler systems, in addition to the requirements of NFPA 13, NFPA 13D and NFPA 13R. The supervision and alarm requirements for sprinkler systems are contained in IBC Section 903.4, whereas IBC Section 903.5 refers to testing and maintenance requirements for sprinkler systems in the *International Fire Code*® (IFC®).

1012.2.2 Fire alarm and detection system. Where a change in occupancy classification occurs or where there is a change of occupancy within a space where there is a different fire protection system threshold requirement in Chapter 9 of the *International Building Code* that requires a fire alarm and detection system to be provided based on the new occupancy in accordance with Chapter 9 of the *International Building Code,* such system shall be provided throughout the area where the *change of occupancy* occurs. Existing alarm notification appliances shall be automatically activated throughout the building. Where the building is not equipped with a fire alarm system, alarm notification appliances shall be provided throughout the area where the *change of occupancy* occurs in accordance with Section 907 of the *International Building Code* as required for new construction.

❖ Fire alarm systems, which typically include manual fire alarm systems and automatic fire detection systems, are required to be installed in accordance with the provisions of IBC Chapter 9 and NFPA 72.

Fire alarm systems are provided in buildings to limit

fire casualties and property losses. Fire alarm systems do this by promptly notifying the occupants of the building of an emergency condition, which increases the time available for evacuation. Similarly, when fire alarm systems are required to be supervised, the fire department will be promptly notified and its response time relative to the onset of the fire incident will be reduced.

Automatic fire detection systems are required under certain conditions to increase the likelihood that fire is detected and occupants are alerted in a timely manner. The detection system is a system of devices and associated hardware that activates the alarm system. The automatic detecting devices are to be smoke detectors unless a condition exists mandating the use of a different type of detector (see IBC Section 907).

1012.3 Interior finish. In areas of the building undergoing the change of occupancy classification, the interior finish of walls and ceilings shall comply with the requirements of the *International Building Code* for the new occupancy classification.

❖ The interior finish of corridor and exit walls and ceilings in the area where the change of occupancy occurs must meet the flame spread and smoke-development limitations and requirements of the IBC for each new occupancy classification. Corridors and exits are defined in IBC Section 202. Flame spread and smoke-developed classification are determined from IBC Table 803.11 and are further based on ASTM E84.

1012.4 Means of egress, general. Hazard categories in regard to life safety and means of egress shall be in accordance with Table 1012.4.

❖ This section requires that hazard categories for means of egress are established in accordance with Table 1012.4. See also the commentary to Table 1012.4.

TABLE 1012.4
MEANS OF EGRESS HAZARD CATEGORIES

RELATIVE HAZARD	OCCUPANCY CLASSIFICATIONS
1 (Highest Hazard)	H
2	I-2, I-3, I-4
3	A, E, I-1, M, R-1, R-2, R-4
4	B, F-1, R-3, S-1
5 (Lowest Hazard)	F-2, S-2, U

❖ As can be seen, the means-of-egress provisions for hazard classification contain five levels. These levels are established based on IBC requirements for different occupancies for means of egress, fire protection and fire-resistance-rated construction.

1012.4.1 Means of egress for change to higher hazard category. When a change of occupancy classification is made to a higher hazard category (lower number) as shown in Table 1012.4, the means of egress shall comply with the requirements of Chapter 10 of the *International Building Code*.

Exceptions:

1. Stairways shall be enclosed in compliance with the applicable provisions of Section 903.1.

2. Existing stairways including handrails and guards complying with the requirements of Chapter 9 shall be permitted for continued use subject to approval of the *code official*.

3. Any stairway replacing an existing stairway within a space where the pitch or slope cannot be reduced because of existing construction shall not be required to comply with the maximum riser height and minimum tread depth requirements.

4. Existing corridor walls constructed on both sides of wood lath and plaster in good condition or $^1/_2$-inch-thick (12.7 mm) gypsum wallboard shall be permitted. Such walls shall either terminate at the underside of a ceiling of equivalent construction or extend to the underside of the floor or roof next above.

5. Existing corridor doorways, transoms and other corridor openings shall comply with the requirements in Sections 805.5.1, 805.5.2 and 805.5.3.

6. Existing dead-end corridors shall comply with the requirements in Section 805.6.

7. An existing operable window with clear opening area no less than 4 square feet (0.38 m^2) and minimum opening height and width of 22 inches (559 mm) and 20 inches (508 mm), respectively, shall be accepted as an emergency escape and rescue opening.

❖ The basic premise here is for the means-of-egress requirements to be the same as required for new construction when the change of occupancy is to a higher (lower number) classification. The exceptions relate to common problem areas where it is practical to fully comply with requirements for new construction. Exception 1 allows the same provision for the limited enclosures of stairways as allowed for Level 3 alterations. Exception 2 allows handrails and guards to meet the limited requirements of Chapter 9. The reader will note that Chapter 9 contains no specific requirements for guards and handrails. However, it must be remembered that the provisions of Chapter 8 also apply to Chapter 9; therefore, the provisions of Sections 805.9 and 805.11 are applicable. Exceptions 3 through 7 give allowances for requirements commonly found in existing construction that make full compliance with the present IBC impractical.

1012.4.2 Means of egress for change of use to equal or lower hazard category. When a change of occupancy classification is made to an equal or lesser hazard category (higher number) as shown in Table 1012.4, existing elements of the means of egress shall comply with the requirements of Section 905 for the new occupancy classification. Newly constructed or configured means of egress shall comply with the requirements of Chapter 10 of the *International Building Code*.

> **Exception:** Any stairway replacing an existing stairway within a space where the pitch or slope cannot be reduced because of existing construction shall not be required to comply with the maximum riser height and minimum tread depth requirements.

❖ Where the change of occupancy is to a lower hazard category (higher number), the code allows for the modifications to the means of egress provided for Level 3 alterations in Chapter 9 (see commentary, Section 905). Detailed requirements are actually provided in Section 805, which are applicable to Level 3 alterations in Chapter 9, as well. Section 905 gives additional requirements regarding means of egress lighting and exit signs, requiring these to be installed in accordance with the IBC.

The exception deals with a practical problem common in existing buildings, where the space provided for the stairway cannot accommodate the shallower pitch of steps required in the IBC.

1012.4.3 Egress capacity. Egress capacity shall meet or exceed the occupant load as specified in the *International Building Code* for the new occupancy.

❖ This provision is consistent throughout the code for all alterations and changes of occupancy. Reduction of egress capacity is strictly prohibited by Section 1001.2 of the IBC.

1012.4.4 Handrails. Existing stairways shall comply with the handrail requirements of Section 805.9 in the area of the change of occupancy classification.

❖ The special provisions of Section 805.9 for handrails are allowed for this category of change of use, which is similar to that allowed for change of occupancy to a higher classification in Exception 2 of Section 1012.4.1, except that the code official's specific approval is required in Section 1012.4.1. The provisions of Section 805.9 boil down to a simple requirement for a single handrail for all stairways with a width less than 66 inches (1676 mm). Stairways wider than 66 inches (1676 mm) require two handrails. The design of the handrails is regulated by the applicable provisions of the IBC.

1012.4.5 Guards. Existing guards shall comply with the requirements in Section 805.11 in the area of the change of occupancy classification.

❖ The provisions of Section 805.11 apply to guards for changes of occupancy in this category. The provisions are the same as that of the IBC; however, guards are only required to comply with Section 805.11 in the area of the change of occupancy.

1012.5 Heights and areas. Hazard categories in regard to height and area shall be in accordance with Table 1012.5.

❖ This section requires that hazard categories for building heights and areas are established in accordance with Table 1012.5. See also the commentary to Table 1012.5.

The provisions for hazard categories and requirements related to change of occupancy classification for height and areas are considerably simpler and more straightforward than the life safety and exits hazard category. Essentially, the change to a higher hazard category requires compliance with height and area limitations for new construction as given by the IBC, and the change to an equal or lesser hazard category allows for the height and area of the existing building to be deemed acceptable.

TABLE 1012.5
HEIGHTS AND AREAS HAZARD CATEGORIES

RELATIVE HAZARD	OCCUPANCY CLASSIFICATIONS
1 (Highest Hazard)	H
2	A-1, A-2, A-3, A-4, I, R-1, R-2, R-4
3	E, F-1, S-1, M
4 (Lowest Hazard)	B, F-2, S-2, A-5, R-3, U

❖ Table 1012.5 groups the relative hazards based on allowable heights and areas given by the IBC.

1012.5.1 Height and area for change to higher hazard category. When a change of occupancy classification is made to a higher hazard category as shown in Table 1012.5, heights and areas of buildings and structures shall comply with the requirements of Chapter 5 of the *International Building Code* for the new occupancy classification.

> **Exception:** For high-rise buildings constructed in compliance with a previously issued permit, the type of construction reduction specified in Section 403.2.1 of the *International Building Code* is permitted. This shall include the reduction for columns. The high-rise building is required to be equipped throughout with an automatic sprinkler system in accordance with Section 903.3.1.1 of the *International Building Code*.

❖ As stated in Section 1012.5.1, the change to a higher hazard category requires compliance with height and area limitations for new construction as given by the IBC, with a single exception regarding Group E, one-story buildings. This approach is consistent with the premise that any alteration to a building cannot increase the hazard in the building. By definition, utilization of a building for a use in which the height or area for the occupancy is exceeded is an increase in the hazard of the building.

The exception addresses the revision to the IBC that prohibited the reduction of ratings on columns for high-rise buildings for Type IA construction under 420 feet and general reductions for Type IA construction over 420 feet in building height. This created a problem for existing buildings that had previously applied these reductions. Without this exception, the code would not recognize those previously complying buildings. For example, a change of occupancy from an office to a retail area would require a complete upgrade in the fire-resistance rating for all the columns in the entire building. This is excessive for small changes in occupancy and often impractical. The exception makes it clear that if the building is protected throughout with an automatic fire sprinkler system, designed to meet NFPA 13 (not 13R), then the column ratings can be what was allowed prior to that revision to the IBC. Additions will need to meet the requirements for new construction, but a change in occupancy of this type should not require the entire building to fall into noncompliance when it was fully compliant when it was built.

1012.5.1.1 Fire wall alternative. In other than Groups H, F-1 and S-1, fire barriers and horizontal assemblies constructed in accordance with Sections 707 and 711, respectively, of the *International Building Code* shall be permitted to be used in lieu of fire walls to subdivide the building into separate buildings for the purpose of complying with the area limitations required for the new occupancy where all of the following conditions are met:

1. The buildings are protected throughout with an automatic sprinkler system in accordance with Section 903.3.1.1 of the *International Fire Code*.

2. The maximum allowable area between fire barriers, horizontal assemblies, or any combination thereof shall not exceed the maximum allowable area determined in accordance with Chapter 5 of the *International Building Code* without an increase allowed for an automatic sprinkler system in accordance with Section 506 of the *International Building Code*.

3. The fire-resistance rating of the fire barriers and horizontal assemblies shall be not less than that specified for fire walls in Table 706.4 of the *International Building Code*.

Exception: Where horizontal assemblies are used to limit the maximum allowable area, the required fire-resistance rating of the horizontal assemblies shall be permitted to be reduced by 1 hour provided the height and number of stories increases allowed for an automatic sprinkler system by Section 504 of the *International Building Code* are not used for the buildings.

❖ This section allows the use of fire barriers and rated horizontal assemblies to create separate buildings for the purpose of applying the height and area requirements. This allowance comes with several criteria, such as requiring an NFPA 13 sprinkler system, ratings in accordance with the fire wall ratings of IBC Table 706.4 and a limitation on the allowable area that does not include a sprinkler increase. This section provides necessary flexibility in meeting the height and area requirements in existing buildings such as old mill complexes where achieving the height and area limitations is very difficult.

1012.5.2 Height and area for change to equal or lesser hazard category. When a change of occupancy classification is made to an equal or lesser hazard category as shown in Table 1012.5, the height and area of the *existing building* shall be deemed acceptable.

❖ When the change of occupancy is to an equal or lesser hazard, the height and area of the existing building is not at issue. This can, however, create an interesting scenario where the new occupancy is housed in a facility with a height or area that would exceed that allowable height and area for new construction for that use. For instance, a Type IIB, Group S-1 building that was constructed to be 17,500 square feet (1626 m^2) could be converted entirely to an Occupancy Group M, even though the maximum allowable area for the Type IIB construction for Group M is 12,500 square feet (1161 m^2).

1012.5.3 Fire barriers. When a change of occupancy classification is made to a higher hazard category as shown in Table 1012.5, fire barriers in separated mixed use buildings shall comply with the fire-resistance requirements of the *International Building Code*.

Exception: Where the fire barriers are required to have a 1-hour fire-resistance rating, existing wood lath and plaster in good condition or existing $^1/_2$-inch-thick (12.7 mm) gypsum wallboard shall be permitted.

❖ This provision is in addition to the provisions of Sections 1012.1.1.1 and 1012.1.1.2. This provision requires fire barriers in compliance with the IBC in a mixed-use building where the proposed change of occupancy is made to a higher hazard category. The exception recognizes the ability of existing plaster and lath construction to provide a 1-hour fire-resistance rating.

1012.6 Exterior wall fire-resistance ratings. Hazard categories in regard to fire-resistance ratings of exterior walls shall be in accordance with Table 1012.6.

❖ The third hazard classification area is the exposure of the exterior walls, relating to the need to protect surrounding buildings from fire conflagration. This section requires that hazard categories for fire exposure of exterior walls are established in accordance with Table 1012.6. See also the commentary to Table 1012.6.

TABLE 1012.6
EXPOSURE OF EXTERIOR WALLS HAZARD CATEGORIES

RELATIVE HAZARD	OCCUPANCY CLASSIFICATION
1 (Highest Hazard)	H
2	F-1, M, S-1
3	A, B, E, I, R
4 (Lowest Hazard)	F-2, S-2, U

❖ Table 1012.6 groups the uses in the relative hazard levels based primarily on the requirements of Section 705, as well as IBC Table 602.

1012.6.1 Exterior wall rating for change of occupancy classification to a higher hazard category. When a change of occupancy classification is made to a higher hazard category as shown in Table 1012.6, exterior walls shall have fire resistance and exterior opening protectives as required by the *International Building Code*.

> **Exception:** A 2-hour fire-resistance rating shall be allowed where the building does not exceed three stories in height and is classified as one of the following groups: A-2 and A-3 with an occupant load of less than 300, B, F, M or S.

❖ Consistent with the general rule throughout this chapter for a change of occupancy classification to a higher hazard category, the exterior wall rating is required to meet the ratings called for by the IBC for new construction. The exception is somewhat academic. These occupancy classifications are never required to have a 3-hour fire-resistance rating on exterior walls.

1012.6.2 Exterior wall rating for change of occupancy classification to an equal or lesser hazard category. When a change of occupancy classification is made to an equal or lesser hazard category as shown in Table 1012.6, existing exterior walls, including openings, shall be accepted.

❖ The exterior wall fire-resistance rating is allowed to remain without change when the change of occupancy classification is to an equal or lesser hazard category. Again, this is in keeping with the philosophy that no change is needed when the hazard to the building is not increased by the change or alteration of the building.

1012.6.3 Opening protectives. Openings in exterior walls shall be protected as required by the *International Building Code*. Where openings in the exterior walls are required to be protected because of their distance from the lot line, the sum of the area of such openings shall not exceed 50 percent of the total area of the wall in each story.

> **Exceptions:**
>
> 1. Where the *International Building Code* permits openings in excess of 50 percent.
> 2. Protected openings shall not be required in buildings of Group R occupancy that do not exceed three stories in height and that are located not less than 3 feet (914 mm) from the lot line.
> 3. Where exterior opening protectives are required, an automatic sprinkler system throughout may be substituted for opening protection.
> 4. Exterior opening protectives are not required when the change of occupancy group is to an equal or lower hazard classification in accordance with Table 1012.6.

❖ At first glance, this provision would seem to contradict the provision in Section 1012.6.2, which states that opening protectives need not change for the change of occupancy to an equal or lesser hazard classification. However, Exception 4 eliminates that apparent contradiction. Still, it should be remembered that this requirement would apply in all cases where change is to a higher hazard category as described in Section 1012.6.1. The significance of this section is that it imposes a maximum area of openings to be 50 percent of the wall area when the fire separation distance requires opening protectives, contrary to the percentages given by IBC Table 705.8.

Exception 1 allows a higher percentage when the separation distance is such that the IBC would otherwise allow a greater distance than 50 percent. Exception 2 relates to the separation distances and requirements allowed by the *International Residential Code*® (IRC®). Exception 3 is consistent with, albeit more liberal than, the trade-off provided for automatic sprinkler systems given by IBC Table 705.8.

1012.7 Enclosure of vertical shafts. Enclosure of vertical shafts shall be in accordance with Sections 1012.7.1 through 1012.7.4.

❖ Section 1012.7 outlines the requirements for the enclosure of vertical shafts in the building.

1012.7.1 Minimum requirements. Vertical shafts shall be designed to meet the *International Building Code* requirements for atriums or the requirements of this section.

❖ Minimum requirements are set forth in Section 1012.7.1, which offers two options for compliance. First, vertical shafts are required to be designed to meet the requirements for atriums as prescribed in IBC Section 404. As an alternative, vertical shafts can comply with the requirements contained in the remaining subsections.

1012.7.2 Stairways. When a change of occupancy classification is made to a higher hazard category as shown in Table 1012.4, interior stairways shall be enclosed as required by the *International Building Code*.

> **Exceptions:**
>
> 1. In other than Group I occupancies, an enclosure shall not be required for openings serving only one adjacent floor and that are not connected with corridors or stairways serving other floors.
> 2. Unenclosed existing stairways need not be enclosed in a continuous vertical shaft if each story is separated from other stories by 1-hour fire-resistance-rated construction or approved wired glass set in steel frames and all exit corridors are sprinklered. The openings between the corridor and the occupant space shall have at least one sprinkler head above the openings on the tenant side. The sprinkler system shall be permitted to be supplied from the domestic water-supply systems, provided the system is of adequate pressure, capacity, and sizing for the combined domestic and sprinkler requirements.
> 3. Existing penetrations of stairway enclosures shall be accepted if they are protected in accordance with the *International Building Code*.

❖ Specific requirements for stairways are found in Section 1012.7.2. When the change of occupancy is to a higher hazard category, as shown in Table 1012.4, compliance with the IBC is required. Three exceptions are provided. Exceptions 1 and 2 provide equivalencies to the requirements for atrium enclosures. Exception 3 allows existing penetrations into the stairway if they are protected in accordance with the IBC.

1012.7.3 Other vertical shafts. Interior vertical shafts other than stairways, including but not limited to elevator hoistways and service and utility shafts, shall be enclosed as required by the *International Building Code* when there is a change of use to a higher hazard category as specified in Table 1012.4.

Exceptions:

1. Existing 1-hour interior shaft enclosures shall be accepted where a higher rating is required.

2. Vertical openings, other than stairways, in buildings of other than Group I occupancy and connecting less than six stories shall not be required to be enclosed if the entire building is provided with an approved automatic sprinkler system.

❖ Section 1012.7.3 outlines compliance requirements for all interior vertical shafts other than stairways. Again, there are minimum requirements when an increase in the hazard category occurs. Two exceptions are provided. Exception 1 is the same rating allowed for atriums, as stated in Section 404.6 of the IBC. Exception 2 is unique to the code and would allow five stories to be connected without a shaft enclosure when the building is sprinklered throughout. Most likely, Exception 2 would encourage the installation of a sprinkler system in buildings that would not have otherwise been required.

1012.7.4 Openings. All openings into existing vertical shaft enclosures shall be protected by fire assemblies having a fire protection rating of not less than 1 hour and shall be maintained self-closing or shall be automatic-closing by actuation of a smoke detector. All other openings shall be fire protected in an approved manner. Existing fusible link-type automatic door-closing devices shall be permitted in all shafts except stairways if the fusible link rating does not exceed 135°F (57°C).

❖ Section 1012.7.4 prescribes a minimum fire protection rating of 1 hour for all openings into vertical shaft enclosures. All openings, such as those created by doors, archways or window-type openings, are to be self-closing or automatic closing by actuation of a smoke detector. Any other openings are required to be protected by an approved method. The use of fusible links is acceptable to be used to control the closing of doors in all vertical shafts, except for stairways. This acceptance is limited to fusible links that have a temperature rating not exceeding 135°F (57°C).

1012.8 Accessibility. *Existing buildings* that undergo a change of group or occupancy classification shall comply with this section.

Exception: Type B dwelling or sleeping units required by Section 1107 of the *International Building Code* are not required to be provided in existing buildings and facilities undergoing a *change of occupancy* in conjunction with less than a Level 3 *alteration*.

❖ This section establishes accessibility requirements for existing buildings that undergo a change of occupancy. There are two thresholds to determine accessibility requirements: partial change of occupancy and complete change of occupancy.

Note that there is an exception provided to Section 1012.8 that essentially states that only a change of occupancy with a Level 3 alteration would be required to provide Type B dwelling units. See the commentary to Section 906.2 for more discussion on the applicability of the Type B accessible units in accordance with IBC Section 1107.

1012.8.1 Partial change in occupancy. Where a portion of the building is changed to a new occupancy classification, any *alteration* shall comply with Sections 705, 806 and 906, as applicable.

❖ Accessibility requirements for partial changes of occupancy must comply with the requirements for alterations in Sections 705, 806 and 906, as applicable. These sections approach the application of accessibility provisions to a facility that is altered by broadly requiring full conformance to new construction, meaning full accessibility is expected. Exceptions are then provided to indicate the conditions under which less than full accessibility is permitted. For more information, see the commentary to Sections 705, 806 and 906.

1012.8.2 Complete change of occupancy. Where an entire building undergoes a *change of occupancy*, it shall comply with Section 1012.8.1 and shall have all of the following accessible features:

1. At least one accessible building entrance.

2. At least one accessible route from an accessible building entrance to *primary function* areas.

3. Signage complying with Section 1111 of the *International Building Code*.

4. Accessible parking, where parking is provided.

5. At least one accessible passenger loading zone, where loading zones are provided.

6. At least one accessible route connecting accessible parking and accessible passenger loading zones to an accessible entrance.

Where it is *technically infeasible* to comply with the new construction standards for any of these requirements for a change of group or occupancy, the above items shall conform to the requirements to the maximum extent technically feasible.

> **Exception:** The accessible features listed in Items 1 through 6 are not required for an accessible route to Type B units.

❖ For a project that involves a complete change of occupancy, full compliance with accessibility requirements is expected and reasonable, except where technical infeasibility can be demonstrated. If full compliance is technically infeasible, the element must be made accessible to the fullest extent that is feasible. This is consistent with the general approach that has always been taken relative to other matters regulated by the code.

In addition to accessibility requirements in the space, an accessible route is required to that space. Six items that make up that accessible route are listed. The intent is to provide the bare minimum to get people from a point of arrival into the building and to the area of the new use. This is not based on any specific provisions of the *Americans with Disabilities Act Accessibility Guidelines* (ADAAG), but parallels the intent of the requirements for the removal of barriers. If the altered area would not be required to be served by an accessible route in new construction, an accessible route would not be required for a change of occupancy (see Section 705.1.13).

If the area undergoing a change of occupancy is being altered and contains a primary function area, the accessible route provisions in Section 705.2 are also applicable. This would not only require an accessible route to the change of occupancy area, but would also include possible upgrades to toilet rooms and drinking fountains that serve the area. See the commentary to Section 705.2 for additional information on the exceptions.

If the change of occupancy results in Type B dwelling units being provided, the exception permits the building to be exempt from providing the additional accessible route requirements listed in this section. This allowance addresses concerns of site impracticality for existing structures. This also reinforces the intent that the inclusion of Type B units is not meant to require elevators when alterations are performed on upper floors in nonelevator buildings (see IBC Section 1107.7). These areas would have been exempted if built new under the Fair Housing Act (FHA) and the IBC, and should continue to be exempted.

Bibliography

The following resource material was used in the preparation of the commentary for this chapter of the code.

FEMA 368, *NEHRP Recommended Provisions for Seismic Regulations for New Buildings and Other Structures, Part 1*. Washington, DC: Federal Emergency Management Agency, 2001.

Chapter 11: Additions

General Comments

An "Addition" is defined in Chapter 2 as "an extension or increase in the floor area, number of stories, or height of a building or structure." Chapter 11 contains the minimum requirements for an addition that is not separated from the existing building by a fire wall.

Purpose

Chapter 11 provides the requirements for additions, which correlate to the requirements for new construction. There are, however, some exceptions that are specifically stated in this chapter.

SECTION 1101
GENERAL

1101.1 Scope. An *addition* to a building or structure shall comply with the *International Codes* as adopted for new construction without requiring the *existing building* or structure to comply with any requirements of those codes or of these provisions, except as required by this chapter. Where an *addition* impacts the *existing building* or structure, that portion shall comply with this code.

❖ This section states the scope of Chapter 11 and indicates that the addition, but not the existing, unaltered portion of the building or structure, must meet the requirements of the current editions of the *International Codes*®. An "Addition" is defined in Chapter 2 as "an extension or increase in floor area, number of stories, or height of a building or structure."

The scope of this chapter would not include a situation where a new building is erected immediately adjacent to an existing building and they are separated by a fire wall, which would create separate buildings and not an addition to the existing structure. The new building must be designed to comply with the technical provisions of the *International Building Code*® (IBC®), not with the provisions of this chapter.

1101.2 Creation or extension of nonconformity. An *addition* shall not create or extend any nonconformity in the *existing building* to which the *addition* is being made with regard to accessibility, structural strength, fire safety, means of egress, or the capacity of mechanical, plumbing, or electrical systems.

❖ If an existing building is noncompliant in any way that may be affected by the addition, or if the addition causes a noncompliant situation, the addition is not allowed to be constructed without correction of the noncompliance. For example, if the existing building has a travel distance that is greater than allowed to an exit and the addition adds to that distance, the addition is not allowed unless the total travel distance is brought to within that which is required by the code. Another example is that an addition that blocks an existing required exit would not be allowed.

An existing structure that is of a type of construction that does not comply with the height and area limitations of IBC Section 503 may not be added to, unless the type of construction is upgraded or the allowable building height and area is increased through the addition of a sprinkler system throughout both the existing building and the addition as allowed by Chapter 5 of the IBC.

1101.3 Other work. Any *repair* or *alteration* work within an *existing building* to which an *addition* is being made shall comply with the applicable requirements for the work as classified in Chapter 5.

❖ This section requires that any existing structure to which an addition is made must comply with the provisions of Chapter 5 for the type of work performed as defined in that chapter. In other words, the provisions of Chapters 7, 8, 9, 10 and 12 would also apply depending on whether the work is a Level 1, 2 or 3 alteration, a change of occupancy or a renovation to a historic building, etc.

SECTION 1102
HEIGHTS AND AREAS

1102.1 Height limitations. No *addition* shall increase the height of an *existing building* beyond that permitted under the applicable provisions of Chapter 5 of the *International Building Code* for new buildings.

❖ See the commentary to Section 1102.2.

1102.2 Area limitations. No *addition* shall increase the area of an *existing building* beyond that permitted under the applicable provisions of Chapter 5 of the *International Building Code* for new buildings unless fire separation as required by the *International Building Code* is provided.

> **Exception:** In-filling of floor openings and nonoccupiable appendages such as elevator and exit stairway shafts shall be permitted beyond that permitted by the *International Building Code*.

❖ The purposes of Sections 1102.1 and 1102.2 are to establish that, with one exception, the height and area of an existing building plus its addition must comply with the requirements for a new building of its same occupancy and type of construction in accordance with Chapter 5 of the IBC.

A building with a proposed addition is to be evaluated based on the type of construction of the existing building or the addition, whichever is the lower type. When reviewing for compliance with IBC Table 506.2, see IBC Section 602.1.1.

Commentary Figure 1102.2(1) shows an example of this situation. If a sprinkler system is not planned for both the existing building and the addition, the proposed addition does not meet the requirements of IBC Table 506.2. The area limitation in IBC Table 506.2 for the nonsprinklered existing construction Type VB, Group F-1 is 8,500 square feet (789.7 m^2) and the total proposed area is 11,000 square feet (1022 m^2). The area evaluation is based on Type VB construction, even though the proposed addition is Type VA construction, because the allowable area in IBC Table 506.2 for Type VB construction is less than the allowable area for Type VA construction. One way to satisfy the requirements of IBC Table 506.2 is to upgrade the type of construction of the existing building to Type VA, which has an allowable area of 14,000 square feet (130 m^2). Another solution is to install an automatic sprinkler system that complies with NFPA 13 throughout the building. The revised allowable area of the building would be 34,000 square feet based on the use of sprinklers. This is more than the proposed building area of 11,000 square feet (1022 m^2).

The exception allows undesirable existing openings in floors that are no longer being utilized to be filled in without resulting in an increase to the building area. Additionally, areas of the building not normally occupied, such as exit stairs and elevator shafts, would not need to be counted into the building area. The logic being that those spaces do not contribute to the fire hazard in the building and some credit can be given for such spaces.

1102.3 Fire protection systems. Existing fire areas increased by the *addition* shall comply with Chapter 9 of the *International Building Code*.

❖ If an addition increases an existing building's fire area or areas to a level that is required to have fire protection systems by IBC Chapter 9, then those fire areas must comply with the new construction requirements, both in the addition and the existing building. The code provides no exceptions to the sprinkler threshold limits given in IBC Chapter 9.

Example: A restaurant is being increased from 4,000 square feet to 6,000 square feet (372 m^2 to 557 m^2), with no intervening fire barriers in the facility. IBC Section 903.2.1.2 requires sprinklers for this Group A-2 occupancy when the fire area is greater than 5,000 square feet (465 m^2). Therefore, with the new addition, the threshold of 5,000 square feet (465 m^2) is exceeded, and sprinklers are required throughout the building.

One area that is somewhat unclear is, where IBC Chapter 9 has a zero square footage threshold for sprinklers such as for Group R occupancies, whether the sprinklers would be required throughout the addition as well as the existing building or simply be limited to the addition.

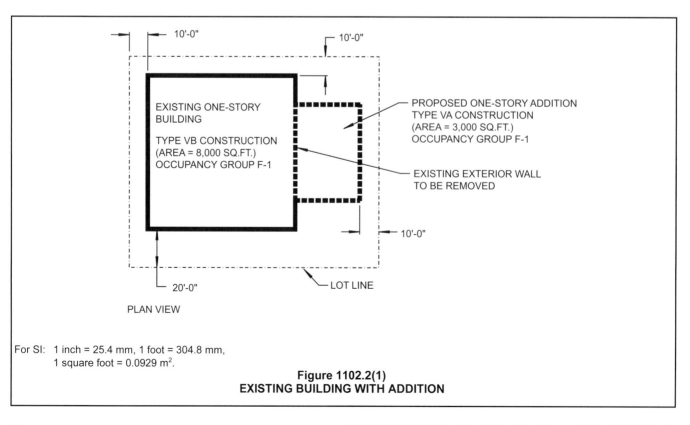

For SI: 1 inch = 25.4 mm, 1 foot = 304.8 mm, 1 square foot = 0.0929 m^2.

Figure 1102.2(1)
EXISTING BUILDING WITH ADDITION

SECTION 1103
STRUCTURAL

[BS] 1103.1 Compliance with the International Building Code. *Additions* to *existing buildings* or structures are new construction and shall comply with the *International Building Code*.

❖ An addition, by its very nature, is new construction so the addition itself must comply with the IBC requirements for new construction.

[BS] 1103.2 Additional gravity loads. Existing structural elements supporting any additional gravity loads as a result of additions shall comply with the *International Building Code*.

Exceptions:

1. Structural elements whose stress is not increased by more than 5 percent.

2. Buildings of Group R occupancy with no more than five dwelling units or sleeping units used solely for residential purposes where the *existing building* and the *addition* comply with the conventional light-frame construction methods of the *International Building Code* or the provisions of the *International Residential Code*.

❖ Unless the existing building was initially designed with future expansion in mind, it is very likely that an addition will impose loads on some portion of the existing structure that exceed the member's capacity and require reinforcement or other mitigation. In any event, where an addition results in added gravity load supported by an existing structure, the affected structural components must be checked to verify that they satisfy the requirements of the IBC for a new structure. Verification for gravity loading indicates compliance with the majority of Chapter 16 of the IBC except for seismic and wind loads.

There are two exceptions to full compliance with the IBC. One exception allows for an increase in gravity loads as long as the existing members affected by the increased loads do not experience an increase in stress of more than 5 percent. The other exception permits smaller Group R buildings and their additions that comply with either the *International Residential Code®* (IRC®) or the conventional light-frame construction provisions of the IBC. This is a prescriptive alternative to the loading provisions also permitted for new construction in IBC Chapter 16. See the limitations on these provisions in the IRC or IBC Chapter 23.

[BS] 1103.3 Lateral force-resisting system. The lateral force-resisting system of *existing buildings* to which additions are made shall comply with Sections 1103.3.1, 1103.3.2 and 1103.3.3.

Exceptions:

1. Buildings of Group R occupancy with no more than five dwelling units or sleeping units used solely for residential purposes where the *existing building* and the *addition* comply with the conventional light-frame construction methods of the *International Building Code* or the provisions of the *International Residential Code*.

2. Any existing lateral load-carrying structural element whose demand-capacity ratio with the addition considered is not more than 10 percent greater than its demand-capacity ratio with the addition ignored shall be permitted to remain unaltered. For purposes of this exception, comparisons of demand-capacity ratios and calculation of design lateral loads, forces and capacities shall account for the cumulative effects of additions and alterations since original construction. For purposes of calculating demand-capacity ratios, the demand shall consider applicable load combinations involving *International Building Code*-level seismic forces in accordance with Section 301.1.4.1.

❖ An addition to an existing building requires the lateral force-resisting system to comply with Section 1103.3.1, if a story is added or the height is increased; Section 1103.3.2, if the addition is horizontal such as an extension of the floor area; or Section 1103.3.3, if there is a voluntary addition of structural elements to the lateral force-resisting system.

There are two exceptions to the previously listed requirements. One exception permits smaller Group R buildings, with not more than five dwelling or sleeping units and their additions, that comply with either the IRC or the conventional light-frame construction provisions of the IBC. Exception 2 permits an addition that is not structurally independent without alteration to the seismic-force-resisting system of the existing structure, provided the demand-capacity ratio is increased no more than 10 percent in any structural element. The demand is determined using the load combinations including wind and seismic effects. By considering the demand-capacity ratio, any decrease in lateral resistance is accounted for as well. This exception is the same as that allowed for additions in Section 402.4 in the prescriptive method.

Previously, this exception was story based versus element based. The intent is that elements triggered for lateral upgrade by Section 1103.3.1 or 1103.3.2 should be exempt based on their individual demand-capacity ratios, not on the overall story shear. A focus on story shear can miss critical individual elements in vertical additions and can be difficult to define in the case of horizontal additions. The *SEAOC Blue Book* and the *Uniform Building Code®* define "Story shear" as "the summation of design lateral forces above the story under consideration."

[BS] 1103.3.1 Vertical addition. Any element of the lateral force-resisting system of an *existing building* subjected to an increase in vertical or lateral loads from the vertical *addition* shall comply with the *International Building Code* wind provisions and the *International Building Code*-level seismic forces specified in Section 301.1.4.1 of this code.

❖ Where the addition of a story or an increase in height imposes additional loads, either vertical or lateral, on portions of the existing lateral force-resisting system, this section requires that those members meet two specific lateral-load requirements: the wind-load requirements of the IBC and the IBC-level seismic provisions as specified in Section 301.1.4.1. Section 301.1.4.1 provides scoping language for using the IBC and ASCE 41 for seismic evaluation and design, which provides alternative options for compliance consistent with the overall goal of the code. Any element not meeting these provisions requires replacement, reinforcement or other measures in order to comply. See the commentary to Section 301.1.4.1 for more information.

[BS] 1103.3.2 Horizontal addition. Where horizontal *additions* are structurally connected to an existing structure, all lateral force-resisting elements of the existing structure affected by such *addition* shall comply with the *International Building Code* wind provisions and the IBC-level seismic forces specified in Section 301.1.4.1 of this code.

❖ A horizontal addition that is isolated from the existing structure is self-supporting and, therefore, has no impact on the existing structure. Where this is not the case, portions of the existing lateral force-resisting system affected by the addition are required to meet two specific lateral load requirements: the wind-load requirements of the IBC and the IBC-level seismic provisions as specified in Section 301.1.4.1. Section 301.1.4.1 provides scoping language for using the IBC and ASCE 41 for seismic evaluation and design, which provides alternative options for compliance consistent with the overall goal of the code. Any element not meeting these provisions requires replacement, reinforcement or other measures in order to comply. See the commentary to Section 301.1.4.1 for more information.

Compliance with the seismic load provisions of Section 301.1.4.1 necessitates consideration of many aspects of the new and existing construction. One of these is that in rigid diaphragm structures, the story shear is distributed according to the relative stiffness of the vertical elements of the lateral force system (rather than merely using the tributary area, as is permitted for flexible diaphragms). This requires consideration of torsional effects, as well as direct shear. This is illustrated in Commentary Figure 1103.3.2, which shows a horizontal addition is likely to affect the center of mass as well as the structure's center of rigidity. It is important, therefore, that the net effect of the resulting torsion on the combined structure be considered. This section simply reiterates what is already a required step in the seismic analysis of any rigid diaphragm structure.

[BS] 1103.3.3 Voluntary addition of structural elements to improve the lateral force-resisting system. Voluntary addition of structural elements to improve the lateral force-resisting system of an *existing building* shall comply with Section 807.6.

❖ Strictly speaking, the definition of "Addition" in Section 202 would not include adding structural elements voluntarily in order to improve a building's resistance to lateral forces. This is more likely to be considered an alteration. For good measure, this section cross references the provision under alterations that apply to voluntary upgrades.

[BS] 1103.4 Snow drift loads. Any structural element of an *existing building* subjected to additional loads from the effects of snow drift as a result of an *addition* shall comply with the *International Building Code*.

Exceptions:

1. Structural elements whose stress is not increased by more than 5 percent.

2. Buildings of Group R occupancy with no more than five dwelling units or sleeping units used solely for residential purposes where the *existing building* and the *addition* comply with the conventional light-frame construction methods of the *International Building Code* or the provisions of the *International Residential Code*.

❖ An addition with a roof level different than that of an adjoining existing roof necessitates the consideration of the potential for drifting snow. Commentary Figure 1103.4 illustrates this condition. If the addition is the portion of the building having the higher roof (side A in Commentary Figure 1103.4) then the potential impact on the existing roof structure (side B) is the greatest. If the addition is the lower portion of the building (side B), then the impact of drifting snow on the existing structure is limited only to those members, if any, of the existing structure that support the new roof. The magnitude of the drift loading that must be included is a function of building geometry. All things being equal, the larger the difference in roof heights, the higher the code-required drift loading.

IBC Chapter 16 requires the use of ASCE 7 procedures for snow drifts. The exceptions in this section of the code are similar to those of Section 1103.2, and

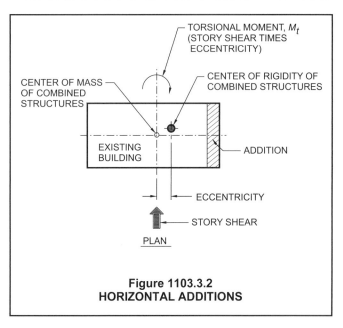

Figure 1103.3.2
HORIZONTAL ADDITIONS

could be construed as already requiring compliance with all the snow-load provisions of the IBC, including snow drifts. For good measure, this section makes it clear that the impact of this commonly occurring condition of existing elements must be considered.

[BS] 1103.5 Flood hazard areas. *Additions* and *foundations* in *flood hazard areas* shall comply with the following requirements:

1. For horizontal *additions* that are structurally interconnected to the *existing building*:

   1.1. If the *addition* and all other proposed work, when combined, constitute *substantial improvement*, the *existing building* and the *addition* shall comply with Section 1612 of the *International Building Code*, or Section R322 of the *International Residential Code*, as applicable.

   1.2. If the *addition* constitutes *substantial improvement*, the *existing building* and the *addition* shall comply with Section 1612 of the *International Building Code*, or Section R322 of the *International Residential Code*, as applicable.

2. For horizontal *additions* that are not structurally interconnected to the *existing building*:

   2.1. The *addition* shall comply with Section 1612 of the *International Building Code*, or Section R322 of the *International Residential Code*, as applicable.

   2.2. If the *addition* and all other proposed work, when combined, constitute *substantial improvement*, the *existing building* and the *addition* shall comply with Section 1612 of the *International Building Code*, or Section R322 of the *International Residential Code*, as applicable.

3. For vertical *additions* and all other proposed work that, when combined, constitute *substantial improvement*, the *existing building* shall comply with Section 1612 of the *International Building Code*, or Section R322 of the *International Residential Code*, as applicable.

4. For a raised or extended foundation, if the foundation work and all other proposed work, when combined, constitute *substantial improvement*, the *existing building* shall comply with Section 1612 of the *International Building Code*, or Section R322 of the *International Residential Code*, as applicable.

5. For a new foundation or replacement foundation, the foundation shall comply with Section 1612 of the *International Building Code* or Section R322 of the *International Residential Code*, as applicable.

❖ The definition of "Flood hazard area" provided in Section 202 is taken from the IBC and establishes where this provision applies. In these areas, additions that separately or in combination with other repairs or alterations are considered to be "substantial improvements" (see the definition of "Substantial improvement" in Section 202) require the entire structure to meet the flood-resistant provisions of IBC Section 1612 or IRC Section R322. Buildings are considered to have undergone substantial improvement when the cost of the improvement is 50 percent or more of its market value before the improvement is undertaken. A horizontal addition that is structurally separated must always comply with the flood-resistant provisions of the IBC or IRC even if it does not qualify as a substantial improvement, and this is merely a restatement of the requirement in Section 1103.1 for additions to comply with the IBC.

IBC Section 1612 and IRC Section R322 address requirements for buildings in designated flood hazard areas. The design and construction are required to be in accordance with ASCE 24, which in turn references the flood loading given in ASCE 7. Through use of these provisions, communities meet a significant portion of the flood plain management regulation requirements necessary to participate in the National

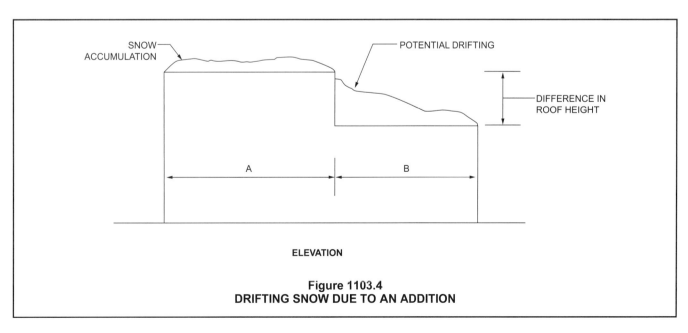

Figure 1103.4
DRIFTING SNOW DUE TO AN ADDITION

Flood Insurance Program (NFIP) (see the definition of "Substantial improvement" in Chapter 2). The requirements of Section 1103.5 correlate to the Federal Emergency Management Agency (FEMA) requirements for additions and existing buildings.

SECTION 1104
SMOKE ALARMS IN OCCUPANCY GROUPS R AND I-1

1104.1 Smoke alarms in existing portions of a building. Where an *addition* is made to a building or structure of a Group R or I-1 occupancy, the *existing building* shall be provided with smoke alarms as required by Section 1103.8 of the *International Fire Code* or Section R314 of the *International Residential Code* as applicable.

❖ Section 1104.1 requires that, for buildings containing Group R and I-1 occupancies, the existing building added to must comply with Section 1103.8 of the *International Fire Code*® (IFC®) or IRC Section R314 (if single family). Note that the addition, in accordance with Section 1101.1, would need to comply with the provisions in IBC Chapter 9 for new buildings pertaining to smoke alarms.

SECTION 1105
ACCESSIBILITY

1105.1 Minimum requirements. Accessibility provisions for new construction shall apply to additions. An addition that affects the accessibility to, or contains an area of, *primary function* shall comply with the requirements of Sections 705, 806 and 906, as applicable.

❖ Additions must comply with new construction. However, an addition is also an alteration to an existing building; therefore, accessible route provisions for existing buildings are applicable (see commentary, Sections 705 and 806).

Example: A new dining area is added to a restaurant. All accessible elements in the addition must be constructed accessible. If the accessible route to the addition is through the existing building, the route must be evaluated to see if elements need to be altered to meet accessibility requirements. In addition, any toilet rooms or drinking fountains that serve the addition would have to be evaluated for accessibility. This would include the route to these elements, the toilet rooms themselves, the fixtures in the toilet room and the drinking fountains.

1105.2 Accessible dwelling units and sleeping units. Where Group I-1, I-2, I-3, R-1, R-2 or R-4 dwelling or sleeping units are being added, the requirements of Section 1107 of the *International Building Code* for accessible units apply only to the quantity of spaces being added.

❖ This section sets forth the rate for providing accessible dwelling or sleeping units in Groups I-1, I-2, I-3, R-1, R-2 and R-4 where such facilities are being added. Assuming that the required number of accessible units is not already provided, the number of accessible units to be incorporated is based on the number of dwelling and sleeping units being added. The technical criteria for accessible units is found in ICC/ANSI A117.1-2010, Chapters 3 through 9.

Example: A nursing home is expanding through an addition that will have 30 sleeping units. Based on that number, IBC Section 1107.5.2.1 would require at least 50 percent to be accessible units. Therefore, 15 sleeping units would be required to be accessible.

1105.3 Type A dwelling or sleeping units. Where more than 20 Group R-2 dwelling or sleeping units are being added, the requirements of Section 1107 of the *International Building Code* for Type A units and Chapter 9 of the *International Building Code* for visible alarms apply only to the quantity of the spaces being added.

❖ This section sets forth the rate for providing Type A dwelling or sleeping units in Group R-2 where such facilities are added. The number of Type A units to be incorporated is based on the number being added. For example, if 24 units are being added to a Group R-2 apartment building, 2 percent of the 24 units being altered would be required to be designed to conform to Type A unit requirements. This would require only one unit to be added. The technical criteria for Type A units is found in ICC/ANSI A117.1-2010, Chapter 10. This section also references visible and audible alarm requirements in IBC Chapter 9. IBC Section 907.5.2.3.4 contains requirements for alarms in Group R-2 units. The visible alarm requirements would only require that such units have the capability to support the notification appliances of the visible alarms.

1105.4 Type B dwelling or sleeping units. Where four or more Group I-1, I-2, R-1, R-2, R-3 or R-4 dwelling or sleeping units are being added, the requirements of Section 1107 of the *International Building Code* for Type B units and Chapter 9 of the *International Building Code* for visible alarms apply only to the quantity of spaces being added.

❖ This section sets forth the rate for providing Type B dwelling or sleeping units in Groups I-1, I-2, I-3, R-1, R-2 and R-4 where such facilities are added to a building. The number of Type B units is based on the number being added. For example, if a nursing home was adding six sleeping units, those six units would be required to meet Type B requirements. The dwelling or sleeping units in the existing building do not need to meet Type B requirements. The technical criteria for Type B units is found in ICC A117.1, Chapter 10. This section also references visible and audible alarm requirements in IBC Chapter 9 for only the sleeping or dwelling units being added. Sleeping accommodations in Groups I-1 and R-1 are required to have visible alarms in accordance with IBC Table 907.5.2.3.2. An addition with 20 new sleeping or dwelling units in a Group R-1 would require four of the units to be equipped with visible alarms. IBC Section 907.5.2.3.3 also contains requirements for alarms in Group R-2 units but focuses on future capabilities of such systems to support visible alarms.

SECTION 1106
ENERGY CONSERVATION

1106.1 Minimum requirements. *Additions* to *existing buildings* shall conform to the energy requirements of the *International Energy Conservation Code* or *International Residential Code* as they relate to new construction.

❖ This section is merely a pointer to the applicable requirements of the *International Energy Conservation Code*® (IECC®) or IRC, as applicable, with regard to energy conservation. No additional requirements are provided here; simply a reminder that requirements exist in other codes for these additions.

Chapter 12:
Historic Buildings

General Comments

This section provides some overall exceptions from code requirements when the building in question has historic value. The most important criterion for the application of this section is that the building must be essentially accredited as being of historic significance by a qualified third party or agency. Usually, this is done by a state or local authority after careful review of the historical value of the building. Most, if not all, states have such authorities, as do many local jurisdictions. The agencies with such authority can be located at the state or local government level or through the local chapter of the American Institute of Architects (AIA). Other considerations include the structural condition of the building (i.e., whether the building is structurally sound), its proposed use, its impact on life safety and how the intent, if not the letter, of the code will be achieved.

Purpose

A large number of existing buildings and structures do not comply with current building code requirements for new construction. Although many of these buildings are potentially salvageable, rehabilitation is often cost prohibitive because they may not be able to comply with all of the requirements for new construction. At the same time, it is necessary to regulate construction in existing buildings that undergo additions, alterations, renovations, extensive repairs or change of occupancy. Such activity represents an opportunity to ensure that new construction complies with current building codes and that existing conditions are maintained, at a minimum, to their current level of compliance or are improved as required. To accomplish this objective, and to make the rehabilitation process easier, this chapter allows for a controlled departure from full compliance with the technical codes, without compromising the minimum standards for fire prevention and life safety features of the rehabilitated building.

SECTION 1201
GENERAL

1201.1 Scope. It is the intent of this chapter to provide means for the preservation of *historic buildings*. Historical buildings shall comply with the provisions of this chapter relating to their *repair*, *alteration*, relocation and *change of occupancy*.

❖ This section establishes that this chapter deals with the requirements for the repair, alteration, relocation and change of occupancy of buildings previously defined by the code as historic. Note that the code would not regard a reconstruction of a historic building as qualifying for treatment under this chapter, even if the building is reconstructed entirely from original, stored materials and assemblies.

[BS] 1201.2 Report. A *historic building* undergoing *repair*, *alteration*, or *change of occupancy* shall be investigated and evaluated. If it is intended that the building meet the requirements of this chapter, a written report shall be prepared and filed with the *code official* by a registered design professional when such a report is necessary in the opinion of the *code official*. Such report shall be in accordance with Chapter 1 and shall identify each required safety feature that is in compliance with this chapter and where compliance with other chapters of these provisions would be damaging to the contributing historic features. For buildings assigned to Seismic Design Category D, E or F, a structural evaluation describing, at a minimum, the vertical and horizontal elements of the lateral force-resisting system and any strengths or weaknesses therein shall be prepared. Additionally, the report shall describe each feature that is not in compliance with these provisions and shall demonstrate how the intent of these provisions is complied with in providing an equivalent level of safety.

❖ This section provides for a report that acts as a technical backup for the code official's judgment. The intent is to be able to retain a record of the decisions and the basis for those decisions for future reference.

The definition of "Historic building" refers to listing/eligibility for national register, state designation, inclusion in a survey or contributing to a national or state historic district. Contributing features may be limited to those features in nominations, historic structure reports or other such documents. This represents a systematized approach for evaluating proposals of equivalent performance (see the definition of "Historic building" in Section 202).

1201.3 Special occupancy exceptions—museums. When a building in Group R-3 is also used for Group A, B, or M purposes such as museum tours, exhibits, and other public assembly activities, or for museums less than 3,000 square feet (279 m^2), the *code official* may determine that the occupancy is Group B when life-safety conditions can be demonstrated in accordance with Section 1201.2. Adequate means of egress in such buildings, which may include a means of maintaining doors in an open position to permit egress, a limit on building occupancy to an occupant load permitted by the means of egress capacity, a limit on occupancy of certain areas or floors, or supervision by a person knowledgeable in the emergency exiting procedures, shall be provided.

❖ This section acknowledges that dwellings used as museums and other public assembly activities are a special case.

The provision allows a small museum or historic house in a museum use to demonstrate life safety using operational controls that would not be permitted in most other uses. This is likely to be effective because of the high degree of supervision that would be necessary for security reasons in this occupancy. In this case, the building occupancy is classified as Occupancy Group B even though it may contain assembly or mercantile uses. It would also permit the occasional use of a residential building for tours and museum use.

[BS] 1201.4 Flood hazard areas. In *flood hazard areas*, if all proposed work, including repairs, work required because of a *change of occupancy*, and *alterations*, constitutes *substantial improvement*, then the *existing building* shall comply with Section 1612 of the *International Building Code*, or Section R322 of the *International Residential Code*, as applicable.

Exception: If an *historic building* will continue to be an *historic building* after the proposed work is completed, then the proposed work is not considered a *substantial improvement*. For the purposes of this exception, an *historic building* is:

1. Listed or preliminarily determined to be eligible for listing in the National Register of Historic Places;
2. Determined by the Secretary of the U.S. Department of Interior to contribute to the historical significance of a registered historic district or a district preliminarily determined to qualify as a historic district; or
3. Designated as historic under a state or local historic preservation program that is approved by the Department of Interior.

❖ In flood areas, historic buildings that meet one of the exceptions are not required to comply with the flood provisions, including the substantial improvement provision, as long as the building continues to be designated as historic. However, if plans to substantially improve a historic building would result in the loss of its designation as a "historic building," the building would be required to be brought into compliance with the requirements for flood hazard areas. The code official may require applicants to consult with the appropriate historic preservation authority to determine whether proposed work will jeopardize a structure's historic designation. To the extent practicable, owners of historic buildings may wish to incorporate flood-resistant provisions when such buildings undergo repairs and alterations. If the work on a historic structure does not cause the loss of its eligibility as a historic building as a result of the proposed work, then the work would not be considered as a substantial improvement; therefore, the building would not have to comply with Section 1612 of the *International Building Code*® (IBC®) or Section R322 of the *International Residential Code*® (IRC®), as applicable.

SECTION 1202
REPAIRS

1202.1 General. Repairs to any portion of an *historic building* or structure shall be permitted with original or like materials and original methods of construction, subject to the provisions of this chapter. Hazardous materials, such as asbestos and lead-based paint, shall not be used where the code for new construction would not permit their use in buildings of similar occupancy, purpose and location.

❖ Repair with like materials may be needed for aesthetic reasons or to meet material conservation needs. It is important to note that original or like materials can be used, but these materials cannot contain substances such as asbestos that are hazardous and would not be permitted under the current building code for new construction.

1202.2 Unsafe conditions. Conditions determined by the *code official* to be *unsafe* shall be remedied. No work shall be required beyond what is required to remedy the *unsafe* conditions.

❖ This section requires the repair of identified unsafe conditions in historic buildings without triggering secondary requirements. It is important to note the term "unsafe" in this section as it is specifically defined in Section 202. It is a more general term than the term "dangerous," which is specific to structural conditions.

1202.3 Relocated buildings. Foundations of relocated *historic buildings* and structures shall comply with the *International Building Code*. Relocated *historic buildings* shall otherwise be considered an *historic building* for the purposes of this code. Relocated *historic buildings* and structures shall be sited so that exterior wall and opening requirements comply with the *International Building Code* or with the compliance alternatives of this code.

❖ The foundation of a relocated building is a new structure; there is no reason why it should not comply with provisions for new construction. However, compared with code provisions that treat a relocated building as a new building, the rest of the building is still treated as a historic structure. In relocation, siting is, to some degree, a new design decision. Relationship of walls and openings to property lines is, therefore, reasonably able to be brought into compliance with the fire separation requirements of the code.

1202.4 Replacement. Replacement of existing or missing features using original materials shall be permitted. Partial replacement for repairs that match the original in configuration, height, and size shall be permitted.

Replacement glazing in hazardous locations shall comply with the safety glazing requirements of Chapter 24 of the *International Building Code*.

Exception: Glass block walls, louvered windows, and jalousies repaired with like materials.

❖ In some cases, historic materials have not been tested for fire resistance or structural performance, but have been shown to perform adequately when in use. An excellent example would be a railing, where spacing, height or end conditions would not meet modern requirements. If sections of a railing are missing, or if repairs are needed, the railing is not required to meet the code provisions for new construction; however, the second paragraph of this section requires upgrading to the safety glazing requirements of IBC Chapter 24 since there is an established hazard and the cost to fabricate a historically correct reproduction of a proper material is negligible.

SECTION 1203
FIRE SAFETY

1203.1 Scope. *Historic buildings* undergoing *alterations*, *changes of occupancy*, or that are moved shall comply with Section 1203.

❖ This section recognizes the unique aspects associated with older historical structures. It reinforces the importance of the means of egress pathway while granting the code official some latitude in accepting some degree of variance in the egress components. (e.g., direction of exit door swing, egress path width and height).

1203.2 General. Every *historic building* that does not conform to the construction requirements specified in this code for the occupancy or use and that constitutes a distinct fire hazard as defined herein shall be provided with an approved automatic fire-extinguishing system as determined appropriate by the *code official*. However, an automatic fire-extinguishing system shall not be used to substitute for, or act as an alternative to, the required number of exits from any *facility*.

❖ Fire-extinguishing systems are effective substitutes for some requirements that are typical for new construction, particularly passive systems such as rated doors and corridors. By slowing or suppressing the development of a fire, a sprinkler system will make passive fire resistance unnecessary. This section also establishes that an automatic sprinkler system is not considered appropriate as a substitution for a reduction in the required number of exits.

1203.3 Means of egress. Existing door openings and corridor and stairway widths less than those specified elsewhere in this code may be approved, provided that, in the opinion of the *code official*, there is sufficient width and height for a person to pass through the opening or traverse the means of egress. When approved by the *code official*, the front or main exit doors need not swing in the direction of the path of exit travel, provided that other approved means of egress having sufficient capacity to serve the total occupant load are provided.

❖ This provision would permit the continuance of a structure no more hazardous than before rehabilitation, with minimum standards of usability. Provisions for new construction would require that exit doors swing in the direction of exit travel for an occupant load exceeding 50.

1203.4 Transoms. In fully sprinklered buildings of Group R-1, R-2 or R-3 occupancy, existing transoms in corridors and other fire-resistance-rated walls may be maintained if fixed in the closed position. A sprinkler shall be installed on each side of the transom.

❖ This permits the retention of nonwired/nonrated glass in historic transoms in rated walls in residential occupancies where protected by an automatic fire sprinkler system throughout the building. This section requires the transoms to be closed and specific protection on both sides of the transoms. This approach basically affords a level of safety equivalent to the passive protection provided by a rated corridor or other rated construction.

1203.5 Interior finishes. The existing finishes of walls and ceilings shall be accepted when it is demonstrated that they are the historic finishes.

❖ While existing wood paneling or other finishes may not meet flame spread requirements for walls or ceilings, this section allows the finishes to remain in place if they can be shown to be historic in nature. If not, there should be no reason not to address the flame spread rating of the materials since the sensitivity of the historic nature of the building is no longer an issue.

1203.6 Stairway enclosure. In buildings of three stories or less, exit enclosure construction shall limit the spread of smoke by the use of tight-fitting doors and solid elements. Such elements are not required to have a fire-resistance rating.

❖ Enclosure of stairways to control smoke would provide an improvement, but permitting the enclosure to be nonrated would allow for use of traditional materials. Example enclosures include plain or wired glass, smoke-activated doors and similar assemblies. This provides for flexibility, but requires at least a minimum level of passive smoke control to protect stairways for exiting.

1203.7 One-hour fire-resistant assemblies. Where 1-hour fire-resistance-rated construction is required by these provisions, it need not be provided, regardless of construction or occupancy, where the existing wall and ceiling finish is wood or metal lath and plaster.

❖ The substitution of standard, old-fashioned lath and plaster for 1-hour-rated wall construction is a well-established alternative and is considered as meeting the intent of the code to provide a safe path for exit.

1203.8 Glazing in fire-resistance-rated systems. Historic glazing materials are permitted in interior walls required to have a 1-hour fire-resistance rating where the opening is provided with approved smoke seals and the area affected is provided with an automatic sprinkler system.

❖ Glazing of interior partitions can be vulnerable because of potential leakage around the edge of the glazing or because of heat-induced glass breakage. This provision addresses both concerns. The sprinkler system should be provided on both sides of the wall containing the glazing.

1203.9 Stairway railings. Grand stairways shall be accepted without complying with the handrail and guard requirements. Existing handrails and guards at all stairways shall be permitted to remain, provided they are not structurally *dangerous*.

❖ New construction requirements for handrail and guardrail height have increased over the years. However, since an ornamental railing is typically difficult to modify, particularly without significant architectural change, the code permits historic railings that are essentially safe to remain. This also would allow handrail profiles that may not meet current grip requirements, provided they are not structurally dangerous.

1203.10 Guards. Guards shall comply with Sections 1203.10.1 and 1203.10.2.

❖ This section establishes that guards in existing historic structures can remain in the configuration in which they were originally constructed. Guards would still have to conform to the restrictions in effect at the time the building was built.

1203.10.1 Height. Existing guards shall comply with the requirements of Section 604.

❖ Section 604 requires that repairs maintain the existing level of protection.

1203.10.2 Guard openings. The spacing between existing intermediate railings or openings in existing ornamental patterns shall be accepted. Missing elements or members of a guard may be replaced in a manner that will preserve the historic appearance of the building or structure.

❖ Requirements for voids in railings and grilles have become more stringent over the years. The spacing of a balustrade or ornamental pattern of a grille is frequently a significant contributor to the building's historic character. This provision enables the retention and repair of these elements.

1203.11 Exit signs. Where exit sign or egress path marking location would damage the historic character of the building, alternative exit signs are permitted with approval of the *code official*. Alternative signs shall identify the exits and egress path.

❖ Signs and exit path identification are still required, but location is permitted to be flexible within the limits of functionality.

1203.12 Automatic fire-extinguishing systems. Every historical building that cannot be made to conform to the construction requirements specified in the *International Building Code* for the occupancy or use and that constitutes a distinct fire hazard shall be deemed to be in compliance if provided with an approved automatic fire-extinguishing system.

Exception: When the *code official* approves an alternative life-safety system.

❖ This term is the generic name for all types of automatic fire-extinguishing systems, including the most common type: the automatic sprinkler system. See IBC Section 904 for requirements for particular alternative automatic fire-extinguishing systems, such as wet-chemical, dry-chemical, foam, carbon dioxide, halon and clean agent systems.

Note that the alternative is specific to the construction requirements (i.e., height and area limitations, construction type and passive fire resistance). Section 1203.2 states that an automatic fire-extinguishing system is not a substitute for the reduction in the number of exits from a facility.

The exception would allow the utilization of an alternative life safety system; however, it would have to first be approved by the code official or the local authority having jurisdiction.

SECTION 1204
ALTERATIONS

1204.1 Accessibility requirements. The provisions of Sections 705, 806 and 906, as applicable, shall apply to facilities designated as historic structures that undergo *alterations*, unless *technically infeasible*. Where compliance with the requirements for accessible routes, entrances or toilet rooms would threaten or destroy the historic significance of the building or *facility*, as determined by the *code official*, the alternative requirements of Sections 1204.1.1 through 1204.1.4 for that element shall be permitted.

Exception: Type B dwelling or sleeping units required by Section 1107 of the *International Building Code* are not required to be provided in historical buildings.

❖ The regulations of individual adopting jurisdictions may provide additional guidance or limitations on the use of this section. The basic Americans with Disabilities Act (ADA) requirements permit exceptions and alternatives for historic buildings. These minimum requirements provide reasonable accommodations for building users.

For this section to be applicable, the building must be registered as historic. Historic buildings are treated much the same as provided for in Sections 705, 806 and 906, in that a historic building that is altered is expected to comply with accessibility requirements, unless technical infeasibility can be demonstrated. However, this section also goes on to acknowledge that the historic character of a building may be adversely affected by strict compliance with accessibility provisions. For example, compliance with door width requirements may necessitate the removal of an existing set of doors that is critical to the historic character of the building. To assist the code official in determining if the required provisions are detrimental to the historic significance, recommended guidelines have been incorporated into Appendix B, Section B101.

This section is intended to exempt such conditions in order to maintain the historic character of the building. Because some degree of accessibility is desired in all facilities, Sections 1204.1.1 through 1204.1.4 allow for alternatives. In addition, there is an exception that would exempt the Type B dwelling or sleeping unit requirements for historical buildings, including those undergoing a Level 3 alteration. This exception is important in that owners will not need to demonstrate that the addition of such units will threaten or destroy the historical character of the building.

In an effort to inform designers and owners of federal regulations that are not typically handled through the code enforcement process, information on the *Americans with Disabilities Act Accessibility Guidelines* (ADAAG) requirements for displays associated with historical buildings have been included in Appendix B, Section B101.5.

1204.1.1 Site arrival points. At least one accessible route from a site arrival point to an *accessible* entrance shall be provided.

❖ Full compliance would require an accessible route from all site arrival points. If this requirement would adversely affect the historical significance of the building, the available alternative is to provide an accessible route from one site arrival point to an accessible entrance.

1204.1.2 Multilevel buildings and facilities. An accessible route from an accessible entrance to public spaces on the level of the accessible entrance shall be provided.

❖ It is not required in building alteration that accessibility to spaces above or below the level of accessible entrance be provided. Full compliance for new construction might require an accessible route to levels above or below, as well as throughout, the entrance level. If this requirement would adversely affect the historical significance of the building, the alternative is to provide an accessible route from the accessible entrance to all spaces open to the public on the entrance level. If elevators are provided, but are not accessible, signage in accordance with IBC Section 1110.2 is required.

1204.1.3 Entrances. At least one main entrance shall be accessible.

Exceptions:

1. If a main entrance cannot be made accessible, an accessible nonpublic entrance that is unlocked while the building is occupied shall be provided; or

2. If a main entrance cannot be made accessible, a locked accessible entrance with a notification system or remote monitoring shall be provided.

❖ Although uniform building access is the rule for new construction, older buildings may have main entrances that are both inaccessible and a significant element of the building's character. In these cases, providing access by an alternative route is deemed to meet the primary intent of the accessibility regulations.

Full compliance would require 60 percent of the public entrances to be accessible, as well as any of the special types of entrances listed in Section 1205. If this requirement would adversely affect the historical significance of the building, only one main entrance is required to be made accessible. If a main entrance cannot be made accessible, then an employee or service entrance may serve as the accessible entrance, provided that it remains unlocked when the building is open. Alternatively, if the entrance is locked, some type of notification system (e.g., doorbell, intercom) or remote monitoring (e.g., security camera) is provided so that a person inside the facility would know to go to that entrance to admit someone. Signage must be provided at accessible and inaccessible entrances in accordance with IBC Sections 1110.1 and 1110.2.

1204.1.4 Toilet and bathing facilities. Where toilet rooms are provided, at least one accessible family or assisted-use toilet room complying with Section 1109.2.1 of the *International Building Code* shall be provided.

❖ If altering the existing facilities to be accessible would adversely affect the historical significance of the building, only one unisex bathroom that complies with the accessible family or assisted-use toilet/bathing requirements in IBC Section 1109.2.1 is required. Signage must be provided at inaccessible toilet rooms in accordance with IBC Section 1110.2.

SECTION 1205
CHANGE OF OCCUPANCY

1205.1 General. *Historic buildings* undergoing a *change of occupancy* shall comply with the applicable provisions of Chapter 10, except as specifically permitted in this chapter. When Chapter 10 requires compliance with specific requirements of Chapter 7, Chapter 8 or Chapter 9 and when those requirements are subject to the exceptions in Section 1202, the same exceptions shall apply to this section.

❖ Occupancy choice ideally minimizes code conflicts and the necessity for change. According to the Secretary of the Interior's *Standards for Treatment of Historic Properties*, "A property will be used as it was historically, or be given a new use that maximizes the retention of distinctive materials, features, spaces and spatial relationships." However, economic pressures may dictate that a structure be used in a different occupancy than that for which it was intended.

1205.2 Building area. The allowable floor area for *historic buildings* undergoing a *change of occupancy* shall be permitted to exceed by 20 percent the allowable areas specified in Chapter 5 of the *International Building Code*.

❖ This section permits an allowable floor area in excess of 20 percent as required by the IBC. This is critical since changing the type of construction to be able to meet the current building area requirements may be very difficult and will likely affect the historical features of the building. This section does not prohibit the use of area increases from automatic sprinkler systems and frontage.

1205.3 Location on property. Historic structures undergoing a change of use to a higher hazard category in accordance with Section 1012.6 may use alternative methods to comply with the fire-resistance and exterior opening protective requirements. Such alternatives shall comply with Section 1201.2.

❖ When changing to a higher hazard category, a building should be evaluated for means to manage the new or increased hazards. Literal code compliance may not be the only way, or the best way, to accomplish this objective in a historic building.

1205.4 Occupancy separation. Required occupancy separations of 1 hour may be omitted when the building is provided with an approved automatic sprinkler system throughout.

❖ Substitution of automatic extinguishing systems for passive protection is established as an acceptable procedure.

1205.5 Roof covering. Regardless of occupancy or use group, roof-covering materials not less than Class C, when tested in accordance with ASTM E108 or UL 790, shall be permitted where a fire-retardant roof covering is required.

❖ Roof covering requirements are a spread-of-fire issue: first, to prevent fire spread to the building in question and, second, to keep burning brands generated by a fire in that building from igniting other buildings in a general conflagration. However, if a limited number of buildings do not meet this requirement, the hazard is not significantly increased.

1205.6 Means of egress. Existing door openings and corridor and stairway widths less than those that would be acceptable for nonhistoric buildings under these provisions shall be approved, provided that, in the opinion of the *code official*, there is sufficient width and height for a person to pass through the opening or traverse the exit and that the capacity of the exit system is adequate for the occupant load, or where other operational controls to limit occupancy are approved by the *code official*.

❖ This provision would permit the continued use of a structure that is no more hazardous than before rehabilitation, with minimum standards of usability.

1205.7 Door swing. When approved by the *code official*, existing front doors need not swing in the direction of exit travel, provided that other approved exits having sufficient capacity to serve the total occupant load are provided.

❖ Provisions for new construction would require that exit doors swing in the direction of exit travel for an occupant load exceeding 50. This section prescribes a method by which existing entry doors can be allowed to swing into the building rather than in the direction of egress travel.

1205.8 Transoms. In corridor walls required by these provisions to be fire-resistance rated, existing transoms may be maintained if fixed in the closed position, and fixed wired glass set in a steel frame or other approved glazing shall be installed on one side of the transom.

Exception: Transoms conforming to Section 1203.4 shall be accepted.

❖ These alternatives would be used primarily by nonresidential occupancies, as residential occupancies would use Section 1203.4.

1205.9 Finishes. Where interior finish materials are required to have a flame spread index of Class C or better, when tested in accordance with ASTM E84 or UL 723, existing nonconforming materials shall be surfaced with approved fire-retardant paint or finish.

Exception: Existing nonconforming materials need not be surfaced with an approved fire-retardant paint or finish where the building is equipped throughout with an automatic sprinkler system installed in accordance with the *International Building Code* and the nonconforming materials can be substantiated as being historic in character.

❖ This section sets a minimum flame spread index for interior finishes in historic structures. As an option, fire-retardant paint can be used to make the current historic materials comply. Fire retardants have the effect of slowing the development and spread of fire. The exception, which requires an automatic sprinkler system, can be used where conditions are unsuitable for the use of intumescent paint or varnish or other fire-retardant finishes.

1205.10 One-hour fire-resistant assemblies. Where 1-hour fire-resistance-rated construction is required by these provisions, it need not be provided, regardless of construction or occupancy, where the existing wall and ceiling finish is wood lath and plaster.

❖ The substitution of standard lath and plaster construction for 1-hour fire-resistance-rated wall construction is a well-established alternative. Pressed tin ceilings are an example of this issue. Some State Historic Preservation Offices (SHPO) may provide additional guidance on this subject and alternatives.

1205.11 Stairways and guards. Existing stairways shall comply with the requirements of these provisions. The *code official* shall grant alternatives for stairways and guards if alternative stairways are found to be acceptable or are judged to meet the intent of these provisions. Existing stairways shall comply with Section 1203.

> **Exception:** For buildings less than 3,000 square feet (279 m^2), existing conditions are permitted to remain at all stairways and guards.

❖ This provision gives an opportunity to analyze the stairways' functionality as an exit, and to alter only those elements that are judged to be unsafe or inadequate. The exception allows existing stairway conditions to remain for smaller buildings, taking into account the shorter time required for egress.

1205.12 Exit signs. The *code official* may accept alternative exit sign locations where such signs would damage the historic character of the building or structure. Such signs shall identify the exits and exit path.

❖ Signs and exit path identification are still required, but location is permitted to be flexible within limits of functionality.

[BS] 1205.13 Exit stair live load. Existing historic stairways in buildings changed to a Group R-1 or R-2 occupancy shall be accepted where it can be shown that the stairway can support a 75-pounds-per-square-foot (366 kg/m^2) live load.

❖ In the case of a building changed to a Group R occupancy, the likelihood is that stairs will be designed to a higher capacity than use would require. To substantiate that likelihood, structural calculations or other data using the structural properties of the existing stair construction could be submitted to the code official for approval.

1205.14 Natural light. When it is determined by the *code official* that compliance with the natural light requirements of Section 1011.1 will lead to loss of historic character or historic materials in the building, the existing level of natural lighting shall be considered acceptable.

❖ At the time the provisions requiring natural light in residential occupancies were enacted, they were a necessary public health reform. At this time, deviations from these requirements would probably not have health consequences.

1205.15 Accessibility requirements. The provisions of Section 1012.8 shall apply to facilities designated as historic structures that undergo a *change of occupancy*, unless *technically infeasible*. Where compliance with the requirements for accessible routes, ramps, entrances, or toilet rooms would threaten or destroy the historic significance of the building or *facility*, as determined by the authority having jurisdiction, the alternative requirements of Sections 1204.1.1 through 1204.1.4 for those elements shall be permitted

> **Exception:** Type B dwelling or sleeping units required by Section 1107 of the *International Building Code* are not required to be provided in historical buildings.

❖ The regulations of individual adopting jurisdictions may provide additional guidance or limitations on the use of this section. The basic ADA requirements permit exceptions and alternatives for historic buildings. These minimum requirements provide reasonable accommodations for building users.

For this section to be applicable, the building must be registered as historic. Historic buildings are treated much the same as provided for in Section 1012.8, in that a historic building that has undergone a change of occupancy is expected to comply with accessibility requirements, unless technical infeasibility can be demonstrated. However, this section also goes on to acknowledge that the historic character of a building may be adversely affected by strict compliance with accessibility provisions. For example, compliance with door width requirements may necessitate the removal of an existing set of doors that is critical to the historic character of the building. To assist the code official in determining if the required provisions are detrimental to the historic significance, recommended guidelines have been incorporated into Appendix B, Section B101.

This section is intended to exempt such conditions in order to maintain the historic character of the building. Because some degree of accessibility is desired in all facilities, Sections 1204.1.1 through 1204.1.4 allow for alternatives. In addition, there is an exception that would exempt the Type B dwelling or sleeping unit requirements for historical buildings, including those undergoing a Level 3 alteration. This exception is important in that owners will not need to demonstrate that the addition of such units will threaten or destroy the historical character of the building.

In an effort to inform designers and owners of federal regulations that are not typically handled through the code enforcement process, information on the ADAAG requirements for displays associated with historical buildings have been included in Appendix B, Section B101.5.

SECTION 1206
STRUCTURAL

[BS] 1206.1 General. *Historic buildings* shall comply with the applicable structural provisions for the work as classified in Chapter 5.

> **Exception:** The *code official* shall be authorized to accept existing floors and approve operational controls that limit the live load on any such floor.

❖ Any proposed work on historic buildings must comply with the structural provisions that apply based on the nature of that work. In other words, the classification of any proposed work is the first consideration. The only exception to this authorizes the code official to accept existing floors and approve operational controls that limit the live load on any such floor. This could be achieved by posting load limit signs in conspicuous locations. Actual loads on floors are typically less than the code-required loads. This section permits consideration of the actual loads as opposed to the probabilistic loads used for new design. Section 1201.2 allows alternative approaches through the evaluation and report.

[BS] 1206.2 Dangerous conditions. Conditions determined by the *code official* to be *dangerous* shall be remedied. No work shall be required beyond what is required to remedy the *dangerous* condition.

❖ The definition of "Dangerous" in Section 202 gives five conditions under which a building, a portion of a building or an individual structural component is considered dangerous. Any building or structural component that is determined to be dangerous presents an unacceptable risk to public safety. Section 115.1 clarifies that a building considered dangerous is deemed to be unsafe, and in order for an unsafe building to remain in place the building must be made safe.

This section clarifies that in a historic building, any structural repair, strengthening or replacement applies only to the specific component(s) or portion(s) necessary to correct the dangerous condition(s).

Bibliography

The following resource materials were used in the preparation of the commentary for this chapter of the code.

2010 *ADA Standard for Accessible Design*. Washington, DC: Department of Justice, September 15, 2010.

36 CFR, Parts 1190 and 1191 Final Rule, *The Americans with Disabilities Act (ADA) Accessibility Guidelines; Architectural Barriers Act (ABA) Accessibility Guidelines*. Washington, DC: Architectural and Transportation Barriers Compliance Board, July 23, 2004.

Standards for Treatment of Historic Properties. Washington, DC: Secretary of the Interior, 1992.

Chapter 13:
Relocated or Moved Buildings

General Comments

Chapter 13 is applicable to any building that is moved or relocated. This also includes relocatable buildings. Relocatable buildings are purposely built with the intention of being used multiple times at different sites. Relocatable buildings are defined in Chapter 2. The relocation of a building will automatically cause an inspection and evaluation process that enables the jurisdiction to determine the level of compliance with the *International Fire Code*® (IFC®) and the *International Property Maintenance Code*® (IPMC®). These two codes, by their scope, are applicable to existing buildings. Note the IFC has provisions that also apply to new construction in addition to existing buildings and facilities. The intent of this reference is to review the applicability of IFC Section 102 to determine the level of compliance necessary. For example, IFC Section 102.1, Item 3 requires that the construction and design provisions of Chapter 11 apply to existing buildings. This is the case regardless of any repair, remodeling, alteration work or change of occupancy occurring (see the IFC and IPMC).

Section 1302 addresses the building location of lot and structural considerations. Note that electrical, plumbing, and heating, ventilating and air conditioning (HVAC) are not addressed for relocated buildings. This means that relocated buildings are not necessarily required to improve the existing electrical, plumbing or HVAC systems beyond what is necessary for the appropriate interface with the public utilities at the new location, unless the code official discovers an unsafe or hazardous condition.

Section 1302.7 contains an inspection requirement that helps to determine the structural integrity of the structure's components and connections. The inspection will help the code official to properly evaluate a building to ensure it is safe for occupancy.

If the project scope involves repairs, alterations or a change of occupancy, in addition to moving or relocating the building, then it will also have to comply with the provisions of the chapters related to repair, alteration or change of occupancy. Any new construction of accessory elements to include decks, stairs, fences or accessory buildings must comply with the applicable code requirements for new construction.

Purpose

A large number of existing buildings and structures would not be able to comply with the current building code requirements for new construction. Although many of these buildings are potentially salvageable, relocation and rehabilitation is often cost prohibitive. It is necessary to regulate construction in existing buildings that undergo additions, alterations, renovations, extensive repairs or change of occupancy to ensure the safety of the public.

SECTION 1301
GENERAL

1301.1 Scope. This chapter provides requirements for relocated or moved structures, including relocatable buildings as defined in Chapter 2.

❖ This section states the scope of Chapter 13 and references the requirements for relocated or moved structures.

1301.2 Conformance. The building shall be safe for human occupancy as determined by the *International Fire Code* and the *International Property Maintenance Code*. Any *repair, alteration,* or *change of occupancy* undertaken within the moved structure shall comply with the requirements of this code applicable to the work being performed. Any field-fabricated elements shall comply with the requirements of the *International Building Code* or the *International Residential Code* as applicable.

❖ Moved structures that are subject to repair, alteration or change of occupancy are required to comply with the provisions of the code for such repair, alteration or change of occupancy. The prescriptive or performance method could potentially be applied to address such repairs, alterations or changes of occupancy. Additionally, it is important to note that the IFC and IPMC have provisions that are also applicable to existing buildings.

SECTION 1302
REQUIREMENTS

1302.1 Location on the lot. The building shall be located on the lot in accordance with the requirements of the *International Building Code* or the *International Residential Code* as applicable.

❖ Exterior walls of buildings may need fire-resistance ratings in accordance with either the *International Building Code*® (IBC®) (see IBC Section 705.5) or the *International Residential Code*® (IRC®) (see IRC Section R302), as applicable. The required rating in either code is based on the fire separation distance and, in accordance with the IBC, the occupancy and type of construction must also be considered. Since the relocated building's exterior wall construction and, therefore, its fire-resistance rating are fixed, the building must be situated so that the exterior wall's fire-resistance rating is in compliance. Alternatively, the fire-resistance ratings and opening protectives of the existing exterior walls and openings can be upgraded as required to meet the IBC for the desired fire separation distance.

[BS] 1302.2 Foundation. The foundation system of relocated buildings shall comply with the *International Building Code* or the *International Residential Code* as applicable.

❖ The foundation for a relocated building is constructed at the building's new location. As new construction, the foundation, as well as the building's connection to it, is therefore required to comply with either the IBC or IRC, as applicable.

[BS] 1302.2.1 Connection to the foundation. The connection of the relocated building to the foundation shall comply with the *International Building Code* or the *International Residential Code* as applicable.

❖ See the commentary to Section 1302.2.

[BS] 1302.3 Wind loads. Buildings shall comply with *International Building Code* or *International Residential Code* wind provisions as applicable.

Exceptions:

1. Detached one- and two-family dwellings and Group U occupancies where wind loads at the new location are not higher than those at the previous location.

2. Structural elements whose stress is not increased by more than 10 percent.

❖ When a building is moved, it can be subjected to different design wind loading at the new location. The structure must be checked, therefore, to verify that it satisfies the wind-load criteria for a new structure using either the IBC or IRC, as applicable. There are two exceptions to full compliance, one of which exempts detached one- and two-family dwellings, as well as Group U occupancies if the design wind loads at the new location are no greater than those at the former location. Exception 2 allows additional wind loads that do not increase stresses in affected structural elements by more than 10 percent. Allowing overstresses of up to 10 percent in existing structural members has been a long-standing rule of thumb used by structural engineers.

[BS] 1302.4 Seismic loads. Buildings shall comply with *International Building Code* or *International Residential Code* seismic provisions at the new location as applicable.

Exceptions:

1. Structures in Seismic Design Categories A and B and detached one- and two-family dwellings in Seismic Design Categories A, B and C where the seismic loads at the new location are not higher than those at the previous location.

2. Structural elements whose stress is not increased by more than 10 percent.

❖ When a building is moved, the structure must satisfy the seismic load criteria for a new structure using either the IBC or IRC, as applicable at the new location. There are two exceptions to full compliance. Exception 1 exempts any building classified as Seismic Design Category A or B, in addition to detached one- and two-family dwellings classified as Seismic Design Category C, provided the design seismic loads at the new location are no greater than those at the former location. See the commentary to Section 301.1.4 for an explanation of seismic design category classification. While the design seismic loads are a function of several variables, with a relocated building that is not also undergoing a change of occupancy, the difference in the seismic load will merely be a function of two of those variables: the mapped spectral accelerations (e.g., S_S) and the site soil coefficients (e.g., F_a). In other words, comparing the design spectral accelerations (e.g., $S_{DS} = 2F_aS_S/3$) calculated at each site provides an indication of the difference in seismic loads. These parameters need to be determined using the IBC and are explained in the commentary to Section 301.1.4 (see IBC Section 1613.3). Exception 2 allows additional seismic loads that do not increase stresses in affected structural elements by more than 10 percent. Allowing overstresses of up to 10 percent in existing structural members has been a long-standing rule of thumb used by structural engineers.

[BS] 1302.5 Snow loads. Structures shall comply with *International Building Code* or *International Residential Code* snow loads as applicable where snow loads at the new location are higher than those at the previous location.

Exception: Structural elements whose stress is not increased by more than 5 percent.

❖ When a building is moved to a location that requires an increased design snow load, the structure must be checked to verify that it satisfies the snow load requirements for a new structure using either the IBC or IRC, as applicable. The exception to full compliance allows additional snow loading that does not increase stresses in affected structural elements by more than 5 percent. Allowing overstresses of up to 5 percent in existing structural members has been a long-standing rule of thumb used by structural engineers.

[BS] 1302.6 Flood hazard areas. If relocated or moved into a flood hazard area, structures shall comply with Section 1612 of the *International Building Code*, or Section R322 of the *International Residential Code*, as applicable.

❖ When a building is moved into a flood hazard area or relocated within a flood hazard area, the structure must satisfy the flood hazard criteria for a new structure under the IBC or the IRC, as applicable.

[BS] 1302.7 Required inspection and repairs. The code official shall be authorized to inspect, or to require approved professionals to inspect at the expense of the owner, the various structural parts of a relocated building to verify that structural components and connections have not sustained structural damage. Any repairs required by the code official as a result of such inspection shall be made prior to the final approval.

❖ In addition to the inspections required by Section 109, this section gives the code official the authority to inspect the structure of a relocated building to verify that structural components and connections have not been damaged. This is important because, in the course of being moved, the building can be subjected to movements and displacements that were never anticipated in the original structural design. In the event more specific expertise is required, the code official may require such inspections to be made by an approved professional at the owner's expense.

Chapter 14:
Performance Compliance Methods

General Comments

Chapter 14 contains provisions for an alternative method of evaluation of additions, alterations or changes of occupancy in existing buildings and structures based on a numerical scoring system involving various safety issues and the degree of code compliance for each issue. This section contains the provisions to be included in the evaluation and the conditions on which that evaluation is based.

Neither the designer nor the code official can physically inspect and evaluate every aspect of an existing building or structure, because many of its features may be concealed in the construction. It is, therefore, necessary to emphasize those items that can be evaluated. There are 21 critically important elements that can be quantified and evaluated to determine the level of safety for an existing building.

This type of analysis provides the designer and the code official with a rational basis for establishing the safety of an existing building or structure without having physical access to every part of the building, documentation of the original design or the construction history of the building.

Purpose

A large number of existing buildings and structures do not comply with current building code requirements for new construction. Although many of these buildings are potentially salvageable, rehabilitation is often cost prohibitive because they may not be able to comply with all the requirements for new construction. At the same time, it is necessary to regulate construction in existing buildings that undergo additions, alterations, renovations, extensive repairs or changes of occupancy. Such activity represents an opportunity to ensure that new construction complies with the current building codes and that existing conditions are maintained, at a minimum, to their current level of compliance or are improved as required. To accomplish this objective, and to make the rehabilitation process easier, this chapter allows for a controlled departure from full compliance with the technical codes, without compromising the minimum standards for fire prevention and life safety features of the rehabilitated building.

SECTION 1401
GENERAL

1401.1 Scope. The provisions of this chapter shall apply to the *alteration, repair, addition* and *change of occupancy* of existing structures, including historic and moved structures, as referenced in Section 301.1.3. The provisions of this chapter are intended to maintain or increase the current degree of public safety, health and general welfare in *existing buildings* while permitting *repair, alteration, addition* and *change of occupancy* without requiring full compliance with Chapters 5 through 13, except where compliance with other provisions of this code is specifically required in this chapter.

❖ This section states the scope of Chapter 14 and references alternative methods of code compliance for alteration, repair, addition and change of occupancy of existing structures. Chapter 14 also describes the responsibilities for maintenance, repairs, compliance with other codes and periodic testing.

1401.1.1 Compliance with other methods. *Alterations, repairs, additions* and *changes of occupancy* to existing structures shall comply with the provisions of this chapter or with one of the methods provided in Section 301.1.

❖ This section references Section 301.1 for the options available to deal with alterations, repairs, additions and changes of occupancy to existing structures. The following briefly describes the options available, other than compliance with this chapter:

1. Subject to the approval of the code official, repairs and alterations can comply with the requirements of the code at the time the building was built.

2. Repairs, alterations, additions and changes of occupancy can comply with the prescriptive compliance method found in Chapter 4. These provisions are duplicated from Sections 3403 through 3411 of the *International Building Code*® (IBC®).

3. Repairs, alterations, additions and changes of occupancy can comply with the proportional approach where upgrades are triggered by the type and extent of the work. These requirements are found in Chapters 5 through 13.

Please note that these options are seperate and distinct and must not be combined in any way.

1401.2 Applicability. Structures existing prior to [**DATE TO BE INSERTED BY THE JURISDICTION. Note: it is recommended that this date coincide with the effective date of building codes within the jurisdiction**], in which there is work involving *addi-*

tions, alterations or *changes of occupancy* shall be made to conform to the requirements of this chapter or the provisions of Chapters 5 through 13. The provisions of Sections 1401.2.1 through 1401.2.5 shall apply to existing occupancies that will continue to be, or are proposed to be, in Groups A, B, E, F, I-2, M, R and S. These provisions shall not apply to buildings with occupancies in Group H or I-1, I-3 or I-4.

❖ The adopting jurisdiction is to insert the desired date for applicability as indicated in this section; therefore, Chapter 14 applies only to structures existing prior to the established date. The date that construction was first regulated through a comprehensive building code in the jurisdiction is recommended because buildings predating any building regulation are often not equipped with the types of systems and features that modern codes require. Newer buildings are more likely to be in closer compliance with contemporary code requirements for new construction. These older buildings are assumed to face more difficulty in achieving a minimum level of life safety and are more likely to need the greater flexibility of the provisions in Chapter 14. The occupancies that qualify for the provisions of Chapter 14 are listed in this section. Chapter 14 does not apply to Groups H and I-1, I-3, or I-4.

1401.2.1 Change in occupancy. Where an *existing building* is changed to a new occupancy classification and this section is applicable, the provisions of this section for the new occupancy shall be used to determine compliance with this code.

❖ When a building undergoes a change of occupancy classification and Chapter 14 is applied, the evaluation method in Chapter 14 must be applied to the new occupancy for determining whether the existing building meets the compliance alternative in the code. This recognizes that it is the proposed conditions and relative hazards that will exist and must be determined to be acceptable. It is also consistent with how changes of occupancy are regulated in the absence of the Chapter 14 alternative.

1401.2.2 Partial change in occupancy. Where a portion of the building is changed to a new occupancy classification and that portion is separated from the remainder of the building with fire barrier or horizontal assemblies having a fire-resistance rating as required by Table 508.4 of the *International Building Code* or Section R317 of the *International Residential Code* for the separate occupancies, or with approved compliance alternatives, the portion changed shall be made to conform to the provisions of this section.

Where a portion of the building is changed to a new occupancy classification and that portion is not separated from the remainder of the building with fire barriers or horizontal assemblies having a fire-resistance rating as required by Table 508.4 of the *International Building Code* or Section R317 of the *International Residential Code* for the separate occupancies, or with approved compliance alternatives, the provisions of this section which apply to each occupancy shall apply to the entire building. Where there are conflicting provisions, those requirements which secure the greater public safety shall apply to the entire building or structure.

❖ Where a portion of the building is changed to a new occupancy classification, the following options may be employed, dependent on the fire separation of the portion of the building from the remainder of the existing building.

Where a portion of the building is changed to a new occupancy classification and that portion is separated from the remainder of the building by a fire barrier that complies with the requirements for new construction, the new occupancy portion must be evaluated with the existing or proposed building design to be in full compliance with the provisions of Chapter 14. The remainder of the existing building must also be evaluated in accordance with Chapter 14. The mandatory safety scores for the new occupancy portion of the building and the existing occupancy are obtained from those listed in Table 1401.8 and are incorporated in the building's final evaluation score (see Table 1401.7).

Where a portion of the building is changed to a new occupancy classification and that portion is not separated by a fire barrier that complies with the requirements for new construction, the provisions of Chapter 14 for each occupancy must apply to the entire building. The requirements offering greater public safety are applied to the entire building. Any proposed or existing building attributes and modifications must be reviewed in light of these requirements. See the examples described in Sections 1401.6.16 and 1401.9.1 for mixed occupancies.

1401.2.3 Additions. *Additions* to *existing buildings* shall comply with the requirements of the *International Building Code*, *International Residential Code*, and this code for new construction. The combined height and area of the *existing building* and the new *addition* shall not exceed the height and area allowed by Chapter 5 of the *International Building Code*. Where a fire wall that complies with Section 706 of the *International Building Code* is provided between the *addition* and the *existing building*, the *addition* shall be considered a separate building.

❖ Additions effectively represent new construction. It is therefore reasonable to mandate that additions comply with code requirements for new construction. There are not assumed to be any existing conditions or other factors that would make full compliance impractical or difficult.

The requirements in this section are applied in the same way as those in Section 1401.1. This section is included in Chapter 14 so that provisions for additions are included for both optional methods of code compliance indicated in Section 1401.1 (see the commentary to Section 1401.1 for a discussion of these options).

The evaluation method in Chapter 14 may be used for an existing building that has been modified by an addition, provided that it meets the applicability requirements of Section 1401.2.

1401.2.4 Alterations and repairs. An *existing building* or portion thereof that does not comply with the requirements of this code for new construction shall not be altered or repaired in such a manner that results in the building being less safe or sanitary than such building is currently. If, in the *alteration* or *repair*, the current level of safety or sanitation is to be reduced, the portion altered or repaired shall conform to the requirements of Chapters 2 through 12 and Chapters 14 through 33 of the *International Building Code.*

❖ An existing building that is altered or repaired may be designed and evaluated in accordance with Chapter 14, provided it meets the applicability provisions of Section 1401.2.

When an existing building is altered or repaired, materials or methods consistent with the original construction must be used. This is described in Chapter 5. The alteration or repair must not cause the building to be less safe or sanitary than it is before the alterations are undertaken. This is true even if the existing condition exceeds the minimum requirements of the code under which it was originally built. Should the alteration or repair cause a reduction in safety or sanitation, the resulting condition must meet the requirements of Chapters 2 through 12 and Chapters 14 through 33 of the IBC. This effectively allows existing conditions that exceed new construction requirements to be reduced to, but not below, new construction requirements.

1401.2.5 Accessibility requirements. Accessibility shall be provided in accordance with Section 410 or 705.

❖ Any building or part of a building that has a change of occupancy must meet the requirements for accessibility in existing buildings found Section 410 or 705. These sections basically require full accessibility in the portion of the building undergoing a change of occupancy and, in addition, may require some work improving the accessible route to that space, and the bathrooms and drinking fountains that serve that space (see commentary, Sections 410 and 705).

1401.3 Acceptance. For *repairs, alterations, additions*, and *changes of occupancy* to *existing buildings* that are evaluated in accordance with this section, compliance with this section shall be accepted by the *code official.*

❖ The *International Codes*® (I-Codes®) regulate safety in existing buildings by establishing the appropriate minimum levels of safety and sanitation deemed necessary for the safe occupancy of buildings. This is accomplished in several different portions of the family of I-Codes®. The *International Fire Code*® (IFC®) addresses fire safety issues in existing buildings and the *International Property Maintenance Code*® (IPMC®) addresses matters of health and sanitation in existing buildings and sites. Codes traditionally give broad authority to the code official to abate unusually hazardous conditions or operations that may be encountered in existing buildings. Although these fire safety, health and sanitation requirements are addressed in different codes, they together make up the package of requirements that represent the minimum conditions any existing building must meet to be considered acceptable for occupancy under the I-Codes. Chapter 14 works in conjunction with the other codes in the family of I-Codes that regulate existing buildings by requiring compliance with those same minimum provisions. Without such references, Chapter 14 would be incomplete as an alternative tool for regulating existing buildings because it would otherwise allow conditions to exist that would not be permitted for any other existing building.

When an owner or designer of an existing building decides to apply Chapter 14 and complies with all the provisions of the chapter, including the applicability requirements in Section 1401.2, the code official must accept for review the proposed work or change of occupancy.

1401.3.1 Hazards. Where the *code official* determines that an unsafe condition exists as provided for in Section 115, such unsafe condition shall be abated in accordance with Section 115.

❖ When the code official finds an unsafe condition in the building that is not being corrected by the proposed work, he or she must order the abatement or correction of the unsafe condition or hazard just as would be ordered in an existing building that is not being renovated, as stipulated by Section 116. This section sets forth a required comprehensive performance objective of abating any condition that is unsafe, insanitary, illegal or of improper occupancy, egress deficient, fire hazardous, poorly maintained or otherwise dangerous to human life or the public welfare. Guidelines for abatement are provided in the code and in Chapter 1 of both the IPMC (see Sections 108 and 110) and the IFC.

1401.3.2 Compliance with other codes. Buildings that are evaluated in accordance with this section shall comply with the *International Fire Code* and *International Property Maintenance Code.*

❖ This section requires an existing building that is subjected to the evaluation scoring process of Section 1401.6 to also comply with the IFC and the IPMC. Those codes provide minimum requirements for health and safety that all existing buildings are expected to meet, regardless of whether there are any changes being made to the building or occupancy. Regardless of an existing building's final safety scores, the requirements of these referenced codes must be followed so occupants are safeguarded from the hazards addressed by those codes. This provision is similar to the requirements of Section 101.4.2.

1401.3.3 Compliance with flood hazard provisions. In *flood hazard areas*, buildings that are evaluated in accordance with this section shall comply with Section 1612 of the *International Building Code*, or Section R322 of the *Interna-*

tional Residential Code, as applicable if the work covered by this section constitutes *substantial improvement*.

❖ Regardless of how compliance with the provisions of the code is evaluated, buildings that are in flood hazard areas must meet the flood-resistant provisions of the IBC or *International Residential Code*® (IRC®), as applicable, if the proposed work is determined to be a substantial improvement.

1401.4 Investigation and evaluation. For proposed work covered by this chapter, the building owner shall cause the *existing building* to be investigated and evaluated in accordance with the provisions of Sections 1401.4 through 1401.9.

❖ This section and the subsequent subsections address what must be done by the owner/designer who elects to employ Chapter 14 for a proposed rehabilitation program and the corresponding action by the code official to assess the program objectively for approval or disapproval. The following actions must be taken:

- Fully investigate the building using both on-site inspections and research of all available building construction documents.
- Evaluate the building for conformance to Sections 1401.5 through 1401.9.
- Perform the required structural analysis of the structure.
- Determine whether the existing building, proposed work or change of occupancy complies with the accessibility provisions of Section 1401.2.5 (see commentary, Section 1401.5).
- Submit to the code official documented results of the investigation and evaluation plus any proposed compliance alternatives.

Thus, the election and implementation of Chapter 14 by the designer or owner is part of a code-compliance option that may be used for existing buildings that the code official has the responsibility to review and assess. A proper and satisfactorily prepared submission by the owner/designer offering qualitative and quantitative data requires review by the code official for approval. If the submission is disapproved, the code official must specifically cite any deficiencies and violations.

The owner is required to have an existing building and any proposed work therein investigated and evaluated for compliance with the 21 parameters of the evaluation process as specified in Section 1401.6.

[BS] 1401.4.1 Structural analysis. The owner shall have a structural analysis of the *existing building* made to determine adequacy of structural systems for the proposed *alteration*, *addition* or *change of occupancy*. The analysis shall demonstrate that the building with the work completed is capable of resisting the loads specified in Chapter 16 of the *International Building Code*.

❖ The owner is required to have a complete structural analysis of the building performed to ensure that it can support the required loads. This requires that all interior loads meet the minimum load requirements of Chapter 16 of the IBC. Through this analysis, existing and altered buildings must be shown to be capable of supporting the expected loading. Any existing exterior member not affected by either interior loading or additional exterior loading need not be evaluated, provided that the structural member has, over a period of time, proven its ability to withstand the forces that normally create stress. Loads imposed on existing structural members by alterations, additions or a change of occupancy must be shown to sustain the requirements of IBC Chapter 16. This structural analysis provides the owner and the code official with reasonable assurance that the building is structurally safe.

1401.4.2 Submittal. The results of the investigation and evaluation as required in Section 1401.4, along with proposed compliance alternatives, shall be submitted to the *code official*.

❖ The results of the investigation, including the structural analysis and evaluation, must be submitted to the code official. If alternative methods, materials or equivalency concepts are proposed, these must also be submitted to the code official for review and approval.

1401.4.3 Determination of compliance. The *code official* shall determine whether the *existing building*, with the proposed *addition*, *alteration*, or *change of occupancy*, complies with the provisions of this section in accordance with the evaluation process in Sections 1401.5 through 1401.9.

❖ When the results of the investigation and evaluation are submitted to the code official, he or she must determine whether the proposed work conforms to the provisions of Chapter 14 and whether the evaluation was performed in accordance with Sections 1401.5 through 1401.9.

1401.5 Evaluation. The evaluation shall be comprised of three categories: fire safety, means of egress, and general safety, as defined in Sections 1401.5.1 through 1401.5.3.

❖ This section and the subsequent subsections address three general areas of safety to be evaluated: fire safety (FS), means of egress (ME) and general safety (GS). Section 1401.6 and the subsequent subsections address 21 safety parameters that reflect on those areas. Each of the 21 safety parameters indicated in Sections 1401.6.1 through 1401.6.21 must be carefully reviewed and assigned a numerical value that signifies the degree of safety influence on the three overall general safety categories. The overall evaluation is divided into three categories or areas, which are defined in Sections 1401.5.1 through 1401.5.3.

1401.5.1 Fire safety. Included within the fire safety category are the structural fire resistance, automatic fire detection, fire alarm, automatic sprinkler system and fire suppression system features of the *facility*.

❖ A partial list of the items used to evaluate fire safety in a building is given in this section.

1401.5.2 Means of egress. Included within the means of egress category are the configuration, characteristics, and support features for means of egress in the facility.

❖ The means-of-egress features evaluated by Chapter 14 fall in the general areas of configuration, characteristics and support features. The specific features include travel distance, dead ends, emergency lighting and exit capacity and number.

1401.5.3 General safety. Included within the general safety category are the fire safety parameters and the means-of-egress parameters.

❖ This category includes every item that is used in either the fire safety or means-of-egress evaluation.

1401.6 Evaluation process. The evaluation process specified herein shall be followed in its entirety to evaluate *existing buildings* in Groups A, B, E, F, M, R, S and U. For existing buildings in Group I-2, the evaluation process specified herein shall be followed and applied to each and every individual smoke compartment. Table 1401.7 shall be utilized for tabulating the results of the evaluation. References to other sections of this code indicate that compliance with those sections is required in order to gain credit in the evaluation herein outlined. In applying this section to a building with mixed occupancies, where the separation between the mixed occupancies does not qualify for any category indicated in Section 1401.6.16, the score for each occupancy shall be determined, and the lower score determined for each section of the evaluation process shall apply to the entire building, or to each smoke compartment for Group I-2 occupancies.

Where the separation between the mixed occupancies qualifies for any category indicated in Section 1401.6.16, the score for each occupancy shall apply to each portion, or smoke compartment of the building based on the occupancy of the space.

❖ This section is the key to understanding the entire evaluation process. The first sentence of this section clearly states that every one of the 21 safety parameters indicated in Sections 1401.6.1 through 1401.6.21 must be evaluated and nothing may be omitted. Although the evaluation process does not specifically evaluate all of the many issues that are regulated by the IBC for new construction, these 21 safety parameters have been determined to be the most critical factors related to the minimum degree of life safety and property protection needed in an existing building. The 21 safety parameters that must be evaluated are:

- Building height (see Section 1401.6.1).
- Building area (see Section 1401.6.2).
- Compartmentation (see Section 1401.6.3).
- Tenant and dwelling unit separations (see Section 1401.6.4).
- Corridor walls (see Section 1401.6.5).
- Vertical openings (see Section 1401.6.6).
- HVAC (heating, ventilating and air-conditioning) systems (see Section 1401.6.7).
- Automatic fire detection (see Section 1401.6.8).
- Fire alarm systems (see Section 1401.6.9).
- Smoke control (see Section 1401.6.10).
- Means-of-egress capacity and number (see Section 1401.6.11).
- Dead ends (see Section 1401.6.12).
- Maximum exit access travel distance to an exit (see Section 1401.6.13).
- Elevator control (see Section 1401.6.14).
- Means-of-egress emergency lighting (see Section 1401.6.15).
- Mixed occupancies (see Section 1401.6.16).
- Automatic sprinklers (see Section 1401.6.17).
- Standpipes (see Section 1401.6.18).
- Incidental uses (see Section 1401.6.19).
- Smoke compartmentation.
- Patient ability, concentration, and attendant to patient ratio (for I-2 occupancies).

The assigning of numerical values to each of the 21 safety parameters establishes a measurable quantity of each parameter's contribution to overall building safety. Some evaluated parameters have a negative influence; others have a positive one. In total, the parameters may or may not result in an acceptable building score. The evaluation will determine whether the existing building has enough positive parameters to overcome the negative parameters, or will indicate the negative factors that must be overcome. Modifications can be made to any aspect of the building that will accrue sufficient additional positive points to achieve the mandatory safety scores in each of the three evaluation categories. In other words, if, for example, the absence of a mixed occupancy separation under Section 1401.6.16.1 yielded a score of -10 (assume, for example, that the evaluation failed by 10 points in the fire safety category), it does not necessarily mean that a fire barrier must be constructed to separate the mixed occupancies. While that would be an acceptable solution, modifications could be made that accrue at least 10 additional points in any one or more of the other parameters in the fire safety category.

After the 21 safety parameters have been evaluated and assigned a numerical value, the values are entered in Table 1401.7. The values must be tabulated to obtain the building score for each of the three evaluation categories. For I-2 occupancies, the evaluation score must be entered for each smoke compartment to obtain a score for each smoke compartment in all three evaluation categories.

Some of the 21 safety parameters listed require mandatory compliance with other sections of the code and establish the foundation for determining a proper evaluation of those parameters, regardless of whether the result is positive or negative. This involves a coordination of mandatory basic requirements with the respective existing building conditions to arrive at the numerical evaluations prescribed in Sections 1401.6.1 through 1401.6.21.

Section 1401.6.16 also addresses how mixed occupancies are handled in the evaluation. To apply this section, it is necessary to understand IBC Section 508. When mixed occupancies in an existing building are not separated by fire barriers or fire walls complying with the requirements for new construction, the entire evaluation must be based on the occupancy with the most restrictive requirements. The evaluation process considers the score for the various occupancies and applies the lowest score to the entire building. When the mixed occupancies are separated by fire barriers in compliance with IBC Section 508.4.4, they are to be evaluated separately and the score for each occupancy will apply to each portion based on its occupancy classification. If there are four different occupancies in the building, the values must be computed for each of the four occupancies. Each occupancy is required to meet the applicable mandatory building score for its occupancy classification. When mixed occupancies are separated by fire walls complying with the requirements of IBC Section 706, separate buildings are created and must be evaluated separately for all 21 safety parameters. See the commentary to Sections 1401.2.2 and 1401.6.16 for a further discussion of the application of the evaluation procedure for mixed occupancies.

1401.6.1 Building height and number of stories. The value for building height and number of stories shall be the lesser value determined by the formula in Section 1401.6.1.1. Section 504 of the *International Building Code* shall be used to determine the allowable height and number of stories of the building. Subtract the actual building height from the allowable height and divide by $12^1/_2$ feet (3810 mm). Enter the height value and its sign (positive or negative) in Table 1401.7 under Safety Parameter 1401.6.1, Building Height, for fire safety, means of egress, and general safety. The maximum score for a building shall be 10.

❖ As a starting point for the actual evaluation process, this section and Section 1401.6.1.1 define, in detail, how to perform the building height and number of stories evaluation. The exact values are to be computed and compared against the mandatory safety scores. The values are not to be rounded. For calculation purposes, two decimal places would be appropriate. It is not necessary to build in any inaccuracy that occurs through the rounding process.

The maximum number of points that can be scored for a building, regardless of the building's allowable height as compared to its actual height, is 10. Building height and number of stories, as well as building area, which is covered in Section 1401.6.2, determine the type of construction classification. Limiting the number of points that can be scored for height limits the weight that classification could otherwise have on the overall evaluation. A low-rise building that is built of a high type of construction could accrue a disproportionately high number of points for height as compared to the number of points that are scored for other safety parameters. The intent is to avoid a situation in which many deficiencies in other critical safety parameters can be overcome simply by a high type of construction classification.

1401.6.1.1 Height formula. The following formulas shall be used in computing the building height value.

$$\text{Height value, feet} = \frac{(AH) - (EBH)}{12.5} \times CF$$

(Equation 14-1)

$$\text{Height value, stories} = (AS - EBS) \times CF$$

(Equation 14-2)

where:

AH = Allowable height in feet (mm) from Section 504 of the *International Building Code*.

EBH = *Existing building* height in feet (mm).

AS = Allowable height in stories from Section 504 of the *International Building Code*.

EBS = *Existing building* height in stories.

CF = 1 if $(AH) - (EBH)$ is positive.

CF = Construction-type factor shown in Table 1401.6.6(2) if $(AH) - (EBH)$ is negative.

Note: Where mixed occupancies are separated and individually evaluated as indicated in Section 1401.6, the values AH, AS, EBH and EBS shall be based on the height of the occupancy being evaluated.

❖ Two height formulas for calculating the score to be entered in Table 1401.7 are given. One formula determines a height value based on the height in feet and the other formula determines a height value in number of stories, based on IBC Tables 504.4 and 504.3, respectively. The denominator in the formula for height in feet, 12.5, represents an average story height in feet. Both formulas must be calculated and the lesser of the two calculated values is the value that must be used in the evaluation.

The actual story height and the overall existing building height are to be directly compared to IBC

Tables 504.4 and 504.3, respectively. That table serves as a datum level that allows for establishing a numerical height value for the existing building by comparing its actual height and its type of construction as represented by a construction factor (*CF*). If the existing building's actual height in feet is less than or equal to the allowable height of IBC Table 504.3, then the construction factor value is 1 (no negative or positive multiplier is factored into the calculation). If the actual height exceeds the allowable height in IBC Table 504.3, the building is not in compliance with that table and represents a safety deficiency. A deficiency rating is assigned dependent on the type of construction. The construction factors to establish this negative value are given in Table 1401.6.6(2). When a building is not in compliance with IBC Tables 504.3 and 504.4, it is considered less safe than a building that does comply with those tables. As a result, deficiency points are assessed. Additional safeguards must be provided to compensate for this condition. This is the primary reason that a different construction factor must be used for buildings not in compliance. The construction factor is the equivalent deficiency that must be overcome by providing additional protection in other areas that are to be evaluated.

Example 1:

A six-story building of Type IB construction is 60 feet (18 288 mm) tall. The building is not sprinklered. It contains a Group B business occupancy.

The allowable story height from IBC Table 504.4 is 11 stories and the allowable height, in feet, from IBC Table 504.3 is 160 feet (48 800 mm).

AH = 160 feet

AS = 11 stories

EBH = 60 feet

EBS = 6 stories

CF = 1 from Table 1401.6.6(2) (because 160 - 60 is a positive number)

The governing building height value is 5, because it is the lesser of the two height values.

Example 2:

A seven-story building of Type IIIB construction is 80 feet (24 400 mm) tall. The building is sprinklered. It contains a Group M mercantile occupancy.

The allowable story height from IBC Table 504.4 is four stories and the allowable height, in feet, from IBC Table 504.3 is 55 feet (16 775 mm).

AH = 55 + 20 (for sprinklers) = 75 feet

AS = 4 + 1 (for sprinklers) = 5 stories

EBS = 7 stories

EBH = 80 feet

CF = 3.5 from Table 1401.6.6(2) (because 75 - 80 is a negative number)

The governing building height value is -7, because it is the lesser of the two height values.

The values in the examples determine the entries for the Summary Sheet of Table 1401.7 for the Safety Parameter of Building Height on line 1401.6.1.

In Example 1, the height value of 5 is entered into the columns for fire safety (FS), means of egress (ME) and general safety (GS).

In Example 2, the height value of -7 is entered into the columns for fire safety (FS), means of egress (ME) and general safety (GS).

The assessed height value parameter is only one of the 21 safety parameters that need to be evaluated. Example 1 shows a positive contribution; however, this may not be enough to result in a sufficient overall building score. Similarly, the negative value in Example 2 may not in itself result in an overall insufficient building score.

1401.6.2 Building area. The value for building area shall be determined by the formula in Section 1401.6.2.2. Section 506 of the *International Building Code* and the formula in Section 1401.6.2.1 shall be used to determine the allowable area of the building. Subtract the actual building area from the allowable area and divide by 1,200 square feet (112 m²). Enter the area value and its sign (positive or negative) in Table 1401.7 under Safety Parameter 1401.6.2, Building Area, for fire safety, means of egress and general safety. In determining the area value, the maximum permitted positive value for area is 50 percent of the fire safety score as listed in Table 1401.8, Mandatory Safety Scores. Group I-2 occupancies shall be scored zero.

❖ In this section, the code user is shown how to calculate the building score for the building area. This section also requires the maximum score of a building in this category. The maximum value is 50 percent of just the fire safety score listed in Table 1401.8 (see the commentary to Section 1401.6.2.2). This maximum score is applicable and equal for the three building score categories—fire safety, means of egress and general safety. The positive score is limited to prevent this one parameter from providing enough points to unjustifiably overcome too many deficiencies in other parameters.

1401.6.2.1 Allowable area formula. The following formula shall be used in computing allowable area:

$$A_a = A_t + (NS \times I_f)$$ (Equation 14-3)

where:

A_a = Allowable building area per story (square feet).

A_t = Tabular allowable area factor (NS, S1, S13R, or SM value, as applicable) in accordance with Table 506.2 of the *International Building Code*.

NS = Tabular allowable area factor in accordance with Table 506.2 of the *International Building Code* for a nonsprinklered building (regardless of whether the building is sprinklered).

I_f = Area factor increase due to frontage as calculated in accordance with Section 506.3 of the *International Building Code*.

❖ The formula used to calculate the allowable area of the building is a direct correlation of the area increases and reductions allowed for new buildings in IBC Section 506 The use of these increases and reductions is addressed in IBC Chapter 5.

1401.6.2.2 Area formula. The following formula shall be used in computing the area value. Determine the area value for each occupancy floor area on a floor-by-floor basis. For each occupancy, choose the minimum area value of the set of values obtained for the particular occupancy.

$$\text{Area value}_i = \frac{\text{Allowable area}_i}{1200 \text{ square feet}} \left[1 - \left(\frac{\text{Actual area}_i}{\text{Allowable area}_i} + \ldots + \frac{\text{Actual area}_n}{\text{Allowable area}_n} \right) \right]$$

(Equation 14-4)

where:

i = Value for an individual separated occupancy on a floor.

n = Number of separated occupancies on a floor.

❖ The area formula provides a numerical value for the actual building area to be entered in Table 1401.7 for the Safety Parameter Building Area on line 1401.6.2. If the area of the existing building is less than the allowable area, it is considered to be safer and the building receives a positive score. If the area of the existing building is larger than the allowable area, it represents a condition that is judged to be less safe and the building receives a negative score.

To use the formula for existing buildings that contain mixed occupancies, an area value must be determined for each story of a building. If a building contains just one occupancy or just one occupancy on a particular floor separated from other floors, the formula simply reduces to the allowable area minus the actual area divided by the constant of 1,200. The formula is also applicable to a story containing several separated occupancies. In such a situation, the formula requires each actual area to be divided by its respective allowable area. The resulting fractions are added together, subtracted from the constant of 1 and multiplied by the ratio of the allowable area of the particular use divided by the constant 1,200. This method of determining area values for several occupancies that are separated on a story is directly comparable to the unity formula in IBC Section 508.2.4.

Example 1:

Commentary Figure 1401.6.2.2(1) illustrates a Group B building of Type IIA construction.

The building is five stories in height, unsprinklered.

The overall building area is 200 feet (60 960 mm) by 200 feet (60 960 mm).

Area value $= \dfrac{75{,}000}{1{,}200}\left[1 - \left(\dfrac{40{,}000}{75{,}000}\right)\right]$

$= 62.5\,(1 - 0.533)$

$= 29.2$

Allowable area = 75,000 square feet

Actual area = 40,000 square feet

In Table 1401.7, enter the following values for the Safety Parameter of Building Area on line 1401.6.2:

Fire safety (FS) = 13.6

Means of egress (ME) = 13.6

General safety (GS) = 13.6

The positive values entered in Table 1401.7 under Safety Parameters 1401.6.2, Building Area (FS = 13.6, ME = 13.6 and GS = 13.6), represent the evaluation for only one of the 21 safety parameters that must be assessed to arrive at the overall building score evaluation of Table 1401.7. As previously described for the building height parameter values under line 1401.6.1 of Table 1401.7, a single parameter of the total 21 safety parameters to be assessed does not, in itself, determine the ultimate acceptable building score.

Example 2:

Commentary Figure 1401.6.2.2(2) illustrates a separated mixed occupancy building (Groups B, M and S-1) of Type IIIB construction. The building is two stories in height, fully sprinklered.

The overall building area is 100 feet (30 480 mm) by 200 feet (60 960 mm).

Actual area for Group M on first story = 20,000 square feet (1858 m^2).

Actual area for Group M on second story = 4,400 square feet (409 m^2).

Actual area for Group B on second story = 10,000 square feet (929 m^2).

Actual area for Group S-1 on second story = 5,600 square feet (520 m^2).

Frontage Increase = 5.6% (from Section 506.2)

Tabular areas from Table 506.2:

Group B = 57,000 square feet (1765 m^2)

Group M = 37,500 square feet (1161 m^2)

Group S-1 = 52,500 square feet (1626 m^2)

PERFORMANCE COMPLIANCE METHODS

Allowable Areas:

    AA = (NS × 0.056) × Tabular Area

    NS for Group B = 19,000 square feet

    AA for Group B = 58,064 square feet

    NS for Group M = 12,500 square feet

    AA for Group M = 38,200 square feet

    NS for Group S-1 = 17,500 square feet

    AA for Group S-1 = 53,480 square feet

Area Values:

For Group M, 1st story

$$\text{Area value} = \frac{38,200}{1,200}\left[1 - \left(\frac{20,000}{38,200}\right)\right]$$

$$= 31.83[1 - 0.52]$$

$$= 15.27$$

For Group B, 2nd story

$$\text{Area value} = \frac{58,064}{1,200}\left[1 - \left(\frac{10,000}{58,064} + \frac{4,400}{38,200} + \frac{5,600}{53,480}\right)\right]$$

$$= 48.39[1 - 0.39]$$

Area value = 29.51

For Group M, 2nd story

$$\text{Area value} = \frac{38,200}{1,200}[1 - 0.39]$$

Area value = 19.42

For Group S-1, 2nd story

$$\text{Area value} = \frac{53,480}{1,200}[1 - 0.39]$$

Area value = 27.19

Since these mixed occupancies are being separated so that one of the categories indicated in Section 1401.6.16 is applicable, a separate score must be computed for each occupancy.

In this example, the area value for the Group B occupancy is calculated to be 29.51, but the maximum area value permitted is 50 percent of the mandatory fire safety score listed in Table 1401.8. For a Group B occupancy, the mandatory fire safety score is 30; therefore, the maximum positive value that can be entered into Table 1401.7 is 15.

When an occupancy is located in more than one story, a separate area value must be calculated for each story. For input into Table 1401.7, the area value to be used is the lesser of all the individual area

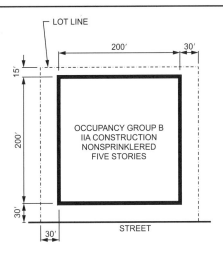

ACTUAL AREA = 200′ × 200′ = 40,000 SQ. F.T.

SP = 0 (FROM IBC SECTION 506.3)
OP = 100% (FROM IBC SECTION 506.2)
ALLOWABLE AREA IN IBC TABLE 506.2 = 37,500 SQ. F.T.

$$AA = \frac{(0 + 100 + 100) \times 37,500}{100} = \frac{200 \times 37,500}{100} = 75,000 \text{ SQ.FT.}$$

For SI: 1 foot = 304.8 mm, 1 square foot = 0.0929 m².

**Figure 1401.6.2.2(1)
ALLOWABLE AREA—SINGLE-OCCUPANCY BUILDING**

values for that occupancy group, but not greater than 50 percent of the mandatory fire safety score. In Commentary Figure 1401.6.2.2(2), there is an area classified as a Group M occupancy on both the first and second stories. The area value for the Group M occupancy on the first story is 15.27, and the area value for the second story is 19.42. Fifty percent of the mandatory fire safety score for Group M is 11.5. The lowest value for all the Group M occupant areas in the building is 11.5, which is 50 percent of the mandatory fire safety score.

The area value for the Group S-1 occupancy portion of the building is 27.19. This value, just as the values for the other occupancies in this example, exceeds the 50-percent maximum permitted by the mandatory fire safety score. The maximum for Group S-1 occupancy is 9.5, 50 percent of 19.

1401.6.3 Compartmentation. Evaluate the compartments created by fire barriers or horizontal assemblies which comply with Sections 1401.6.3.1 and 1401.6.3.2 and which are exclusive of the wall elements considered under Sections 1401.6.4 and 1401.6.5. Conforming compartments shall be figured as the net area and do not include shafts, chases, stairways, walls, or columns. Using Table 1401.6.3, determine the appropriate compartmentation value (CV) and enter that value into Table 1401.7 under Safety Parameter 1401.6.3, Compartmentation, for fire safety, means of egress, and general safety.

❖ This section establishes and evaluates the compartments contained in an existing building by the effectiveness of the enclosing fire barrier walls and fire-resistant floor/ceiling assemblies. Larger compartments are considered to be a greater safety risk than smaller compartments because the entire compartment is assumed to be involved when a fire incident occurs in the compartment and, therefore, a single fire incident affects a greater portion of the building at one time.

Fire barriers must comply with Sections 1401.6.3.1 and 1401.6.3.2. Fire barriers are exclusive of the other separations or enclosures that are evaluated in Sections 1401.6.4 and 1401.6.5.

The evaluation of the compartments contained in an existing building is a linear function allowing inter-

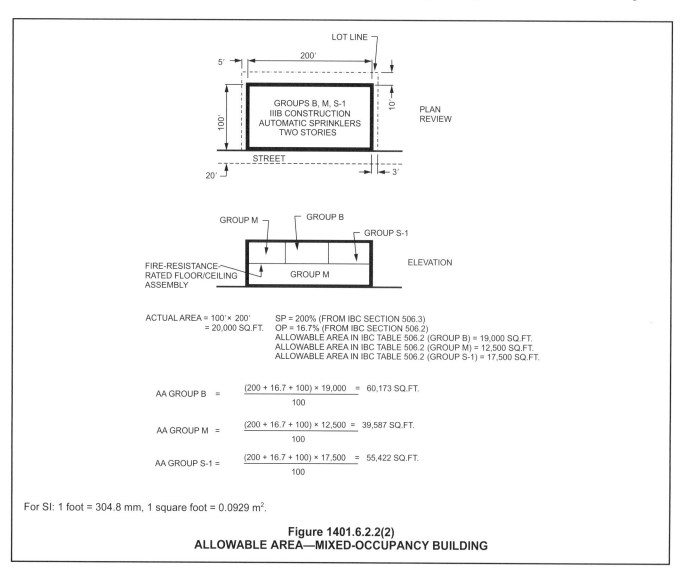

Figure 1401.6.2.2(2)
ALLOWABLE AREA—MIXED-OCCUPANCY BUILDING

polation between the various categories. This approach allows the compartmentation value to increase or decrease consistent with the actual changes in compartment sizes. Such an adjustment removes the previously built-in bias against smaller-sized buildings. Higher compartmentation values are assigned to buildings with smaller compartments.

TABLE 1401.6.3. See below.

❖ Table 1401.6.3 contains compartment values based on the group and the square footage of individual compartments. The value increases as the size of the compartment decreases in order to recognize the safety benefit of fire-resistance-rated compartmentation.

If the existing floor construction inherently complies with Section 1401.6.3.2, creating compartments by constructing fire barriers in accordance with Section 1401.6.3.1 could be a cost-effective way to increase the building's safety score.

1401.6.3.1 Wall construction. A wall used to create separate compartments shall be a fire barrier conforming to Section 707 of the *International Building Code* with a fire-resistance rating of not less than 2 hours. Where the building is not divided into more than one compartment, the compartment size shall be taken as the total floor area on all floors. Where there is more than one compartment within a story, each compartmented area on such story shall be provided with a horizontal exit conforming to Section 1026 of the *International Building Code*. The fire door serving as the horizontal exit between compartments shall be so installed, fitted, and gasketed that such fire door will provide a substantial barrier to the passage of smoke.

❖ This section states that the walls determining the boundary of the compartment need to have fire-barrier ratings of not less than 2 hours. These assemblies must be constructed in accordance with IBC Section 707. For an existing building, this may need to be evaluated by both analyzing available plans and on-site investigations with a professional engineering determination of the required 2-hour fire-resistance rating. If the fire-resistance rating is less than 2 hours or cannot be reasonably determined, such walls should not be considered as creating a compartment. The entire story of an existing building must then be considered the compartment. If 2-hour-rated floor/ceiling assemblies are not present, the compartment size becomes the total area of all stories combined.

Opening protection and continuity requirements of IBC Section 707 must be followed to maintain the integrity of the fire-resistance-rated wall assemblies and thus the compartments. Horizontal exits and their fire doors must comply with IBC Section 1025.

The evaluation of an existing door condition requires an investigation of available engineering data and on-site inspections. Any uncertainty as to the performance of such doors may result in the openings being considered unprotected, which results in a larger assumed compartment area or requires modification of the door to meet the current requirements of IBC Section 1025.3.

1401.6.3.2 Floor/ceiling construction. A floor/ceiling assembly used to create compartments shall conform to Section 711 of the *International Building Code* and shall have a fire-resistance rating of not less than 2 hours.

❖ The building features that create the horizontal boundaries of the compartment must provide effective fire resistance between floors. The existing floor/ceiling assemblies must be rated for 2 hours and need to be tight against exterior walls. Penetrations in the floor/ceiling assemblies must be protected in accordance with IBC Section 714 to maintain their fire-resistant integrity. The floor/ceiling assemblies must conform to all of the requirements of IBC Section 711 to create the level of compartmentation required for this evaluation parameter.

1401.6.4 Tenant and dwelling unit separations. Evaluate the fire-resistance rating of floors and walls separating tenants, including dwelling units, and not evaluated under Sections 1401.6.3 and 1401.6.5. Group I-2 occupancies shall evaluate the rating of the separations between patient sleeping rooms.

Under the categories and occupancies in Table 1401.6.4, determine the appropriate value and enter that value in Table 1401.7 under Safety Parameter 1401.6.4, Tenant and Dwelling Unit Separation, for fire safety, means of egress, and general safety.

❖ This parameter is used to evaluate partitions in an existing building other than those used for the creation of compartments in Section 1401.6.3 or the enclosure of corridors in Section 1401.6.5. This sec-

TABLE 1401.6.3
COMPARTMENTATION VALUES

OCCUPANCY	CATEGORIES				
	a Compartment size equal to or greater than 15,000 square feet	b Compartment size of 10,000 square feet	c Compartment size of 7,500 square feet	d Compartment size of 5,000 square feet	e Compartment size of 2,500 square feet or less
A-1, A-3	0	6	10	14	18
A-2	0	4	10	14	18
A-4, B, E, S-2	0	5	10	15	20
F, M, R, S-1	0	4	10	16	22

For SI: 1 square foot = 0.0929 m².

tion examines the level of separation between tenant spaces and dwelling units. The listed categories specifically reference IBC Sections 707, 708 and 711 not only for fire-resistance ratings, but also for continuity and opening-protection purposes. Further credit is provided for existing buildings that have a 2-hour-rated separation between adjacent tenant spaces or dwelling units, which exceeds the requirement for new construction.

**TABLE 1401.6.4
SEPARATION VALUES**

OCCUPANCY	CATEGORIES				
	a	b	c	d	e
A-1	0	0	0	0	1
A-2	-5	-3	0	1	3
R	-4	-2	0	2	4
A-3, A-4, B, E, F, M, S-1	-4	-3	0	2	4
I-2	0	1	2	3	4
S-2	-5	-2	0	2	4

❖ Table 1401.6.4 provides values for tenant space and dwelling unit separations. The rationale for considering a nonfire-resistance-rated or incomplete separation as a safety deficiency is that even though the assembly may have some limited fire resistance, it cannot be assumed to provide the level of fire performance expected of a fully complying assembly. Buildings containing Group R occupancies have separation values based on the assumption that dwelling unit separations are more critical than tenant separations in other occupancies.

1401.6.4.1 Categories. The categories for tenant and dwelling unit separations are:

1. Category a—No fire partitions; incomplete fire partitions; no doors; doors not self-closing or automatic-closing.

2. Category b—Fire partitions or floor assemblies with less than 1-hour fire-resistance ratings or not constructed in accordance with Section 708 or 711 of the *International Building Code*, respectively.

3. Category c—Fire partitions with 1-hour or greater fire-resistance ratings constructed in accordance with Section 708 of the *International Building Code* and floor assemblies with 1-hour but less than 2-hour fire-resistance ratings constructed in accordance with Section 711 of the *International Building Code* or with only one tenant within the floor area.

4. Category d—Fire barriers with 1-hour but less than 2-hour fire-resistance ratings constructed in accordance with Section 707 of the *International Building Code* and floor assemblies with 2-hour or greater fire-resistance ratings constructed in accordance with Section 711 of the *International Building Code*.

5. Category e—Fire barriers and floor assemblies with 2-hour or greater fire-resistance ratings and constructed in accordance with Sections 707 and 711 of the *International Building Code*, respectively.

❖ Tenant space and dwelling unit separations are categorized by the partitions being evaluated. The values of each category are listed in Table 1401.6.4 by occupancy classifications. Typical illustrations of the types of partitions are shown in Commentary Figures 1401.6.4.1(1) and 1401.6.4.1(2). The listed catego-

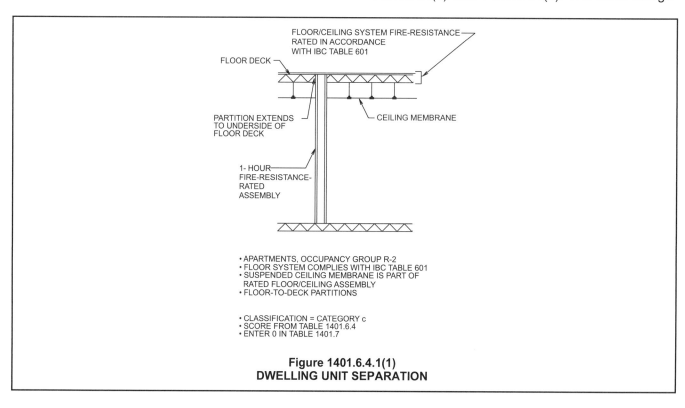

**Figure 1401.6.4.1(1)
DWELLING UNIT SEPARATION**

ries provide a graduated level of separation when compared to new construction requirements.

Category a addresses the situation where there is no separation or there are gaps in the separation provided between tenant spaces or dwelling units.

Category b accounts for those tenant spaces or dwelling units that are separated from one another with less than a 1-hour rating.

Category c represents what is required for new construction for tenant space and dwelling unit separation. An existing building meeting this level of compliance gains no benefit or penalty and, therefore, the separation value is zero.

Category d is given additional credit because the tenant space or dwelling unit separation has a fire-resistance rating that marginally exceeds the minimum required for new construction. Additionally, the walls are required to meet the fire barrier requirements of IBC Section 707.

Category e provides increased credit for existing buildings that have walls and floor/ceiling assemblies with fire-resistance ratings exceeding the new construction requirement by at least 1 full hour. This increased level of fire-resistant separation in an existing building is credited accordingly with high separation values.

1401.6.5 Corridor walls. Evaluate the fire-resistance rating and degree of completeness of walls which create corridors serving the floor and that are constructed in accordance with Section 1020 of the *International Building Code*. This evaluation shall not include the wall elements considered under Sections 1401.6.3 and 1401.6.4. Under the categories and groups in Table 1401.6.5, determine the appropriate value and enter that value into Table 1401.7 under Safety Parameter 1401.6.5, Corridor Walls, for fire safety, means of egress, and general safety.

❖ Corridor walls are evaluated as fire partitions possessing an adequate fire-resistance rating and completeness to restrict the spread of fire into the corridor. Various categories require compliance with prescribed requirements in IBC Sections 708 and 1018. Existing corridor walls require investigation and analysis to determine the equivalency to code requirements. Commentary Figures 1401.6.5.1(1) and 1401.6.5.1(2) illustrate various corridor wall values. Corridor walls contrast with the compartmentation and tenant dwelling unit separations of Sections 1401.6.3 and 1401.6.4 by requiring an appropriate fire-resistance rating and continuity (see commentary, IBC Section 708.4). The corridor wall evaluations do not include partitions required to establish a compartment (see Section 1401.6.3) or tenant and dwelling unit separations (see Section 1401.6.4). If a corridor wall serves as more than one of these elements, for example where it is both a corridor wall and it defines a compartment under Section 1401.6.3 and a corridor under this subsection, it may be evaluated under one parameter or the other at the designer's option, but not both.

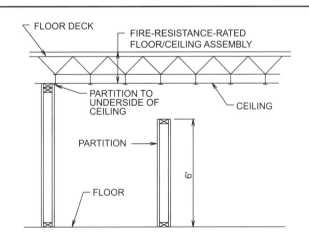

- FOOD COURT RESTAURANTS, OCCUPANCY GROUP A-3
- SOME PARTITIONS EXTEND TO CEILING OF FIRE-RESISTANCE-RATED FLOOR/CEILING ASSEMBLY
- NO CLOSERS ON DOORS BETWEEN ADJACENT TENANT SPACES
- SOME 6' PRIVACY PARTITIONS (PARTIAL PARTITIONS)

- CLASSIFICATION = CATEGORY a
- SCORE FROM TABLE 1401.6.4
- ENTER -4 IN TABLE 1401.7

For SI: 1 foot = 304.8 mm.

**Figure 1401.6.4.1(2)
TENANT SEPARATION**

PERFORMANCE COMPLIANCE METHODS

**TABLE 1401.6.5
CORRIDOR WALL VALUES**

OCCUPANCY	CATEGORIES			
	a	b	c[a]	d[a]
A-1	-10	-4	0	2
A-2	-30	-12	0	2
A-3, F, M, R, S-1	-7	-3	0	2
A-4, B, E, S-2	-5	-2	0	5
I-2	-10	0	1	2

a. Corridors not providing at least one-half the exit access travel distance for all occupants on a floor shall use Category b.

❖ Table 1401.6.5 assigns values to the different uses based on the relative degree of fire resistance and smoke resistance of corridor walls. Since corridors are enclosed (confined) spaces subject to the rapid buildup of smoke and heat, a degree of protection is necessary to minimize this hazard to the occupants egressing the building. The table reflects this emphasis through substantial negative scores in buildings without properly enclosed corridors. Both Section 1401.6.5 and this table place an emphasis on corridor walls, which differs from that expressed elsewhere in the code. For example, Chapter 10 of the IBC does not require corridors to be provided except in Group I-2 occupancies. The emphasis here is that when corridors are already present, they must provide basic protection. The table is an assessment of the relative risk represented by the corridor.

Note a to the table further controls the application of Categories c and d to existing buildings with certain means-of-egress arrangements. Although an existing building may have corridors with significant fire-resistance ratings and protected openings, little credit can be granted when the building occupants are protected for only a short period of time. Very short corridors in large floor plans or corridors located in just one tenant space of a floor plan provide a benefit to only a portion of the occupant load for a small portion of the overall exit access travel and, thus, do not accrue any positive points. Unless rated corridors are available for all occupants of that particular floor or they provide a protected path of travel for at least one-half of the occupants' overall travel length, negative corridor wall values are assigned. In existing buildings with very short or limited-use corridors, the code user is directed to use Category b, regardless of the corridors' fire-resistance ratings and opening protectives.

1401.6.5.1 Categories. The categories for corridor walls are:

1. Category a—No fire partitions; incomplete fire partitions; no doors; or doors not self-closing.

2. Category b—Less than 1-hour fire-resistance rating or not constructed in accordance with Section 708.4 of the *International Building Code*.

3. Category c—1-hour to less than 2-hour fire-resistance rating, with doors conforming to Section 716 of the *International Building Code* or without corridors as permitted by Section 1020 of the *International Building Code*.

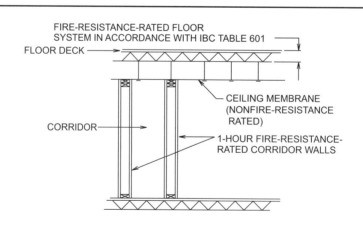

For SI: 1 foot = 304.8 mm.

**Figure 1401.6.5.1(1)
CORRIDOR WALL VALUES—CATEGORY A**

4. Category d—2-hour or greater fire-resistance rating, with doors conforming to Section 716 of the *International Building Code*.

❖ Corridor walls are categorized by the partitions being evaluated. The values of each category are listed in Table 1401.6.5 by occupancy classification.

1401.6.6 Vertical openings. Evaluate the fire-resistance rating of interior exit stairways or ramps, hoistways, escalator openings, and other shaft enclosures within the building, and openings between two or more floors. Table 1401.6.6(1) contains the appropriate protection values. Multiply that value by the construction-type factor found in Table 1401.6.6(2). Enter the vertical opening value and its sign (positive or negative) in Table 1401.7 under Safety Parameter 1401.6.6, Vertical Openings, for fire safety, means of egress, and general safety. If the structure is a one-story building or if all the unenclosed vertical openings within the building conform to the requirements of Section 713 of the *International Building Code*, enter a value of 2. The maximum positive value for this requirement shall be 2.

❖ Vertical openings are used to evaluate the fire-resistance ratings of openings between floors of a building and between shaft enclosures, such as stairs, elevator hoistways and escalator openings. This section also gives the formula for determining the score to be entered in Table 1401.7. This section recognizes the inherent safety of one-story buildings and of buildings where all vertical openings are protected in accordance with one of the methods in Section 713 by allowing the maximum value of 2 for these instances.

TABLE 1401.6.6(1)
VERTICAL OPENING PROTECTION VALUE

PROTECTION	VALUE
None (unprotected opening)	-2 times number of floors connected
Less than 1 hour	-1 times number of floors connected
1 to less than 2 hours	1
2 hours or more	2

❖ Table 1401.6.6(1) assigns relative protection values based on the fire-resistance ratings of the vertical openings in a building. The lower the fire-resistance rating, the greater the hazard to the rest of the building. The table also reflects the varying levels of impact to an existing building based on the number of stories that are connected by unprotected openings. The greater the number of floors that are interconnected by unprotected openings, the greater the number of negative points assessed in the existing building's evaluation scores. The closer the building comes to meeting new construction requirements for shaft protection, the greater the number of positive points that can be accrued. Noncomplying, unenclosed vertical openings consistently show up as contributing factors in unsuccessful fires. Consequently, a substantial number of points are at risk in this parameter, which is intended to be an incentive to bring noncomplying situations into compliance with new construction requirements.

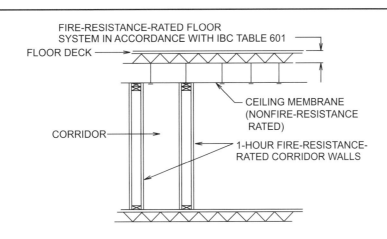

- OCCUPANCY GROUP B
- NO CLOSERS ON DOORS
- FLOOR SYSTEM COMPLIES WITH IBC TABLE 601
- SUSPENDED CEILING MEMBRANE IS NONFIRE-RESISTANCE RATED
- CORRIDOR WALLS ARE 1-HR FIRE-RESISTANCE RATED

NOTE: THESE CONDITIONS PRODUCE A CATEGORY a BECAUSE OF BOTH INCOMPLETE FIRE PARTITIONS AND NO SELF-CLOSING DOORS

- CLASSIFICATION = CATEGORY a
- SCORE FROM TABLE 1401.6.5 = -5
- ENTER -5 IN TABLE 1401.7

Figure 1401.6.5.1(2)
CORRIDOR WALL VALUES—CATEGORY B

**TABLE 1401.6.6(2)
CONSTRUCTION-TYPE FACTOR**

FACTOR	TYPE OF CONSTRUCTION								
	IA	IB	IIA	IIB	IIIA	IIIB	IV	VA	VB
	1.2	1.5	2.2	3.5	2.5	3.5	2.3	3.3	7

❖ Relative values for each type of construction are assigned in Table 1401.6.6(2). These represent the relative degree for fire hazard of each type of construction when compared to other types of construction. Similar factors were considered in the original development of IBC tables for height and area limitations. When one building has two different opening circumstances that individually result in different values, the lower value must be used.

1401.6.6.1 Vertical opening formula. The following formula shall be used in computing vertical opening value.

$$VO = PV \times CF \qquad \textbf{(Equation 14-5)}$$

where:

VO = Vertical opening value.

PV = Protection value from Table 1401.6.6.(1).

CF = Construction-type factor from Table 1401.6.6.(2).

❖ See the commentary to Section 1401.6.6.

1401.6.7 HVAC systems. Evaluate the ability of the HVAC system to resist the movement of smoke and fire beyond the point of origin. Under the categories in Section 1401.6.7.1, determine the appropriate value and enter that value into Table 1401.7 under Safety Parameter 1401.6.7, HVAC Systems, for fire safety, means of egress, and general safety. Facilities in Group I-2 occupancies meeting Categories a, b or c shall be considered to fail the evaluation.

❖ This section evaluates the HVAC system's potential for resisting the movement and spread of fire and smoke. This section does not address HVAC systems that are used exclusively for smoke control in the building. The systems evaluated in this section are those that use either supply air, return air or exhaust air. For example, a typical building might have supply air ducts or shafts, return air ducts or shafts, toilet exhaust ducts or shafts, and kitchen exhaust ducts or shafts, all of which are considered HVAC systems. All systems in the building are evaluated and the lowest score obtained by any of the systems is the score that must be assigned to the entire building.

These provisions include two other safety aspects that are also applicable for new construction: plenums and air movement in egress elements, such as exit access corridors and exit stairways. These factors can significantly affect the relative safety of the occupants of the existing building in a fire condition. In some cases, these safety aspects can be more important than just the number of stories connected by an HVAC system.

1401.6.7.1 Categories. The categories for HVAC systems are:

1. Category a—Plenums not in accordance with Section 602 of the *International Mechanical Code*. -10 points.

2. Category b—Air movement in egress elements not in accordance with Section 1018.5 of the *International Building Code*. -5 points.

3. Category c—Both Categories a and b are applicable. -15 points.

4. Category d—Compliance of the HVAC system with Section 1020.5 of the *International Building Code* and Section 602 of the *International Mechanical Code*. 0 points.

5. Category e—Systems serving one story; or a central boiler/chiller system without ductwork connecting two or more stories. +5 points.

❖ The five categories that must be used in the evaluation process are defined in this section, along with their applicable values. The applicable value is entered in Table 1401.7 for the Safety Parameter of HVAC Systems on line 1401.6.7. These values are not occupancy sensitive, since the spread of fire is dependent on the HVAC system present in the existing building, not on its occupancy classification.

Category a requires a value of -10 for existing buildings that contain plenums not in compliance with the requirements of Section 602 of the *International Mechanical Code*® (IMC®). Locations of plenums in the existing buildings, the materials they are built of relative to the building's type of construction and the materials exposed to plenum air must all be evaluated.

Category b corresponds to existing buildings that have exit access corridors or exit stairways that are used to supply, return or exhaust air, or for other ventilation purposes. That type of layout puts the existing building's occupants at greater risk and must, therefore, be penalized with a value of -5.

Category c is applicable when an existing building has both noncomplying plenums and corridors or stairways used for air movement. A value of -15 must be assigned to such a building because the movement of smoke and fire throughout would pose an even greater hazard to the occupants.

Category d represents the base value of zero. A newly constructed building is required to meet all the provisions of IBC Section 1018.5 and IMC Section 602. An existing building that complies with these provisions is meeting this same minimum compliance level and, therefore, is neither penalized nor given benefit for that compliance.

Category e specifies a value of 5 points for HVAC systems that serve only one story of a building. The hazards of a fire spreading laterally through a story of a building via the HVAC system are minimal; therefore, the code assigns a positive value. A boiler/chiller system also does not lend itself to fire spread as long as there is no air movement in ducts.

Example 1:

A Group R-2 apartment building is six stories in height. Each apartment has its own HVAC equipment located in the dwelling unit. Bathrooms are ventilated by fans connected to a central exhaust shaft; kitchen exhaust hoods connect to a central exhaust shaft serving all floors. Corridors have a direct outside-supply air system on each floor with no exhaust. The HVAC systems are classified as Category b because the corridors are being used as the makeup air source for the bathroom and kitchen exhausts. A score of -5 is, therefore, entered in Table 1401.7 for the Safety Parameter for the HVAC Systems on line 1401.6.7.

Example 2:

A Group S-2 low-hazard storage occupancy is located in a one-story building. The building is preengineered steel Type IIB construction, and has a suspended gypsum board ceiling. Gypsum board has also been attached to the underside of the ceiling joists to serve as the upper membrane of a return air plenum. The plenum also contains plastic fire sprinkler piping for the automatic fire suppression system that is installed throughout the building.

The plenum is classified as noncombustible in accordance with Section 602 of the IMC. As long as the plastic fire sprinkler piping meets the optical density and flame spread limits, the plenum is in compliance with the code requirements for new construction. Therefore, the HVAC system will be classified as Category d. A value of zero is, therefore, entered in Table 1401.7 for the Safety Parameter of the HVAC Systems on line 1401.6.7.

1401.6.8 Automatic fire detection. Evaluate the smoke detection capability based on the location and operation of automatic fire detectors in accordance with Section 907 of the *International Building Code* and the *International Mechanical Code*. Under the categories and occupancies in Table 1401.6.8, determine the appropriate value and enter that value into Table 1401.7 under Safety Parameter 1401.6.8, Automatic Fire Detection, for fire safety, means of egress, and general safety. Facilities in Group I-2 occupancies meeting Category a, b or c shall be considered to fail the evaluation.

❖ This section considers the use of smoke detectors in a building. To receive credit for the smoke detectors, they must be connected to audible alarms and installed in accordance with IBC Section 907 and IMC Section 606.

TABLE 1401.6.8
AUTOMATIC FIRE DETECTION VALUES

OCCUPANCY	CATEGORIES					
	a	b	c	d	e	f
A-1, A-3, F, M, R, S-1	-10	-5	0	2	6	—
A-2	-25	-5	0	5	9	—
A-4, B, E, S-2	-4	-2	0	4	8	—
I-2	NP	NP	NP	4	5	2

❖ Table 1401.6.8 assigns values for each occupancy. The use of detectors increases the safety in a building by providing early warning of a fire condition to occupants. The lack of detectors and the associated early warning is considered less than optimum for safety in occupancies with high population densities and high combustible loads. Large deficiency points are accrued if adequate detection systems are not provided.

Example:

A four-story Group R-1 hotel has corridors and an elevator lobby on each story. Smoke detectors are installed throughout the corridors, closets, rooms and elevator lobbies. Single-station detectors are installed in the guestrooms. There are smoke detectors in the HVAC return air system and a fire alarm system is provided. An analysis of these building characteristics results in the selection of Category d. To be classified in Category e, guestrooms must have detectors connected to the building's emergency electrical system and annunciated by each room at a constantly attended location, such as the front desk. Additionally, the fire alarm system must be capable of being manually activated by the front desk when a smoke detector operates.

Category d value from Table 1401.6.8 = 2.

Enter 2 in Table 1401.7 for the Safety Parameter 1401.6.8 of Automatic Fire Detection on line 1401.6.8.

1401.6.8.1 Categories. The categories for automatic fire detection are:

1. Category a—None.

2. Category b—Existing smoke detectors in HVAC systems and maintained in accordance with the *International Fire Code*.

3. Category c—Smoke detectors in HVAC systems. The detectors are installed in accordance with the requirements for new buildings in the *International Mechanical Code*.

4. Category d—Smoke detectors throughout all floor areas other than individual sleeping units, tenant spaces and dwelling units.

5. Category e—Smoke detectors installed throughout the floor area.

6. Category f—Smoke detectors in corridors only.

❖ The categories are based on the location and completeness of the smoke detection system.

The categories represent a graduation in the levels of smoke detection that ranges from no detectors in Category a to full detection throughout all fire area spaces in Category e.

Category a is a facility that has no automatic smoke detection system.

Category b acknowledges that the IMC currently requires HVAC system detectors in more locations than may have been required in less contemporary model codes. This category, therefore, assumes that there are some smoke detectors in an existing building's HVAC system, but not to the extent required by the IMC for new construction. If the limited HVAC system detectors are maintained in accordance with the IFC, the detection system qualifies as Category b.

Category c addresses existing buildings that have upgraded HVAC systems with duct detectors installed in accordance with the IMC. The detection values for Category c are zero, since this is the level of protection required for new construction.

Category d is the classification for existing buildings that have full detection coverage throughout the public and common use spaces.

Category e has smoke detectors throughout floor areas in compliance with the provisions for new construction.

1401.6.9 Fire alarm systems. Evaluate the capability of the fire alarm system in accordance with Section 907 of the *International Building Code*. Under the categories and occupancies in Table 1401.6.9, determine the appropriate value and enter that value into Table 1401.7 under Safety Parameter 1401.6.9, Fire Alarm System, for fire safety, means of egress, and general safety.

❖ This section evaluates the capabilities of building fire alarm systems that are separate from the automatic fire detection system evaluated in Section 1401.6.8. A fire alarm system that is manually operated or activated by smoke detectors or sprinkler water-flow devices alerts the occupants to a fire condition. The fire alarm system will notify the occupants with visible or audible alarms so they may begin to take appropriate action. These systems are of particular importance in assembly, business, educational or residential occupancies, which can have large numbers of occupants in rooms with concentrated seating or people who are sleeping.

TABLE 1401.6.9
FIRE ALARM SYSTEM VALUES

OCCUPANCY	CATEGORIES			
	a	b[a]	c	d
A-1, A-2, A-3, A-4, B, E, R	-10	-5	0	5
F, M, S	0	5	10	15
I-2	-4	1	2	5

a. For buildings equipped throughout with an automatic sprinkler system, add 2 points for activation by a sprinkler water-flow device.

❖ Table 1401.6.9 gives values for each occupancy and type of fire alarm system provided. It reflects the idea that the presence of an alarm system in a building usually creates a safer condition for the occupants when compared to a building without a fire alarm system.

Deficiency points are assigned to occupancies without a fire alarm system but that have a high occupant load or number of sleeping occupants, such as Groups A-1, A-2, A-3, A-4, B, E and R, which are included in Categories a and b.

Example:

A one-story building with an assembly occupancy has a complete manual fire alarm system, a voice alarm system, a public address system and a fire command station that does not contain status indicators and controls for the air-handling system. The fire command station does not have emergency power or lighting system controls. It also does not have a fire department communication panel. As a result, the building is classified in Category c.

Category c value from Table 1401.6.9 = 0.

Enter 0 in Table 1401.7 for the Safety Parameter of Fire Alarm System on line 1401.6.9.

1401.6.9.1 Categories. The categories for fire alarm systems are:

1. Category a—None.

2. Category b—Fire alarm system with manual fire alarm boxes in accordance with Section 907.4 of the *International Building Code* and alarm notification appliances in accordance with Section 907.5.2 of the *International Building Code*.

3. Category c—Fire alarm system in accordance with Section 907 of the *International Building Code*.

4. Category d—Category c plus a required emergency voice/alarm communications system and a fire command station that conforms to Section 911 of the *International Building Code* and contains the emergency

voice/alarm communications system controls, fire department communication system controls, and any other controls specified in Section 911 of the *International Building Code* where those systems are provided.

❖ These categories are defined by the fire alarm system provided in the existing building.

Category a means that there is no fire alarm system in the building or it does not conform to all the requirements of IBC Section 907. This includes a system that does not have a secondary power supply in accordance with IBC Section 907 or one where the zones of a floor exceed 22,500 square feet (2090 m^2), as noted in IBC Section 907.6.3.

Category b applies when the manual fire alarm boxes comply with IBC Section 907.4.2 and NFPA 72. The alarm-notification appliances, specifically the audible alarms, are in accordance with IBC Section 907.5.2.1. Location, height and color of the manual fire alarm boxes must be specifically evaluated for compliance, along with the sound levels of the audible alarms when compared to the normal sound levels in the existing building.

Category c requires that the fire alarm system comply with IBC Section 907 and NFPA 72. This category indicates that a complete fire alarm system is present in the existing building and that the system complies with all the requirements for new construction.

Category d is applicable for a building that is provided with a fire command center for fire department operations that complies with IBC Section 911. This is a type of fire command station that is only required in new construction for high-rise buildings (see IBC Section 403.4.6). IBC Section 911 may be used when evaluating any building that complies with its performance requirements. This section specifies the fire command operations required, including the following:

- Voice/alarm communication system panels.
- Fire department communication panels.
- Fire detection and alarm system annunciator panels.
- An annunciator for visually indicating floor locations of elevators and whether they are operational.
- Status indicators and controls for air-handling systems.
- Controls for unlocking all stairway doors simultaneously.
- Sprinkler valve and water-flow detector display panels.
- Emergency and standby power status indicators.
- Elevator fire recall switch in accordance with ASME A17.1.

1401.6.10 Smoke control. Evaluate the ability of a natural or mechanical venting, exhaust, or pressurization system to control the movement of smoke from a fire. Under the categories and occupancies in Table 1401.6.10, determine the appropriate value and enter that value into Table 1401.7 under Safety Parameter 1401.6.10, Smoke Control, for means of egress and general safety.

❖ This section is used to evaluate characteristics that could limit smoke migration in the building, including operable windows, mechanical exhaust systems, pressurized stairways or smokeproof enclosures.

TABLE 1401.6.10
SMOKE CONTROL VALUES

OCCUPANCY	CATEGORIES					
	a	b	c	d	e	f
A-1, A-2, A-3	0	1	2	3	6	6
A-4, E	0	0	0	1	3	5
B, M, R	0	2^a	3^a	3^a	3^a	4^a
F, S	0	2^a	2^a	3^a	3^a	3^a
I-2	-4	0	0	0	3	0

a. This value shall be 0 if compliance with Category d or e in Section 1401.6.8.1 has not been obtained.

❖ Table 1401.6.10 assigns values for each occupancy and category of smoke control in the building. The table represents the relative benefits to the building occupants of various levels of smoke control methods provided in the building.

There are no negative points accrued under any circumstances, but zero positive points are indicated if no smoke control or only limited smoke control is provided. The table also indicates the value of smoke control in the exit stairs of a building and of automatic sprinkler systems as a means of limiting the volume of smoke production from a fire. Note a of the table assigns zero credit points for Groups B, M, R, F and S unless the building complies with Category d or e in Section 1401.6.8.1, which are the automatic fire detection system requirements.

Example 1:

A Group A-3 church sanctuary is located in a one-story building. It has no operable windows and no smoke control system.

Category a is applicable and the value from Table 1401.6.10 = 0.

Enter 0 in Table 1401.7 for the Safety Parameter of Smoke Control on line 1401.6.10, for only means of egress (ME) and general safety (GS) parameters. No entry is made under the fire safety (FS) parameter.

Example 2:

A three-story Group E high school has operable windows throughout the building. The stairways are interior without windows or a pressurization system.

Category b is applicable and the value from Table 1401.6.10 = 0.

Enter 0 in Table 1401.7 under Safety Parameter 1401.6.10, Smoke Control, for means of egress (ME) and general safety (GS) parameters.

Example 3:

A two-story Group A-2 nightclub is sprinklered throughout and has a smoke control system that meets the requirements for Category e.

Category e is applicable and the value from Table 1401.6.10 = 6.

Enter 6 in Table 1401.7 under Safety Parameter 1401.6.10, Smoke Control, for means of egress (ME) and general safety (GS) parameters.

Example 4:

A six-story Group B office building has three stairways: one is a smokeproof enclosure conforming to Section 909.20; one is pressurized in accordance with Section 909.20.5; and one has operable exterior windows. The building has smoke detectors throughout all floor spaces and fire areas.

Category f is applicable and the value from Table 1401.6.10 = 4.

Enter 4 in Table 1401.7 under Safety Parameter 1401.6.10, Smoke Control, for means of egress (ME) and general safety (GS).

If detectors are omitted from some of the offices of the buildings, then a score of 0 must be entered in Table 1401.7. This is determined from Note a in Table 1401.6.10.

1401.6.10.1 Categories. The categories for smoke control are:

1. Category a—None.

2. Category b—The building is equipped throughout with an automatic sprinkler system. Openings are provided in exterior walls at the rate of 20 square feet (1.86 m^2) per 50 linear feet (15 240 mm) of exterior wall in each story and distributed around the building perimeter at intervals not exceeding 50 feet (15 240 mm). Such openings shall be readily openable from the inside without a key or separate tool and shall be provided with ready access thereto. In lieu of operable openings, clearly and permanently marked tempered glass panels shall be used.

3. Category c—One enclosed exit stairway, with ready access thereto, from each occupied floor of the building. The stairway has operable exterior windows, and the building has openings in accordance with Category b.

4. Category d—One smokeproof enclosure and the building has openings in accordance with Category b.

5. Category e—The building is equipped throughout with an automatic sprinkler system. Each floor area is provided with a mechanical air-handling system designed to accomplish smoke containment. Return and exhaust air shall be moved directly to the outside without recirculation to other floor areas of the building under fire conditions. The system shall exhaust not less than six air changes per hour from the floor area. Supply air by mechanical means to the floor area is not required. Containment of smoke shall be considered as confining smoke to the floor area involved without migration to other floor areas. Any other tested and approved design that will adequately accomplish smoke containment is permitted.

6. Category f—Each stairway shall be one of the following: a smokeproof enclosure in accordance with Section 1023.11 of the *International Building Code*; pressurized in accordance with Section 909.20.5 of the *International Building Code*; or shall have operable exterior windows.

❖ The six categories to be evaluated are compared to the occupancies to determine the score to be entered in Table 1401.7 for the Safety Parameter of Smoke Control on line 1401.6.10.

Category a means there is no method of controlling smoke in the building.

Category b means the existing building is sprinklered and there are exterior windows that can be readily opened without the use of keys or tools. This category recognizes the benefit of manual venting capability taken from the high-rise provisions of the IBC for new construction, which include automatic sprinkler protection. Operable panels or windows in the exterior walls must be provided at the rate of 20 square feet per 50 lineal feet (1.85 m^2 per 15 240 mm) of exterior wall in each story. The openings must be distributed around the perimeter of the building at intervals not more than 50 feet (15 240 mm) between windows. A story with very high ceilings may have a strip of windows located well above the floor with latches or controls that are not easily reachable. Such openings would not meet Category b standards, even if they are provided and distributed at the required rate.

Category c requires at least one enclosed exit stairway to have operable windows that open to the exterior. Additionally, the building must have operable windows complying with the size and spacing requirements of Category b.

Category d requires a minimum of one smokeproof enclosure and operable windows in the building. These operable windows are the same as those addressed in Category b. By definition, a smokeproof enclosure refers to an enclosed interior exit stairway that conforms to IBC Section 1009 as designated in IBC Section 1022.

Section 1022.9 requires stairways to be protected from smoke by one of two methods: a smokeproof enclosure or stairway pressurization design (see IBC Section 909.20.5).

Additional requirements that must be considered when evaluating a smokeproof enclosure are: access (see IBC Section 909.20.1), construction (see IBC Section 909.20.2), ventilating equipment (see IBC Section 909.20.6) and standby power (see IBC Section 909.20.6.2).

Category e recognizes the use of a mechanical smoke control system designed in accordance with the provisions of the code in a fully sprinklered building. Each fire area in the building must have a mechanical smoke control system. Return and exhaust air from the system must be discharged directly to the exterior to achieve the necessary level of protection. A specific air change requirement is provided, independent of the fire area's volume or size. Note that a smoke control system designed in accordance with IBC Section 909 is also considered acceptable.

Category f recognizes the merits of smokeproof enclosures, pressurized stairways and stairways with operable exterior windows. To be classified in this category, all stairways in the building must comply with the requirements of any one, or a combination of, the three types of stairways. It is possible for a single building to have a smokeproof enclosure, a pressurized stairway and a stairway with operable exterior windows.

Both the categories and the occupancy of the building are used in determining the score that will be entered in Table 1401.7. This score is determined from Table 1401.6.10. Note a in Table 1401.6.10 can have a significant impact on the scores. The note states that, even if the building has some level of smoke control, it is to receive no credit if it does not have an automatic fire detection system complying with Category d or e in Section 1401.6.8.1.

1401.6.11 Means of egress capacity and number. Evaluate the means of egress capacity and the number of exits available to the building occupants. In applying this section, the means of egress are required to conform to the following sections of the *International Building Code:* 1003.7, 1004, 1005, 1006, 1007, 1016.2, 1026.1, 1028.2, 1028.5, 1029.2, 1029.3, 1029.4 and 1030. The number of exits credited is the number that is available to each occupant of the area being evaluated. Existing fire escapes shall be accepted as a component in the means of egress when conforming to Section 405.

Under the categories and occupancies in Table 1401.6.11, determine the appropriate value and enter that value into Table 1401.7 under Safety Parameter 1401.6.11, Means of Egress Capacity, for means of egress and general safety.

❖ This section addresses the exit capacity and number of existing exits available to the building occupants. Before a building can be evaluated in this category, and Chapter 14 in general, it must comply with the listed sections of the IBC.

The means of egress is required to conform to the following IBC sections for new construction before evaluation:

- IBC Section 1003.7 establishes that elevators, escalators and moving walks cannot be considered as part of the means of egress.

- IBC Section 1004 establishes the minimum number of occupants the exit facilities must accommodate.
- IBC Section 1005.3 defines the capacity of the means of egress by identifying a minimum width of egress component per occupant. This is used to calculate the total capacity of the means-of-egress component.
- IBC Section 1006.2.1 establishes requirements for common path of egress travel.
- IBC Section 1007.1 establishes requirements for the remoteness of exit doors and means-of-egress doors.
- IBC Section 1006 establishes requirements for minimum number of exits.
- IBC Sections 1027.1, 1027.2 and 1027.5 establish requirements related to the exit discharge.
- IBC Sections 1028.2, 1028.3 and 1028.4 establish requirements related to exits and lobbies for assembly occupancies.
- IBC Section 1030 establishes requirements for emergency escape and rescue.

TABLE 1401.6.11
MEANS OF EGRESS VALUES[a]

OCCUPANCY	CATEGORIES				
	a	b	c	d	e
A-1, A-2, A-3, A-4, E, I-2	-10	0	2	8	10
M	-3	0	1	2	4
B, F, S	-1	0	0	0	0
R	-3	0	0	0	0

a. The values indicated are for buildings six stories or less in height. For buildings over six stories above grade plane, add an additional -10 points.

❖ Table 1401.6.11 assigns values for each occupancy and category from Section 1401.6.11.1. The table gives credit for providing additional numbers of exits and additional exit capacity beyond the minimum required for new construction.

This additional evaluation contributes to the overall assessment of the building's safety and may offset other safety deficiencies [see Commentary Figures 1401.6.11(1) and 1401.6.11(2)]. Note a of the table provides a significant penalty for high-rise buildings that use fire escapes as part of the means of egress. In this case, if the building is seven stories or greater in height and uses fire escapes, -10 points must be added to the values already listed for Category a buildings.

PERFORMANCE COMPLIANCE METHODS

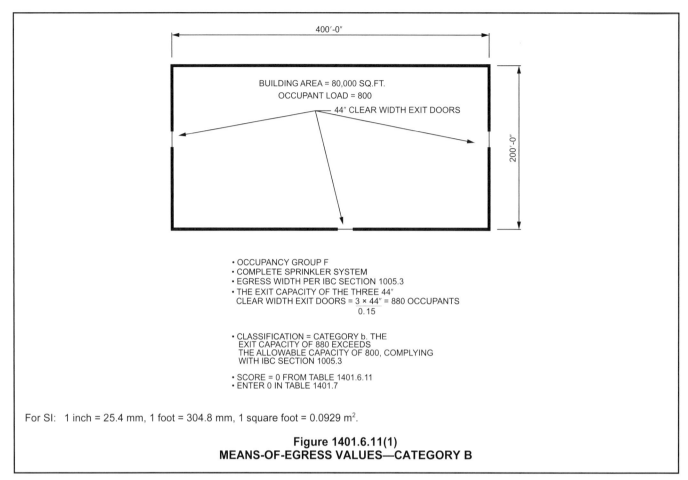

Figure 1401.6.11(1)
MEANS-OF-EGRESS VALUES—CATEGORY B

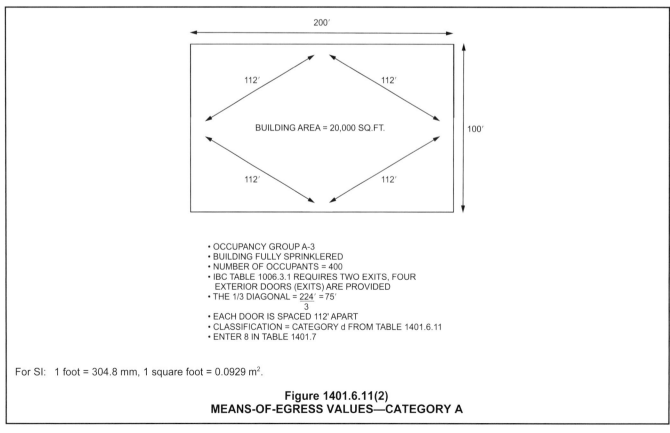

Figure 1401.6.11(2)
MEANS-OF-EGRESS VALUES—CATEGORY A

1401.6.11.1 Categories. The categories for means-of-egress capacity and number of exits are:

1. Category a—Compliance with the minimum required means-of-egress capacity or number of exits is achieved through the use of a fire escape in accordance with Section 405.

2. Category b—Capacity of the means of egress complies with Section 1005 of the *International Building Code*, and the number of exits complies with the minimum number required by Section 1006 of the *International Building Code*.

3. Category c—Capacity of the means of egress is equal to or exceeds 125 percent of the required means-of-egress capacity, the means of egress complies with the minimum required width dimensions specified in the *International Building Code*, and the number of exits complies with the minimum number required by Section 1006 of the *International Building Code*.

4. Category d—The number of exits provided exceeds the number of exits required by Section 1006 of the *International Building Code*. Exits shall be located a distance apart from each other equal to not less than that specified in Section 1007 of the *International Building Code*.

5. Category e—The area being evaluated meets both Categories c and d.

❖ Five categories must be considered. These categories and the occupancy of the building will determine the score entered in Table 1401.7.

Category a is applicable to buildings that comply with either the means-of-egress capacity or the number of exits, including the use of fire escapes. Although the code allows fire escapes to be used as an egress element in existing buildings, this is the least desirable option in any type of building. The code requires a building using fire escapes to use Category a and its corresponding negative points.

Category b is used for buildings that meet the minimum requirements of IBC Section 1004 for capacity of means of egress and IBC Section 1021 for minimum number of exits and continuity.

Category c is used for buildings that exceed the requirements for new construction. This category provides small positive points for existing buildings that meet all of the following requirements:

- The capacity of all means-of-egress components is greater than 125 percent of the required capacity.

- All means-of-egress components comply with the minimum required widths [e.g., 32-inch (814 mm) clear for doors, corridors that are 44 inches (1118 mm) wide and stairways].

- The minimum number of exits is provided based on the number of occupants on each floor level.

By providing oversized egress capacity and minimum egress width requirements, an existing building has added safety, which is rewarded with a small number of positive points.

Category d simply relates to buildings where more exits are provided than required by IBC Section 1021, which contributes a positive factor to the building's exit capacity.

Before credit can be given to an existing building with a greater number of exits, the exits must be evaluated for their remoteness and independence from each other. All of the exits must be located at least one-half the length of the diagonal from each other before Category d can be used. If the building is equipped throughout with an automatic sprinkler system in accordance with NFPA 13 or 13R, the exits can be located one-third the length of the diagonal from each other.

Category e provides additional credit for existing buildings that have the characteristics of both Categories c and d. Such a building, which has oversized capacities in its egress elements, minimum required clear widths for the egress components and additional exits that are all remotely located from one another, is considered to have the highest degree of safety; however, this is true for only some of the applicable occupancies. In Group B, F, S and R occupancies, any additional capacity or increased number of exits is not a significant factor. No positive points are awarded for this life-safety issue for these occupancies.

1401.6.12 Dead ends. In spaces required to be served by more than one means of egress, evaluate the length of the exit access travel path in which the building occupants are confined to a single path of travel. Under the categories and occupancies in Table 1401.6.12, determine the appropriate value and enter that value into Table 1401.7 under Safety Parameter 1401.6.12, Dead Ends, for means of egress and general safety.

❖ This section is used to evaluate dead-end exit access conditions within the building. This section uses the terminology "confined to a single path of travel." Another way to illustrate the meaning of this is "a single direction of travel to reach an exit." A corridor is typically of a width that provides effectively only one path in which a building occupant can travel in one of two directions to reach exits. A dead end may be a single path, but the key feature of a dead end is that only one direction is available to reach an exit. When building occupants have only one direction of travel available, a potentially hazardous condition is created because they may become trapped if the direction of travel to the exit is blocked by fire or smoke.

This section addresses only dead ends that are a component of exit access travel, which is "that portion of a means of egress that leads to an entrance to an exit." Although a room containing only one egress door has only one way out, it is not considered a dead-end condition. The code takes this condition into account in establishing the limitations under which a room is allowed to have only one means of egress. Dead ends are a concern in passageways and corridors where occupants may not realize that the corridor ends and they may have to retrace their steps in order to remain on a path toward an exit.

**TABLE 1401.6.12
DEAD-END VALUES**

OCCUPANCY	CATEGORIES[a]			
	a	b	c	d
A-1, A-3, A-4, B, F, M, R, S	-2	0	2	-4
A-2, E	-2	0	2	-4
I-2	-2	0	2	-6

a. For dead-end distances between categories, the dead-end value shall be obtained by linear interpolation.

❖ The table reflects the relative degree of hazard associated with dead-end passageways and corridors. This is shown by the deficiency points that apply where a dead end exceeds 35 or 70 feet (10 668 or 21 336 mm) in an occupancy with a relatively high occupant load, such as an assembly occupancy. Note a allows for the interpolation for actual dead-end lengths between the distances specified in the categories. For example, if a building contains a corridor that has a dead-end length of 10 feet (3048 mm) at each end beyond the exits, the value of 1 is used; this is the midpoint between Categories b and c.

1401.6.12.1 Categories. The categories for dead ends are:

1. Category a—Dead end of 35 feet (10 670 mm) in nonsprinklered buildings or 70 feet (21 340 mm) in sprinklered buildings.

2. Category b—Dead end of 20 feet (6096 mm); or 50 feet (15 240 mm) in Group B in accordance with Section 1020.4, Exception 2, of the *International Building Code*.

3. Category c—No dead ends; or ratio of length to width (l/w) is less than 2.5:1.

4. Category d—Dead ends exceeding Category a.

❖ This section defines the categories for dead-end exit access conditions.

Category a allows dead-end conditions of up to 35 feet (10 668 mm) for existing buildings that are not fully sprinklered and up to 70 feet (21 336 mm) for existing buildings that are fully sprinklered. These distances correspond to the absolute maximum dead-end lengths allowed in existing buildings by the IFC. Dead ends greater than these distances are considered an unsafe condition because they are not allowed by the IFC to occur in an existing building. It would, therefore, be inconsistent to allow greater dead-end lengths under this evaluation method. Since these lengths far exceed the allowable lengths permitted for new construction, negative values are associated with this category.

Category b is the classification for buildings that comply with the requirements for new construction. This zero-based category provides no extra credit for complying with new construction requirements.

Category c represents conditions that exceed the requirements for new construction. If there are no dead-end corridors or the "corridor" is more of a "space," Category c can be used. In Category c, the length-to-width ratio of 2.5 to define a space/corridor is the same as Exception 3 in IBC Section 1020.4. Such a space allows the building user a more circular route and full view of the space and, therefore, does not represent as great a potential hazard as a classic dead-end corridor.

These categories and occupancies of the building are used in Table 1401.6.12 to determine the score to be entered in Table 1401.7 for the Safety Parameter of Dead Ends on line 1401.6.12 [see Commentary Figures 1401.6.12.1(1), 1401.6.12.1(2) and 1401.6.12.1(3)].

1401.6.13 Maximum exit access travel distance to an exit. Evaluate the length of exit access travel to an approved exit. Determine the appropriate points in accordance with the following equation and enter that value into Table 1401.7 under Safety Parameter 1401.6.13, Maximum Exit Access Travel Distance for means of egress and general safety. The maximum allowable exit access travel distance shall be determined in accordance with Section 1016.1 of the *International Building Code*.

$$\text{Points} = 20 \times \frac{\text{Maximum allowable travel distance} - \text{Maximum actual travel distance}}{\text{Maximum allowable travel distance}}$$

(Equation 14-6)

❖ The length of exit access travel distance is evaluated in comparison to the travel distance allowed by IBC Table 1017.2 for a particular occupancy. The exit access travel distance is measured from the most remote point in the building to the nearest entrance to an exit. Total exit access travel distance in IBC Table 1017.2 includes travel within a room or space plus any exit access corridor travel to the exit. Some modifications to the IBC Table 1017.2 requirements are listed for certain occupancies and buildings as listed in the notes to that table.

To determine the value to be assigned for maximum travel distances to an exit, the specified equation must be used. This equation allows a graduated scale to be used to evaluate compliance of an existing situation with new construction travel distance requirements. With the equation, a more definite evaluation can occur with a broader range of scores. Any existing building having overall travel distances less than those specified in IBC Table 1017.2 will achieve a positive credit. Existing buildings with travel distances greater than those allowed for new construction will be assigned negative points based on the extent the travel distance exceeds the allowable limit.

For example, an existing Group B business building has a travel distance of 150 feet (45 720 mm). The distance is measured from the most remote corner of a partitioned office, down a corridor and to an exit door. IBC Table 1017.2 permits an unsprinklered Group B building to have a travel distance of 200 feet (60 960 mm). The equation yields the following:

$$\text{Points} = 20\left(\frac{200 \text{ feet} - 150 \text{ feet}}{200 \text{ feet}}\right)$$

$$= 5.0$$

See Commentary Figures 1401.6.13(1), 1401.6.13(2) and 1401.6.13(3) for other examples.

1401.6.14 Elevator control. Evaluate the passenger elevator equipment and controls that are available to the fire department to reach all occupied floors. Emergency recall and in-car operation of elevators shall be provided in accordance with the *International Fire Code*. Under the categories and occupancies in Table 1401.6.14, determine the appropriate value and enter that value into Table 1401.7 under Safety Parameter 1401.6.14, Elevator Control, for fire safety, means of egress and general safety. The values shall be zero for a single-story building.

❖ The availability of elevators in the building and their incapability in an emergency are evaluated. This section addresses four different categories of elevators in existing buildings. This section does not address the requirements of IBC Chapter 30. Access must be provided to all occupied floors by passenger elevators for the purposes of this evaluation. Freight elevators cannot be considered, since they may be in locations not readily accessible for fire department use. Emergency elevator operation must comply with the IFC for either a Phase I emergency recall or Phase II emergency in-car operation. One-story buildings, including those with mezzanines served by an elevator, are awarded zero value.

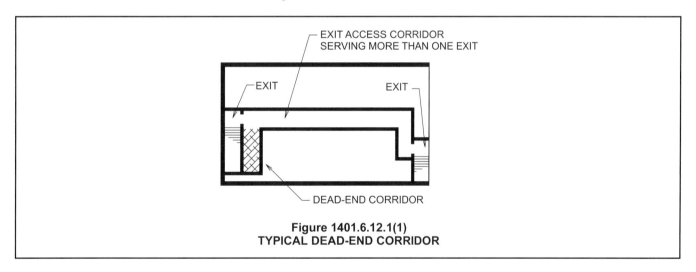

**Figure 1401.6.12.1(1)
TYPICAL DEAD-END CORRIDOR**

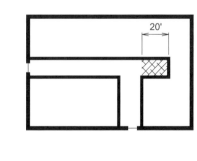

- OCCUPANCY GROUP B WITH MOVABLE PARTITIONS AND FURNITURE 6'-6" HIGH
- DEAD-END PASSAGEWAY IS 20'-0"
- CLASSIFICATION = CATEGORY b
- VALUE = 0 FROM TABLE 1401.6.12
- ENTER 0 IN TABLE 1401.7

For SI: 1 inch = 25.4 mm, 1 foot = 304.8 mm.

**Figure 1401.6.12.1(2)
DEAD-END VALUES—CATEGORY B**

PERFORMANCE COMPLIANCE METHODS

TABLE 1401.6.14
ELEVATOR CONTROL VALUES

ELEVATOR TRAVEL	CATEGORIES			
	a	b	c	d
Less than 25 feet of travel above or below the primary level of elevator access for emergency fire-fighting or rescue personnel	-2	0	0	+2
Travel of 25 feet or more above or below the primary level of elevator access for emergency fire-fighting or rescue personnel	-4	NP	0	+4

For SI: 1 foot = 304.8 mm.
NP = Not permitted.

❖ The table assigns values based on the types of controls the elevators have, and the distance the elevators must travel to reach the floors they serve. The 25-foot (7620 mm) threshold of elevator travel is based on the IFC requirements and ASME A17.1. Elevator travel distance is based on the level where the elevator is accessed by the fire department. Usually, this is the ground floor or grade-level floor and corresponds to the level where fire command stations or fire alarm system annunciator panels are located. Any elevator that travels 25 feet (7620 mm) or more must always be provided with Phase I and II recall capabilities.

Example:

Assume a Group B business occupancy, three stories in height where the elevator lobbies are equipped

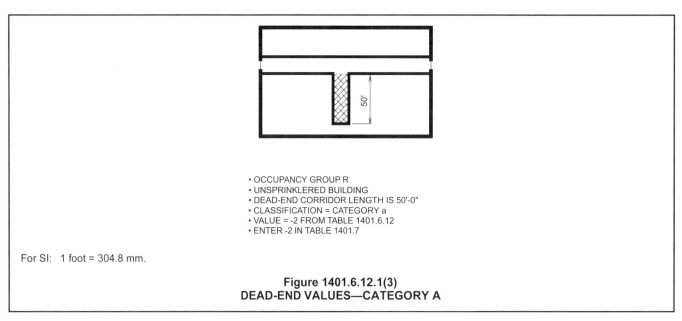

For SI: 1 foot = 304.8 mm.

**Figure 1401.6.12.1(3)
DEAD-END VALUES—CATEGORY A**

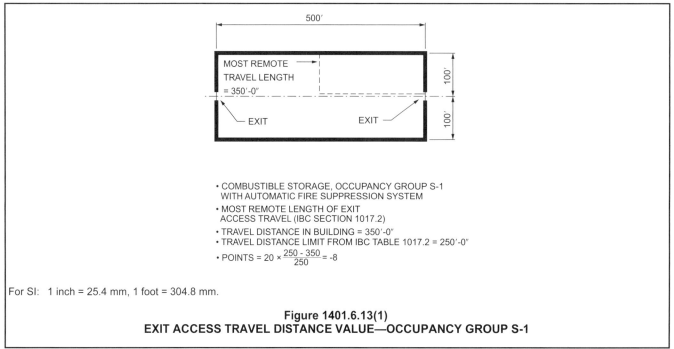

For SI: 1 inch = 25.4 mm, 1 foot = 304.8 mm.

**Figure 1401.6.13(1)
EXIT ACCESS TRAVEL DISTANCE VALUE—OCCUPANCY GROUP S-1**

with smoke detectors that recall the elevator to the main floor, or to an alternative floor if the main floor detector is activated. The elevator travels more than 25 feet (7620 mm) above the main floor.

This building must be placed in Category c. Because the elevator travels more than 25 feet (7620 mm) above the main floor, it must be brought into compliance with the IFC, which requires Phase I or II recall. When the elevators are brought into compliance with the IFC, they are placed automatically in Category c. The value from Table 1401.6.14 is zero; therefore, zero must be entered in Table 1401.7 for the Safety Parameter of Elevator Control on line 1401.6.14.

1401.6.14.1 Categories. The categories for elevator controls are:

1. Category a—No elevator.

2. Category b—Any elevator without Phase I emergency recall operation and Phase II emergency in-car operation.

3. Category c—All elevators with Phase I emergency recall operation and Phase II emergency in-car operation as required by the *International Fire Code*.

4. Category d—All meet Category c; or Category b where permitted to be without Phase I emergency recall operation and Phase II emergency in-car operation; and at least one elevator that complies with new construction requirements serves all occupied floors.

❖ The categories that a building may be placed in range from buildings with no elevators to buildings with elevators complying with new construction requirements and total elevator recall capability.

Category a is applicable to buildings with no elevators. Without an elevator in a multiple-story building, fire department personnel are required to use the stairs for rescuing people and accessing fire floors. Category a is assigned a negative value.

Category b is applicable to buildings in which elevators are present but have no recall controls. Without recall or fire department control, the elevators are allowed vertical travel distances of less than 25 feet (7620 mm). This approach is consistent with the IFC, which requires controls for elevators reaching 25 feet (7620 mm) or more.

Category c is applicable to buildings with elevators that have automatic recall as required by the IFC.

Category d is used for buildings when the elevator

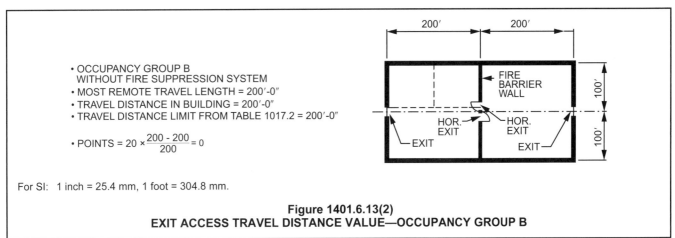

For SI: 1 inch = 25.4 mm, 1 foot = 304.8 mm.

**Figure 1401.6.13(2)
EXIT ACCESS TRAVEL DISTANCE VALUE—OCCUPANCY GROUP B**

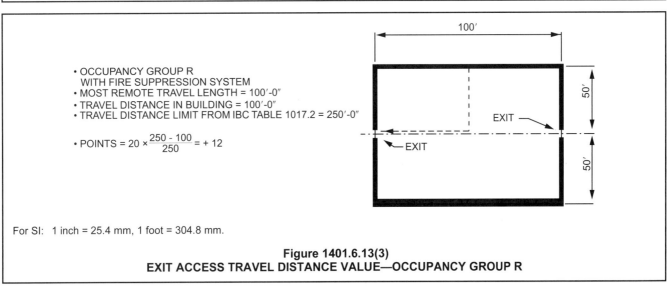

For SI: 1 inch = 25.4 mm, 1 foot = 304.8 mm.

**Figure 1401.6.13(3)
EXIT ACCESS TRAVEL DISTANCE VALUE—OCCUPANCY GROUP R**

controls comply with either Category b or c and at least one of the elevators in the existing building serves all occupied floor levels. The controls must also comply with all requirements for new construction. This category recognizes the benefits of having all elevators with Phase I emergency recall and Phase II emergency in-call operation controls, as well as having a newly constructed elevator with all the features to facilitate fire-fighting and rescue operations.

Based on the controls provided, or the lack of elevators, the appropriate category is determined. This category and the distance the elevator travels are then used along with Table 1401.6.14 to determine the value score the building will receive for this item. The value is then entered in Table 1401.7.

1401.6.15 Means-of-egress emergency lighting. Evaluate the presence of and reliability of means-of-egress emergency lighting. Under the categories and occupancies in Table 1401.6.15, determine the appropriate value and enter that value into Table 1401.7 under Safety Parameter 1401.6.15, Means-of-Egress Emergency Lighting, for means of egress and general safety.

❖ Lighting throughout the entire means of egress is evaluated in this section. Illumination of the means of egress is essential in a building during normal occupancy. During an emergency, illumination becomes even more important, because occupants may be under more stress when seeking to evacuate the building and visibility may be reduced by the buildup of smoke from a fire. The relative value of means-of-egress lighting is represented by the positive points assigned by Table 1401.6.15 when reliability of egress lighting is provided beyond the minimum required by the code.

TABLE 1401.6.15
MEANS-OF-EGRESS EMERGENCY LIGHTING VALUES

NUMBER OF EXITS REQUIRED BY SECTION 1015 OF THE INTERNATIONAL BUILDING CODE	CATEGORIES		
	a	b	c
Two or more exits	NP	0	4
Minimum of one exit	0	1	1

NP = Not permitted.

❖ This table is used to assess the relative risk to the occupants of the building when egress lighting is not adequate. The table reinforces the idea that when a building is required to have two or more exits, lighting has more significance. The lighting values are less critical in smaller buildings where only one exit is required because the hazard of a substantial occupant load locating and using the means of egress is minimal.

Example:

A Group A church building with a number of classrooms that will accommodate up to 60 people has at least 1 footcandle (11 lux) of illumination at the floor level throughout the entire means of egress. The means-of-egress lighting in the sanctuary, corridors and stairways is provided with battery back-up power that is sized to provide lighting for 1 hour. The classroom lighting is wired to the main switch panel in the building and has no emergency power source.

The lighting must be placed in Category a because the classrooms are part of the means of egress (exit access) and the lack of emergency power does not comply with IBC Section 1006.3. The classroom will accommodate up to 60 occupants; therefore, in accordance with IBC Section 1006.3.1 and IBC Table 1006.3.1, the rooms are required to have two means of egress. The lighting must be provided with an emergency power source in accordance with IBC Section 1006.3. The value determined from Table 1401.6.15 is "NP." Emergency power must be added to the lighting in the classrooms. The resulting Category b has a value of 0, which should be entered in Table 1401.7.

1401.6.15.1 Categories. The categories for means-of-egress emergency lighting are:

1. Category a—Means-of-egress lighting and exit signs not provided with emergency power in accordance with Section 2702 of the *International Building Code*.

2. Category b—Means-of-egress lighting and exit signs provided with emergency power in accordance with Section 2702 of the *International Building Code*.

3. Category c—Emergency power provided to means-of-egress lighting and exit signs, which provides protection in the event of power failure to the site or building.

❖ There are three categories of egress lighting. These categories are consistent with the requirements established by the IFC.

Category a is applicable when there is no emergency power source provided for the lighting in the means of egress. Without an emergency power source to ensure continued illumination of the means of egress, the means-of-egress lighting and exit signs would not provide as reliable a level of safety to the building occupants. For buildings in which only one exit is required, this category represents the zero-based criteria, which is the level required for new construction.

Category b is applicable to buildings that are provided with an emergency power source for means-of-egress illumination. For buildings in which a minimum of two exits are required, this category represents the zero-based criteria, which is the level required for new construction.

Category c is applicable when emergency power is provided for means-of-egress lighting and exit signs in excess of the minimum requirements for new construction. The emergency power requirements for new construction assume a power failure occurs in the building or somewhere in the building site. It does not assume that the power failure occurs at the source of power to the site (i.e., from the provider). If the emergency power provides full protection to the site or building during power failure, Category c is applicable. Campus-type complexes or buildings that require extra security may have a power plant available to provide complete backup power for indefinite periods. This will qualify for the positive points of this category.

These categories, along with the number of required exits, are used to determine the appropriate value from Table 1401.6.15. The value determined from Table 1401.6.15 is then entered in Table 1401.7.

1401.6.16 Mixed occupancies. Where a building has two or more occupancies that are not in the same occupancy classification, the separation between the mixed occupancies shall be evaluated in accordance with this section. Where there is no separation between the mixed occupancies or the separation between mixed occupancies does not qualify for any of the categories indicated in Section 1401.6.16.1, the building shall be evaluated as indicated in Section 1401.6, and the value for mixed occupancies shall be zero. Under the categories and occupancies in Table 1401.6.16, determine the appropriate value and enter that value into Table 1401.7 under Safety Parameter 1401.6.16, Mixed Occupancies, for fire safety and general safety. For buildings without mixed occupancies, the value shall be zero. Facilities in Group I-2 occupancies meeting Category a shall be considered to fail the evaluation.

❖ This section is used to evaluate mixed occupancies in an existing building and whether the method of separating mixed occupancies conforms to the requirements of IBC Section 508. If IBC Section 508 is not completely understood, see the associated commentary before proceeding with the evaluation of the existing building in this section. Also, refer to the commentary to Section 1401.6.

This section is applicable only to separated mixed occupancies. If a building is a single occupancy, the applicable value for this section is zero. If an existing mixed-occupancy building has no fire-resistance-rated separation between the different uses, or the separation is fire-resistance rated for less than 1 hour, the applicable value is also zero. The building must also be evaluated in accordance with Section 1401.6. The zero-based category for this section is equivalent to full compliance with IBC Section 508, which is the requirement for new construction.

TABLE 1401.6.16
MIXED OCCUPANCY VALUES[a]

OCCUPANCY	CATEGORIES		
	a	b	c
A-1, A-2, R	-10	0	10
A-3, A-4, B, E, F, M, S	-5	0	5
I-2	NP	0	5

NP = not permitted.

a. For fire-resistance ratings between categories, the value shall be obtained by linear interpolation.

❖ This table addresses the relative risk of a building in or close to compliance with the provisions for separated mixed occupancies. Where mixed occupancies are not separated from each other, the risk from hazards is greater in high-density occupancies, such as Groups A-1 and A-2. This risk is also greater in residential occupancies, because occupants may be sleeping and may not be fully alert. For this reason, inadequate separation is given greater negative values. In buildings with lower occupant loads, and where the occupants are alert, the risks are relatively lower.

Note a permits linear interpolation of the corresponding values in each category. For example, an unsprinklered, Group B/Group F-2 mixed occupancy building is separated with a 1-hour fire-resistance-rated assembly. From IBC Table 508.4, the required fire-resistance rating of the separation between a Group B and a Group F-2 occupancy for Category b is 3 hours. Since the rating that is provided is only 1 hour, an interpolation halfway between Category a and Category b can be used. The resulting points are -2.5.

Example 1:

A three-story building has a Group M mercantile occupancy on the first floor and the upper two stories are Group R residential occupancies. The building is Type VB construction, 9,000 square feet (836 m^2) per floor. It is fully sprinklered, and only 25 percent of the perimeter is accessible. Although Type VB construction is not required to be protected, the Group M occupancy on the first floor is separated from the Group R occupancy on the second floor with a 2-hour fire-resistance-rated floor/ceiling assembly. In this case, IBC Table 508.4 requires a 1-hour fire-resistance rating because the building is equipped throughout with an automatic sprinkler system in accordance with IBC Section 903. The 2-hour rating provided is twice that required; therefore, this building qualifies for Category c. A value of 10 points is assigned to the residential portion of the building, and a value of 5 points is assigned to the mercantile portion (see Section 1401.6 for commentary about mixed uses).

Example 2:

A four-story building is Type IIB construction and is fully sprinklered. It has 15,000 square feet (1394 m^2) per floor with a 25-percent open perimeter. The building is to be a Group R-1 hotel, except for a portion of the first floor, which will be used as a Group A-2 nightclub, occupying 1,000 square feet (93 m^2). Although the building is unprotected Type IIB construction, the floor/ceiling assemblies of the first and second floors are 3 hours and the fire barrier assembly at the first floor between the Group R-1 occupancy and the Group A-2 occupancy is also 3 hours. Considering these building characteristics, the following determinations are made:

Because of the presence of an automatic sprinkler system, IBC Table 508.4 requires a fire-resistance rating of 1 hour. Although this building is required to have a 1-hour separation, it has a 3-hour separation, which is more than double the minimum. This building is classified as Category c. A value of 10 points is awarded to both occupancies.

1401.6.16.1 Categories. The categories for mixed occupancies are:

1. Category a—Occupancies separated by minimum 1-hour fire barriers or minimum 1-hour horizontal assemblies, or both.

2. Category b—Separations between occupancies in accordance with Section 508.4 of the *International Building Code*.

3. Category c—Separations between occupancies having a fire-resistance rating of not less than twice that required by Section 508.4 of the *International Building Code*.

❖ This section addresses three different conditions:

Category a is applicable when the rating between mixed occupancies is less than specified in IBC Section 508.4 but is a fire barrier with a minimum 1-hour fire-resistance rating.

Category b is applicable when the separation between occupancies is in accordance with IBC Section 508.4 for fire-resistance ratings.

Category c is applicable when the fire-resistance rating that separates occupancies in an existing building is no less than twice the rating required by IBC Section 508.4. Category c gives bonus points or increased credits to existing buildings that have higher fire-resistance ratings.

Section 1401 does not require full compliance with all of IBC Section 508.4. Section 1401.6.16 evaluates whether the existing building's compliance with IBC Section 508.4 provides relative safety within the building. If the building does comply with IBC Section 508.4, Section 1401.6.16 acknowledges that there is no basis for assigning negative points, nor is there any basis for assigning positive points for safety; therefore, a neutral zero value is assigned. The appropriate value is entered in Table 1401.7 for the Safety Parameter of Mixed Occupancies on line 1401.

1401.6.17 Automatic sprinklers. Evaluate the ability to suppress a fire based on the installation of an automatic sprinkler system in accordance with Section 903.3.1.1 of the *International Building Code*. "Required sprinklers" shall be based on the requirements of this code. Under the categories and occupancies in Table 1401.6.17, determine the appropriate value and enter that value into Table 1401.7 under Safety Parameter 1401.6.17, Automatic Sprinklers, for fire safety, means of egress divided by 2, and general safety. High-rise buildings defined in Chapter 2 of the *International Building Code* that undergo a *change of occupancy* to Group R shall be equipped throughout with an automatic sprinkler system in accordance with Section 403 of the *International Building Code* and Chapter 9 of the *International Building Code*. Facilities in Group I-2 occupancies meeting Category a, b, c or f shall be considered to fail the evaluation.

❖ These provisions are used to determine the amount of credit that can be applied to the evaluation for the installation of an automatic sprinkler system in an existing building.

The value of sprinkler protection is reflected in the additional credit available in this parameter over and above the basic credits that are provided for building height in Section 1401.6.1, building area in Section 1401.6.2 and maximum travel distance to an exit in Section 1401.6.13. Throughout the code, additional credits are offered for the use of an automatic sprinkler system.

The evaluation of sprinklers in an existing building is based on whether an automatic sprinkler system is both required and installed. The criteria used to determine when an automatic sprinkler system is required are tied to the same requirements for new construction in IBC Section 903.2. The thresholds listed in IBC Sections 903.2.1 through 903.2.12 must be used to evaluate whether those characteristics and occupancies are present and whether a sprinkler system would be required for new construction.

The exceptions to IBC Section 903; the requirements for stories (including basements) without openings of IBC Section 903.2.11.1; and the suppression requirements for covered mall buildings, high-rise buildings, public garages and unlimited area buildings must also be evaluated. The determination of whether sprinklers are required in a building or a portion of a building must be done to correctly determine the category applicable to the existing building.

These guidelines for sprinklers allow for a more equitable evaluation of the contribution of that feature to overall building safety. This factor encourages the installation of automatic sprinkler systems in existing buildings by providing substantial negative and positive points.

The values for sprinklers are given in Table 1401.6.17. The appropriate credit values from Table 1401.6.17 are entered in Table 1401.7 for the Safety Parameter of Automatic Sprinklers on line 1401.6.17, for both fire safety (FS) and general safety (GS), but only one-half the value is entered under the means of egress (ME). The one-half credit for egress is allowed because some credits for sprinklers are incorporated into the parameters for means-of-egress capacity (see Section 1401.6.11), dead ends (see Section 1401.6.12) and maximum travel distance to an exit (see Section 1401.6.13).

**TABLE 1401.6.17
SPRINKLER SYSTEM VALUES**

OCCUPANCY	CATEGORIES					
	a[a]	b[a]	c	d	e	f
A-1, A-3, F, M, R, S-1	-6	-3	0	2	4	6
A-2	-4	-2	0	1	2	4
A-4, B, E, S-2	-12	-6	0	3	6	12
I-2	NP	NP	NP	8	10	NP

NP = not permitted.

a. These options cannot be taken if Category a in Section 1401.6.18 is used.

❖ This table lists the credit values for the respective categories of Section 1401.6.17.1, based on the occupancy being evaluated in the existing building. The assembly occupancies containing large combustible fuel loads with large occupant loads are included with those occupancies containing large fuel loads. These occupancies represent buildings where experience has shown that adequate sprinkler systems save lives and where an on-site sprinkler system is necessary to supplement the local fire department capabilities.

Group A-2 buildings are contained in a separate line in the table for determining sprinkler system values. This occupancy with its densely packed high occupant loads, ill-defined seating and aisle arrangements, and facilities services must be separately evaluated to adequately match its hazards with sprinkler system requirements. Churches and schools, which have high occupant loads but low fuel loads, are combined with other low fuel-load occupancies in the last line item in the table. Category c is the zero-based category. The categories to the left of this column contain negative values that define buildings and occupancies that would be required to be sprinklered as new construction, but are not sprinklered or are provided with inadequate sprinkler systems. The three categories to the right of this column contain positive values that represent buildings and occupancies with sprinkler systems that are deemed adequate or comply with current standards and requirements.

1401.6.17.1 Categories. The categories for automatic sprinkler system protection are:

1. Category a—Sprinklers are required throughout; sprinkler protection is not provided or the sprinkler system design is not adequate for the hazard protected in accordance with Section 903 of the *International Building Code*.

2. Category b—Sprinklers are required in a portion of the building; sprinkler protection is not provided or the sprinkler system design is not adequate for the hazard protected in accordance with Section 903 of the *International Building Code*.

3. Category c—Sprinklers are not required; none are provided.

4. Category d—Sprinklers are required in a portion of the building; sprinklers are provided in such portion; the system is one that complied with the code at the time of installation and is maintained and supervised in accordance with Section 903 of the *International Building Code*.

5. Category e—Sprinklers are required throughout; sprinklers are provided throughout in accordance with Chapter 9 of the *International Building Code*.

6. Category f—Sprinklers are not required throughout; sprinklers are provided throughout in accordance with Chapter 9 of the *International Building Code*.

❖ Six categories are defined in this section for evaluating the automatic sprinkler system in an existing building. These categories address all aspects, from existing buildings that are unsprinklered but are required to be sprinklered if new construction, to existing buildings that are sprinklered and are not required to be sprinklered if new construction. These categories reduce the impact of providing sprinkler protection required by Chapter 9 of the IBC to be sprinklered in new construction. This greater range of points increases the flexibility in the use of this evaluation method.

Category a includes buildings or occupancies that would be required by the new construction criteria of IBC Section 903 to be sprinklered throughout the building. Category a buildings are provided with either no sprinkler system, or one that is inadequate and does not provide the required level of protection. To evaluate the existing sprinkler system, a trained fire protection engineer should evaluate the system's design against the applicable referenced standards. This is not the same type of consideration needed in Category d, where the existing system actually meets all of the requirements of any earlier edition of the applicable referenced standards. This category is considered the lowest acceptable level of compliance and is, therefore, associated with the largest negative values in Table 1401.6.17.

Category b is applicable when only a portion of the existing building is required by the provisions of IBC Section 903 to be sprinklered. For example, a multiple-story building may have two of its stories qualify as windowless stories that require sprinklers, or a mercantile building may have one fire area exceeding 12,000 square feet (1115 m^2) and require sprinklers. In both cases, the buildings are without an automatic sprinkler system, or the sprinkler systems that are in place are inadequate and do not comply with the technical provisions of the code. Since only portions of the buildings are required to be sprinklered, the negative sprinkler system values for this category in Table 1401.6.17 are exactly half of the Category a values.

Category c is the zero-based category for this safety parameter. The existing building is not required by IBC Section 903 to be sprinklered and no sprinkler system is provided. Such buildings are neither penalized nor rewarded by Table 1401.6.17.

Category d is similar to Category b because only a portion of the existing building is required by IBC Section 903 to be sprinklered. In this case, there is a sprinkler system in place in that portion of the building and the system was designed at the time of its installation to comply with the requirements of an earlier edition of IBC Section 903. An example of this is a sprinkler system designed to comply with the hydraulic design criteria of the 1989 edition of NFPA 13 for an ordinary Group 2 hazard. As long as this existing sprinkler system is still properly maintained and supervised, the building qualifies for the small positive values listed in Table 1401.6.17.

Category e addresses existing buildings that are required by the provisions of IBC Chapter 9 to be sprinklered throughout, and are protected throughout by a properly designed, installed and supervised sprinkler system. Although this category more closely represents the requirements for new construction, moderate positive values are awarded by Table 1401.6.17. The evaluation rewards existing buildings that are sprinklered with higher points.

Category f is the highest level of protection and is rewarded with maximum positive points in the accompanying table. Existing buildings, which are not required by IBC Section 903 to be sprinklered but are voluntarily provided with a fully designed, installed and supervised system, provide an added level of protection that justifies substantial bonus points that may ultimately determine whether an existing building meets its mandatory safety scores.

1401.6.18 Standpipes. Evaluate the ability to initiate attack on a fire by a making supply of water available readily through the installation of standpipes in accordance with Section 905 of the *International Building Code*. "Required Standpipes" shall be based on the requirements of the *International Building Code*. Under the categories and occupancies in Table 1401.6.18, determine the appropriate value and enter that value into Table 1401.7 under Safety Parameter 1401.6.18, Standpipes, for fire safety, means of egress, and general safety.

❖ These provisions are used to determine the amount of credit that can be applied to the evaluation for the installation of a standpipe system in an existing building. Standpipes are provided as a tool for use by fire fighters to aid their fire-fighting operations. The general consensus is that they are not recommended or intended for use by untrained building occupants.

The evaluation of standpipe systems in an existing building, much like that for automatic sprinker systems, is based on whether a standpipe system is both required and installed. The criteria used to determine when a standpipe system is required are tied to the same requirements for new construction in IBC Section 905.3. The thresholds listed in IBC Sections 905.3.1 through 905.3.8 must be used to evaluate whether those characteristics and occupancies are present and whether a standpipe system would be required for new construction. These parameters for standpipes encourage the installation of standpipe systems in existing buildings by providing substantial negative and positive points.

The values for standpipe systems are given in Table 1401.6.18. The appropriate values from Table 1401.6.18 are entered in Table 1401.7 for the Safety Parameter of Standpipes on line 1401.6.18, for fire safety (FS), means of egress (ME) and general safety (GS).

**TABLE 1401.6.18
STANDPIPE SYSTEM VALUES**

OCCUPANCY	CATEGORIES			
	a[a]	b	c	d
A-1, A-3, F, M, R, S-1	-6	0	4	6
A-2	-4	0	2	4
A-4, B, E, S-2	-12	0	6	12
I-2	-2	0	1	2

a. This option cannot be taken if Category a or Category b in Section 1401.6.17 is used.

❖ This table lists the credit values for the respective categories of Section 1401.6.18.1, based on the occupancy being evaluated in the existing building. The grouping of occupancies is the same as provided for the sprinkler system parameter since the need for and value of a standpipe system, like sprinkler systems, is a function of the likelihood and potential severity of a fire that the various occupancies present.

Category a contains negative values that define buildings and occupancies that would be required to have a standpipe system as new construction but are not provided with one, or are provided with a standpipe system that is not designed and installed in accordance with the design and installation requirements of IBC Section 905 for new construction. Category b is the zero-based category.

Categories c and d contain positive values that represent buildings and occupancies with standpipe systems that are deemed adequate or comply with current design standards and installation requirements.

1401.6.18.1 Standpipe categories. The categories for standpipe systems are:

1. Category a—Standpipes are required; standpipe is not provided or the standpipe system design is not in compliance with Section 905.3 of the *International Building Code*.

2. Category b—Standpipes are not required; none are provided.

3. Category c—Standpipes are required; standpipes are provided in accordance with Section 905 of the *International Building Code*.

4. Category d—Standpipes are not required; standpipes are provided in accordance with Section 905 of the *International Building Code*.

❖ Four categories are defined in this section for evaluating the standpipe system in an existing building. These categories address all aspects from existing buildings that do not have standpipes but would be required to have standpipes if new construction, to existing buildings that have standpipes and would not be required to have standpipes if new construction.

Category a is applicable to buildings or occupancies that would be required by the new construction criteria to be provided with a Class I, II or III standpipe system. Category a buildings are provided with either no standpipe system, or one that is inadequate and does not comply with the applicable design and installation criteria of IBC Section 905. To evaluate the existing standpipe system, a trained fire protection engineer should evaluate the system's design against the applicable referenced standards. This category is considered the lowest acceptable level of compliance and is, therefore, associated with the largest negative values in Table 1401.6.18. Category a is not an option in unsprinklered or inadequately sprinklered buildings that would be required to be partially or fully sprinklered as new construction. This is reflected in Note a to Table 1401.6.18, which indicates that Category a cannot be used if Category a or b is used in the automatic sprinkler parameter in Section 1401.6.17. The effect of this provision is that the absence of both sprinklers and standpipes is unacceptable in buildings that would be required to provide both sprinklers and standpipes. One or the other system must be provided in order to complete this evaluation method. If only one of the required systems is provided, negative points will be accrued in the parameter for the other system, but it is still possible to have an overall passing score for the building. This reflects the concept that, while both systems are required for new construction, it is permissible to only have one in an existing building because the absence of the other system will have positive attributes or features that offset the negative points accrued by its absence.

Category b is the zero-based category for this safety parameter. The existing building is not required by IBC Section 905 to have a standpipe system and no standpipe system is provided. Such buildings are neither penalized nor rewarded by Table 1401.6.18.

Category c is applicable to buildings that are required by IBC Section 905 to have a standpipe system, and are provided with the class standpipe system that would be required for new construction and is designed and installed in accordance with IBC Section 905.2. Although this category more closely represents the requirements for new construction, moderate positive values are awarded by Table 1401.6.18. The evaluation rewards existing buildings that have a contemporary, reliable standpipe system with higher points.

Category d is the highest level of protection and is rewarded with maximum positive points in the accompanying table. Existing buildings, which are not required by IBC Section 905 to have a standpipe system but are voluntarily provided with a fully designed and installed system, provide an added level of protection that justifies substantial bonus points.

1401.6.19 Incidental uses. Evaluate the protection of incidental uses in accordance with Section 509.4.2 of the *International Building Code*. Do not include those where this code requires automatic sprinkler systems throughout the building including covered and open mall buildings, high-rise buildings, public garages and unlimited area buildings. Assign the lowest score from Table 1401.6.19 for the building or floor area being evaluated and enter that value into Table 1401.7 under Safety Parameter 1401.6.19, Incidental Uses, for fire safety, means of egress and general safety. If there are no specific occupancy areas in the building or floor area being evaluated, the value shall be zero.

❖ This section includes an evaluation system for the separation/protection requirements indicated for incidental uses in an existing building. This evaluation is based on IBC Table 509.4.2. If the building designer chooses to separate/protect the small rooms or areas in an existing building in accordance with this table, the building is classified according to its main use. Because some existing buildings may have been designed in this fashion, an evaluation procedure has been added to account for the level of protection provided. The lowest score must be assigned to the building or fire area for the specific use areas. For example, an existing Group E high school has three laboratory classrooms. If two of the classrooms are protected with limited area sprinkler systems and separated with construction capable of resisting the passage of smoke, and the third room is not protected in the same manner, the lowest score for all the classrooms is determined by the single unprotected classroom. If the building designer chooses to treat the specific occupancy areas as separate occupancies, the provisions of Section 1401.9.1 are applicable and a zero is inserted in Table 1401.7 under Safety Parameter 1401.6.19, Incidental Use Areas.

TABLE 1401.6.19. See page 14-34.

❖ This table provides a matrix of characteristics arranged in a format of rows and columns for determining values for incidental uses. The values are inserted into the building score sheet. The left-hand column of the table is arranged for the separation/protection specified by IBC Table 509 for a particular occupancy room or area. IBC Table 509 must first be consulted to determine the level of separation/protection required by the code for new construction. This same entry is then found in the left-hand column of Table 1401.6.19. The top row represents the actual level of protection that is provided in the existing building. The corresponding use area value is read and then inserted into Table 1401.7 under Safety Parameter 1401.6.19, Incidental Use Areas. Values of zero are assigned to all arrangements that represent compliance with the requirements for new construction. Negative values are assigned based on the degree of noncompliance with the requirements for new construction.

1401.6.20 Smoke compartmentation. Evaluate the smoke compartments for compliance with Section 407.5 of the *International Building Code*. Under the categories and occupancies in Table 1401.6.20, determine the appropriate smoke compartmentation value (SCV) and enter that value into Table 1401.7 under Safety Parameter 1401.6.20, Smoke Compartmentation, for fire safety, means of egress and general safety. Facilities in Group I-2 occupancies meeting Category b or c shall be considered to fail the evaluation.

❖ This section addresses the issue of smoke compartments, which are required for new construction in Group I-2. This provides scores to fully evaluate Group I-2 occupancies under this alternative method.

TABLE 1401.6.20
SMOKE COMPARTMENTATION VALUES

OCCUPANCY	CATEGORIES[a]		
	a	b	c
A, B, E, F, M, R and S	0	0	0
I-2	0	NP	NP

For SI: 1 square foot = 0.093 m².
NP = Not permitted.
a. For areas between categories, the smoke compartmentation value shall be obtained by linear interpolation.

❖ This table lists the credit values for the respective categories in the area of smoke compartments of Section 1401.6.20.1, based on the occupancy being evaluated in the existing building. The grouping of occupancies in the table is simply I-2, and all of the rest (A, B, E, F, M, R and S). The only occupancy that requires smoke compartments is Group I-2, and therefore this table establishes that a qualifying existing Group I-2 occupancy must have smoke compartments that are a maximum of 22,500 square feet. The score is 0 for those qualifying existing Group I-2 occupancies, and not permitted for Group I-2 occupancies with larger smoke compartments.

1401.6.20.1 Categories. Categories for smoke compartment size are:

1. Category a—Smoke compartment size is equal to or less than 22,500 square feet (2092 m²).

2. Category b—Smoke compartment size is greater than 22,500 square feet (2092 m²).

3. Category c—Smoke compartments are not provided.

❖ Three categories are provided in this section for evaluating smoke compartments in Group I-2 facilities. For Group I-2, if the smoke compartment falls in Category b (smoke compartment too large) or Category c (no smoke compartment), then the smoke compartment fails this evaluation.

1401.6.21 Patient ability, concentration, smoke compartment location and ratio to attendant. In I-2 occupancies, the ability of patients, their concentration and ratio to attendants shall be evaluated and applied in accordance with this section. Evaluate each smoke compartment using the categories in Sections 1401.6.21.1, 1401.6.21.2 and 1401.6.21.3 and enter the value in Table 1401.8. To determine the safety factor, multiply the three values together, if the product is 9 or greater, compliance has failed.

❖ This section addresses the issue of patient safety in a smoke compartment in Group I-2 facilities. There are three factors for patient safety included: 1. Patient ability (Section 1401.6.21.1); 2. Patient concentration (Section 1401.6.21.2); and 3. Attendant-to-patient ratio (Section 1401.6.21.3). Basically, these considerations address patient safety from the standpoint of the ability of patients to be brought to safety in the event of an emergency. These three factors are entered into Table 1401.7 for an overall score for the I-2 smoke compartment being evaluated.

This section also establishes a mandatory safety score for these three evaluation categories alone. The rule is that if the product of the three categories when multiplied together is 9 or greater, then the smoke compartment fails compliance. Observing the values for each of these factors will show that the value for Category c in each factor is 3. Therefore, if

TABLE 1401.6.19
INCIDENTAL USE AREA VALUES

PROTECTION REQUIRED BY TABLE 509 OF THE *INTERNATIONAL BUILDING CODE*	PROTECTION PROVIDED						
	None	1 hour	AS	AS with CRS	1 hour and AS	2 hours	2 hours and AS
2 hours and AS	-4	-3	-2	-2	-1	-2	0
2 hours, or 1 hour and AS	-3	-2	-1	-1	0	0	0
1 hour and AS	-3	-2	-1	-1	0	-1	0
1 hour	-1	0	-1	-1	0	0	0
1 hour, or AS with CRS	-1	0	-1	-1	0	0	0
AS with CRS	-1	-1	-1	-1	0	-1	0
1 hour or AS	-1	0	0	0	0	0	0

AS = Automatic sprinkler system;
CRS = Construction capable of resisting the passage of smoke (see IBC Section 509.4.2 of the *International Building Code*).
Note: For Table 1401.7, see page 75 of the code or page 14-36 of this volume.

the category for any two of the three factors is Category c, with the third factor being Category a, then the product of those factors would be 9, and compliance has failed. This means that for any compartment, only one of these three factors can be Category c, and when that occurs, one of the other factors must be Category a. By requiring a maximum multiple of 9, the code is requiring some level of trade-off on these categories.

1401.6.21.1 Patient ability for self-preservation. Evaluate the ability of the patients for self-preservation in each smoke compartment in an emergency. Under the categories and occupancies in Table 1401.6.21.1 determine the appropriate value and enter that value in Table 1401.7 under Safety Parameter 1401.6.21.1, Patient Ability for Self-preservation, for means of egress and general safety.

❖ See the commentary to Section 1401.6.21.

TABLE 1401.6.21.1
PATIENT ABILITY VALUES

OCCUPANCY	CATEGORIES		
	a	b	c
I-2	1	2	3

1401.6.21.1.1 Categories. The categories for patient ability for self-preservation are:

1. Category a—(mobile) Patients are capable of self-preservation without assistance.
2. Category c—(not mobile) Patients rely on assistance for evacuation or relocation.
3. Category d—(not movable) Patients cannot be evacuated or relocated.

❖ This section establishes the different categories for scoring patient ability in a Group I-2 smoke compartment.

1401.6.21.2 Patient concentration. Evaluate the concentration of patients in each smoke compartment under Section 1401.6.21.2. Under the categories and occupancies in Table 1401.6.21.2 determine the appropriate value and enter that value in Table 1401.7 under Safety Parameter 1401.6.21.2, Patient Concentration, for means of egress and general safety.

❖ See the commentary to Section 1401.6.21.

TABLE 1401.6.21.2
PATIENT CONCENTRATION VALUES

OCCUPANCY	CATEGORIES		
	a	b	c
I-2	1	2	3

1401.6.21.3 Attendant-to-patient ratio. Evaluate the attendant-to-patient ratio for each compartment under Section 1401.6.21.3. Under the categories and occupancies in Table 1401.6.21.3 determine the appropriate value and enter that value in Table 1401.7 under Safety Parameter 1401.6.21.3, Attendant-to-patient Ratio, for means of egress and general safety.

❖ See the commentary to Section 1401.6.21.

1401.6.21.3.1 Categories. The categories for attendant-to-patient concentrations are:

1. Category a—attendant-to-patient concentrations is 1:5.
2. Category b—attendant-to-patient concentrations is 1:6 to 1:10.
3. Category c—attendant-to-patient concentrations is greater than 1:10 or no patients.

❖ This section establishes the different categories for scoring attendant-to-patient ratio in a Group I-2 smoke compartment.

TABLE 1401.6.21.3
ATTENDANT-TO-PATIENT RATIO VALUES

OCCUPANCY	CATEGORIES		
	a	b	c
I-2	1	2	3

1401.7 Building score. After determining the appropriate data from Section 1401.6, enter those data in Table 1401.7 and total the building score.

❖ This section is the tally sheet for all of the 21 safety parameters evaluated in Sections 1401.6.1 through 1401.6.19, which determine the building's overall safety profile for fire safety, means of egress and general safety.

This section also directs the data and values of the 21 safety parameters of Sections 1401.6.1 through 1401.6.19 to be entered in Table 1401.7 for totaling the building scores.

TABLE 1401.7. See page 14-36.

❖ Table 1401.7 is the summary sheet containing all the relative attributes of the building. The summary sheet also contains a complete listing of the 21 safety parameters that have been evaluated. The upper portion of the summary sheet serves as a guide to the user to catalog and highlight existing building elements that relate to the 21 safety parameters and to the evaluation. The lower portion of the summary sheet is used to record the results of the 21 safety parameters that have been evaluated. These are added to produce the building score total values for fire safety, means of egress and general safety.

PERFORMANCE COMPLIANCE METHODS

**TABLE 1401.7
SUMMARY SHEET-BUILDING CODE**

Existing occupancy _____   Proposed occupancy _____

Year building was constructed _____   Number of stories _____ Height in feet _____

Type of construction _____   Area per floor _____

Percentage of open perimeter increase _____%

Completely suppressed: Yes _____ No _____   Corridor wall rating _____

                                                                         Type: _____

Compartmentation: Yes _____ No _____   Required door closers: Yes _____ No _____

Fire-resistance rating of vertical opening enclosures _____

Type of HVAC system _____ , serving number of floors _____

Automatic fire detection: Yes _____ No _____   Type and location _____

Fire alarm system: Yes _____ No _____   Type _____

Smoke control: Yes _____ No _____   Type _____

Adequate exit routes: Yes _____ No _____   Dead ends: _____ Yes _____ No _____

Maximum exit access travel distance _____   Elevator controls: Yes _____ No _____

Means of egress emergency lighting: Yes ___ No ___   Mixed occupancies: Yes _____ No _____

Standpipes   Yes _____ No _____   Patient ability for self-preservation _____

Incidental use   Yes _____ No _____   Patient concentration _____

Smoke compartmentation less
   than 22,500 sq. feet (2092 m²) Yes ___ No ___   Attendant-to-patient ratio _____

SAFETY PARAMETERS	FIRE SAFETY (FS)	MEANS OF EGRESS (ME)	GENERAL SAFETY (GS)
1401.6.1 Building Height 1401.6.2 Building Area 1401.6.3 Compartmentation			
1401.6.4 Tenant and Dwelling Unit Separations 1401.6.5 Corridor Walls 1401.6.6 Vertical Openings			
1401.6.7 HVAC Systems 1401.6.8 Automatic Fire Detection 1401.6.9 Fire Alarm System			
1401.6.10 Smoke control 1401.6.11 Means of Egress 1401.6.12 Dead ends	* * * * * * * * * * * *		
1401.6.13 Maximum Exit Access Travel Distance 1401.6.14 Elevator Control 1401.6.15 Means of Egress Emergency Lighting	* * * * * * * *		
1401.6.16 Mixed Occupancies 1401.6.17 Automatic Sprinklers 1401.6.18 Standpipes 1401.6.19 Incidental Use 1401.6.20 Smoke compartmentation 1401.6.21.1 Patient ability for self-preservation 1401.6.21.2 Patient concentration 1401.6.21.3 Attendant-to-patient Ratio	 * * * * * * * * * * * *	* * * * ÷2 =	
Building score—total value			

* * * *No applicable value to be inserted.

PERFORMANCE COMPLIANCE METHODS

1401.8 Safety scores. The values in Table 1401.8 are the required mandatory safety scores for the evaluation process listed in Section 1401.6.

❖ This section lists the minimum scores for fire safety, means of egress and general safety that must be obtained from the evaluation of the 21 safety parameters to be acceptable as a building meeting the code's objectives for public safety and health. This section summarizes the mandatory values of Table 1401.8 that must be met from the evaluation program.

TABLE 1401.8
MANDATORY SAFETY SCORES[a]

OCCUPANCY	FIRE SAFETY (MFS)	MEANS OF EGRESS (MME)	GENERAL SAFETY (MGS)
A-1	20	31	31
A-2	21	32	32
A-3	22	33	33
A-4, E	29	40	40
B	30	40	40
F	24	34	34
I-2	19	34	34
M	23	40	40
R	21	38	38
S-1	19	29	29
S-2	29	39	39

a. MFS = Mandatory Fire Safety.
MME = Mandatory Means of Egress.
MGS = Mandatory General Safety.

❖ The table lists the minimum mandatory safety scores for various occupancies in the areas of fire safety, means of egress and general safety. The mandatory safety values are based on the scores considered to provide an overall acceptable level of safety in an existing building on which approval of the alterations, repairs, change of occupancy or addition can be based. This is the zero-based concept. The scores have been determined as representing one level of compliance higher than the code's minimum requirements for new construction. The mandatory safety scores are consistent with the idea of establishing an equivalent level of safety, even though the existing building is evaluated only for the 21 safety parameters.

1401.9 Evaluation of building safety. The mandatory safety score in Table 1401.8 shall be subtracted from the building score in Table 1401.7 for each category. Where the final score for any category equals zero or more, the building is in compliance with the requirements of this section for that category. Where the final score for any category is less than zero, the building is not in compliance with the requirements of this section.

❖ The sections and table that follow are the final steps in the evaluation process. This section also discusses how mixed occupancies must be treated during the final step of the evaluation process.

This section compares the three building scores from Table 1401.7 to the three mandatory safety scores from Table 1401.8. If the values in all three categories from Table 1401.7 exceed the corresponding mandatory safety scores in Table 1401.8, the building passes and is in compliance with the code. If the score in any one category is less than the mandatory safety score, the building is deemed to have failed and additional measures must be taken to bring the scores to a point that will at least equal the mandatory safety scores.

TABLE 1401.9. See below.

❖ The sections and table that follow are the final steps in the evaluation process. This section also discusses how mixed occupancies must be treated during the final step of the evaluation process.

This section compares the three building scores from Table 1401.7 to the three mandatory safety scores from Table 1401.8. If the values in all three categories from Table 1401.7 exceed the corresponding mandatory safety scores in Table 1401.8, the building passes and is in compliance with the code. If the score in any one category is less than the mandatory safety score, the building is deemed to have failed and additional measures must be taken to bring the scores to a point that will at least equal the mandatory safety scores.

Table 1401.9 shows in simple equations whether a building passes the evaluation by subtracting the mandatory safety score from Table 1401.7. This is done for each of the three general categories of evaluation: fire safety, means of egress and general safety. If the difference for each category is zero or greater, the existing building passes and is considered to comply with the code objectives for public safety.

TABLE 1401.9
EVALUATION FORMULAS[a]

FORMULA	T1401.7	T1401.8		SCORE	PASS	FAIL
FS − MFS ≥ 0	_____ (FS) −	_____ (MFS)	=	_____	_____	_____
ME − MME ≥ 0	_____ (ME) −	_____ (MME)	=	_____	_____	_____
GS − MGS ≥ 0	_____ (GS) −	_____ (MGS)	=	_____	_____	_____

a. FS = Fire Safety.　　　　　　　　　　　　　　MFS = Mandatory Fire Safety.
ME = Means of Egress.　　　　　　　　　　　MME = Mandatory Means of Egress.
GS = General Safety.　　　　　　　　　　　　MGS = Mandatory Means of Safety.

Example:

A Group M mercantile occupancy receives evaluations for the 21 safety parameters that result in total building scores in Table 1401.7 as follows:

Fire safety (FS) = 23

Means of egress (ME) = 41

General safety (GS) = 42

These scores are compared to the mandatory safety scores for a Group M occupancy from Table 1401.8 (see Commentary Figure 1401.9).

Mandatory fire safety (MFS) = 23

Mandatory means of egress (MME) = 40

Mandatory general safety (MGS) = 40

The mandatory score is subtracted from the building score.

Conclusion:

The building is acceptable; all of the final safety scores are zero or greater. Each category must individually have a building score equal to or greater than the respective mandatory safety score for the building to pass the overall evaluation.

Table 1401.9
EVALUATION FORMULAS[a]

Formula	Table 1401.7	Table 1401.8	Score	Pass	Fail
FS-MFS ≥ 0	23 (FS)	- 23 (MFS) =	0	X	—
ME-MME ≥ 0	41 (ME)	- 40 (MME) =	1	X	—
GS-MGS ≥ 0	42 (GS)	- 40 (MGS) =	1	X	—

Note a.
FS = Fire Safety MFS = Mandatory Fire Safety
ME = Means of Egress MME = Mandatory Means of Egress
GS = General Safety MGS = Mandatory General Safety

Figure 1401.9
EXAMPLE OF EVALUATION FORMULAS

1401.9.1 Mixed occupancies. For mixed occupancies, the following provisions shall apply:

1. Where the separation between mixed occupancies does not qualify for any category indicated in Section 1401.6.16, the mandatory safety scores for the occupancy with the lowest general safety score in Table 1401.8 shall be utilized. (See Section 1401.6.)

2. Where the separation between mixed occupancies qualifies for any category indicated in Section 1401.6.16, the mandatory safety scores for each occupancy shall be placed against the evaluation scores for the appropriate occupancy.

❖ This section explains how to determine whether a mixed occupancy building passes or fails the process of evaluation. It restates the information in Sections 1401.6 and 1401.6.16. The mixed occupancy evaluation is based on the requirements of IBC Section 508. The two procedures described are:

Procedure 1:

For nonseparated occupancies in accordance with IBC Section 508, or for occupancies that are separated with a fire-resistance rating of less than 1 hour, mandatory safety scores for the occupancy with the lowest general safety score of Table 1401.8 apply.

Procedure 2:

For separated occupancies in accordance with IBC Section 508.4, or one of the categories listed in Section 1401.6.16, mandatory safety scores for each occupancy are compared to the evaluation scores for the appropriate occupancy. The total building score of Table 1401.7 is computed for each appropriate occupancy.

This section does not address a third option for mixed occupancies, and that is the separation of multiple occupancies within a single building or structure with one or more fire walls. Likewise, this option is not addressed in IBC Section 508. A fire wall creates separate and independent buildings; therefore, a separate evaluation must be done for each building that is created by fire walls.

Example:

A 40-foot (12 192 mm), three-story building has an open perimeter of 25 percent. It is Type IIB unprotected construction, unsprinklered. The building has 9,000 square feet (836 m^2) per floor. A Group A-2 assembly is on the first floor, with Group B business occupancies on the second and third floors. Although the building is unprotected Type IIB construction, there is a 2-hour fire-resistance-rated floor/ceiling assembly separating the first and second floors. An analysis of this building will result in the following: In accordance with IBC Section 508.4 and IBC Table 508.4, the mixed occupancies are separated by the required fire-rated assembly of 2 hours.

This building qualifies for Category b in Section 1401.6.16.1. A separate summary sheet for tabulating the building score in Table 1401.7 has to be completed for each of the two occupancies. The values of the building scores for fire safety, means of egress and general safety for the Group M mercantile occupancy and the Group R residential occupancy must be calculated. The mercantile building scores are compared to the mandatory safety scores from Table 1401.8 of 23, 40 and 40, and the residential building scores are compared to the mandatory safety scores from Table 1401.8 of 21, 38 and 38. If any one of the three building scores from either of the two occupancies does not equal or exceed the applicable safety score, the entire building fails the evaluation. For the building to pass, the safety parameters that are less than the mandatory safety score must be upgraded until a passing score is achieved.

Bibliography

The following resource materials were used in the preparation of the commentary for this chapter of the code.

36 CFR, Parts 1190 and 1191, Draft, *The Americans with Disabilities Act Accessibility Guidelines* (ADAAG). Washington, DC: Architectural and Transportation Barriers Compliance Board, April 2, 2002.

42 USC 3601-88, *Fair Housing Amendments Act* (FHAA). Washington, DC: United States Code, 1988.

Accessibility and Egress for People with Physical Disabilities. Falls Church, VA: CABO Board for the Coordination of the Model Codes Report, October 5, 1993.

"Final Report, Recommendations for a New ADAAG." Washington DC: ADAAG Review Federal Advisory Committee, September 30, 1996.

Rehabilitation Guidelines/1980. Washington, DC: National Institute of Building Science for the US Department of Housing and Urban Development, 1980.

1. Guideline for Setting and Adopting Standards for Building Rehabilitation;
2. Guideline for Approval of Building Rehabilitation;
3. Statutory Guideline for Building Rehabilitation;
4. Guideline for Managing Official Liability Associated with Building Rehabilitation;
5. Egress Guideline for Residential Rehabilitation;
6. Electrical Guideline for Residential Rehabilitation;
7. Plumbing DWV Guideline for Residential Rehabilitation; and
8. Guideline on Fire Ratings of Archaic Materials and Assemblies.

Chapter 15: Construction Safeguards

General Comments

The building construction process involves a number of known and unanticipated hazards. Chapter 15 establishes specific regulations in order to minimize risk to the public and adjacent property. Some construction failures have resulted during the initial stages of grading, excavation and demolition. During these early stages, poorly designed and installed sheeting and shoring have resulted in ditch and embankment cave-ins. Also, inadequate underpinning of adjoining existing structures and the careless removal of existing structures have produced construction failures.

The most critical period in building construction related to the physical safety of those on the job site is during the ongoing process when all building components have not yet been completed. Compounding this incomplete state is the use of dangerous construction methods, materials and equipment.

The importance of reasonable precautions is evidenced by the federal Occupational Safety and Health Act (OSHA), and related state and local regulations.

Purpose

The purpose of Chapter 15 is to cite safety requirements during construction or demolition of buildings and structures. These requirements are intended to protect the public from injury and adjoining property from damage.

SECTION 1501
GENERAL

[BG] 1501.1 Scope. The provisions of this chapte. shall govern safety during construction that is under the jurisdiction of this code and the protection of adjacent public and private properties.

❖ The fundamental rationale behind this chapter is to establish that reasonable safety precautions are to be provided during the construction or demolition process to protect the public from injury and adjoining property from damage. This chapter does not address projects not associated with a building or structure. Section 101.2 establishes the scope as applicable to buildings or structures. For example, the widening or deepening of an existing river for flood control purposes or a similar project would not be regulated by this chapter.

[BG] 1501.2 Storage and placement. Construction equipment and materials shall be stored and placed so as not to endanger the public, the workers or adjoining property for the duration of the construction project.

❖ This section requires that construction materials and equipment must be located and protected pursuant to the governing provisions of this chapter so that the public and adjoining property are safeguarded at all times during construction or demolition processes.

[BG] 1501.3 Alterations, repairs, and additions. Required exits, existing structural elements, fire protection devices, and sanitary safeguards shall be maintained at all times during *alterations, repairs,* or *additions* to any building or structure.

Exceptions:

1. When such required elements or devices are being altered or repaired, adequate substitute provisions shall be made.

2. When the *existing building* is not occupied.

❖ Demolition and construction operations must not create a hazard for the occupants of a building during an alteration or addition. As such, the existing fire protection, means of egress elements and safety systems must remain in place and be functional. However, two exceptions are provided in this section that permit the fire protection, means of egress elements and safety systems to be changed, modified or removed. The first option provides an alternative method that must be approved by the code official. The second option is simply to have an unoccupied building, which means that residents of a dwelling unit or employees of a business are not to be present during the construction or demolition process.

[BG] 1501.4 Manner of removal. Waste materials shall be removed in a manner which prevents injury or damage to persons, adjoining properties, and public rights-of-way.

❖ Safe and sanitary procedures for the removal of building construction and demolition waste must be provided. The method of waste removal must be controlled such that debris will not pose a hazard, eyesore or nuisance to the public and neighboring properties. Examples of acceptable practices include: evidence that a professional disposal service will haul away the debris; limiting areas used for storing and handling demolished materials; enclosing storage areas so that only authorized personnel can gain access; establishing routes that waste removal vehicles are permitted to use; covering or tarping debris to prevent flying objects; providing fully enclosed chutes to control falling objects and dust; and scheduling waste removal when an adjoining property and the public will be least exposed to unusual and possibly dangerous situations caused by fumes, noise, dust and unfamiliar events involved during a demolition process. Note that the accepted practice of waste removal is subject to the approval of the code official. A growing number of local jurisdictions as part of their sustainability, or green, programs limit the amount of construction and demolition debris that can be sent to landfills.

[BG] 1501.5 Fire safety during construction. Fire safety during construction shall comply with the applicable requirements of the *International Building Code* and the applicable provisions of Chapter 33 of the *International Fire Code*.

❖ This requirement correlates Section 3302 of the *International Building Code*® (IBC®) addressing "Construction Safeguards," and the construction safeguard requirements of this chapter with the *International Fire Code*® (IFC®) requirements on fire safety during construction and demolition. Chapter 33 of the IFC, "Fire Safety During Construction and Demolition," has all the fire safety requirements provided in one place for fire-safe operations during construction and demolition.

[BS] 1501.6 Protection of pedestrians. Pedestrians shall be protected during construction and demolition activities as required by Sections 1501.6.1 through 1501.6.7 and Table 1501.6. Signs shall be provided to direct pedestrian traffic.

❖ Safeguards are required to be in place during construction or demolition operations in accordance with Table 1501.6 and this chapter. In addition, since construction operations alter the familiar setting and path of travel, it is necessary to provide some form of visible directional sign to lead the public toward safety and away from potential hazards.

TABLE 1501.6. See below.

❖ Table 1501.6 establishes the type of protection required based on the location to overhead hazards and distance in relation to ground hazards. Once the type of protection is determined from the table, the applicable code section, such as Section 1501.6.3 for construction railings, Section 1501.6.4 for barriers and Section 1501.6.5 for covered walkways, must be followed.

[BS] 1501.6.1 Walkways. A walkway shall be provided for pedestrian travel in front of every construction and demolition site unless the applicable governing authority authorizes the sidewalk to be fenced or closed. Walkways shall be of sufficient width to accommodate the pedestrian traffic, but in no case shall they be less than 4 feet (1219 mm) in width. Walkways shall be provided with a durable walking surface. Walkways shall be accessible in accordance with Chapter 11 of the *International Building Code* and shall be designed to support all imposed loads and in no case shall the design live load be less than 150 pounds per square foot (psf) (7.2 kN/m^2).

❖ Construction operations must not narrow or impede the normal flow of pedestrian traffic along a walkway by the placement of a fence or other enclosure. Note that the applicable governing authority must approve any construction operations that will cause the narrowing or impedance of a walkway, such as the sidewalk. If a walkway is narrowed or enclosed, the construction of another or wider walkway is required for all pedestrians. The walkway must be able to handle the normal anticipated flow of pedestrian traffic and must not be less than the minimum 4-foot (1219 mm) width. Even though such walkways may be temporary installations, they may still need to be usable by persons with disabilities where such walkways must provide an accessible route in accordance with IBC Chapter 11. In addition, the walkway must be a stable surface that is capable of supporting all imposed loads. The minimum design load of the walkway must not be less than 150 pounds per square foot (psf) (7.2 kN/m^2 Pa).

[BS] 1501.6.2 Directional barricades. Pedestrian traffic shall be protected by a directional barricade where the walkway extends into the street. The directional barricade shall be of sufficient size and construction to direct vehicular traffic away from the pedestrian path.

❖ Similar to the local, federal and state guidelines that govern the provisions to protect employees while doing road work, this section establishes that a barrier must be erected. The barrier must be capable of redirecting and regulating the flow of vehicular traffic where constructed or projected into a street. An example of how to regulate traffic is to provide the necessary warnings, notice of caution and instructions to the operators of motor vehicles with signs

[BS] TABLE 1501.6
PROTECTION OF PEDESTRIANS

HEIGHT OF CONSTRUCTION	DISTANCE OF CONSTRUCTION TO LOT LINE	TYPE OF PROTECTION REQUIRED
8 feet or less	Less than 5 feet	Construction railings
	5 feet or more	None
More than 8 feet	Less than 5 feet	Barrier and covered walkway
	5 feet or more, but not more than one-fourth the height of construction	Barrier and covered walkway
	5 feet or more, but between one-fourth and one-half the height of construction	Barrier
	5 feet or more, but exceeding one-half the height of construction	None

For SI: 1 foot = 304.8 mm.

and colors that are common to the local, federal and state guidelines. Note that the intent of this visual barrier is to keep pedestrians out of vehicular traffic and prevent vehicles from encroaching on or using the pedestrian walkway as a roadway.

[BS] 1501.6.3 Construction railings. Construction railings shall be at least 42 inches (1067 mm) in height and shall be sufficient to direct pedestrians around construction areas.

❖ A barrier consisting of a horizontal rail and supports must be constructed with a minimum height of 42 inches (1067 mm). In addition, this barrier must be capable of controlling the flow of pedestrian traffic by routing travel away from the hazards associated with a construction area. Where construction railings are located where guards would be required by IBC Section 1015.1, such railings would also need to comply with IBC Section 1015. Even where guards are not required, the design standards for guards could be useful in designing construction railings.

[BS] 1501.6.4 Barriers. Barriers shall be a minimum of 8 feet (2438 mm) in height and shall be placed on the side of the walkway nearest the construction. Barriers shall extend the entire length of the construction site. Openings in such barriers shall be protected by doors which are normally kept closed.

❖ When a barrier is required by Table 1501.6, it must be constructed to impede, separate and obstruct passage of pedestrians onto a construction site. The structure must be a minimum 8 feet (2438 mm) in height and located continuously along the walkway on the side where construction activities are being done. Note that openings, such as doors or gates, are permitted as long as the only time they are open is during use by authorized personnel. As such, when the doors or gates are not being used they must remain closed.

[BS] 1501.6.4.1 Barrier design. Barriers shall be designed to resist loads required in Chapter 16 of the *International Building Code* unless constructed as follows:

1. Barriers shall be provided with 2 × 4 top and bottom plates.
2. The barrier material shall be a minimum of $^3/_4$ inch (19.1 mm) boards or $^1/_4$ inch (6.4 mm) wood structural use panels.
3. Wood structural use panels shall be bonded with an adhesive identical to that for exterior wood structural use panels.
4. Wood structural use panels $^1/_4$ inch (6.4 mm) or $^1/_{16}$ inch (1.6 mm) in thickness shall have studs spaced not more than 2 feet (610 mm) on center.
5. Wood structural use panels $^3/_8$ inch (9.5 mm) or $^1/_2$ inch (12.7 mm) in thickness shall have studs spaced not more than 4 feet (1219 mm) on center, provided a 2-inch by 4-inch (51 mm by 102 mm) stiffener is placed horizontally at the mid-height where the stud spacing exceeds 2 feet (610 mm) on center.
6. Wood structural use panels $^5/_8$ inch (15.9 mm) or thicker shall not span over 8 feet (2438 mm).

❖ This section establishes two methods that can be used to design barriers. The first method establishes that the barrier can be designed using the same materials and assembly instructions as indicated in Items 1 through 6. The second method provides for a barrier that will resist the loads imposed on it to be part of a designed assembly. The loads, which need to be resisted, are the same as those established in IBC Chapter 16.

[BS] 1501.6.5 Covered walkways. Covered walkways shall have a minimum clear height of 8 feet (2438 mm) as measured from the floor surface to the canopy overhead. Adequate lighting shall be provided at all times. Covered walkways shall be designed to support all imposed loads. In no case shall the design live load be less than 150 psf (7.2 kN/m^2) for the entire structure.

Exception: Roofs and supporting structures of covered walkways for new, light-frame construction not exceeding two stories above grade plane are permitted to be designed for a live load of 75 psf (3.6 kN/m^2) or the loads imposed on them, whichever is greater. In lieu of such designs, the roof and supporting structure of a covered walkway are permitted to be constructed as follows:

1. Footings shall be continuous 2 × 6 members.
2. Posts not less than 4 × 6 shall be provided on both sides of the roof and spaced not more than 12 feet (3658 mm) on center.
3. Stringers not less than 4 × 12 shall be placed on edge upon the posts.
4. Joists resting on the stringers shall be at least 2 × 8 and shall be spaced not more than 2 feet (610 mm) on center.
5. The deck shall be planks at least 2 inches (51 mm) thick or wood structural panels with an exterior exposure durability classification at least $^{23}/_{32}$ inch (18.3 mm) thick nailed to the joists.
6. Each post shall be knee-braced to joists and stringers by 2 × 4 minimum members 4 feet (1219 mm) long.
7. A 2 × 4 minimum curb shall be set on edge along the outside edge of the deck.

❖ Walkways with a covering that extends out over the pedestrian path of travel, pursuant to Table 1501.6, must be at least 8 feet (2438 mm) in height. The measurement must be taken from the top of the walking surface vertically upward to the underside of the roof covering. The walkway must maintain a satisfactory level of light to the space that will be equal to or better than that provided prior to the installation of the protective covering. The covered walkway must also be structurally capable of resisting the design loads established in IBC Chapter 16, but not less than 150 psf (7.2 kN/m^2) for the live load.

Note that this section provides two alternative design options. One option permits the covered walkway to resist the design loads established in IBC Chapter 16, but not less than the minimum 75-psf (3.6 kN/m^2) live load if nearby construction does not exceed two stories in height and the construction materials are lightweight, such as light-frame construction. The second option is to construct the covered walkway in accordance with the seven criteria set forth in this section.

[BS] 1501.6.6 Repair, maintenance and removal. Pedestrian protection required by Section 1501.6 shall be maintained in place and kept in good order for the entire length of time pedestrians may be endangered. The owner or the owner's agent, upon the completion of the construction activity, shall immediately remove walkways, debris and other obstructions and leave such public property in as good a condition as it was before such work was commenced.

❖ Any safeguards required by this chapter must be kept in good functional condition for the entire duration of the construction or demolition activity so that the public will not be placed in harm's way. Once a building or structure is occupiable and the site is properly graded, the protection must be removed. As such, public property that was affected by the construction activity must be restored to or left in the condition that existed prior to the work.

[BS] 1501.6.7 Adjacent to excavations. Every excavation on a site located 5 feet (1524 mm) or less from the street lot line shall be enclosed with a barrier not less than 6 feet (1829 mm) high. Where located more than 5 feet (1524 mm) from the street lot line, a barrier shall be erected when required by the *code official*. Barriers shall be of adequate strength to resist wind pressure as specified in Chapter 16 of the *International Building Code*.

❖ This section establishes that whenever excavation is to take place less than 5 feet (1524 mm) from the edge of a roadway, a barrier must be erected with a minimum height of 6 feet (1829 mm). If the excavation is located greater than 5 feet (1524 mm) from the edge of the roadway, the code official must evaluate the level of hazard for the public and the necessary precautions to take. As such, the code official has the authority to order the construction of a structural barrier. Note that any barrier erected must maintain other provisions of the code such that it is capable of handling and resisting the design wind loads denoted in IBC Chapter 16.

1501.7 Facilities required. Sanitary facilities shall be provided during construction or demolition activities in accordance with the *International Plumbing Code*.

❖ Construction employees must have plumbing facilities available during the construction or demolition process of a building. The facilities must conform to the requirements set forth in the *International Plumbing Code*® (IPC®).

SECTION 1502
PROTECTION OF ADJOINING PROPERTY

[BS] 1502.1 Protection required. Adjoining public and private property shall be protected from damage during construction and demolition work. Protection must be provided for footings, foundations, party walls, chimneys, skylights and roofs. Provisions shall be made to control water runoff and erosion during construction or demolition activities. The person making or causing an excavation to be made shall provide written notice to the owners of adjoining buildings advising them that the excavation is to be made and that the adjoining buildings should be protected. Said notification shall be delivered not less than 10 days prior to the scheduled starting date of the excavation.

❖ This section emphasizes the need to protect all existing public and private property bordering proposed construction or demolition operations. The term "property" only alludes to existing buildings. As such, any building element or system must be provided with a safeguard that will limit the damage that could be caused from the processes involved to the equipment and materials used. Additionally, soil erosion and land disbursement control must be provided to prevent spillage and spread of disturbed soil debris. The owner or owner's agent has the responsibility to provide a written notice 10 days in advance for any demolition or construction activities that may warrant bordering lots to be protected from damage.

SECTION 1503
TEMPORARY USE OF STREETS, ALLEYS AND PUBLIC PROPERTY

[BG] 1503.1 Storage and handling of materials. The temporary use of streets or public property for the storage or handling of materials or equipment required for construction or demolition, and the protection provided to the public shall comply with the provisions of the applicable governing authority and this chapter.

❖ In addition to applicable regulations, this chapter establishes procedures for the storage of construction materials and equipment to protect access to public safety equipment, utilities and public transportation facilities.

[BG] 1503.2 Obstructions. Construction materials and equipment shall not be placed or stored so as to obstruct access to fire hydrants, standpipes, fire or police alarm boxes, catch basins or manholes, nor shall such material or equipment be located within 20 feet (6.1 m) of a street intersection, or placed so as to obstruct normal observations of traffic signals or to hinder the use of public transit loading platforms.

❖ This section indicates that precautions for material and equipment storage and placement must be provided so as not to block or obstruct access to fire hydrants, standpipes (including fire department siamese connections for sprinklers and standpipes), fire or police alarm boxes, utility boxes and meters, catch basins or manholes, or any other vital facility with a function that contributes to the health, safety and welfare of the public. Also, storage must not be placed within 20 feet (6096 mm) of a street intersection if it obstructs the normal observation of traffic signals or hinders the use of any mass-transit loading platforms, such as sidewalk bus stops, taxi waiting areas, etc.

[BG] 1503.3 Utility fixtures. Building materials, fences, sheds or any obstruction of any kind shall not be placed so as to obstruct free approach to any fire hydrant, fire department connection, utility pole, manhole, fire alarm box, or catch basin, or so as to interfere with the passage of water in the gutter. Protection against damage shall be provided to such utility fixtures during the progress of the work, but sight of them shall not be obstructed.

❖ Utility fixtures, such as those used by electrical, telephone, water, gas or sewer companies, as well as fire protection devices, must not be hidden from view and not be blocked from access and use. When construction operations are near a utility fixture, precautions must be provided so as not to cause damage. Note that the utility companies and the authority having jurisdiction, including the fire department, may have specific requirements and guidelines to follow to limit the possibility of damage and obstruction.

SECTION 1504
FIRE EXTINGUISHERS

[F] 1504.1 Where required. All structures under construction, *alteration*, or demolition shall be provided with not less than one approved portable fire extinguisher in accordance with Section 906 of the *International Fire Code* and sized for not less than ordinary hazard as follows:

1. At each stairway on all floor levels where combustible materials have accumulated.
2. In every storage and construction shed.
3. Additional portable fire extinguishers shall be provided where special hazards exist including, but not limited to, the storage and use of flammable and combustible liquids.

❖ A means to provide fire protection during the construction and demolition process is required. As such, this section indicates provisions for portable fire extinguishers. In addition to the provisions of this section, the regulations of IFC Section 906 must be maintained.

[F] 1504.2 Fire hazards. The provisions of this code and of the *International Fire Code* shall be strictly observed to safeguard against all fire hazards attendant upon construction operations.

❖ Methods, procedures and construction materials each contribute, in some way, to creating a fire hazard. Therefore, in addition to the provisions of the code, the IFC must be used to regulate the proper means of safety that must be provided for a building being demolished, altered or constructed.

SECTION 1505
MEANS OF EGRESS

[BE] 1505.1 Stairways required. Where a building has been constructed to a building height of 50 feet (15 240 mm) or four stories, or where an *existing building* exceeding 50 feet (15 240 mm) in building height is altered, at least one temporary lighted stairway shall be provided unless one or more of the permanent stairways are erected as the construction progresses.

❖ Temporary stairways must be constructed in accordance with IBC Section 1009, including riser, tread, guard and handrail requirements.

[F] 1505.2 Maintenance of means of egress. Required means of egress shall be maintained at all times during construction, demolition, remodeling or *alterations* and *additions* to any building.

Exception: Approved temporary means of egress systems and facilities.

❖ Access to all existing required exits must remain unobstructed and usable by any occupants in the existing building while construction or demolition work is being done. Note that, depending on the extent of work, an exit may become obstructed. Therefore, an alternative exit must be provided if the blocked exit is a required exit.

SECTION 1506
STANDPIPE SYSTEMS

[F] 1506.1 Where required. In buildings required to have standpipes by Section 905.3.1 of the *International Building Code*, not less than one standpipe shall be provided for use during construction. Such standpipes shall be installed prior to construction exceeding 40 feet (12 192 mm) in height above the lowest level of fire department vehicle access. Such standpipe shall be provided with fire department hose connections at accessible locations adjacent to usable stairways. Such standpipes shall be extended as construction progresses to within one floor of the highest point of construction having secured decking or flooring.

❖ The scope of this section is to provide fire safety procedures during the construction operation in accordance with the code and the IFC. Standpipes that are required by IFC Section 905 and are to be a permanent part of the building must be installed and remain functional as construction progresses. Functional standpipes are required so that fire-fighting capability is available at all times within a reasonable proximity to all potential fire locations. Note that standpipes must be in place prior to a building or structure under construction exceeding 40 feet (12 192 mm) in height and thereafter as it progresses to its completed height. During construction, the standpipe must be operational no lower than one floor below the highest point of construction.

[F] 1506.2 Buildings being demolished. Where a building or portion of a building is being demolished and a standpipe is existing within such a building, such standpipe shall be maintained in an operable condition so as to be available for use by the fire department. Such standpipe shall be demolished with the building but shall not be demolished more than one floor below the floor being demolished.

❖ Standpipes in buildings under demolition must remain in service during the demolition process so that firefighting capability is maintained. Similar to buildings under construction, the standpipe is to remain in operation up to one floor level below the floor being demolished. Note that the purpose of this section is to establish the requirement for fire safety procedures during the demolition process in accordance with the code and the IFC.

[F] 1506.3 Detailed requirements. Standpipes shall be installed in accordance with the provisions of Chapter 9 of the *International Building Code*.

> **Exception:** Standpipes shall be either temporary or permanent in nature, and with or without a water supply, provided that such standpipes conform to the requirements of Section 905 of the *International Building Code* as to capacity, outlets and materials.

❖ During the construction or demolition process, standpipes must be provided in a temporary or permanent location so that fire fighters will have a sufficient means to supply water to their connection. These standpipes must either be connected to a permanent water supply source or have the capabilities of being connected to one in accordance with IBC Section 1509.1. Note that all standpipes must comply with IBC Section 905 and must be installed and provided in a building pursuant to the regulations of IBC Chapter 9.

SECTION 1507
AUTOMATIC SPRINKLER SYSTEM

[F] 1507.1 Completion before occupancy. In portions of a building where an automatic sprinkler system is required by this code, it shall be unlawful to occupy those portions of the building until the automatic sprinkler system installation has been tested and approved, except as provided in Section 110.3.

❖ A certificate of occupancy must not be issued by the code official if a required automatic sprinkler system is not approved. However, a temporary occupancy may be granted at the discretion of the code official.

[F] 1507.2 Operation of valves. Operation of sprinkler control valves shall be permitted only by properly authorized personnel and shall be accompanied by notification of duly designated parties. When the sprinkler protection is being regularly turned off and on to facilitate connection of newly completed segments, the sprinkler control valves shall be checked at the end of each work period to ascertain that protection is in service.

❖ The scope of this section is to provide fire safety procedures during construction operations in accordance with the code and the IFC. Note that a sprinkler system must remain operable unless work is being done on it. In such a case, the water must only be shut off by authorized personnel coupled with the notification of the proper authorities so that a form of check and balance is achieved to make certain the water is turned back on.

SECTION 1508
ACCESSIBILITY

[BE] 1508.1 Construction sites. Structures, sites, and equipment directly associated with the actual process of construction, including but not limited to scaffolding, bridging, material hoists, material storage, or construction trailers are not required to be accessible.

❖ This section exempts structures directly associated with the construction process because the need for accessibility on a continuous or regular basis in those circumstances is unlikely to arise. Note that all structures that may be involved during a construction project are not exempt, only those specifically involved in the actual process of construction. For example, if mobile units are brought into house classrooms during a school addition project, those mobile units would have to comply with accessibility provisions. This is consistent with the exception permitted for new construction in IBC Section 1103.2.5.

SECTION 1509
WATER SUPPLY FOR FIRE PROTECTION

[F] 1509.1 When required. An approved water supply for fire protection, either temporary or permanent, shall be made available as soon as combustible material arrives on the site.

❖ Where a standpipe system is present during the construction process, pursuant to Section 1506.1, a water source must be readily available to provide firefighting capabilities at all times.

Chapter 16:
Referenced Standards

General Comments

Chapter 16 contains a comprehensive list of all standards that are referenced in the code. It is organized in a manner that makes it easy to locate specific document references.

This chapter lists the standards that are referenced in various sections of this document. The standards are listed herein by the promulgating agency of the standard; the standard identification; the date and title; and the section or sections of this document that reference the standard. The application of the referenced standards shall be as specified in Section 102.4. It is important to understand that not every document related to building design and construction is qualified to be a "referenced standard." The International Code Council® (ICC®) has adopted a criterion that standards referenced in the *International Codes*® and standards intended for adoption into the *International Codes* must meet in order to qualify as a referenced standard. The policy is summarized as follows:

- Code references: The scope and application of the standard must be clearly identified in the code text.
- Standard content: The standard must be written in mandatory language and appropriate for the subject covered. The standard shall not have the effect of requiring proprietary materials or prescribing a proprietary testing agency.
- Standard promulgation: The standard must be readily available and developed and maintained in a consensus process, such as ASTM or ANSI.

It should be noted that the ICC Code Development Procedures, of which the standards policy is a part, are updated periodically. A copy of the latest version can be obtained from the ICC offices.

Once a standard is incorporated into the code through the code development process, it becomes an enforceable part of the code. When the code is adopted by a jurisdiction, the standard is also a part of that jurisdiction's adopted code. It is for this reason that the criteria were developed. Compliance with this policy provides that documents incorporated into the code are, among others, developed through the use of the consensus process, written in mandatory language and do not mandate the use of proprietary materials or agencies. The requirement for a standard to be developed through a consensus process is vital, as it means that the standard will be representative of the most current body of available knowledge on the subject as determined by a broad spectrum of interested or affected parties without dominance by any single interest group. A true consensus process has many attributes, including but not limited to:

- An open process that has formal (published) procedures that allow for the consideration of all viewpoints;
- A definitive review period that allows for the standard to be updated or revised;
- A process of notification to all interested parties; and
- An appeals process.

Many available documents related to design, installation and construction, though useful, are not "standards" and are not appropriate for reference in the code. Often, these documents are developed or written with the intention of being used for regulatory purposes and are unsuitable for use as a regulation due to extensive use of recommendations, advisory comments and nonmandatory terms. Typical examples of such documents include installation instructions, guidelines and practices.

The objective of ICC's standards policy is to provide regulations that are clear, concise and enforceable—thus, the requirement for standards to be written in mandatory language. This requirement is not intended to mean that a standard cannot contain informational or explanatory material that will aid the user of the standard in its application. When it is the desire of the standard's promulgating agency for such material to be included, however, the information must appear in a nonmandatory location, such as an annex or appendix, and be clearly identified as not being part of the standard.

Overall, standards referenced by the code must be authoritative, relevant, up to date and, most important, reasonable and enforceable. Standards that comply with ICC's standards policy fulfill these expectations.

Purpose

As a performance-oriented code, the code contains numerous references to documents that are used to regulate materials and methods of construction. The references to these documents within the code text consist of the promulgating agency's acronym and its publication designation (e.g., ASME A17.1) and a further indication that the document being referenced is the one that is listed in Chapter 16. Chapter 16 contains all the information that is necessary to identify the specific referenced document. Included is the following information on a document's promulgating agency (see Figure 16):

- The promulgating agency (i.e. the agency's title);

REFERENCED STANDARDS

- The promulgating agency's acronym; and
- The promulgating agency's address.

For example, a reference to an ASME standard within the code indicates that the document is promulgated by the American Society of Mechanical Engineers (ASME), which is located in New York City. This chapter lists the standards agencies alphabetically for ease of identification. This chapter also includes the following information on the referenced document itself (see Figure 16):

- The document's publication designation;
- The document's edition year;
- The document's title;
- Any addenda or revisions to the document that are applicable; and
- Every section of the code in which the document is referred.

For example, a reference to ASME A17.1 indicates that this document can be found in Chapter 16 under the heading ASME. The specific standards designation is A17.1/CSA B44. For convenience, these designations are listed in alphanumeric order. Chapter 16 identifies that ASME A17.1/CSA B44 is titled Safety Code for Elevators and Escalators, the applicable edition (i.e., its year of publication) is 2007 and is referenced in numerous sections of the code.

This chapter also indicates when a document has been discontinued or replaced by its promulgating agency. When a document is replaced by a different one, a note will appear to tell the user the designation and title of the new document.

The key aspect of the manner in which standards are referenced by the code is that a specific edition of a specific standard is clearly identified. In this manner, the requirements necessary for compliance can be readily determined. The basis for code compliance is, therefore, established and available on an equal basis to the code official, contractor, designer and owner.

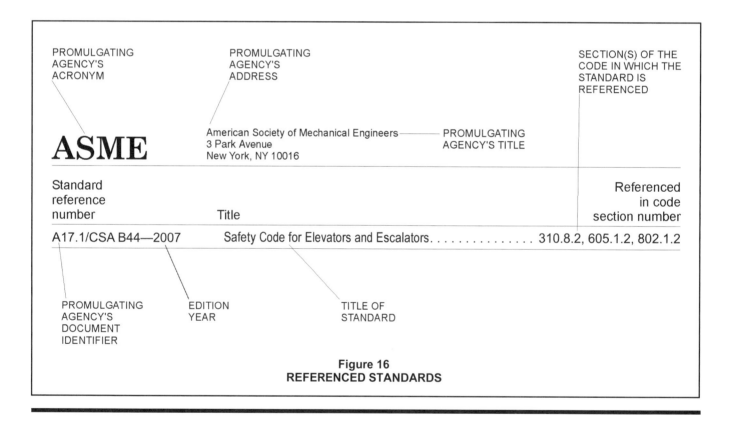

Figure 16
REFERENCED STANDARDS

REFERENCED STANDARDS

This chapter lists the standards that are referenced in various sections of this document. The standards are listed herein by the promulgating agency of the standard, the standard identification, the effective date and title, and the section or sections of this document that reference the standard. The application of the referenced standards shall be as specified in Section 102.4.

ASCE/SEI
American Society of Civil Engineers
Structural Engineering Institute
1801 Alexander Bell Drive
Reston, VA 20191-4400

Standard reference number	Title	Referenced in code section number
7—10	Minimum Design Loads for Buildings and Other Structures with Supplement No. 1	301.1.4.1, 403.4, 403.9, 807.5
41—13	Seismic Evaluation and Retrofit of Existing Buildings	301.1.4, 301.1.4.1, Table 301.1.4.1 301.1.4.2, Table 301.1.4.2, 402.4, Table 402.4, 403.4, 404.2.1, Table 404.2.1, 404.2.3, 407.4

ASHRAE
American Society of Heating, Refrigerating and Air Conditioning Engineers
1791 Tullie Circle, NE
Atlanta, GA 30329

Standard reference number	Title	Referenced in code section number
62.1—2013	Ventilation for Acceptable Indoor Air Quality	809.2

ASME
American Society of Mechanical Engineers
3 Park Avenue
New York, NY 10016

Standard reference number	Title	Referenced in code section number
ASME A17.1/ CSA B44—2013	Safety Code for Elevators and Escalators	410.8.2, 705.1.2, 902.1.2
A17.3—2008	Safety Code for Existing Elevators and Escalators	902.1.2
A18.1—2008	Safety Standard for Platform Lifts and Stairway Chair Lifts	410.8.3, 705.1.3

ASTM
ASTM International
100 Barr Harbor Drive
West Conshohocken, PA 19428-2959

Standard reference number	Title	Referenced in code section number
C 94/C94M—13	Specification for Ready-mixed Concrete	109.3.1
E 84—13A	Test Method for Surface Burning Characteristics of Building Materials	1205.9

REFERENCED STANDARDS

ASTM—continued

Standard	Title	Referenced in code section number
E 108—11	Standard Test Methods for Fire Tests of Roof Coverings	1205.5
E 136—2012	Test Method for Behavior of Materials in a Vertical Tube Furnace at 750°C	202
F 2006—10	Standard Safety Specification for Window Fall Prevention Devices for Non Emergency Escape (Egress) and Rescue (Ingress) Windows	406.2, 702.4
F 2090—10	Standard Specification for Window Fall Prevention Devices with Emergency (Egress) Release Mechanisms	406.2, 702.4, 705.5

ICC

International Code Council, Inc.
500 New Jersey Avenue, NW, 6th Floor
Washington, DC 20001

Standard reference number	Title	Referenced in code section number
IBC—15	International Building Code®	101.4.1, 106.2.2, 109.3.3, 109.3.8, 110.2, 202, 301.1, 301.1.4, 301.1.4.1, 301.1.4.2 401.2.3, 402.1, 402.2, 402.3, 402.3.1, 402.4, 403.1, 403.2, 403.3, 403.3.1, 403.4, 403.4.1, 403.8, 403.9, 404.2.1, 404.2.3, 404.3, 404.4, 404.5, 406.3, 407.1, 407.3, 407.4, 408.3, 410.4, 410.4.2, 410.6, 410.8.1, 410.8.4, 410.8.6, 410.8.5, 410.8.7, 410.8.8, 410.8.10, 410.8.14, 410.9, 410.9.3, 410.9.4, 501.3, 601.3, 602.3, 606.1, 606.2.1, 606.2.2.1, 606.2.3, 606.2.3, 606.2.4 701.2, 701.3, 702.1, 702.2, 702.3, 702.4, 702.5, 702.6, 705.1, 705.1.1, 705.1.4, 705.1.7, 705.1.8, 705.1.9, 706.1, 706.3, 706.3.2, 707.2, 707.3.1, 707.3.2 801.3, 802.1, 803.2.1, 803.2.3, 803.3, 803.4, 803.5.2, 803.6, 804.1.1, 804.2, 804.2.2, 804.2.3, 804.2.4, 804.3, 805.3.1, 805.3.1.2.1, 805.4.3, 805.5, 805.6, 805.7.1, 805.8.1, 805.9.2, 805.10.1.1, 805.10.1.2, 805.10.1.3, 805.10.2, 805.11.2, 806.2, 806.3, 806.4, 806.5, 807.2, 807.4, 807.5, 807.6 904.1.2, 904.1.3, 904.2, 904.2.1, 904.2.2, 905.2, 905.3, 906.2, 907.4, 907.4.2 1001.3, 1002.1, 1002.2, 1007.1, 1007.2, 1007.3.1, 1011.1, 1012.1.1.1, 1012.1.1.2, 1012.2.1, 1012.2.2, 1012.3, 1012.4.1, 1012.4.2, 1012.4.3, 1012.5.1, 1012.5.1.1, 1012.5.3, 1012.6.1, 1012.6.3, 1012.7.1, 1012.7.2, 1012.7.3, 1012.8, 1012.8.2 1102.1, 1102.2, 1102.3, 1103.1, 1103.2, 1103.3, 1103.3.1, 1103.3.2, 1103.4, 1103.5, 1201.4, 1202.3, 1202.4, 1203.12, 1204.1, 1204.1.4, 1205.2, 1205.9, 1205.15, 1301.2, 1302.1, 1302.2, 1302.2.1, 1302.3, 1302.4, 1302.5, 1302.6, 1401.2.2, 1401.2.3, 1401.2.4, 1401.3.3, 1401.4.1, 1401.6.1, 1401.6.1.1, 1401.6.2, 1401.6.2.1, 1401.6.3.1, 1401.6.3.2, 1401.6.4.1, 1401.6.5.1, 1401.6.6, 1401.6.7.1, 1401.6.8, 1401.6.9, 1401.6.9.1, 1401.6.10, 1401.6.10.1, 1401.6.11, 1401.6.11.1, 1401.6.12.1, 1401.6.13, 1401.6.15.1, 1401.6.16.1, 1401.6.17, 1401.6.17.1, 1401.6.18, 1401.6.18.1, 1401.6.19, Table 1401.6.19, 1501.5, 1501.6.1, 1501.6.4.1, 1501.6.7, 1506.1, 1506.3
ICC A117.1—09	Accessible and Usable Buildings and Facilities	410.8.2, 410.8.3, 410.8.10, 705.1.2, 705.1.3
ICC 300—12	ICC Standard on Bleachers, Folding and Telescopic Seating and Grandstands	401.1
IECC—15	International Energy Conservation Code®	301.2, 702.6, 708.1, 811.1, 908.1
IFC—15	International Fire Code®	101.4.2, 301.1.1, 301.2, 402.5, 403.10, 803.2.1, 803.2.3, 804.4.1.1, 804.4.1.2, 804.4.1.3, 804.4.1.4, 804.4.1.5, 804.4.1.6, 804.4.1.7, 804.4.3, 1012.5.1.1, 1104.1, 1301.2, 1401.3.2, 1401.6.8.1, 1401.6.14, 1401.6.14.1, 1501.5, 1504.1, 1504.2
IFGC—15	International Fuel Gas Code®	301.2, 702.6.1
IMC—15	International Mechanical Code®	301.2, 702.6, 809.1, 902.1.1, 902.2.1, 1009.1, 1401.6.7.1, 1401.6.8, 1401.6.8.1
IPC—15	International Plumbing Code®	301.2, 609.1, 702.6, 810.1, 1010.1, 1010.2, 1010.3, 1010.5, 1501.7
IPMC—15	International Property Maintenance Code®	101.4.2, 301.2, 1301.2, 1401.3.2
IRC—15	International Residential Code®	101.4.1, 301.2, 402.2, 403.2, 404.5, 408.3, 602.3, 701.3, 702.5, 706.2, 707.2, 707.4, 707.5, 708.1 807.4, 808.3, 811.1, 907.4, 908.1, 1103.2, 1103.3, 1103.4, 1104.1, 1106.1, 1201.4, 1301.2, 1302.1, 1302.2, 1302.2.1, 1302.3, 1302.4, 1302.6, 1302.5, 1401.2.2, 1401.2.3, 1401.3.3

REFERENCED STANDARDS

NFPA

National Fire Protection Agency
1 Batterymarch Park
Quincy, MA 02269-9101

Standard reference number	Title	Referenced in code section number
NFPA 13R—13	Standard for the Installation of Sprinkler Systems in Residential Occupancies up to and Including Four Stories in Height	804.2.5
NFPA 70—14	National Electrical Code	107.3, 301.2, 607.1.1, 607.1.2, 607.1.3, 607.1.4, 607.1.5, 808.1, 808.3.4, 808.3.7, 1008.1, 1008.2, 1008.3, 1008.4
NFPA 72—13	National Fire Alarm and Signaling Code	804.2.5, 804.4
NFPA 99—15	Health Care Facilities Code	607.1.4
NFPA 101—15	Life Safety Code	805.2

UL

UL LLC
333 Pfingsten Road
Northbrook, IL 60062

Standard reference number	Title	Referenced in code section number
723—08	Standard for Test for Surface Burning Characteristics of Building Materials with Revisions Through September 2010	1205.9
790—04	Standard Test Methods for Fire Tests of Roof Coverings with Revisions through October 2008	1205.5

Appendix A: Guidelines for the Seismic Retrofit of Existing Buildings

Chapter A1:
Seismic Strengthening Provisions for Unreinforced Masonry Bearing Wall Buildings

[Commentary portions of Chapters A1 through A5 are produced by the Structural Engineers Association of California (SEAOC). This commentary is jointly copyrighted by SEAOC and ICC.]

General

Appendix Chapter A1, along with the rest of Appendix A of the code, is referenced by the *International Codes®* in two ways:

- Section 301.1.4.2 allows the use of Appendix A for qualifying buildings when "reduced IBC level seismic forces" are triggered.
- Section 807.6 allows the use of Appendix A for voluntary seismic retrofit.

The provisions in Appendix Chapter A1 require the selection of one of two procedures—either the general procedure or the special procedure for the analysis and seismic strengthening of unreinforced masonry (URM) buildings. The special procedure is based on research on the behavior of URM buildings at their limit state and on the experience of engineers who have upgraded buildings in the Los Angeles hazard reduction program; use of the special procedure is expected for most buildings. The general procedure is a procedure for buildings with rigid diaphragms that do not qualify for the special procedure; it is a combination of a conventional code approach and some portions of the special procedure. The commentary discusses the general and special procedures individually and in detail.

URM bearing wall buildings have shown poor performance in past earthquakes, including the following significant U.S. events: 1868 Hayward, California; 1906 San Francisco, California; 1925 Santa Barbara, California; 1933 Long Beach, California; 1935 Helena, Montana; 1952 Kern County, California; 1964 Olympia, Washington; 1971 San Fernando, California; 1983 Coalinga, California; 1987 Whittier, California; 1989 Loma Prieta, California; 1994 Northridge, California and 2001 Nisqually, Washington.

The loss of lives and property damage in future earthquakes will be heaviest in buildings that exist today, especially buildings of URM construction. The short-term impact of improved knowledge, codes and design practices will be limited to those relatively few recent buildings that take advantage of this improved knowledge. Thus, the dominant policy issues posed by earthquakes involve not new but existing buildings, particularly those structures that have obvious weaknesses and do not comply with the general intent and necessary requirements of current regulations. The issue before the public and the profession is determining how to set standards for these noncompliant existing buildings that are consistent with both the desire for safety and the limited resources available to achieve improved safety.

It is generally accepted that the intensity of earthquakes reasonably expected to occur in the moderate and high seismic zones of the United States would be sufficient to cause buildings with minimal seismic resistance characteristics to be seriously damaged, causing injury or death to occupants or passers-by.

It is reasonable, where a real seismic risk exists, to take steps to significantly reduce it. The objective of Appendix A1 is the reduction or elimination of seismic risk associated with URM buildings.

The main goal is to reduce life safety risks as best as possible with the available resources. The efforts are directed toward ensuring a coherent load path for lateral loads, reduction of out-of-plane wall failures, reduction of loss of support for floors and roofs and reduction of falling parapets or ornamentation. Application of this chapter will decrease the probability of loss of life, but this cannot be prevented. Inherent to the procedure is the acceptability of some major and irreparable damage as long as there is a decrease in the likelihood of falling building elements or loss of support for horizontal framing and a reduction in serious injury and loss of life.

The goal of this recommended strengthening is lower than the goal set for new construction. Appendix Chapter A1 recognizes that the economic difficulty of strengthening existing buildings necessitates reliance on building components with seismic performance characteristics that are less than ideal.

SECTION A101
PURPOSE

[BS] A101.1 Purpose. The purpose of this chapter is to promote public safety and welfare by reducing the risk of death or injury that may result from the effects of earthquakes on existing unreinforced masonry bearing wall buildings.

The provisions of this chapter are intended as minimum standards for structural seismic resistance, and are established primarily to reduce the risk of life loss or injury. Compliance with these provisions will not necessarily prevent loss of life or injury, or prevent earthquake damage to rehabilitated buildings.

❖ Appendix Chapter A1 is intended to reduce risks associated with the seismic performance of URM buildings. It represents minimum standards required to reduce risk of life loss or injury. The objective is accomplished by reducing the possibility of damage and the extent of damage. However, compliance with this chapter will not necessarily prevent earthquake damage to rehabilitated buildings. Risk of life loss or injury is significantly reduced but not eliminated by compliance with this chapter.

The risk reduction applies to the overall average performance of rehabilitated URM buildings. An individual building may have damage levels above or below the average, depending on its structural characteristics and the local ground motion. If it is desired to further limit damage to any specific building or preserve its post-earthquake function, the engineer should consider additional measures for strengthening the building.

SECTION A102
SCOPE

[BS] A102.1 General. The provisions of this chapter shall apply to all existing buildings having at least one unreinforced masonry bearing wall. The elements regulated by this chapter shall be determined in accordance with Table A1-A. Except as provided herein, other structural provisions of the building code shall apply. This chapter does not apply to the *alteration* of existing electrical, plumbing, mechanical or fire safety systems.

❖ This chapter is applicable to all existing buildings that have at least one URM wall or URM bearing wall. Section A103 defines a URM wall and URM bearing wall. URM walls constructed within concrete or steel frames are not bearing walls. These frame buildings with URM infill are considered as potentially hazardous, but they are not covered by this chapter.

Table A1-A lists elements of buildings that are regulated by this chapter. This listing of elements is related to observed damage or partial collapse of these elements and spectral intensity of ground shaking. Separation of unbraced parapets has been reported for cities distant from a large magnitude earthquake source. The measure of spectral velocity, S_{D1}, is used instead of short period spectral acceleration, S_{Ds}, because S_{D1} more accurately reflects the seismic behavior of unreinforced masonry buildings with flexible diaphragms and, to a lesser extent, the other items listed in Table A1-A. The upper bound value of S_{D1} given in Table A1-A corresponds to descriptions of Seismic Design Categories B, C and D.

Appendix Chapter A1 does not regulate nonstructural systems.

[BS] A102.2 Essential and hazardous facilities. The provisions of this chapter shall not apply to the strengthening of buildings in Risk Category III or IV. Such buildings shall be strengthened to meet the requirements of the *International Building Code* for new buildings of the same risk category or other such criteria approved by the *code official*.

❖ The degree of earthquake risk reduction anticipated in Appendix Chapter A1 is not considered acceptable for IBC Risk Categories III and IV. Buildings in these risk categories require special detailing considerations, increased seismic loading and limitations on seismic force-resisting systems.

SECTION A103
DEFINITIONS

For the purpose of this chapter, the applicable definitions in the building code shall also apply.

❖ In general, the definitions given in this section are self-explanatory. The commentary expands on several of the definitions that are critical to the use of Appendix Chapter A1.

[BS] COLLAR JOINT. The vertical space between adjacent wythes. A collar joint may contain mortar or grout.

❖ The collar joint is the space between wythes; space that may be empty or may be filled with mortar or grout. Although the condition of the collar joint may not be critical for in-plane wall shear stress, the procedure for determining allowable stress from in-plane shear tests takes into account the probable effect of mortar in the collar joint. The condition of the joint is of greater importance for out-of-plane forces where it is necessary to have the wythes of the wall act integrally. A visual examination of the collar joint to determine its mortar coverage is required to allow the use of the increased height-to-thickness ratios given in Table A1-B for buildings with crosswalls in all stories.

[BS] CROSSWALL. A new or existing wall that meets the requirements of Section A111.3 and the definition of Section A111.3. A crosswall is not a shear wall.

❖ A crosswall is a light-frame wall sheathed with new or existing materials. Light-frame walls with shear panels have a desirable hysteretic behavior and can, by their coupling strength, stiffen the diaphragm and reduce diaphragm deflections. They function as energy dissipaters when connecting flexible diaphragms or a flexible diaphragm to grade within the span of the diaphragm. They act similar to shear walls to the extent that they diminish the displacement of a floor or roof relative to the building base, but are not true shear walls. Their in-plane stiffness is not comparable with shear walls of masonry, concrete or lateral load-resisting elements of structural steel. Moment-resisting frames may also be designed as crosswalls.

[BS] CROSSWALL SHEAR CAPACITY. The unit shear value times the length of the crosswall, $v_c L_c$.

❖ The unit shear capacities are listed in Tables A1-D and A1-E for existing and new construction, respectively. For crosswalls with openings, the minimum length of the sum of the individual crosswalls along a line through the openings should be used to determine the total length of crosswall, L_c.

[BS] DIAPHRAGM EDGE. The intersection of the horizontal diaphragm and a shear wall.

[BS] DIAPHRAGM SHEAR CAPACITY. The unit shear value times the depth of the diaphragm, $v_u D$.

[BS] FLEXIBLE DIAPHRAGM. A diaphragm of wood or untopped metal deck construction.

❖ The definition for "Flexible diaphragm" is required for Section A111.1, which determines where the special procedure can be used. For further explanation, refer to the commentary to the definition of "Rigid diaphragm."

[BS] NORMAL WALL. A wall perpendicular to the direction of seismic forces.

❖ A normal wall is a URM bearing wall with loads perpendicular to the wall. When the earthquake loads are parallel to an URM bearing wall, the wall is considered a shear wall.

[BS] OPEN FRONT. An exterior building wall line without vertical elements of the lateral force-resisting system in one or more stories.

❖ "Open front" is a term for the side of a building that does not have a shear wall, frame or braced frame at the exterior wall line. The open front may be at the first story or at all stories. The open front is defined as "on one side only." The definition is intended to be a limitation. Open-front buildings on street corners must have lateral load-resisting elements, such as shear walls installed on one of the street-front sides.

[BS] POINTING. The process of removal of deteriorated mortar from between masonry units and placement of new mortar. Also known as repointing or tuckpointing for purposes of this chapter.

❖ Pointing is equivalent to repointing. The recommended procedure for pointing is described in Section A106.3.3.9.

[BS] REPOINTING. See "Pointing."

[BS] RIGID DIAPHRAGM. A diaphragm of concrete construction.

❖ Whether a diaphragm should be considered flexible or rigid depends on the relationship of diaphragm in-plane shear stiffness to masonry wall in-plane stiffness. Concrete diaphragms, including older concrete diaphragms that are not "reinforced" based on current code definitions, are typically considered rigid based on stiffness. The engineer must consider the relative stiffness and strength limit states of other materials, such as gypsum concrete roofs, to determine a rational analysis procedure for the existing structural materials. An alternative procedure would be to use the IBC definition of a "Flexible diaphragm," which is based on the stiffness of only the diaphragm and the shear walls below the diaphragm level. Appendix Chapter A1 does not consider semirigid diaphragms.

[BS] TUCKPOINTING. See "Pointing."

[BS] UNREINFORCED MASONRY. Includes burned clay, concrete or sand-lime brick; hollow clay or concrete block; plain concrete; and hollow clay tile. These materials shall comply with the requirements of Section A106 as applicable.

[BS] UNREINFORCED MASONRY BEARING WALL. A URM wall that provides the vertical support for the reaction of floor or roof-framing members.

[BS] UNREINFORCED MASONRY (URM) WALL. A masonry wall that relies on the tensile strength of masonry units, mortar and grout in resisting design loads, and in which the area of reinforcement is less than 25 percent of the minimum ratio required by the building code for reinforced masonry.

❖ A URM wall is a wall constructed of solid clay or concrete units, or a wall constructed of hollow units, either clay or concrete. If the wall is constructed of hollow units, the net area of the masonry wall must be used for in-plane shear calculations. The allowable height-to-thickness ratios given in Table A1-B are not affected by the solidity of the masonry unit. Only the height-to-thickness ratio affects the stability of the wall. The weight per square foot of the wall is not a parameter for out-of-plane wall stability.

The in-plane bed joint shear strength or tensile splitting strength of the masonry must be determined by testing. Field stone or adobe masonry does not have a reliable test method for determination of in-plane shear strength. The engineer should, with the concurrence of the code official, determine strength limit state materials properties.

Appendix Chapter A1 has been developed for URM. The tensile strength of bed joints is considered as zero. A wall is considered unreinforced if the amount of reinforcing is less than 25 percent of the minimum amount that would be required by the IBC. The intention is to exclude engineered systems that have small but definite amounts of horizontal and vertical reinforcing.

In most cases, a wall will either contain reinforcing at or near the IBC requirements, or it will have no reinforcing at all. The reinforcing must be both vertical and horizontal. Steel straps laid along bed joints exist in some older types of construction; these should not be considered horizontally reinforced since these straps could actually weaken the bed joint.

[BS] YIELD STORY DRIFT. The lateral displacement of one level relative to the level above or below at which yield stress is first developed in a frame member.

❖ This term refers to the lateral displacement of one level relative to the level above or below at which yield stress is first developed in a frame member of moment frames acting as crosswalls.

SECTION A104
SYMBOLS AND NOTATIONS

For the purpose of this chapter, the following notations supplement the applicable symbols and notations in the building code.

a_n = Diameter of core multiplied by its length or the area of the side of a square prism.

A = Cross-sectional area of unreinforced masonry pier or wall, square inches (10^{-6} m^2).

A_b = Total area of the bed joints above and below the test specimen for each in-place shear test, square inches (10^{-6} m^2).

D = In-plane width dimension of pier, inches (10^{-3} m), or depth of diaphragm, feet (m).

DCR = Demand-capacity ratio specified in Section A111.4.2.

f'_m = Compressive strength of masonry.

f_{sp} = Tensile-splitting strength of masonry.

F_{wx} = Force applied to a wall at level x, pounds (N).

H = Least clear height of opening on either side of a pier, inches (10^{-3} m).

h/t = Height-to-thickness ratio of URM wall. Height, h, is measured between wall anchorage levels and/or slab-on-grade.

L = Span of diaphragm between shear walls, or span between shear wall and open front, feet (m).

L_c = Length of crosswall, feet (m).

L_i = Effective span for an open-front building specified in Section A111.8, feet (m).

P = Applied force as determined by standard test method of ASTM C496 or ASTM E519, pounds (N).

P_D = Superimposed dead load at the location under consideration, pounds (kN). For determination of the rocking shear capacity, dead load at the top of the pier under consideration shall be used.

P_{D+L} = Press resulting from the dead plus actual live load in place at the time of testing, pounds per square inch (kPa).

P_w = Weight of wall, pounds (N).

R = Response modification factor for Ordinary plain masonry shear walls in Bearing Wall System from Table 12.2-1 of ASCE 7, where $R = 1.5$.

S_{DS} = Design spectral acceleration at short period, in g units.

S_{D1} = Design spectral acceleration at 1-second period, in g units.

v_a = The shear strength of any URM pier, $v_m A/1.5$ pounds (N).

v_c = Unit shear capacity value for a crosswall sheathed with any of the materials given in Table A1-D or A1-E, pounds per foot (N/m).

v_m = Shear strength of unreinforced masonry, pounds per square inch (kPa).

V_{aa} = The shear strength of any URM pier or wall, pounds (N).

V_{ca} = Total shear capacity of crosswalls in the direction of analysis immediately above the diaphragm level being investigated, $v_c L_c$, pounds (N).

V_{cb} = Total shear capacity of crosswalls in the direction of analysis immediately below the diaphragm level being investigated, $v_c L_c$, pounds (N).

V_p = Shear force assigned to a pier on the basis of its relative shear rigidity, pounds (N).

V_r = Pier rocking shear capacity of any URM wall or wall pier, pounds (N).

v_t = Mortar shear strength as specified in Section A106.3.3.5, pounds per square inch (kPa).

V_{test} = Load at incipient cracking for each in-place shear test performed in accordance with Section A106.3.3.1, pounds (kN).

v_{to} = Mortar shear test values as specified in Section A106.3.3.5, pounds per square inch (kPa).

v_u = Unit shear capacity value for a diaphragm sheathed with any of the materials given in Table A1-D or A1-E, pounds per foot (N/m).

V_{wx} = Total shear force resisted by a shear wall at the level under consideration, pounds (N).

W = Total seismic dead load as defined in the building code, pounds (N).

W_d = Total dead load tributary to a diaphragm level, pounds (N).

W_w = Total dead load of a URM wall above the level under consideration or above an open-front building, pounds (N).

W_{wx} = Dead load of a URM wall assigned to level x halfway above and below the level under consideration, pounds (N).

$\Sigma v_u D$ = Sum of diaphragm shear capacities of both ends of the diaphragm, pounds (N).

$\Sigma\Sigma v_u D$ = For diaphragms coupled with crosswalls, $v_u D$ includes the sum of shear capacities of both ends of diaphragms coupled at and above the level under consideration, pounds (N).

ΣW_d = Total dead load of all the diaphragms at and above the level under consideration, pounds (N).

❖ Selected definitions are given to clarify the use of the following symbols:

P_D = The tributary dead load at the top of the pier under consideration. For uniformly loaded bearing walls, it consists of the dead load on a width of wall extending between the centers of the openings adjacent to an interior pier, from the wall edge to the center of the adjacent opening for the exterior piers or from wall edge to wall edge. For use in Equations A1-4, A1-5 and A1-6, P_D is the superimposed dead load at the test location.

P_{D+L} = Stress resulting from the dead load plus actual live load in place at the time of testing, in pounds per square inch. For use in Equation A1-3, the quantity P_{D+L} is the stress obtained by dividing the axial load on the wall by the wall area, not the area of the test bed joints.

V_{test} = Load at incipient cracking for the in-place shear test described in Section A106.3.3.1, Item 1. This description of the in-place shear test has more detail than the obsolete UBC 21-6.

$\Sigma v_u D$ = The sum of the diaphragm shear capacities of both ends of the diaphragm that are parallel to the direction of the lateral load.

$\Sigma\Sigma v_u D$ = The sum of the Σv_{uD} values for all of the coupled diaphragms at and above the level under consideration.

SECTION A105
GENERAL REQUIREMENTS

[BS] A105.1 General. The seismic force-resisting system specified in this chapter shall comply with the building code, except as modified herein.

❖ Chapter 16 of the *International Building Code*® (IBC®) provides the basic requirements for a seismic force-resisting system. Distribution of lateral loads shall consider the relative rigidity of horizontal diaphragms. Flexible diaphragms have inadequate relative stiffness to redistribute force between those elements that are considered shear walls by the special procedure. Diaphragms in buildings that use the general procedure are capable of redistributing the lateral loads, causing increased forces due to torsional response and coupling the weight at each story level with the shear walls.

[BS] A105.2 Alterations and repairs. Alterations and repairs required to meet the provisions of this chapter shall comply with applicable structural requirements of the building code unless specifically provided for in this chapter.

❖ The IBC applies unless explicitly excluded or modified herein.

[BS] A105.3 Requirements for plans. The following construction information shall be included in the plans required by this chapter:

1. Dimensioned floor and roof plans showing existing walls and the size and spacing of floor and roof-framing members and sheathing materials. The plans shall indicate all existing and new crosswalls and shear walls and their materials of construction. The location of these walls and their openings shall be fully dimensioned and drawn to scale on the plans.

2. Dimensioned wall elevations showing openings, piers, wall classes as defined in Section A106.3.3.8, thickness, heights, wall shear test locations, cracks or damaged portions requiring repairs, the general condition of the mortar joints, and if and where pointing is required. Where the exterior face is veneer, the type of veneer, its thickness and its bonding and/or ties to the structural wall masonry shall also be noted.

3. The type of interior wall and ceiling materials, and framing.

4. The extent and type of existing wall anchorage to floors and roof when used in the design.

5. The extent and type of parapet corrections that were previously performed, if any.

6. *Repair* details, if any, of cracked or damaged unreinforced masonry walls required to resist forces specified in this chapter.

7. All other plans, sections and details necessary to delineate required retrofit construction.

8. The design procedure used shall be stated on both the plans and the permit application.

9. Details of the anchor prequalification program required by Section A107.5.3, if used, including location and results of all tests.

❖ These are minimum requirements for the plans to be submitted. Although this may seem like an administrative requirement, this section is needed to ensure that a thorough investigation of the building has been made and is shown on required plans that are submitted to the code official.

The plans should include openings and their dimensions. Each diaphragm should be investigated at several locations. The lay-up of the sheathing should be observed and properly described in the plans. However, the continuity of the finish floor sheathing under partitions need not be verified.

Section A107.5.3, referenced by Item 9, is based on the requirements originally written for the *Uniform Code for Building Conservation* [and later as *Uniform Building Code* (UBC) Standard 21-7] on the subject of direct tension testing of existing anchors and new bolts, torque testing of new bolts and prequalification testing for bolts and other types of anchors. Prequalification testing may be used to justify capacities for new bolts that differ from the ones specified in Table A1-E; however, the tension and shear capacities of new bolts determined by prequalification testing may not exceed 1.5 times the values specified in Table A1-E.

[BS] A105.4 Structural observation, testing and inspection. Structural observation, in accordance with Section 1708 of the *International Building Code*, shall be required for all structures in which seismic retrofit is being performed in accordance with this chapter. Structural observation shall include visual observation of work for conformance with the approved construction documents and confirmation of existing conditions assumed during design.

Structural testing and inspection for new construction materials shall be in accordance with the building code, except as modified by this chapter.

❖ The construction phase of the retrofit program will usually uncover unforeseen conditions. The engineer or architect responsible for the design must provide a more extensive observation of this type of project than that required for new construction. This is the reason that structural observation is required for all structures undergoing a seismic retrofit.

SECTION A106
MATERIALS REQUIREMENTS

[BS] A106.1 General. Materials permitted by this chapter, including their appropriate strength design values and those existing configurations of materials specified herein, may be used to meet the requirements of this chapter.

❖ Existing materials, as described in Table A1-D; new materials, as described in Table A1-E; and masonry materials, when tested as described in this section or categorized by f'_m, may be used as a part of the seismic force-resisting structural system. New materials permitted by the IBC may be used to supplement the strength of existing materials.

[BS] A106.2 Existing materials. Existing materials used as part of the required vertical load-carrying or lateral force-resisting system shall be in sound condition, or shall be repaired or removed and replaced with new materials. All other unreinforced masonry materials shall comply with the following requirements:

1. The lay-up of the masonry units shall comply with Section A106.3.2, and the quality of bond between the units has been verified to the satisfaction of the building official;

2. Concrete masonry units are verified to be load-bearing units complying with ASTM C90 or such other standard as is acceptable to the building official; and

3. The compressive strength of plain concrete walls shall be determined based on cores taken from each class of concrete wall. The location and number of tests shall be the same as those prescribed for tensile-splitting strength tests in Sections A106.3.3.3 and A106.3.3.4, or in Section A108.1.

The use of materials not specified herein or in Section A108.1 shall be based on substantiating research data or engineering judgment, with the approval of the building official.

❖ ASTM C90 can be used for description of dimensions of load-bearing hollow concrete units. Face shell thickness of hollow units is needed for calculation of net bedded cross section; traditionally, only face shells are mortared in ungrouted hollow concrete units. Face shell thickness for nominal thickness of the unit is: 6 inches nominal, 1 inch; 8 inches nominal, $1^1/_4$ inches; 10 inches nominal, $1^3/_8$ inches; 12 inches nominal, $1^1/_2$ inches.

[BS] A106.3 Existing unreinforced masonry.

[BS] A106.3.1 General. Unreinforced masonry walls used to carry vertical loads or seismic forces parallel and perpendicular to the wall plane shall be tested as specified in this section. All masonry that does not meet the minimum standards established by this chapter shall be removed and replaced with new materials, or alternatively, shall have its structural functions replaced with new materials and shall be anchored to supporting elements.

❖ It is essential for the engineer to make a proper evaluation of the existing masonry. Allowable shear values for use in the analysis of shear walls and allowable height-to-thickness values for normal walls that are not analyzed for out-of-plane loading are based on the assumption of adequate strengths and lay-up of the masonry. Nonconforming masonry must be removed or treated as veneer.

[BS] A106.3.2 Lay-up of walls.

[BS] A106.3.2.1 Multiwythe solid brick. The facing and backing shall be bonded so that not less than 10 percent of the exposed face area is composed of solid headers extending not less than 4 inches (102 mm) into the backing. The clear distance between adjacent full-length headers shall not exceed 24 inches (610 mm) vertically or horizontally. Where the backing consists of two or more wythes, the headers shall extend not less than 4 inches (102 mm) into the most distant wythe, or the backing wythes shall be bonded together with separate headers with their area and spacing conforming to the foregoing. Wythes of walls not bonded as described above shall be considered veneer. Veneer wythes shall not be included in the effective thickness used in calculating the height-to-thickness ratio and the shear capacity of the wall.

Exception: Where S_{D1} is not more than 0.3, veneer wythes anchored as specified in the building code and made composite with backup masonry may be used for calculation of the effective thickness.

❖ The concern is for the integrity of the multiwythe wall acting as a whole in resisting out-of-plane forces. "Facing" means the wythe at the face of the wall, whether interior or exterior; "backing" means the inner wythes that are not normally visible. "Lay-up" means the pattern of masonry units in each wythe and in the interlocking of wythes. The primary requirement is that all wythes be adequately interlocked by header bricks. The quantity of mortar in the collar joints between wythes is also important, having an effect on the allowable height-to-thickness ratio for the wall as specified in the notes to Table A1-B.

The section exception to the requirements of the common lay-up of multiwythe walls is permitted for lower seismic hazard zones. This exception is applicable where S_{D1} as defined and mapped in the IBC is 0.3g or less. The exception in this section permits veneer wythes with anchorage as specified by the IBC and made composite with the backing masonry to be considered as a part of the structural wall. The wythes may be bonded by a combination of additional ties or the combination of grout and ties and should be checked to determine if they can transfer the calculated stress between wythes.

[BS] A106.3.2.2 Grouted or ungrouted hollow concrete or clay block and structural hollow clay tile. Grouted or ungrouted hollow concrete or clay block and structural hollow clay tile shall be laid in a running bond pattern.

[BS] A106.3.2.3 Other lay-up patterns. Lay-up patterns other than those specified in Sections A106.3.2.1 and A106.3.2.2 above are allowed if their performance can be justified.

[BS] A106.3.3 Testing of masonry.

[BS] A106.3.3.1 Mortar tests. The quality of mortar in all masonry walls shall be determined by performing in-place shear tests in accordance with the following:

1. The bed joints of the outer wythe of the masonry shall be tested in shear by laterally displacing a single brick relative to the adjacent bricks in the same wythe. The head joint opposite the loaded end of the test brick shall be carefully excavated and cleared. The brick adjacent to the loaded end of the test brick shall be carefully removed by sawing or drilling and excavating to provide space for a hydraulic ram and steel loading blocks. Steel blocks, the size of the end of the brick, shall be used on each end of the ram to distribute the load to the brick. The blocks shall not contact the mortar joints. The load shall be applied horizontally, in the plane of the wythe. The load recorded at first movement of the test brick as indicated by spalling of the face of the mortar bed joints is V_{test} in Equation A1-3.

2. Alternative procedures for testing shall be used where in-place testing is not practical because of crushing or other failure mode of the masonry unit (see Section A106.3.3.2).

❖ **Item 1.** Mortar tests are performed by shearing of the mortar joints above and below the test masonry unit. The test load is recorded at the first sign of movement of the test brick. This can be detected when sand grains are detaching from the mortar joint. The user can refer to ASTM C1531 for a procedure for conducting shear strength tests.

Item 2. Many masonry walls in the United States have been constructed with mortars that have shear strength such that the mortar shear strength test is not possible (e.g., hollow masonry). The failure may be the bearing on the end of the masonry units at the jack. Where this occurs, an alternative testing method described in Section 106.3.3.2 determines tensile splitting strength by the standard ASTM procedure.

[BS] A106.3.3.2 Alternative procedures for testing masonry. The tensile-splitting strength of existing masonry, f_{sp}, or the prism strength of existing masonry, f'_m, may be determined in accordance with one of the following procedures:

1. Wythes of solid masonry units shall be tested by sampling the masonry by drilled cores of not less than 8 inches (203 mm) in diameter. A bed joint intersection with a head joint shall be in the center of the core. The tensile-splitting strength of these cores should be determined by the standard test method of ASTM C496. The core should be placed in the test apparatus with the bed joint 45 degrees (0.79 rad) from the horizontal. The tensile-splitting strength should be determined by the following equation:

$$f_{sp} = \frac{2P}{\pi a_n} \quad \text{(Equation A1-1)}$$

2. Hollow unit masonry constructed of through-the-wall units shall be tested by sampling the masonry by a sawn square prism of not less than 18 inches square (11 613 mm²). The tensile-splitting strength should be determined by the standard test method of ASTM E519. The diagonal of the prism should be placed in a vertical position. The tensile-splitting strength should be determined by the following equation:

$$f_{sp} = \frac{0.494P}{a_n} \quad \text{(Equation A1-2)}$$

3. An alternative to material testing is estimation of the f'_m of the existing masonry. This alternative should be limited to recently constructed masonry. The determination of f'_m requires that the unit correspond to a specification of the unit by an ASTM standard and classification of the mortar by type.

❖ These procedures determine the tensile splitting strength of masonry, f_{sp}. The tensile splitting strength of masonry adjusted for the effects of axial load, as used in Equation A1-5, is used to determine the unreinforced masonry shear strength, v_m. The pier shear strength, V_m, is calculated using $v_m A_n/1.5$, which is the average strength in shear for a rectangular cross section.

Recently constructed masonry (post-1950) may be categorized by unit strength and mortar mix according to Tables 1 and 2 of TMS 602-11/ACI 530.1-11/ASCE 6-11. These tables may be used to estimate the compressive strength of the masonry. The estimate will be a conservative prism strength, which may be used to calculate peak shear stress. The shear strength of a pier is calculated by Equation A1-20.

The shear strength is an expected shear strength; no load factors or capacity reduction factors are used in Appendix Chapter A1.

[BS] A106.3.3.3 Location of tests. The shear tests shall be taken at locations representative of the mortar conditions throughout the entire building, taking into account variations in workmanship at different building height levels, variations in weathering of the exterior surfaces, and variations in the condition of the interior surfaces due to deterioration caused by leaks and condensation of water and/or by the deleterious effects of other substances contained within the building. The exact test locations shall be determined at the building site by the engineer or architect in responsible charge of the structural design work. An accurate record of all such tests and their locations in the building shall be recorded, and these results shall be submitted to the building department for approval as part of the structural analysis.

❖ The test locations should be distributed over the wall surface to account for variability in construction, environmental influences and weathering. The exact location of each test shall be determined at the site by the engineer. Relatively small areas that are pointed due to local deterioration should not be used for test areas since pointed areas will produce higher values that can skew the test results.

[BS] A106.3.3.4 Number of tests. The minimum number of tests per class shall be as follows:

1. At each of both the first and top stories, not less than two tests per wall or line of wall elements providing a common line of resistance to lateral forces.

2. At each of all other stories, not less than one test per wall or line of wall elements providing a common line of resistance to lateral forces.

3. In any case, not less than one test per 1,500 square feet (139.4 m^2) of wall surface and not less than a total of eight tests.

❖ Numbers and locations are specified to ensure a representative sample of existing masonry, considering that the walls were not necessarily built at the same time and with the same workmanship. Workmanship may vary in a given wall, often being poorer near the top, and may not have been subjected to the same environmental influences. These and other conditions may make it desirable to divide walls into classes according to their relative overall quality (see Section A106.3.3.8). The number of tests in each class of wall should be established to provide a body of data that is adequate to establish a 20-percent value, v_t.

A common line of resistance is defined as a set of one or more masonry walls at an edge of a diaphragm. The wall may have offsets in that common line. A minimum of eight tests for a building is specified. This would be the minimum number for a single-story building. The area of 1,500 square feet (139 m^2) is gross area, including wall openings.

[BS] A106.3.3.5 Minimum quality of mortar.

1. Mortar shear test values, v_{to}, in pounds per square inch (kPa) shall be obtained for each in-place shear test in accordance with the following equation:

   $$v_{to} = (V_{test}/A_b) - P_{D+L} \quad \textbf{(Equation A1-3)}$$

2. Individual unreinforced masonry walls with v_{to} consistently less than 30 pounds per square inch (207 kPa) shall be entirely pointed prior to retesting.

3. The mortar shear strength, v_t, is the value in pounds per square inch (kPa) that is exceeded by 80 percent of the mortar shear test values, v_{to}.

4. Unreinforced masonry with mortar shear strength, v_t, less than 30 pounds per square inch (207 kPa) shall be removed, pointed and retested or shall have its structural function replaced, and shall be anchored to supporting elements in accordance with Sections A106.3.1 and A113.8. When existing mortar in any wythe is pointed to increase its shear strength and is retested, the condition of the mortar in the adjacent bed joints of the inner wythe or wythes and the opposite outer wythe shall be examined for extent of deterioration. The shear strength of any wall class shall be no greater than that of the weakest wythe of that class.

❖ A mortar shear test value, v_{to}, is obtained for each test location by dividing the test force, V_{test}, by the sum of the areas of the upper and lower surfaces of the brick, and subtracting the axial stress in the wall at the time of testing.

Axial stress is defined to include actual live load where it is significant; ordinarily the use of the dead weight of the superimposed masonry is close enough. A set of values of v_{to} is assembled for the whole building.

When an individual wall has values of v_{to} that are consistently below 30 pounds per square inch (psi) (207 kPa), the wall may be upgraded by pointing. Where this is done, the whole wall must be pointed. Then the upgraded wall is retested and improved values of v_{to} will be obtained. These improved values of v_{to} may be used in place of the original values for that wall, and a new set of values of v_{to} is assembled for calculation of v_{to}, since this wall now represents a class of masonry.

Once a set of values of v_{to} is obtained for the building or a class of masonry, the mortar shear strength, v_t, for the building or the class of masonry is determined as the value of v_{to} that is exceeded by 80 percent of the values of v_{to}.

When v_t is less than 30 psi (207 kPa) for the building or a common line of resistance, the masonry in the building, or line of resistance, must be either removed or upgraded by repointing, and then retested to verify that the mortar meets the minimum shear strength value of 30 psi (207 kPa). Alternatively, the wall's structural function shall be replaced and the masonry is to be anchored in accordance with code requirements.

[BS] A106.3.3.6 Minimum quality of masonry.

1. The minimum average value of tensile-splitting strength determined by Equation A1-1 or A1-2 shall be 50 pounds per square inch (344.7 kPa). The minimum value of f'_m determined by categorization of the masonry units and mortar should be 1,000 pounds per square inch (6895 kPa).

2. Individual unreinforced masonry walls with average tensile-splitting strength of less than 50 pounds per square inch (344.7 kPa) shall be entirely pointed prior to retesting.

3. Hollow unit unreinforced masonry walls with estimated prism compressive strength of less than 1,000 pounds per square inch (6895 kPa) shall be grouted to increase the average net area compressive strength.

❖ A minimum quality of an assemblage of units and mortar is required in order to have confidence in the determination of strength limit values. Hollow unit masonry can have its tensile splitting strength increased by pointing. Hollow unit masonry commonly has only the face shell supported on a mortar bed joint. Hollow unit masonry may also be grouted to increase its net area and tensile splitting strength.

[BS] A106.3.3.7 Collar joints. The collar joints shall be inspected at the test locations during each in-place shear test, and estimates of the percentage of adjacent wythe surfaces that are covered with mortar shall be reported along with the results of the in-place shear tests.

❖ When a masonry unit is removed for the in-place shear test, the collar joint between the exterior wythe and the interior wythe is exposed. The percentage of mortar coverage of the collar joint is estimated and reported with the results of the in-place shear testing. Fifty percent of the collar joint must be filled to meet the requirements of the notes to Table A1-B. Table A1-B specifies the allowable height-thickness ratios of URM walls.

[BS] A106.3.3.8 Unreinforced masonry classes. Existing unreinforced masonry shall be categorized into one or more classes based on shear strength, quality of construction, state of *repair*, deterioration and weathering. A class shall be characterized by the allowable masonry shear stress determined in accordance with Section A108.2. Classes shall be defined for whole walls, not for small areas of masonry within a wall.

❖ This section allows categorization of masonry into classes. A single wall may be defined as a class. The provision that 80 percent of the tests made in this wall exceed the test value v_{to} would require a minimum of five tests in that wall. A three-story or higher building will have a minimum of five tests in any wall. If categorization of a wall or walls as a class is contemplated in a building of fewer than three stories, additional tests will be required in each of the walls or line of walls.

[BS] A106.3.3.9 Pointing. Deteriorated mortar joints in unreinforced masonry walls shall be pointed in accordance with the following requirements:

1. **Joint preparation.** The deteriorated mortar shall be cut out by means of a toothing chisel or nonimpact power tool to a depth at which sound mortar is reached but not less than $^3/_4$ inch (19.1 mm). Care shall be taken not to damage the brick edges. After cutting is complete, all loose material shall be removed with a brush, air stream or water stream.

2. **Mortar preparation.** The mortar mix shall be proportioned as required by the registered design professional. The pointing mortar shall be prehydrated by first thoroughly mixing all ingredients dry and then mixing again, adding only enough water to produce a damp workable mix which will retain its form when pressed into a ball. The mortar shall be kept in a damp condition for $1^1/_2$ hours; then sufficient water shall be added to bring it to a consistency that is somewhat drier than conventional masonry mortar.

3. **Packing.** The joint into which the mortar is to be packed shall be damp but without freestanding water. The mortar shall be tightly packed into the joint in layers not exceeding $^1/_4$ inch (6.4 mm) in depth until it is filled; then it shall be tooled to a smooth surface to match the original profile.

Nothing shall prevent pointing of any deteriorated masonry wall joints before testing is performed in accordance with Section A106.3.3, except as required in Section A107.1.

❖ Pointing is maintenance work. Old mortars are subject to deterioration and, if so, they need to be pointed. Mortar joints may be pointed before testing to increase the shear test values, but this pointing must be performed with special inspection as required by Section A107.1.

A source for a detailed description of pointing of brick masonry joints is Technical Note 46 (2005). Alternative repointing procedures, such as those described in Preservation Brief 2, may also be appropriate subject to the code official's approval. Note it is very important that the deteriorated mortar be removed carefully so as to minimize damage to the existing materials that remain. The requirements of this section are based on those contained in UBC Standard 21-8, which is no longer maintained and therefore no longer referenced. The technical content of UBC 21-8 has been inserted into this section in its entirety (with minor editorial changes), with the only exception being the removal of a requirement in UBC Standard 21-8 for Type N or Type S pointing mortar. Selection of the mortar herein is left to the registered design professional.

APPENDIX A

SECTION A107
QUALITY CONTROL

[BS] A107.1 Pointing. Preparation and mortar pointing shall be performed with special inspection.

Exception: At the discretion of the building official, incidental pointing may be performed without special inspection.

❖ Preparation and pointing of mortar joints shall have special inspection. Section A106.3.3.9 provides guidance for the pointing work. The exception allows the code official to allow pointing of small, deteriorated areas at his or her discretion without special inspection.

[BS] A107.2 Masonry shear tests. In-place masonry shear tests shall comply with Section A106.3.3.1. Testing of masonry for determination of tensile-splitting strength shall comply with Section A106.3.3.2.

❖ See the commentary to Section A106.

[BS] A107.3 Existing wall anchors. Existing wall anchors used as all or part of the required tension anchors shall be tested in pullout according to Section A107.5.1. The minimum number of anchors tested shall be four per floor, with two tests at walls with joists framing into the wall and two tests at walls with joists parallel to the wall, but not less than 10 percent of the total number of existing tension anchors at each level.

❖ Refer to the commentary to Section A108.5 for a description of the testing procedure and interpretation of test results.

[BS] A107.4 New bolts. All new embedded bolts shall be subject to periodic special inspection in accordance with the building code, prior to placement of the bolt and grout or adhesive in the drilled hole. Five percent of all bolts that do not extend through the wall shall be subject to a direct-tension test, and an additional 20 percent shall be tested using a calibrated torque wrench. Testing shall be performed in accordance with Section A107.5. New bolts that extend through the wall with steel plates on the far side of the wall need not be tested.

Exception: Special inspection in accordance with the building code may be provided during installation of new anchors in lieu of testing.

All new embedded bolts resisting tension forces or a combination of tension and shear forces shall be subject to periodic special inspection in accordance with the building code, prior to placement of the bolt and grout or adhesive in the drilled hole. Five percent of all bolts resisting tension forces shall be subject to a direct-tension test, and an additional 20 percent shall be tested using a calibrated torque wrench. Testing shall be performed in accordance with Section A107.5. New through-bolts need not be tested.

[BS] A107.5 Tests of anchors in unreinforced masonry walls. Tests of anchors in unreinforced masonry walls shall be in accordance with Sections A107.5.1 through A107.5.4.

❖ The requirements of Section A107.5 are based on the requirements contained in UBC 21-7, which is no longer maintained and therefore no longer referenced herein. The technical content of UBC 21-7 has been inserted into this section in its entirety (with minor editorial changes).

[BS] A107.5.1 Direct tension testing of existing anchors and new bolts. The test apparatus shall be supported by the masonry wall. The distance between the anchor and the test apparatus support shall be not less than one-half the wall thickness for existing anchors and 75 percent of the embedment for new embedded bolts. Existing wall anchors shall be given a preload of 300 pounds (1335 N) prior to establishing a datum for recording elongation. The tension test load reported shall be recorded at $1/8$ inch (3.2 mm) relative movement between the existing anchor and the adjacent masonry surface. New embedded tension bolts shall be subject to a direct tension load of not less than 2.5 times the design load but not less than 1,500 pounds (6672 N) for five minutes (10-percent deviation).

[BS] A107.5.2 Torque testing of new bolts. Bolts embedded in unreinforced masonry walls shall be tested using a torque-calibrated wrench to the following minimum torques:

$1/2$-inch-diameter (12.7 mm) bolts: 40 foot pounds (54.2 N-m).

$5/8$-inch-diameter (15.9 mm) bolts: 50 foot pounds (67.8 N-m).

$3/4$-inch-diameter (19.1 mm) bolts: 60 foot pounds (81.3 N-m).

[BS] A107.5.3 Prequalification test for bolts and other types of anchors. This section is applicable when it is desired to use tension or shear values for anchors greater than those permitted by Table A1-E. The direct-tension test procedure set forth in Section A107.5.1 for existing anchors shall be used to determine the allowable tension values for new embedded through bolts, except that no preload is required. Bolts shall be installed in the same manner and using the same materials as will be used in the actual construction. A minimum of five tests for each bolt size and type shall be performed for each class of masonry in which they are proposed to be used. The allowable tension values for such anchors shall be the lesser of the average ultimate load divided by a safety factor of 5.0 or the average load at which $1/8$ inch (3.2 mm) elongation occurs for each size and type of bolt and class of masonry.

The test procedure for prequalification of shear bolts shall comply with ASTM E488 or another approved procedure.

The allowable values determined in this manner shall be permitted to exceed those set forth in Table A1-E.

[BS] A107.5.4 Reports. Results of all tests shall be reported. The report shall include the test results as related to anchor size and type, orientation of loading, details of the anchor installation and embedment, wall thickness and joist orientation.

SECTION A108
DESIGN STRENGTHS

[BS] A108.1 Values.

1. Strength values for existing materials are given in Table A1-D and for new materials in Table A1-E.

2. Capacity reduction factors need not be used.

3. The use of new materials not specified herein shall be based on substantiating research data or engineering judgment, with the approval of the building official.

❖ The materials resistance values used in Appendix Chapter A1 are strength values, but are not used with capacity reduction factors. Strength values obtained by experimental testing and in-place testing of URM are expected strength values. The strength values obtained by the ABK (Agbabian, Barnes and Kariotis) dynamic testing of ductile materials, such as sheathed diaphragms, are strength limit values.

Design of new elements intended to supplement the existing material should be designed using strength procedures, but not with capacity reduction factors.

When designing new elements, special attention should be paid to the relative stiffness of the existing structural elements and the new elements. For example, if new shear walls or braced frames supplement the strength of a URM wall, the loading should not be reduced by an increased R-factor. However, if the existing structural system is strengthened so that the system's ductility is increased, a higher R-factor could be justified. Such an example would be where new concrete shear walls are added and are designed to carry the full lateral load, and URM piers are adequate to accommodate the expected sum of linear and nonlinear displacements without shear failure.

[BS] A108.2 Masonry shear strength. The unreinforced masonry shear strength, v_m, shall be determined for each masonry class from one of the following equations:

1. The unreinforced masonry shear strength, v_m, shall be determined by Equation A1-4 when the mortar shear strength has been determined by Section A106.3.3.1.

$$v_m = 0.56 v_t + \frac{0.75 P_D}{A} \quad \textbf{(Equation A1-4)}$$

The mortar shear strength values, v_t, shall be determined in accordance with Section A106.3.3.

2. The unreinforced masonry shear, v_m, shall be determined by Equation A1-5 when tensile-splitting strength has been determined in accordance with Section A 106.3.3.2, Item 1 or 2.

$$v_m = 0.8 f_{sp} + 0.5 \frac{P_D}{A} \quad \textbf{(Equation A1-5)}$$

3. When f'_m has been estimated by categorization of the units and mortar in accordance with Section 2105.1 of the *International Building Code*, the unreinforced masonry shear strength, v_m, shall not exceed 200 pounds per square inch (1380 kPa) or the lesser of the following:

   a) $2.5 \sqrt{f'_m}$ or

   b) 200 psi or

   c) $v + 0.75 \frac{P_D}{A}$ **(Equation A1-6)**

For SI: 1 psi = 6.895 kPa.

where:

v = 62.5 psi (430 kPa) for running bond masonry not grouted solid.

v = 100 psi (690 kPa) for running bond masonry grouted solid.

v = 25 psi (170 kPa) for stack bond grouted solid.

❖ **Item 1.** The correlation of v_t and v_m was obtained by physical testing performed by the ABK joint venture. Equation A1-4 is an empirical formula.

Item 2. Equation A1-5 is a theoretical formula adjusted for a probable coefficient of variation of the test data.

Item 3. Equation A1-6 is similar to the shear values given in the IBC. They are allowable strength values adjusted upward by about 1.7.

[BS] A108.3 Masonry compression. Where any increase in dead plus live compression stress occurs, the compression stress in unreinforced masonry shall not exceed 300 pounds per square inch (2070 kPa).

❖ This maximum allowable stress is a conservative estimate of the compressive strength of minimum acceptable masonry. No strength increase for seismic loading is permitted.

[BS] A108.4 Masonry tension. Unreinforced masonry shall be assumed to have no tensile capacity.

❖ Masonry is assumed to have no tensile strength. The overturning forces should therefore be fully resisted by the gravity resisting moment.

[BS] A108.5 Existing tension anchors. The resistance values of the existing anchors shall be the average of the tension tests of existing anchors having the same wall thickness and joist orientation.

❖ Each variation of wall anchor types used should be tested, in accordance with Section A107.3, considering wall thickness, wall framing conditions, masonry quality and other factors affecting the anchor. It should be noted that wall anchor capacity is dependent on its embedment in the masonry and attachment to the roof or floor framing.

[BS] A108.6 Foundations. For existing foundations, new total dead loads may be increased over the existing dead load by 25 percent. New total dead load plus live load plus seismic forces may be increased over the existing dead load plus live load by 50 percent. Higher values may be justified only in conjunction with a geotechnical investigation.

❖ Foundation loads may be increased over existing loads for several reasons. First, consolidation of foundation materials has decreased the probability of settlement due to added loads. Second, the dynamic loading of soils by earthquakes can be tolerated by foundation materials without additional settlement.

Third, in many cases, the dead load added by seismic strengthening is usually relatively insignificant compared to the dead load from the existing URM-bearing wall.

The restrictions may be illustrated by the following example:

Given original dead load = 100

Added dead load = 20

Live load = 50

Seismic = 30

The new dead load is 100 + 20 = 120.

According to the first sentence in Section A108.6, new dead load may be increased over the existing dead load by 25 percent: in the example, 120 is less than 1.25 x 100 = 125; therefore, the added dead load is acceptable. According to the second sentence, new dead plus live plus seismic may be increased over the existing dead load plus live load by 50 percent. In the example, new dead plus live plus seismic is 120 + 50 + 30 = 200, which is less than 1.5 x (100 + 50) = 225; therefore the added dead load is acceptable. If the dead load exceeds the specified limits, higher values may be used if justified by a geotechnical engineer.

Every structure should adequately resist overturning effects. This is less important for existing unreinforced masonry elements since their lateral load-carrying capacity is limited by their rocking capacity. However, where new elements are added, overturning can be a controlling issue. In these cases, the new elements should be designed to resist overturning in general accordance with the IBC.

SECTION A109
ANALYSIS AND DESIGN PROCEDURE

[BS] A109.1 General. The elements of buildings hereby required to be analyzed are specified in Table A1-A.

❖ Appendix Chapter A1 uses analysis and design procedures that are specific for URM buildings. The chapter is for analysis and retrofit for existing buildings. The elements of the masonry buildings that are regulated by Appendix Chapter A1 are given in Table A1-A. This table exempts certain items from regulation in low-to-moderate seismic hazard regions.

[BS] A109.2 Selection of procedure. Buildings with rigid diaphragms shall be analyzed by the general procedure of Section A110, which is based on the building code. Buildings with flexible diaphragms shall be analyzed by the general procedure or, when applicable, may be analyzed by the special procedure of Section A111.

❖ URM buildings with concrete floors and roofs (rigid diaphragms) must be analyzed using the general procedure. The general procedure follows the IBC in that the analysis base shear is calculated using the total mass of the building. The rigid diaphragms couple the shear walls of each level with the weight of the story level and normal walls. URM buildings with flexible diaphragms have an entirely different dynamic response characteristic. The mass of the floor, roofs and normal walls is coupled with the shear wall parallel to the direction of the seismic loading by a shear beam element that has a fundamental elastic period in the range where dynamic response is related to T, the elastic period of the diaphragm. This shear beam has significant ductility potential and the coupling of the story and normal wall mass with the shear wall will be limited to the shear capacity of the diaphragm at each story level. Section A111.1 provides the limits a building must satisfy in order for the special procedure to be used.

SECTION A110
GENERAL PROCEDURE

[BS] A110.1 Minimum design lateral forces. Buildings shall be analyzed to resist minimum lateral forces assumed to act nonconcurrently in the direction of each of the main axes of the structure in accordance with the following:

$$V = \frac{0.75 S_{DS} W}{R} \quad \textbf{(Equation A1-7)}$$

❖ The base shear of the URM building with rigid diaphragms is given by Equation A1-7. This formula is recognizable as being related to the seismic design formula in the IBC. It is a simplified form of the usual base shear formula.

The fundamental elastic period of the building is taken as being in the constant acceleration range, i.e., S_{DS}. The base shear is at the maximum value for the seismic hazard zone.

This base shear is reduced by 25 percent to be consistent with the policy that the analysis forces for existing buildings are three-fourths of the design forces specified for new buildings. The base shear is also divided by an R-value of 1.5. This is consistent with the IBC for bearing wall systems having ordinary plain (unreinforced) masonry shear walls. It is recommended that this base shear be applied in a uniform loading pattern on a multidegree of freedom model, if the shear wall has substantial perforations at each story level. The multidegree of freedom model is a shear-yielding model, not a flexural beam model. Story yield mechanisms may be only a single floor level if the yield mechanism is in-plane shear in the piers. If rocking of the piers is the yield mechanism, it is a possibility that yield mechanisms may form at several story levels. If the wall is basically solid, a triangular loading should be applied, as per the IBC. The engineer should make preliminary calculations as to probable story strengths and yield mechanisms to determine a rational distribution of the base shear over the height of a multistory URM shear wall.

Global overturning is not a critical issue for URM shear walls. Flexural tensile forces due to global overturning cannot be carried downward without a tension load path. Story height overturning moments cause a modification of the axial load on the pier, but equilibrium requires that the reduction in vertical pier loading be equal to the increase in vertical pier loading on other adjacent piers. The in-plane shear capacity and rocking shear capacity of individual piers is affected by pier vertical loading, but the total interstory shear capacity is not significantly changed.

[BS] A110.2 Lateral forces on elements of structures. Parts and portions of a structure not covered in Section A110.3 shall be analyzed and designed per the current building code, using force levels defined in Section A110.1.

Exceptions:

1. Unreinforced masonry walls for which height-to-thickness ratios do not exceed ratios set forth in Table A1-B need not be analyzed for out-of-plane loading. Unreinforced masonry walls that exceed the allowable *h/t* ratios of Table A1-B shall be braced according to Section A113.5.

2. Parapets complying with Section A113.6 need not be analyzed for out-of-plane loading.

3. Where walls are to be anchored to flexible floor and roof diaphragms, the anchorage shall be in accordance with Section A113.1.

❖ The failure of building elements, such as parapets and portions of walls, represents a major source of risk posed by URM buildings. All vulnerable elements should be identified and analyzed.

Diaphragms are analyzed for the loadings calculated by the distribution of base shear at the level of consideration. The allowable shear values of existing and new materials applied to existing materials given in Tables A1-D and A1-E are applicable to the general and special procedures. These provisions require the same force factors on parts and portions as are required for a new building.

Although not specifically stated in buildings being analyzed by the general procedure and having flexible diaphragms and crosswalls conforming to Section A111.3, use of the crosswalls to provide partial lateral support for the diaphragm loading has been used by some engineers and permitted by several jurisdictions.

Exception 1 states that normal walls of unreinforced masonry need not be analyzed for flexural capacity but may be deemed adequate if the height-to-thickness ratios specified in Table A1-B are not exceeded. Use of this table is very conservative for rigid diaphragms. The rigid diaphragm does not amplify the story level acceleration like a flexible diaphragm. In addition to this effect, the overburden/wall-weight ratio used in the development of Table A1-B uses the weight of wood floors and only one additional story for multistory buildings. The probable vertical load of concrete floors and roofs would greatly increase the acceptable height-to-thickness ratio for normal walls.

Exception 2 states that parapets that have height-to-thickness ratios less than specified in Section A113.6 need not be analyzed. Parapets with height-to-thickness ratios in excess of this limit shall be braced.

The special wall anchorage requirements of Exception 3 for flexible diaphragms are based on the ABK research performed in the 1980s. The wall anchorage force specified in Section A113.1.3 (0.9 S_{DS}) is higher than the anchorage force requirements for new masonry construction with flexible diaphragms of the IBC (in accordance with Section 12.11.2.1 of ASCE 7-10) for flexible diaphragms with a span less than 125 feet (38 100 mm).

The provisions of the IBC do apply for wall anchorage with rigid diaphragms (the exception only applies to flexible diaphragms). It is appropriate to make this distinction, since the ABK testing program did not cover buildings with rigid diaphragms.

[BS] A110.3 In-plane loading of URM shear walls and frames. Vertical lateral load-resisting elements shall be analyzed in accordance with Section A112.

❖ Once the story shear forces are determined by the methods of the general procedure, the analysis of unreinforced masonry shear walls follows the same procedures as specified for the special procedure. The masonry is tested as prescribed by Section A106; the allowable shear stress is determined in accordance with Section A108; and the piers in the shear wall are analyzed in accordance with Section A112.2.3.

[BS] A110.4 Redundancy and overstrength factors. Any redundancy or overstrength factors contained in the building code may be taken as unity. The vertical component of earthquake load (E_v) may be taken as zero.

❖ No redundancy or overstrength factors are required to be used, as the forces and the capacities are at strength limit states. The lack of redundancy is generally not an issue for buildings of unreinforced masonry construction. In addition, the overstrength factors of the wood-framed elements that couple the dynamic response of the rigid masonry walls are already included in the provisions.

SECTION A111
SPECIAL PROCEDURE

[BS] A111.1 Limits for the application of this procedure. The special procedures of this section may be applied only to buildings having the following characteristics:

1. Flexible diaphragms at all levels above the base of the structure.

2. Vertical elements of the lateral force-resisting system consisting predominantly of masonry or concrete shear walls.

3. Except for single-story buildings with an open front on one side only, a minimum of two lines of vertical elements of the lateral force-resisting system parallel to

each axis of the building (see Section A111.8 for open-front buildings).

❖ **Item 1.** The building must have flexible diaphragms at all story levels to qualify for this procedure, and the seismic loading used in this section is exclusively for flexible diaphragms. The definition of "Rigid diaphragm" given in Section A103 is purposely very narrow. Investigation of existing URM wall buildings will find materials that do not fit into this narrow definition. Such a building would have to be modeled with the unique stiffness characteristics of the materials. Stiffness degradation should be assessed if a linear elastic analysis is made. It is outside the scope of Appendix Chapter A1 to provide guidance for analysis of unique buildings. The engineer and the code official should make the decisions for an analysis procedure specific to a unique structural system.

Item 2. The vertical seismic force-resisting system should have stiffness such that the relative stiffness ratio of the diaphragms and shear walls assumed for this procedure is maintained. A steel or reinforced concrete moment frame does not have the story stiffness of a reinforced concrete or masonry shear wall designed for the same lateral loading. A moment frame may be adequate to act as a shear wall in the middle portion of a flexible diaphragm or at an open front of an unreinforced masonry building. Section A111.3.5 has special limits on yield drift of a moment frame used as a crosswall. Section A111.6.4 requires use of the same design forces as for a shear wall and further limits the elastic drift of this moment frame. Use of an *R*-factor in the design of the moment frame is prohibited.

Item 3. A minimum of two lines of vertical seismic force-resisting elements on each axis of the building is required. Section A111.1, Item 2 requires that each of these elements be "predominantly of masonry or concrete shear walls." If structural steel or reinforced concrete bracing is used for one or both of these vertical elements, a relative rigidity check of the bracing versus a shear wall should be made to confirm that the concept of a "rigid wall-flexible diaphragm" response to earthquake is probable.

[BS] A111.2 Lateral forces on elements of structures. With the exception of the provisions in Sections A111.4 through A111.7, elements of structures shall comply with Sections A110.2 through A110.4.

❖ This section specifies that floor and roof diaphragms are only checked for strength limit state and stiffness. The requirement for the check is given in Section A111.4. In addition, requirements for diaphragm shear, transfer shear walls and out-of-plane forces are given in Sections A111.5, A111.6 and A111.7, respectively. Other elements of structures are analyzed as prescribed by the IBC, but the forces are those given in Section A110, and the material's resistance values are those given in Section A108. These design strengths are calculated using the principle of load and resistance factor design (LRFD) (strength) methods. However, load is not factored and capacity reduction factors are not used for resistance values.

[BS] A111.3 Crosswalls. Crosswalls shall meet the requirements of this section.

[BS] A111.3.1 Crosswall definition. A crosswall is a wood-framed wall sheathed with any of the materials described in Table A1-D or A1-E or other system as defined in Section A111.3.5. Crosswalls shall be spaced no more than 40 feet (12 192 mm) on center measured perpendicular to the direction of consideration, and shall be placed in each story of the building. Crosswalls shall extend the full story height between diaphragms.

Exceptions:

1. Crosswalls need not be provided at all levels when used in accordance with Section A111.4.2, Item 4.

2. Existing crosswalls need not be continuous below a wood diaphragm at or within 4 feet (1219 mm) of grade, provided:

    2.1. Shear connections and anchorage requirements of Section A111.5 are satisfied at all edges of the diaphragm.

    2.2. Crosswalls with total shear capacity of $0.5S_{D1}\Sigma W_d$ interconnect the diaphragm to the foundation.

    2.3. The demand-capacity ratio of the diaphragm between the crosswalls that are continuous to their foundations does not exceed 2.5, calculated as follows:

$$DCR = \frac{(2.1S_{D1}W_d + V_{ca})}{2v_u D}$$

(Equation A1-8)

❖ The definition of a "Crosswall" is given and fully defined in this commentary in Sections A103 and A111.3.1. A crosswall is a light-frame sheathed wall parallel to the direction of earthquake loading. A crosswall is not considered as a shear wall in that it need not be designed for the tributary loads of the flexible diaphragm or diaphragms. The crosswall decreases the displacement of the center of the diaphragm relative to the shear walls and provides damping of the response of the diaphragm to earthquake shaking.

The spacing of the crosswalls along the diaphragm span length is limited to 40 feet (12 192 mm). This restriction is intended to limit the significant higher mode participation of the diaphragm between crosswalls. In general, if crosswalls are used to increase the allowable height-to-thickness ratio of normal walls, the crosswalls must be in each story of the building. Existing and new crosswalls must extend the full story height between diaphragm levels.

Exception 1 allows the use of crosswalls that extend only between the roof diaphragm and a floor diaphragm to couple these into a combined diaphragm for its demand capacity ratio (*DCR*) analysis.

Exception 2 allows a special condition that commonly occurs in residential occupancies. Often in this occupancy, many crosswalls interconnect the roof and floors. However, the first floor is commonly constructed over a crawl space without crosswalls. This exception allows the use of special crosswalls spaced as far apart as 40 feet (12 192 mm) to stiffen the lowest diaphragm. The crosswalls must have adequate strength to limit the deformation of the diaphragm at the special crosswalls. The loading of the first floor diaphragm is its tributary load and the entire capacity of the crosswalls that are above the first floor.

[BS] A111.3.2 Crosswall shear capacity. Within any 40 feet (12 192 mm) measured along the span of the diaphragm, the sum of the crosswall shear capacities shall be at least 30 percent of the diaphragm shear capacity of the strongest diaphragm at or above the level under consideration.

❖ The stiffness of sheathed systems such as flexible diaphragms and crosswalls is directly related to their peak strength; therefore, the minimum strength of a crosswall is related to the strength of the diaphragm that it is restraining. The minimum strength of a crosswall must be 30 percent of the strength of the diaphragm shear capacity of the stronger diaphragm at or above the level under consideration.

[BS] A111.3.3 Existing crosswalls. Existing crosswalls shall have a maximum height-to-length ratio between openings of 1.5 to 1. Existing crosswall connections to diaphragms need not be investigated as long as the crosswall extends to the framing of the diaphragms above and below.

❖ Existing crosswalls generally are perforated by door openings. The effective length of the crosswall used for calculations of the required crosswall strength cannot include portions that have a height-to-length ratio less than 1.5 to 1. If the openings were 9-foot-high (2743 mm) doors, a section between the door or doors and the end of the wall of less than 6 feet (1828 mm) cannot be used for the calculation of crosswall capacity.

The capacities of the connections of the existing crosswall to the diaphragms above or below the wall need not be investigated if the crosswall extends to the diaphragm level above. If the crosswall only extends to ceiling joists that are separate from the floor or roof framing above, then continuity of the crosswall must be provided. Note b of Table A1-D limits the total capacity to 900 pounds per foot (13.1 kN/m), regardless of the combined capacity of the existing materials on the crosswall. This limitation was provided in lieu of an investigation of the capacity of the connection of the crosswall to the diaphragm.

[BS] A111.3.4 New crosswalls. New crosswall connections to the diaphragm shall develop the crosswall shear capacity. New crosswalls shall have the capacity to resist an overturning moment equal to the crosswall shear capacity times the story height. Crosswall overturning moments need not be cumulative over more than two stories.

❖ Connections of new crosswalls to diaphragms shall be designed by the applicable strength design sections of the IBC. This includes the design of a vertical connection of the end of the crosswall to the foundation of the crosswall or support below the level of the crosswall. This connection is required to protect the floor framing from wall overturning effects. The dead load of the floor or roof framing supported by the crosswall cannot be used to resist the calculated overturning due to lateral forces. However, the calculated force at the end of the crosswall need not use an overturning moment accumulated from more than two stories, including the level of calculation.

[BS] A111.3.5 Other crosswall systems. Other systems, such as moment-resisting frames, may be used as crosswalls provided that the yield story drift does not exceed 1 inch (25 mm) in any story.

❖ Other systems may be used as crosswalls. A steel frame may be more desirable than a crosswall in a commercial space. Such a frame should be a moment frame rather than a braced frame, so as to have the damping effect of a crosswall rather than the rigid bracing effect of a shear wall. The prescription of yield story drift not to exceed 1 inch (25 mm) is intended to provide stiffness, an inelastic threshold and load deformation behavior similar to that of existing crosswalls.

The frame used as a crosswall is intended to yield, and the frame sections at which yielding will occur must be braced to building elements that can provide adequate restraint against lateral buckling.

Considering the cost of adding a frame, regardless of the design criteria, it may be better to provide a moment frame designed as a shear wall. When used as a shear wall, the moment frame is designed to resist the loading of the tributary area of the diaphragm or diaphragms, and its drift under this loading is limited by Section A111.6.4 to 0.015 times the story height.

[BS] A111.4 Wood diaphragms.

[BS] A111.4.1 Acceptable diaphragm span. A diaphragm is acceptable if the point (L, DCR) on Figure A1-1 falls within Region 1, 2 or 3.

❖ Conventional diaphragm analysis is not required by the special procedure. The strength and stiffness of a flexible diaphragm are acceptable if its span length and DCR fall in Region 1, 2 or 3 of Figure A1-1. Figure A1-1 is a plot of the estimated dynamic displacement of the center of the diaphragm of 5 inches (127 mm) measured relative to the shear walls that are shaking the diaphragm. The ground motions used for these calculations were scaled to the ATC-3 5-percent damped response spectrum having 0.40g effective peak acceleration and 12 inches per second (305 mm/s) peak ground velocity. The analysis of the DCR

of diaphragms is only required in IBC seismic areas where S_{D1} is 0.20g or greater.

[BS] A111.4.2 Demand-capacity ratios. Demand-capacity ratios shall be calculated for the diaphragm at any level according to the following formulas:

1. For a diaphragm without qualifying crosswalls at levels immediately above or below:

$$DCR = 2.1 S_{D1} W_d / \Sigma v_u D \quad \textbf{(Equation A1-9)}$$

2. For a diaphragm in a single-story building with qualifying crosswalls, or for a roof diaphragm coupled by crosswalls to the diaphragm directly below:

$$DCR = 2.1 S_{D1} W_d / \Sigma v_u D + V_{cb} \quad \textbf{(Equation A1-10)}$$

3. For diaphragms in a multistory building with qualifying crosswalls in all levels:

$$DCR = 2.1 S_{D1} \Sigma W_d / (\Sigma \Sigma v_u D + V_{cb})$$

$$\textbf{(Equation A1-11)}$$

   DCR shall be calculated at each level for the set of diaphragms at and above the level under consideration. In addition, the roof diaphragm shall also meet the requirements of Equation A1-10.

4. For a roof diaphragm and the diaphragm directly below, if coupled by crosswalls:

$$DCR = 2.1 S_{D1} \Sigma W_d / \Sigma \Sigma v_u D \quad \textbf{(Equation A1-12)}$$

❖ All of the *DCR* formulas assume that the resistance of each end of the diaphragm is nearly equal. If a large inequality exists, the total capacity of the diaphragm should be adjusted. A conservative assumption would be to use twice the capacity of the end with the least depth. The analysis of a flexible diaphragm for a *DCR* assumes that the diaphragm shape in any span is relatively rectangular. Buildings with plan irregularities but parallel walls can be analyzed by this method. The span of the diaphragm is the distance between shear walls when ties or collectors develop the load transfer to the shear wall.

*DCR*s are calculated for the following four conditions:

Item 1. If there are no qualifying crosswalls above or below the diaphragm level analyzed, the total lateral load tributary to the diaphragm, multiplied by a lateral load coefficient and spectral response factor, is divided by the total shear capacity of both ends of the diaphragm.

Item 2. The crosswalls in a single-story building are also usually an additional load path between the roof diaphragm and the ground. The capacity of both ends of the diaphragm and the crosswalls below the diaphragm are added together to determine the capacity of the diaphragm.

Item 3. If the building has crosswalls in all levels of the building, the calculation of *DCR* begins with the roof level and is made at each story level. The sum of the diaphragm loads above the level under consideration is used as the demand. The total capacity of all the diaphragms above the level under consideration and the capacity of the crosswalls below the diaphragm level analyzed are the combined capacity. The crosswalls at each level have a minimum capacity of 30 percent of the diaphragm capacity above and cause a nearly common displacement of the diaphragms relative to the shear walls.

Item 4. A special treatment of the roof diaphragm is provided in this paragraph. Roof diaphragms are generally more flexible than adjacent floors. This provision allows the calculation of a *DCR* for the combination of a roof and the adjacent floor when only these two diaphragms are coupled by crosswalls. Crosswalls may or may not exist in the stories below the floor adjacent to the roof diaphragm.

[BS] A111.4.3 Chords. An analysis for diaphragm flexure need not be made, and chords need not be provided.

❖ Diaphragms are checked using the demand-capacity formulas and Figure A1-1. The conventional analysis for diaphragm chords is not part of the special procedure and the analysis is explicitly excluded in the provisions. The reason is that there is not a significant flexural response in flexible wood diaphragms; the primary effect is shear yielding in the end zones. The source of flexural capacity of the diaphragm is the continuity inherent in the diaphragm construction. Since significant life-threatening damage related to lack of diaphragm chords has not been observed and virtually every diaphragm has some form of edge restraint and/or internal continuity capable of providing adequate flexural capacity, no rules for formal analysis of chords are included in Appendix Chapter A1.

[BS] A111.4.4 Collectors. An analysis of diaphragm collector forces shall be made for the transfer of diaphragm edge shears into vertical elements of the lateral force-resisting system. Collector forces may be resisted by new or existing elements.

❖ The transfer of shear forces at the edge of a diaphragm to the shear resisting elements at that edge may require a collector at the edge of the diaphragm for distribution of the diaphragm shear to the elements of the shear wall. If the shear wall is the full depth of the diaphragm, no need for a collector would exist as the uniformly distributed shear load of the diaphragm would be uniformly resisted along the length of the shear wall. If the shear wall strength and stiffness were at one end of that line of resistance, the collector would accumulate load from the diaphragm and act as a tie to the shear wall element. The calculated collector force may be resisted by existing or new elements. The relative rigidity of new and existing elements should be considered in apportioning the loads to the various elements.

A pier adjacent to a corner of the two intersecting

shear walls is much stiffer than the usual pier analysis indicates. The effect of the flange, which is the other wall around the corner, is significant when the flange is in tension. The stiffness of this pier may cause the spandrel to have a significant collector stress. This should be considered in the design of the collector. Because of this increased collector force, a minimum required spacing of the first shear bolt from the corner is specified in Section A113.1.4 to provide an attachment of the collector immediately adjacent to the corner.

[BS] A111.4.5 Diaphragm openings.

1. Diaphragm forces at corners of openings shall be investigated and shall be developed into the diaphragm by new or existing materials.

2. In addition to the demand-capacity ratios of Section A111.4.2, the demand-capacity ratio of the portion of the diaphragm adjacent to an opening shall be calculated using the opening dimension as the span.

3. Where an opening occurs in the end quarter of the diaphragm span, the calculation of $v_u D$ for the demand-capacity ratio shall be based on the net depth of the diaphragm.

❖ **Item 1.** The analysis of a diaphragm for its *DCR* assumes a rectangular shape. The presence of a large opening can increase the flexibility of the diaphragm and cause localized stresses. The diaphragm shear is assumed to be an average shear stress and this assumption implies that the uniform shear in the diaphragm adjacent to the opening must be transferred by collectors to the portion of the diaphragm that is continuous past the opening. Section A111.4.5 describes both the stiffness and strength check; the reduction in stiffness caused by an opening is checked as indicated in Item 3. If the strength of the existing diaphragm with the opening is adequate for the strength analysis, the transfer of uniform shear to the remaining section of the diaphragm could be assumed to be accomplished by the diaphragm itself, and the diaphragm edge adjacent to the opening could be assumed to have no shear stress.

Item 2. When the length of an opening is a significant part of the diaphragm span length and it is in the center half of the diaphragm, the *DCR* of the diaphragm that is continuous shall be checked. The demand is that part of the diaphragm load, W_d, within the length of opening and the span is the length of the opening measured parallel to the diaphragm span.

Item 3. Openings within the length of the diaphragm significantly affect its stiffness. The reduction of stiffness is approximated by using a capacity of the diaphragm that is calculated for the net depth rather than the full depth.

[BS] A111.5 Diaphragm shear transfer. Diaphragms shall be connected to shear walls with connections capable of developing the diaphragm-loading tributary to the shear wall given by the lesser of the following formulas:

$$V = 1.2 S_{D1} C_p W_d \qquad \text{(Equation A1-13)}$$

using the C_p values in Table A1-C, or

$$V = v_u D \qquad \text{(Equation A1-14)}$$

❖ The calculation of the capacity of the shear connection of the diaphragm to the shear wall is based on the coefficients for spectral acceleration, S_{D1} for the IBC, and a C_p factor that is based on the materials used for construction of the diaphragm. The values of the C_p range from 0.5 to 0.75. The least value is for a single layer of sheathing. The greatest value of C_p is for multiple layers of sheathing. A single layer of plywood with all edges supported on the framing or blocked is considered equivalent to a double-layer system. Steel decks without fill material have C_p values between those of single- and double-sheathed diaphragms. The stronger and stiffer diaphragms amplify the ground motions transmitted to the ends of the diaphragm by the shear walls more than single-sheathed systems and are assigned a high C_p coefficient.

The evaluation of the required capacity of the shear connection of a diaphragm to the shear wall is not a calculation of the reaction of a beam with a loading that may be variable within its span length. The required capacity of the connection is one-half of the total dead load of the diaphragm and normal walls, multiplied by the spectral response and construction coefficients. The analysis of diaphragms by the special procedure accepts nonlinear behavior of the diaphragm and limits the required capacity of the shear connection to the capacity of the diaphragm. The check for the acceptable *DCR* of the diaphragm is for displacement control. The limitation of the shear connection capacity to the shear capacity of the diaphragm provides a connection strength that causes shear yielding in the diaphragm.

[BS] A111.6 Shear walls (In-plane loading).

[BS] A111.6.1 Wall story force. The wall story force distributed to a shear wall at any diaphragm level shall be the lesser value calculated as:

$$F_{wx} = 0.8 S_{D1}(W_{wx} + W_d/2) \qquad \text{(Equation A1-15)}$$

but need not exceed

$$F_{wx} = 0.8 S_{D1} W_{wx} + v_u D \qquad \text{(Equation A1-16)}$$

❖ The special procedure differs from the general procedure in that the loading of the shear wall is determined by the diaphragm capacity rather than an arbitrary distribution of a base shear to the levels of the diaphragm. If there are no crosswalls, the story force of a diaphragm level is equal to the spectral response coefficient, S_{D1}, times the tributary weight of the wall at the level of calculation, plus one-half of the dead load of the diaphragm and normal walls at the level of consideration. However, the story force need not exceed the spectral response coefficient times the tributary weight of the shear wall, plus the shear capacity of the diaphragm at the level of consideration.

[BS] A111.6.2 Wall story shear. The wall story shear shall be the sum of the wall story forces at and above the level of consideration.

$$V_{wx} = \Sigma F_{wx} \qquad \text{(Equation A1-17)}$$

❖ The design shear load in a shear wall at any level is the total of the wall story forces calculated at the stories above by Equations A1-15 and A1-16. The least value of calculated story shear is to be used for analysis of the existing shear wall.

[BS] A111.6.3 Shear wall analysis. Shear walls shall comply with Section A112.

❖ The forces calculated in this section are used to analyze the existing shear walls by the procedures specified in Section A112.

[BS] A111.6.4 Moment frames. Moment frames used in place of shear walls shall be designed as required by the building code, except that the forces shall be as specified in Section A111.6.1, and the story drift ratio shall be limited to 0.015, except as further limited by Section A112.4.2.

❖ Moment frames may be used as shear walls in locations where a shear wall does not exist. A common use for moment frames is at the open front of a commercial building. The loading of the moment frame is calculated as if it were an existing shear wall. The interstory drift of the moment frame is limited to 0.015 of the story height. If the moment frame is used in a line of resistance with a URM wall, more severe limitations on interstory drift are prescribed by Section A112.4.2 and the moment frame must be designed to take 100 percent of the load along that line.

[BS] A111.7 Out-of-plane forces-unreinforced masonry walls.

[BS] A111.7.1 Allowable unreinforced masonry wall height-to-thickness ratios. The provisions of Section A110.2 are applicable, except the allowable height-to-thickness ratios given in Table A1-B shall be determined from Figure A1-1 as follows:

1. In Region 1, height-to-thickness ratios for buildings with crosswalls may be used if qualifying crosswalls are present in all stories.

2. In Region 2, height-to-thickness ratios for buildings with crosswalls may be used whether or not qualifying crosswalls are present.

3. In Region 3, height-to-thickness ratios for "all other buildings" shall be used whether or not qualifying crosswalls are present.

❖ URM walls need not be analyzed for forces normal to the wall surface. Instead of a flexural analysis, acceptable height-to-thickness ratios for the URM walls are specified. Walls conforming to these height-to-thickness ratios are stable due to dynamic rocking stability, not because of the tensile capacity of the bed joints. The concept of dynamic stability assumes that the bed joints are cracked at the wall anchorage levels and near the center of the wall between wall anchorage levels. If the height-to-thickness ratio of the URM wall exceeds these prescribed ratios, the wall must be braced to increase its stability.

The allowable height-to-thickness ratios are given in Table A1-B. The height-to-thickness ratios were determined by data obtained by dynamic testing of full size walls. The walls at the uppermost story level are most vulnerable because there is no overburden. Walls at the first story of the buildings have less vulnerability as the unamplified ground motion shakes at one end of the wall. The increased ratios for buildings with crosswalls are due to the effect of damping by the crosswalls of the diaphragms. These height-to-thickness ratios, with the exception of buildings with crosswalls, are applicable to buildings analyzed by either the general or special procedure.

Buildings with crosswalls analyzed by the special procedure may have larger height-to-thickness ratios if several special limitations are met:

Item 1. In Region 1, the increased height-to-thickness ratios are applicable only to buildings with crosswalls in all story levels and if the following conditions are met:

- The tested mortar shear strength shall not be less than 100 psi (690 kPa) or where the visual examination of the collar joints indicates 50 percent or more of the collar joint is filled with mortar and the tested mortar shear strength is 60 psi (414 mm) or more.

- The separation of the building from an adjacent building shall not be less than 5 inches (127 mm) (see Section A113.10).

Item 2. A special condition exists when the *DCR* of the diaphragms falls in Region 2 of Figure A1-1. In this special condition, the height-to-thickness ratios for crosswalls may be used even though qualifying crosswalls are not present.

Item 3. This special provision allowing ratios of buildings "with crosswalls" when no crosswalls are present is due to the presence of hysteretic damping of diaphragms that have *DCR*s in excess of 2.5.

[BS] A111.7.2 Walls with diaphragms in different regions. When diaphragms above and below the wall under consideration have demand-capacity ratios in different regions of Figure A1-1, the lesser height-to-thickness ratio shall be used.

❖ The *DCR* of both the upper and lower diaphragms that shake the out-of-plane wall shall be checked. If the *DCR* of either diaphragm falls into Region 3 of Figure A1-1, the height-to-thickness ratios for "all other" buildings shall be used even if qualifying crosswalls are not present. The *DCR* of both diaphragms must fall within Region 2 for the height-to-thickness ratios for "with crosswalls" to be used, even though crosswalls are not present. In the calculation of DCR for determination of height-to-thickness ratios, the capability of crosswalls to reduce diaphragm lengths or acting as lines to resist diaphragm shear is excluded.

[BS] A111.8 Open-front design procedure. A single-story building with an open front on one side and crosswalls parallel to the open front may be designed by the following procedure:

1. Effective diaphragm span, L_i, for use in Figure A1-1 shall be determined in accordance with the following formula:

    $L_i = 2[(W_w/W_d)L + L]$ **(Equation A1-18)**

2. Diaphragm demand-capacity ratio shall be calculated as:

    $DCR = 2.1 S_{D1}(W_d + W_w)/[(v_u D) + V_{cb}]$
    (Equation A1-19)

❖ It should be noted that "open-front" has an ordinary meaning, referring to a building face that is open, i.e., free from significant solid walls, usually at one story. Most commonly, an open-front building is the ground floor of a commercial building. Buildings on corners of blocks often have open fronts on two sides.

If a moment frame complying with Section A111.6.4 is placed in the open front, then the building meets the requirement of Section A111.1, Item 3 for two lines of vertical elements in the direction parallel to the front; therefore, the open-front procedure is not applicable.

Under the exception of Section A111.1, Item 3, the open front of a single-story building is permitted to remain open if the building satisfies the requirements of the open-front procedure in this section and the building contains crosswalls.

The open-front procedure is permitted only in one direction: if there are two open fronts, one of them must be provided with a braced frame or shear wall.

In case the open-front procedure is tried but cannot be satisfied, the solution would be to provide a braced frame or a moment frame in the open front. The open-front procedure uses Equation A1-18 to determine an effective span, L_i; the distance from the nearest shear wall to the open front is the length L. An additional length is calculated in order to account for the weight of masonry at the open front. This weight, which is a concentrated load at the end of the diaphragm, is assumed to be distributed uniformly along a length, $(W_w/W_d) L$. The augmented length is $[(W_w/W_d) L + L]$. This length is then doubled to estimate the span, L_i, of an equivalent diaphragm that has a deflected shape similar to a shear beam spanning between two shear walls.

The diaphragm DCR is calculated by Equation A1-19, using the spectral response coefficient, the sum of the diaphragm and normal wall dead load, the weight of the wall at the open front and the shear capacity of one end of the diaphragm and the required crosswall. This analysis determines the capacity of the crosswall that is required to control the displacement at the end of the diaphragm at the open front.

The required crosswall need not be right in the open front, but may be at some distance back within the building. It should also be noted that a moment frame or other system might be used in place of a crosswall in accordance with Section A111.3.5. Where the crosswall or other system is at some distance back from the open front, the analysis performed to determine the maximum distance that is permitted becomes an iterative process. It is similar to the analysis described above, except that a crosswall capacity is not included in the calculation of Equation A1-19. The maximum distance is L_i. A trial value of L is used in Equation A1-18, and the corresponding L_i and DCR are used to check the diaphragm. The process is repeated until an acceptable value of L is obtained.

SECTION A112
ANALYSIS AND DESIGN

[BS] A112.1 General. The following requirements are applicable to both the general procedure and the special procedure for analyzing vertical elements of the lateral force-resisting system.

❖ The requirements of this section are applicable to both the general and special procedures. The two procedures will determine different in-plane shear loading for the shear walls for buildings of equal floor area due to the difference in overall building weight and the coupling of the weight of normal walls with the shear walls.

[BS] A112.2 Existing unreinforced masonry walls.

[BS] A112.2.1 Flexural rigidity. Flexural components of deflection may be neglected in determining the rigidity of an unreinforced masonry wall.

❖ The relative rigidity of piers in a wall may be determined on the basis of net area and height as in a shear beam. Wall piers rarely have span-depth ratios that exceed the ratio where flexural stress-strain relationships exist. Use of shear deformations only for calculation of relative rigidity is an appropriate procedure.

[BS] A112.2.2 Shear walls with openings. Wall piers shall be analyzed according to the following procedure, which is diagrammed in Figure A1-2.

1. For any pier,

    1.1. The pier shear capacity shall be calculated as:

    $V_a = v_m A/1.5$ **(Equation A1-20)**

1.2. The pier rocking shear capacity shall be calculated as:

$$V_r = 0.9P_D D/H \quad \text{(Equation A1-21)}$$

2. The wall piers at any level are acceptable if they comply with one of the following modes of behavior:

2.1. Rocking controlled mode. When the pier rocking shear capacity is less than the pier shear capacity, i.e., $V_r < V_a$ for each pier in a level, forces in the wall at that level, V_{wx}, shall be distributed to each pier in proportion to $P_D D/H$.

For the wall at that level:

$$0.7V_{wx} < \Sigma V_r \quad \text{(Equation A1-22)}$$

2.2. Shear controlled mode. Where the pier shear capacity is less than the pier rocking capacity, i.e., $V_a < V_r$ in at least one pier in a level, forces in the wall at the level, V_{wx}, shall be distributed to each pier in proportion to D/H.

For each pier at that level:

$$V_p < V_a \quad \text{(Equation A1-23)}$$

and

$$V_p < V_r \quad \text{(Equation A1-24)}$$

If $V_p < V_a$ for each pier and $V_p > V_r$ for one or more piers, such piers shall be omitted from the analysis, and the procedure shall be repeated for the remaining piers, unless the wall is strengthened and reanalyzed.

3. Masonry pier tension stress. Unreinforced masonry wall piers need not be analyzed for tension stress.

❖ This section is for the analysis of a URM wall with windows and/or openings. The in-plane strength of the wall is the strength of the piers between the windows and doors. The pier shear capacity is its allowable shear, v_m, times the net area of the pier, A, and adjusted from peak shear stress to average shear stress. This adjusted shear strength is the average shear as used in the design of reinforced masonry and reinforced concrete. A second method of calculating the in-plane shear capacity is given by Equation A1-21. This rocking shear capacity assumes that the bed joints at the top and bottom of the pier have cracked and the axial load on the pier acting eccentrically on the displaced pier provides the restoring shear or rocking shear capacity. The effective lever arm of the axial loads on the ends of the pier is about 90 percent of the pier width. This restoring force is a strength limit state.

Rocking controlled mode. When the rocking shear capacity of each individual pier is less than that pier's shear capacity, the pier will rotate, or rock, as a rigid block and have a stable dynamic behavior. In this case, the capacity of the wall consisting of a series of piers is the sum of the rocking shear capacities of each pier, and each pier is loaded with a portion of the total base shear. The rocking shear capacity, P_D (D/H), is used as the relative rigidity of each pier.

Shear controlled mode. When the axial load on any wall pier is large enough to prevent bed joint cracking prior to the formation of a diagonal shear crack, a shear failure in this pier may occur at a small displacement. In this case, the load is proportioned to the piers by the relative shear rigidity. Flexural stiffness is not included in the relative rigidity calculations as the common pier shape is such that flexural deformations are those related to deep beams.

When the pier shear capacity, V_a, is less than the pier rocking capacity, V_r, in even one pier, the shear force in the wall, V_w, is distributed to the piers in accordance with their relative rigidity, D/H. The distributed load to a pier, V_p, must be less than V_a, and if $V_p < V_a$, the shear resistance of the wall piers is adequate. If $V_p < V_a$ for each pier and $V_p > V_r$ for one or more piers, then an incompatible pier capacity exists. The piers where $V_p > V_r$ are omitted from the analysis and the procedure is repeated using only the piers where $V_p > V_a$. This procedure is shown in a flow diagram in Figure A1-2.

Masonry pier tension stress. Flexural moments and tensile stresses are not calculated for unreinforced masonry piers. Flexural cracking may occur on bed joints due to out-of-plane flexure, but the pier system can be stable when subjected to the dynamic loading of an earthquake.

[BS] A112.2.3 Shear walls without openings. Shear walls without openings shall be analyzed the same as for walls with openings, except that V_r shall be calculated as follows:

$$V_r = 0.9(P_D + 0.5P_w)D/H \quad \text{(Equation A1-25)}$$

❖ These walls shall be analyzed as a single pier for in-plane shear capacity. The rocking shear computation uses one-half of the wall weight as the restoring force because the product of P_D is one-half of the wall length, D.

[BS] A112.3 Plywood-sheathed shear walls. Plywood-sheathed shear walls may be used to resist lateral forces for buildings with flexible diaphragms analyzed according to provisions of Section A111. Plywood-sheathed shear walls may not be used to share lateral forces with other materials along the same line of resistance.

❖ Sheathed light-frame shear walls may be used as shear walls in buildings with flexible diaphragms in conformity with the restrictions in the IBC. These sheathed walls cannot share lateral forces with other materials in the line of the sheathed light-frame walls.

[BS] A112.4 Combinations of vertical elements.

[BS] A112.4.1 Lateral-force distribution. Lateral forces shall be distributed among the vertical-resisting elements in proportion to their relative rigidities, except that moment-resisting frames shall comply with Section A112.4.2.

❖ The distribution of the lateral force should be based on realistic calculations of stiffness. The compressive modulus of unreinforced brick has been determined by testing to have values vary from 100,000 psi (690 MPa) to 500,000 psi (345 Pa). Effective moduli of reinforced concrete and masonry shear walls vary from 10 to 20 percent of uncracked moduli. A special case is given in Section 112.4.2 for assignment of lateral loads, regardless of relative stiffness.

[BS] A112.4.2 Moment-resisting frames. Moment-resisting frames shall not be used with an unreinforced masonry wall in a single line of resistance unless the wall has piers that have adequate shear capacity to sustain rocking in accordance with Section A112.2.2. The frames shall be designed in accordance with the building code to carry 100 percent of the lateral forces tributary to that line of resistance, as determined from Equation A1-7. The story drift ratio shall be limited to 0.0075.

❖ Use of a combination of moment frame and URM shear wall in the same line of resistance is prohibited unless the piers in the wall have a rocking shear capacity that exceeds the pier shear capacity (see Equations A1-20 and A1-21). If this condition is met, a moment frame may be used with 100 percent of the load assigned to it. The minimum stiffness of the moment frame shall be twice that specified for other moment frames that are used to resist lateral loads (see Section A111.6.4). This section should not be interpreted to be applicable to a moment frame used in the open front of a URM building where the piers do not have significant stiffness. These masonry columns may support lintels that have a substantial load. The engineer should assess the stability of the masonry-bearing surface when the design drift of the open front occurs.

SECTION A113
DETAILED SYSTEM DESIGN REQUIREMENTS

[BS] A113.1 Wall anchorage.

[BS] A113.1.1 Anchor locations. Unreinforced masonry walls shall be anchored at the roof and floor levels as required in Section A110.2. Ceilings of plaster or similar materials, when not attached directly to roof or floor framing and where abutting masonry walls, shall either be anchored to the walls at a maximum spacing of 6 feet (1829 mm), or be removed.

❖ URM walls are anchored to all roofs and floors, and to ceilings in certain cases. If the ceiling is forced to displace with the roof diaphragm by existing wood trussing, the anchor should also be at the ceiling level. The URM wall is forced to remain vertical within the depth of the truss and the flexural crack will occur at the ceiling level. A depth of truss that triggers ceiling anchorage has not been defined. Engineering judgment should be used to determine an acceptable depth of the roof trusses not having anchorage at the ceiling level.

Ceiling systems of substantial weight act as loading on the walls. The transfer of the ceiling weight to the shear wall could be made by the diaphragm above and its shear connection, or by an independent shear connection and diaphragm at the ceiling level.

[BS] A113.1.2 Anchor requirements. Anchors shall consist of bolts installed through the wall as specified in Table A1-E, or an approved equivalent at a maximum anchor spacing of 6 feet (1829 mm). All wall anchors shall be secured to the joists to develop the required forces.

❖ When existing wall anchors have been tested to confirm the adequacy of the embedded end, the connection of the existing anchor to the wood framing shall be analyzed by rational engineering methods for its adequacy.

[BS] A113.1.3 Minimum wall anchorage. Anchorage of masonry walls to each floor or roof shall resist a minimum force determined as $0.9S_{DS}$ times the tributary weight or 200 pounds per linear foot (2920 N/m), whichever is greater, acting normal to the wall at the level of the floor or roof. Existing wall anchors, if used, must meet the requirements of this chapter or must be upgraded.

[BS] A113.1.4 Anchors at corners. At the roof and floor levels, both shear and tension anchors shall be provided within 2 feet (610 mm) horizontally from the inside of the corners of the walls.

❖ The end pier in a system of piers has additional stiffness due to the effect of the intersecting wall acting as a flange. The placement of the first shear anchor within 2 feet (610 mm) of the end of the wall aids in transferring the shear load from the diaphragm to the corner pier.

[BS] A113.2 Diaphragm shear transfer. Bolts transmitting shear forces shall have a maximum bolt spacing of 6 feet (1829 mm) and shall have nuts installed over malleable iron or plate washers when bearing on wood, and heavy-cut washers when bearing on steel.

[BS] A113.3 Collectors. Collector elements shall be provided that are capable of transferring the seismic forces originating in other portions of the building to the element providing the resistance to those forces.

❖ The transfer of shear forces at the edge of a diaphragm to the shear resisting elements at that edge may require a collector at the edge of the diaphragm for distribution of the diaphragm shear to the elements of the shear wall. If the shear wall is the full depth of the diaphragm, no need for a collector would exist, as the uniformly distributed shear load of the diaphragm would be uniformly resisted along the length of the shear wall. If the shear wall strength and stiffness were at one end of that line of resistance, the collector will accumulate load from the diaphragm and act as a tie to the shear wall element. The calculated collector force may be resisted by existing or new elements. The relative rigidity of new and existing elements should be considered in apportioning loads to the various elements.

A pier adjacent to a corner of two intersecting shear walls is much stiffer than the usual pier analysis indicates. The effect of the flange, which is the other wall around the corner, is significant when the flange is in tension. The stiffness of this pier may cause the spandrel to have a significant collector stress. Because of this increased collector force, a minimum required spacing of the first shear bolt from the corner is specified to minimize the collector stress in the spandrel above the opening nearest to the building corner.

[BS] A113.4 Ties and continuity. Ties and continuity shall conform to the requirements of the building code.

[BS] A113.5 Wall bracing.

[BS] A113.5.1 General. Where a wall height-to-thickness ratio exceeds the specified limits, the wall may be laterally supported by vertical bracing members per Section A113.5.2 or by reducing the wall height by bracing per Section A113.5.3.

❖ Normal walls with excessive height-to-thickness ratios may be supported by an additional line of wall anchors that reduce the unbraced height or by vertical members that support the wall when cracked.

[BS] A113.5.2 Vertical bracing members. Vertical bracing members shall be attached to floor and roof construction for their design loads independently of required wall anchors. Horizontal spacing of vertical bracing members shall not exceed one-half of the unsupported height of the wall or 10 feet (3048 mm). Deflection of such bracing members at design loads shall not exceed one-tenth of the wall thickness.

❖ The vertical member shall be attached to the diaphragms. The reaction shall not be transferred into the URM wall and to its anchorage system. The horizontal spacing of the wall braces is limited to a distance equal to one-half of the wall height in order to provide a uniformly distributed resistance to out-of-plane displacements. This bracing does not prevent tension cracking of the URM wall. The brace should be anchored to the wall with a wall anchor at the center of the wall height as a minimum. The design loading of the wall brace is determined by the requirements of the IBC, and the stiffness of the wall brace must be sufficient to meet the deflection limitations.

[BS] A113.5.3 Intermediate wall bracing. The wall height may be reduced by bracing elements connected to the floor or roof. Horizontal spacing of the bracing elements and wall anchors shall be as required by design, but shall not exceed 6 feet (1829 mm) on center. Bracing elements shall be detailed to minimize the horizontal displacement of the wall by the vertical displacement of the floor or roof.

❖ Braces that extend between floors or roofs can reduce the unbraced height of the URM wall. A horizontal girt anchored on the face of the wall can be used to distribute the wall anchor force to braces that are spaced greater than 6 feet (1829 mm). Vertical displacement of the anchorage of the brace to the floor and roof will cause an outward displacement or bending stress in the URM wall. The deflection of the brace anchorage point caused by live loading shall be investigated and minimized.

[BS] A113.6 Parapets. Parapets and exterior wall appendages not conforming to this chapter shall be removed, or stabilized or braced to ensure that the parapets and appendages remain in their original positions.

The maximum height of an unbraced unreinforced masonry parapet above the lower of either the level of tension anchors or the roof sheathing shall not exceed the height-to-thickness ratio shown in Table A1-F. If the required parapet height exceeds this maximum height, a bracing system designed for the forces determined in accordance with the building code shall support the top of the parapet. Parapet corrective work must be performed in conjunction with the installation of tension roof anchors.

The minimum height of a parapet above any wall anchor shall be 12 inches (305 mm).

> **Exception:** If a reinforced concrete beam is provided at the top of the wall, the minimum height above the wall anchor may be 6 inches (152 mm).

❖ A maximum height of unbraced parapets above the wall anchorage points is specified. This height of unreinforced wall will be dynamically stable, but may slide laterally by the diaphragm response. A minimum height of parapet above the wall anchorage is specified, as its tension capacity is dependent on the axial stress in the wall at the level of anchorage. An exception is provided for a minimum height of 6 inches (152 mm). This exception is necessary as it was specified for standard details used for parapet risk reduction prior to the development of comprehensive requirements for earthquake risk reduction. It is not a recommended detail, and reinforced concrete beams may be slid laterally on the top of the wall by anchorage forces. However, failure of the anchorage system has not been observed.

[BS] A113.7 Veneer.

1. Veneer shall be anchored with approved anchor ties conforming to the required design capacity specified in the building code and shall be placed at a maximum spacing of 24 inches (610 mm) with a maximum supported area of 4 square feet (0.372 m^2).

   > **Exception:** Existing anchor ties for attaching brick veneer to brick backing may be acceptable, provided the ties are in good condition and conform to the following minimum size and material requirements.
   >
   > Existing veneer anchor ties may be considered adequate if they are of corrugated galvanized iron strips not less than 1 inch (25 mm) in width, 8 inches (203 mm) in length and $^1/_{16}$ inch (1.6 mm) in thickness, or the equivalent.

2. The location and condition of existing veneer anchor ties shall be verified as follows:

   2.1. An approved testing laboratory shall verify the location and spacing of the ties and shall submit a report to the building official for approval as part of the structural analysis.

   2.2. The veneer in a selected area shall be removed to expose a representative sample of ties (not less than four) for inspection by the building official.

❖ Unbonded brick veneer and veneer poorly bonded with angled bricks or sporadic flat metal ties to the URM structural wall have been found to detach from the structural wall when shaken with 20 to 30 percent of gravity ground motion. The exception in this section was taken from the 1930 *Los Angeles Building Code*.

[BS] A113.8 Nonstructural masonry walls. Unreinforced masonry walls that carry no design vertical or lateral loads and that are not required by the design to be part of the lateral force-resisting system shall be adequately anchored to new or existing supporting elements. The anchors and elements shall be designed for the out-of-plane forces specified in the building code. The height- or length-to-thickness ratio between such supporting elements for such walls shall not exceed nine.

❖ A nonstructural wall must be isolated from the structure on three sides. If the wall is not isolated from the structure, it must be considered as a shear wall and/or bearing masonry wall. The allowable height-to-thickness ratio is reduced from that allowed for a one-story bearing wall because vertical loading increases the stability of an unreinforced masonry wall. The height-to-thickness ratio may not exceed 9 to 1 or the limitations of Section A113.5.3.

[BS] A113.9 Truss and beam supports. Where trusses and beams other than rafters or joists are supported on masonry, independent secondary columns shall be installed to support vertical loads of the roof or floor members.

Exception: Secondary supports are not required where S_{D1} is less than 0.3g.

❖ Conformance of the URM wall with the height-to-thickness limitations does not imply that these walls will be uncracked and deform as flexural beams. It is highly likely that a crack will occur at each anchorage point and in the center part of the wall where the design level of intensity of shaking occurs. These blocks of walls will rotate on these cracks but will be dynamically stable. The rotation angle between a block above and below a wall anchorage could be greater than 5 degrees. A truss or beam that has an unyielding bearing surface can load the edge of its bearing on the masonry and cause spalling of the bearing surface. The secondary support serves the same purpose as shoring that would be installed after the earthquake. A foundation is not required for the secondary support, and the secondary support loads need not be accumulative in a multistory building. The exception to the requirement for secondary support of vertical loads of roof or floor members is permitted for lower seismic hazard zones. The exception is applicable where S_{D1}, as defined and mapped in the IBC, is 0.3g or less.

[BS] A113.10 Adjacent buildings. Where elements of adjacent buildings do not have a separation of at least 5 inches (127 mm), the allowable height-to-thickness ratios for "all other buildings" per Table A1-B shall be used in the direction of consideration.

❖ The increases in the allowable height-to-thickness ratios of URM walls when crosswalls are present in each story were predicated on allowing the diaphragms to have dynamic displacements that cause nonlinear displacements in the crosswalls and the diaphragm or in the diaphragm alone. Figure A1-1 was plotted by nonlinear analysis using ground motions scaled to 0.4g effective peak acceleration. The acceptable demand-capacity-span boundary is equivalent to 5 inches (127 mm) of displacement of the center of the diaphragm relative to the shear walls at the end of the span. If an adjacent building restricts the diaphragm displacement, the increase in the allowable height-to-thickness ratio is not applicable.

SECTION A114
WALLS OF UNBURNED CLAY, ADOBE OR STONE MASONRY

[BS] A114.1 General. Walls of unburned clay, adobe or stone masonry construction shall conform to the following:

1. Walls of unburned clay, adobe or stone masonry shall not exceed a height- or length-to-thickness ratio specified in Table A1-G.

2. Adobe may be allowed a maximum value of 9 pounds per square inch (62.1 kPa) for shear unless higher values are justified by test.

3. Mortar for repointing may be of the same soil composition and stabilization as the brick, in lieu of cement-mortar.

Bibliography

The following resource materials were used in the preparation of the commentary for this chapter of the code.

ABK-84, *Methodology for Mitigation of Seismic Hazards in Existing Unreinforced Masonry Buildings: The Methodology.* Topical Report 08, National Science

APPENDIX A

Foundation, Contract No. NSF-C-PFR78-19200. Washington, DC: Applied Science and Research Applications, 1984.

ABK-86, Guidelines for the Evaluation of Historic Brick Masonry Buildings in Earthquake Hazard Zones. ABK, A Joint Venture, Funded by the Department of Parks and Recreation of the State of California and the National Park Service, United States Government, January 1986.

ATC-78, *Tentative Provisions for the Development of Seismic Regulations for Buildings*, Report ATC 3-16. Redwood City, CA: Applied Technology Council, 1978.

ATC-87, *Evaluating the Seismic Resistance of Existing Buildings*, Report ATC-14. Redwood City, CA: Applied Technology Council, 1987.

ICBO-1988 through 1997, *Uniform Code for Building Conservation.* Whittier, CA: International Conference of Building Officials, 1988–1997.

ICBO-01, *Guidelines for the Seismic Retrofit of Existing, Buildings.* Whittier, CA: International Conference of Building Officials, 2006.

Preservation Brief 2, *Repointing Mortar Joints in Historic Masonry Buildings.* Robert C. Mack, FAIA, and John P. Speweik for the National Park Service, United States Government, October 1998.

TN 46, *Technical Note on Maintenance of Brick Masonry*, pp. 3-5, *Mortar Joint Repair.* Reston, VA: The Brick Industry Association, December 2005.

[BS] TABLE A1-A
ELEMENTS REGULATED BY THIS CHAPTER

BUILDING ELEMENTS	S_{D1}			
	$\geq 0.067g < 0.133g$	$\geq 0.133g < 0.20g$	$\geq 0.20g < 0.30g$	$> 0.30g$
Parapets	X	X	X	X
Walls, anchorage	X	X	X	X
Walls, h/t ratios		X	X	X
Walls, in-plane shear		X	X	X
Diaphragms[a]			X	X
Diaphragms, shear transfer[b]		X	X	X
Diaphragms, demand-capacity ratios[b]			X	X

a. Applies only to buildings designed according to the general procedures of Section A110.
b. Applies only to buildings designed according to the special procedures of Section A111.

[BS] TABLE A1-B
ALLOWABLE VALUE OF HEIGHT-TO-THICKNESS RATIO OF UNREINFORCED MASONRY WALLS

WALL TYPES	$0.13g \leq S_{D1} < 0.25g$	$0.25g \leq S_{D1} < 0.4g$	$S_{D1} \geq 0.4g$ BUILDINGS WITH CROSSWALLS[a]	$S_{D1} > 0.4g$ ALL OTHER BUILDINGS
Walls of one-story buildings	20	16	16[b,c]	13
First-story wall of multistory building	20	18	16	15
Walls in top story of multistory building	14	14	14[b,c]	9
All other walls	20	16	16	13

a. Applies to the special procedures of Section A111 only. See Section A111.7 for other restrictions.
b. This value of height-to-thickness ratio may be used only where mortar shear tests establish a tested mortar shear strength, v_t, of not less than 100 pounds per square inch (690 kPa). This value may also be used where the tested mortar shear strength is not less than 60 pounds per square inch (414 kPa), and where a visual examination of the collar joint indicates not less than 50-percent mortar coverage.
c. Where a visual examination of the collar joint indicates not less than 50-percent mortar coverage, and the tested mortar shear strength, v_t, is greater than 30 pounds per square inch (207 kPa) but less than 60 pounds per square inch (414 kPa), the allowable height-to-thickness ratio may be determined by linear interpolation between the larger and smaller ratios in direct proportion to the tested mortar shear strength.

APPENDIX A

[BS] TABLE A1-C
HORIZONTAL FORCE FACTOR, C_p

CONFIGURATION OF MATERIALS	C_p
Roofs with straight or diagonal sheathing and roofing applied directly to the sheathing, or floors with straight tongue-and-groove sheathing.	0.50
Diaphragms with double or mulitple layers of boards with edges offset, and blocked plywood systems.	0.75
Diaphragms of metal deck without topping:	
Minimal welding or mechanical attachment.	0.6
Welded or mechanically attached for seismic resistance.	0.68

[BS] TABLE A1-D
STRENGTH VALUES FOR EXISTING MATERIALS

EXISTING MATERIALS OR CONFIGURATION OF MATERIALS[a]		STRENGTH VALUES
		x 14.594 for N/m
Horizontal diaphragms	Roofs with straight sheathing and roofing applied directly to the sheathing.	300 lbs. per ft. for seismic shear
	Roofs with diagonal sheathing and roofing applied directly to the sheathing.	750 lbs. per ft. for seismic shear
	Floors with straight tongue-and-groove sheathing.	300 lbs. per ft. for seismic shear
	Floors with straight sheathing and finished wood flooring with board edges offset or perpendicular.	1,500 lbs. per ft. for seismic shear
	Floors with diagonal sheathing and finished wood flooring.	1,800 lbs. per ft. for seismic shear
	Metal deck welded with minimal welding.[c]	1,800 lbs. per ft. for seismic shear
	Metal deck welded for seismic resistance.[d]	3,000 lbs. per ft. for seismic shear
Crosswalls[b]	Plaster on wood or metal lath.	600 lbs. per ft. for seismic shear
	Plaster on gypsum lath.	550 lbs. per ft. for seismic shear
	Gypsum wallboard, unblocked edges.	200 lbs. per ft. for seismic shear
	Gypsum wallboard, blocked edges.	400 lbs. per ft. for seismic shear
Existing footing, wood framing, structural steel, reinforcing steel	Plain concrete footings.	$f'_c = 1{,}500$ psi (10.34 MPa) unless otherwise shown by tests
	Douglas fir wood.	Same as D.F. No. 1
	Reinforcing steel.	$F_y = 40{,}000$ psi (124.1 N/mm^2) maximum
	Structural steel.	$F_y = 33{,}000$ psi (137.9 N/mm^2) maximum

For SI: 1 inch = 25.4 mm, 1 square inch = 645.16 mm^2, 1 pound = 4.4 N.

a. Material must be sound and in good condition.
b. Shear values of these materials may be combined, except the total combined value should not exceed 900 pounds per foot (4380 N/m).
c. Minimum 22-gage steel deck with welds to supports satisfying the standards of the Steel Deck Institute.
d. Minimum 22-gage steel deck with $^3/_4$ φ plug welds at an average spacing not exceeding 8 inches (203 mm) and with sidelap welds appropriate for the deck span.

[BS] TABLE A1-E
STRENGTH VALUES OF NEW MATERIALS USED IN CONJUNCTION WITH EXISTING CONSTRUCTION

NEW MATERIALS OR CONFIGURATION OF MATERIALS		STRENGTH VALUES
Horizontal diaphragms	Plywood sheathing applied directly over existing straight sheathing with ends of plywood sheets bearing on joists or rafters and edges of plywood located on center of individual sheathing boards.	675 lbs. per ft.
Crosswalls	Plywood sheathing applied directly over wood studs; no value should be given to plywood applied over existing plaster or wood sheathing.	1.2 times the value specified in the current building code.
	Drywall or plaster applied directly over wood studs.	The value specified in the current building code.
	Drywall or plaster applied to sheathing over existing wood studs.	50 percent of the value specified in the current building code.
Tension bolts[e]	Bolts extending entirely through unreinforced masonry wall secured with bearing plates on far side of a three-wythe-minimum wall with at least 30 square inches of area.[b,c]	5,400 lbs. per bolt. 2,700 lbs. for two-wythe walls.
Shear bolts[e]	Bolts embedded a minimum of 8 inches into unreinforced masonry walls; bolts should be centered in $2^1/_2$-inch-diameter holes with dry-pack or nonshrink grout around the circumference of the bolt.	The value for plain masonry specified for solid masonry in the current building code; no value larger than those given for $^3/_4$-inch bolts should be used.
Combined tension and shear bolts	Through-bolts—bolts meeting the requirements for shear and for tension bolts.[b,c]	Tension—same as for tension bolts. Shear—same as for shear bolts.
	Embedded bolts—bolts extending to the exterior face of the wall with a $2^1/_2$-inch round plate under the head and drilled at an angle of $22^1/_2$ degrees to the horizontal; installed as specified for shear bolts.[a,b,c]	Tension—3,600 lbs. per bolt. Shear—same as for shear bolts.
Infilled walls	Reinforced masonry infilled openings in existing unreinforced masonry walls; provide keys or dowels to match reinforcing.	Same as values specified for unreinforced masonry walls.
Reinforced masonry[d]	Masonry piers and walls reinforced per the current building code.	The value specified in the current building code for strength design.
Reinforced concrete[d]	Concrete footings, walls and piers reinforced as specified in the current building code.	The value specified in the current building code for strength design.

For SI: 1 inch = 25.4 mm, 1 square inch = 645.16 mm^2, 1 pound = 4.4 N.

a. Embedded bolts to be tested as specified in Section A107.4.
b. Bolts to be $^1/_2$ inch minimum in diameter.
c. Drilling for bolts and dowels shall be done with an electric rotary drill; impact tools should not be used for drilling holes or tightening anchors and shear bolt nuts.
d. No load factors or capacity reduction factor shall be used.
e. Other bolt sizes, values and installation methods may be used, provided a testing program is conducted in accordance with Section A107.5.3. The strength value shall be determined by multiplying the calculated allowable value, determined in accordance with Section A107.5.3, by 3.0, and the usable value shall be limited to a maximum of 1.5 times the value given in the table. Bolt spacing shall not exceed 6 feet on center and shall be not less than 12 inches on center.

[BS] TABLE A1-F
MAXIMUM ALLOWABLE HEIGHT-TO-THICKNESS RATIOS FOR PARAPETS

	S_{D1}		
	$0.13_g \leq S_{D1} < 0.25_g$	$0.25_g \leq S_{D1} < 0.4_g$	$S_{D1} \geq 0.4_g$
Maximum allowable height-to-thickness ratios	2.5	2.5	1.5

[BS] TABLE A1-G
MAXIMUM HEIGHT-TO-THICKNESS RATIOS FOR ADOBE OR STONE WALLS

	S_{D1}		
	$0.13_g \leq S_{D1} < 0.25_g$	$0.25_g \leq S_{D1} < 0.4_g$	$S_{D1} \geq 0.4_g$
One-story buildings	12	10	8
Two-story buildings			
First story	14	11	9
Second story	12	10	8

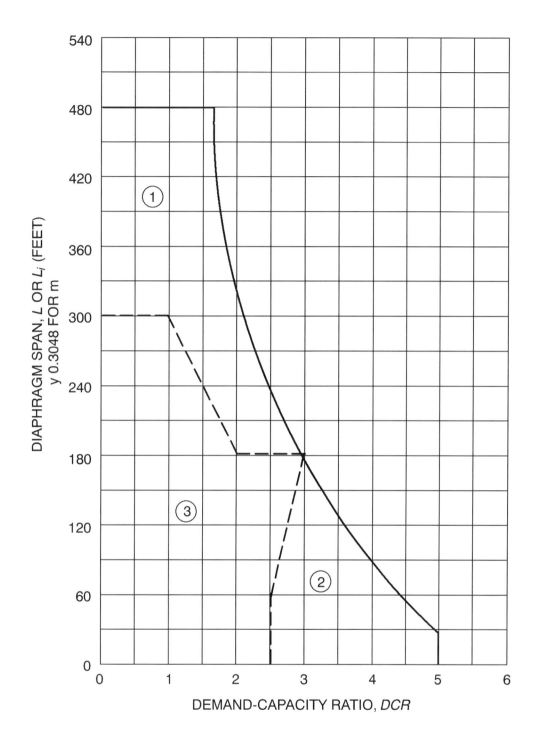

1. Region of demand-capacity ratios where crosswalls may be used to increase h/t ratios.
2. Region of demand-capacity ratios where h/t ratios of "buildings with crosswalls" may be used, whether or not crosswalls are present.
3. Region of demand-capacity ratios where h/t ratios of "all other buildings" shall be used, whether or not crosswalls are present.

[BS] FIGURE A1-1
ACCEPTABLE DIAPHRAGM SPAN

APPENDIX A

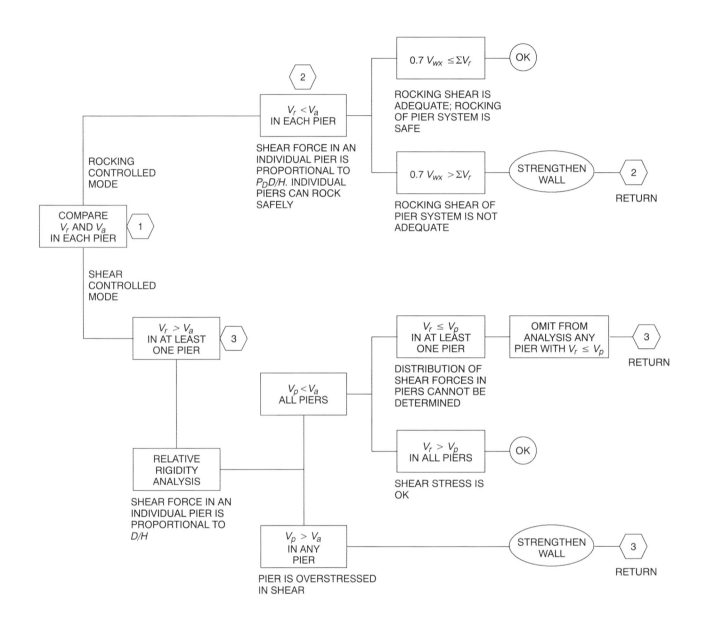

V_a = Allowable shear strength of a pier.
V_p = Shear force assigned to a pier on the basis of a relative shear rigidity analysis.
V_r = Rocking shear capacity of pier.
V_{wx} = Total shear force resisted by the wall.
ΣV_r = Rocking shear capacity of all piers in the wall.

[BS] FIGURE A1-2
ANALYSIS OF URM WALL IN-PLANE SHEAR FORCES

Chapter A2:
Earthquake Hazard Reduction in Existing Reinforced Concrete and Reinforced Masonry Wall Buildings with Flexible Diaphragms

SECTION A201
PURPOSE

[BS] A201.1 Purpose. The purpose of this chapter is to promote public safety and welfare by reducing the risk of death or injury that may result from the effects of earthquakes on reinforced concrete and reinforced masonry wall buildings with flexible diaphragms. Based on past earthquakes, these buildings have been categorized as being potentially hazardous and prone to significant damage, including possible collapse in a moderate to major earthquake. The provisions of this chapter are minimum standards for structural seismic resistance established primarily to reduce the risk of life loss or injury on both subject and adjacent properties. These provisions will not necessarily prevent loss of life or injury, or prevent earthquake damage to an *existing building* that complies with these standards.

❖ Appendix Chapter A2, along with the rest of Appendix A of the code, is referenced by the *International Codes®* in two ways:

  • Section 301.1.4.2 allows the use of Appendix A for qualifying buildings when "reduced IBC level seismic forces" are triggered.

  • Section 807.6 allows the use of Appendix A for voluntary seismic retrofit.

The provisions of this chapter apply to buildings designed prior to the adoption of the 1997 *Uniform Building Code* (UBC) or a code based on the 1997 NEHRP *Recommended Provisions for Seismic Regulations for New Buildings and Other Structures,* such as the 2000 *International Building Code®* (IBC®). They address deficiencies that have been found to be direct threats to life safety during prior earthquakes. These deficiencies first became apparent during the 1971 San Fernando, California earthquake when a large number of tilt-up buildings partially collapsed. Further significant damage was observed in tilt-ups and concrete block wall buildings with flexible diaphragms in subsequent significant U.S. events, including the 1987 Whittier Narrows, 1989 Loma Prieta and 1994 Northridge earthquakes. These deficiencies are as follows:

  • A commonly used method of connecting wall panels to roofs and floors for out-of-plane loading was to nail the diaphragms to wood ledgers and bolt the wood ledgers to the walls. This method of connection allowed the wall panels to separate from the roof or floor by failure of the ledger in cross-grain bending, nail pullout from the ledger or by tear-out of the nails through the edge of the wood structural panel.

  • When used, wall-to-roof and wall-to-floor anchors have often included eccentric, flexible sheet metal hardware that is susceptible to buckling and fracture at stress concentrations such as those caused by holes for connectors to wood. Poor installation, including misalignment of anchors, created additional stresses.

  • Out-of-plane forces have caused distress at girder-to-pilaster connections. The pilaster provides an edge support to the wall panel, which concentrates a substantial portion of the out-of-plane structural response to this connection. Commonly, the anchor bolts connecting girders to the top of the pilasters did not have adequate edge distance or confinement in the concrete. Typically, past damage included spalling of the inside face of the concrete pilasters and loss of support for the girders. The closely spaced ties required by the IBC at the top of the pilasters did not prevent this kind of damage.

  • The lack of continuous ties across the full depth of the diaphragm allowed cross-grain tension failures of the framing members at continuous joints in the wood structural panel sheathing. This occurred in the interior of the roof or floor diaphragms. Lack of tension ties at glulam hinge locations also led to loss of support for suspended girders.

  • Walls at reentrant corners or short architectural fin walls not designed to resist tributary loads and/or provided with collectors with adequate collector connections also experienced disproportionate damage.

Voluntary upgrade provisions for buildings designed with provisions predating the 1976 UBC

APPENDIX A

were developed in Los Angeles after the 1987 Whittier Narrows earthquake (Division 91, City of Los Angeles), and were then made mandatory for buildings designed to pre-1976 building codes in the City of Los Angeles after the 1994 Northridge earthquake. Other California jurisdictions have adopted similar provisions as either mandatory or voluntary ordinances.

This chapter requires:

- Positive anchorage of all reinforced concrete and masonry walls to develop the out-of-plane wall loads into the floor and roof diaphragms and to prevent splitting of the ledger or nails pulling through the edge of the sheathing.
- Provision for collectors and collector connections for walls.

Where large earthquakes have previously occurred and portions of the existing wall anchorage system may have been damaged due to deficiencies in the original design, it may be appropriate to further investigate and prioritize retrofits for these structures. Deteriorated and modified structures must also be investigated to determine the effect on the building's ability to withstand strong ground motions.

Other possible deficiencies, such as inadequate diaphragm strength, inadequate diaphragm stiffness or inadequate reinforcement at diaphragm openings, have not been reported as causing collapse or partial collapse, and are not addressed in this chapter because the level of risk associated with them is generally lower. (It is assumed that structural diaphragms exist in the structures covered by this chapter. That is, diaphragms such as gypsum or other proprietary components with minimal lateral load capacity are not used.) Although reinforced concrete and masonry wall buildings with metal deck diaphragms are within the scope of this chapter, they have been relatively rare in areas that have experienced earthquakes, and consequently post-earthquake observations and lessons learned from these configurations are very limited. It is possible that vulnerabilities associated with metal deck diaphragms will be observed in future earthquakes.

Similarly, the analysis of the walls for either in-plane or out-of-plane forces is not part of this chapter. Although walls are designed for smaller forces for out-of-plane bending than the forces used for wall anchor design, performance of reinforced walls in past earthquakes, provided that they have not been weakened by the introduction of large openings, has been good. Some engineers are concerned that walls consisting primarily of narrow piers without proper detailing may experience substantial damage caused by in-plane loading in a major earthquake. Thus, an engineer considering items beyond the scope of this chapter may want to evaluate and possibly strengthen walls consisting primarily of truck doors or large windows for in-plane shear and out-of-plane bending. Additionally, walls constructed using building codes without seismic requirements but located in regions with significant seismicity may represent an additional risk not addressed by the provisions.

This chapter was originally part of the *Guidelines for the Seismic Rehabilitation of Existing Buildings* (GSREB) (ICBO, 2001) and was intended to serve as a voluntary or mandatory retrofit ordinance. The chapter may be used for retrofits triggered by alterations to the building, occupancy changes at the building or repairs triggered by significant damage in an earthquake. However, in such cases, it is necessary to use other provisions (e.g., the provisions of the current edition of the IBC with reduced forces) to evaluate members not within the scope of this chapter (e.g., diaphragms and walls).

SECTION A202
SCOPE

[BS] A202.1 Scope. The provisions of this chapter shall apply to wall anchorage systems that resist out-of-plane forces and to collectors in existing reinforced concrete or reinforced masonry buildings with flexible diaphragms. Wall anchorage systems that were designed and constructed in accordance with the 1997 *Uniform Building Code*, 1999 *BOCA National Building Code*, 1999 *Standard Building Code* or the 2000 and subsequent editions of the *International Building Code* shall be deemed to comply with these provisions.

❖ The scope includes all reinforced concrete and reinforced masonry buildings with flexible diaphragms. Concrete tilt-up buildings are a subset of this type of structure. In terms of earthquake performance, there is little or no difference between buildings with precast concrete walls or other heavy walls, assuming the buildings have similar diaphragm spans, wall heights and wall reinforcing. The key is the transfer of forces associated with out-of-plane earthquake forces on these walls into the flexible diaphragms of the building.

The predecessor to Chapter A2, Division 91 of the *Los Angeles Building Code*, was developed for single-story reinforced concrete or masonry shear wall buildings with roofs sheathed with wood-based materials that are considered "flexible" diaphragms. Buildings with unreinforced walls are the subject of Chapter A1 and are not within the scope of this chapter.

The scope only includes buildings that were designed under a building code in effect prior to the adoption of the seismic provisions of the 1997 UBC, 1999 BOCA *National Building Code* (NBC), 1999 *Standard Building Code* (SBC) or the 2000 and subsequent editions of the IBC [changes based on observations in the 1994 Northridge earthquake were incorporated into both the 1997 UBC and 1997 *National Earthquake Hazards Reduction Program* (NEHRP)]. Structures that were designed according to these more recent building codes are presumed to have adequate designs and need not be strengthened. Past performance has indicated that partial and full collapse of these structures built with the deficient seismic design provisions of the past are likely at peak ground accelerations exceeding 0.20g. (This observation is based on past earthquakes with short to moderate duration. Ground shaking of less severity but longer duration could also result in collapses.)

The scope of this chapter includes only elements of the wall anchorage system and collectors and their connections. The wall anchorage system is defined in the IBC as including those elements in the diaphragm that are required to develop wall anchorage forces, such as wall anchors, struts, subdiaphragms, cross-ties and continuity ties.

SECTION A203
DEFINITIONS

[BS] A203.1 Definitions. For the purpose of this chapter, the applicable definitions listed in Chapters 16, 19, 21, 22 and 23 of the *International Building Code* and the following shall apply:

[BS] FLEXIBLE DIAPHRAGMS. Roofs and floors including, but not limited to, those sheathed with plywood, wood decking (1-by or 2-by) or metal decks without concrete topping slabs.

❖ The definition of "Flexible diaphragms" in this chapter is nearly identical to that in Section 12.3 of ASCE 7-10. It is not intended that materials such as gypsum, with minimal structural capacity, be included.

SECTION A204
SYMBOLS AND NOTATIONS

[BS] A204.1 General. For the purpose of this chapter, the applicable symbols and notations in the *International Building Code* shall apply.

❖ All symbols and notations used in this chapter are the same as those used in the IBC and ASCE/SEI 7.

SECTION A205
GENERAL REQUIREMENTS

[BS] A205.1 General. The seismic-resisting elements specified in this chapter shall comply with provisions of Section 1613 of the *International Building Code*, except as modified herein.

❖ It is only intended that the wall anchorage system, collectors and collector connections comply with the requirements of the IBC, as modified herein. Other structural components are considered lower priority in terms of reducing the potential for building collapse for these types of structures. As discussed in the commentary to Section A202, conditions where other components could be evaluated are identified.

[BS] A205.2 Alterations and repairs. Alterations and repairs required to meet the provisions of this chapter shall comply with applicable structural requirements of the building code unless specifically modified in this chapter.

[BS] A205.3 Requirements for plans. The plans shall accurately reflect the results of the engineering investigation and design and shall show all pertinent dimensions and sizes for plan review and construction. The following shall be provided:

1. Floor plans and roof plans shall show existing framing construction, diaphragm construction, proposed wall anchors, cross-ties and collectors. Existing nailing, anchors, cross-ties and collectors shall also be shown on the plans if they are considered part of the lateral force-resisting systems.

2. At elevations where there are alterations or damage, details shall show roof and floor heights, dimensions of openings, location and extent of existing damage and proposed *repair*.

3. Typical wall panel details and sections with panel thickness, height, pilasters and location of anchors shall be provided.

4. Details shall include existing and new anchors and the method of developing anchor forces into the diaphragm framing, existing and/or new cross-ties, and existing and/or new or improved support of roof and floor girders at pilasters or walls.

5. The basis for design and the building code used for the design shall be stated on the plans.

[BS] A205.4 Structural observation, testing and inspection. Structural observation, in accordance with Section 1709 of the *International Building Code*, shall be required for all structures in which seismic retrofit is being performed in accordance with this chapter. Structural observation shall include visual observation of work for conformance to the approved construction documents and confirmation of existing conditions assumed during design.

Structural testing and inspection for new construction materials shall be in accordance with the building code, except as modified by this chapter.

❖ The provisions of this chapter are intended to address deficiencies that represent direct threats to life safety. The primary concern is providing an adequate wall anchorage system. Lack of adequate collectors or collector connections at reentrant corners or internal walls can also be significant deficiencies that can lead to localized collapses. If the engineer finds that complete load paths are absent (e.g., some shear walls have no structural connections for force transfer from the roof to the shear walls), or if modifications or deterioration have introduced weaknesses, the requirements of the IBC for complete load paths (with reduced force levels) should be applied.

SECTION A206
ANALYSIS AND DESIGN

[BS] A206.1 Reinforced concrete and reinforced masonry wall anchorage. Concrete and masonry walls shall be anchored to all floors and roofs that provide lateral support for the wall. The anchorage shall provide a positive direct connection between the wall and floor or roof construction capable of resisting 75 percent of the horizontal forces specified in Section 1613 of the *International Building Code*.

❖ This chapter specifies that wall anchorage systems in existing buildings be evaluated using a force level of 75 percent of the design force required for new buildings, as it has historically. The wall anchorage system is defined in Section 12.11.2 of ASCE 7-10 as including the elements in the diaphragm required to develop wall anchorage forces, such as: wall anchors, struts, subdiaphragms, cross-ties and continuity ties. For design of wall anchors, ρ may be taken as 1.0. Note also that although wall anchors are connections between the roof diaphragm and the walls, it is not required that they be designed for 25-percent higher loads as required for buildings with reentrant corners (horizontal irregularity 2 in Table 12.3-1 of ASCE 7-10) as required in Section 12.3.3.4 of ASCE 7-10.

Once it is decided that retrofit of the wall anchorage system will be undertaken, additional risk reduction can often be achieved in a cost-effective manner through the use of stronger hardware that meets current IBC requirements. The engineer may want to discuss the possibility of such voluntary enhancements with the building owner.

[BS] A206.2 Special requirements for wall anchorage systems. The steel elements of the wall anchorage system shall be designed in accordance with the building code without the use of the 1.33 short duration allowable stress increase when using allowable stress design.

Wall anchors shall be provided to resist out-of-plane forces, independent of existing shear anchors.

Exception: Existing cast-in-place shear anchors are allowed to be used as wall anchors if the tie element can be readily attached to the anchors, and if the engineer or architect can establish tension values for the existing anchors through the use of approved as-built plans or testing and through analysis showing that the bolts are capable of resisting the total shear load (including dead load) while being acted upon by the maximum tension force due to an earthquake. Criteria for analysis and testing shall be determined by the building official.

Expansion anchors are only allowed with special inspection and approved testing for seismic loading.

Attaching the edge of plywood sheathing to steel ledgers is not considered compliant with the positive anchoring requirements of this chapter. Attaching the edge of steel decks to steel ledgers is not considered as providing the positive anchorage of this chapter unless testing and/or analysis are performed to establish shear values for the attachment perpendicular to the edge of the deck. Where steel decking is used as a wall anchor system, the existing connections shall be subject to field verification and the new connections shall be subject to special inspection.

❖ The wall anchorage forces are dependent on the span and stiffness of the diaphragm and the weight, thickness, height and flexibility of the walls. In a design earthquake, the probable dynamic forces will exceed the design force. Recent studies of strong motion records in buildings with flexible diaphragms show amplification at the roof diaphragm level of two to four times the peak ground acceleration. (Higher amplifications typically occur at lower ground accelerations where lower damping and no inelastic behavior of the diaphragm occurs.) Such amplifications of peak ground acceleration are expected for structures that have diaphragm periods corresponding with the peak of the response spectrum.

Although it is not recommended that wall anchors be designed to yield when subjected to major earthquakes (yielding would result in large deformations that could cause damage to diaphragm nailing at the ledgers for wood diaphragms), it is still recommended that the designer provide wall anchors that will yield rather than fail in a brittle manner at reduced sections (e.g., twisted straps or at steel straps with stress concentrations, such as those caused by holes in steel components).

Structural steel, which has a known ductile behavior and a small coefficient of variation in strength, has the lowest ratio of yield stress to allowable stress of all of the materials used in the wall anchor system. The material-specific factor of 1.4 in Section 12.11.2.2.2 of ASCE 7-10 was first introduced in the 1997 UBC for strength design, which adjusts the strength of the steel to meet the expected anchorage loading and, thus, maintains consistency in the elastic capacity of all of the materials used in the wall anchorage system. The steel components of the anchor hardware are designed using this load factor. (The embedment of the anchor bolts and bolts and nails in wood members are evaluated without the 1.4 factor for steel.)

It is also important to consider the stiffness of the steel and other materials in the anchorage. For example, the wall anchors should be substantially stiffer than the ledger nails adjacent to a wall anchor in order to prevent the nails from resisting substantial load and causing failure of the wood ledger in cross-grain bending. Clearly, stiff wall anchors and minimal bolt slip (e.g., no oversized holes) will help form a good wall anchorage system.

Existing ledger bolts embedded in the concrete or masonry wall may be used as a part of the wall anchorage system if their tension values can be established by testing or through analysis. The size and spacing of these anchor bolts must be shown on the approved plans and verified by site investigation. A strength design analysis must show that these embedded anchor bolts are capable of resisting the factored tension load in combination with the factored combination of earthquake and gravity-induced shear loads in both the horizontal and vertical directions.

Use of new expansion anchors is only allowed if they are installed with special inspection, and if they meet the appropriate criteria [such as those anchors evaluated in an ICC Evaluation Services® (ICC-ES®) report] indicating the conditions for which the anchors are permitted to resist seismic forces.

Wall anchorage that relies on the tear-out strength (shot pins pulling through the edge) of the wood structural panel for attachment to a steel ledger is not considered compliant with the provisions of this chapter. This is because such connectors at the edge of the wood structural panel sheets are loaded with a combination of shear and tension (in-plane shear and tear-out) during an earthquake, resulting in local demand that is much higher than the original design capacity. In addition, tear-out values of connectors adjacent to the edge of the wood structural panel sheet are not specified in the IBC.

Welding steel decking to a steel ledger angle is not considered positive anchorage, as required by this chapter, when flutes are parallel to the wall in accordance with Section 12.11.2.2.4 of ASCE 7-10. The reasons are the same as for wood structural panel attachment, and also because the flutes have substantial flexibility. When flutes are perpendicular to the wall, it should be demonstrated through testing or calculation that the metal decking and ledger angle can be used for wall anchorage without significant displacement normal to the wall surface. Calculations should consider that the welds are also simultaneously loaded in shear due to global diaphragm behavior. The 1.4 material-specific factor should be used in calculations.

[BS] A206.3 Development of anchor loads into the diaphragm. Development of anchor loads into roof and floor diaphragms shall comply with Section 1613 of the *International Building Code* using horizontal forces that are 75 percent of those used for new construction.

> **Exception:** If continuously tied girders are present, the maximum spacing of the continuity ties is the greater of the girder spacing or 24 feet (7315 mm).

In wood diaphragms, anchorage shall not be accomplished by use of toenails or nails subject to withdrawal. Wood ledgers, top plates or framing shall not be used in cross-grain bending or cross-grain tension. The continuous ties required in Section 1613 of the *International Building Code* shall be in addition to the diaphragm sheathing.

Lengths of development of anchor loads in wood diaphragms shall be based on existing field nailing of the sheathing unless existing edge nailing is positively identified on the original construction plans or at the site.

❖ This section presents the requirements of IBC Chapter 16 and Section 12.11.2.2.1 of ASCE 7-10 for development of wall anchorage loading. Continuous ties or struts between diaphragm chords are required to distribute wall anchorage forces into the body of the diaphragm. Subdiaphragms created with subdiaphragm chords at their boundaries may be used to transmit the anchorage forces to the main cross-ties. Note that although subdiaphragms are useful tools for helping to define rules for developing wall anchor forces into the diaphragms, the engineer should not really expect the force transfer to occur in a manner consistent with the subdiaphragm assumptions.

The span-depth ratio of subdiaphragms shall not exceed 2.5 to 1. This allowable span-depth ratio is substantially less than the 4.0 value permitted prior to the 1997 UBC. This limit is an indirect attempt to limit subdiaphragm shear. The combination of global diaphragm shear (from orthogonal directions) and subdiaphragm shears can exceed the actual diaphragm capacity, though this effect is commonly ignored by designers even in new structures. There was some discussion by the chapter authors about relaxing this requirement to allow a subdiaphragm ratio of 3:1 for existing buildings since subdiaphragms with 8-foot by 24-foot (2438 mm by 7315 mm) dimensions used to be common. However, this was a minority opinion and the requirement for retrofitting subdiaphragms with a length-to-depth ratio of more than 2.5:1 is a requirement of this chapter. For cases when retrofits are being performed on a voluntary basis and subdiaphragm ratios exceed 2.5:1 by a small amount, the engineer may want to consider omitting correction of this deficiency, since it is a lower priority item than other items included in this chapter.

The engineer should consider the existing wood structural panel layout when selecting locations for wall anchors. If members without edge nailing are used as part of the wall anchorage system, the reduced capacity for field nailing must be considered or new nailing must be provided. This includes locations where the wood structural panel is staggered and the subpurlin in the first bay has edge nailing, but the subpurlin in the second bay does not, or vice versa. If the diaphragm nailing is unknown (no design drawings exist), the roofing should be removed locally in a number of locations to determine edge nailing unless field nailing is assumed. As nailing will probably vary over the length of the diaphragm (zone nailing), the nailing should be verified in a number of regions. If there are indications of poor construction quality (e.g., nail spacing erratic, nails overdriven), removal of the roofing to verify nailing at wall anchor locations is recommended.

Note that the engineer should consider whether additional nailing into 2x members to increase either wall anchor transfer or diaphragm shear capacity will split the members. Generally, nail spacing (new and existing combined) should not be less than that permitted in 2x members in new diaphragms. It is sometimes difficult to tighten existing nail spacing without triggering the provisions associated with closely spaced nailing. For example, spacing at 6 inches (152 mm) cannot be decreased to 4 inches (102 mm) without placing one nail between two existing nails, effectively producing a 3-inch (76 mm) spacing. Nailing from strap anchors on top of the 2x members increases the likelihood of splitting. If straps are applied on top of plywood, new diaphragm nailing at the same location may be omitted. Splitting of short 2x members (e.g., blocking) is more of a concern than longer 8-foot-long (2438 mm) members typical of panelized roofs. Predrilling of holes can be used to reduce the tendency for splitting. Staples may be used to enhance diaphragm capacity without splitting, although relative flexibility must be considered by using data from testing.

In previous editions a minimum member size of 3 inches (76 mm) nominal was required for wood members used to develop anchorage forces into the diaphragms. Since a minimum member size requirement does not exist in ASCE 7, this was considered inappropriate. Instead, whatever size can be shown adequate through calculation is allowed. It is difficult or impossible to make 2x members work due to edge distance requirements and typical construction tolerances, notwithstanding that it may appear adequate through calculation. An existing 2-inch (51 mm) nominal member can be supplemented with other 2-inch (51 mm) nominal member(s) if the additional member(s) are internailed. It is necessary to provide sufficient internailing between the new and existing member for force transfer. In addition, if no new nailing into the diaphragm is provided, the stresses due to the eccentric loading from the diaphragm nails in the existing member only must be considered. It is common practice to glue the top of the new member to the diaphragm in order to achieve stress transfer from the diaphragm to the new 2x member. However, the strength of the glue should not be relied upon in calculations unless it can be demonstrated that the glue does not lose its strength or become more brittle with age.

[BS] A206.4 Anchorage at pilasters. Anchorage at pilasters shall be designed for the tributary wall-anchoring load per Section A206.1, considering the wall as a two-way slab. The edges of the two-way slab shall be considered fixed when there is continuity at pilasters and shall be considered pinned at roof and floor. The pilasters or the walls immediately adjacent to the pilasters shall be anchored directly to the roof framing such that the existing vertical anchor bolts at the top of the pilasters are bypassed without permitting tension or shear failure at the top of the pilasters.

Exception: If existing vertical anchor bolts at the top of the pilasters are used for the anchorage, additional exterior confinement shall be provided as required to resist the total anchorage force.

The minimum anchorage force at a floor or roof between the pilasters shall be that specified in Section A206.1.

❖ It has been past practice to anchor walls to the floor and the roof diaphragms for tributary wall loading, assuming the wall panel spans between adjacent floors or between the floor and roof. However, when pilasters are present in the walls, the stiffening effect of the pilasters on the deformed shape of the wall-pilaster system must be taken into account. This is consistent with Section 12.11.2.2.7 of ASCE 7-10. Since the relative stiffness of the walls and pilasters may vary, a rational method is required to calculate the wall panel load carried by the pilaster. The reaction of a pilaster at the diaphragm level, resulting from the analysis of the wall-pilaster system, is the load applied directly to the girder through its existing anchorage to the top of the pilaster.

Three concerns must be addressed for existing wall anchors at pilasters: 1. The existing bolted connection to the girder must have adequate strength for the pilaster reaction; 2. The hardware must be strong enough (including the net section); and 3. The shear strength of the existing anchors, with consideration of the edge distance of the embedded bolts in the top of the pilaster, must be adequate.

An inadequate bolted connection to the girder can be enhanced by additional bolts. Side plates can be extended by either welding new plates to the existing plates, or by removing the existing bolts, providing a second longer plate with additional holes over each plate, and reinstalling the existing, plus new bolts.

Inadequate hardware strength can be addressed by replacing the existing anchors with new, stiff wall anchors designed to take the entire load. The stiffness of such a bypass system must be checked to verify that it is stiffer under loading than the existing system. If the new anchors are not significantly stiffer than the existing system, the existing system should be disconnected (e.g., remove bolts from girder).

If the shear strength of the existing bolts is not adequate due to their edge distance, additional exterior confinement at the top of the pilaster must be provided if the bolts are to be relied on to resist loads.

A single-story wall panel with pilasters at each end is supported at four edges by combination of the pilasters, the roof diaphragm and the slab-on-grade. Typically, the support at the slab-on-grade should be considered as a support without moment restraint. (The panel could also be modeled with lateral restraint at both the slab-on-grade and the footing in cases where the elevations of the two are substantially different.) The support condition of the panel at the pilaster can be a continuous span, a single span having moment restraint or a simple support. The moment restraint depends on whether continuity of the panel through the pilaster is shown on the drawings or whether an open joint exists at one side of the pilaster. The support condition at the roof level is typically considered as a support without moment restraint. A textbook or standard for the design of slabs supported on four edges may be used to determine reactions at each edge of the panel. These references consider the aspect ratio of the panel and the rotational restraint at each panel edge. Analyses can be complicated by the presence of openings and parapets. In such cases, approximate, conservative, simplifying assumptions should be used. A logical approach regarding tributary area for the design of pilaster wall anchorage is included in Figure C108.1 of the 1999 Structural Engineers Association of California (SEAOC) *Recommended Lateral Force Requirements and Commentary (Bluebook)*. In any case, without an exact solution that considers pilaster to wall stiffness and different wall anchor stiffnesses, some conservatism is recommended.

The stiffening effect of pilasters should be considered even when framing members are not directly supported on the pilasters (e.g., purlins). Their presence will cause the out-of-plane loading at the diaphragm level adjacent to the pilaster to exceed the "average" anchorage loading. If possible, direct connection from the pilaster to nearby members is desirable.

Wall anchors at pilaster locations should be located close enough to the top of the pilaster to avoid overstressing the wall panels in bending from the higher wall anchor forces associated with the pilaster. The desire to avoid bending in the wall should be balanced with the fact that the roof framing members experience less bending when the wall anchors are close to the roof sheathing where the wall anchor load is transferred to the diaphragm. Also, tying them to the wall near the level of the roof provides some flexibility, negating, to an extent, the stiffening effect of the pilaster.

Although it is appropriate to increase the tributary wall anchor forces at pilasters to consider the pilasters' stiffness, design forces for adjacent wall anchors (i.e., subpurlins) should not be reduced.

[BS] A206.5 Symmetry. Symmetry of wall anchorage and continuity connectors about the minor axis of the framing member is required.

> **Exception:** Eccentricity may be allowed when it can be shown that all components of forces are positively resisted. The resistance must be supported by calculations or tests.

❖ Symmetry of wall anchorage connectors that are placed on the vertical sides of the framing members (i.e., wall anchors that would generate bending about minor axis of the framing members) is required unless stresses caused by the eccentricity are accounted for in the design. This is consistent with Section 12.11.2.2.6 of ASCE 7-10. In the past, engineers often used allowable hardware values in the manufacturers' catalogs without consideration of member capacity, and specifically the stresses caused by eccentric connections. The values in the hardware catalog typically were developed by testing the hardware on a steel jig, and thus did not consider the eccentricity stresses in the wood members. It is very common to find single-sided wall anchor connections in existing buildings. This is especially a problem for subpurlins, but may be a problem for purlins and beams.

The flexural stresses about the minor axis caused by unsymmetrical connectors must be combined with stresses caused by gravity loading. Eccentricity between the line of loading on the connector and the centerline of the resisting member must be resisted by providing supports normal to the line of loading or by a resisting couple. The reactions at these supports must be determined by calculations or testing.

The engineer should be aware of the fact that eccentricity can arise either through design or because of bolt or strap misalignment (i.e., poor installation). Placing an anchor on each vertical face of the framing member at each connection is an easy method of providing symmetry. For a condition with an existing single-sided wall anchor, a second new wall anchor can be added, but the new anchor should closely match the existing anchor in terms of stiffness and strength unless the analysis considers the rigidity and strength (see Section A206.6).

Where two new anchors are provided, an offset of anchors along the length of the framing member may be required to avoid fastener interference for anchors that use screws or nails rather than through bolts.

[BS] A206.6 Combination of anchor types. New anchors used in combination on a single framing member shall be of compatible behavior and stiffness.

❖ The sharing of the load on a single framing member between wall anchors of different types requires an analysis of their relative stiffness. Stiffness must consider the type of anchor bolt used. Testing of J bolts has found that the smooth portion can exhibit minimal bond and pull-out may occur by straightening of the J bend. Thus J bolts tend to be inferior to headed bolts. Similarly, expansion anchors may slip prior to failure, resulting in unequal load distribution if used in combination with cast-in-place anchors or through bolts. If no data exists for calculation of relative stiffness of existing hardware and/or anchor bolts, load sharing between different types of wall anchors is not permitted.

Older hardware is often more flexible than new hardware, and it may be more practical to remove existing hardware or ignore it and provide two identical pieces of new hardware, one on each side of the member.

[BS] A206.7 Anchorage at interior walls. Existing interior reinforced concrete or reinforced masonry walls that extend to the floor above or to the roof diaphragm shall be anchored for out-of-plane forces per Sections A206.1 and A206.3. Walls extending through the roof diaphragm shall be anchored for out-of-plane forces on both sides, and continuity ties shall be spliced across or continuous through the interior wall to provide diaphragm continuity.

❖ This section requires interior walls to be provided with wall anchor systems. In Case 1, where the wall does not project through the diaphragm and the framing is continuous, it may only be necessary to provide wall anchorage on one side. For Case 2, in diaphragms interrupted by walls, it is necessary to provide wall anchors on both sides. The intent is that every wall be anchored into the framing and that the loads be distributed into the diaphragm for seismic loading in each direction. For Case 2, this requires either subdiaphragms on both sides of the wall or continuity ties through the wall.

[BS] A206.8 Collectors. If collectors are not present at reentrant corners or interior shear walls, they shall be provided. Existing or new collectors shall be designed for the capacity required to develop into the diaphragm a force equal to the lesser of the rocking or shear capacity of the reentrant wall or the tributary shear based on 75 percent of the horizontal forces specified in Chapter 16 of the *International Building Code*. The capacity of the collector need not exceed the capacity of the diaphragm to deliver loads to the collector. A connection shall be provided from the collector to the reentrant wall to transfer the full collector force (load). If a truss or beam other than a rafter or purlin is supported by the reentrant wall or by a column integral with the reentrant wall, then an independent secondary column is required to support the roof or floor members whenever rocking or shear capacity of the reentrant wall is less than the tributary shear.

❖ At reentrant corners (and fin walls), the concrete or masonry return walls may have been ignored in the diaphragm analysis performed in the original design. Because of their in-plane stiffness, these walls will function as diaphragm supports and may attract very large loads based on tributary area. Although lack of collectors at interior masonry or concrete walls (or braced frames) is not common, the walls may attract larger loads by tributary area than they were designed for due to changes in IBC provisions (i.e., increase in base shear or introduction of Ω). Even for narrow, relatively flexible walls, deflection of the diaphragm spanning between end walls of the building may not be compatible with the deformation of the top of these reentrant or interior masonry or concrete walls. This deformation incompatibility may result in diaphragm damage and/or separation of the diaphragm from the return wall. As an alternative to using a deflection compatibility approach, a simple comparison of strengths or capacities is recommended. The objective of the comparison is to find the minimum load needed to adequately tie the return wall to the diaphragm. The capacities of the diaphragm system or the return wall could be estimated as follows for a single-story structure:

1. Rocking capacity of a wall: The actual rocking capacity, V_R, of the wall shall be calculated from an equilibrium equation, using all the dead loads on the walls, including the weight of the wall, taken about an extreme edge of the footing. Other factors such as soil weight and contributions from continuous footings should be included, as underestimating the overturning resistance in this context is not conservative. Load factors specified in the IBC that provide the maximum resistance shall be applied.

2. Shear capacity of the wall: V_n shall be the nominal shear strength of the wall provided by the concrete or masonry and shear reinforcement without any strength reduction factor.

3. Maximum force that can be delivered by the diaphragm: $V_d = (v)(L_1 + L_2)$ where:

   L_1, L_2 = Depth of the diaphragm measured on each side of the return or interior wall.

   v = Allowable diaphragm shear value from IBC Chapter 23 times a factor of 3. The 2003 NEHRP may be used for strength design values for diagonal sheathing.

If the collector and collector connection capacity exceed each of the three capacities discussed above, then together they will not represent the weak link in the structure, and no strengthening is required. It should be noted that the consequences of failure of whichever element above has the weakest capacity should be considered even if such analysis is not required by this chapter. For instance, shear failure of the wall may not be acceptable.

The engineer should consider providing a redundant, independent vertical support beneath major framing members, such as girders or trusses supported by ends of return or interior walls, if these walls are limited in capacity (when either $V > V_n$ or $V > V_R$) and $V > V_d$, where V is the demand on the wall (strength design). The failure mode could be shear in the reinforced concrete or reinforced masonry wall, or a dynamic response that causes rocking in the return wall. A wall that fails in-plane shear may lose its capacity to support vertical loads. A wall that rocks on its foundation may cause damage to the bearing surface providing support for the member framing onto the return wall.

Provision of a redundant support for smaller members such as purlins, rafters or subpurlins is not as important. It is likely that loss of support for such minor roof framing members will not result in significant collapse or threat to life safety.

Note that for design purposes, the three capacities identified above (rocking capacity of wall, shear capacity of wall and diaphragm capacity) are compared with the collector and collector connection capacity without reduction. The 0.75 factor is applied to code design forces only.

[BS] A206.9 Mezzanines. Existing mezzanines relying on reinforced concrete or reinforced masonry walls for vertical and/or lateral support shall be anchored to the walls for the tributary mezzanine load. Walls depending on the mezzanine for lateral support shall be anchored per Sections A206.1, A206.2 and A206.3.

Exception: Existing mezzanines that have independent lateral and vertical support need not be anchored to the walls.

❖ Existing buildings commonly have mezzanines that are used for office or light storage. Mezzanines may be dependent on the wall and pilasters for lateral support. In these configurations, the seismic load of a mezzanine is additive to the seismic load of the walls for calculation of the anchorage force at the top of the wall. This combined loading should be used for calculation of wall anchorage forces at the roof diaphragm level above and below (if applicable) the mezzanine. An alternative would be to provide a lateral load-resisting system for the mezzanine and use the mezzanine for support of the adjacent walls. In either of these cases, the mezzanine must be anchored to the wall. If the mezzanine is isolated from the wall in accordance with the building separation requirement of the IBC and has independent lateral and vertical support, no consideration of the anchorage of the mezzanine is required. This is what is meant by "independent lateral and vertical support" in the exception.

This chapter does not require that the flexural strength of the walls and/or pilasters be analyzed for lateral loads from the mezzanine. However, it is recommended that consideration be given to the existing strength of these elements before making a decision about the alternatives of providing lateral bracing below the existing mezzanine, or using the wall to transmit the seismic loading of the mezzanine to the roof and slab-on-grade levels. This is especially true if the mezzanine was added after the original construction. In past earthquakes, horizontal flexural cracks and bowing attributable to mezzanine loading on walls have been observed.

SECTION A207
MATERIALS OF CONSTRUCTION

[BS] A207.1 Materials. All materials permitted by the building code, including their appropriate strength or allowable stresses, may be used to meet the requirements of this chapter.

❖ This section generally gives permission to use any material allowed by the IBC as applicable to the situation.

Bibliography

The following resource materials were used in the preparation of the commentary for this chapter of the code.

Loma Prieta Earthquake Reconnaissance Report. EERI Spectra Vol. 6 Supplement, May 1990.

NEHRP, *Recommended Provisions for Seismic Regulation of New Buildings and Other Structures (FEMA 302). Part 1: Provisions.* Washington, DC: Federal Emergency Management Agency, 1997.

NEHRP, *Recommended Provisions for Seismic Regulation of New Buildings and Other Structures (FEMA 450-1/2003). Part 1: Provisions.* Washington, DC: Federal Emergency Management Agency, 2004.

Northridge Earthquake Reconnaissance Report, Vol. 2. EERI Spectra Vol. Supplement C to Vol. 11, January 1996.

SEAOC, *Recommended Lateral Force Requirements and Commentary.* Sacramento, CA: Structural Engineers Association of California, Seismology Committee, 1999.

The 1987 Whittier Earthquake. EERI Spectra Vol. 4, No. 1, Feb. 1988; and Vol. 4, No. 2, May 1988.

Chapter A3:
Prescriptive Provisions for Seismic Strengthening of Cripple Walls and Sill Plate Anchorage of Light, Wood-frame Residential Buildings

SECTION A301
GENERAL

❖ Appendix Chapter A3, along with the rest of Appendix A, is referenced by the *International Codes®* in two ways:

- Section 301.1.4.2 allows the use of Appendix A for qualifying buildings when "reduced IBC level seismic forces" are triggered.
- Section 807.6 allows the use of Appendix A for voluntary seismic retrofit.

History has shown that light, wood-frame residential buildings with specific structural weaknesses in their original construction are susceptible to severe damage from earthquakes. The most common structural weaknesses are: 1. Absence of proper connection between the exterior walls and the foundation (i.e., anchor bolts); 2. Inadequate bracing of cripple walls between the foundation and first floor (i.e., lack of structural sheathing); 3. Discontinuous or inadequate foundations below the exterior walls; and 4. A perimeter foundation system that is constructed of weak or deteriorated unreinforced masonry or stone.

Unreinforced masonry chimneys and poorly reinforced masonry chimneys are also a common source of damage and, in some cases, pose a life safety threat. Chimneys are not included in the scope of this document since reduction of chimney vulnerability through pointing of mortar and bracing is not typically considered cost effective, particularly if the risks to life can be controlled by other means. For example, the Applied Technology Council (ATC) recommends adding plywood above the ceiling framing to reduce the chances of falling masonry from penetrating through the ceiling (ATC, 2002). Curtailing the occupancy and frequent use of property within the falling radius of chimneys could also be an effective way of minimizing the risk of casualties. The ATC recommends replacing upper portions of damaged chimneys with light-frame construction rather than diagonal bracing. Otherwise, masonry bearing wall chimneys should be replaced with light-frame construction with or without veneer [Los Angeles Department of Building and Safety (LA), 2001]. Generally, the cost of replacing existing chimneys is comparable to the cost of repairing or replacing a damaged chimney after earthquakes, so in many cases it is arguably best to forego chimney retrofits and instead take steps to reduce the exposure of occupants and neighbors and to anticipate the possibility of chimney replacement after future earthquakes.

Water heaters that are not restrained from toppling in earthquakes also pose significant risks of water and fire damage that are not addressed herein [National Institute of Standards and Technology (NIST), 1997]. Existing premanufactured restraint systems are available at many hardware stores and have been stamped as preapproved by a government agency for limited use. They are both economical and effective for water heaters typically located adjacent to wood-frame walls [California Division of State Architect (DSA), 2002]. However, atypical water heater bracing installations require engineered designs.

After the Loma Prieta, California, earthquake of 1989, the average cost to repair dwellings that suffered damage due to the above-stated weaknesses ranged from $25,000 to $30,000 in 1990 dollars (Gallagher, 1990), and would be much more today. Some dwellings with these deficiencies have been total losses. Experience indicates that the average cost for a licensed contractor to install sill bolts and cripple wall bracing to undamaged dwellings ranges from $2,500 to $5,000 in 1998 dollars [California Seismic Safety Commission (CSSC), 2002]. The total cost will also vary, depending on accessibility in the crawl space area. If dwelling owners elect to perform the strengthening work themselves, costs can be substantially lower. The cost effectiveness of correcting these weaknesses becomes more favorable by considering the potential costs of emergency shelters, interim housing and lost employee productivity while damaged dwellings are being repaired [Association of Bay Area Governments (ABAG), 2000] (Comerio, 1996). Some insurance companies offer retrofit discounts that can lower earthquake insurance premiums and insured losses [California Earthquake Authority (CEA), 2002]. Some cities also offer expedited and/or reduced plan check fees and, in some cases, financial assistance that can reduce the total project cost to strengthen these deficiencies.

The purpose of this chapter is to provide minimum

standards addressing the four primary structural weaknesses stated above. This chapter is intended to encourage and facilitate seismic strengthening of conventional dwellings and to provide standardized methods for performing this work. The provisions are written in a prescriptive format to eliminate the need, in most cases, for engineering design and to encourage the direct use of the chapter by building owners and contractors. In practice, the engineering fees involved in designing retrofits for small single-family dwellings can be substantial compared with the construction cost and can discourage owners from undertaking the work. Contractors who perform the work, however, are generally not licensed to provide engineering design services. Homeowners, meanwhile, might not fully recognize the distinction between an engineer's role and a contractor's. Therefore, prescriptive retrofit provisions are expected to be cost effective, to help ensure a complete scope of work for the contractor and to provide a measure of consumer protection for the owner.

Retrofit training for contractors has been periodically offered by the Federal Emergency Management Agency (FEMA) and the Association of Bay Area Governments (ABAG). By adopting and enforcing the provisions in this chapter, jurisdictions can further help protect the interests of homeowners and establish a standard for retrofit practice. Note that some jurisdictions (e.g., the cities of Los Angeles, San Leandro, Santa Barbara, Berkeley, Oakland and San Mateo) have prepared drawings with approved details to further assist in retrofitting dwellings. For regions outside of the western U.S., the Institute for Building and Home Safety (IBHS) offers a retrofit guide.

Wood-frame buildings with continuous concrete foundations in good condition, anchorages meeting the requirements of this chapter and wood structural panel sheathing or diagonal sheathing around the entire building perimeter, including the cripple walls anchored to the foundation, have performed well in past earthquakes. The provisions of this chapter need not be applied to these buildings.

"Prescriptive" means these provisions apply to specific conditions and must be used in precisely the manner described. "Prescriptive" also means "determined in advance," without the need for case-specific analysis or design. Through the use of these provisions and the accompanying details, the dwelling owner or contractor can develop plans without the services of a design professional (civil engineer, structural engineer or architect); however, the use of other materials, proprietary systems or methods not shown by the figures and details in this chapter may require the services of a registered design professional.

[BS] A301.1 Purpose. The provisions of this chapter are intended to promote public safety and welfare by reducing the risk of earthquake-induced damage to existing wood-frame residential buildings. The requirements contained in this chapter are prescriptive minimum standards intended to improve the seismic performance of residential buildings; however, they will not necessarily prevent earthquake damage.

This chapter sets standards for strengthening that may be approved by the *code official* without requiring plans or calculations prepared by a registered design professional. The provisions of this chapter are not intended to prevent the use of any material or method of construction not prescribed herein. The *code official* may require that construction documents for strengthening using alternative materials or methods be prepared by a registered design professional.

❖ In contrast to other earthquake retrofit guidelines and codes, the provisions of this chapter are not strictly designed for life safety protection. These provisions are, in fact, expected to reduce property damage, reduce the number of uninhabitable dwellings and avoid increased public assistance expenditures to repair damage and provide temporary housing after an earthquake (ABAG, 1999).

These provisions do not guarantee that strengthened structures will not be damaged; however, it is anticipated that costly damage to the vulnerable parts of dwellings below the first floor will be greatly reduced. According to the Consortium of Universities for Research in Earthquake Engineering (CUREE): "Cripple walls retrofitted to these provisions are generally expected to meet a life safety performance objective" (CUREE, 2002). However, other parts of the building may exceed or perform worse than this objective.

Analysis and design of strengthening for other structural or nonstructural components not addressed by these provisions must be performed in accordance with Section A301.3.

[BS] A301.2 Scope. The provisions of this chapter apply to residential buildings of light-frame wood construction containing one or more of the structural weaknesses specified in Section A303.

Exception: The provisions of this chapter do not apply to the buildings, or elements thereof, listed below. These buildings or elements require analysis by a registered design professional in accordance with Section A301.3 to determine appropriate strengthening:

1. Group R-1, R-2 or R-4 occupancies with more than four dwelling units.

2. Buildings with a lateral force-resisting system using poles or columns embedded in the ground.

3. Cripple walls that exceed 4 feet (1219 mm) in height.

4. Buildings exceeding three stories in height and any three-story building with cripple wall studs exceeding 14 inches (356 mm) in height.

5. Buildings where the *code official* determines that conditions exist that are beyond the scope of the prescriptive requirements of this chapter.

6. Buildings or portions thereof constructed on concrete slabs on grade.

❖ The provisions are intended to deal with specific structural weaknesses; therefore, the work that is being done is structural in nature and generally requires submitting plans and obtaining a building permit. The weaknesses that these provisions address are listed in Section A303. It should be clearly understood that the application of these provisions is limited because they are not suitable to use for strengthening hotels, motels or large multiunit apartments, which are typically larger structures that require engineered retrofits by licensed design professionals. Also excluded are dwellings built with cripple walls with studs taller than 4 feet (1219 mm) in any location; dwellings that have columns or poles embedded in the ground as their foundation system; and buildings exceeding three stories or any three-story building with cripple wall studs exceeding 14 inches (356 mm) in height. Each of these types of buildings presents unique conditions that preclude the use of prescriptive criteria to strengthen them.

The 4-foot (1219 mm) cripple wall stud height limits the amount of overturning in braced crippled walls with lengths defined in Figure A3-10. When the height of the wall studs exceeds 4 feet (1219 mm), the owner will need to have the bracing designed by an engineer or architect. The 4-foot (1219 mm) height is also intended to limit the provisions to level or very low-slope sites. Bracing and anchorage of dwellings on moderate- to high-slope sites are beyond the scope of this chapter.

Conventional construction provisions Section 2308.2.1 of the *International Building Code*® (IBC®) state that cripple walls with studs exceeding 14 inches (356 mm) in height are to be considered as first-story walls for the purpose of determining bracing. However, the bracing layout requirements in these provisions (see Figure A3-10) are provided according to the number of stories above the cripple wall.

The code official is also allowed to exclude other residential buildings to which these provisions would otherwise apply if they have vertical or horizontal irregularities or other features not considered by these prescriptive standards. Very few dwellings are rectangular in plan. U-, T- or L-shaped plans are not necessarily problems in conventional construction with wood diaphragms. However, split-level dwellings, hillside dwellings or structures where the upper-level exterior walls are horizontally offset from the line of the lower story exterior walls may be determined by the code official to be beyond the scope of these provisions. It is recommended that the owner or contractor consult with the authority having jurisdiction to determine whether these provisions can be applied.

Observations after past earthquakes have shown that cripple walls with substantially varying heights, such as those with sloping or stepped foundations, suffer more damage than cripple walls of constant height (SEAOSC, 2002). Where the height of cripple walls varies significantly, such as the case of hillside homes with stepped foundation systems, seismic strengthening should be determined by a registered design professional. It should be noted that hillside homes are beyond the scope of this prescriptive chapter. In general, shorter sections of cripple walls along the same direction are stiffer and tend to resist more horizontal seismic forces than taller ones. For buildings with more than a 2:1 height ratio between the tallest and shortest cripple wall bracing panels, additional anchors may be prudent. Additional information on hillside homes can be found at http://peer.berkeley.edu/publications/peer_reports/reports_2000/0003.pdf.

Also excluded from these provisions are buildings or portions of buildings constructed on concrete slabs-on-grade since these dwellings do not have cripple stud walls. While these buildings may have wall anchorage deficiencies, retrofit of these structures for installation of sill bolts would require removal of wall finishes and may not be as cost effective as retrofitting buildings with crawl spaces. While sliding of dwellings on slab-on-grade foundations has occasionally occurred in past earthquakes, it has not caused widespread economic and habitation losses. Dwellings with stem walls (reinforced concrete or masonry foundation walls that project above the ground to the underside of the first floor framing), however, will experience substantial damage if the dwelling slides off the foundation, resulting in partial or full collapse.

A majority of dwellings constructed in California prior to 1950 were unanchored. The UBC did not begin to specify anchorage until its 1946 edition (SEAOC, 1995); however, most local governments did not uniformly adopt such model codes immediately after their publication until the late 1970s. Further, some dwellings were constructed with inadequate cripple wall bracing in the 1970s and even later, particularly where building codes were not enforced. In the past, dwellings often used horizontal wood siding or stucco as wall sheathing material. Owners of older dwellings should examine the exterior walls from within the crawl space under the first floor to determine if the sill plates are adequately bolted to the foundation with bolt size and spacing that comply with Table A3-A. They should also determine whether reliable wall sheathing such as wood structural panels or diagonal sheathing already exists. Straight sheathing (where 1x boards are oriented horizontally) typically does not qualify as having adequate strength. Plaster or stucco finishes that are not underlain with wood structural panels or diagonal sheathing may also perform poorly. The structural performance of these finishes, while initially strong, often cracks and quickly loses strength under earthquake motion.

Note that most dwellings constructed after 1950 were nominally anchored to their foundations; however, this anchorage is often considered inadequate when compared to current requirements in terms of size, spacing and the use of plate washers. This nominal anchorage was done simply to hold the mudsill in place during construction; it was not specifically designed to resist horizontal forces due to wind and earthquakes. Any existing anchors, if relied on, should be inspected to ensure they have not lost capacity due to significant corrosion or wood splitting or that they were not originally installed in large oversized holes in sill plates that will allow sliding or uneven load distribution. In many areas, it is customary to ignore the capacity of these existing anchors as the relative cost of additional anchors is minimal. In high seismic regions, cripple wall bracing now considered inadequate was in common use in the 1970s and 1980s.

Previous versions of the code required that these provisions be applied to light-framed wood structures in Seismic Design Categories C, D and E. These requirements were deleted in the 2012 edition. While it is likely that these provisions would only be applied to structures in these seismic design categories, it is plausible that they could also be used for buildings in Seismic Design Category B. It should be noted that this appendix chapter is used either to satisfy triggered upgrades via Section 301.1.4.2 or for a voluntary seismic upgrade allowed under Section 807.6. For both reasons, the scoping of the appendix chapter should not restrict the use of upgrades to certain seismic design categories. The sections of the code where seismic upgrade triggers occur, such as Section 606.2.2, 807.5 or 907.4, do not distinguish between seismic design categories, so neither should this chapter.

[BS] A301.3 Alternative design procedures. The details and prescriptive provisions herein are not intended to be the only acceptable strengthening methods permitted. Alternative details and methods shall be permitted to be used where approved by the code official. Approval of alternatives shall be based on a demonstration that the method or material used is at least equivalent in terms of strength, deflection and capacity to that provided by the prescriptive methods and materials.

Where analysis by a registered design professional is required, such analysis shall be in accordance with all requirements of the building code, except that the seismic forces may be taken as 75 percent of those specified in the building code.

❖ "Prescriptive" means these provisions apply to specific conditions and must be used in precisely the manner described. "Prescriptive" also means "determined in advance," without the need for case-specific analysis or design. Through the use of these provisions and the accompanying details, the dwelling owner or contractor can develop plans without the services of a design professional (civil engineer, structural engineer or architect); however, the use of other materials, proprietary systems or methods not shown by the figures and details in this chapter may require the services of a registered design professional. This section is not intended to allow an owner or contractor to bypass the requirements of the engineer of record's details.

This section intentionally omits the commonly used statement that the design must comply with all the requirements of the IBC because complete code compliance is often not feasible with respect to existing buildings. For example, design professionals should not be expected to rigorously address issues such as the stiffness variations in existing flooring systems due to differences in the type, thickness or composition of the flooring. However, any strengthening designed by the design professional should at least be equivalent in terms of strength and deflection to that provided by the prescriptive methods. The specific reference to "test data" was deleted from the 2012 edition. The code typically allows alternatives, and test data is not always required. Reasonable alternatives can also be justified on the basis of engineering analysis, proprietary data, conventional methods or other consensus documents often derived from test data, but not necessarily based on that data directly.

The 75-percent factor applied to IBC design forces accounts in a general way for differences between current design criteria and the less conservative criteria that were likely in effect when the dwelling was originally constructed. Historically, existing buildings have been granted a 25-percent reduction in forces—in part to encourage reasonable and cost-effective strengthening measures and to account for the presumed reduced remaining life of the structure compared to that for new structures.

Where the geometry and construction of a residential building falls within the scope of this chapter, but existing building conditions do not permit the use of those prescriptive fastening requirements provided as part of this chapter, it is acceptable to use alternative fastening requirements provided that they are: documented in sufficient detail to permit plan check and shown by calculation to have a capacity not less than the equivalent connection requirements in this chapter. It is not necessary to provide an engineering design for the overall building. This approach is intended to encourage retrofitting by permitting the construction of equivalent strengthening measures to the provisions of this chapter without requiring the expense of a full engineered design.

SECTION A302
DEFINITIONS

For the purpose of this chapter, in addition to the applicable definitions in the building code, certain additional terms are defined as follows:

❖ Throughout this chapter there are references to "the building code." This term generally refers to the current edition of the IBC or the jurisdiction's governing code.

[BS] ADHESIVE ANCHOR. An assembly consisting of a threaded rod, washer, nut, and chemical adhesive approved by the *code official* for installation in existing concrete or masonry.

[BS] CRIPPLE WALL. A wood-frame stud wall extending from the top of the foundation to the underside of the lowest floor framing.

[BS] EXPANSION ANCHOR. An approved post-installed anchor, inserted into a predrilled hole in existing concrete or masonry, that transfers loads to or from the concrete or masonry by direct bearing or friction or both.

[BS] PERIMETER FOUNDATION. A foundation system that is located under the exterior walls of a building.

[BS] SNUG-TIGHT. As tight as an individual can torque a nut on a bolt by hand, using a wrench with a 10-inch-long (254 mm) handle, and the point at which the full surface of the plate washer is contacting the wood member and slightly indenting the wood surface.

[BS] WOOD STRUCTURAL PANEL. A panel manufactured from veneers, wood strands or wafers or a combination of veneer and wood strands or wafers bonded together with waterproof synthetic resins or other suitable bonding systems. Examples of wood structural panels are:

Composite panels. A wood structural panel that is comprised of wood veneer and reconstituted wood-based material and bonded together with waterproof adhesive;

Oriented strand board (OSB). A mat-formed wood structural panel comprised of thin rectangular wood strands arranged in cross-aligned layers with surface layers normally arranged in the long panel direction and bonded with waterproof adhesive; or

Plywood. A wood structural panel comprised of plies of wood veneer arranged in cross-aligned layers. The plies are bonded with waterproof adhesive that cures on application of heat and pressure.

SECTION A303
STRUCTURAL WEAKNESSES

[BS] A303.1 General. For the purposes of this chapter, any of the following conditions shall be deemed a structural weakness:

1. Sill plates or floor framing that are supported directly on the ground without a foundation system that conforms to the building code.

2. A perimeter foundation system that is constructed only of wood posts supported on isolated pad footings.

3. Perimeter foundation systems that are not continuous.

   Exceptions:
   1. Existing single-story exterior walls not exceeding 10 feet (3048 mm) in length, forming an extension of floor area beyond the line of an existing continuous perimeter foundation.

   2. Porches, storage rooms and similar spaces not containing fuel-burning appliances.

4. A perimeter foundation system that is constructed of unreinforced masonry or stone.

5. Sill plates that are not connected to the foundation or that are connected with less than what is required by the building code.

   Exception: Where approved by the *code official*, connections of a sill plate to the foundation made with other than sill bolts may be accepted if the capacity of the connection is equivalent to that required by the building code.

6. Cripple walls that are not braced in accordance with the requirements of Section A304.4 and Table A3-A, or cripple walls not braced with diagonal sheathing or wood structural panels in accordance with the building code.

❖ This section provides criteria that allow owners and contractors to evaluate dwellings and identify the major areas of potential weakness below the first floor. The structural weaknesses listed in this section might not be the only weaknesses that may lead to structural damage when the building is subjected to earthquake forces; however, they represent conditions that can be addressed by prescriptive, nonengineered provisions that are most common and most cost effective to strengthen.

Sill Plates or Floor Framing without Approved Foundation System. Some older dwellings do not have a foundation system. Instead, the wall sill plate, and much of the floor framing, are supported directly on the ground. When subjected to earthquake-induced lateral and vertical forces, these structures can easily move because they are not adequately anchored. This movement can result in a variety of structural and nonstructural damage, including broken gas and utility lines. Damage to these utility lines can lead to fires. Further, these structures are highly susceptible to biologic growth and insect infestation due to inadequate separation between wood and earth.

Post and Pier Foundation Systems. Many dwellings have continuous cripple walls and foundation systems around their perimeter, but are supported on wood posts and isolated concrete pad footings (called a "post-and-pier" system) under the interior of the dwelling. Such a system is not generally assumed deficient since the continuous perimeter foundation system should provide ample resistance; however, if a post and pier system is directly below any perimeter walls, it is considered a structural weakness since they generally provide little horizontal resistance. This deficient foundation system is found in some older dwellings, including Victorian-era structures and buildings in areas where soil moisture is high.

Providing diagonal bracing members between the posts does not generally solve the problem. Each post would then need to be adequately connected to a competent foundation system. Typically, the existing footing pads are too small to make the necessary connections and lack the required strength and stiffness to resist lateral earthquake forces. Simply providing bracing between the posts only moves the point of failure from the top of the post to the bottom of the post at the pad footing. For these buildings, bracing, anchorage and additional footings may be required.

Some post-and-pier-type structures may be considered as historic buildings. Care needs to be taken when performing strengthening work so the historic fabric of the building is not adversely impacted. Many jurisdictions have adopted specific requirements for historic buildings, such as those in Chapter 12. If a dwelling utilizes this type of system and might be considered a historic building, consult with the building department.

Noncontinuous Perimeter Foundation System. Noncontinuous perimeter foundations are another deficiency often found in older residential structures. There are many variations of partial foundation systems, and some do not necessarily represent a significant weakness. When applying these provisions to an existing building, the intent is to reduce the potential for damage to habitable portions of structures. Therefore, the standards include two exceptions to the requirement for continuous perimeter foundations.

> **Exception 1:** Observations after past earthquakes have shown that failures of relatively small enclosed spaces tend to be localized and may not result in the remainder of the dwelling being uninhabitable. Consequently, small extensions or appendages do not require addition of a perimeter footing.
>
> **Exception 2:** This exception addresses nonhabitable spaces (i.e., porches, storage rooms and other similar areas), the failure of which is unlikely to result in the structure becoming uninhabitable. Failures of these structures rarely impact the habitability of the structure although they may temporarily limit access or egress. Storage rooms, in this exception, include those areas adjacent to the dwelling and accessible from the outside of the building only, and storage buildings or areas that are separate from the main structure. Attached rooms with water heaters or furnaces are not exempt since damage to these areas could result in fires.

Unreinforced Masonry Perimeter Foundation. A perimeter foundation constructed of unreinforced masonry is assumed to lack the necessary deformation capacity to resist earthquake forces and potential building displacements that often result from earthquakes. These systems are common in many older dwellings built before codes were adopted in high seismic regions and may also exist in newer dwellings where codes have not been enforced. Section A304.2.2 requires analysis of unreinforced masonry foundations by either an architect or an engineer.

Inadequate Sill Plate Anchorage. While sliding between an unanchored sill plate and the foundation can occur, it is one of the more uncommon sources of damage. Regardless, it is still important that sill plate anchorage be present in order to complete the lateral force path. Bracing cripple walls without bolting the sill plate to the foundation simply moves the weak link to the interface of the sill and foundation. Compliance with either Tables A3-A and B or the IBC is deemed acceptable. Care should also be exercised when depending on existing anchors to meet Table A3-A and A3-B. Existing anchors often have minimal capacity to resist sliding where deteriorated or installed in oversize holes. In addition, anchors are often missing nuts and washers.

> **Exception:** Because details common in new construction are often impractical in retrofits, the provisions recognize alternative means of anchorage, including proprietary products. These proprietary products are especially useful in conditions where limited clearance prevents installation of bolts through the sill plate in the traditional fashion. Generally, cripple wall studs need to be 24 inches (610 mm) or taller to install anchor bolts with a normal drill. In some cases, drills with right-angle attachments can be used in more confined spaces.

Inadequate Cripple Wall Bracing. Past earthquakes have shown that the most common cause of major damage in dwellings is poorly braced cripple walls.

A common and very weak type of cripple wall can be found in older buildings constructed with horizontal exterior wood siding. This type of siding and its nailing are not adequate to resist earthquake forces associated with nearby moderate or major earthquakes.

Recent earthquakes have also shown that "let-in" diagonal bracing does not adequately brace cripple walls (LA, 1994). Let-in braces are usually single, nominal 1-inch (25 mm) thick and placed in a notch cut into the face of the stud or placed on the face of the stud. Let-in braces are no longer permitted as an acceptable bracing method in the IBC for buildings located in regions where strong earthquakes are expected to occur.

Let-in braces should not be confused with diagonal wood sheathing. Diagonal wood sheathing, which is acceptable by these provisions, is composed of individual boards nominally 1-inch (25 mm) thick, laid diagonally across the face of the entire stud wall. These boards are laid next to one another covering the entire width and length of the wall extending from the top plate to the sill plate. If the cripple walls are covered with diagonal sheathing, the wall is assumed to be adequately braced, provided the boards are nailed to each stud they cross and to the top and bottom plates. Adequate nailing consists of three 8-penny nails at each stud and the ends. If the boards or the studs are split, or if the end nails are too close to the ends of the sheathing, this system can be deficient.

The most effective cripple wall bracing system that significantly reduces the risk of damage is wood structural panel sheathing [i.e., plywood or oriented strand board (OSB)] fully nailed around the sheet perimeters to each stud and to the top and bottom plates. It is critical that the structural sheathing be nailed along all edges; failure to nail the sheathing to the top plate or the sill plate can result in premature failure. If the dwelling has plywood sheathing as an exterior finish, check for nails spaced no more than 6 inches (152 mm) apart along all the edges of each sheet. If adequate nailing is not present, comply with the nailing requirements of this chapter.

Exterior plywood siding with vertical grooves (referred to as T1-11) can have another serious deficiency. At the edges where two adjacent panels adjoin, each panel must be nailed to the wall stud with a separate row of nails. These sheets have "lips" so that they overlap at the joints. A common, improper construction practice is providing only one row of nails through both sheets (at the overlap). This creates a weakness since the plywood thickness is only one-half of its normal thickness at the overlap and only half the number of nails are provided. Such practice led to failures in past earthquakes. In all cases where nailing is exposed to the elements, it is recommended that hot-dip galvanized nails be used.

Dwellings with existing Portland cement plaster (stucco) as the exterior finish may also be at risk since they often lack the toughness or durability to resist repeated accelerations. While stucco has been a recognized bracing material for a number of years, it is only as good as the connection of the lath to the studs and plates. Where cracks develop quickly in the stucco, these connections quickly degrade and cause loss of wall strength. Many dwellings with stucco applied directly over the studs without plywood or diagonal sheathing under the stucco experienced serious damage in the 1994 Northridge earthquake (LA, 1994) (NAHB, 1994). In high seismic regions, this failure most often was caused by inadequate attachment of the lath to the bottom sill plate (LA, 1994).

SECTION A304
STRENGTHENING REQUIREMENTS

[BS] A304.1 General.

[BS] A304.1.1 Scope. The structural weaknesses noted in Section A303 shall be strengthened in accordance with the requirements of this section. Strengthening work may include both new construction and alteration of existing construction. Except as provided herein, all strengthening work and materials shall comply with the applicable provisions of the building code.

❖ Use of materials, proprietary systems or methods not shown by the figures and details in this chapter requires the services of a design professional.

Where a dwelling has an unusual or irregular configuration or unusual features, the services of an engineer or architect to design a strengthening program utilizing the alternative procedures of Section A301.3 is required. The code official may require a predesign special inspection as described in Section A304.5 to determine which portions of the work require the services of a design professional. Alternative means of compliance are possible and addressed in Section A301.3.

Section A303 defines the most common structural weaknesses associated with these structures while Section A304 provides the means to "prescriptively" strengthen them. If a particular condition is not addressed by Section A304 then alternative design procedures may be used as defined in Section A301.3.

[BS] A304.1.2 Condition of existing wood materials. All existing wood materials that will be a part of the strengthening work (sills, studs, sheathing, etc.) shall be in a sound condition and free from defects that substantially reduce the capacity of the member. Any wood material found to contain fungus infection shall be removed and replaced with new material. Any wood material found to be infested with insects or to have been infested with insects shall be strengthened or replaced with new materials to provide a net dimension of sound wood at least equal to its undamaged original dimension.

❖ Wood decay, commonly known as "dry rot," can occur to wood framing exposed to dampness or to water leakage. Termite infestation is another cause of damage to wood members. All buildings being strengthened where damage is suspected should have a thorough inspection, but only those elements affected by retrofit need to be checked for the purposes of this chapter. If not repaired, dry rot and pest damage can weaken sill plates, studs and wood siding, and have a substantial adverse effect on a building's response to earthquakes.

Even dwellings without dry rot or termite damage may have weaknesses because of poor construction quality. For example, insufficient nailing of plywood, OSB or diagonal sheathing will result in a structure that is unable to resist the forces imposed during earthquakes. Simply repairing the weaknesses may not be adequate if the condition of the existing wood-framing members to be utilized is in doubt. It is recommended that the exposed wood be thoroughly inspected to ensure that all wood that is part of the strengthening work is in good condition. Existing wood members showing evidence of wood decay commonly referred to as dry rot, or evidence of insect infestation, must be removed and replaced. Wood decay can be found by probing the wood members with a sharp object, like a knife or awl. If the probe easily penetrates the wood, the member likely has experienced decay and is weakened. Sound wood will be difficult to probe. It is important to perform probing and visual observations prior to construction and to be alert for the presence of wood decay or insect infestation during construction as well. Concealed wood decay may be encountered when drilling for new sill anchors. If the drill suddenly moves through the wood, a pocket of wood decay has likely been encountered. The portion of the sill plate containing the decayed wood will need to be cut out and replaced with a new piece of sill plate. The new sill plate must be anchored in accordance with Tables A3-A and A3-B. When replacing pieces of sill plate, pressure-treated lumber will need to be used to protect the new member from future decay.

Wood decay occurs where wood is made continually or repeatedly damp by a leaking plumbing pipe or by repeated saturation and drying from an exterior wall that leaks during rains. Simply removing and replacing damaged wood will not necessarily prevent the decay from recurring. It is important to find the cause of the leak, repair it and allow the remaining wood to dry.

Insect infestation, on the other hand, stops damaging the wood once the infestation has been stopped. Consequently, a member that has been significantly damaged by insects does not need to be removed if it can be strengthened. The easiest method of strengthening is to add a new member next to the damaged member. Unfortunately, there are no clear guidelines to indicate when insect damage requires strengthening. This determination must be based on judgment gained through experience.

Prior to removing sill plates or studs for repair caused by fungal or infestation damage, temporary shoring may be installed. The design of this shoring must be carefully planned by a qualified design professional and installed by a contractor, both having shoring experience.

[BS] A304.1.3 Floor joists not parallel to foundations. Floor joists framed perpendicular or at an angle to perimeter foundations shall be restrained either by an existing nominal 2-inch-wide (51 mm) continuous rim joist or by a nominal 2-inch-wide (51 mm) full-depth block between alternate joists in one-and two-story buildings, and between each joist in three-story buildings. Existing blocking for multistory buildings must occur at each joist space above a braced cripple wall panel.

Existing connections at the top and bottom edges of an existing rim joist or blocking need not be verified in one-story buildings. In multistory buildings, the existing top edge connection need not be verified; however, the bottom edge connection to either the foundation sill plate or the top plate of a cripple wall shall be verified. The minimum existing bottom edge connection shall consist of 8d toenails spaced 6 inches (152 mm) apart for a continuous rim joist, or three 8d toenails per block. When this minimum bottom edge-connection is not present or cannot be verified, a supplemental connection installed as shown in Figure A3-8A or A3-8C shall be provided.

Where an existing continuous rim joist or the minimum existing blocking does not occur, new $^3/_4$-inch (19.1 mm) or $^{23}/_{32}$-inch (18 mm) wood structural panel blocking installed tightly between floor joists and nailed as shown in Figure A3-9 shall be provided at the inside face of the cripple wall. In lieu of wood structural panel blocking, tight fitting, full-depth 2-inch (51 mm) blocking may be used. New blocking may be omitted where it will interfere with vents or plumbing that penetrates the wall.

❖ For these strengthening procedures to be effective, there must be a continuous horizontal load path from the exterior walls to the foundation. Where floor joists are perpendicular to a cripple wall, or frame into the cripple wall at an angle, existing rim joists (or blocking) need to be connected to either the foundation sill plate (if there are no cripple stud walls) or the top plate of the cripple wall. When reviewing existing construction, if there is a connection between the rim joist and the plate that meets the nailing requirements of this section, the connection may be considered adequate for this link in the load path. Where these connections do not exist, new connections must be made. Rim joists will need to be toe-nailed with 8-penny $2^1/_2$-inch-long (64 mm) common nails, spaced 6 inches (152 mm) apart, through the joist into the plate. Blocking will need to be toe-nailed. Use of proprietary products for these connections may be easier and more effective than toe-nailing. When approved by the code official, these connections may be made by using products with current evaluation reports by an independent testing authority.

Because the forces in a single-story structure are relatively small, it is not necessary to verify these connections if the blocking or rim joists are present. In multistory buildings, the connections between the foundation and the blocking or rim joists must be verified. When these requirements are not met or cannot be verified, the provisions of this chapter apply.

In some cases, existing construction might not include a rim joist or blocking. In other cases, the members are smaller in width than a nominal 2 inches (51 mm) [$1^1/_2$ inches (38 mm)]. In these cases, a new nominal 2-inch (51 mm) wide full-depth joist or blocking or one of the methods described in the chapter may be used to provide the load path from the floor to the sill plate or cripple wall. Figures A3-8A, A3-8B and A3-8C assume that an existing "end" or "rim" joist is present and provides the necessary information by which to connect this "rim" or "end" joint to the cripple wall or mud sill. Where this 2x "rim" or "end" joist is not present, solid 2x blocking can be added to achieve the same result. Figure A3-9 utilizes plywood as an alternative to the addition of 2x solid blocking. In addition to providing a load path link, the rim joist or blocking provides rotational restraint for the ends of the floor joists. Note that these figures use the term "end joist" and "rim joist" interchangeably.

[BS] A304.1.4 Floor joists parallel to foundations. Where existing floor joists are parallel to the perimeter foundations, the end joist shall be located over the foundation and, except for required ventilation openings, shall be continuous and in continuous contact with the foundation sill plate or the top plate of the cripple wall. Existing connections at the top and bottom edges of the end joist need not be verified in one-story buildings. In multistory buildings, the existing top edge connection of the end joist need not be verified; however, the bottom edge connection to either the foundation sill plate or the top plate of a cripple wall shall be verified. The minimum bottom edge connection shall be 8d toenails spaced 6 inches (152 mm) apart. If this minimum bottom edge connection is not present or cannot be verified, a supplemental connection installed as shown in Figure A3-8B, A3-8C or A3-9 shall be provided.

❖ Where floor joists are parallel to a cripple wall, the same load path concept applies as with joists perpendicular to foundations. In this condition, the end floor joist must occur over the foundation wall or cripple wall and be connected. If this member is not connected to the plate, it will need to be toe-nailed with 8-penny common nails spaced 6 inches (152 mm) apart or with equivalent approved hardware. This connection need only be verified for multistory buildings, for which seismic forces are larger. If an end joist in a multistory building is not connected to the sill plate on top of the foundation, or this connection cannot be determined, the end joist may be connected to the sill plate with sheet metal angles (proprietary hardware is available). Where clearances do not permit installation of this angle, an alternative method using $^3/_4$-inch (19 mm) plywood attached to the foundation plate or cripple wall top plate and to the underside of the flooring as shown in Figure A3-9 may be used. Other alternatives are provided in Figure A3-8B or A3-8C. Figures A3-8B and A3-8C require a framing clip with a minimum horizontal capacity of 450 pounds (2002 N) where an existing rim joist or blocking nailing cannot be verified.

[BS] A304.2 Foundations.

[BS] A304.2.1 New perimeter foundations. New perimeter foundations shall be provided for structures with the structural weaknesses noted in Items 1 and 2 of Section A303. Soil investigations or geotechnical studies are not required for this work unless the building is located in a special study zone as designated by the *code official* or other authority having jurisdiction.

[BS] A304.2.2 Evaluation of existing foundations. Partial perimeter foundations or unreinforced masonry foundations shall be evaluated by a registered design professional for the force levels specified in Section A301.3. Test reports or other substantiating data to determine existing foundation material strengths shall be submitted to the *code official*. Where approved by the *code official*, these existing foundation systems may be strengthened in accordance with the recommendations included with the evaluation in lieu of being replaced.

Exception: In lieu of testing existing foundations to determine material strengths, and where approved by the *code official*, a new nonperimeter foundation system designed for the forces specified in Section A301.3 may be used to resist lateral forces from perimeter walls. A registered design professional shall confirm the ability of the existing diaphragm to transfer seismic forces to the new nonperimeter foundations.

❖ It might not be economical to replace existing partial perimeter foundations or unreinforced masonry foundations. In order to determine if existing foundation systems are adequate, an engineer or an architect should evaluate both the condition of the system and its ability to resist the prescribed forces. This analysis would be limited to the foundation system only. If other strengthening is to be performed, it must either be designed by a registered design professional or comply with the prescriptive provisions of this chapter. As specified in the exception, new interior foundation walls (nonperimeter foundations) can be used to resist lateral forces coming from the perimeter of the building. The design professional responsible for the new nonperimeter foundation should, however, check the floor diaphragm as part of this offset load path to ensure exterior lateral loads can be resisted at these interior locations.

[BS] A304.2.3 Details for new perimeter foundations. All new perimeter foundations shall be continuous and constructed according to either Figure A3-1 or A3-2. All new construction materials shall comply with the requirements of building code. Where approved by the *code official*, the existing clearance between existing floor joists or girders and existing grade below the floor need not comply with the building code.

Exception: Where designed by a registered design professional and approved by the *code official*, partial perimeter foundations may be used in lieu of a continuous perimeter foundation.

❖ The first three weaknesses listed in Section A303 involve buildings without complete foundation systems. These conditions will be resolved by installing a new concrete or masonry foundation system around the perimeter of the dwelling. It is the intent of these provisions that all new foundations meet the current minimum standards of the IBC.

The IBC has specific provisions for minimum clearance under the structure for both access and ventilation. Occasionally, existing construction does not provide the clearances that are required today. These provisions are not intended to require current code clearance when a new foundation must be installed. Excavating or raising the building would be extremely difficult and costly and would not improve its resistance to earthquakes. If substantial fungal or insect infestation has occurred in the past, the owner may want to consider measures to prevent future damage. In some cases, remedial work will be required in accordance with Section A304.1.2.

An existing partial concrete foundation (weakness Type 3 in Section A303) may be replaced, or it may be evaluated by a design professional to determine if it can perform in a manner equivalent to a continuous foundation. An existing unreinforced masonry or stone foundation (weakness Type 4) may be replaced with a new foundation that complies with the IBC, or it may be evaluated in accordance with Section A304.2.2. Replacement might be uneconomical or aesthetically displeasing.

A well-maintained unreinforced masonry foundation might be adequate to support a building for normal vertical loads, but its strength and ability to brace the building during earthquakes should be evaluated. Where anchor bolts are being added between the foundation sill plate and foundation, the shear stress in the plane of the foundation wall is generally small and acceptable. The foundation, however, must be adequate to resist these anchorage forces without local failure at the anchor. Unless retrofit design includes the addition of tie-down anchors, significant spanning of the foundation to carry these tension loads is not necessary. An evaluation of the foundation should primarily consider the ability of the foundation to develop anchorage loads. This may be evaluated by installing and tension testing anchors in representative locations. Where poor materials or foundation deterioration result in a foundation that is not able to develop these local anchorage loads, it will be necessary to replace, augment or otherwise modify the existing foundation. One approach that has been used in deteriorated masonry foundations is the removal of a section of foundation approximately 1 foot long by 1 foot high (305 mm by 305 mm) at each anchor bolt location. Concrete is then cast back in around each anchor bolt, providing a concrete key into the deteriorated foundation. Another approach is to use longer anchors.

Deteriorated foundations are often common in certain locations and the result of local geographical climate or soil conditions. Local building jurisdictions are often aware and even familiar with similar problems and may provide guidance for repair or strengthening. The code user is encouraged to contact the local building department to see if there is knowledge of similar concerns and whether any additional guidance or requirements apply.

An evaluation of a foundation should also seek to identify signs of foundation settlement or other movement. Where this is the case, recommendations from a geotechnical engineer for stabilization of the foundation may be needed before anchor bolts can be installed.

If an existing unreinforced masonry foundation is assumed not to resist earthquake forces, a new foundation bracing system must be provided that is independent of the existing foundation. The new system must be designed to resist all the earthquake forces from the building occurring at the foundation level. In this case, unreinforced masonry foundations do not require analysis and strengthening when an alternative foundation system is used.

Older existing concrete foundations are often not reinforced and commonly have cracks due to shrinkage or long-term differential settlement. Even new footings have shrinkage cracks. Common locations for these cracks are at corners and near changes in footing height or thickness, as well as more uniformly distributed over the length of the footing. Typical shrinkage cracks in footings are straight and vertical and have uniform narrow width. Isolated cracks less than $1/_8$ inch (3.2 mm) in width can be assumed not to significantly diminish the strength of the foundation (ATC, 2002). Where significant cracking or settlement of the foundation is present, the foundation shall by evaluated by an engineer. Additionally, a geotechnical engineer may be required for recommendations regarding stabilization of the foundation.

[BS] A304.2.4 New concrete foundations. New concrete foundations shall have a minimum compressive strength of 2,500 pounds per square inch (17.24 MPa) at 28 days.

[BS] A304.2.5 New hollow-unit masonry foundations. New hollow-unit masonry foundations shall be solidly grouted. The grout shall have minimum compressive strength of 2,000 pounds per square inch (13.79 MPa). Mortar shall be Type M or S.

[BS] A304.2.6 New sill plates. Where new sill plates are used in conjunction with new foundations, they shall be minimum 2× nominal thickness and shall be preservative-treated wood or naturally durable wood permitted by the building code for similar applications, and shall be marked or branded by an approved agency. Fasteners in contact with preservative-treated wood shall be hot-dip galvanized or other material permitted by the building code for similar applications. Anchors, that attach a preservative-treated sill plate to the foundation, shall be permitted to be of mechanically deposited zinc-coated steel with coating weights in accordance with ASTM B695, Class 55 minimum. Metal framing anchors in contact with preservative-treated wood shall be galvanized in accordance with ASTM A653 with a G 185 coating.

[BS] A304.3 Foundation sill plate anchorage.

[BS] A304.3.1 Existing perimeter foundations. Where the building has an existing continuous perimeter foundation, all perimeter wall sill plates shall be anchored to the foundation with adhesive anchors or expansion anchors in accordance with Table A3-A.

Anchors shall be installed in accordance with Figure A3-3, with the plate washer installed between the nut and the sill plate. The nut shall be tightened to a snug-tight condition after curing is complete for adhesive anchors and after expansion wedge engagement for expansion anchors. All anchors shall be installed in accordance with manufacturer's recommendations. Where existing conditions prevent anchor installations through the sill plate, this connection shall be made in accordance with Figure A3-4A, A3-4B or A3-4C. The spacing of these alternate connections shall comply with the maximum spacing requirements of Table A3-A. Expansion anchors shall not be used where the installation causes surface cracking of the foundation wall at the locations of the anchor.

❖ The provisions for connecting existing sill plates to existing foundations maintain the older code requirements for $^1/_2$-inch-diameter (12.7 mm) anchors spaced a maximum of 6 feet (1829 mm) apart for one-story buildings. Two- and three-story buildings need progressively more anchors because their height and added weight result in larger forces to be resisted. Five-eighth-inch-diameter (15.9 mm) anchors, which are more common for new construction, are required for three-story buildings and can be substituted for $^1/_2$-inch-diameter (12.7 mm) expansion anchors or adhesive anchors.

Expansion anchors and adhesive anchors are acceptable for connecting to existing concrete. These connectors, due to their shorter length of embedment into the concrete, have lower capacity in concrete than anchor bolts that are cast in place when the foundation is placed. However, even with the required 4-inch (102 mm) embedment, these connectors will have the same capacity in the wood sill plate, which is the weakest link in this connection. Consequently, properly installed expansion or adhesive anchors can provide the same resistance against sliding as cast-in-place anchors. An expansion anchor is effective when the hole is drilled to the correct size and is relatively clean, and when the bolt is properly tightened to set the expanding portion of the assembly in accordance with manufacturers' specifications. For expansion anchors to be fully effective, the foundation material must be able to engage the expansion portion without cracking. Where subsequent cracking indicates conditions of poor quality concrete or masonry during installation or limited edge distance, expansion anchors should not be used. If cracks are observed during installation, installation should be stopped and an anchor should be installed at a new location at least 1 foot (305 mm) away. If the problem continues, adhesive or screw-type anchors should be used instead. All anchors must be installed at least 9 inches (229 mm) away from the end of the sill plate in order to be effective.

Adhesive anchors are allowed for all types of foundations, but are required where existing concrete is in poor condition or when the installation of expansion anchors causes cracking of the concrete. An adhesive anchor utilizes a threaded rod with an epoxy-type substance to set the anchor. Adhesive anchors are most effective when the hole is the correct size and is clean. Concrete dust must be removed in accordance with manufacturers' specifications. A clean hole is more critical for adhesive anchors than for expansion-type anchors.

Some adhesives are viscous enough for use in horizontal holes, but others are too fluid and tend to drain out of the holes before setting. To avoid this problem, consult the current product evaluation report, such as ICC Evaluation Services® (ICC-ES®) reports on anchor systems, as well as manufacturers' instructions before purchasing.

These provisions require square steel plate washers to be placed between the nuts and the sill plate. Table A3-A provides the size of the plate washers required.

The provisions call for the nut to be tightened to a "snug-tight condition" after epoxy curing is complete or after the nut has been tightened to set an expansion anchor. Tightening the nut to set the expansion anchors and tightening the nut to connect the sill plate to the foundation are separate operations. The setting requirements of expansion anchors vary according to the bolt used. The specific bolt manufacturer's procedures must be closely followed to ensure that the bolt is properly set and is capable of transmitting forces into the foundation. Because these procedures vary, the provisions only address how tight the nut should be after an expansion anchor has been properly set or the adhesive has set. If the nuts are not tight against the washer plates, the anchors may not function as intended. The nut should be tightened to the point at which the full surface of the plate washer is in contact with the wood member and slightly indents the wood surface. This section also gives the code official the authority to spot test the nut tightness during the required inspection. Since adhesive anchors require a larger diameter hole to be drilled through the sill and into the foundation than the diameter of the bolt, these provisions require the annular space between the hole and the bolt to be filled with additional adhesive to help ensure there is no relative movement between the mudsill and anchor.

Figures A3-4A, A3-4B and A3-4C show side plates with adhesive or expansion anchors into the foundation and lag bolts into the narrow face of the existing wood sill plate. Proprietary systems with ICC-ES reports or other independent test reports may be used with code official approval in lieu of the connections shown if they provide an equal or greater capacity.

Figure A3-4C also shows the condition of a battered footing. This type of slanted face footing will require that a wood shim installed between the steel plate and the wood sill plate must be shaped so the steel plate will have full contact against the shim when the lag screws are tightened. Further, a beveled washer under the head of the lag screw is needed to ensure that it bears fully on the steel plate. It is recommended that the shim be nailed to the sill plate (in addition to the lag screws), but the nailing must not split the shim. Predrilling of holes at $^1/_{16}$ inch (1.6 mm) less than the nail diameter may be necessary. Alternative details, including proprietary anchors, such as the Simpson UFP10, may be easier and faster to install and should be acceptable in principle to the code official. Discuss potential alternatives with the building department or consider hiring a registered design professional to prepare an alternative for unique conditions.

[BS] A304.3.2 Placement of anchors. Anchors shall be placed within 12 inches (305 mm), but not less than 9 inches (229 mm), from the ends of sill plates and shall be placed in the center of the stud space closest to the required spacing. New sill plates may be installed in pieces where necessary because of existing conditions. For lengths of sill plates 12 feet (3658 mm) or greater, anchors shall be spaced along the sill plate as specified in Table A3-A. For other lengths of sill plate, anchor placement shall be in accordance with Table A3-B.

> **Exception:** Where physical obstructions such as fireplaces, plumbing or heating ducts interfere with the placement of an anchor, the anchor shall be placed as close to the obstruction as possible, but not less than 9 inches (229 mm) from the end of the plate. Center-to-center spacing of the anchors shall be reduced as necessary to provide the minimum total number of anchors required based on the full length of the wall. Center-to-center spacing shall be not less than 12 inches (305 mm).

❖ Careful attention needs to be given to the proper location and spacing of sill bolts. In order to ensure that the sills are properly connected, this section not only specifies the minimum spacing, but also limits the placement of bolts at the ends of pieces of a sill plate. These provisions differ from those in the IBC in requiring the bolts to be placed no closer than 9 inches (229 mm) from the end of the sill plate. When bolts are placed closer than 9 inches (229 mm) to the end of a plate, there is a potential for that bolt to split the sill from the bolt hole to the end of the plate as the bolt is loaded from earthquake forces. When the bolt is placed more than 12 inches (305 mm) from the end of the piece, there is a tendency for the end of the plate to lift due to overturning forces on the wall (CUREE, 2002). Placing the bolt between 9 and 12 inches (229 mm and 305 mm) from the end will minimize both tendencies. Some judgment may be needed where placement of the anchor in this very small region is precluded by existing conditions.

These provisions also address the fact that existing sill plates may be installed in short pieces, either where the foundation wall steps or where new pieces of a sill plate must be installed to replace sections damaged by wood decay or insect infestation. Therefore, the provisions specify a minimum number of bolts for various lengths of sill plate.

It will not always be possible to install sill bolts at the exact spacing. There are many existing elements that can interfere with their placement, such as a fireplace, plumbing or mechanical ducts. The provisions of this chapter have taken these field situations into account and allow that where physical obstructions exist, the bolts may be omitted. However, the spacing of the remaining bolts needs to be adjusted so that the same total number of bolts is installed as though the obstruction did not exist. It is recommended that if possible, the bolts with close spacing should coincide with the sheathing locations.

[BS] A304.3.3 New perimeter foundations. Sill plates for new perimeter foundations shall be anchored in accordance with Table A3-A and as shown in Figure A3-1 or A3-2.

❖ Table A3-A specifies the size, type and spacing of anchors, which varies depending on the number of stories above the cripple wall. The anchor bolts should be centered as much as possible on the sill plate while having sufficient edge distance in the concrete or masonry wall to preclude premature failure.

[BS] A304.4 Cripple wall bracing.

[BS] A304.4.1 General. Exterior cripple walls not exceeding 4 feet (1219 mm) in height shall be permitted to be specified by the prescriptive bracing method in Section A304.4. Cripple walls over 4 feet (1219 mm) in height require analysis by a registered design professional in accordance with Section A301.3.

❖ When bracing a cripple wall, consideration must be given to providing adequate resistance to both the horizontal forces and the tendency for uplifting the ends of the wall. Any wall panel that is subject to earthquake forces has a tendency to lift up at an end, as well as slide. This uplift can be resisted by one of two methods. In new construction, a "hold-down" anchor consisting of a heavy gauge metal angle is anchored or attached to a stud and also anchored into the concrete foundation. The use of hold-downs, however, would require the existing foundation system to resist significant vertical tension and compression forces, which might not be possible without associated damage. As such, the method used in this chapter is based on increasing the length-to-height ratio of the cripple wall bracing panels. By making the panels longer, more weight from the walls and floor above can be engaged to resist these uplift forces. The basic proportion required is that the minimum length of the braced cripple wall panel be at least two times its height [Residential Retrofit and Repair Committee (RRR), 1992]. In addition, longer panels are needed as the number of stories above the wall increases. This is simply because a taller building imposes larger horizontal forces on the braced cripple wall panel.

Where floor joists run parallel to the wall, the cripple wall element will not engage as much weight, so hold-downs at the ends of nonbearing walls may be prudent, especially for multistory buildings. Hold-downs, however, should only be used where the existing foundation is of adequate quality to locally develop the expected tension forces and where the foundation is of sufficient size to provide enough dead load capacity. Where cripple wall bracing panels can be connected at corners of buildings and installed in combined panel lengths longer than the minimum defined in Figure A3-10, the potential for wall overturning can be reduced. In addition, continuity provided by rim joists, plates and floor framing tends to create appreciable stability at the tops of the cripple walls that offsets wall overturning.

To stay within the limits of these prescriptive methods, a maximum of 4 feet (1219 mm) for the height of the cripple wall was established to limit overturning effects. When the height of the cripple wall exceeds 4 feet (1219 mm), the dwelling owner will need to have the bracing designed by a registered design professional.

[BS] A304.4.1.1 Sheathing installation requirements. Wood structural panel sheathing shall be not less than $^{15}/_{32}$-inch (12 mm) thick and shall be installed in accordance with Figure A3-5 or A3-6. All individual pieces of wood structural panels shall be nailed with 8d common nails spaced 4 inches (102 mm) on center at all edges and 12 inches (305 mm) on center at each intermediate support with not less than two nails for each stud. Nails shall be driven so that their heads are flush with the surface of the sheathing and shall penetrate the supporting member a minimum of $1^{1}/_{2}$ inches (38 mm). When a nail fractures the surface, it shall be left in place and not counted as part of the required nailing. A new 8d nail shall be located within 2 inches (51 mm) of the discounted nail and be hand-driven flush with the sheathing surface. Where the installation involves horizontal joints, those joints shall occur over nominal 2-inch by 4-inch (51 mm by 102 mm) blocking installed with the nominal 4-inch (102 mm) dimension against the face of the plywood.

Vertical joints at adjoining pieces of wood structural panels shall be centered on studs such that there is a minimum $^{1}/_{8}$ inch (3.2 mm) between the panels. Where required edge distances cannot be maintained because of the width of the existing stud, a new stud shall be added adjacent to the existing studs and connected in accordance with Figure A3-7.

❖ Plywood with a minimum thickness of $^{15}/_{23}$ inch (12 mm) with at least five plies or layers of wood is prescribed as the required sheathing because of observations of ruptured $^{3}/_{8}$-inch-thick (10 mm) (3 ply) plywood panels documented in the Modified Mercalli Intensity (MMI) VIII and IX intensity areas caused by the Northridge earthquake (LA, 1994).

Even though the provisions accept existing diagonal wood sheathing (see Section A303) as sheathing material for existing construction, it is no longer cost effective for strengthening weak cripple walls. The omission of this material was not based on its ability to resist lateral forces, as it has performed well in past earthquakes and high winds (LA, 1992). Instead, it was based on cost considerations and practicality, since this type of sheathing is more time consuming to install and more expensive than wood structural panels. Proprietary bracing methods may also be used when approved by the code official.

The most important component of wood structural sheathing is proper nailing. To prevent splitting of existing wood framing, 8-penny common nails are considered optimum for 2x material. If splitting of studs is observed, predrilling of holes is recommended. Predrilled holes should have about three-fourths of the diameter of the nail. Nail guns tend to produce less splitting than hand nailing. Minimum edge distance for nails should be maintained for plywood, the wood studs and top and sill plates to prevent splitting or premature nail failure. With this size nail, 4-inch (102 mm) spacing provides adequate capacity with the minimum bracing length permitted by Table A3-A and Figure A3-10. Further, using larger nails or closer spacing would, by comparison of capacity, require larger diameter sill anchors or closer spacing of the anchors than specified in Section A304.3.2.

In the code, the nail edge distance requirement was decreased from $^{1}/_{2}$ inch to $^{3}/_{8}$ inch. A $^{3}/_{8}$-inch edge distance in the plywood or OSB is considered adequate for this application. The engineer should be aware that in existing buildings with 2x members, satisfying edge distances for both the sheathing and framing may not be possible. The engineer should consider slanting the nails or predrilling to protect the existing framing.

When plywood is installed on the inside face of cripple walls with an exterior surface of stucco, care must be used to prevent damage to the stucco.

If a nail gun is used, the operator must make sure that the nail heads do not penetrate the surface of the plywood. Local variations in the density of the backing (new or existing wood-framing members) can create situations where it will be difficult to maintain consistent nail penetration. The use of a flush head attachment on a nailing gun will usually prevent overdriving. When a nail head fractures the plywood surface, the amount of force that this particular connection is capable of resisting is reduced significantly. It becomes much easier for the nail head to pull through the sheathing material. Whenever a nail head fractures the surface of the sheathing, the nail must be discounted.

When purchasing structural sheathing, one of the structural grades must be stamped on the sheets used. For installation in a raised floor space subject to moisture, exterior grade plywood should be used. While OSB board is also permitted within these provisions, it is generally not recommended where moisture can be present. The user is encouraged to refer to the IBC for more information.

[BS] A304.4.2 Distribution and amount of bracing. See Table A3-A and Figure A3-10 for the distribution and amount of bracing required for each wall line. Each braced panel length must be at least two times the height of the cripple stud. Where the minimum amount of bracing prescribed in Table A3-A cannot be installed along any walls, the bracing must be designed in accordance with Section A301.3.

Exception: Where physical obstructions such as fireplaces, plumbing or heating ducts interfere with the placement of cripple wall bracing, the bracing shall then be placed as close to the obstruction as possible. The total amount of bracing required shall not be reduced because of obstructions.

❖ Table A3-A and Figure A3-10 note the location and distribution of bracing panels based on the number of stories and cripple stud height. The amount of bracing described is for each wall line. Bracing panels are required at or near each end of each wall line.

[BS] A304.4.3 Stud space ventilation. When bracing materials are installed on the interior face of studs forming an enclosed space between the new bracing and the existing exterior finish, each braced stud space must be ventilated. Adequate ventilation and access for future inspection shall be provided by drilling one 2-inch to 3-inch-diameter (51 mm to 76 mm) round hole through the sheathing, nearly centered between each stud at the top and bottom of the cripple wall. Such holes should be spaced a minimum of 1 inch (25 mm) clear from the sill or top plates. In stud spaces containing sill bolts, the hole shall be located on the center line of the sill bolt but not closer than 1 inch (25 mm) clear from the nailing edge of the sheathing. When existing blocking occurs within the stud space, additional ventilation holes shall be placed above and below the blocking, or the existing block shall be removed and a new nominal 2-inch by 4-inch (51 mm by 102 mm) block shall be installed with the nominal 4-inch (102 mm) dimension against the face of the plywood. For stud heights less than 18 inches (457 mm), only one ventilation hole need be provided.

❖ The most common form of cripple wall bracing will be to add sheathing to the interior face of the cripple wall from within the crawl space. When this is done, a closed space is created between each stud that does not allow natural ventilation. This can result in a buildup of moisture that may lead to wood decay. In order to protect these concealed spaces from decay, 2- to 3-inch diameter (51 mm to 76 mm) ventilation holes must be provided at the top and bottom of each stud space (see Figure A3-7). These ventilation holes will allow the free movement of air within the stud space, thereby minimizing the risk of future decay. When 2x horizontal blocking is needed in the stud space to provide backing for panel joint nailing, it must be installed with the wide face oriented vertically flush with the face of the stud on which the sheathing is being installed. This can be easily accomplished using commercially available fence rail hardware at each end to attach the block to the studs. This will eliminate blockage of ventilation inside the stud space. If the blocking is installed such that it blocks ventilation inside the stud space, ventilation holes must be provided above and below the blocking to ventilate all portions of the stud space.

Ventilation holes should be round wherever possible. Hole-cutting tools are available to cut the properly sized hole. Square holes, or other shapes with sharp corners or notches, can result in high concentrations of stress when the panel is loaded.

[BS] A304.4.4 Existing underfloor ventilation. Existing underfloor ventilation shall not be reduced without providing equivalent new ventilation as close to the existing ventilation as possible. Braced panels may include underfloor ventilation openings when the height of the opening, measured from the top of the foundation wall to the top of the opening, does not exceed 25 percent of the height of the cripple stud wall; however, the length of the panel shall be increased a distance equal to the length of the opening or one stud space minimum. Where an opening exceeds 25 percent of the cripple wall height, braced panels shall not be located where the opening occurs. See Figure A3-7.

Exception: For homes with a post and pier foundation system where a new continuous perimeter foundation system is being installed, new ventilation shall be provided in accordance with the building code.

❖ Air circulation under the floor protects the framing from wood decay.

[BS] A304.5 Quality control. All work shall be subject to inspection by the *code official* including, but not limited to:

1. Placement and installation of new adhesive or expansion anchors installed in existing foundations. Special inspection is not required for adhesive anchors installed in existing foundations regulated by the prescriptive provisions of this chapter.

2. Installation and nailing of new cripple wall bracing.

3. Any work may be subject to special inspection when required by the *code official* in accordance with the building code.

❖ Strengthening work is only as good as the quality of the construction. In most jurisdictions, the strengthening work required by this chapter will require building permits and inspections. Prior to requesting a permit, the owner or contractor should survey and determine all existing conditions, dimensions and other considerations significant to the retrofit or repair work. A plan should be prepared showing the location of the proposed sheathing and spacing of anchors. The drawings should differentiate between new and existing components. Because of the nature of the work being performed and the materials being used, there are some additional inspections that need to be performed that are not specified in the IBC for new construction.

Placement and Installation of New Adhesive and Expansion Anchors. The code official must approve the use of expansion and adhesive anchors. Code officials often use evaluation reports from agencies such as the ICC-ES as a guide for the products they approve. These reports set allowable design values based on testing and specify requirements for construction quality control. They often call for special inspection, especially for bolts that might be subject to tension forces. Special inspection generally involves inspection of the work while it is being performed, as opposed to when it is complete, and, in some cases, actual testing of the anchor. It is also performed by qualified individuals retained by the owner.

For the purposes of this appendix chapter, special inspection is not required because the anchors in question are intended to act primarily in shear, not in tension; these anchors tend to work fairly well in shear whether or not they are installed perfectly. The waiving of special inspection, thus, represents a justifiable cost savings. While special inspection is not typically required, the code official may still require verification of proper installation in accordance with Section A304.3.1.

In lieu of special inspection, it is recommended that a post-installation torque test for expansion anchors be done together with the inspection for anchor spacing, end distance and a spot check to make sure the nuts are properly tightened. Usually, this inspection would be performed after the anchors were installed and before the cripple wall sheathing is placed.

Both expansion and adhesive anchors must be approved, as stated in the definitions in Section A302. This means that they generally must have a valid evaluation report from agencies such as the ICC-ES or an approved equivalent independent test report. Normally, adhesive anchors require continuous inspection during their installation as a part of their approval for use. Continuous inspection checks that the hole is the correct diameter and depth and is sufficiently clean prior to placing the epoxy material. The purpose of all the checks is to ensure that each anchor will attain the tension strength allowed by its evaluation report.

Installation and Nailing of New Cripple Wall Bracing. It is important to make sure that the connections of the wall bracing panels are installed correctly and completely. The bracing serves no purpose if the connections do not engage the proper framing members or, in the case of wood sheathing, are overdriven. The nailing of the new structural elements should be inspected to ensure that they meet the minimum requirements of this chapter.

Work May be Subject to Special Inspection. The code official may require a special inspection where conditions on a particular job site make inspections difficult. Retrofit and strengthening work often involve unusual conditions or proprietary components that require additional levels of quality control. In order to address these problems, the code official is allowed to require that a special inspector verify work.

Most dwellings have some features that will not conform to the conditions and strengthening provisions used in this chapter. To avoid situations where those existing conditions either preclude the use of these prescriptive provisions or where other complications may occur that would make their application to a specific building difficult, the code official is encouraged to perform a predesign inspection. The cost of this inspection, however, may be in addition to the normal permit fee. Such an inspection might not be necessary if the owner provides adequate drawings supplemented by photographs to permit adequate review by the code official.

The purpose of a predesign inspection is to notify the owner or contractor of problems that may need the services of design professionals. It is not intended to be a consulting service to the owner. Typically, this inspection should focus on the following issues:

1. Areas where obstructions in the crawl space along exterior walls might prevent installation of adequate lengths of bracing.

2. Areas that may be questionable with respect to insect or fungal damage to wood members to be used in the strengthening.

3. Foundations that may be questionable or too weak to be effectively used for anchoring sill plates.

4. Tests of nut snug tightness on sill anchors.

5. Inadequate rim joist or blocking conditions or unusual framing conditions that would require alterations to the prescriptive details indicated.

6. Other concerns that the owner or contractor believes will preclude the use of the prescriptive details or methods described in this chapter.

Phasing:

With the publication of the 2012 code, the former section on phasing (Section A304.6) was removed, as it was believed to be both unnecessary and costly. First, work may always be phased or sequenced at the discretion of the code official. The current provision is intended to ensure that the greatest seismic risks are reduced before other work is performed. But the probability that a significant earthquake occurs during the project is extremely low, even if the work is phased. Since this chapter is triggered by other provisions in the body of the code, the seismic improvements it requires will already be part of the project scope and should not inadvertently be delayed or left incomplete. By requiring a specific sequence of work, the provision could unnecessarily restrict an owner from completing work in the least disruptive or expensive way.

[BS] A304.5.1 Nails. All nails specified in this chapter shall be common wire nails of the following diameters and lengths: 8d nails shall be 0.131 inch by $2^1/_2$ inches. 10d nails shall be 0.148 inch by 3 inches. 12d nails shall be 0.148 inch by $3^1/_4$ inches. 16d nails shall be 0.162 inch by $3^1/_2$ inches. Nails used to attach metal framing connectors directly to wood members shall be as specified by the connector manufacturer in an approved report.

❖ The provisions of this chapter make many references to nails, but do not always indicate "common wire nails." Many different types of nails are sold at lumber supply yards and hardware stores, so choosing the appropriate type of nail may be confusing. Further, the substitution of inadequate nails is commonplace, where gage and length are not specified. This section clarifies what is required.

Bibliography

The following resource materials were used in the preparation of the commentary for this chapter of the code.

ABAG 99, *Preventing the Nightmare—Designing a Model Program to Encourage Owners of Homes and Apartments to Do Earthquake Retrofits*. Association of Bay Area Governments, 1999.

ABAG 2000, *Preventing the Nightmare—Post-earthquake Housing Issue Papers*. Association of Bay Area Governments, 2000.

ATC 50, *Simplified Seismic Assessment of Detached, Single-family, Wood-frame Dwellings*. Applied Technology Council, 2002.

ATC 50-1, *Seismic Retrofit Guidelines for Detached, Single-family Wood-frame Dwellings*. Applied Technology Council, 2002.

ATC 50-2, *Safer at Home in Earthquakes: A Proposed Earthquake Safety Program*. Applied Technology Council, 2002.

CEA 2002, *Earthquake Insurance Retrofit Premium Discounts, Rates and Premiums*. Policy Information, California Earthquake Authority, http://www.earthquakeauthority.com/rates/rates_premiums.html#top, 2002.

Comerio, Mary G., John D. Landis, Cathurin J. Fispo and Juan Pablo Manzan. *Residential Earthquake Recovery—Improving California's Post-disaster Rebuilding Policies and Programs*. Berkeley, CA: California Policy Seminar, 1996.

CSSC-98, *Homeowner's Guide to Earthquake Safety*. California Seismic Safety Commission, 2002.

CUREE 2002, *Recommendations for Earthquake Resistance in the Design and Construction of Wood-frame Buildings*. Consortium of Universities for Research in Earthquake Engineering, 2002 Final Review Draft.

DSA 2002, *Guidelines for Earthquake Bracing of Residential Water Heaters. DSA Review and Acceptance Procedures—Manufactured Earthquake Bracing Systems for Residential Water Heaters*. Division of the State Architect, Department of General Services, State of California, 2002.

FEMA 1995, *Seismic Retrofit Training for Building Contractors and Building Inspectors—Participant Handbook*. Wiss Janney, Elstner Assoc., EMW-92-C3852, 1995.

Gallagher, Ron. *Mitigation of Principal Earthquake Hazards in Wood-frame Dwellings and Mobile Dwellings*. California Department of Insurance, June 1990.

IBHS 1999, *Is Your Home Protected from Earthquake Disaster—A Homeowner's Guide to Earthquake Retrofit*. Institute for Building and Home Safety, 1999.

LA 1994, *Prescriptive Provisions for Seismic Strengthening of Light, Wood-frame, Residential Buildings*. Los Angeles: Department of Building and Safety, Joint City of Los Angeles/SEAOSC Subcommittee on Cripple Wall Buildings, 1994.

LA 2001, *Reconstruction and Replacement of Earthquake Damaged Masonry Chimneys*. Los Angeles: Department of Building & Safety, 2001.

NAHB 1994, *Assessment of Damage to Residential Buildings Caused by the Northridge Earthquake*. National Association of Home Builders, 1994.

NIST 1997, *Fire Hazards and Mitigation Measures Associated with Seismic Damage of Water Heaters*. Mroz and Soong, National Institute of Standards and Technology, 1997.

RRR 1992, *Prescriptive Provisions for Voluntary Seismic Strengthening of Light, Wood-frame Residential Buildings, (with Commentary)*. Residential Retrofit and Repair Committee Jointly sponsored by California Office of Emergency Services, California Building Officials, American Institute of Architects—California Council, Structural Engineers Association of California, State Historical Building Safety Board, available from California Seismic Safety Commission, 1994.

SEAOC 1995, *History of Conventional Construction*. Structural Engineers Association of California, 1995.

SEAOSC 2002, *Seismic Retrofit of Existing Buildings - Retrofit of Hillside Dwellings*. Seminar Notes, Nels Roselund, *Prescriptive Provisions for Seismic Strengthening of Cripple Walls*. Ben Schmid, June 2002.

SL 2002, City of San Leandro, CA *Retrofit Plan Set*, http://www.sanleandro.org/depts/cd/bldg/retrofit/default.asp, San Leandro, 2002.

[BS] TABLE A3-A
SILL PLATE ANCHORAGE AND CRIPPLE WALL BRACING

NUMBER OF STORIES ABOVE CRIPPLE WALLS	MINIMUM SILL PLATE CONNECTION AND MAXIMUM SPACING[a, b]	AMOUNT OF BRACING FOR EACH WALL LINE[c,d,e]	
		A Combination of Exterior Walls Finished with Portland Cement Plaster and Roofing Using Clay Tile or Concrete Tile Weighing More than 6 psf (287 N/m²)	All Other Conditions
One story	$1/2$ inch (12.7 mm) spaced 6 feet, 0 inch (1829 mm) center-to-center with washer plate	Each end and not less than 50 percent of the wall length	Each end and not less than 40 percent of the wall length
Two stories	$1/2$ inch (12.7 mm) spaced 4 feet, 0 inch (1219 mm) center-to-center with washer plate; or $5/8$ inch (15.9 mm) spaced 6 feet, 0 inch (1829 mm) center-to-center with washer plate	Each end and not less than 70 percent of the wall length	Each end and not less than 50 percent of the wall length
Three stories	$5/8$ inch (15.9 mm) spaced 4 feet, 0 inch (1219 mm) center-to-center with washer plate	100 percent of the wall length[f]	Each end and not less than 80 percent of the wall length[f]

a. Sill plate anchors shall be adhesive anchors or expansion anchors in accordance with Section A304.3.1.
b. All washer plates shall be 3 inches by 3 inches by 0.229 inch minimum. The hole in the plate washer is permitted to be diagonally slotted with a width of up to $3/16$ inch larger than the bolt diameter and a slot length not to exceed $1 3/4$ inches, provided a standard cut washer is placed between the plate washer and the nut.
c. See Figure A3-10 for braced panel layout.
d. Braced panels at ends of walls shall be located as near to the end as possible.
e. All panels along a wall shall be nearly equal in length and shall be nearly equal in spacing along the length of the wall.
f. The minimum required underfloor ventilation openings are permitted in accordance with Section A304.4.4.

[BS] TABLE A3-B
SILL PLATE ANCHORAGE FOR VARIOUS LENGTHS OF SILL PLATE[a,b]

NUMBER OF STORIES	LENGTHS OF SILL PLATE		
	Less than 12 feet (3658 mm) to 6 feet (1829 mm)	Less than 6 feet (1829 mm) to 30 inches (762 mm)	Less than 30 inches (762 mm)[c]
One story	Three connections	Two connections	One connection
Two stories	Four connections for $1/2$-inch (12.7 mm) anchors or bolts or three connections for $5/8$-inch (15.9 mm) anchors or bolts	Two connections	One connection
Three stories	Four connections	Two connections	One connection

a. Connections shall be either adhesive anchors or expansion anchors.
b. See Section A304.3.2 for minimum end distances.
c. Connections shall be placed as near to the center of the length of plate as possible.

APPENDIX A

NUMBER OF STORIES	MINIMUM FOUNDATION DIMENSIONS					MINIMUM FOUNDATION REINFORCING	
	W	F	$D^{a, b, c}$	T	H	VERTICAL REINFORCING	
						Single-pour wall and footing	Footing placed separate from wall
1	12 inches (305 mm)	6 inches (152 mm)	12 inches (305 mm)	6 inches (152 mm)	≤ 24 inches (610 mm)	#4 @ 48 inches (1219 mm) on center	#4 @ 32 inches (813 mm) on center
2	15 inches (381 mm)	7 inches (178 mm)	18 inches (457 mm)	8 inches (203 mm)	≥ 36 inches (914 mm)	#4 @ 48 inches (1219 mm) on center	#4 @ 32 inches (813 mm) on center
3	18 inches (457 mm)	8 inches (203 mm)	24 inches (610 mm)	10 inches (254 mm)	≥ 36 inches (914 mm)	#4 @ 48 inches (1219 mm) on center	#4 @ 18 inches (457 mm) on center

a. Where frost conditions occur, the minimum depth shall extend below the frost line.
b. The ground surface along the interior side of the foundation may be excavated to the elevation of the top of the footing.
c. Where the soil is designated as expansive, the foundation depth and reinforcement shall be approved by the code official.

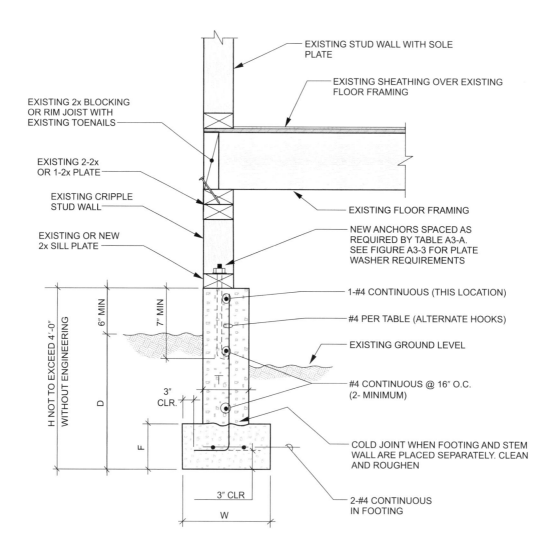

For SI: 1 inch = 25.4 mm, 1 foot = 304.8 mm.

[BS] FIGURE A3-1
NEW REINFORCED CONCRETE FOUNDATION SYSTEM

APPENDIX A

NUMBER OF STORIES	MINIMUM FOUNDATION DIMENSIONS					MINIMUM FOUNDATION REINFORCING	
	W	F	D[a, b, c]	T	H	VERTICAL REINFORCING	HORIZONTAL REINFORCING
1	12 inches (305 mm)	6 inches (152 mm)	12 inches (305 mm)	6 inches (152 mm)	≤ 24 inches (610 mm)	#4 @ 24 inches (610 mm) on center	#4 continuous at top of stem wall
2	15 inches (381 mm)	7 inches (178 mm)	18 inches (457 mm)	8 inches (203 mm)	≥ 24 inches (610 mm)	#4 @ 24 inches (610 mm) on center	#4 @ 16 inches (406 mm) on center
3	18 inches (457 mm	8 inches (203 mm)	24 inches (610 mm)	10 inches (254 mm)	≥ 36 inches (914 mm)	#4 @ 24 inches (610 mm) on center	#4 @ 16 inches (406 mm) on center

a. Where frost conditions occur, the minimum depth shall extend below the frost line.
b. The ground surface along the interior side of the foundation may be excavated to the elevation of the top of the footing.
c. Where the soil is designated as expansive, the foundation depth and reinforcement shall be approved by the code official.

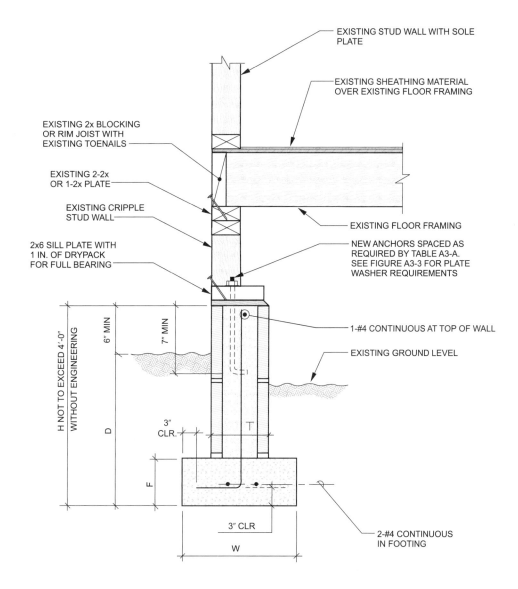

For SI: 1 inch = 25.4 mm, 1 foot = 304.8 mm.

[BS] FIGURE A3-2
NEW MASONRY CONCRETE FOUNDATION

APPENDIX A

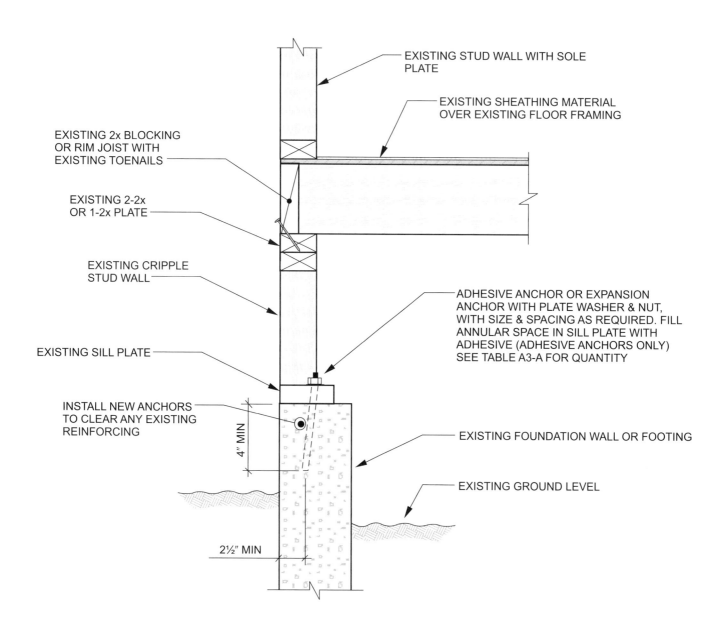

For SI: 1 inch = 25.4 mm.
NOTES:
1. Plate washers shall comply with the following:

   $^1/_2$-inch anchor or bolt—3 inches × 3 inches × 0.229 inch minimum.

   $^5/_8$-inch anchor or bolt—3 inches × 3 inches × 0.229 inch minimum.

   A diagonal slot in the plate washer is permitted in accordance with Table A3-A, Footnote b.
2. See Figure A3-5 or A3-6 for cripple wall bracing.

**[BS] FIGURE A3-3
SILL PLATE BOLTING TO EXISTING FOUNDATION**

APPENDIX A

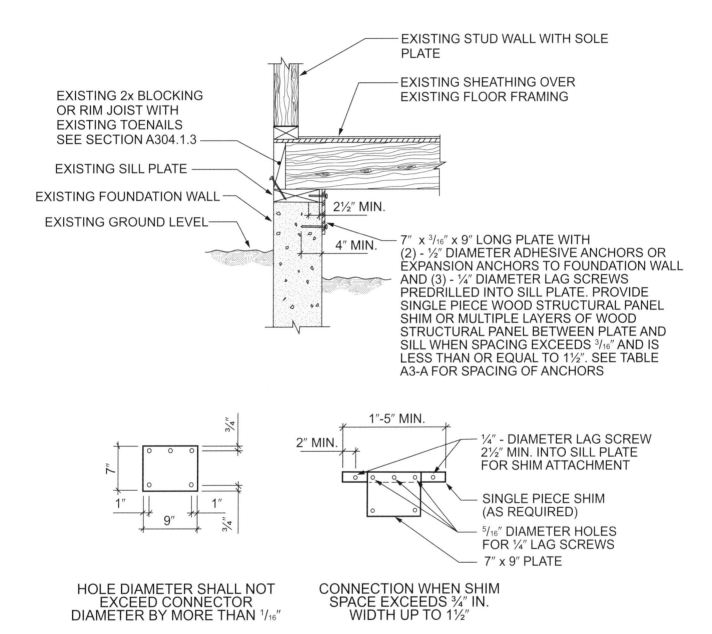

For SI: 1 inch = 25.4 mm, 1 foot = 304.8 mm.
NOTES:
1. If shim space exceeds 1 1/2 inches, alternate details will be required.
2. Where required, single piece shim shall be naturally durable wood or preservative-treated wood. If perservative-treated wood is used, it shall be isolated from the foundation system with a moisture barrier.

[BS] FIGURE A3-4A
ALTERNATE SILL PLATE ANCHORING IN EXISTING FOUNDATION—
WITHOUT CRIPPLE WALLS AND FLOOR FRAMING NOT PARALLEL TO FOUNDATIONS

APPENDIX A

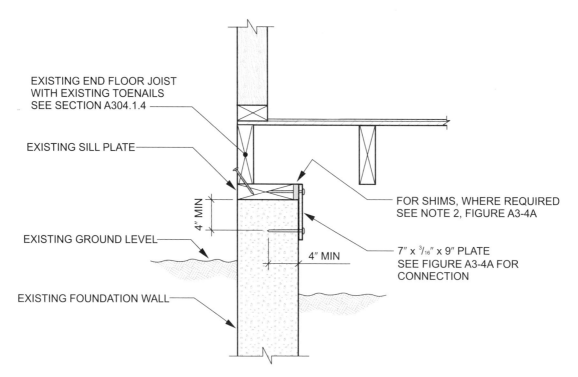

For SI: 1 inch = 25.4 mm.

[BS] FIGURE A3-4B
ALTERNATE SILL PLATE ANCHOR TO EXISTING FOUNDATION WITHOUT CRIPPLE
WALL AND FLOOR FRAMING PARALLEL TO FOUNDATIONS

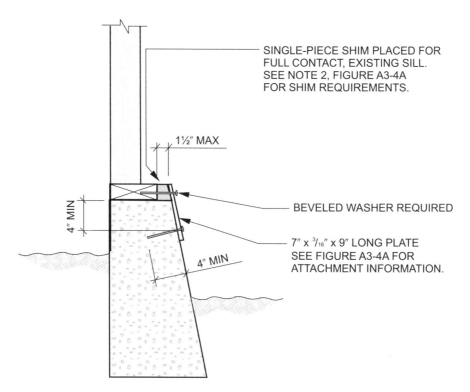

For SI: 1 inch = 25.4 mm.

[BS] FIGURE A3-4C
SILL PLATE ANCHORING TO EXISTING FOUNDATION—ALTERNATE CONNECTION FOR BATTERED FOOTING

APPENDIX A

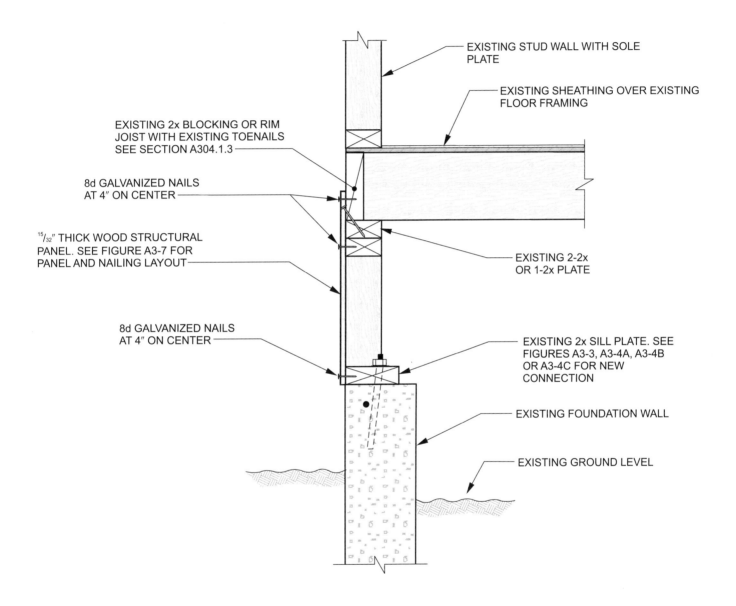

For SI: 1 inch = 25.4 mm.
NOTE: See Figure A3-3 for sill plate anchoring.

[BS] FIGURE A3-5
CRIPPLE WALL BRACING WITH NEW WOOD STRUCTURAL PANEL ON EXTERIOR FACE OF CRIPPLE STUDS

APPENDIX A

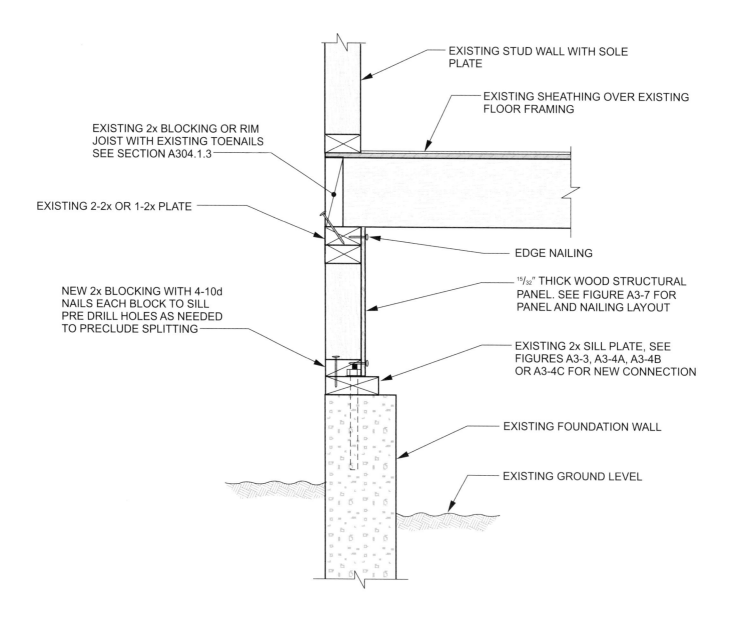

For SI: 1 inch = 25.4 mm.

[BS] FIGURE A3-6
CRIPPLE WALL BRACING WITH WOOD STRUCTURAL PANEL ON INTERIOR FACE OF CRIPPLE STUDS

APPENDIX A

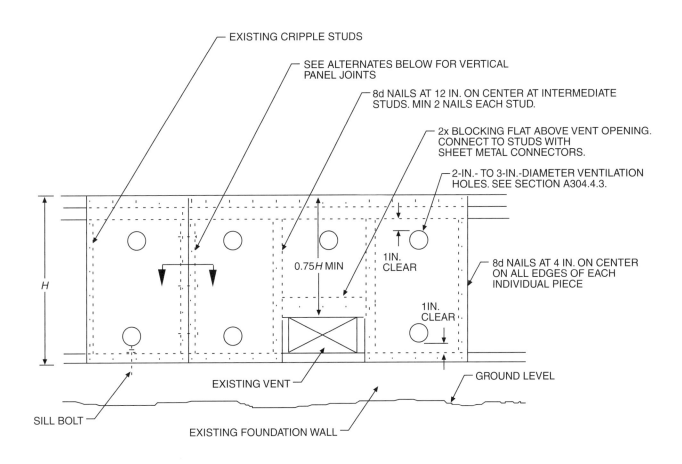

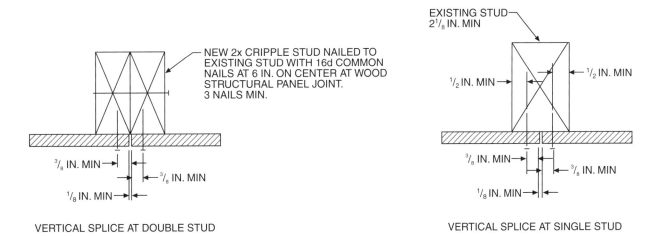

VERTICAL SPLICE AT DOUBLE STUD

VERTICAL SPLICE AT SINGLE STUD

For SI: 1 inch = 25.4 mm.

[BS] FIGURE A3-7
PARTIAL CRIPPLE STUD WALL ELEVATION

APPENDIX A

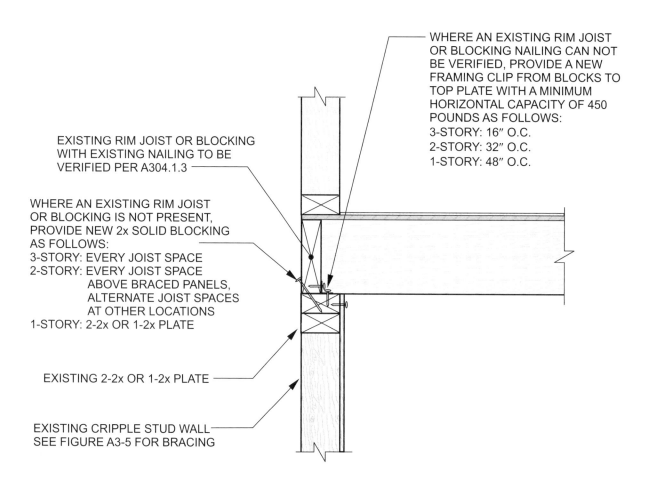

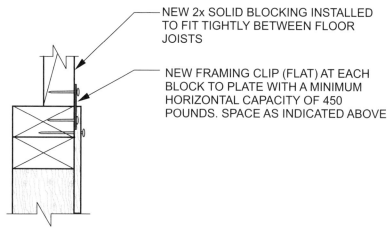

ALTERNATE DETAIL FOR FLUSH CONDITION

For SI: 1 inch = 25.4 mm, 1 pound = 4.4 N.
NOTE: See manufacturing instructions for nail sizes associated with metal framing clips.

[BS] FIGURE A3-8A
TYPICAL FLOOR TO CRIPPLE WALL CONNECTION (FLOOR JOISTS NOT PARALLEL TO FOUNDATIONS)

APPENDIX A

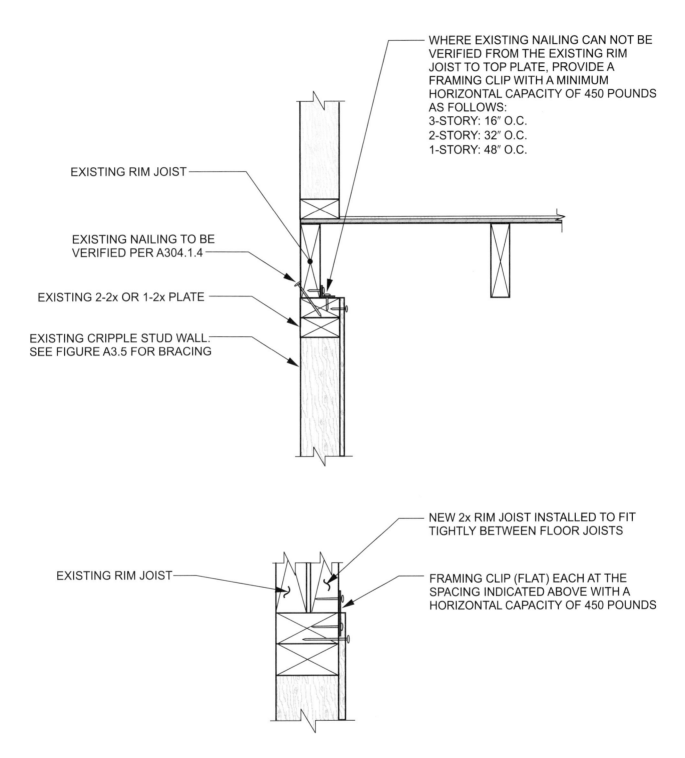

For SI: 1 inch = 25.4 mm, 1 pound = 4.4 N.
NOTE: See manufacturing instructions for nail sizes associated with metal framing clips.

**[BS] FIGURE A3-8B
TYPICAL FLOOR TO CRIPPLE WALL CONNECTION (FLOOR JOISTS PARALLEL TO FOUNDATIONS)**

APPENDIX A

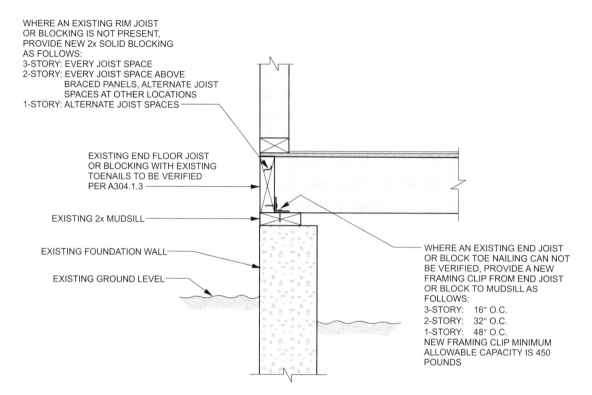

FLOOR JOISTS NOT PARALLEL TO FOUNDATIONS

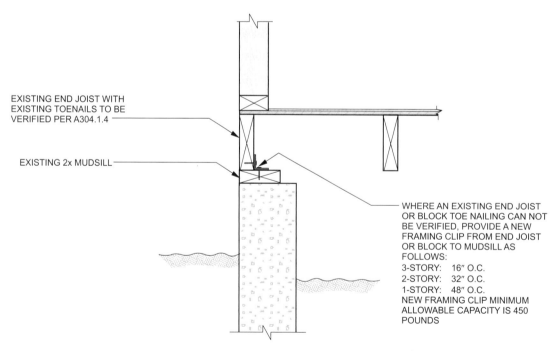

FLOOR JOISTS PARALLEL TO FOUNDATIONS

For SI: 1 inch = 25.4 mm.

NOTES:
1. See Section A304.3 for sill plate anchorage.
2. See manufacturing instructions for nail sizes associated with metal framing clips.

**[BS] FIGURE A3-8C
TYPICAL FLOOR TO MUDSILL CONNECTIONS**

APPENDIX A

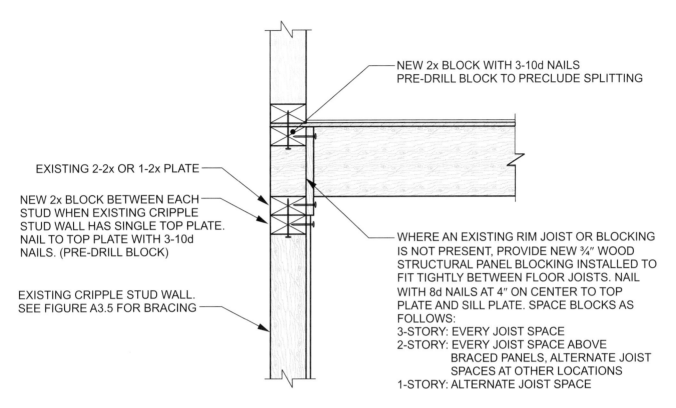

FLOOR JOISTS NOT PARALLEL TO FOUNDATION

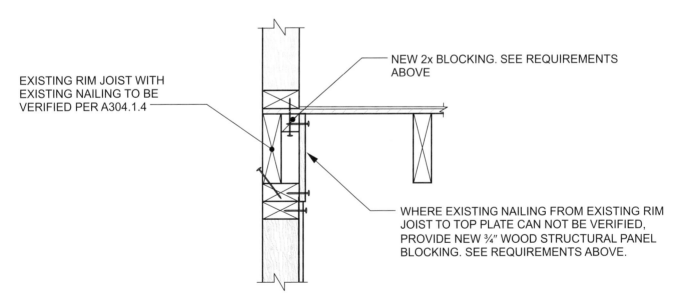

FLOOR JOISTS PARALLEL TO FOUNDATION

For SI: 1 inch = 25.4 mm, 1 pound = 4.4N.
NOTE: See Section A304.4 for cripple wall bracing.

**[BS] FIGURE A3-9
ALTERNATE FLOOR FRAMING TO CRIPPLE WALL CONNECTION**

APPENDIX A

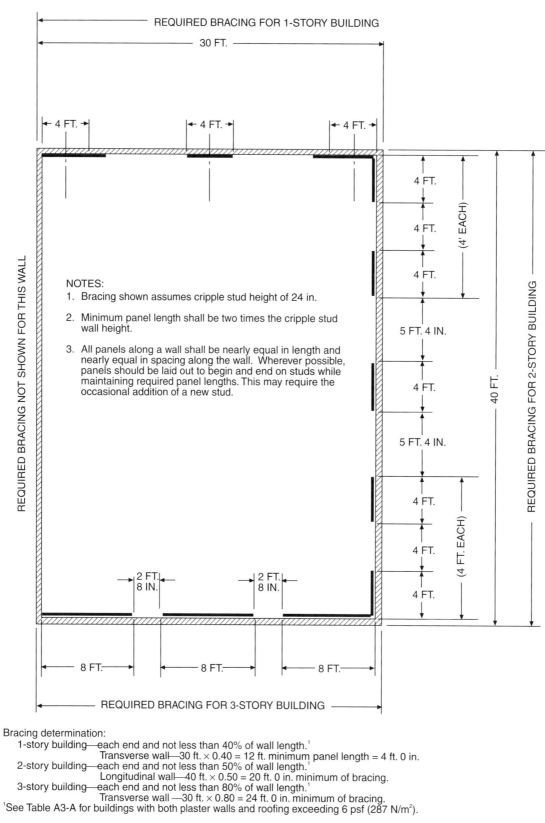

Bracing determination:
   1-story building—each end and not less than 40% of wall length.[1]
       Transverse wall—30 ft. × 0.40 = 12 ft. minimum panel length = 4 ft. 0 in.
   2-story building—each end and not less than 50% of wall length.[1]
       Longitudinal wall—40 ft. × 0.50 = 20 ft. 0 in. minimum of bracing.
   3-story building—each end and not less than 80% of wall length.[1]
       Transverse wall —30 ft. × 0.80 = 24 ft. 0 in. minimum of bracing.
[1] See Table A3-A for buildings with both plaster walls and roofing exceeding 6 psf (287 N/m^2).

For SI: 1 inch = 25.4 mm, 1 foot = 304.8 mm.

[BS] FIGURE A3-10
FLOOR PLAN-CRIPPLE WALL BRACING LAYOUT

Chapter A4:
Earthquake Risk Reduction in Wood-framed Residential Buildings with Soft, Weak or Open Front Walls

❖ Appendix Chapter A4, along with the rest of Appendix A of the code, is referenced by the *International Codes®* in two ways:

- Section 301.1.4.2 allows the use of Appendix A for qualifying buildings when "reduced IBC level seismic forces" are triggered.
- Section 807.6 allows the use of Appendix A for voluntary seismic retrofit.

This chapter has its origins in documents developed for the City of Los Angeles [Los Angeles (LA), 1999] and the City of Fremont, California, after the 1994 Northridge earthquake. Later, those provisions were extended by the Structural Engineers Association of California (SEAOC) to become Chapter 4 of the 2000 edition of *Guidelines for the Seismic Retrofit of Existing Buildings* (GSREB), published in 2001 by the International Conference of Building Officials (ICBO). Chapter 4 of the GSREB was reprinted in its entirety as Appendix Chapter A4 of the 2003 code. The 2006 through 2015 editions made many clarifications and revisions, most proposed by SEAOC.

For convenience, this commentary uses the acronym "SWOF" to refer to buildings, wall lines or structural conditions characterized by soft, weak or open-front (SWOF) walls.

SECTION A401
GENERAL

[BS] A401.1 Purpose. The purpose of this chapter is to promote public welfare and safety by reducing the risk of death or injury that may result from the effects of earthquakes on existing wood-frame, multiunit residential buildings. The ground motions of past earthquakes have caused the loss of human life, personal injury and property damage in these types of buildings. This chapter creates minimum standards to strengthen the more vulnerable portions of these structures. When fully followed, these minimum standards will improve the performance of these buildings but will not necessarily prevent all earthquake-related damage.

❖ The provisions in this chapter are intended to reduce earthquake risks by preventing concentrations of drift and structural damage in the vulnerable first stories of typical wood-framed SWOF buildings. These retrofit provisions are not intended to make the entire structure conform to the current code. Instead, Appendix Chapter A4 provisions are intended to "reduce the risk" by addressing the principal seismic weaknesses of a SWOF building with significantly less design effort, construction cost and tenant disruption. Risk reduction does not take a comprehensive approach to life safety, does not aim to protect property or function and is not necessarily equivalent to new construction under the *International Building Code®* (IBC®). For many owners, tenants and jurisdictions, however, this risk-reduction approach represents an acceptable trade-off.

Risk reduction identifies and improves the structure's more vulnerable portions, often leaving the rest of the building untouched. The SWOF condition itself is considered by far the most vulnerable attribute of a SWOF building, and retrofit of the SWOF wall line can be expected to substantially reduce the likelihood of excessive drift or collapse. Indeed, if the SWOF condition is the building's only serious structural deficiency, proper retrofit of the SWOF condition may achieve the benefits of a full life safety retrofit. Other examples of risk-reduction retrofit schemes include the strengthening and bolting of unbraced cripple walls in a wood-framed house without consideration of the upper stories, and the bracing of parapets in unreinforced masonry (URM) buildings without consideration of the remaining URM walls.

The question of performance objectives and the suitability of Appendix Chapter A4 have been debated when jurisdictions in areas of high seismicity have shifted the focus of seismic programs from basic safety to community resilience. If a program's objective is to minimize the loss of housing stock so as to limit the need for post-earthquake emergency shelters and to facilitate recovery, Appendix Chapter A4 may not facilitate that objective. These issues are discussed in an Applied Technology Council (ATC) report for the City and County of San Francisco (ATC, 2009) and are among the topics studied by the ATC 71-1 project. Nevertheless, that debate is about retrofit mandates and ordinances. For work triggered by codes for individual buildings, the objective remains safety based, and Appendix Chapter A4 is deemed equivalent to the other methods prescribed in Appendix Chapter A1.

As with any building or retrofit code, these provisions are formulated for certain typical conditions, and the intended performance is expected to be achieved by the great majority of the buildings to which the provisions apply. The performance of any specific building, however, might be better or worse than that intended by the provisions.

Section A401.1 refers to past earthquakes. Multistory SWOF buildings have shown unacceptable performance, including collapse and consequent loss of life, in the 1971 San Fernando, 1978 Santa Barbara, 1989 Loma Prieta and 1994 Northridge earthquakes. For more information on their past performance, see Harris et al. (1990), (1994), Mendes (1995) and Holmes and Somers (1996). For more information on analysis and retrofit approaches, see Vukazich (1998), Los Angeles Department of Building and Safety (LADBS) (1999), Rutherford and Chekene (2000), Federal Emergency Management Agency (FEMA) FEMA 547 (2006), Harris et al. (2009), Searer et al. (2009) and FEMA P-807 (2012).

[BS] A401.2 Scope. The provisions of this chapter shall apply to all existing Occupancy Group R-1 and R-2 buildings of wood construction or portions thereof where the structure has a soft, weak, or open-front wall line, and there exists one or more stories above.

❖ Section A401.2 describes the construction, occupancy and deficiencies for which this chapter is intended. The original intent of this section was that it apply only to voluntary retrofits in areas of relatively high seismicity (that is, Seismic Design Categories C, D and E but not F, which is for special occupancies outside the scope of the chapter). This appendix is currently written to be applied to SWOF buildings regardless of seismicity, and is now referenced from the body of the code (Section 301.1.4.2) as a design alternative for Seismic Risk Category I or II.

The provisions refer only to residential occupancies. Nevertheless, Appendix Chapter A4 is also intended to apply to certain buildings classified as "mixed occupancies," such as apartment buildings (Group R-2) with substantial ground floor private parking (Group U), restaurant (Group A-2), retail (Group B or M), public parking (Group S-2) or other incidental uses. In addition, these provisions are expected to apply to structurally similar cases, such as a building originally built as an apartment house but that has upper stories that were converted to office suites (Group B). Nevertheless, use of these provisions for nonresidential occupancies might require the concurrence of the code official. The important consideration is that the provisions of Appendix Chapter A4 presume that the diaphragm just above the critical SWOF story is relatively rigid and strong enough to distribute some lateral loads by rotation. This condition can be reasonably expected in wood-framed residential buildings, particularly buildings with floors composed of diagonal sheathing, plywood sheathing, two layers of perpendicular straight sheathing and/or acoustical concrete topping. Buildings for which a relatively stiff diaphragm cannot be reasonably presumed, regardless of occupancy, warrant additional attention by the designer and code official.

The most common SWOF conditions occur when the use of the building's first story differs from that of its upper stories, and particularly where the first-story use requires uninterrupted open areas and large openings in exterior walls, resulting in fewer and shorter walls to resist lateral loads. In general, the provisions in this chapter were conceived to address deficiencies in a building type with "tuck-under parking," so called because a ground-floor parking area is "tucked" under the upper stories. In California, many such structures were built in the 1960s and 1970s without considering earthquake loads or with sheathing materials that would not be allowed by the current code.

Structural characteristics of a typical wood-framed SWOF building of this era include:

- Perimeter wood-stud walls sheathed on the exterior face with stucco (sometimes over gypsum wallboard or plywood) and on the interior face with gypsum wallboard (drywall).
- Interior stud-framed partitions sheathed with drywall on both sides.
- No hold-down hardware.
- Floors of wood joists with wood sheathing, sometimes with lightweight concrete topping. Where parking is provided at the ground floor, the soffit sheathing and/or second floor construction would often be of a type sufficient for the code-required fire separation.
- Wood joist roof framing with wood sheathing.
- At the ground-floor parking or open area, steel pipe columns or wood posts supporting steel or wood beams and girders.

This chapter addresses only SWOF buildings that have wood-framed walls in the critical story.

The provision refers to SWOF wall lines "as defined in this chapter." The definitions in Section A402 differ somewhat from similar definitions of vertical irregularities in Table 12.3-2 of ASCE 7-10.

Appendix Chapter A4 is not intended to apply to one-story buildings because the principal risk from SWOF buildings is collapse of the first story under the weight of upper stories. Likewise, it does not apply to two-story houses over stucco-braced cripple walls because single-family houses are not in Occupancy Group R-1 or R-2 (such a structure might be covered by Appendix Chapter A3).

Though not explicitly stated, a multistory building might be exempted if the portion of the building with the SWOF condition is distinct from the rest of the building. That is, "stories above" may be reasonably understood to mean "stories above and influenced by the SWOF condition." For example, a three-story building with a one-story parking wing might be exempt if the SWOF condition occurs only in the one-story portion of the building. Such an interpretation requires engineering judgment and would likely be subject to the approval of the code official.

SECTION A402
DEFINITIONS

Notwithstanding the applicable definitions, symbols and notations in the building code, the following definitions shall apply for the purposes of this chapter:

❖ In general, "the building code" means the codes, standards, regulations and interpretations in effect in a specific jurisdiction. States and local jurisdictions typically adopt, and sometimes modify, a model code such as the *International Building Code*® (IBC®).

[BS] ASPECT RATIO. The span-width ratio for horizontal diaphragms and the height-length ratio for shear walls.

❖ "Vertical diaphragms" should be understood to mean "shear walls."

[BS] GROUND FLOOR. Any floor whose elevation is immediately accessible from an adjacent grade by vehicles or pedestrians. The ground floor portion of the structure does not include any floor that is completely below adjacent grades.

[BS] NONCONFORMING STRUCTURAL MATERIALS. Wall bracing materials other than wood structural panels or diagonal sheathing.

❖ This term is used only to define the scope of this chapter in Section A401.2. By this definition, some materials currently permitted for new construction by model codes such as the IBC are designated as nonconforming.

The Los Angeles and Fremont provisions for SWOF buildings (Los Angeles, 1999; Fremont) define "nonconforming structural materials" as any that are no longer permitted for new construction or any with design values (allowable shear or aspect ratio) that have been reduced since construction. Those standards are, therefore, more restrictive than the provisions given in this chapter.

For purposes of identifying SWOF conditions in Section A401.2, wood structural panels and diagonal sheathing are given more credit than other bracing methods such as let-in bracing, stucco (Portland cement plaster) or straight sheathing. Though deemed "conforming," it is possible that plywood or diagonal sheathing, even in buildings from the 1970s and 1980s, will not meet current post-Northridge requirements for new construction (for example, in terms of material specifications, hold-downs or nailing).

[BS] OPEN-FRONT WALL LINE. An exterior wall line, without vertical elements of the lateral force-resisting system, that requires tributary seismic forces to be resisted by diaphragm rotation or excessive cantilever beyond parallel lines of shear walls. Diaphragms that cantilever more than 25 percent of the distance between lines of lateral force-resisting elements from which the diaphragm cantilevers shall be considered excessive. Exterior exit balconies of 6 feet (1829 mm) or less in width shall not be considered excessive cantilevers.

❖ "Without vertical elements of the lateral force-resisting system" may be understood to mean wall lines without vertical elements (i.e., walls) sufficient to resist tributary lateral loads in the absence of diaphragm rotation. If there are insufficient elements of the seismic force-resisting system along a given diaphragm edge, then the wall line along that edge must meet the cantilever criteria of this definition or be considered "open."

More important than the solid wall length, though harder to calculate, is the reliance on diaphragm rotation or diaphragm cantilever. The open-front condition is largely a proxy for torsional irregularity, defined in Table 12.3-1 of ASCE 7-10. Torsional irregularity is often considered to be a concern only in rigid diaphragm structures. However, research has shown that torsion, or diaphragm rotation, contributes significantly to excessive drift along the open wall line, even in buildings with wood diaphragms [Consortium of Universities for Research in Earthquake Engineering (CUREE), 2002]. In lieu of complicated and potentially inaccurate drift calculations for a semirigid diaphragm that would be needed to determine torsional irregularity, the provisions use the more prescriptive definition of "Open front wall line."

Commentary Figure A402 illustrates the 25-percent calculation, which compares the cantilever length, c, to the backspan length, b. If c/b exceeds 0.25, the cantilever is considered excessive, and Line A is considered an open-front wall line. The 25-percent value is derived from judgment for consistency with previous code provisions, not from testing, analysis or performance statistics. See Section A403.8 and its commentary for additional discussion of cantilevered diaphragms.

Exit balconies are excluded because, in typical configurations, the outside edge of the balcony does not carry substantial gravity loads from the roof or stories above.

[BS] RETROFIT. An improvement of the lateral force-resisting system by *alteration* of existing structural elements or *addition* of new structural elements.

[BS] SOFT WALL LINE. A wall line whose lateral stiffness is less than that required by story drift limitations or deformation compatibility requirements of this chapter. In lieu of analysis, a soft wall line may be defined as a wall line in a story where the story stiffness is less than 70 percent of the story above for the direction under consideration.

❖ The definition calls for a calculation of story drift, but provides an alternative approach that compares stiffness in adjacent stories. The 70-percent value provided for the alternative approach is similar to the definition of soft story irregularity in Table 12.3-2 of ASCE 7-10; it does not make the distinction that ASCE 7 makes between "soft story" and "extreme soft story" irregularities.

It is generally difficult to make a meaningful or reliable calculation of story drifts for buildings for which seismic bracing in the upper stories is provided by interior plaster, gypsum wallboard and exterior wall finishes that are not part of an engineered seismic force-resisting system. These materials generally provide great stiffness until their strength is exceeded, at which time their stiffnesses can deteriorate rapidly. Consequently, it will likely be easier to perform the soft-story calculation based on a comparative approach.

While the definition is for an individual wall line, the calculation is generally performed for an entire story. According to the definition, if the story is soft in a given direction, then each wall line in that direction is considered a soft wall line. It is probably more appropriate, however, to consider the sum of individual wall lines and to compare their total stiffness with the corresponding stiffness in the story above.

If actual story stiffness is to be calculated, it should include contributions from nonbearing walls and partitions (even those of nonconforming structural materials). Also, walls with hold-downs will be stiffer than walls subject to rocking or uplift. Section D.7.5 of the CUREE publication W-30b (Cobeen et al., 2004) provides information on reduced strength and stiffness for shear walls without hold-downs to restrain uplift. Experimental research by Ni and Karacabeyli (2000) also provides information regarding the response of structural panel walls with and without hold-downs.

[BS] STORY. A story as defined by the building code, including any basement or underfloor space of a building with cripple walls exceeding 4 feet (1219 mm) in height.

[BS] STORY STRENGTH. The total strength of all seismic-resisting elements sharing the same story shear in the direction under consideration.

❖ It is sometimes preferable to calculate element capacity based on strength values. For purposes of comparing the strengths of adjacent stories, however, allowable stresses may be used.

[BS] WALL LINE. Any length of wall along a principal axis of the building used to provide resistance to lateral loads. Parallel wall lines separated by less than 4 feet (1219 mm) shall be considered one wall line for the distribution of loads.

❖ "Separated by less than 4 feet (1219 mm)" refers to out-of-plane separation; for example, where part of a wall line is recessed for architectural purposes. A given wall line might consist of several individual noncontiguous shear wall panels separated in-plane by large door or window openings or by other architectural elements.

[BS] WEAK WALL LINE. A wall line in a story where the story strength is less than 80 percent of the story above in the direction under consideration.

❖ As discussed in the commentary to "Soft wall line," it is generally difficult to make a very precise calculation of story strength for upper stories of these buildings. The 80-percent value is consistent with Table 12.3-2 of ASCE 7-10, but Appendix Chapter A4 does not make the same distinction as ASCE 7 between "weak story" and "extreme weak story" irregularities.

See the commentary to "Soft wall line" for discussion of wall capacities with and without hold-downs.

Where story strength is calculated, it should include contributions from nonbearing walls and partitions (even those of nonconforming structural materials).

Another potential shortcoming of this definition is that it does not account for different force levels in adjacent stories. In a two-story building, the first-story shear is substantially higher than the second-story shear; thus, even if the two stories have the same strength, the first story will be critical. Therefore, where the lower story is just marginally strong enough to avoid classification as "weak," a comparison of demand-to-capacity ratios in adjacent stories is useful. If the lower story is clearly critical and its seismic force-resisting system does not provide ample ductility, the engineer might consider voluntarily classifying the story as "weak," regardless of the definition.

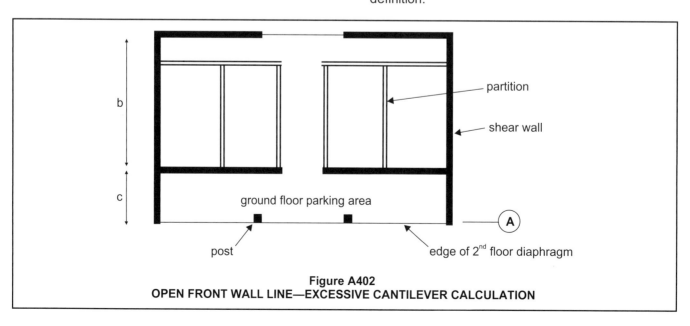

Figure A402
OPEN FRONT WALL LINE—EXCESSIVE CANTILEVER CALCULATION

SECTION A403
ANALYSIS AND DESIGN

[BS] A403.1 General. All modifications required by the provisions in this chapter shall be designed in accordance with the *International Building Code* provisions for new construction, except as modified by this chapter.

> **Exception:** Buildings for which the prescriptive measures provided in Section A404 apply and are used.

No *alteration* of the existing lateral force-resisting system or vertical load-carrying system shall reduce the strength or stiffness of the existing structure, unless the altered structure would remain in conformance to the building code and this chapter.

❖ This section mandates that any modifications required by these provisions, principally new structural elements, must be made in accordance with IBC requirements for new construction unless specifically modified by this chapter. Further, reference to "the building code" throughout Appendix Chapter A4 should be understood to mean the provisions for new construction.

If the building qualifies for the prescriptive measures in Section A404, then the requirements of Section A403—including the weak story limits, drift limits, etc.—are waived. The requirements of Sections A405, A406 and A407 would still apply.

[BS] A403.2 Scope of analysis. This chapter requires the *alteration*, *repair*, replacement or *addition* of structural elements and their connections to meet the strength and stiffness requirements herein. The lateral-load-path analysis shall include the resisting elements and connections from the wood diaphragm immediately above any soft, weak or open-front wall lines to the foundation soil interface or to the uppermost story of a podium structure comprised of steel, masonry, or concrete structural systems that supports the upper, wood-framed structure. Stories above the uppermost story with a soft, weak, or open-front wall line shall be considered in the analysis but need not be modified. The lateral-load-path analysis for added structural elements shall also include evaluation of the allowable soil-bearing and lateral pressures in accordance with the building code. Where any portion of a building within the scope of this chapter is constructed on or into a slope steeper than one unit vertical in three units horizontal (33-percent slope), the lateral force-resisting system at and below the base level diaphragm shall be analyzed for the effects of concentrated lateral forces at the base caused by this hillside condition.

> **Exception:** When an open-front, weak or soft wall line exists because of parking at the ground floor of a two-story building and the parking area is less than 20 percent of the ground floor area, then only the wall lines in the open, weak or soft directions of the enclosed parking area need comply with the provisions of this chapter.

❖ This section distinguishes Appendix Chapter A4's risk-reduction objective from a full life-safety objective. Using this chapter, the only structural elements that need to be considered for retrofit are those between the diaphragm immediately above the SWOF story and the foundation soil. Stories above that level need not be checked, though their mass and stiffness must still be included in the lateral analysis of the building and their role in global building response considered.

Further, if the wood-framed SWOF structure is supported by a steel, concrete or masonry podium structure, that podium structure need not be checked. The wording of this section is unclear as to whether the diaphragm strength itself needs to be checked or designed. Section A403.8 clarifies that it does not.

Top stories are exempt for the same reason that one-story buildings were exempted in Section A401.2. These provisions focus on the specific risks posed by the collapse of SWOF stories with substantial mass above. Racking or excessive drift in a one-story open-front building or in the upper stories of a multistory building is of lesser concern.

Soil stresses need only be checked where they are affected by added structural elements, such as walls, diagonal braces, frame columns, posts or other structural elements that are added as part of the retrofit. Where allowable soil stresses are exceeded due to these added elements, foundation elements may need to be supplemented as well.

The exception further reduces the scope of Appendix Chapter A4 for SWOF deficiencies that are judged to be remediable by local measures. Again, the risk reduction approach acknowledges that when the most critical deficiencies are easily identified and remedied, comprehensive analysis of the entire structure should not be necessary.

The exception is intended to cover a large subclass of SWOF buildings with minimal tuck-under parking. While the provision specifically mentions "parking," it could be reasonably applied to similar unoccupied spaces used for storage or other purposes. The 20-percent value is derived from judgment, not from testing, analysis or performance statistics. Use of the term "enclosed parking area" is intended to indicate a defined area bounded in plan by walls, partitions or the edge of the building; it does not mean that the parking (or other nonoccupiable) area must be physically enclosed.

The 20-percent value is to be based on floor area. All nonoccupiable areas (as opposed to just designated parking area) between the SWOF wall line and the nearest parallel wall line should be counted as the parking area. Commentary Figure A403.2 shows the type of common SWOF building for which this exception was intended. Other configurations are also eligible for the exception. For complex floor layouts, designation and calculation of the parking area might require some judgment.

With reference to Commentary Figure A403.2, Line A is assumed to be a SWOF wall line. Line B is the nearest parallel wall line. If $A_p/(L_x L_y)$ is less than 0.20, then only wall Lines A and B need to be checked and potentially strengthened or stiffened.

The intention of this exception is only to exempt

wall Lines C, 1 and 2 from drift and strength requirements. These wall lines must still be included in the structural model used to derive forces and deflections for Lines A and B. In considering Lines A and B, the load path from the second floor diaphragm to the soil-foundation interface must still be checked in accordance with Section A403.2.

This exception does not apply if the prescriptive provisions of Section A404 are used.

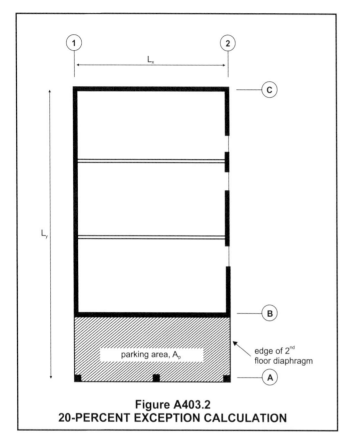

Figure A403.2
20-PERCENT EXCEPTION CALCULATION

[BS] A403.3 Design base shear and design parameters. The design base shear in a given direction shall be permitted to be 75 percent of the value required for similar new construction in accordance with the building code. The value of R used in the design of the strengthening of any story shall not exceed the lowest value of R used in the same direction at any story above. The system overstrength factor, Δ_0, and the deflection amplification factor, C_d, shall be not less than the largest respective value corresponding to the R factor being used in the direction under consideration.

Exceptions:

1. For structures assigned to Seismic Design Category B, values of R, Δ_0 and C_d shall be permitted to be based on the seismic force-resisting system being used to achieve the required strengthening.

2. For structures assigned to Seismic Design Category C or D, values of R, Δ_0 and C_d shall be permitted to be based on the seismic force-resisting system being used to achieve the required strengthening, provided that when the strengthening is complete, the strengthened structure will not have an extreme weak story irregularity defined as Type 5b in ASCE 7 Table 12.3-2.

3. For structures assigned to Seismic Design Category E, values of R, Δ_0 and C_d shall be permitted to be based on the seismic force-resisting system being used to achieve the required strengthening, provided that when the strengthening is complete, the strengthened structure will not have an extreme soft story, a weak story, or an extreme weak story irregularity defined, respectively, as Types 1b, 5a and 5b in ASCE 7 Table 12.3-2.

❖ The 75-percent value represents a common allowance for existing buildings with precedent in FEMA 178 (FEMA, 1992). FEMA 178 specified a design base shear of 85 percent (for short-period buildings) or 67 percent (for longer period buildings) of that required for new construction. At the time, the reduction was understood to be related to the difference between a "mean" earthquake and a "mean + one-sigma" earthquake. It is not clear whether that relationship still holds for the seismicity estimates and design parameters in current codes. Nevertheless, a reduction of about 25 percent remains traditional and is consistent with the concept of "reduced IBC level seismic forces" in Section 301.1.4.2 of the code.

Appendix Section A403.3 requires that the value of R, Ω_0 and C_d used the in strengthening of any story be used for similar new construction. These values would be those in Table 12.2-1 of ASCE 7-10. There are three exceptions to this requirement:

Exception 1 is specifically for Seismic Design Category B. It allows the use of design coefficients for the seismic force-resisting system used.

Exception 2 is specifically for Seismic Design Categories C and D. The design coefficients may be those of the seismic force-resisting system used if the strengthened structure does not have an extreme weak story irregularity.

Exception 3 is specifically for Seismic Design Category E. The seismic design coefficients may be those of the seismic force-resisting system used if the strengthened structure does not have an extreme soft story, a weak story or an extreme weak story irregularity. This escalation of conditions and requirements is based on the seismic design category (seismic demand) of the location and probable duration of ground shaking.

It is recommended that the design professional consider the strength of the upper stories, as required by the provisions of Section A403.2. Increasing the first-story strength and stiffness also increases the base shear that can be transmitted to the floors above. If the retrofitted strength of the first story exceeds the strength of the story above, the increased demand on the second story may propagate the expected failure up to the next level. The user is referred to FEMA P-807 (FEMA, 2012) for further information.

[BS] A403.4 Story drift limitations. The calculated story drift for each retrofitted story shall not exceed the allowable deformation compatible with all vertical load-resisting elements and 0.025 times the story height. The calculated story drift shall not be reduced by the effects of horizontal diaphragm stiffness but shall be increased when these effects produce rotation. Drift calculations shall be in accordance with the building code.

❖ The IBC is based on ASCE 7, and "the calculated story drift" means the design story drift as defined in Section 12.8.6 of ASCE 7-10, based on center of mass deflections, d_{xe}, calculated from a linear elastic analysis using the forces prescribed in Section A403.3. $P\Delta$ effects must be included in the analysis as required by Section A403.5 (by reference to corresponding Section 12.8.7 of ASCE 7-10).

"Each retrofitted story" means each story subject to consideration, according to Section A403.2. Drifts need not be checked at stories above the SWOF conditions.

The limiting drift ratio of 0.025 was specified in the 2000 GSREB to match the limit for low- and mid-rise buildings in Section 1630.10.2 of the 1997 *Uniform Building Code* (UBC). Table 12.12-1 of ASCE 7-10 also limits drift ratios in most new low-rise buildings in Risk Categories I and II to 0.025.

Many existing SWOF buildings also have wood posts or steel pipe columns along the open side. For these, the limit of 0.025 is judged appropriate unless their condition is poor (decayed or corroded) or their connections are unable to accommodate that much deformation.

The provision regarding "horizontal diaphragm stiffness" requires that diaphragms be modeled as stiff or (semi) rigid relative to the walls, so that the effects of actual and accidental torsion can be conservatively estimated. The (semi) rigid diaphragm assumption is supported by testing that found that the effects of a wood diaphragm on an open-front structure were sufficiently approximated by modeling the diaphragm as a rotating rigid body (CUREE, 2002).

The provision to add, but not subtract, the effects of diaphragm rotation is a conservative requirement similar to those in Sections 12.8.6 and 12.12.1 of ASCE 7-10 for buildings subject to significant torsion.

Unless conservative assumptions regarding materials and details are made, stiffness calculations for elements not shown on plans will likely require destructive investigation per Section A405.3. This can be disruptive and expensive. The relative stiffness of existing materials can be estimated from procedures given in references such as ASCE 41 (2013) or CUREE (2002). However, while a given reference is likely to be internally consistent regarding its assumptions and definitions, different references might vary from one another. Therefore, the same reference should be used to derive the properties of all relevant materials whenever possible.

For purposes of comparing the stiffness of adjacent stories, any reasonable material and fixity assumptions are acceptable, as long as similar assumptions are made for the two stories being compared and the calculation is not given credit for undue precision. For simplicity, elastic (i.e., uncracked, nondegraded) stiffnesses may be used, and partitions and shear walls may be assumed fixed at the ground and floor levels with respect to in-plane rotation. Section 12.8.6 of ASCE 7-10 indicates that, where torsional deflections are significant, they should be included in story drift. Where a flexible diaphragm analysis is used, it is more practical to check drift at each shear wall line. Where a rigid or semirigid diaphragm analysis is used, peak drifts that include torsion should be used.

[BS] A403.4.1 Pole structures. The effects of rotation and soil stiffness shall be included in the calculated story drift where lateral loads are resisted by vertical elements whose required depth of embedment is determined by pole formulas. The coefficient of subgrade reaction used in deflection calculations shall be based on a geotechnical investigation conducted in accordance with the building code.

❖ The "rotation" referred to in this paragraph is the rotation of an embedded pole bearing laterally against soil, either with or without restraint at the ground surface by a slab or rigid pavement. The provision is intended to address pipe columns or posts supporting gravity loads along the building's open side. Pole systems are especially flexible and vulnerable to loss of vertical capacity. This provision applies mostly to analysis of the existing structure; in the retrofitted structure, the lateral stiffness contribution of posts is generally negligible compared to the stiffness of new wall, frame or brace elements and to the wall elements above the SWOF story. The coefficient of subgrade reaction refers to the passive stiffness of soil reacting against movement or rotation of the embedded post.

[BS] A403.5 Deformation compatibility and P Δ effects. The requirements of the building code shall apply, except as modified herein. All structural framing elements and their connections not required by design to be part of the lateral force-resisting system shall be designed and/or detailed to be adequate to maintain support of expected gravity loads when subjected to the expected deformations caused by seismic forces. Increased demand due to P Δ effects and story sidesway stability shall be considered in retrofit stories that rely on the strength and stiffness of cantilever columns for lateral resistance.

❖ Sections 12.1.2 and 12.12.5 of ASCE 7-10 require deformation compatibility checks for structural elements that are not part of the seismic force-resisting system (but are part of the gravity system), such as posts, pipe columns and wood-framed walls deformed out of plane. Inelasticity is permitted when members and connections are checked for deformation compatibility as long as the structure maintains its resistance to gravity loads.

In existing SWOF buildings, the most critical gravity

support elements are probably the posts along the open front because they often have weak flexural connections and because drift is greatest along that wall line. The design check should rule out any failure of the columns, base plates or connection hardware that would compromise stability.

[BS] A403.6 Ties and continuity. All parts of the structure included in the scope of Section A403.2 shall be interconnected as required by the building code.

❖ The applicable code provisions are in Sections 12.1.3 and 12.1.4 of ASCE 7-10.

As noted in ASCE 7, the connection forces prescribed by this provision do not apply to the overall design of the seismic force-resisting system. That is, they need not be added to the design forces prescribed by Section A403.3 for analysis and design of the seismic force-resisting system.

The minimum connection strengths required by this provision are prescriptive and are derived from judgment, not from testing, analysis or performance statistics.

[BS] A403.7 Collector elements. Collector elements shall be provided that can transfer the seismic forces originating in other portions of the building to the elements within the scope of Section A403.2 that provide resistance to those forces.

❖ This provision is similar to Section 12.10.2 of ASCE 7-10. "Other portions" does not mean elements outside the scope of Section A403.2; this terminology remains from past model code provisions for new construction (and is still present in ASCE 7-10). The intent of Section A403.7 is merely to ensure adequate collectors in the load path described in Section A403.2; that is, from the diaphragm above the SWOF condition to the existing or added components of the seismic force-resisting system.

The provision does not specify means for determining design forces for collectors. Acceptable design forces may be derived from formulas such as Equation 12.10-1 of ASCE 7-10, with the applied forces F_i determined from Section A403.3. Provisions in Section A403.6 set a minimum on this design force.

[BS] A403.8 Horizontal diaphragms. The strength of an existing horizontal diaphragm sheathed with wood structural panels or diagonal sheathing need not be investigated unless the diaphragm is required to transfer lateral forces from vertical elements of the seismic force-resisting system above the diaphragm to elements below the diaphragm because of an offset in placement of the elements.

Rotational effects shall be accounted for when asymmetric wall stiffness increases shear demands.

❖ Floor diaphragms are considered capable of rotating as rigid bodies without substantial distortion. They are therefore judged capable of transmitting forces if the seismic force-resisting system is not symmetric in plan or if the system has elements on only three sides.

"Rotational effects shall be accounted for when asymmetric wall stiffness increases shear demands" refers to torsional effects that increase the forces and drifts imposed on the seismic force-resisting system. The requirement to account for rotational effects on shear walls should apply as well to existing or new elements of other seismic force-resisting systems.

[BS] A403.9 Wood-framed shear walls. Wood-framed shear walls shall have strength and stiffness sufficient to resist the seismic loads and shall conform to the requirements of this section.

❖ Section A403.9 addresses only wood-framed shear walls. Other structural elements added for retrofit are to be evaluated against the demands of Sections A403.3 and A403.4, and detailed according to the IBC for new construction.

[BS] A403.9.1 Gypsum or cement plaster products. Gypsum or cement plaster products shall not be used to provide lateral resistance in a soft or weak story or in a story with an open-front wall line, whether or not new elements are added to mitigate the soft, weak or open-front condition.

❖ "Gypsum" refers primarily to gypsum wallboard but also includes traditional gypsum plaster and gypsum lath (sometimes called "buttonboard"). "Cement plaster" refers to plaster with Portland cement, commonly known as "stucco."

Even after retrofit, the SWOF story may experience significant ductility demands during intense ground shaking. The intent of this provision is to preclude existing gypsum and Portland cement plaster materials from being considered in the strength or stiffness of the retrofitted SWOF story.

[BS] A403.9.2 Wood structural panels.

[BS] A403.9.2.1 Drift limit. Wood structural panel shear walls shall meet the story drift limitation of Section A403.4. Conformance to the story drift limitation shall be determined by approved testing or calculation. Individual shear panels shall be permitted to exceed the maximum aspect ratio, provided the allowable story drift and allowable shear capacities are not exceeded.

❖ The intent of this section is to ensure that the drift in the strengthened story does not exceed reasonable limits.

[BS] A403.9.2.2 Openings. Shear walls are permitted to be designed for continuity around openings in accordance with the building code. Blocking and steel strapping shall be provided at corners of the openings to transfer forces from discontinuous boundary elements into adjoining panel elements. Alternatively, perforated shear wall provisions of the building code are permitted to be used.

❖ In the 2006 IBC, the applicable code provisions were those in Section 2305.8.2 for perforated shear walls. Section 2305 had been substantially revised by 2009. The intent of these provisions is to ensure performance that matches the characteristics assumed for analysis and design, principally that the shear wall acts as a unit, without substantial deformations or stress concentrations at openings. Commentary, background and design examples regarding perforated shear walls are available in Section 12.3 of the

2003 National Earthquake Hazards Reduction Program (NEHRP) commentary (BSSC, 2004).

[BS] A403.9.3 Hold-down connectors.

[BS] A403.9.3.1 Expansion anchors in tension. Expansion anchors that provide tension strength by friction resistance shall not be used to connect hold-down devices to existing concrete or masonry elements.

❖ This provision prohibits friction-based expansion anchors for tension loads because of poor performance in the 1994 Northridge earthquake. Since then, approval criteria have been made more restrictive and the design of some expansion anchors has been modified. Any anchor type approved for cyclic tension loads should be acceptable in this application.

[BS] A403.9.3.2 Required depth of embedment. The required depth of embedment or edge distance for the anchor used in the hold-down connector shall be provided in the concrete or masonry below any plain concrete slab unless satisfactory evidence is submitted to the *code official* that shows that the concrete slab and footings are of monolithic construction.

❖ "Monolithic construction" generally means concrete placed continuously (at the same time). A slab placed over the top of a previously placed footing will not be monolithic unless the top surface of the footing was cleaned/roughened, bent reinforcing dowels connect the two pours and there is no "visqueen" vapor barrier laid between the two pours. A slab with nominal reinforcing or wire mesh provided only for crack control should be considered plain concrete.

SECTION A404
PRESCRIPTIVE MEASURES FOR WEAK STORY

[BS] A404.1 Limitation. These prescriptive measures shall apply only to two-story buildings and only when deemed appropriate by the *code official*. These prescriptive measures rely on rotation of the second floor diaphragm to distribute the seismic load between the side and rear walls of the ground floor open area. In the absence of an existing floor diaphragm of wood structural panel or diagonal sheathing, a new wood structural panel diaphragm of minimum thickness of $^3/_4$ inch (19.1 mm) and with 10d common nails at 6 inches (152 mm) on center shall be applied.

❖ This section is not limited to weak story conditions. Any structure that meets the qualifications is eligible for these prescriptive measures. For buildings that qualify, Section A404 is intended as an alternative to Section A403. The requirements of Sections A405, A406 and A407 still apply. The main purpose of this prescriptive alternative is to reduce the cost of engineered evaluation and design. It is expected that Section A404 could be applied by a qualified contractor without the assistance of an engineer.

The provisions of Section A404 are derived from judgment, not from testing, analysis or performance statistics. No studies or tests have been performed to confirm that these prescriptive measures will result in the same performance with the same reliability as designs engineered to the provisions of Section A403. Nevertheless, for the buildings that qualify, these prescriptive measures are expected to achieve the same risk reduction objective.

Section A404 is intended for a two-story building with an uncomplicated footprint and a SWOF wall line on only one side. Commentary Figure A404.1 illustrates two such buildings in plan. In the bottom part of the figure, only the portion of the side wall between Lines A and B is considered to be enclosing the parking/open area, and only that portion should be counted toward the required wall length under these prescriptive measures.

The provision does not identify any detail requirements for the existing wood diaphragm. Instead, the provision appears to presume that the presence of a conforming material ensures acceptable sheathing thickness, material grade, condition and nailing.

Where a new diaphragm is required, it may be applied over the existing sheathing or to the underside of the second floor joists. Application to the underside of the floor joists will often be less disruptive.

[BS] A404.1.1 Additional conditions. To qualify for these prescriptive measures, the following additional conditions need to be satisfied by the retrofitted structure:

1. Diaphragm aspect ratio L/W is less than 0.67, where W is the diaphragm dimension parallel to the soft, weak or open-front wall line and L is the distance in the orthogonal direction between that wall line and the rear wall of the ground floor open area.

2. Minimum length of side shear walls = 20 feet (6096 mm).

3. Minimum length of rear shear wall = three-fourths of the total rear wall length.

4. No plan or vertical irregularities other than a soft, weak or open-front wall line.

5. Roofing weight less than or equal to 5 pounds per square foot (240 N/m^2).

6. Aspect ratio of the full second floor diaphragm meets the requirements of the building code for new construction.

❖ The ratios and shear wall lengths in this provision are derived from judgment, not from testing, analysis or performance statistics.

The six conditions need not be satisfied by the existing building as long as they are satisfied by the retrofit. In other words, wall elements may be added and openings may be modified to meet the specified criteria. Whether new or existing, only walls that comply with Section A404.2.3 should be counted for the purposes of satisfying these conditions.

L and W are the principal dimensions of the diaphragm bounded by the shear walls immediately adjacent to the SWOF wall line. The intention of the aspect ratio limit is to restrict the prescriptive provisions to buildings that are not prone to substantial torsion. Where the critical diaphragm is the entire second floor, as in the top part of Commentary Figure A404.1, the preferred condition has the SWOF wall line along the long side of the building so that the center of mass and the center of rigidity are close together and torsion is limited. If the open wall line were along the short side, the three-sided structure would be subject to high torsion.

The bottom part of Commentary Figure A404.1 illustrates a potential shortcoming of this simple screening criterion. If L/W for this building were greater than 0.67, it would not qualify for the prescriptive measures of Section A404. Yet, this building might still be less vulnerable to torsion than the building in the top part of the figure. In the bottom part of the figure, the critical condition is the demand on the wall along Line B due to torsional response.

For the building in the bottom part of Commentary Figure A404.1, the side shear walls include only the portions between Line A and Line B. The portion between Line B and Line C is not counted toward the required 20 feet (6096 mm). Reasonable designs that make use of wall lengths not immediately adjacent to the open area (that is, lengths between Lines B and C) are possible, but should be subject to engineered design. For these prescriptive measures, it is appropriate to restrict certain design choices in order to ensure reliable outcomes in the absence of engineering calculations.

Plan and vertical structural irregularities are described in Tables 12.3-1 and 12.3-2 of ASCE 7-10.

The roofing weight limit of 5 pounds per square foot (240 N/m^2) is intended to rule out unusually heavy structures to which the prescriptive provisions should not apply. The limit is not intended to rule out the fairly common condition of two layers of asphalt roofing.

[BS] A404.2 Minimum required retrofit.

❖ For buildings that qualify, Section A404 is intended as an alternative to Section A403. The requirements of Sections A405, A406 and A407 still apply.

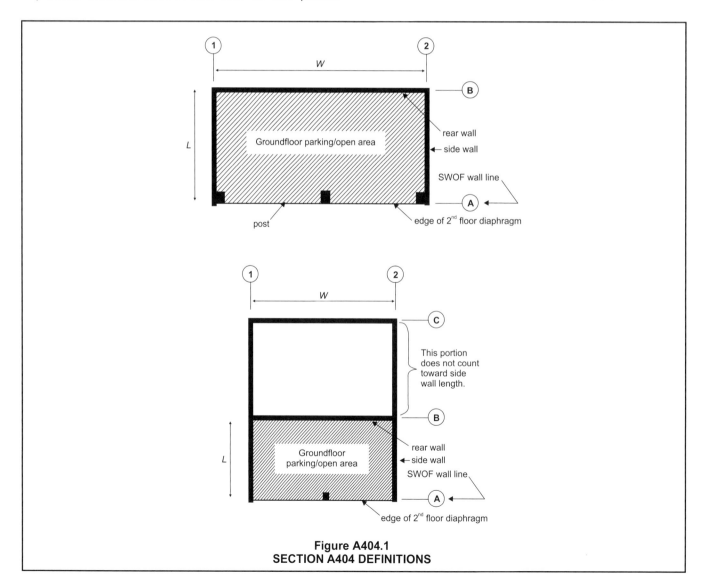

Figure A404.1
SECTION A404 DEFINITIONS

[BS] A404.2.1 Anchor size and spacing. The anchor size and spacing shall be a minimum of $^3/_4$ inch (19.1 mm) in diameter at 32 inches (813 mm) on center. Where existing anchors are inadequate, supplemental or alternative approved connectors (such as new steel plates bolted to the side of the foundation and nailed to the sill) shall be used.

❖ Anchors as specified in this provision must be provided only along the length of the side and rear walls.

"Inadequate" refers principally to anchor size and spacing, but judgment should be applied where existing anchors indicate poor workmanship (e.g., oversized holes in sill plates, insufficient edge distance or embedment) or poor condition (e.g., corrosion, spalling).

The provision specifies $^3/_4$-inch-diameter (19 mm) anchors. Smaller anchors are sometimes necessary for proper edge distance in narrow sill plates or concrete curbs. These smaller anchor sizes may be used if the code official approves and the designer demonstrates that the proposed anchor layout, which may combine new and existing anchors of different sizes, provides capacity equivalent to $^3/_4$-inch (19 mm) anchors at 32 inches (813 mm) on center. If the new anchors are substantially stiffer than the existing anchors, then the existing anchors should not be counted toward the required capacity.

Embedment must be into the footing (see commentary, Section A403.9.3.2).

The provision allows approved steel plate connectors. These are commonly used for retrofit of cripple walls where limited access prevents installation of new anchors vertically through the sill plate.

Sill plate damage is best prevented by proper design, location and installation of hold-downs. Plate washers are recommended near ends of shear walls.

Installation of expansion anchors can crack weak or deteriorated concrete. Adhesive, undercut or threaded screw-type anchors might be useful in these conditions. An adhesive anchor may be used to resist shear and/or tension loads; adhesive anchors use a structural adhesive (e.g., epoxy) to secure the metal fastener in a predrilled hole in hardened concrete or masonry. An undercut anchor, without grout or adhesive, may also be used to resist shear and/or tension; when tightened, an undercut anchor engages the sides and undercut surfaces of a specially predrilled hole in hardened concrete or masonry. Screw-type anchors self-thread a predrilled hole in hardened concrete and resist forces through the mechanical interlock of the anchor threads with the concrete.

[BS] A404.2.2 Connection to floor above. Shear wall top plates shall be connected to blocking or rim joist at upper floor with a minimum of 18-gage galvanized steel angle clips $4^1/_2$ inches (114 mm) long with 12-8d nails spaced no farther than 16 inches (406 mm) on center, or by equivalent shear transfer methods.

[BS] A404.2.3 Shear wall sheathing. The shear wall sheathing shall be a minimum of $^{15}/_{32}$ inch (11.9 mm) 5-Ply Structural I with 10d nails at 4 inches (102 mm) on center at edges and 12 inches (305 mm) on center at field; blocked all edges with 3 by 4 board or larger. Where existing sill plates are less than 3-by thick, place flat 2-by on top of sill between studs, with flat 18-gage galvanized steel clips $4^1/_2$ inches (114 mm) long with 12-8d nails or $^3/_8$-inch-diameter (9.5 mm) lags through blocking for shear transfer to sill plate. Stagger nailing from wall sheathing between existing sill and new blocking. Anchor new blocking to foundation as specified above.

❖ The requirements for $^{15}/_{32}$-inch (11.9 mm) five-ply sheathing and 3x framing and blocking are consistent with changes made to building codes after the 1994 Northridge earthquake. For additional discussion of their background, see Harder (1994) and SEAOC (1999), Appendix F.

The provision makes no allowance for existing $^3/_8$-inch (10 mm) sheathing. For these prescriptive measures, the specified sheathing is required without exception. If the existing sheathing is $^1/_2$ inch (12.7 mm), nail size and spacing must be verified; the nailing must be sufficient to develop the capacity equivalent to that required by the provision. If the existing sheathing or nailing does not meet the prescriptive requirements, the engineered approach of Section A403 may be used to take advantage of their strength contribution. If the required $^{15}/_{32}$-inch (11.9 mm) sheathing is applied to studs already sheathed with lesser material, care should be taken not to split the existing wall studs with new nailing. Doubled studs, predrilled holes or lag screws may be beneficial in this regard.

Sheathing applied to the inside face of existing studs should be provided with ventilation and inspection holes.

[BS] A404.2.4 Shear wall hold-downs. Shear walls shall be provided with hold-down anchors at each end. Two hold-down anchors are required at intersecting corners. Hold-downs shall be approved connectors with a minimum $^5/_8$-inch-diameter (15.9 mm) threaded rod or other approved anchor with a minimum allowable load of 4,000 pounds (17.8 kN). Anchor embedment in concrete shall be not less than 5 inches (127 mm). Tie-rod systems shall be not less than $^5/_8$ inch (15.9 mm) in diameter unless using high-strength cable. High-strength cable elongation shall not exceed $^5/_8$ inch (15.9 mm) under a 4,000 pound (17.8 kN) axial load.

❖ While the design provisions of Section A403 are generally waived, applicable portions of the hold-down provisions of Section A403.9.3 should be applied in addition to those given here.

The hold-down capacity might be limited by available concrete strength or edge distance. It is acceptable to use two 2,000-pound (8.9 kN) hold-downs, one on each side of the post, in lieu of a single 4,000-pound (17.8 kN) hold-down.

Posts to which hold-downs are connected should be sufficient to prevent failures due to reductions in cross section at bolt or connector holes. Posts built up from multiple 2x or 3x members are acceptable if adequately connected, but 4x or larger posts are preferred.

SECTION A405
MATERIALS OF CONSTRUCTION

[BS] A405.1 New materials. New materials shall meet the requirements of the *International Building Code*, except where allowed by this chapter.

[BS] A405.2 Allowable foundation and lateral pressures. The use of default values from the building code for continuous and isolated concrete spread footings shall be permitted. For soil that supports embedded vertical elements, Section A403.4.1 shall apply.

❖ See the commentary to Section A403.2 for additional discussion of soil-bearing pressures and the commentary to Section A403.4.1 for additional discussion of the coefficient of subgrade reaction.

[BS] A405.3 Existing materials. The physical condition, strengths, and stiffnesses of existing building materials shall be taken into account in any analysis required by this chapter. The verification of existing materials conditions and their conformance to these requirements shall be made by physical observation, material testing or record drawings as determined by the registered design professional subject to the approval of the *code official*.

❖ This section requires that existing materials be taken into account in terms of condition, strengths and stiffness when addressed in any analysis required by Appendix Chapter A4. The provision is intended to ensure that load path elements will not fail prematurely due to deterioration or errors/omissions in the original construction (e.g., "shiner" nails, misplaced anchors or missing hold-downs). This section allows archaic materials (e.g., an unreinforced brick foundation) if they can be shown by analysis or testing to work within the context of this chapter. The user is referred to Appendix D of FEMA P-807 (FEMA, 2012) for a summary of component testing (monotonic or backbone load-deflection curves) of commonly used materials for existing construction, retrofit and combinations thereof.

[BS] A405.3.1 Wood-structural-panel shear walls.

[BS] A405.3.1.1 Existing nails. When the required calculations rely on design values for common nails or surfaced dry lumber, their use in construction shall be verified by exposure.

❖ This section is intended to ensure that assumptions made in the analysis phase are reasonable. If an engineer is going to assume that the lumber installed during original construction was dry or that common nails were used, these items should be verified prior to analysis.

[BS] A405.3.1.2 Existing plywood. When verification of the existing plywood is by use of record drawings alone, plywood shall be assumed to be of three plies.

❖ Visual verification usually requires no more than a small core sample. If a sample is not taken and reliance is simply on the design drawing, then a conservative assumption of three plies is required to be made.

[BS] A405.3.2 Existing wood framing. Wood framing is permitted to use the design stresses specified in the building code under which the building was constructed or other stress criteria approved by the *code official*.

❖ Wood design stresses have changed significantly over the years; for the purposes of this chapter, it is acceptable to use the original design stresses.

[BS] A405.3.3 Existing structural steel. All existing structural steel shall be permitted to be assumed to comply with ASTM A36. Existing pipe or tube columns shall be assumed to be of minimum wall thickness unless verified by testing or exposure.

❖ This section is intended to ensure that conservative assumptions regarding the strength of steel elements are used; if higher strengths are needed, the physical characteristics of the steel must be verified.

[BS] A405.3.4 Existing concrete. All existing concrete footings shall be permitted to be assumed to be plain concrete with a compressive strength of 2,000 pounds per square inch (13.8 MPa). Existing concrete compressive strength taken greater than 2,000 pounds per square inch (13.8 MPa) shall be verified by testing, record drawings or department records.

❖ This section is intended to ensure that conservative assumptions regarding the strength of concrete elements are used; if higher strengths are needed, the physical characteristics of the concrete must be verified.

[BS] A405.3.5 Existing sill plate anchorage. The analysis of existing cast-in-place anchors shall be permitted to assume proper anchor embedment for purposes of evaluating shear resistance to lateral loads.

❖ This provision should be understood to mean that evaluation of existing cast-in-place anchors for shear resistance may reasonably assume anchor embedment sufficient to develop the capacity of the anchor, as limited by the wood sill plate.

SECTION A406
INFORMATION REQUIRED TO BE ON THE PLANS

[BS] A406.1 General. The plans shall show all information necessary for plan review and for construction and shall accurately reflect the results of the engineering investigation and design. The plans shall contain a note that states that this retrofit was designed in compliance with the criteria of this chapter.

❖ The requirements of this section are in addition to any other documentation required by the IBC or the code official.

Documentation requirements regarding calculated member capacities, design forces or results of engineering analysis may be waived when the prescriptive measures of Section A404 are used.

This provision is intended to ensure that the designer has accounted for existing conditions likely to compromise the seismic load path or interfere with the intended details. An investigation should verify representative load path details. Specific conditions might require additional field investigation and detailing during the construction phase.

[BS] A406.2 Existing construction. The plans shall show existing diaphragm and shear wall sheathing and framing materials; fastener type and spacing; diaphragm and shear wall connections; continuity ties; and collector elements. The plans shall also show the portion of the existing materials that needs verification during construction.

❖ Details of existing construction need only be shown for the areas within the scope of the retrofit as indicated by Section A403.2.

The provisions in this chapter that address verification include Sections A405.3 (general condition assessment), A404.1.1 (diaphragm type), A404.2.1 (anchorage), A405.3.1 (wood shear wall construction), A405.3.3 (steel tube thickness) and A405.3.4 (concrete strength).

[BS] A406.3 New construction.

[BS] A406.3.1 Foundation plan elements. The foundation plan shall include the size, type, location and spacing of all anchor bolts with the required depth of embedment, edge and end distance; the location and size of all shear walls and all columns for braced frames or moment frames; referenced details for the connection of shear walls, braced frames or moment-resisting frames to their footing; and referenced sections for any grade beams and footings.

[BS] A406.3.2 Framing plan elements. The framing plan shall include the length, location and material of shear walls; the location and material of frames; references on details for the column-to-beam connectors, beam-to-wall connections and shear transfers at floor and roof diaphragms; and the required nailing and length for wall top plate splices.

[BS] A406.3.3 Shear wall schedule, notes and details. Shear walls shall have a referenced schedule on the plans that includes the correct shear wall capacity in pounds per foot (N/m); the required fastener type, length, gage and head size; and a complete specification for the sheathing material and its thickness. The schedule shall also show the required location of 3-inch (76 mm) nominal or two 2-inch (51 mm) nominal edge members; the spacing of shear transfer elements such as framing anchors or added sill plate nails; the required holddown with its bolt, screw or nail sizes; and the dimensions, lumber grade and species of the attached framing member.

Notes shall show required edge distance for fasteners on structural wood panels and framing members; required flush nailing at the plywood surface; limits of mechanical penetrations; and the sill plate material assumed in the design. The limits of mechanical penetrations shall also be detailed showing the maximum notching and drilled hole sizes.

❖ See the commentary to Section A406.1 regarding documentation of member capacities on plans.

This provision is intended to document new and modified elements. The extensive list indicates the importance of quality control for shear walls.

"Complete specification" means enough information to ensure selection of the correct product by the builder and to allow inspection by the authority having jurisdiction. Reference to a material standard and a panel grade designation is generally sufficient (for example, "United States Voluntary Product Standard PS 1-95, Structural I").

[BS] A406.3.4 General notes. General notes shall show the requirements for material testing, special inspection and structural observation.

❖ In accordance with Section A403.1, the testing and inspection requirements are the same as those for new construction. For example, see IBC Chapter 17.

SECTION A407
QUALITY CONTROL

[BS] A407.1 Structural observation, testing and inspection. Structural observation, in accordance with Section 1709 of the *International Building Code*, shall be required for all structures in which seismic retrofit is being performed in accordance with this chapter. Structural observation shall include visual observation of work for conformance to the approved construction documents and confirmation of existing conditions assumed during design.

Structural testing and inspection for new construction materials shall be in accordance with the building code, except as modified by this chapter.

❖ In accordance with Section A403.1, the structural observation requirements are generally the same as those for new construction. For example, see IBC Section 1708. However, retrofit projects routinely involve unanticipated conditions. The designer and builder should make allowances for such conditions when developing a project schedule and quality control plan.

Bibliography

The following resource materials were used in the preparation of the commentary for this chapter of the code.

BSSC, *NEHRP Recommended Provisions for Seismic Regulations for New Buildings and Other Structures (FEMA 450-2/2003 Edition), Part 2: Commentary*. Washington, DC: Building Seismic Safety Council, 2004.

Cobeen, K., J. Russell, and D.J. Dolan., *Recommendations for Earthquake Resistance in the Design and Construction of Woodframe Buildings.* (CUREE Publication No. W-30b), 2004.

CUREE, *Recommendations for Earthquake Resistance in the Design and Construction of Woodframe Buildings, Part I—Recommendations and Part II—Topical Discussions* (Final Review Draft). Richmond, CA: Consortium of Universities for Research in Earthquake Engineering, November 15, 2002.

FEMA 547/ICSSC-7, *Techniques for the Seismic Rehabilitation of Existing Buildings.* Washington, DC: Federal Emergency Management Agency, 2006.

Harder, Robert. *City of Los Angeles Department of Building and Safety and Structural Engineers Association of Southern California Joint Task Force: Wood Frame Construction Code Recommendations.* Proceedings: 63rd Annual Convention (1994 SEAOC Convention): Structural Engineers Association of California, 1994.

Harris, S.K., C. Scawthorn, and J. Egan, *Damage in the Marina District of San Francisco in the October 17, 1989 Loma Prieta Earthquake.* Tokyo: Proceedings of the 8th Japan Earthquake Engineering Symposium, 1990.

Holmes, W. and P. Somers, editors. "Northridge Earthquake of January 17, 1994." Reconnaissance Report, Volume 2. Supplement C to Volume 11. *Earthquake Spectra*, January 1996.

LADBS, *Multistory Wood Frame Buildings with Nonconforming Structural Materials and Tuck-under Parking Features: Summary of Wood Light-frame Building Subcommittee (Draft version 5.7.99).* LADBS/ SCEC/SEAOSC Ground Motion Joint Task Force Committee, March 1999.

Mendes, S. *Lessons Learned from Four Earthquake Damaged Multi-story Type V Structures.* Proceedings: 64th Annual Convention. Indian Wells, CA: 1995 SEAOC Convention, 1995.

Ni, Chun and Erol, Karacabeyli, *Effect of Overturning Restraint on Performance of Shear Wall.* WCTC Conf., 2000.

Rutherford and Chekene. *Seismic Rehabilitation of Three Model Buildings with Tuckunder Parking: Engineering Assumptions and Cost Information.* San Jose, CA: Department of Housing and Office of Emergency Services, May 2000.

Searer, G.R, J. Valancius and K.E. Cobeen, *Soft/Weak Story Problems and Solutions for Residential Structures.* Applied Technology Council and Structural Engineering Institute Conference on Improving the Seismic Performance of Existing Buildings and Other Structures. San Francisco, CA: December 2009.

Seismology Committee, SEAOC, *Recommended Lateral Force Requirements and Commentary, Seventh Edition.* Sacramento, CA: Structural Engineers Association of California, 1999.

Vukazich, S. *The Apartment Owner's Guide to Earthquake Safety.* San Jose, CA: Residential Seismic Safety Program, August 1998.

FEMA P-807, Seismic Evaluation and Retrofit of Multi-Unit Wood Frame Buildings with Weak First Stories. Washington, DC: Federal Emergency Management Agency, 2012.

Chapter A5:
Earthquake Hazard Reduction in Existing Concrete Buildings

Purpose

Appendix Chapter A5, along with the rest of Appendix A of the code, is referenced by *International Codes*® in two ways:

- Section 301.1.4.2 allows the use of Appendix A for qualifying buildings when "reduced IBC level seismic forces" are triggered.
- Section 807.6 allows the use of Appendix A for voluntary seismic retrofit.

Appendix Chapter A5 addresses structures where the gravity (and seismic) load-resisting system consists of bearing walls, columns and beams of reinforced cast-in-place concrete construction. The ability of these structures to dissipate seismic energy is strongly dependent on the detailing of the reinforcing. Older reinforced concrete buildings with inadequate detailing can experience sudden, brittle failures, as opposed to controlled local damage. Past performance has shown that strong ground shaking can cause these buildings to be heavily damaged or collapse, causing serious injury or death to the occupants or passers-by.

Code-required detailing for reinforced concrete frame seismic force-resisting systems ("space frame-ductile moment resisting") changed in the 1967 *Uniform Building Code* (UBC). The effectiveness of these design and detailing revisions was confirmed by damage surveys made after the 1971 San Fernando earthquake. The 1967 UBC changes were only applicable to UBC "zones" 2 and 3 in a seismic hazard mapping that has little relationship with current *International Building Code*® (IBC®) hazard mapping. Though there were few changes in detailing requirements going from the 1967 UBC to the 1976 UBC, the seismic loading requirements of the 1976 UBC were significantly higher. Buildings designed in accordance with the 1976 UBC can therefore be expected to have superior performance, and are selected as the benchmark building in Section A502.

Because attention to detailing was typically not a priority prior to 1967, caution is recommended for accepting what appears to be adequate detailing on contract documents for pre-1967 buildings without verification by destructive and nondestructive methods. The adoption dates in Section A502.1 [the 1992 BOCA National Building Code of Canada (NBC)], 1993 Standard Building Code (SBC) and 1976 UBC) give reasonable confidence that "special reinforced concrete moment frames" will have the detailing required for such structures.

Substantial damage and collapse of concrete buildings with inadequate detailing was observed in the San Fernando earthquake of 1971 (U.S. Department of Commerce, 1973), the Mexico City earthquake of 1985 (EERI, 1988), the Whittier Narrows earthquake of 1987 (EERI, 1988), the Loma Prieta earthquake of 1989 (EERI, 1990), the Northridge earthquake of 1994 (EERI, 1996), the Kobe (Japan) earthquake of 1995 (EERI, 1995), the Izmit (Turkey) earthquake of 1999 (EERI, 2000) and the Chi-Chi (Taiwan) earthquake of 1999 (EERI, 2001). Also, more recent earthquakes such as the 2009 earthquake in l'Aquila, Italy (Progettazione Sismica, 2009) and the 2011 Christchurch earthquake in New Zealand caused substantial damage to concrete buildings. Further, a disproportionate loss of life was associated with the collapse of a limited number of concrete buildings with inadequate detailing because the occupancy of these so-called nonductile buildings tends to be above average, the relative mass (weight) of floor and roof construction is several times greater than steel frame and light-framed buildings, and stiffness and strength degradation of reinforced concrete systems significantly affects the fragility of the gravity load-carrying system. Stiffness degradation (i.e., cracking of the concrete) may result in concentration of nonlinear displacements in a few stories. This may exacerbate shear strength degradation in plastic hinge zones and result in loss of vertical support in walls and columns. Therefore, the consensus among code officials and engineers is that existing concrete buildings covered by this chapter can pose a significant seismic risk.

The impact of improved knowledge, codes and design practices will be limited to those relatively few recent buildings that take advantage of this improved knowledge. Thus, the dominant policy issues posed by earthquakes involve not new but existing buildings, particularly those structures that have obvious weaknesses and do not comply with the general intent and necessary requirements of current regulations. The issue before the public and the professionals is how to set standards for these noncompliant existing buildings consistent with both the desire for safety and the limited resources available to achieve improved safety.

It is reasonable to take steps to significantly reduce this risk. The objective of Appendix Chapter A5 is the reduction or mitigation of risk to the greatest extent practicable. Application of these provisions will decrease the probability of loss of life, but without a guarantee that loss of life will be prevented. Also, the possibility of major, perhaps irreparable, damage should be accepted as long as there is a decrease in the likelihood of collapse. The provisions in this chapter were derived

APPENDIX A

from several sources. The three-tiered approach is based on ASCE 41, sponsored by the Federal Emergency Management Agency (FEMA), which specifies methodologies to analyze existing concrete buildings. These provisions further draw on the provisions for analysis of existing reinforced concrete buildings developed by the City of Los Angeles (Chapter 95 of the *City of Los Angeles Building Code*) in response to California Senate Bill 547. Finally, this chapter builds on ongoing activities of the Structural Engineers Association of California's (SEAOC) Existing Building Committee related to the analysis of nonductile concrete buildings, such as Appendix Chapter A5 of the *Guidelines for Seismic Retrofit of Existing Buildings* (GSREB) (ICC, 2001).

The performance goal of Appendix Chapter A5 is lower than the performance goal set for new construction since the force level used for Tier 2 and Tier 3 analyses is 75 percent of current code. Use of a reduced seismic design force corresponds to a "reduced rehabilitation objective (Section 2.2.3.1 of ASCE 41-13). It should be recognized that the economic costs and difficulty of strengthening existing buildings necessitates reliance on building components with seismic performance characteristics that are less than ideal.

Analysis of concrete moment-resisting frame buildings with unreinforced masonry infill is no longer covered in Appendix Chapter A5.

SECTION A501
PURPOSE

[BS] A501.1 Purpose. The purpose of this chapter is to promote public safety and welfare by reducing the risk of death or injury that may result from the effects of earthquakes on concrete buildings and concrete frame buildings.

The provisions of this chapter are intended as minimum standards for structural seismic resistance, and are established primarily to reduce the risk of life loss or injury. Compliance with the provisions in this chapter will not necessarily prevent loss of life or injury or prevent earthquake damage to the rehabilitated buildings.

❖ The purpose of this chapter is to identify and reduce the likelihood of significant earthquake damage and collapse of existing concrete buildings designed and constructed prior to the adoption of the building codes listed in Section A502.

SECTION A502
SCOPE

[BS] A502.1 Scope. The provisions of this chapter shall apply to all buildings having concrete floors or roofs supported by reinforced concrete walls or by concrete frames and columns. This chapter shall not apply to buildings with roof diaphragms that are defined as flexible diaphragms by the building code, and shall not apply to concrete frame buildings with masonry infilled walls.

Buildings that were designed and constructed in accordance with the seismic provisions of the 1993 *BOCA National Building Code*, the 1994 *Standard Building Code*, the 1976 *Uniform Building Code*, the 2000 *International Building Code* or later editions of these codes shall be deemed to comply with these provisions, unless the seismicity of the region has increased since the design of the building.

Exception: This chapter shall not apply to buildings assigned to Risk Category IV.

❖ The exception in the first paragraph of Section A502 about concrete buildings "with roof diaphragms that are defined as flexible diaphragms" is related to a definition in Section 12.3.1.1 of ASCE 7-10: "Diaphragms of untopped steel decking or wood structural panels are permitted to be idealized as 'flexible' for concrete shear wall buildings." Note that this exception is not applicable for diaphragms of concrete moment frames. This chapter does not cover concrete buildings with flexible diaphragms such as untopped steel decking or wood structural panels.

Starting with the 2009 edition of the code, analysis of concrete moment-resisting framed buildings with unreinforced masonry infill is no longer covered in Appendix Chapter A5. The complex behavior of these buildings does not fit the three-tiered approach adopted in this chapter.

The concept of "benchmark building" used in ASCE 41 is adopted to limit the scope of this chapter. While benchmark buildings need not proceed with further evaluation, the design professional should demonstrate that the building is compliant with the benchmark document. Benchmark buildings are deemed likely to have improved performance compared to prebenchmark buildings and to not require strengthening "unless the mapped seismicity of the region has increased since the design of the building." This provision enables the authority having jurisdiction to decide whether newer buildings designed in accordance with the seismic provisions of more recent building codes comply. Unintentionally, this exception could be misinterpreted to nearly eliminate the concept of the benchmark building, and the design professional is encouraged to verify what the local interpretation of these provisions is before judging a building compliant based on the building code used for its design.

Chapter A5 is intended to improve a building's performance with respect to safety but not necessarily with respect to post-earthquake functionality or recovery. As such, it is not applicable to buildings assigned to Risk Category IV. Previously, this exemption had been applied to Seismic Design Category A because the perceived low seismicity made these provisions unnecessary, but that limitation has been removed from this section and Chapter 3.

SECTION A503
GENERAL REQUIREMENTS

[BS] A503.1 General. This chapter provides a three-tiered procedure to evaluate the need for *seismic rehabilitation* of existing concrete buildings. The evaluation shall show that the *existing building* is in compliance with the appropriate part of the evaluation procedure as described in Sections A505, A506 and A507, or shall be modified to conform to the respective acceptance criteria. This chapter does not preclude a building from being evaluated or modified to conform to the acceptance criteria using other well-established procedures, based on rational methods of analysis in accordance with principles of mechanics and approved by the authority having jurisdiction.

❖ The term "compliant" (compliance) is used in Chapter 3. Determination of "compliance" by the requirements of the appendix chapters is defined in Section 301.1.4.2. As a result of the use of reduced IBC level seismic forces, achieving compliance through Tier 1, 2 or 3 analysis has a significant impact on the costs of the resulting repair and retrofit requirements.

If the building fails acceptance by Tier 1 evaluation, the engineer may reanalyze the structural system by Tier 2 procedures. If the building fails acceptance by the Tier 2 procedure, the Tier 3 procedure may be used for acceptance of the existing structural system. If the building cannot be shown to work by any of these methods, the building shall be strengthened so that it meets the acceptance criteria of one or more of the three tiers.

[BS] A503.2 Properties of cast-in-place materials. Except where specifically permitted herein, the stress-strain relationship of concrete and reinforcement shall be determined from published data or by testing. All available information, including building plans, original calculations and design criteria, site observations, testing and records of typical materials and construction practices prevalent at the time of construction, shall be considered when determining material properties.

For Tier 3 analysis, lower-bound and expected material properties shall be established in accordance with Section 10.2 of ASCE 41.

❖ All available data relating to the design and detailing of the existing building shall be used in the analysis. If data is unavailable or limited, the procedures of ASCE 41-13, Section 4.2.3, Default Properties shall be used for Tier 1 procedures.

The material properties used in Tier 2 procedures shall be nominal values. When sufficient information regarding the material properties is not available in the original design and construction documents, material properties shall be determined by sampling and testing as specified in Section 10.2 of ASCE 41-13. Nominal material values used in Tier 2 analyses should be reduced by the capacity reduction factors of ACI 318, in accordance with Section A506.5.2.

[BS] A503.3 Structural observation, testing and inspection. Structural observation, in accordance with Section 1709 of the *International Building Code,* shall be required for all structures in which seismic retrofit is being performed in accordance with this chapter. Structural observation shall include visual observation of work for conformance to the approved construction documents and confirmation of existing conditions assumed during design.

Structural testing and inspection for new construction materials shall be in accordance with the building code, except as modified by this chapter.

❖ Similar to new construction, structural observation is critical for quality control during construction of any retrofit of existing buildings. The structural observation required by Section A503.3 will help ensure that the construction conforms to the approved construction documents. In addition, structural observation can also provide a unique opportunity for confirmation of structural details that may affect the modeling and design assumptions.

Structural testing and inspection for new construction materials shall be in accordance with the building code.

SECTION A504
SITE GROUND MOTION

[BS] A504.1 Site ground motion for Tier 1 analysis. The earthquake loading used for the determination of demand on elements of the structure shall correspond to that required by ASCE 41 Chapter 4.

❖ The seismic force used for Tier 1 analyses may be higher than that used for Tier 3 analyses to limit the possibility that a building with marginal strength and detailing will be determined compliant with Appendix Chapter A5.

[BS] A504.2 Site ground motion for Tier 2 analysis. The earthquake loading used for the determination of demand on elements and the structure shall conform to 75 percent of that required by the building code.

❖ The earthquake loading used for determination of the demand on elements for a Tier 2 analysis is reduced to 75 percent of that required by current code. This earthquake loading is increased by the importance factor of Table 1.5-2 of ASCE 7-10, when applicable. This reduction in earthquake demand is consistent with Chapter 2 of ASCE 41-13. The 0.75 factor is consistent with historical retrofit requirements, is identical to that used in ASCE 31-03 for Tier 3 analyses and near identical to that used for FEMA 178/ June 1992 NEHRP *Handbook for the Seismic Evaluation of Existing Buildings* analyses.

[BS] A504.3 Site ground motion for Tier 3 analysis. The site ground motion shall be an elastic design response spectrum prepared in conformance to the building code but having spectral acceleration values equal to 75 percent of the code design response spectrum. The spectral acceleration values shall be increased by the occupancy importance factor when required by the building code.

❖ The loading criterion for a Tier 3 analysis is an elastic design response spectrum compatible with mapped spectral acceleration and velocity parameters as modified by site soil class factors. The spectral acceleration values shall be reduced to 75 percent of the design-basis ground motion similar to that permitted for a Tier 2 analysis. The spectral acceleration values are increased by occupancy importance factors related to the building's risk category when applicable.

SECTION A505
TIER 1 ANALYSIS PROCEDURE

[BS] A505.1 General. Structures conforming to the requirements of the ASCE 41 Chapter 4, Screening Phase, are permitted to be shown to be in conformance to this chapter by submission of a report to the building official, as described in this section.

❖ This section provides a means for conducting simplified and rapid evaluations of existing concrete buildings. This section is not intended to be used for concrete-framed buildings with masonry infill. A few "quick checks" require analysis as part of the checklist process. The analysis is based on the simplified equivalent lateral force procedure of Section 4.5 of ASCE 41, using BSE-1E.

[BS] A505.2 Evaluation report. The registered design professional shall prepare a report summarizing the analysis conducted in compliance with this section. As a minimum, the report shall include the following items:

1. Building description.
2. Site inspection summary.
3. Summary of reviewed record documents.
4. Earthquake design data used for the evaluation of the building.
5. Completed checklists.
6. Quick-check analysis calculations.
7. Summary of deficiencies.

❖ The evaluation report is intended to summarize the evaluation performed and facilitate review by the code official. The report should state clearly that only compliant items exist in the building and as such no rehabilitation measures are necessary, or noncompliant items do exist in the building and what subsequent steps are to be followed by the design professional to address these items. As noted in Section A503.1, the design professional has the option of performing subsequent analyses (Tiers 2 or 3) to show that the noncompliant items are acceptable or, if not acceptable, to pursue a mitigation effort to correct the deficiencies noted from the Tier 1 analysis.

When a subsequent analysis is performed, the report required under this section is intended to be used as supporting documentation for limiting the analysis to only the noncompliant items.

SECTION A506
TIER 2 ANALYSIS PROCEDURE

[BS] A506.1 General. A Tier 2 analysis includes an analysis using the following linear methods: Static or equivalent lateral force procedures. A linear dynamic analysis may be used to determine the distribution of the base shear over the height of the structure. The analysis, as a minimum, shall address all potential deficiencies identified in Tier 1, using procedures specified in this section.

If a Tier 2 analysis identifies a nonconforming condition, such condition shall be modified to conform to the acceptance criteria. Alternatively, the design professional may choose to perform a Tier 3 analysis to verify the adequacy of the structure.

❖ The Tier 2 linear analysis procedure is the equivalent lateral force procedure described in Section 12.8 of ASCE 7-10. This procedure uses a single "global" response modification factor, R, for the entire structure. This approach differs from the linear static procedure analysis in ASCE 41, which employs a "component-based" ductility related factor, m, in checking the acceptability of the component evaluated.

The R factor for Tier 2 analyses shall be selected based on the type of existing seismic force-resisting system given in Table 12.2-1 of ASCE 7; ordinary reinforced concrete moment frame, $R = 3$; ordinary reinforced concrete shear walls, $R = 4$: ordinary reinforced shear walls in building frame, $R = 5$. The Ω_o and C_d factors associated with the R factors shall be used when appropriate. Classification of seismic force-resisting systems as other than "ordinary" for selection of an R-factor from Table 12.2-1 of ASCE 7 is only allowed as specified in Sections 301.1.4.1, Item 1, and 301.1.4.2 of the code. It is unlikely that an existing building will meet all the provisions of the current code to qualify as a system other than "ordinary."

Acceptance of the building using the Tier 2 procedure is based on having the flexural strength, axial strength, shear strength and connection strength that are prescribed by referenced standards given in the IBC for resistance to the reduced seismic loading. These referenced standards include ACI 318 and ASCE 7.

[BS] A506.2 Limitations. A Tier 2 analysis procedure may be used if:

1. There is no in-plane offset in the lateral force-resisting system.
2. There is no out-of-plane offset in the lateral force-resisting system.
3. There is no torsional irregularity present in any story. A torsional irregularity may be deemed to exist in a story when the maximum story drift, computed including

accidental torsion, at one end of the structure transverse to an axis is more than 1.2 times the average of the story drifts at the two ends of the structure.

4. There is no weak story irregularity at any floor level on any axis of the building. A weak story is one in which the story strength is less than 80 percent of that in the story above. The story strength is the total strength of all seismic-resisting elements sharing the story shear for the direction under consideration.

Exception: Static or equivalent lateral force procedures shall not be used if:

1. The building is more than 100 feet (30 480 mm) in height.
2. The building has a vertical mass or stiffness irregularity (soft story). Mass irregularity shall be considered to exist where the effective mass of any story is more than 150 percent of the effective mass of any adjacent story. A soft story is one in which the lateral stiffness is less than 70 percent of that in the story above or less than 80 percent of the average stiffness of the three stories above.
3. The building has a vertical geometric irregularity. Vertical geometric irregularity shall be considered to exist where the horizontal dimension of the lateral force-resisting system in any story is more than 130 percent of that in an adjacent story.
4. The building has a nonorthogonal lateral force-resisting system.

❖ These limits for use of equivalent lateral force design procedures are similar to those limitations given in Table 12-6 and in Section 12.7.3 of ASCE 7. Linear static procedures are most applicable to buildings with regular geometries and prescribed distributions of stiffness, strength and mass.

For buildings with plan irregularities, soft stories, or weak stories, the inelastic ductility demand may significantly differ from the result of linear analyses; therefore, Tier 3 analysis procedures shall be used. For buildings with vertical irregularities or significant higher mode response, linear dynamic procedures may provide a more accurate vertical distribution of seismic demand as compared to the linear static procedure.

[BS] A506.3 Analysis procedure. A structural analysis shall be performed for all structures in accordance with the requirements of the building code, except as modified in Section A506. The response modification factor, R, shall be selected based on the type of seismic force-resisting system employed and shall comply with the requirements of Section 301.1.4.1.

❖ The stiffness of any component shall be calculated using the values given in Table 10-5 of ASCE 41-13. The building period calculated using a percentage of the gross stiffness to account for effective stiffness may exceed and is permitted to exceed the upper limit of calculated period given in the IBC. To determine the effective stiffness of a flat plate or flat slab, the design professional shall refer to available research data. The empirical limits on elastic fundamental period given in the IBC are not appropriate for analysis of the diversity of structural systems found in existing buildings.

[BS] A506.3.1 Mathematical model. The three-dimensional mathematical model of the physical structure shall represent the spatial distribution of mass and stiffness of the structure to an extent that is adequate for the calculation of the significant features of its distribution of lateral forces. All concrete and masonry elements shall be included in the model of the physical structure.

Exception: Concrete or masonry partitions that are isolated from the concrete frame members and the floor above.

Cast-in-place reinforced concrete floors with span-to-depth ratios less than three-to-one may be assumed to be rigid diaphragms. Other floors, including floors constructed of precast elements with or without a reinforced concrete topping, shall be analyzed in conformance to the building code to determine if they must be considered semi-rigid diaphragms. The effective in-plane stiffness of the diaphragm, including effects of cracking and discontinuity between precast elements, shall be considered. Parking structures that have ramps rather than a single floor level shall be modeled as having mass appropriately distributed on each ramp. The lateral stiffness of the ramp may be calculated as having properties based on the uncracked cross section of the slab exclusive of beams and girders.

❖ The mathematical model of the building must be three dimensional (3-D) and include all concrete elements that will be deformed by lateral forces. The exception allows isolated partitions to be omitted from the 3-D model. The partitions must be isolated from the structure at the top and along both vertical edges by joints adequate to accommodate expected interstory drifts. Section 12.7.3 of ASCE 7-10 describes the elements that must be included in the 3-D model. This 3-D model shall be used for analysis on each axis of the building for compliance with applicable subsections of Section 12.5 of ASCE 7-10. Cast-in-place reinforced concrete floor systems may be assumed to act as rigid diaphragms if in conformance with the specified span-to-depth ratio of Section 12.3.1.1 of ASCE 7-10. Other floor systems must be analyzed in conformance with Section 12.3.1.3 of ASCE 7-10 to confirm that the floor diaphragm is not flexible. If the diaphragm is determined to be not flexible, the horizontal flexibility of the diaphragm must be included in the 3-D model. This definition of when a diaphragm is rigid or semirigid may be related to its orthogonal axis; the definition may be different along each axis of a rectangular building. In this case, use of a plate element as representative of the orthogonal horizontal stiffness of the diaphragm may simplify the analysis.

Parking structures that consist of ramps in lieu of horizontal floor levels shall have the mass distributed along the ramp length to determine their expected in-plane deformations. The lateral stiffness of the ramps shall be the uncracked stiffness without stiffening effects of edge beams and girders. Columns that are restrained by floor beams at each adjacent ramp level may have critical shear stress induced by the floor beams and should be carefully considered in the model.

[BS] A506.3.2 Component stiffness. Component stiffness shall be calculated based on the approximate values shown in ASCE 41 Table 10-5.

❖ The ranges of approximate percentages of gross stiffness given in ASCE 41–reduce the element's gross section properties to an effective stiffness property. Cracking of the concrete in the elastic stress range and additional cracking in nonlinear behavior range must be considered. Effective stiffness properties are required for consideration of the relative rigidities of reinforced concrete elements and for linear dynamic procedures.

[BS] A506.4 Design, detailing requirements and structural component load effects. The design and detailing of new components of the seismic force-resisting system shall comply with the requirements of the *International Building Code*, unless specifically modified herein.

❖ The design and detailing of new elements of a supplemental seismic force-resisting system shall conform to the requirements of ACI 318 Chapter 18 and as modified by IBC Chapter 19.

[BS] A506.5 Acceptance criteria. The calculated strength of a member shall be not less than the load effects on that member.

❖ The strength of elements calculated in conformance with the IBC shall equal or exceed the load effects determined in conformance with the IBC.

[BS] A506.5.1 Load combinations. For load and resistance factor design (strength design), structures and all portions thereof shall resist the most critical effects from the combinations of factored loads prescribed in the building code.

Exception: For concrete beams and columns, the shear effect shall be determined based on the most critical load combinations prescribed in the building code. The shear load effect, because of seismic forces, shall be multiplied by a factor of C_d, but combined shear load effect need not be greater than V_e, as calculated in accordance with Equation A5-4. M_{pr1} and M_{pr2} are the end moments, assumed to be in the same direction (clockwise or counter clockwise), based on steel tensile stress being equal to $1.25 f_y$, where f_y is the specified yield strength.

$$V_e = \frac{M_{pr1} + M_{pr2}}{L} \pm \frac{W_g}{2}$$ (Equation A5-1)

where:

W_g = Total gravity loads on the beam

❖ The critical load combinations for strength design or load and resistance factor design of the IBC shall be used. The exception in this section requires that the shear demands from seismic forces be multiplied by C_d, but need not exceed the value of V_e calculated using Equation A5-1. The C_d factor is used in this section to calculate the post-yield moments. C_d is the appropriate magnifier as the maximum moment occurs when the plastic hinge is at its maximum curvature which occurs at the point of maximum drift. It should be noted that the plastic moments and V_e will generally place an upper bound on the design shear force.

The end moments in Equation A5-1 are based on steel tensile stress of 125 percent of specified yield stress. The total gravity load may be calculated using the lower of the unfactored dead and reduced live load, or the gravity component of the factored load combination with earthquake loads as applicable for new construction.

This requirement is intended to minimize the probability of a shear failure in components with flexural and axial loadings prior to attaining the nonlinear displacement assumed by the response reduction factor in the calculation of the seismic loading. The commentary to Chapter 4 of the 2003 NEHRP (FEMA 450) for Figures C4.2-1 and C4.2-3 shows the relationship of nonlinear displacement and nonlinear flexural moments in systems or elements. The elastic moment increases the strength limit moment (ultimate) as the nonlinear curvature of the plastic hinge increases. These post-yield moments can be calculated by assuming a strain of yield strain x C_D at the tension face and a strain of 0.003 at the compression face. Equation A5-1 simplifies the determination of the probable moments by limiting the increase to 125 percent of yield moment.

[BS] A506.5.2 Determination of the strength of members. The strength of a member shall be determined by multiplying the nominal strength of the member by a strength reduction factor, ϕ. The nominal strength of the member shall be determined in accordance with the building code.

❖ Strength of members shall be determined by reducing the nominal strength by strength reduction factors (ϕ) specified in ACI 318. ACI 318 is the referenced standard given in the IBC for design of reinforced concrete.

SECTION A507
TIER 3 ANALYSIS PROCEDURE

[BS] A507.1 General. A Tier 3 evaluation shall be performed using the Nonlinear Static Procedure or Nonlinear Dynamic Procedure of Section 10.3.1.2.2 of ASCE 41. The general assumptions and requirements of Section 10.3 of ASCE 41, excluding those for concrete frames with infills, shall be used in the evaluation. Reduced *International Building Code* level site-ground motions in accordance with Section A504.3 are permitted for this evaluation. Structures meeting the ASCE 41 Life Safety (LS) acceptance criteria shall be deemed to comply with this chapter. If a Tier 3 analysis identifies nonconforming conditions, such conditions shall be modified to conform to the acceptance criteria.

❖ The Tier 3 evaluation procedure is specified as a nonlinear procedure corresponding to ASCE 41. It is intended that life safety acceptance criteria be used. Use of linear static procedures is not permitted for Tier 3 analyses. The Tier 2 procedure uses the code equivalent lateral force procedure in lieu of the linear static procedures. If the building is shown noncompliant by a Tier 2 analysis, it would not be expected to be shown compliant by another linear analysis.

Engineers should familiarize themselves with ASCE 41 prior to beginning a Tier 3 analysis. ASCE 41 provides commentary in lieu of commentary in this document.

The procedure for Tier 3 analyses is described in Section 7.4.3 of ASCE 41.

The mathematical model of the building shall be subjected to monotonically increasing loads to 150 percent of target displacement, The target displacement shall be amplified for horizontal torsion, if applicable, in accordance with Section 7.2.3.2.2 of ASCE 41.

Acceptance criteria for nonlinear procedures shall be as prescribed in Section 7.5.3.2 of ASCE 41. Primary and secondary components shall have expected deformation capacities not less than the maximum deformation demands calculated at the target displacement. Component deformation limits are as described in Chapter 10 of ASCE 41 for the life safety performance level.

Force-controlled actions of primary and secondary components shall have strengths not less than the maximum design force.

For buildings with negative post-yield stiffness with target displacement greater than the displacement at maximum base shear, the maximum strength ratio, $\mu_{strength}$, shall be determined by Equation 7-32 of ASCE 41. If $\mu_{strength}$ exceeds μ_{max}, the nonlinear dynamic procedure shall be used to show compliance with reduced-level seismic forces. In lieu of this nonlinear dynamic procedure analysis, the lateral force-resisting system and/or gravity support system of this building may be deemed to be noncompliant and modified to conform to the acceptance criteria.

Bibliography

The following resource materials were used in the preparation of the commentary for this chapter of the code.

California Earthquake of February 9, 1971. San Fernando, CA: U.S. Department of Commerce National Oceanic and Atmospheric Administration,1973.

"Chi-Chi, Taiwan, Earthquake." *EERI Spectra*, Supplement to Vol. 17, April 2001 and Vol. 17, No. 4, November 2001.

"Earthquake Hazard Reduction in Existing Reinforced Concrete Buildings and Concrete Frame Buildings with Masonry Fill." City of Los Angeles Building Code—Volume 2, 2008.

EERI, *The Hyogo-Ken Nanbu Earthquake Great Hanshin Earthquake Disaster January 17, 1995*. Preliminary Reconnaissance Report, February 1995.

Guidelines for the Seismic Retrofit of Existing Buildings, International Code Council, Washington, DC: 2001.

"Izmut Turkey Earthquake, 1999." *EERI Spectra*. Supplement A to Vol. 16, December 2000.

"Loma Prieta Earthquake, 1989." *EERI Spectra*. Vol. 6 Supplement, May 1990.

"Mexico City Earthquake, 1985." *EERI Spectra*. Vol. 4. No. 3, August 1988

NEHRP *Handbook for the Seismic Evaluation of Existing Buildings (FEMA 178)*. Washington, DC: Federal Emergency Management Agency, 1992.

NEHRP, *Recommended Provisions and Commentary for Seismic Regulations for New Buildings and Other Structures (FEMA 450)*. Washington, DC: Federal Emergency Management Agency, 2003.

"Northridge Earthquake, 1994" *EERI Spectra*. Vol. 4. No. 3, 1994

Progettazione Sismica, Special Issue on the April 6th 2009 l'Aquila earthquake. IUSS Press, Pavia, Vol. 1, No. 3, 2009.

"Whittier Earthquake, 1987." *EERI Spectra*. Vol. 4. No.1, February 1988 and Vol. 4 No. 2, May 1988.

Chapter A6:
Referenced Standards

ASCE/SEI
American Society of Civil Engineers
Structural Engineering Institute
1801 Alexander Bell Drive
Reston, VA 20191-4400

Standard reference number	Title	Referenced in code section number
7—10	Minimum Design Loads for Buildings and Other Structures with Supplement No. 1	A104, A403.3
41—13	Seismic Rehabilitation of Existing Buildings	A503.2, A504.1, A505.1, A506.3.2, A507.1

ASTM
ASTM International
100 Barr Harbor Drive
West Conshohocken, PA 19428-2959

Standard reference number	Title	Referenced in code section number
A36/A36M-08	Specification for Carbon Structural Steel	A405.3.3
A 653/A653M—11	Standard Specification for Steel Sheet, Zinc Coated (Galvanized) or Zinc-Iron Alloy-Coated (Galvannealed) by Hot-Dip Process	A304.2.6
B 695—04(2009)	Standard Specification for Coating of Zinc Mechanically Deposited on Iron And Steel	A304.2.6
C 496—96/C496M-11	Standard Test Method for Splitting Tensile Strength of Cylindrical Concrete Specimens	A104, A106.3.3.2
E 488-10	Test Method for Strength of Anchors in Concrete and Masonry Elements	A107.5.3
E 519/E519M—2010	Standard Test Method for Diagonal Tension (Shear) in Masonry Assemblages	A104, A106.3.3.2

ICC
International Code Council
500 New Jersey Avenue, NW, 6th Floor
Washington, DC 20001

Standard reference number	Title	Referenced in code section number
BNBC—93	BOCA National Building Code®	A502
BNBC—96	BOCA National Building Code®	A502
BNBC—99	BOCA National Building Code®	A202
IBC—00	International Building Code®	A202.1, A502.1
IBC—03	International Building Code®	A202.1, A502.1
IBC—06	International Building Code®	A202.1, A502.1
IBC—09	International Building Code®	A202.1, A502.1
IBC—12	International Building Code®	A202.1, A502.1
IBC—15	International Building Code®	A102.2, A108.2, A202.1, A203, A206.3, A206.9, A403.1, A405.1, A407.1, A502.1, A503.3, A506.4,
SBC—94	Standard Building Code®	A502
SBC—97	Standard Building Code®	A502
SBC—99	Standard Building Code®	A202, A502
UBC—76	Uniform Building Code®	A502
UBC—97	Uniform Building Code®	A202, A502

Appendix B: Supplementary Accessibility Requirements for Existing Buildings and Facilities

The provisions contained in this appendix are not mandatory unless specifically referenced in the adopting ordinance.

General Comments

As stated in Section 101.6, the provisions of the appendices do not apply unless specifically adopted.

Chapter 11 of the *International Building Code®* (IBC®) contains provisions that set forth requirements for accessibility to buildings and their associated sites and facilities for people with physical disabilities. Sections 410, 605, 705, 806, 906, 1006, 1012.1.4, 1012.8, 1105, 1204.1 and 1205.15 in the code address accessibility provisions and alternatives permitted in existing buildings. Appendix B was added to address accessibility in construction for items that are not typically enforceable through the traditional building code enforcement process.

Section B101 deals with historical facilities that are registered through the National Historic Preservation Act (NHPA) or a similar state or local law. If such a facility undergoes an alteration or change of occupancy, it must comply with accessibility regulations as specified in Sections 1204 and 1205.15. If the required alterations would adversely affect some historic aspect of the structure, alternatives are available. Before these alternatives can be utilized, the owner must go through a process to evaluate the requirements and implications of each. Section B101.3 describes the process if the building is registered through the NHPA. Section B101.4 describes the process if the building is registered through a state or local registration law. Section B101.5 provides alternatives for displays in historical buildings.

Section B102 lists criteria specific to fixed transportation facilities and stations, such as train stations.

Section B103 contains provisions related to communication features for altered or added dwelling or sleeping units.

Section B104 lists the standards referenced in this appendix.

Purpose

Appendix B includes scoping requirements found in the *Americans with Disabilities Act Accessibility Guidelines* (ADAAG) that are not in the accessibility provisions in the IBC or the code. Items in this appendix deal with topics not typically addressed in building codes. For example, Section B101 specifies the process to follow regarding alterations or change of occupancy in historically registered buildings.

This appendix is being included in the code for the convenience of building owners and designers who are required to comply with ADAAG, and for the benefit of jurisdictions that may wish to go beyond the traditional boundary of the code and formally adopt these additional ADAAG requirements. The requirements in the code indicate what is required and enforceable by the code official. The appendix includes items that are outside the scope of code enforcement, but must be complied with in addition to the code requirements to satisfy ADAAG regulations.

SECTION B101
QUALIFIED HISTORICAL BUILDINGS AND FACILITIES

B101.1 General. Qualified historical buildings and facilities shall comply with Sections B101.2 through B101.5.

❖ Special considerations are given to registered historic buildings (see commentary, Sections 410.9, 1204 and 1205.15).

B101.2 Qualified historic buildings and facilities. These procedures shall apply to buildings and facilities designated as historic structures that undergo alterations or a *change of occupancy*.

❖ This section provides a process to evaluate historical registered buildings in the event where accessibility requirements may adversely affect the historical significance of the building.

B101.3 Qualified historic buildings and facilities subject to Section 106 of the National Historic Preservation Act. Where an *alteration* or *change of occupancy* is undertaken to a qualified *historic building* or facility that is subject to Section 106 of the National Historic Preservation Act, the federal agency with jurisdiction over the undertaking shall follow the Section 106 process. Where the state historic preservation officer or Advisory Council on Historic Preservation determines that compliance with the requirements for accessible routes, ramps, entrances, or toilet facilities would threaten or destroy the historic significance of the building or facility, the alternative requirements of Section 410.9 for that element are permitted.

❖ If a facility is subject to Section 106 of the NHPA, this section outlines the procedure to follow with the federal agency involved to allow the facility to utilize the alternatives offered in Section 410.9. This information is repeated in Sections 1204 and 1205.15.

B101.4 Qualified historic buildings and facilities not subject to Section 106 of the National Historic Preservation Act. Where an *alteration* or *change of occupancy* is undertaken to a qualified *historic building* or facility that is not subject to Section 106 of the National Historic Preservation Act, and the entity undertaking the alterations believes that compliance with the requirements for accessible routes, ramps, entrances, or toilet facilities would threaten or destroy the historic significance of the building or facility, the entity shall consult with the state historic preservation officer. Where the state historic preservation officer determines that compliance with the accessibility requirements for accessible routes, ramps, entrances, or toilet facilities would threaten or destroy the historical significance of the building or facility, the alternative requirements of Section 410.9 for that element are permitted.

❖ If a facility is not subject to Section 106 of the NHPA, this section outlines the procedure to follow with the state historic preservation officer to allow the facility to utilize the alternatives offered in Section 410.9. This information is repeated in Sections 1204 and 1205.15.

B101.4.1 Consultation with interested persons. Interested persons shall be invited to participate in the consultation process, including state or local accessibility officials, individuals with disabilities, and organizations representing individuals with disabilities.

❖ Some type of public forum or meeting must be held to allow for the input of interested parties that may wish to utilize the facilities offered in the historic building.

B101.4.2 Certified local government historic preservation programs. Where the state historic preservation officer has delegated the consultation responsibility for purposes of this section to a local government historic preservation program that has been certified in accordance with Section 101 of the National Historic Preservation Act of 1966 [(16 U.S.C. 470a(c)] and implementing regulations (36 CFR 61.5), the responsibility shall be permitted to be carried out by the appropriate local government body or official.

❖ Where the state has certified a local group in accordance with the NHPA to oversee activity in a historic district or area, that group can take on the role of the state historic preservation officer.

B101.5 Displays. In qualified historic buildings and facilities where alternative requirements of Section 1105 are permitted, displays and written information shall be located where they can be seen by a seated person. Exhibits and signs displayed horizontally shall be 44 inches (1120 mm) maximum above the floor.

❖ Displays, exhibits and signs in historic buildings must be located so that a person in a wheelchair can look at them.

SECTION B102
FIXED TRANSPORTATION FACILITIES AND STATIONS

B102.1 General. Existing fixed transportation facilities and stations shall comply with Section B102.2.

❖ Fixed transportation facilities and stations, such as train or light rail stations, provide access to public transportation and are considered an essential service; therefore, accessibility in these facilities is of special concern. Requirements for new construction can be found in Appendix E of the IBC.

B102.2 Existing facilities—key stations. Rapid rail, light rail, commuter rail, intercity rail, high-speed rail and other fixed guideway systems, altered stations, and intercity rail and key stations, as defined under criteria established by the Department of Transportation in Subpart C of 49 CFR Part 37, shall comply with Sections B102.2.1 through B102.2.3.

❖ Existing guideway system facilities are required to meet a lesser degree of accessibility specified in Sections B102.2.1 through B102.2.3.

B102.2.1 Accessible route. At least one accessible route from an accessible entrance to those areas necessary for use of the transportation system shall be provided. The accessible route shall include the features specified in Appendix E109.2 of the *International Building Code*, except that escalators shall comply with *International Building Code* Section 3004.2.2. Where technical unfeasibility in existing stations requires the accessible route to lead from the public way to a paid area of the transit system, an accessible fare collection machine complying with *International Building Code* Appendix E109.2.3 shall be provided along such accessible route.

❖ At least one accessible route is required throughout passenger areas in an existing facility. Existing escalators in subway stations are not required to meet the criteria for a 32-inch (813 mm) minimum clear width in IBC Section 3004.2.2. If it is technically infeasible to provide an accessible route from the entrance to a ticket sales area, some type of fare-taking machinery may be provided along the accessible route to the train platform.

B102.2.2 Platform and vehicle floor coordination. Station platforms shall be positioned to coordinate with vehicles in accordance with applicable provisions of 36 CFR Part 1192. Low-level platforms shall be 8 inches (250 mm) minimum above top of rail.

Exception: Where vehicles are boarded from sidewalks or street-level, low-level platforms shall be permitted to be less than 8 inches (250 mm).

❖ In existing facilities, the vertical separation between the platform and the cars is slightly less restrictive than in new construction. The gap between the platform and the cars is only measured at accessible cars. Retrofitted cars may have an even larger variance. Where train cars cannot meet the vertical separation and gap requirements, alternatives such as lifts and ramps may be utilized.

B102.2.3 Direct connections. New direct connections to commercial, retail, or residential facilities shall, to the maximum extent feasible, have an accessible route complying with Section 705.2 from the point of connection to boarding platforms and transportation system elements used by the public. Any elements provided to facilitate future direct connections shall be on an accessible route connecting boarding platforms and transportation system elements used by the public.

❖ New buildings that provide a direct connection to the existing transit systems must also have an accessible route to the transit system. The exceptions in Section 705.2 are applicable. This information is repeated in Section 410.7.

SECTION B103
DWELLING UNITS AND SLEEPING UNITS

B103.1 Communication features. Where dwelling units and sleeping units are altered or added, the requirements of Section E104.3 of the *International Building Code* shall apply only to the units being altered or added until the number of units with accessible communication features complies with the minimum number required for new construction.

❖ Communication features are required in a limited number of rooms in transient lodging and jails (see IBC Appendix E, Section E104.2). The intent is to allow for special features for people with hearing impairments in those facilities.

Where transient lodging dwelling and sleeping units are altered or added, only these units are counted to determine the number of units that will require accessible communication features. Once the entire building meets the number required for new construction, no additional accessible communication features are required. This is consistent with the intent in the code expressed for when Accessible and Type A units are required in existing buildings.

SECTION B104
REFERENCED STANDARDS

Y3.H626 2P National Historic Preservation J101.2, 43/933 Act of 1966, as amended J101.3, 3rd Edition, Washington, DC: J101.3.2 US Government Printing Office, 1993.

49 CFR Part 37.43 (c), Alteration of Transportation Facilities by Public Entities, Department of Transportation, 400 7th Street SW, Room 8102, Washington, DC 20590-0001.

APPENDIX C: Guidelines for the Wind Retrofit of Existing Buildings

Chapter C1: Gable End Retrofit for High-wind Areas

The provisions contained in this appendix are not mandatory unless specifically referenced in the adopting ordinance.

General Comments

The provisions in Appendix Chapter C1 are focused on the retrofit of gable ends to increase the resistance to the effects of wind for existing structures. This is one of two appendix chapters focused on providing prescriptive methods to increase wind resistance for existing structures. Appendix Chapter C2 addresses roof deck fastening requirements.

The content of both Appendix Chapters C1 and C2 is similar in concept to that provided in Appendix A chapters dealing with seismic retrofit requirements. It is anticipated that, over time, additional retrofit methods will be provided in this appendix as new Appendix C chapters. These retrofits are voluntary and, as such, may or may not meet the requirements of new construction. However, these voluntary measures will serve to better protect the public and reduce damage from high-wind events.

Because most of America's buildings located in hurricane-prone regions were not built to today's building code standards, there is significant value added to the code if the retrofitting of buildings could be facilitated by the provisions of prescriptive means. This would inherently reduce the cost of retrofitting. The need for structural retrofitting has been highlighted in the recent spate of hurricanes and the insurance crises that have developed in the coastal high-wind areas of a number of states because of older buildings that do not meet current building code structural requirements. Clearly, it is in the best interests of the public health, welfare and safety to facilitate retrofitting. Given the importance of retrofitting to the public, retrofitting of buildings should be encouraged and facilitated by removing as many impediments as possible. The code can actually facilitate and encourage retrofitting by providing prescriptive means. Such methods should encourage, facilitate and reduce the cost of improving America's building stock.

Purpose

The purpose of Appendix Chapter C1 is to provide prescriptive means for retrofitting gable ends to resist high winds. These provisions will facilitate the retrofitting of gable ends without requiring site-specific engineering for common applications, thus removing some of the obstacles that might impede this important retrofit in hurricane-prone regions. Gable end failures are one of the most common types of structural failures observed in hurricanes. They have been documented in most major hurricanes and in many weaker hurricanes.

The provisions are intended to be a prescriptive approach to reduce retrofitting costs, facilitate retrofitting, minimize the need for engineering and facilitate code review and inspection. Appendix Chapter C1 provides standardized off-the-shelf methods that can be readily approved and easily inspected by building department personnel. Building departments can thus become creditable third-party resources for authenticating retrofitting just as they do for other structural issues of buildings.

The provisions were added as an appendix for several reasons. The first being the ease in adding future Appendix C chapters. Further, by grouping retrofit measures into a separate chapter, users will find them and perhaps even use the chapter as a catalog of potential retrofit measures. Additionally, grouping voluntary measures into a separate chapter (separate from mandatory measures) reduces confusion in the application of the code.

Currently, Appendix Chapter C1 is broken into four major sections

- C101 General. Provides the purpose, eligibility criterion and applicability of the *International Building Code®* (IBC®) and *International Residential Code®* (IRC®).

- C102 Definitions. Provides definitions relative to the application of the appendix chapter.

- C103 Materials of Construction. Provides the parameters for the materials that can be used for the retrofit of gable ends.

- C104 Retrofitting Gable End Walls to Enhance Wind Resistance. Provides the methodologies for retrofitting the gable ends.

SECTION C101
GENERAL

[BS] C101.1 Purpose. This chapter provides prescriptive methods for partial structural retrofit of an existing building to increase its resistance to out-of-plane wind loads. It is intended for voluntary use and for reference by mitigation programs. The provisions of this chapter do not necessarily satisfy requirements for new construction. Unless specifically cited, the provisions of this chapter do not necessarily satisfy requirements for structural improvements triggered by addition, alteration, repair, change of occupancy, building relocation or other circumstances.

[BS] C101.2 Eligible buildings and gable end walls. The provisions of this chapter are applicable only to buildings that meet the following eligibility requirements:

1. The building is not more than three stories tall, from adjacent grade to the bottom plate of each gable end wall being retrofitted with this chapter.
2. The building is classified as Occupancy Group R3 or is within the scope of the *International Residential Code*.
3. The structure includes one or more wood-framed gable end walls, either conventionally framed or metal-plate-connected.

In addition, the provisions of this chapter are applicable only to gable end walls that meet the following eligibility requirements:

4. Each gable end wall has or shall be provided with studs or vertical webs spaced 24 inches (610 mm) on center maximum.
5. Each gable end wall has a maximum height of 16 feet (4877 mm).

[BS] C101.3 Compliance. Eligible gable end walls in eligible buildings may be retrofitted with this chapter. All other modifications required for conformance with this chapter shall be designed and constructed in accordance with the *International Building Code* or *International Residential Code* provisions for new construction, except as specifically provided for by this chapter.

SECTION C102
DEFINITIONS

The following words and terms shall, for the purposes of this chapter, have the meanings shown herein.

[BS] ANCHOR BLOCK. A piece of lumber secured to horizontal braces and filling the gap between existing framing members for the purpose of restraining horizontal braces from movement perpendicular to the framing members.

[BS] COMPRESSION BLOCK. A piece of lumber used to restrain in the compression mode (force directed towards the interior of the attic) an existing or retrofit stud. It is attached to a horizontal brace and bears directly against the existing or retrofit stud.

[BS] CONVENTIONALLY FRAMED GABLE END. A gable end framed with studs whose faces are perpendicular to the gable end wall.

[BS] GABLE END FRAME. A factory or site-fabricated frame, installed as a complete assembly that incorporates vertical webs with their faces parallel to the plane of the frame.

[BS] HORIZONTAL BRACE. A piece of lumber used to restrain both compression and tension loads applied by a retrofit stud. It is typically installed horizontally on the top of attic floor framing members (truss bottom chords or ceiling joists) or on the bottom of pitched roof framing members (truss top chord or rafters).

[BS] HURRICANE TIES. Manufactured metal connectors designed to provide uplift and lateral restraint for roof framing members.

[BS] NAIL PLATE. A manufactured metal plate made of galvanized steel with factory-punched holes for fasteners. A nail plate may have the geometry of a strap.

[BS] RETROFIT. The voluntary process of strengthening or improving buildings or structures, or individual components of buildings or structures for the purpose of making existing conditions better serve the purpose for which they were originally intended or the purpose that current building codes intend.

[BS] RETROFIT STUD. A lumber member used to structurally supplement an existing gable end wall stud or gable end frame web.

[BS] STUD-TO-PLATE CONNECTOR. A manufactured metal connector designed to connect studs to plates.

SECTION C103
MATERIALS OF CONSTRUCTION

[BS] C103.1 Existing materials. All existing wood materials that will be part of the retrofitting work (trusses, rafters, ceiling joists, top plates, wall studs, etc.) shall be in sound condition and free from defects or damage that substantially reduces the load-carrying capacity of the member. Any wood materials found to be damaged or deteriorated shall be strengthened or replaced with new materials to provide a net dimension of sound wood equivalent to its undamaged original dimensions.

[BS] C103.2 New materials. All new materials shall comply with the standards for those materials as specified in the *International Building Code* or the *International Residential Code*.

[BS] C103.3. Material specifications for retrofits. Materials for retrofitting gable end walls shall comply with Table C103.3.

[BS] C103.4 Twists in straps. Straps shall be permitted to be twisted or bent where they transition between framing members or connection points. Straps shall be bent only once at a given location though it is permissible that they be bent or twisted at multiple locations along their length.

APPENDIX C

[BS] C103.5 Fasteners. Fasteners shall meet the requirements of Table C103.5, Sections C103.5.1 and C103.5.2, and shall be permitted to be screws or nails meeting the minimum length requirement shown in the figures and specified in the tables of this appendix. Fastener spacing shall meet the requirements of Section C103.5.3.

[BS] C103.5.1 Screws. Unless otherwise indicated in the appendix, screw sizes and lengths shall be in accordance with Table C103.5. Permissible screws include deck screws and wood screws. Screws shall have at least 1 inch (25 mm) of thread. Fine threaded screws or drywall screws shall not be permitted. Select the largest possible diameter screw such that the shank adjacent to the head fits through the hole in the strap.

[BS] C103.5.2 Nails. Unless otherwise indicated in this appendix, nail sizes and lengths shall be in accordance with Table C103.5.

[BS] C103.5.3 General fastener spacing. Fastener spacing for shear connections of lumber-to-lumber shall meet the requirements shown in Figure C103.6.3 and the following conditions.

[BS] TABLE C103.3
MATERIAL SPECIFICATIONS FOR RETROFITS

COMPONENT	MINIMUM SIZE OR THICKNESS	MINIMUM MATERIAL GRADE	MINIMUM CAPACITY
Anchor blocks, compression blocks, and horizontal braces	2 x 4 nominal lumber	#2 Spruce-Pine-Fir or better	N/A
Nail plates	20 gage thickness 8d minimum nail holes	Galvanized sheet steel	N/A
Retrofit studs	2 x 4 nominal lumber	#2 Spruce-Pine-Fir or better	N/A
Gusset angle	14 gage thickness	Galvanized sheet steel	350 pounds uplift and lateral load
Stud-to-plate connector	20 gage thickness	Galvanized sheet steel	500 pounds uplift
Metal plate connectors, straps, and anchors	20 gage thickness	Galvanized sheet steel	N/A

For SI: 1 pound = 4.4 N.

N/A = Not applicable

a. Metal plate connectors, nail plates, stud-to-plate connectors, straps and anchors shall be products *approved* for connecting wood-to-wood or wood-to-concrete as appropriate.

[BS] TABLE C103.5
NAIL AND SCREW REQUIREMENTS

FASTENER TYPE	MINIMUM SHANK DIAMETER	MINIMUM HEAD DIAMETER	MINIMUM FASTENER LENGTH
#8 screws	N/A	0.28 inches	1-1/4 inches
8d common nails	0.131 inches	0.28 inches	2-1/2 inches
10d common nails	0.148 inches	0.28 inches	3 inches

For SI: 1 inch = 25.4 mm.

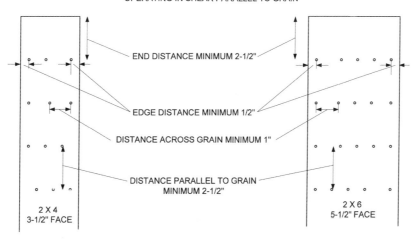

[BS] FIGURE C103.6.3
FASTENER SPACINGS FOR LUMBER-TO-LUMBER CONNECTIONS OPERATING IN SHEAR PARALLEL TO GRAIN

APPENDIX C

[BS] C103.5.3.1 General fastener spacing. Fastener spacing shall meet the following conditions except as provided for in Section C103.5.3.

The distance between fasteners and the edge of lumber that is less than $3^1/_2$ inches deep (89 mm) in the direction of the fastener length shall be a minimum of $^3/_4$ inch (19.1 mm).

1. The distance between fasteners and the edge of lumber that is more than 2 inches (51 mm) thick in the direction of the fastener length shall be a minimum of $^1/_2$ inch (12.7 mm).

2. The distance between a fastener and the end of lumber shall be a minimum of $2^1/_2$ inches (64 mm).

3. The distance between fasteners parallel to the grain (center-to-center) shall be a minimum of $2^1/_2$ inches (64 mm).

4. The distance between fasteners perpendicular to the grain (center-to-center) in lumber that is less than $3^1/_2$ inches (89 mm) deep in the direction of the fastener length shall be 1 inch (25 mm).

5. The distance between fasteners perpendicular to the grain (center-to-center) in lumber that is more than 2 inches (51 mm) thick in the direction of the fastener length shall be $^1/_2$ inch (12.7 mm).

[BS] C103.5.3.2 Wood-to-wood connections of two members each 2 inches or less in thickness. Wood-to-wood connections fastener spacing shall meet the following conditions.

1. The distance between fasteners parallel to grain (center-to-center) shall be a minimum of $2^1/_2$ inches (64 mm).

2. The distance between fasteners across grain (center-to-center) shall be a minimum of 1 inch (25 mm).

3. For wood-to-wood connections of lumber at right angles, fasteners shall be spaced a minimum of $2^1/_2$ inches (64 mm) parallel to the grain and 1 inch (25 mm) perpendicular to the grain in any direction.

[BS] C103.5.3.3 Metal connectors for wood-to-wood connections. Metal connectors for wood-to-wood connections shall meet the following conditions.

1. Fastener spacing to edge or ends of lumber shall be as dictated by the prefabricated holes in the connectors and the connectors shall be installed in a configuration that is similar to that shown by the connector manufacturer.

2. Fasteners in $1^1/_4$-inch-wide (32 mm) metal straps that are installed on the narrow face of lumber shall be a minimum $^1/_4$ inch (6.4 mm) from either edge of the lumber. Consistent with Section C103.5.3.1, fasteners shall be permitted to be spaced according to the fastener holes fabricated into the strap.

3. Fasteners in metal nail plates shall be spaced a minimum of $^1/_2$ inch (12.7 mm) perpendicular to grain and a minimum of $1^1/_2$ inches (38 mm) parallel to grain.

SECTION C104
RETROFITTING GABLE END WALLS TO ENHANCE WIND RESISTANCE

[BS] C104.1 General. These prescriptive methods of retrofitting are intended to increase the resistance of existing gable end construction for out-of-plane wind loads resulting from high-wind events. The ceiling diaphragm shall be comprised of minimum $^1/_2$-inch-thick (12.7 mm) gypsum board, minimum nominal $^3/_8$-inch-thick (9.5 mm) wood structural panels, or plaster. An overview isometric drawing of one type of gable end retrofit to improve wind resistance is shown in Figure C104.1.1.

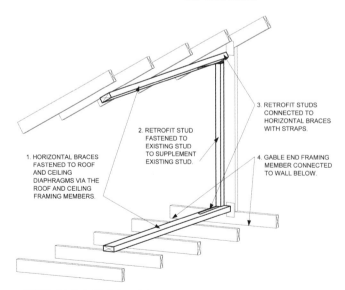

THIS FIGURE SHOWS A TRUSS GABLE END.
THE METHODOLOGY FOR A CONVENTIONALLY FRAMED GABLE END IS SIMILAR.
THE NUMBERS INDICATE A TYPICAL SEQUENCE OF INSTALLATION.
IN ORDER TO SHOW STRAPS COMPRESSION BLOCKS ARE NOT SHOWN.

**[BS]FIGURE C104.1.1
BASIC GABLE END RETROFIT METHODOLOGY**

[BS] C104.2 Horizontal braces. Horizontal braces shall be installed perpendicular to the roof and ceiling framing members at the location of each existing gable end stud greater than 3 feet (91 cm) in length. Unless it is adjacent to an omitted horizontal brace location, horizontal braces shall be minimum 2x4 dimensional lumber as defined in Section C103.3. A single horizontal brace is required at the top and bottom of each gable end stud for Retrofit Configuration A, B, or C. Two horizontal braces are required at the top and bottom of each gable end stud for Retrofit Configuration D. Maximum heights of gable end wall studs and associated retrofit studs for each Retrofit Configuration shall not exceed the values listed in Table C104.2. Horizontal braces shall be oriented with their wide faces across the roof or ceiling framing members, be fastened to a minimum of three framing members, and extend at least 6 feet (183 cm) measured perpendicularly from the gable end plus $2^1/_2$ inches (64 mm) beyond the last top chord or bottom chord member (rafter or ceiling joist) from the gable end as shown in Figures C104.2(1), C104.2(2), C104.2(3) and C104.2(4).

[BS] C104.2.1 Existing gable end studs. If the spacing of existing vertical gable end studs is greater than 24 inches (64 mm), a new stud and corresponding horizontal braces shall be installed such that the maximum spacing between existing and added studs shall be no greater than 24 inches (64 mm). Additional gable end wall studs shall not be required at locations where their length would be 3 feet (914 mm) or less. Each end of each required new stud shall be attached to the existing roofing framing members (truss top chord or rafter and truss bottom chord or ceiling joist) using a minimum of two 3-inch (76 mm) toenail fasteners (#8 wood screws or 10d nails) and a metal connector with minimum uplift capacity of 175 pounds (778 N), or nail plates with a minimum of four $1^1/_4$-inch-long (32 mm) fasteners (No. 8 wood screws or 8d nails).

[BS] C104.2.2 Main method of installation. Each horizontal brace shall be fastened to each existing roof or ceiling member that it crosses using three 3-inch-long (76 mm) fasteners (No. 8 wood screws or 10d nails) as indicated in Figure C104.2(1) and Figure C104.2(3) for trusses and Figure C104.2(2) and Figure C104.2(4) for conventionally framed gable end walls. Alternative methods for providing horizontal bracing of the gable end studs as provided in Sections C104.2.3 through C104.2.9 shall be permitted.

[BS] C104.2.3 Omitted horizontal brace. Where conditions exist that prevent installation in accordance with Section C104.2.2, horizontal braces shall be permitted to be omitted for height limitations corresponding to Retrofit Configurations A and B as defined in Table C104.2 provided installation is as indicated in Figure C104.2.3 and provided all of the following conditions are met. This method is not permitted for Retrofit Configurations C or D.

1. There shall be at least two horizontal braces on each side of an omitted horizontal brace or at least one horizontal brace if it is the end horizontal brace. Omitted horizontal braces must be separated by at least two horizontal braces even if that location is composed of two retrofit studs and two horizontal braces.

2. Horizontal braces adjacent to the omitted horizontal brace shall be 2x6 lumber, shall butt against the existing studs, and shall be fastened to each existing roof or ceiling member crossed using three 3-inch-long (76 mm) fasteners (No. 8 wood screws or 10d nails). For Retrofit Configuration B, four fasteners shall be required on at least one of the connections between the horizontal brace and the existing roof and ceiling framing members. Fasteners shall be spaced a minimum of $3/_4$ inch (19.1 mm) from the edges of the horizontal braces and a minimum of $1^3/_4$ inch (44 mm) from adjacent fasteners.

3. Where the existing studs on each side of an omitted horizontal brace have their wide face perpendicular to the gable end wall, the retrofit studs at those locations

[BS] TABLE C104.2
STUD LENGTH LIMITATIONS BASED ON EXPOSURE AND DESIGN WIND SPEED

EXPOSURE CATEGORY	MAXIMUM 3-SEC GUST BASIC WIND SPEED[a]	MAXIMUM HEIGHT OF GABLE END RETROFIT STUD[b]			
C	140	8'-0"	11'-3"	14'-9"	16'-0"
C	150	7'-6"	10'-6"	13'-6"	16'-0"
C	165	7'-0"	10'-0"	12'-3"	16'-0"
C	180	7'-0"	10'-0"	12'-3"	16'-0"
C	190	6'-6"	8'-9"	11'-0"	16'-0"
B	140	8'-0"	12'-3"	16'-0"	N/R[c]
B	150	8'-0"	11'-3"	14'-9"	16'-0"
B	165	8'-0"	11'-3"	14'-9"	16'-0"
B	180	7'-6"	10'-6"	13'-6"	16'-0"
B	190	7'-0"	10'-0"	12'-3"	16'-0"
	Retrofit Configuration	A	B	C	D

For SI: 1 inch = 25.4 mm, 1 foot = 304.8 mm.
a. Interpolation between given wind speeds is not permitted.
b. Existing gable end studs less than or equal to 3 feet 0 inches in height shall not require retrofitting.
c. N/R = Not Required. Configuration C is acceptable to 16 inches 0 maximum height.

APPENDIX C

and the retrofit stud at the omitted horizontal brace locations shall extend a minimum of $3^3/_4$ inches (95 mm) beyond the interior edge of the existing studs for both Retrofit Configurations A and B. The edges of the three retrofit studs facing towards the interior of the attic shall be aligned such that they are the same distance from the gable end wall.

4. Retrofit studs shall be fastened to existing studs in accordance with Section C104.3.

5. Retrofit studs adjacent to the omitted horizontal brace shall be fastened to the horizontal brace using straps in accordance with Table C104.4.1 consistent with the size of the retrofit stud. The method applicable to Table C104.4.2 is not permitted.

6. A strong back made of minimum of 2x8 lumber shall be placed parallel to the gable end and shall be located on and span between horizontal braces on the two sides of the omitted horizontal brace and shall extend beyond each horizontal brace by a minimum of $2^1/_2$ inches (64 mm). The strong back shall be butted to the three retrofit studs. The strong back shall be attached to each of the horizontal braces on which it rests with five 3-inch-long (76 mm) fasteners (#8 screws or 8d nails). The fasteners shall have a minimum $^3/_4$-inch (19.1 mm) edge distance and a minimum $2^1/_2$-inch (64 mm) spacing between fasteners. Additional compression blocks shall not be required at locations where a strong back butts against a retrofit stud.

7. The retrofit stud at the location of the omitted horizontal braces shall be fastened to the strong back using a connector with minimum uplift capacity of 800 pounds (3559 N) and installed such that this capacity is oriented in the direction perpendicular to the gable end wall.

8. The use of shortened horizontal braces using the alternative method of Section C104.2.5 is not permitted for horizontal braces adjacent to the omitted horizontal braces.

9. Horizontal braces shall be permitted to be interrupted in accordance with Section C104.2.8.

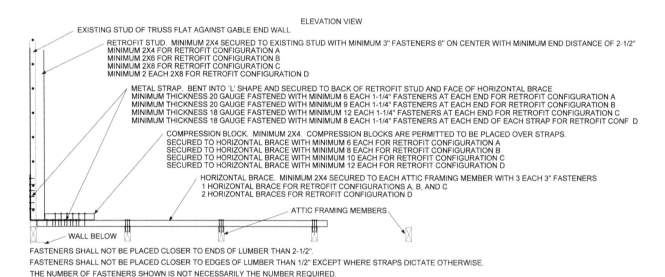

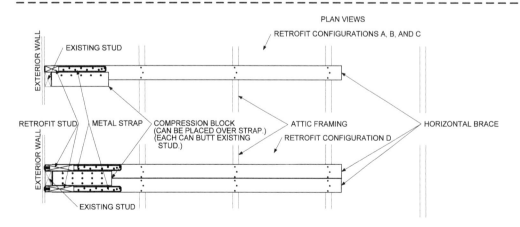

For SI: 1 inch = 25.4 mm.

**[BS] FIGURE C104.2(1)
TRUSS FRAMED GABLE END**

APPENDIX C

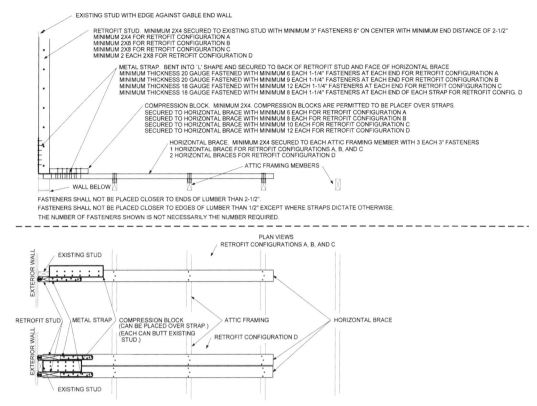

For SI: 1 inch = 25.4 mm.

[BS] FIGURE C104.2(2)
CONVENTIONALLY FRAMED GABLE END L-BENT STRAP

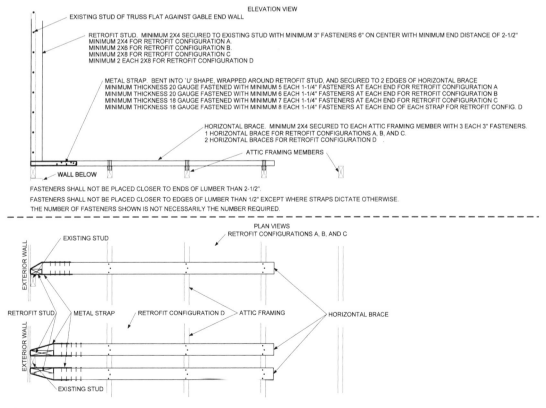

For SI: 1 inch = 25.4 mm.

[BS] FIGURE C104.2(3)
TRUSS FRAMED GABLE END U-BENT STRAP

APPENDIX C

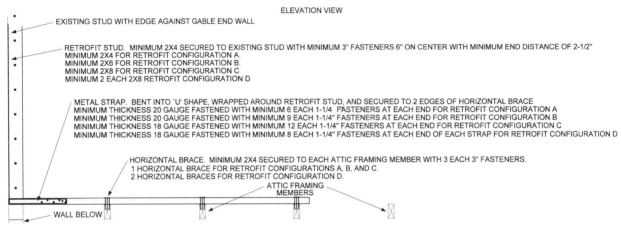

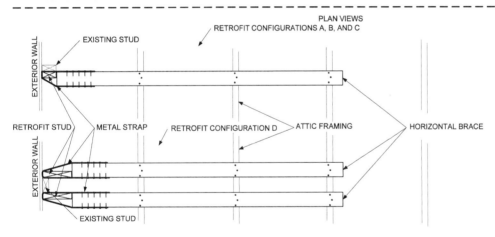

For SI: 1 inch = 25.4 mm.

[BS] FIGURE C104.2(4)
CONVENTIONALLY FRAMED GABLE END U-BENT STRAP

APPENDIX C

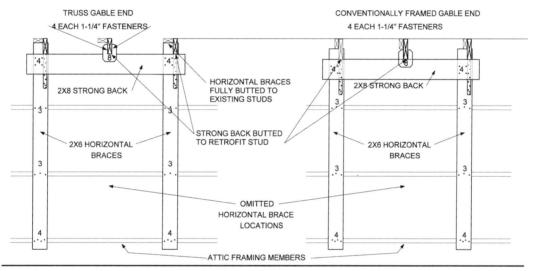

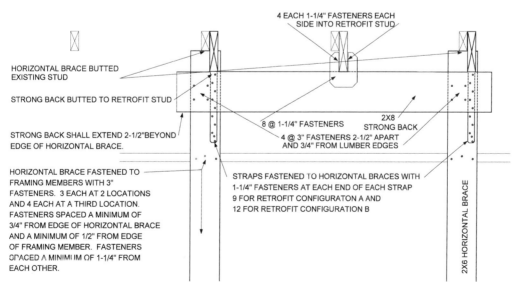

[BS]FIGURE C104.2.3
OMITTED HORIZONTAL BRACE

APPENDIX C

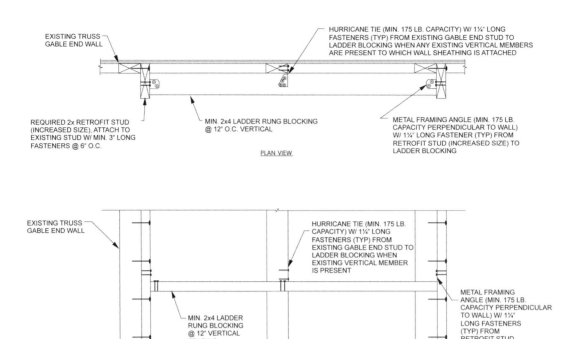

For SI: 1 inch = 25.4 mm; 1 pound = 4.4 N.

**[BS] FIGURE C104.2.4(1)
LADDER BRACING FOR OMITTED RETROFIT STUD (GABLE END FRAME)**

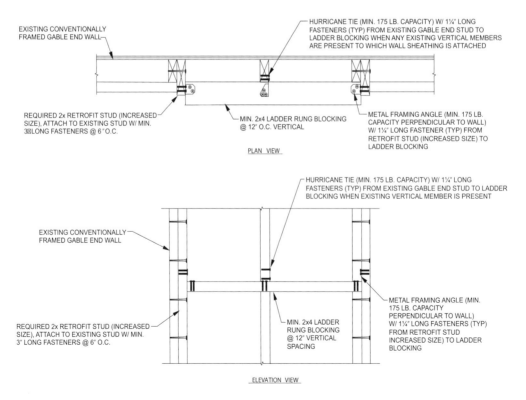

For SI: 1 inch = 25.4 mm; 1 pound = 4.4 N.

**[BS] FIGURE C104.2.4(2)
LADDER BRACING FOR OMITTED RETROFIT STUD (CONVENTIONALLY FRAMED GABLE END)**

[BS] C104.2.4 Omitted horizontal brace and retrofit stud. Where conditions exist that prevent installation in accordance with Section C104.2.2 or C104.2.3, then retrofit studs and horizontal braces shall be permitted to be omitted from those locations by installation of ladder assemblies for Retrofit Configurations A and B as defined in Table C104.2 provided all of the following conditions are met. This method is not permitted for Retrofit Configurations C or D.

1. No more than two ladder assemblies are permitted on a single gable end.

2. There shall be at least two retrofit studs and horizontal brace assemblies on either side of the locations where the retrofit studs and horizontal bracing members are omitted (no two ladder braces bearing on a single retrofit stud).

3. Where the existing studs on each side of an omitted horizontal brace have their wide face parallel to the gable end wall the retrofit studs at those locations and the retrofit stud at the omitted horizontal brace locations shall be 2x6 lumber for Retrofit Configuration A and 2x8 lumber for Retrofit Configuration B.

4. Horizontal braces adjacent to the omitted horizontal brace shall be 2x6 lumber and be fastened to each existing roof or ceiling member crossed using three 3-inch-long (76 mm) fasteners (#8 wood screws or 10d nails) as indicated in Figures C104.2(1) and C104.2(3) for gable end frames and Figures C104.2(2) and C104.2(4) for conventionally framed gable end walls. For Retrofit Configuration B, four fasteners shall be required on at least one of the connections between the horizontal brace and the existing roof and ceiling framing members.

5. Ladder rungs shall be provided across the location of the omitted retrofit studs as indicated in Figure C104.2.4(1) for gable end frames and Figure C104.2.4(2) for conventionally framed gable end walls.

6. Ladder rungs shall be minimum 2x4 lumber oriented with their wide face horizontal and spaced a maximum of 16 inches (41 cm) on center vertically.

7. Where ladder rungs cross wall framing members they shall be connected to the wall framing members with a metal connector with a minimum capacity of 175 pounds (778 N) in the direction perpendicular to the gable end wall.

8. Notching of the ladder rungs shall not be permitted unless the net depth of the framing member is a minimum of $3^1/_2$ inches (89 mm).

[BS] C104.2.5 Short horizontal brace. Where conditions exist that prevent installation in accordance with Section C104.2.2, C104.2.3 or C104.2.4, the horizontal braces shall be permitted to be shortened provided installation is as indicated in Figure C104.2.5 and all of the following conditions are met.

1. The horizontal brace shall be installed across a minimum of two framing spaces, extend a minimum of 4 feet (1220 mm) from the gable end wall plus $2^1/_2$ inches (64 mm) beyond the farthest roof or ceiling framing member from the gable end, and be fastened to each existing framing member with three 3-inch-long (76 mm) fasteners (#8 wood screws or 10d nails).

2. An anchor block shall be fastened to the side of the horizontal brace in the second framing space from the gable end wall as shown in Figure C104.2.5. The anchor block lumber shall have a minimum edge thickness of $1^1/_2$ inches (38 mm) and the depth shall be at a minimum the depth of the existing roof or ceiling framing member. Six 3-inch-long (76 mm) fasteners (#8 wood screws or 10d nails) shall be used to fasten the anchor block to the side of the horizontal brace.

3. The anchor block shall extend into the space between the roof or ceiling framing members a minimum of one-half the depth of the existing-framing members at the location where the anchor block is installed. The anchor block shall be installed tightly between the existing framing members such that the gap at either end shall not exceed $1/_8$ inch (3.2 mm).

4. The use of omitted horizontal braces using the method of Section C104.2.3 adjacent to a short horizontal brace as defined in this section is not permitted.

[BS] C104.2.6 Installation of horizontal braces onto webs of trusses. Where existing conditions preclude installation of horizontal braces on truss top or bottom chords they shall be permitted to be installed on truss webs provided all of the following conditions are met.

1. Horizontal braces shall be installed as close to the top or bottom chords as practical without altering the truss or any of its components and not more than three times the depth of the truss member to which it would ordinarily be attached.

2. A racking block, comprised of an anchor block meeting the definition of anchor block of Section C102 or comprised of minimum $^{15}/_{32}$-inch (12 mm) plywood or $^7/_{16}$-inch (11.1 mm) Oriented Strand Board (OSB), shall be fastened to the horizontal brace in the second framing space from the gable end wall. The racking block shall extend toward the roof or ceiling diaphragm so that the edge of the racking block closest to the diaphragm is within $^1/_2$ the depth of the existing framing member from the diaphragm surface. The racking block shall be attached to horizontal braces using six fasteners (No. 8 wood screws or 10d nails) of sufficient length to provide $1^1/_2$ inches (38 mm) of penetration into the horizontal brace.

3. Racking blocks shall be permitted to be fastened to any face or edge of horizontal braces between each web or truss vertical posts to which a horizontal brace is attached. Racking blocks shall be permitted to be on alternate sides of horizontal braces. Racking blocks shall be installed tightly between the lumber of truss members or truss plates such that the gap at either end shall be a maximum of $^1/_8$ inch (3.2 mm).

APPENDIX C

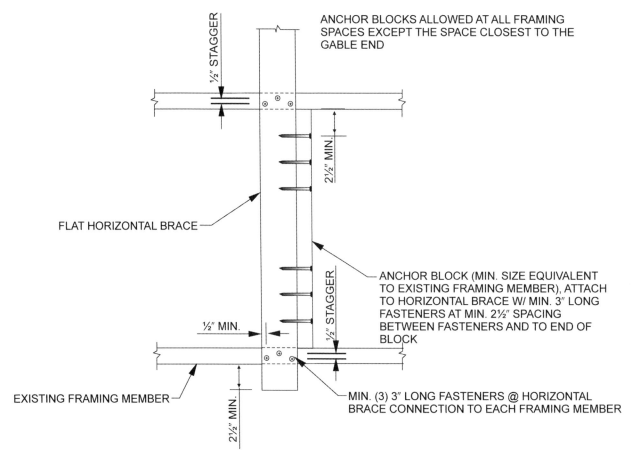

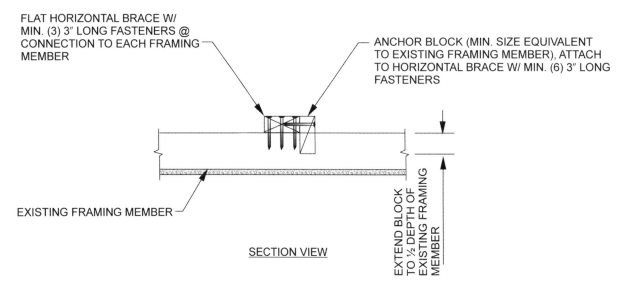

For SI: 1 inch = 25.4 mm.

**[BS] FIGURE C104.2.5
ANCHOR BLOCK INSTALLATION**

[BS] C104.2.7 Alternative method of installation of horizontal braces at truss ridges. Where conditions exist that limit or restrict installation of horizontal braces near the peak of the roof, ridge ties shall be added to provide support for the required horizontal brace. The top of additional ridge tie members shall be installed a maximum of 16 inches (406 mm) below the existing ridge line or 4 inches (102 mm) below impediments. A minimum 2x4 member shall be used for each ridge tie and fastening shall consist of two 3-inch-long (76 mm) wood screws, four 3-inch-long (76 mm) 10d nails or two $3^1/_2$-inch-long (89 mm) 16d nails driven through and clinched at each top chord or web member intersected by the ridge tie as illustrated in Figure C104.2.7.

[BS] C104.2.8 Interrupted horizontal braces. Where conditions exist that prevent the installation of a continuous horizontal brace then horizontal braces shall be permitted to be interrupted using the methods shown in Figures C104.2.8(1), C104.2.8(2), and C104.2.8(3). For interruptions that occur in the attic framing space closest to the gable end, nine 3-inch (76 mm) fasteners shall be used to connect each section of the interrupted horizontal braces. For interruptions that occur in the second attic space from the gable end, six 3-inch (76 mm) fasteners shall be used to connect each section of the interrupted horizontal braces. For interruptions that occur in the attic framing space farthest from the gable end, three 3-inch (76 mm) fasteners shall be used to connect each section of the interrupted horizontal braces. Horizontal braces shall be continued far enough to allow connections to three existing roof framing members as shown in Figure C104.2.8(1), C104.2.8(2) or C104.2.8(3). Fasteners shall be spaced in accordance with Section C103.5.3. Horizontal braces shall be the same width and depth as required for an uninterrupted member.

[BS] C104.2.9 Piggyback gable end frames. Piggyback gable end frames (gable end frames built in two sections one above the other) shall be permitted to be retrofitted if either of the following cases is true:

1. The existing studs in both the upper gable end frames and the lower gable end frames to which wall sheathing, panel siding, or other wall covering are attached are sufficiently in line that retrofit studs can be installed and connections made between the two with retrofit stud(s).

2. Existing studs in the upper frame are not sufficiently in line with the studs in the frame below and the existing studs in the upper frame are 3 feet (91 cm) or shorter.

For Condition 1 both the lower stud and the upper stud shall be retrofitted using the methods of Section C104.2. For Condition 2 the retrofit stud shall be connected to the lower studs using the methods of Section C104.2 and be continuous from the bottom horizontal brace to the top horizontal brace. No connection is required between the retrofit stud and the upper stud. In both conditions the bottom chord of the piggyback truss section shall be fastened to each retrofit stud using a connector with minimum axial capacity of 175 pounds (778 N).

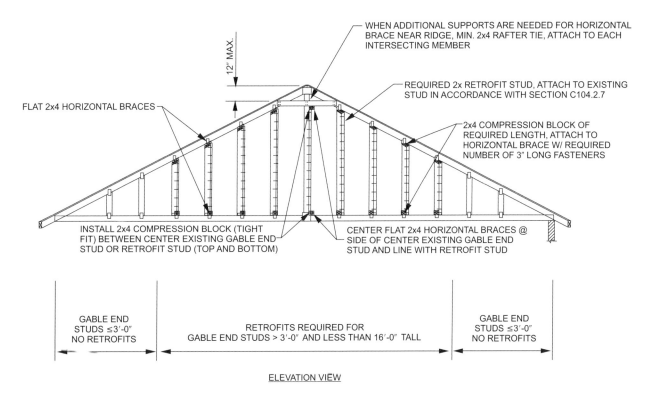

For SI: 1 inch = 25.4 mm; 1 foot = 304.8 mm.

**[BS] FIGURE C104.2.7
DETAIL OF RETROFIT TIE INSTALLATION**

APPENDIX C

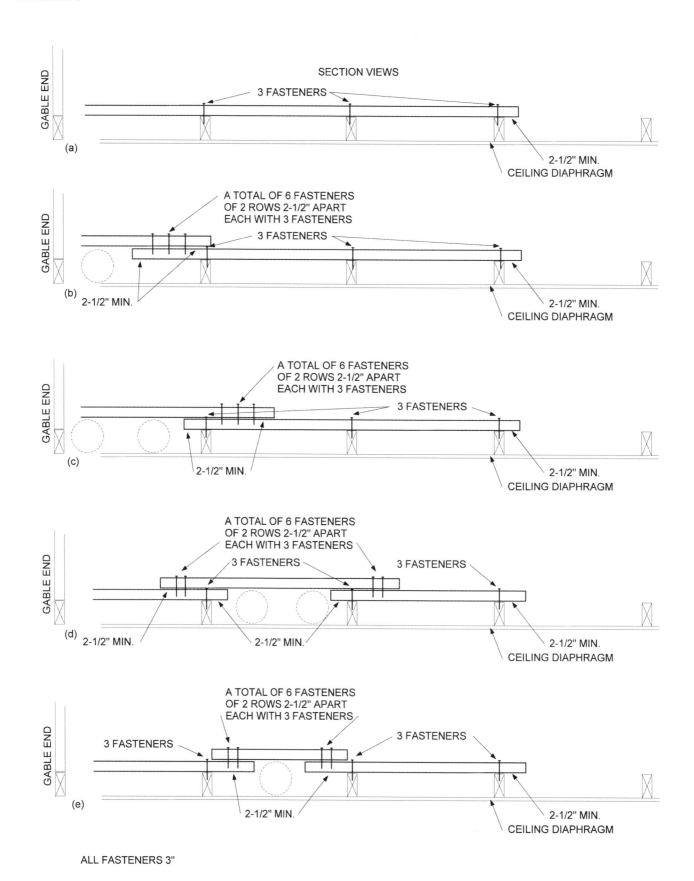

ALL FASTENERS 3"

For SI: 1 inch = 25.4 mm.

[BS] FIGURE C104.2.8(1)
SPLICED HORIZONTAL BRACES

APPENDIX C

SECTION VIEWS

(f) GABLE END — 3 FASTENERS — A TOTAL OF 6 FASTENERS OF 2 ROWS 2-1/2" APART EACH WITH 3 FASTENERS — 3 FASTENERS — 2-1/2" MIN. — 2-1/2" MIN. CEILING DIAPHRAGM

(g) GABLE END — 3 FASTENERS — A TOTAL OF 6 FASTENERS OF 2 ROWS 2-1/2" APART EACH WITH 3 FASTENERS — 3 FASTENERS — 2-1/2" MIN. — 2-1/2" MIN. CEILING DIAPHRAGM

(h) GABLE END — 2-1/2" MIN — A TOTAL OF 6 FASTENERS OF 2 ROWS 2-1/2" APART EACH WITH 3 FASTENERS — 3 FASTENERS — 3 FASTENERS — 2-1/2" MIN. — 2-1/2" MIN. — 2-1/2" MIN. CEILING DIAPHRAGM

(i) GABLE END — 3 FASTENERS — 3 FASTENERS — 3 FASTENERS — 2-1/2" MIN. — 2-1/2" MIN. — 2-1/2" MIN. CEILING DIAPHRAGM

(j) GABLE END — 3 FASTENERS — 3 FASTENERS — 3 FASTENERS — 2-1/2" MIN. — 2-1/2" MIN. CEILING DIAPHRAGM

ALL FASTENERS 3"

For SI: 1 inch = 25.4 mm.

[BS] FIGURE C104.2.8(2)
SPLICED HORIZONTAL BRACES

APPENDIX C

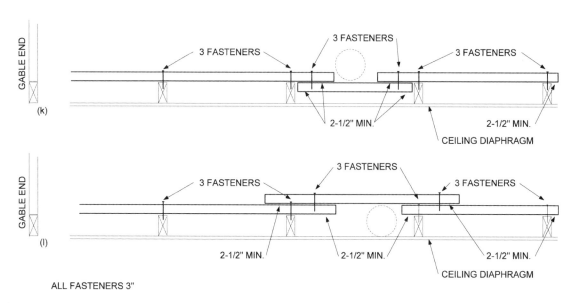

For SI: 1 inch = 25.4 mm.

[BS] FIGURE C104.2.8(3)
SPLICED HORIZONTAL BRACES

[BS] C104.3 Retrofit studs. Retrofit studs shall be installed in accordance with Section C104.3.1 using one of the five methods of Sections C104.3.2, C104.3.3, C104.3.4, C104.3.5, or C104.3.6. Figure C104.3 shows these methods of installation. For the Retrofit Configuration obtained from Table C104.2, the size of retrofit studs shall be as indicated in Table C104.4.1 or Table C104.4.2. Retrofit studs shall extend from the top of the lower horizontal brace to the bottom of the upper horizontal brace except that a maximum gap of $^1/_8$ inch (3.2 mm) is permitted at the bottom and $^1/_2$ inch (12.7 mm) at the top. Where wall sheathing, panel siding, or other wall covering is fastened to a conventionally framed gable end, retrofit studs shall be applied in accordance with Section C104.2.1.

[BS] C104.3.1 Fastening. Where nail plates are not used, retrofit studs shall be attached to existing studs using 3-inch (76 mm) fasteners at a maximum of 6 inches (152 mm) on center but no closer than $2^1/_2$ inches (64 mm) on center with fasteners no closer to ends of members than $2^1/_2$ inches (64 mm).

[BS] C104.3.2 Method #1: Face-to-edge or face-to-face method. Retrofit studs shall be installed immediately adjacent to existing gable end wall studs as indicated in Figure C104.3(a). The retrofit studs shall overlap the edge or side of the existing stud by a minimum of $1^1/_4$ inches (32 mm). Fasteners shall be installed as specified in Section C104.3.1.

[BS] C104.3.3 Method #2: Face-to-face offset method. Retrofit studs shall be installed against the face of existing studs as indicated in Figure C104.3(b) such that the faces overlap a minimum of $1^1/_2$ inches (38 mm) and the edge distance to fasteners is no less than $^3/_4$ inch (19.1 mm). Fasteners shall be installed as specified in Section C104.3.1.

[BS] C104.3.4 Method #3: Butted retrofit stud method. Provided that all of the following fastening conditions are met, retrofit studs shall be permitted to be butted by their edge to existing studs with the addition of nail plates as indicated in Figure C104.3(c) and Figure C104.3.4.

1. The narrow edge of retrofit studs shall be installed against the narrow or the wide face of existing studs.

2. A minimum of two nail plates shall be used.

3. Fasteners used to secure nail plates to studs shall be a minimum $1^1/_4$ inches (32 mm) long (#8 wood screws or 8d nails).

4. Fasteners placed in nail plates shall have a minimum end distance of $2^1/_2$ inches (64 mm) for both studs and a maximum end distance of 6 inches (152 mm) from the ends of the shorter stud.

5. Fasteners shall have a minimum $^1/_2$-inch (12.7 mm) edge distance. Fasteners shall be placed a maximum of $1^1/_2$ inches (38 mm) from the abutting vertical edges of existing studs and retrofit studs.

6. There shall be at least three fasteners through nail plates into all existing and retrofit studs to which the nail plate is attached.

7. Nail plates with three fasteners onto a single existing or retrofit stud shall be spaced a maximum of 15 inches (38 cm) on center.

8. Nail plates with more than three fasteners onto a single existing or retrofit stud shall be spaced a maximum of 20 inches (51 cm) on center.

9. Fasteners used to secure nail plates shall be spaced vertically a minimum of $1^1/_2$ inches (38 mm) on center. Staggered fasteners used to secure nail plates shall be spaced horizontally a minimum of $^1/_2$ inch (12.7 mm).

APPENDIX C

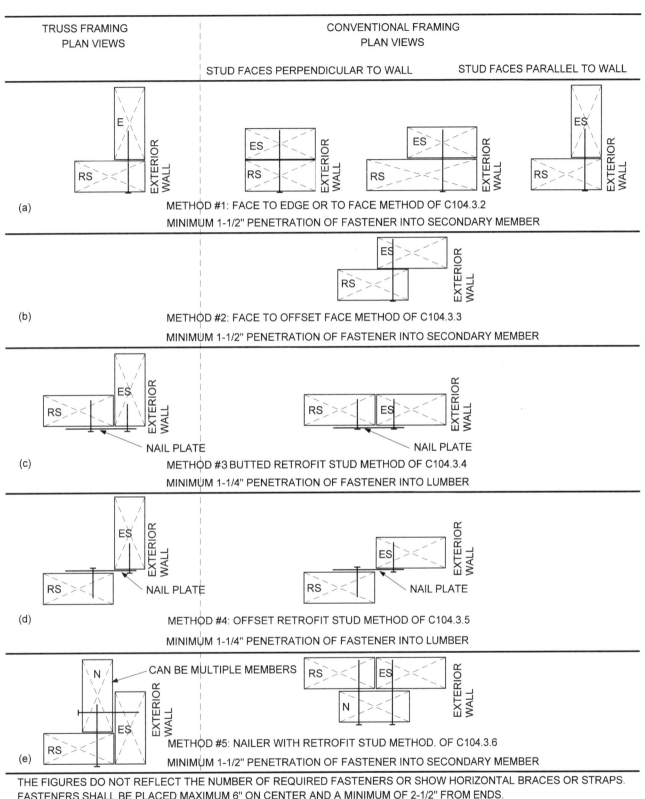

[BS] FIGURE C104.3
METHOD OF INSTALLING RETROFIT STUDS

APPENDIX C

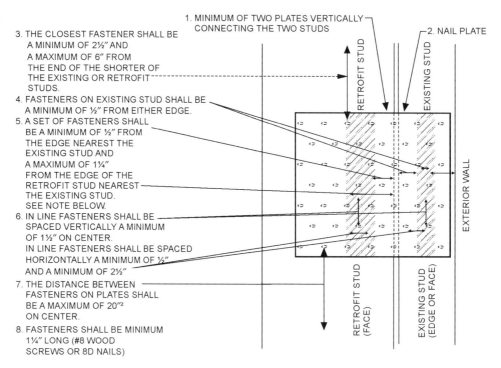

For SI: 1 inch = 25.4 mm.

**[BS]FIGURE C104.3.4
NAIL PLATE FASTENING**

[BS] C104.3.5 Method #4: Offset retrofit stud method. Retrofit studs may be offset from existing studs by use of nail plates as shown in Figure C104.3(d) such that the vertical corner of a retrofit stud shall align with the vertical corner of an existing stud as indicated in Figure C104.3(d) and Figure C104.3.4, and the fastening conditions of Section C104.3.4 are met.

[BS] C104.3.6 Method #5: Nailer with retrofit stud method. Retrofit studs and existing studs shall be permitted to be connected using noncontinuous 2x4 nailers as indicated in Figure C104.3(e) provided the following conditions are met.

1. Both the existing stud and the retrofit stud shall be butted to nailers and both shall be fastened to the nailer with 3-inch-long (76 mm) fasteners (#8 wood screws or 8d nails). Fasteners connecting each stud to the nailer shall be a spaced 6 inches (152 mm) o.c.

2. Fasteners into nailers from any direction shall be offset vertically by a minimum of $2^1/_2$ inches (64 mm).

3. Fasteners into nailers shall be a minimum of $2^1/_2$ inches (64 mm) but not more than 6 inches (152 mm) from the end of the shorter of the existing stud and retrofit stud to which they are fastened.

[BS] C104.3.7 Reduced depth of retrofit studs. Retrofit studs may be reduced in depth by notching, tapering, or other methods at any number of locations along their length provided that all of the following conditions are met:

1. Retrofit studs to be reduced in depth shall be sized such that the remaining minimum depth of member at the location of the notch (including cross-cut kerfs) shall be not less than that required by Table C104.4.1 or C104.4.2.

2. Reduced in-depth retrofit stud shall not be spliced within 12 inches (30 cm) of the location of notches. Splice members shall not be notched.

3. The vertical extent of notches shall not exceed 12 inches (30 cm) as measured at the depth of location of reduced depth.

4. A reduced in-depth retrofit stud member shall be fastened to the side of the existing gable end wall studs in accordance with Section C104.3.1. Two additional 3-inch (76 mm) fasteners (#8 wood screws or 10d nails) shall be installed on each side of notches in addition to those required by Section C104.3.1.

APPENDIX C

[BS] C104.3.8 Retrofit stud splices. Retrofit studs greater than 8 feet (244 cm) in height may be field spliced in accordance with Figure C104.3.8.

[BS] C104.4 Connection between horizontal braces and retrofit studs. Connections between horizontal braces and retrofit studs shall comply with Section C104.4.1 or C104.4.2. Each retrofit stud shall be connected to the top and bottom horizontal brace members with a minimum 20-gage $1^1/_4$-inch-wide (32 mm) flat or coil metal strap with prepunched holes for fasteners. Straps shall be fastened with $1^1/_4$-inch-long (32 mm) fasteners (#8 wood screws or 8d nails) with the number of fasteners as indicated in Tables C104.4.1 and C104.4.2. Fasteners shall be no closer to the end of lumber than $2^1/_2$ inches (64 mm).

[BS] C104.4.1 L-bent strap method. Retrofit studs shall be connected to horizontal braces or to strong backs in accordance with Figure C104.2(1), C104.2(2) or C104.2.3, and shall comply with the following conditions.

1. A strap shall be applied to the edges of a retrofit stud nearest the gable end wall and to the face of horizontal braces using at each end of the strap the number of fasteners specified in Table C104.4.1. Straps shall be long

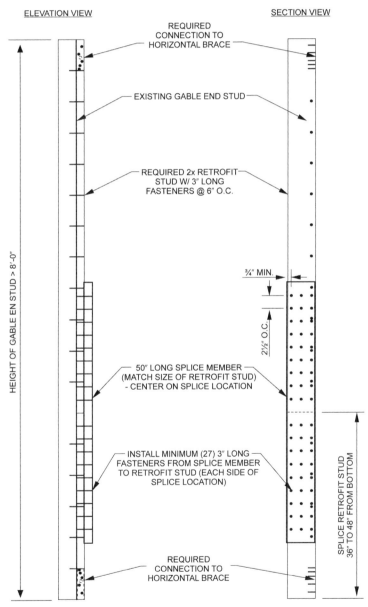

For SI: 1 inch = 25.4 mm; 1 foot = 304.8 mm.

**[BS] FIGURE C104.3.8
RETROFIT STUD SPLICES**

enough so that each strap extends sufficient distance onto the vertical face of the retrofit stud that the fastener closest to the ends of the studs is a minimum of $2^1/_2$ inches (64 mm) from the end of the stud. Straps shall be permitted to be twisted to accommodate the transition between the tops of retrofit studs and horizontal bracings following roof pitches.

2. Compression blocks shall be installed on the horizontal braces directly against either the existing vertical gable end wall stud or the retrofit stud. Figure C104.2(1) (trusses) and Figure C104.2(2) (conventionally framed) show the installation of the compression block against the existing vertical gable end wall stud with the strap from the retrofit stud running beside the compression block. Compression blocks shall be permitted to be placed over straps. Compression blocks shall be fastened to the horizontal braces with at least the minimum number of 3-inch-long (76 mm) fasteners (#8 wood screws or 10d nails) specified in Table C104.4.1. End and edge distances for fasteners shall be in accordance with Section C103.5.3.

[BS] C104.4.2 U-bent strap method. Retrofit studs shall be connected to horizontal braces in accordance with Figure C104.2(3) or C104.2(4), shall be limited to Retrofit Configurations A and B as defined in Table C104.2, and shall comply with the following conditions.

1. Straps of sufficient length to meet the requirements for the number of fasteners in accordance with Table C104.4.2 and meet the end distance requirements of Section C103.5.3 shall be shaped around retrofit studs and fastened to the edges of horizontal braces. Straps shall wrap the back edge of the retrofit stud snugly with a maximum gap of $^1/_4$ inch (6.4 mm). Rounded bends of straps shall be permitted. One fastener shall be installed that connects each strap to the side of the associated retrofit stud.

2. The horizontal brace shall butt snugly against the retrofit stud with a maximum gap of $^1/_4$ inch (6.4 mm).

3. Straps shall be permitted to be twisted to accommodate the transition between the tops of retrofit studs and horizontal braces that follow the roof pitch.

[BS] C104.5 Connection of gable end wall to wall below. The bottom chords or bottom members of wood-framed gable end walls shall be attached to the wall below using one of the methods prescribed in Sections C104.5.1 or C104.5.2. The particular method chosen shall correspond to the framing system and type of wall construction encountered.

[BS] C104.5.1 Gable end frame. The bottom chords of the gable end frame shall be attached to the wall below using gusset angles. A minimum of two fasteners shall be installed into the bottom chord. The gusset angles shall be installed throughout the portion of the gable end where the gable end wall height is greater than 3 feet (91 cm) at the spacing specified in Table C104.5.1. Connection to the wall below shall be by one of the methods listed below:

1. For a wood-frame wall below, a minimum of two fasteners shall be installed. The fasteners shall be of the same diameter and style specified by the gusset angle manufacturer and sufficient length to extend through the double top plate of the wall below.

2. For a concrete or masonry wall below without a sill plate, the type and number of fasteners into the wall shall be consistent with the gusset angle manufacturer's specifications for fasteners installed in concrete or masonry.

[BS] TABLE C104.4.1
ELEMENT SIZING AND SPACING FOR L-BENT RETROFIT METHOD

RETROFIT ELEMENTS	RETROFIT CONFIGURATION			
	A	B	C	D
Minimum size and number of Horizontal Braces	2 x 4	2 x 4	2 x 4	2 each 2 x 4
Minimum size and number of Retrofit Studs	2 x 4	2 x 6	2 x 8	2 each 2 x 8
Minimum number of fasteners connecting each end of straps to Retrofit Studs or to Horizontal Braces #8 screws or 10d nails $1^1/_4$" long	6	9	12	8 on each strap
Minimum number of fasteners to connect Compression Blocks to Horizontal Braces #8 screws or 10d nails 3" long	6	8	10	12

For SI: 1 inch = 25.4 mm, 1 foot = 304.8 mm.

[BS] TABLE C104.4.2
ELEMENT SIZING AND SPACING FOR U-BENT RETROFIT METHOD

RETROFIT ELEMENTS	RETROFIT CONFIGURATION			
	A	B	C	D
Minimum size and number of Horizontal Braces	2 x 4	2 x 4	2 x 4	2 each 2 x 4
Minimum size and number of Retrofit Studs	2 x 4	2 x 6	2 x 8	2 each 2 x 8
Minimum number of fasteners connecting Straps to each edge of Horizontal Braces #8 screws or 10d nails $1^1/_4$" long	6	7	7	6 on each side of strap

For SI: 1 inch = 25.4 mm, 1 foot = 304.8 mm.

3. For a concrete or masonry wall below with a 2x sill plate, the fasteners into the wall below shall be of the diameter and style specified by the gusset angle manufacturer for concrete or masonry connections; but, long enough to pass through the wood sill plate and provide the required embedment into the concrete or masonry below. Alternatively, the gusset angle can be anchored to the sill plate using four each $1^1/_2$-inch-long (38 mm) fasteners of the same type as specified by the gusset angle manufacturer for wood connections, provided that the sill plate is anchored to the wall on each side of the gusset angle by a $^1/_4$-inch-diameter (6.4 mm) masonry screw with $2^3/_4$ inches (70 mm) of embedment into the concrete or masonry wall. A $^1/_4$-inch (6.4 mm) washer shall be placed under the heads of the masonry screws.

[BS] C104.5.2 Conventionally framed gable end wall. Each stud in a conventionally framed gable end wall, throughout the length of the gable end wall where the wall height is greater than 3 feet (.91 m), shall be attached to the bottom or sill plate using a stud to plate connector with minimum uplift capacity of 175 pounds (778 N). The bottom or sill plate shall then be connected to the wall below using one of the methods listed below:

1. For a wood frame wall below, the sill or bottom plate shall be connected to the top plate of the wall below using $^1/_4$-inch-diameter (6.4 mm) lag bolt fasteners of sufficient length to penetrate the bottom plate of the upper gable end wall and extend through the bottom top plate of the wall below. A washer sized for the diameter of the lag bolt shall be placed under the head of each lag bolt. The fasteners shall be installed at the spacing indicated in Table C104.5.2.

2. For a concrete or masonry wall below, the sill or bottom plate shall be connected to the concrete or masonry wall below using $^1/_4$-inch-diameter (6.4 mm) concrete or masonry screws of sufficient length to provide $2^3/_4$ inches (70 mm) of embedment into the top of the concrete or masonry wall. A washer sized for the diameter of the lag bolt shall be placed under the head of each lag bolt. The fasteners shall be installed at the spacing indicated in Table C104.5.2.

[BS] TABLE C104.5.1
SPACING OF GUSSET ANGLES

EXPOSURE CATEGORY	BASIC WIND SPEED (mph)	SPACING OF GUSSET ANGLES (inches)
C	140	38
C	150	32
C	165	28
C	180	24
C	190	20
B	140	48
B	150	40
B	165	36
B	180	30
B	190	26

For SI: 1 inch = 25.4 mm, 1 mile per hour = 0.45 m/s.

[BS] TABLE C104.5.2
SPACING OF LAG OR MASONRY SCREWS USED TO CONNECT SILL PLATE OF GABLE END WALL TO TOP OF THE WALL BELOW

EXPOSURE CATEGORY	BASIC WIND SPEED (mph)	SPACING OF LAG OR MASONRY SCREWS (inches)
C	140	19
C	150	16
C	165	14
C	180	14
C	190	10
B	140	24
B	150	20
B	165	18
B	180	15
B	190	13

For SI: 1 inch = 25.4 mm, 1 mile per hour = 0.45 m/s.

Chapter C2:
Roof Deck Fastening for High-wind Areas

The provisions contained in this appendix are not mandatory unless specifically referenced in the adopting ordinance.

General Comments

The provisions in Appendix Chapter C2 are focused on the fastening of roof decks in high-wind areas. This is one of two appendix chapters focused on providing prescriptive methods to increase wind resistance for existing structures. Appendix Chapter C1 addresses the retrofit of gable ends to resist wind. This applicability is limited to buildings that fall in the scope of the *International Residential Code*® (IRC®) or are assigned to lower risk levels. Buildings assigned to higher risk level buildings would require more rigorous upgrades than this appendix can provide.

The content of both Appendix Chapters C1 and C2 is similar in concept to that provided in Appendix A chapters dealing with seismic retrofit requirements. It is anticipated that, over time, additional retrofit methods will be provided in this appendix as new Appendix C chapters. These retrofits are voluntary and, as such, may or may not meet the requirements of new construction. However, these voluntary measures will serve to better protect the public and reduce damage from high-wind events.

Because most of America's buildings located in hurricane-prone regions were not built to today's building code standards, there is significant value added to the code if the retrofitting of buildings could be facilitated by the provisions of prescriptive means. This would inherently reduce the cost of retrofitting. The need for structural retrofitting has been highlighted in the recent spate of hurricanes and the insurance crises that have developed in the coastal high-wind areas of a number of states because of older buildings that do not meet current building code structural requirements. Clearly, it is in the best interests of public health, welfare and safety to facilitate retrofitting. Given the importance of retrofitting to the public, retrofitting of buildings should be encouraged and facilitated by removing as many impediments as possible. The code can actually facilitate and encourage retrofitting by providing prescriptive means. Such methods should encourage, facilitate and reduce the cost of improving America's building stock.

Purpose

This appendix chapter is intended to provide a prescriptive solution for compliance with Section 707.3.2. These retrofits are voluntary, and as such, may or may not meet the requirements of new construction. However, these voluntary measures will serve to better protect the public and reduce damage from high-wind events.

Section 707.3.2 requires the roof deck to be evaluated and remedial action taken when insufficient or deteriorated connections are found. However, it gives little guidance on making the required determination or providing the required corrections. Ordinarily, one would turn to the requirements for new construction. However, blindly applying the same fastening requirements where fasteners already exist could potentially compromise performance because of damage to roof panels or framing members. The assumption is that there is an optimum spacing of existing and new fasteners that is a function of the number and type of existing connectors.

Adding fasteners where fasteners already exist is different than installing fasteners in new construction because of the greater potential for damaging sheathing or framing members. To date, the code only addresses nailing schedules for new installations without providing any guidance for retrofit nailing. The goal is that by adding additional fastener strengths, the code will at least approach current fastening requirements in order to approach the same performance level.

This appendix chapter is fairly brief and primarily provides guidance on fastening for sawn lumber, wood plank roofs or wood structural panel roofs. The nail spacings shown in Table C202.1.2 are derived from research conducted in the 1990s at Clemson University, tempered by the requirements for roof sheathing attachment for high winds in the *Wood Frame Construction Manual* and SSTD 10-99. Smaller diameter fasteners, such as staples, damage framing members less than larger diameter fasteners and provide significantly lower uplift resistance. Consequently, in these situations supplemental fasteners can be installed at typical new construction spacing without concern for splitting the structural members.

APPENDIX C

SECTION C201
GENERAL

[BS] C201.1 Purpose. This chapter provides prescriptive methods for partial structural retrofit of an existing building to increase its resistance to wind loads. It is intended for voluntary use where the ultimate design wind speed, V_{ult}, determined in accordance with Figure 1609.3(1) of the *International Building Code* exceeds 130 mph (58 m/s) and for reference by mitigation programs. The provisions of this chapter do not necessarily satisfy requirements for new construction. Unless specifically cited, the provisions of this chapter do not necessarily satisfy requirements for structural improvements triggered by addition, alteration, repair, change of occupancy, building relocation or other circumstances.

[BS] C201.2 Eligible conditions. The provisions of this chapter are applicable only to buildings that meet the following eligibility requirements:

1. Buildings assigned to Risk Category I or II in accordance with *International Building Code* Table 1604.5; or

2. Buildings within the scope of the *International Residential Code*.

SECTION C202
ROOF DECK ATTACHMENT FOR WOOD ROOFS

[BS] C202.1 Roof decking attachment for one- and two-family dwellings. For one- and two-family dwellings, fastening shall be in accordance with Section C202.1.1 or C202.1.2 as appropriate for the existing construction. The diameter of 8d nails shall be a minimum of 0.131 inch (3 mm) and the length shall be a minimum of $2^{1}/_{4}$ inches (57 mm) to qualify for the provisions of this section for existing nails regardless of head shape or head diameter.

[BS] C202.1.1 Sawn lumber or wood plank roofs. Roof decking consisting of sawn lumber or wood planks up to 12 inches (30 cm) wide and secured with at least two nails (minimum size 8d) to each roof framing member it crosses shall be deemed to be sufficiently connected. Sawn lumber or wood plank decking secured with smaller fasteners than 8d nails or with fewer than two nails (minimum size 8d) to each framing member it crosses shall be deemed sufficiently connected if fasteners are added such that two clipped head, round head, or ring shank nails (minimum size 8d) are in place on each framing member the nail crosses.

[BS] C202.1.2 Wood structural panel roofs. For roof decking consisting of wood structural panels, fasteners and spacings required in Table C202.1.2 are deemed to comply with the requirements of Section 707.3 of the *International Existing Building Code*.

Supplemental fasteners as required by Table C202.1.2 shall be 8d ring shank nails with round heads and the following minimum dimensions:

1. 0.113-inch-nominal (3 mm) shank diameter.

2. Ring diameter a minimum of 0.012 inch (0.3 mm) greater than shank diameter.

3. 16 to 20 rings per inch.

4. A minimum 0.280-inch (7 mm) full round head diameter.

5. Ring shank to extend a minimum of $1^{1}/_{2}$ inches (38 mm) from the tip of the nail.

6. Minimum $2^{1}/_{4}$-inch (57 mm) nail length.

[BS] TABLE C202.1.2
SUPPLEMENT FASTENERS AT PANEL EDGES AND INTERMEDIATE FRAMING

EXISTING FASTENERS	EXISTING FASTENER SPACING (EDGE OR INTERMEDIATE SUPPORTS)	MAXIMUM SUPPLEMENTAL FASTENER SPACING FOR 130 MPH < V_{ult} ≤ 140 MPH	MAXIMUM SUPPLEMENTAL FASTENER SPACING FOR INTERIOR ZONE[c] LOCATIONS FOR MPH V_{ult} > 140 MPH AND EDGE ZONES NOT COVERED BY THE COLUMN TO THE RIGHT	EDGE ZONE[d] FOR V_{ult} > 160 MPH AND EXPOSURE C, OR V_{ult} > 180 MPH AND EXPOSURE B
Staples or 6d	Any	6″ o.c.[b]	6″ o.c.[b]	4″ o.c.[b] at panel edges and 4″ o.c.[b] at intermediate supports
8d clipped head or round head smooth shank	6″ o.c. or less	None necessary	None necessary along edges of panels but 6″ o.c.[b] at intermediate supports of panel	4″ o.c.[a] at panel edges and 4″ o.c.[a] at intermediate supports
8d clipped head or round head ring shank	6″ o.c. or less	None necessary	None necessary	4″ o.c.[a] at panel edges and 4″ o.c.[a] at intermediate supports
8d clipped head or round head smooth shank	Greater than 6″ o.c.	6″ o.c.[a]	6″ o.c.[a] along panel edges and 6″ o.c.[b] at intermediate supports of panel	4″ o.c.[a] at panel edges and 4″ o.c.[a] at intermediate supports
8d clipped head or round head ring shank	Greater than 6″ o.c.	6″ o.c.[a]	6″ o.c.[a]	4″ o.c.[a] at panel edges and 4″ o.c.[a] at intermediate supports

For SI: 1 inch = 25.4 mm; 1 mile per hour = 0.447 m/s.

a. Maximum spacing determined based on existing fasteners and supplemental fasteners.
b. Maximum spacing determined based on supplemental fasteners only.
c. Interior zone = sheathing that is not located within 4 feet of the perimeter edge of the roof or within 4 feet of each side of a ridge.
d. Edge zone = sheathing that is located within 4 feet of the perimeter edge of the roof and within 4 feet of each side of a ridge.

RESOURCE A

GUIDELINES ON FIRE RATINGS OF ARCHAIC MATERIALS AND ASSEMBLIES

Introduction

The *International Existing Building Code®* (IEBC®) is a comprehensive code with the goal of addressing all aspects of work taking place in existing buildings and providing user-friendly methods and tools for regulation and improvement of such buildings. This resource document is included within the cover of the IEBC with that goal in mind and as a step towards accomplishing that goal.

In the process of *repair* and *alteration* of existing buildings, based on the nature and the extent of the work, the IEBC might require certain upgrades in the fire-resistance rating of building elements, at which time it becomes critical for the designers and the code officials to be able to determine the fire-resistance rating of the *existing building* elements as part of the overall evaluation for the assessment of the need for improvements. This resource document provides a guideline for such an evaluation for fire-resistance ratings of archaic materials that are not typically found in the modern model building codes.

Resource A is only a guideline and is not intended to be a document for specific adoption as it is not written in the format or language of ICC's *International Codes* and is not subject to the code development process.

PURPOSE

The *Guideline on Fire Ratings of Archaic Materials and Assemblies* focuses upon the fire-related performance of archaic construction. "Archaic" encompasses construction typical of an earlier time, generally prior to 1950. "Fire-related performance" includes fire resistance, flame spread, smoke production and degree of combustibility.

The purpose of this guideline is to update the information which was available at the time of original construction, for use by architects, engineers and code officials when evaluating the fire safety of a rehabilitation project. In addition, information relevant to the evaluation of general classes of materials and types of construction is presented for those cases when documentation of the fire performance of a particular archaic material or assembly cannot be found.

It has been assumed that the building materials and their fastening, joining and incorporation into the building structure are sound mechanically. Therefore, some determination must be made that the original manufacture, the original construction practice, and the rigors of aging and use have not weakened the building. This assessment can often be difficult because process and quality control was not good in many industries, and variations among locally available raw materials and manufacturing techniques often resulted in a product which varied widely in its strength and durability. The properties of iron and steel, for example, varied widely, depending on the mill and the process used.

There is nothing inherently inferior about archaic materials or construction techniques. The pressures that promote fundamental change are most often economic or technological matters not necessarily related to concerns for safety. The high cost of labor made wood lath and plaster uneconomical. The high cost of land and the congestion of the cities provided the impetus for high-rise construction. Improved technology made it possible. The difficulty with archaic materials is not a question of suitability, but familiarity.

Code requirements for the fire performance of key building elements (e.g., walls, floor/ceiling assemblies, doors, shaft enclosures) are stated in performance terms: hours of fire resistance. It matters not whether these elements were built in 1908 or 1980, only that they provide the required degree of fire resistance. The level of performance will be defined by the local community, primarily through the enactment of a building or rehabilitation code. This guideline is only a tool to help evaluate the various building elements, regardless of what the level of performance is required to be.

The problem with archaic materials is simply that documentation of their fire performance is not readily available. The application of engineering judgment is more difficult because building officials may not be familiar with the materials or construction method involved. As a result, either a full-scale fire test is required or the archaic construction in question removed and replaced. Both alternatives are time consuming and wasteful.

This guideline and the accompanying appendix are designed to help fill this information void. By providing the necessary documentation, there will be a firm basis for the continued acceptance of archaic materials and assemblies.

1
FIRE-RELATED PERFORMANCE OF ARCHAIC MATERIALS AND ASSEMBLIES

1.1 FIRE PERFORMANCE MEASURES

This guideline does not specify the level of performance required for the various building components. These requirements are controlled by the building occupancy and use and are set forth in the local building or rehabilitation code.

The fire resistance of a given building element is established by subjecting a sample of the assembly to a "standard" fire test which follows a "standard" time-temperature curve. This test method has changed little since the 1920s. The test results tabulated in the Appendix have been adjusted to reflect current test methods.

The current model building codes cite other fire-related properties not always tested for in earlier years: flame spread, smoke production, and degree of combustibility. However, they can generally be assumed to fall within well defined values because the principal combustible component of archaic materials is cellulose. Smoke production is more important today because of the increased use of plastics. However, the early flame spread tests, developed in the early 1940s, also included a test for smoke production.

"Plastics," one of the most important classes of contemporary materials, were not found in the review of archaic materials. If plastics are to be used in a rehabilitated building, they should be evaluated by contemporary standards. Information and documentation of their fire-related properties and performance is widely available.

Flame spread, smoke production and degree of combustibility are discussed in detail below. Test results for eight common species of lumber, published in an Underwriter's Laboratories' report (104), are noted in the following table:

TUNNEL TEST RESULTS FOR EIGHT SPECIES OF LUMBER

SPECIES OF LUMBER	FLAME SPREAD	FUEL CONTRIBUTED	SMOKE DEVELOPED
Western White Pine	75	50-60	50
Northern White Pine	120-215	120-140	60-65
Ponderosa Pine	80-215	120-135	100-110
Yellow Pine	180-190	130-145	275-305
Red Gum	140-155	125-175	40-60
Yellow Birch	105-110	100-105	45-65
Douglas Fir	65-100	50-80	10-100

Flame Spread

The flame spread of interior finishes is most often measured by the ASTM E84 "tunnel test." This test measures how far and how fast the flames spread across the surface of the test sample. The resulting flame spread rating (FSR) is expressed as a number on a continuous scale where cement-asbestos board is 0 and red oak is 100. (Materials with a flame spread greater than red oak have an FSR greater than 100.) The scale is divided into distinct groups or classes. The most commonly used flame spread classifications are: Class I or A*, with a 0-25 FSR; Class II or B, with a 26-75 FSR; and Class III or C, with a 76-200 FSR. The *NFPA Life Safety Code* also has a Class D (201-500 FSR) and Class E (over 500 FSR) interior finish.

These classifications are typically used in modern building codes to restrict the rate of fire spread. Only the first three classifications are normally permitted, though not all classes of materials can be used in all places throughout a building. For example, the interior finish of building materials used in exits or in corridors leading to exits is more strictly regulated than materials used within private dwelling units.

In general, inorganic archaic materials (e.g., bricks or tile) can be expected to be in Class I. Materials of whole wood are mostly Class II. Whole wood is defined as wood used in the same form as sawn from the tree. This is in contrast to the contemporary reconstituted wood products such as plywood, fiberboard, hardboard, or particle board. If the organic archaic material is not whole wood, the flame spread classification could be well over 200 and thus would be particularly unsuited for use in exits and other critical locations in a building. Some plywoods and various wood fiberboards have flame spreads over 200. Although they can be treated with fire retardants to reduce their flame spread, it would be advisable to assume that all such products have a flame spread over 200 unless there is information to the contrary.

Smoke Production

The evaluation of smoke density is part of the ASTM E84 tunnel test. For the eight species of lumber shown in the table above, the highest levels are 275-305 for Yellow Pine, but most of the others are less smoky than red oak which has an index of 100. The advent of plastics caused substantial increases in the smoke density values measured by the tunnel test. The ensuing limitation of the smoke production for wall and ceiling materials by the model building codes has been a reaction to the introduction of plastic materials. In general, cellulosic materials fall in the 50-300 range of smoke density which is below the general limitation of 450 adopted by many codes.

Degree of Combustibility

The model building codes tend to define "noncombustibility" on the basis of having passed ASTM E136 or if the material is totally inorganic. The acceptance of gypsum wallboard as noncombustible is based on limiting paper thickness to not over $1/8$ inch and a 0-50 flame spread rating by ASTM E84. At times there were provisions to define a Class I or A material (0-25 FSR) as noncombustible, but this is not currently recognized by most model building codes.

If there is any doubt whether or not an archaic material is noncombustible, it would be appropriate to send out samples for evaluation. If an archaic material is determined to be noncombustible according to ASTM E136, it can be expected that it will not contribute fuel to the fire.

* Some codes are Roman numerals, others use letters.

1.2
COMBUSTIBLE CONSTRUCTION TYPES

One of the earliest forms of timber construction used exterior load-bearing masonry walls with columns and/or wooden walls supporting wooden beams and floors in the interior of the building. This form of construction, often called "mill" or "heavy timber" construction, has approximately 1 hour fire resistance. The exterior walls will generally contain the fire within the building.

With the development of dimensional lumber, there was a switch from heavy timber to "balloon frame" construction. The balloon frame uses load-bearing exterior wooden walls which have long timbers often extending from foundation to roof. When longer lumber became scarce, another form of construction, "platform" framing, replaced the balloon framing. The difference between the two systems is significant because platform framing is automatically fire-blocked at every floor while balloon framing commonly has concealed spaces that extend unblocked from basement to attic. The architect, engineer, and *code official* must be alert to the details of construction and the ease with which fire can spread in concealed spaces.

2
BUILDING EVALUATION

A given rehabilitation project will most likely go through several stages. The preliminary evaluation process involves the designer in surveying the prospective building. The fire resistance of *existing building* materials and construction systems is identified; potential problems are noted for closer study. The final evaluation phase includes: developing design solutions to upgrade the fire resistance of building elements, if necessary; preparing working drawings and specifications; and the securing of the necessary code approvals.

2.1
PRELIMINARY EVALUATION

A preliminary evaluation should begin with a building survey to determine the existing materials, the general arrangement of the structure and the use of the occupied spaces, and the details of construction. The designer needs to know "what is there" before a decision can be reached about what to keep and what to remove during the rehabilitation process. This preliminary evaluation should be as detailed as necessary to make initial plans. The fire-related properties need to be determined from the applicable building or rehabilitation code, and the materials and assemblies existing in the building then need to be evaluated for these properties. Two work sheets are shown below to facilitate the preliminary evaluation.

Two possible sources of information helpful in the preliminary evaluation are the original building plans and the building code in effect at the time of original construction. Plans may be on file with the local building department or in the offices of the original designers (e.g., architect, engineer) or their successors. If plans are available, the investigator should verify that the building was actually constructed as called for in the plans, as well as incorporate any later alterations or changes to the building. Earlier editions of the local building code should be on file with the building official. The code in effect at the time of construction will contain fire performance criteria. While this is no guarantee that the required performance was actually provided, it does give the investigator some guidance as to the level of performance which may be expected. Under some code administration and enforcement systems, the code in effect at the time of construction also defines the level of performance that must be provided at the time of rehabilitation.

Figure 1 illustrates one method for organizing preliminary field notes. Space is provided for the materials, dimensions, and condition of the principal building elements. Each floor of the structure should be visited and the appropriate information obtained. In practice, there will often be identical materials and construction on every floor, but the exception may be of vital importance. A schematic diagram should be prepared of each floor showing the layout of exits and hallways and indicating where each element described in the field notes fits into the structure as a whole. The exact arrangement of interior walls within apartments is of secondary importance from a fire safety point of view and need not be shown on the drawings unless these walls are required by code to have a fire resistance rating.

The location of stairways and elevators should be clearly marked on the drawings. All exterior means of escape (e.g., fire escapes) should be identified.[1]

The following notes explain the entries in Figure 1.

<u>Exterior Bearing Walls</u>: Many old buildings utilize heavily constructed walls to support the floor/ceiling assemblies at the exterior of the building. There may be columns and/or interior bearing walls within the structure, but the exterior walls are an important factor in assessing the fire safety of a building.

The field investigator should note how the floor/ceiling assemblies are supported at the exterior of the building. If columns are incorporated in the exterior walls, the walls may be considered nonbearing.

<u>Interior Bearing Walls</u>: It may be difficult to determine whether or not an interior wall is load bearing, but the field investigator should attempt to make this determination. At a later stage of the rehabilitation process, this question will need to be determined exactly. Therefore, the field notes should be as accurate as possible.

<u>Exterior Nonbearing Walls</u>: The fire resistance of the exterior walls is important for two reasons. These walls (both bearing and nonbearing) are depended upon to: a) contain a fire within the building of origin; or b) keep an exterior fire *outside* the building. It is therefore important to indicate on the drawings where any openings are located as well as the materials and construction of all doors or shutters. The drawings

---

1. Problems providing adequate exiting are discussed at length in the *Egress Guideline for Residential Rehabilitation*.

should indicate the presence of wired glass, its thickness and framing, and identify the materials used for windows and door frames. The protection of openings adjacent to exterior means of escape (e.g., exterior stairways, fire escapes) is particularly important. The ground floor drawing should locate the building on the property and indicate the precise distances to adjacent buildings.

Interior Nonbearing Walls (Partitions): A partition is a "wall that extends from floor to ceiling and subdivides space within any story of a building." (48) Figure 1 has two categories (A & B) for Interior Nonbearing Walls (Partitions) which can be used for different walls, such as hallway walls as compared to inter-apartment walls. Under some circumstances there may be only one type of wall construction; in others, three or more types of wall construction may occur.

The field investigator should be alert for differences in function as well as in materials and construction details. In general, the details within apartments are not as important as the major exit paths and exit stairways. The preliminary field investigation should attempt to determine the thickness of all walls. A term introduced below called "thickness design" will depend on an accurate ($\pm\ ^1/_4$ inch) determination. Even though this initial field survey is called "preliminary," the data generated should be as accurate and complete as possible.

The field investigator should note the exact location from which observations are recorded. For instance, if a hole is found through a wall enclosing an exit stairway which allows a cataloguing of the construction details, the field investigation notes should reflect the location of the "find." At the preliminary stage it is not necessary to core every wall; the interior details of construction can usually be determined at some location.

Structural Frame: There may or may not be a complete skeletal frame, but usually there are columns, beams, trusses, or other like elements. The dimensions and spacing of the structural elements should be measured and indicated on the drawings. For instance, if there are 10-inch square columns located on a 30-foot square grid throughout the building, this should be noted. The structural material and cover or protective materials should be identified wherever possible. The thickness of the cover materials should be determined to an accuracy of $\pm\ ^1/_4$ inch. As discussed above, the preliminary field survey usually relies on accidental openings in the cover materials rather than a systematic coring technique.

Floor/Ceiling Structural Systems: The span between supports should be measured. If possible, a sketch of the cross-section of the system should be made. If there is no location where accidental damage has opened the floor/ceiling construction to visual inspection, it is necessary to make such an opening. An evaluation of the fire resistance of a floor/ceiling assembly requires detailed knowledge of the materials and their arrangement. Special attention should be paid to the cover on structural steel elements and the condition of suspended ceilings and similar membranes.

Roofs: The preliminary field survey of the roof system is initially concerned with water-tightness. However, once it is apparent that the roof is sound for ordinary use and can be retained in the rehabilitated building, it becomes necessary to evaluate the fire performance. The field investigator must measure the thickness and identify the types of materials which have been used. Be aware that there may be several layers of roof materials.

**FIGURE 1
PRELIMINARY EVALUATION FIELD NOTES**

BUILDING ELEMENT		MATERIALS	THICKNESS	CONDITION	NOTES
Exterior Bearing Walls					
Interior Bearing Walls					
Exterior Nonbearing Walls					
Interior Nonbearing Walls or Partitions:	A				
	B				
Structural Frame: Columns					
Beams					
Other					
Floor/Ceiling Structural System Spanning					
Roofs					
Doors (including frame and hardware): a) Enclosed vertical exitway					
b) Enclosed horizontal exitway					
c) Other					

Doors: Doors to stairways and hallways represent some of the most important fire elements to be considered within a building. The uses of the spaces separated largely controls the level of fire performance necessary. Walls and doors enclosing stairways or elevator shafts would normally require a higher level of performance than between the bedroom and bath. The various uses are differentiated in Figure 1.

Careful measurements of the thickness of door panels must be made, and the type of core material within each door must be determined. It should be noted whether doors have self-closing devices; the general operation of the doors should be checked. The latch should engage and the door should fit tightly in the frame. The hinges should be in good condition. If glass is used in the doors, it should be identified as either plain glass or wired glass mounted in either a wood or steel frame.

Materials: The field investigator should be able to identify ordinary building materials. In situations where an unfamiliar material is found, a sample should be obtained. This sample should measure at least 10 cubic inches so that an ASTM E136 fire test can be conducted to determine if it is combustible.

Thickness: The thickness of all materials should be measured accurately since, under certain circumstances, the level of fire resistance is very sensitive to the material thickness.

Condition: The method of attaching the various layers and facings to one another or to the supporting structural element should be noted under the appropriate building element. The "secureness" of the attachment and the general condition of the layers and facings should be noted here.

Notes: The "Notes" column can be used for many purposes, but it might be a good idea to make specific references to other field notes or drawings.

After the building survey is completed, the data collected must be analyzed. A suggested work sheet for organizing this information is given below as Figure 2.

The required fire resistance and flame spread for each building element are normally established by the local building or rehabilitation code. The fire performance of the existing materials and assemblies should then be estimated, using one of the techniques described below. If the fire performance of the *existing building* element(s) is equal to or greater than that required, the materials and assemblies may remain. If the fire performance is less than required, then corrective measures must be taken.

The most common methods of upgrading the level of protection are to either remove and replace the *existing building* element(s) or to *repair* and upgrade the existing materials and assemblies. Other fire protection measures, such as automatic sprinklers or detection and alarm systems, also could be con-

FIGURE 2
PRELIMINARY EVALUATION WORKSHEET

BUILDING ELEMENT		REQUIRED FIRE RESISTANCE	REQUIRED FLAME SPREAD	ESTIMATED FIRE RESISTANCE	ESTIMATED FLAME SPREAD	METHOD OF UPGRADING	ESTIMATED UPGRADED PROTECTION	NOTES
Exterior Bearing Walls								
Interior Bearing Walls								
Exterior Nonbearing Walls								
Interior Nonbearing Walls or Partitions:	A							
	B							
Structural Frame: Columns								
Beams								
Other								
Floor/Ceiling Structural System Spanning								
Roofs								
Doors (including frame and hardware): a) Enclosed vertical exitway								
b) Enclosed horizontal exitway								
c) Others								

sidered, though they are beyond the scope of this guideline. If the upgraded protection is still less than that required or deemed to be acceptable, additional corrective measures must be taken. This process must continue until an acceptable level of performance is obtained.

2.2
FIRE RESISTANCE OF EXISTING BUILDING ELEMENTS

The fire resistance of the *existing building* elements can be estimated from the tables and histograms contained in the Appendix. The Appendix is organized first by type of building element: walls, columns, floor/ceiling assemblies, beams, and doors. Within each building element, the tables are organized by type of construction (e.g., masonry, metal, wood frame), and then further divided by minimum dimensions or thickness of the building element.

A histogram precedes every table that has 10 or more entries. The X-axis measures fire resistance in hours; the Y-axis shows the number of entries in that table having a given level of fire resistance. The histograms also contain the location of each entry within that table for easy cross-referencing.

The histograms, because they are keyed to the tables, can speed the preliminary investigation. For example, Table 1.3.2, *Wood Frame Walls 4" to Less Than 6" Thick*, contains 96 entries. Rather than study each table entry, the histogram shows that every wall assembly listed in that table has a fire resistance of less than 2 hours. If the building code required the wall to have 2 hours fire resistance, the designer, with a minimum of effort, is made aware of a problem that requires closer study.

Suppose the code had only required a wall of 1 hour fire resistance. The histogram shows far fewer complying elements (19) than noncomplying ones (77). If the existing assembly is not one of the 19 complying entries, there is a strong possibility the existing assembly is deficient. The histograms can also be used in the converse situation. If the existing assembly is not one of the smaller number of entries with a lower than required fire resistance, there is a strong possibility the existing assembly will be acceptable.

At some point, the *existing building* component or assembly must be located within the tables. Otherwise, the fire resistance must be determined through one of the other techniques presented in the guideline. Locating the building component in the Appendix Tables not only guarantees the accuracy of the fire resistance rating, but also provides a source of documentation for the building official.

2.3
EFFECTS OF PENETRATIONS IN FIRE RESISTANT ASSEMBLIES

There are often many features in existing walls or floor/ceiling assemblies which were not included in the original certification or fire testing. The most common examples are pipes and utility wires passed through holes poked through an assembly. During the life of the building, many penetrations are added, and by the time a building is ready for rehabilitation it is not sufficient to just consider the fire resistance of the assembly as originally constructed. It is necessary to consider all penetrations and their relative impact upon fire performance. For instance, the fire resistance of the corridor wall may be less important than the effect of plain glass doors or transoms. In fact, doors are the most important single class of penetrations.

A fully developed fire generates substantial quantities of heat and excess gaseous fuel capable of penetrating any holes which might be present in the walls or ceiling of the fire compartment. In general, this leads to a severe degradation of the fire resistance of those building elements and to a greater potential for fire spread. This is particularly applicable to penetrations located high in a compartment where the positive pressure of the fire can force the unburned gases through the penetration.

Penetrations in a wall or floor/ceiling assembly will generally completely negate the barrier qualities of the assembly and will lead to rapid spread of fire and smoke to adjacent spaces. In order to maintain compartmentation of a building and impede the spread of fire and smoke, penetrations in rated assemblies need to be protected. The upper half of walls are similar to the floor/ceiling assembly in that a positive pressure can reasonably be expected in the top of the room, and this will push hot and/or burning gases through the penetration unless the penetration is completely sealed.

Building codes require doors installed in fire resistive walls to resist the passage of fire for a specified period of time. If the door to a fully involved room is not closed, a large plume of fire will typically escape through the doorway, preventing anyone from using the space outside the door while allowing the fire to spread. This is why door closers are so important. Glass in doors and transoms can be expected to rapidly shatter unless constructed of listed or approved wire glass in a steel frame. As with other building elements, penetrations or nonrated portions of doors and transoms must be upgraded or otherwise protected.

Table 5.1 in Section V of the Appendix contains 41 entries of doors mounted in sound tightfitting frames. Part 3.4 below outlines one procedure for evaluating and possibly upgrading existing doors.

3
FINAL EVALUATION AND DESIGN SOLUTION

The final evaluation begins after the rehabilitation project has reached the final design stage and the choice is made to keep certain archaic materials and assemblies in the rehabilitated building. The final evaluation process is essentially a more refined and detailed version of the preliminary evaluation. The specific fire resistance and flame spread requirements are determined for the project. This may involve local building and fire officials reviewing the preliminary evaluation as depicted in Figures 1 and 2 and the field drawings and notes. When necessary, provisions must be made to upgrade *existing building* elements to provide the required level of fire performance.

There are several approaches to design solutions that can make possible the continued use of archaic materials and assemblies in the rehabilitated structure. The simplest case occurs when the materials and assembly in question are found within the Appendix Tables and the fire performance properties satisfy code requirements. Other approaches must be used, though, if the assembly cannot be found within the Appendix or the fire performance needs to be upgraded. These approaches have been grouped into two classes: experimental and theoretical.

3.1
THE EXPERIMENTAL APPROACH

If a material or assembly found in a building is not listed in the Appendix Tables, there are several other ways to evaluate fire performance. One approach is to conduct the appropriate fire test(s) and thereby determine the fire-related properties directly. There are a number of laboratories in the United States which routinely conduct the various fire tests. A current list can be obtained by writing the Center for Fire Research, National Bureau of Standards, Washington, D.C. 20234.

The contract with any of these testing laboratories should require their observation of specimen preparation as well as the testing of the specimen. A complete description of where and how the specimen was obtained from the building, the transportation of the specimen, and its preparation for testing should be noted in detail so that the building official can be satisfied that the fire test is representative of the actual use.

The test report should describe the fire test procedure and the response of the material or assembly. The laboratory usually submits a cover letter with the report to describe the provisions of the fire test that were satisfied by the material or assembly under investigation. A building official will generally require this cover letter, but will also read the report to confirm that the material or assembly complies with the code requirements. Local code officials should be involved in all phases of the testing process.

The experimental approach can be costly and time consuming because specimens must be taken from the building and transported to the testing laboratory. When a load bearing assembly has continuous reinforcement, the test specimen must be removed from the building, transported, and tested in one piece. However, when the fire performance cannot be determined by other means, there may be no alternative to a full-scale test.

A "nonstandard" small-scale test can be used in special cases. Sample sizes need only be 10-25 square feet (0.93-2.3 m^2), while full-scale tests require test samples of either 100 or 180 square feet (9.3 or 17 m^2) in size. This small-scale test is best suited for testing nonload-bearing assemblies against thermal transmission only.

3.2
THE THEORETICAL APPROACH

There will be instances when materials and assemblies in a building undergoing rehabilitation cannot be found in the Appendix Tables. Even where test results are available for more or less similar construction, the proper classification may not be immediately apparent. Variations in dimensions, loading conditions, materials, or workmanship may markedly affect the performance of the individual building elements, and the extent of such a possible effect cannot be evaluated from the tables.

Theoretical methods being developed offer an alternative to the full-scale fire tests discussed above. For example, Section 4302(b) of the 1979 edition of the *Uniform Building Code* specifically allows an engineering design for fire resistance in lieu of conducting full-scale tests. These techniques draw upon computer simulation and mathematical modeling, thermodynamics, heat-flow analysis, and materials science to predict the fire performance of building materials and assemblies.

One theoretical method, known as the "Ten Rules of Fire Endurance Ratings," was published by T. Z. Harmathy in the May, 1965 edition of *Fire Technology*. (35) Harmathy's Rules provide a foundation for extending the data within the Appendix Tables to analyze or upgrade current as well as archaic building materials or assemblies.

HARMATHY'S TEN RULES

Rule 1: The "thermal"[1] fire endurance of a construction consisting of a number of parallel layers is greater than the sum of the "thermal" fire endurances characteristic of the individual layers when exposed separately to fire.

The minimum performance of an untested assembly can be estimated if the fire endurance of the individual components is known. Though the exact rating of the assembly cannot be stated, the endurance of the assembly is greater than the sum of the endurance of the components.

When a building assembly or component is found to be deficient, the fire endurance can be upgraded by providing a protective membrane. This membrane could be a new layer of brick, plaster, or drywall. The fire endurance of this membrane is called the "finish rating." Appendix Tables 1.5.1 and 1.5.2 contain the finish ratings for the most commonly employed materials. (See also the notes to Rule 2).

The test criteria for the finish rating is the same as for the thermal fire endurance of the total assembly: average temperature increases of 250°F (121°C) above ambient or 325°F (163°C) above ambient at any one place with the membrane being exposed to the fire. The temperature is measured at the interface of the assembly and the protective membrane.

Rule 2. The fire endurance of a construction does not decrease with the addition of further layers.

---

1. The "thermal" fire endurance is the time at which the average temperature on the unexposed side of a construction exceeds its initial value by 250° when the other side is exposed to the "standard" fire specified by ASTM Test Method E-19.

Harmathy notes that this rule is a consequence of the previous rule. Its validity follows from the fact that the additional layers increase both the resistance to heat flow and the heat capacity of the construction. This, in turn, reduces the rate of temperature rise at the unexposed surface.

This rule is not just restricted to "thermal" performance but affects the other fire test criteria: direct flame passage, cotton waste ignition, and load bearing performance. This means that certain restrictions must be imposed on the materials to be added and on the loading conditions. One restriction is that a new layer, if applied to the exposed surface, must not produce additional thermal stresses in the construction, i.e., its thermal expansion characteristics must be similar to those of the adjacent layer. Each new layer must also be capable of contributing enough additional strength to the assembly to sustain the added dead load. If this requirement is not fulfilled, the allowable live load must be reduced by an amount equal to the weight of the new layer. Because of these limitations, this rule should not be applied without careful consideration.

Particular care must be taken if the material added is a good thermal insulator. Properly located, the added insulation could improve the "thermal" performance of the assembly. Improperly located, the insulation could block necessary thermal transmission through the assembly, thereby subjecting the structural elements to greater temperatures for longer periods of time, and could cause premature structural failure of the supporting members.

Rule 3: The fire endurance of constructions containing continuous air gaps or cavities is greater than the fire endurance of similar constructions of the same weight, but containing no air gaps or cavities.

By providing for voids in a construction, additional resistances are produced in the path of heat flow. Numerical heat flow analyses indicate that a 10 to 15 percent increase in fire endurance can be achieved by creating an air gap at the midplane of a brick wall. Since the gross volume is also increased by the presence of voids, the air gaps and cavities have a beneficial effect on stability as well. However, constructions containing combustible materials within an air gap may be regarded as exceptions to this rule because of the possible development of burning in the gap.

There are numerous examples of this rule in the tables. For instance:

Table 1.1.4; Item W-8-M-82: Cored concrete masonry, nominal 8 inch thick wall with one unit in wall thickness and with 62 percent minimum of solid material in each unit, load bearing (80 PSI). Fire endurance: $2^1/_2$ hours.

Table 1.1.5; Item W-10-M-11: Cored concrete mansonry, nominal 10 inch thick wall with two units in wall thickness and a 2-inch (51 mm) air space, load bearing (80 PSI). The units are essentially the same as item W-8-M-82. Fire endurance: $3^1/_2$ hours.

These walls show 1 hour greater fire endurance by the addition of the 2-inch (51 mm) air space.

Rule 4: The farther an air gap or cavity is located from the exposed surface, the more beneficial is its effect on the fire endurance.

Radiation dominates the heat transfer across an air gap or cavity, and it is markedly higher where the temperature is higher.

The air gap or cavity is thus a poor insulator if it is located in a region which attains high temperatures during fire exposure.

Some of the clay tile designs take advantage of these factors. The double cell design, for instance, ensures that there is a cavity near the unexposed face. Some floor/ceiling assemblies have air gaps or cavities near the top surface and these enhance their thermal performance.

Rule 5: The fire endurance of a construction cannot be increased by increasing the thickness of a completely enclosed air layer.

Harmathy notes that there is evidence that if the thickness of the air layer is larger than about $^1/_2$ inch (12.7 mm), the heat transfer through the air layer depends only on the temperature of the bounding surfaces, and is practically independent of the distance between them. This rule is not applicable if the air layer is not completely enclosed, i.e., if there is a possibility of fresh air entering the gap at an appreciable rate.

Rule 6: Layers of materials of low thermal conductivity are better utilized on that side of the construction on which fire is more likely to happen.

As in Rule 4, the reason lies in the heat transfer process, though the conductivity of the solid is much less dependent on the ambient temperature of the materials. The low thermal conductor creates a substantial temperature differential to be established across its thickness under transient heat flow conditions. This rule may not be applicable to materials undergoing physico-chemical changes accompanied by significant heat absorption or heat evolution.

Rule 7: The fire endurance of asymmetrical constructions depends on the direction of heat flow.

This rule is a consequence of Rules 4 and 6, as well as other factors. This rule is useful in determining the relative protection of corridors and walls enclosing an exit stairway from the surrounding spaces. In addition, there are often situations where a fire is more likely, or potentially more severe, from one side or the other.

Rule 8: The presence of moisture, if it does not result in explosive spalling, increases the fire endurance.

The flow of heat into an assembly is greatly hindered by the release and evaporation of the moisture found within cementitious materials such as gypsum, portland cement, or magnesium oxychloride. Harmathy has shown that the gain in fire endurance may be as high as 8 percent for each percent (by volume) of moisture in the construction. It is the moisture chemically bound within the construction material at the time of manufacture or processing that leads to increased fire endurance. There is no direct relationship between the relative humidity of the air in the pores of the material and the increase in fire endurance.

Under certain conditions there may be explosive spalling of low permeability cementitious materials such as dense concrete. In general, one can assume that extremely old concrete has developed enough minor cracking that this factor should not be significant.

Rule 9: Load-supporting elements, such as beams, girders and joists, yield higher fire endurances when subjected to fire endurance tests as parts of floor, roof, or ceiling assemblies than they would when tested separately.

One of the fire endurance test criteria is the ability of a load-supporting element to carry its design load. The element will be deemed to have failed when the load can no longer be supported.

Failure usually results for two reasons. Some materials, particularly steel and other metals, lose much of their structural strength at elevated temperatures. Physical deflection of the supporting element, due to decreased strength or thermal expansion, causes a redistribution of the load forces and stresses throughout the element. Structural failure often results because the supporting element is not designed to carry the redistributed load.

Roof, floor, and ceiling assemblies have primary (e.g., beams) and secondary (e.g., floor joists) structural members. Since the primary load-supporting elements span the largest distances, their deflection becomes significant at a stage when the strength of the secondary members (including the roof or floor surface) is hardly affected by the heat. As the secondary members follow the deflection of the primary load-supporting element, an increasingly larger portion of the load is transferred to the secondary members.

When load-supporting elements are tested separately, the imposed load is constant and equal to the design load throughout the test. By definition, no distribution of the load is possible because the element is being tested by itself. Without any other structural members to which the load could be transferred, the individual elements cannot yield a higher fire endurance than they do when tested as parts of a floor, roof or ceiling assembly.

Rule 10: The load-supporting elements (beams, girders, joists, etc.) of a floor, roof, or ceiling assembly can be replaced by such other load-supporting elements which, when tested separately, yielded fire endurances not less than that of the assembly.

This rule depends on Rule 9 for its validity. A beam or girder, if capable of yielding a certain performance when tested separately, will yield an equally good or better performance when it forms a part of a floor, roof, or ceiling assembly. It must be emphasized that the supporting element of one assembly must not be replaced by the supporting element of another assembly if the performance of this latter element is not known from a separate (beam) test. Because of the load-reducing effect of the secondary elements that results from a test performed on an assembly, the performance of the supporting element alone cannot be evaluated by simple arithmetic. This rule also indicates the advantage of performing separate fire tests on primary load-supporting elements.

ILLUSTRATION OF HARMATHY'S RULES

Harmathy provided one schematic figure which illustrated his Rules.[1] It should be useful as a quick reference to assist in applying his Rules.

EXAMPLE APPLICATION OF HARMATHY'S RULES

The following examples, based in whole or in part upon those presented in Harmathy's paper (35), show how the Rules can be applied to practical cases.

Example 1

Problem

A contractor would like to keep a partition which consists of a $3^{3}/_{4}$ inch (95 mm) thick layer of red clay brick, a $1^{1}/_{4}$ inch (32 mm) thick layer of plywood, and a $^{3}/_{8}$ inch (9.5 mm) thick layer of gypsum wallboard, at a location where 2-hour fire endurance is required. Is this assembly capable of providing a 2-hour protection?

Solution

(1) This partition does not appear in the Appendix Tables.

(2) Bricks of this thickness yield fire endurances of approximately 75 minutes (Table 1.1.2, Item W-4-M-2).

(3) The $1^{1}/_{4}$ inch (32 mm) thick plywood has a finish rating of 30 minutes.

(4) The $^{3}/_{8}$ inch (9.5 mm) gypsum wallboard has a finish rating of 10 minutes.

(5) Using the recommended values from the tables and applying Rule 1, the fire endurance (FI) of the assembly is larger than the sum of the individual layers, or

$$FI > 75 + 30 + 10 = 115 \text{ minutes}$$

Discussion

This example illustrates how the Appendix Tables can be utilized to determine the fire resistance of assemblies not explicitly listed.

Example 2

Problem

(1) A number of buildings to be rehabilitated have the same type of roof slab which is supported with different structural elements.

(2) The designer and contractor would like to determine whether or not this roof slab is capable of yielding a 2-hour fire endurance. According to a rigorous interpretation of ASTM E119, however, only the roof assembly, including the roof slab as well as the cover and the supporting elements, can be subjected to a fire test.

---

1. Reproduced from the May 1065 *Fire Technology* (Vol. 1, No. 2). Copyright National Fire Protection Association, Boston. Reproduced by permission.

Therefore, a fire endurance classification cannot be issued for the slabs separately.

(3) The designer and contractor believe this slab will yield a 2-hour fire endurance even without the cover, and any beam of at least 2-hour fire endurance will provide satisfactory support. Is it possible to obtain a classification for the slab separately?

Solution

(1) The answer to the question is yes.

(2) According to Rule 10 it is not contrary to common sense to test and classify roofs and supporting elements separately. Furthermore, according to Rule 2, if the roof slabs actually yield a 2-hour fire endurance, the endurance of an assembly, including the slabs, cannot be less than 2 hours.

(3) The recommended procedure would be to review the tables to see if the slab appears as part of any tested roof or floor/ceiling assembly. The supporting system can be regarded as separate from the slab specimen, and the fire endurance of the assembly listed in the table is at least the fire endurance of the slab. There would have to be an adjustment for the weight of the roof cover in the allowable load if the test specimen did not contain a cover.

(4) The supporting structure or element would have to have at least a 2-hour fire endurance when tested separately.

Discussion

If the tables did not include tests on assemblies which contained the slab, one procedure would be to assemble the roof slabs on any convenient supporting system (not regarded as part of the specimen) and to subject them to a load which, besides the usually required superimposed load, includes some allowances for the weight of the cover.

Example 3

Problem

A steel-joisted floor and ceiling assembly is known to have yielded a fire endurance of 1 hour and 35 minutes. At a certain location, a 2-hour endurance is required. What is the most economical way of increasing the fire endurance by at least 25 minutes?

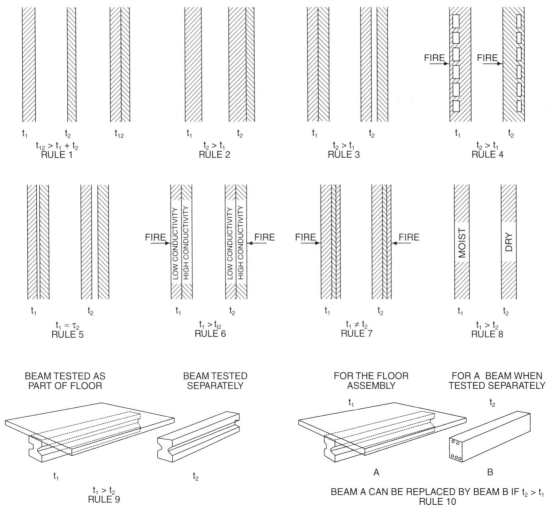

Diagrammatic illustration of ten rules.
t = fire endurance

Solution

(1) The most effective technique would be to increase the ceiling plaster thickness. Existing coats of paint would have to be removed and the surface properly prepared before the new plaster could be applied. Other materials (e.g., gypsum wallboard) could also be considered.

(2) There may be other techniques based on other principles, but an examination of the drawings would be necessary.

Discussion

(1) The additional plaster has at least three effects:

    a) The layer of plaster is increased and thus there is a gain of fire endurance (Rule 1).

    b) There is a gain due to shifting the air gap farther from the exposed surface (Rule 4).

    c) There is more moisture in the path of heat flow to the structural elements (Rules 7 and 8).

(2) The increase in fire endurance would be at least as large as that of the finish rating for the added thickness of plaster. The combined effects in (1) above would further increase this by a factor of 2 or more, depending upon the geometry of the assembly.

Example 4

Problem

The fire endurance of item W-10-M-1 in Table 1.1.5 is 4 hours. This wall consists of two $3^3/_4$ inch (95 mm) thick layers of structural tiles separated by a 2-inch (51 mm) air gap and $^3/_4$ inch (19 mm) portland cement plaster or stucco on both sides. If the actual wall in the building is identical to item W-10-M-1 except that it has a 4-inch (102 mm) air gap, can the fire endurance be estimated at 5 hours?

Solution

The answer to the question is no for the reasons contained in Rule 5.

Example 5

Problem

In order to increase the insulating value of its precast roof slabs, a company has decided to use two layers of different concretes. The lower layer of the slabs, where the strength of the concrete is immaterial (all the tensile load is carried by the steel reinforcement), would be made with a concrete of low strength but good insulating value. The upper layer, where the concrete is supposed to carry the compressive load, would remain the original high strength, high thermal conductivity concrete. How will the fire endurance of the slabs be affected by the change?

Solution

The effect on the thermal fire endurance is beneficial:

(1) The total resistance to heat flow of the new slabs has been increased due to the replacement of a layer of high thermal conductivity by one of low conductivity.

(2) The layer of low conductivity is on the side more likely to be exposed to fire, where it is more effectively utilized according to Rule 6. The layer of low thermal conductivity also provides better protection for the steel reinforcement, thereby extending the time before reaching the temperature at which the creep of steel becomes significant.

3.3
"THICKNESS DESIGN" STRATEGY

The "thickness design" strategy is based upon Harmathy's Rules 1 and 2. This design approach can be used when the construction materials have been identified and measured, but the specific assembly cannot be located within the tables. The tables should be surveyed again for thinner walls of like material and construction detail that have yielded the desired or greater fire endurance. If such an assembly can be found, then the thicker walls in the building have more than enough fire resistance. The thickness of the walls thus becomes the principal concern.

This approach can also be used for floor/ceiling assemblies, except that the thickness of the cover[1] and the slab become the central concern. The fire resistance of the untested assembly will be at least the fire resistance of an assembly listed in the table having a similar design but with less cover and/or thinner slabs. For other structural elements (e.g., beams and columns), the element listed in the table must also be of a similar design but with less cover thickness.

3.4
EVALUATION OF DOORS

A separate section on doors has been included because the process for evaluation presented below differs from those suggested previously for other building elements. The impact of unprotected openings or penetrations in fire resistant assemblies has been detailed in Part 2.3 above. It is sufficient to note here that openings left unprotected will likely lead to failure of the barrier under actual fire conditions.

For other types of building elements (e.g., beams, columns), the Appendix Tables can be used to establish a minimum level of fire performance. The benefit to rehabilitation is that the need for a full-scale fire test is then eliminated. For doors, however, this cannot be done. The data contained in Appendix Table 5.1, Resistance of Doors to Fire Exposure, can only provide guidance as to whether a successful fire test is even feasible.

For example, a door required to have 1 hour fire resistance is noted in the tables as providing only 5 minutes. The likelihood of achieving the required 1 hour, even if the door is upgraded, is remote. The ultimate need for replacement of the doors is reasonably clear, and the expense and time needed for testing can be saved. However, if the performance documented in the table is near or in excess of what is being required, then a fire test should be conducted. The test docu-

---

1. Cover: the protective layer or membrane of material which slows the flow of heat to the structural elements.

mentation can then be used as evidence of compliance with the required level of performance.

The table entries cannot be used as the sole proof of performance of the door in question because there are too many unknown variables which could measurably affect fire performance. The wood may have dried over the years; coats of flammable varnish could have been added. Minor deviations in the internal construction of a door can result in significant differences in performance. Methods of securing inserts in panel doors can vary. The major non-destructive method of analysis, an x-ray, often cannot provide the necessary detail. It is for these, and similar reasons, that a fire test is still felt to be necessary.

It is often possible to upgrade the fire performance of an existing door. Sometimes, "as is" and modified doors are evaluated in a single series of tests when failure of the unmodified door is expected. Because doors upgraded after an initial failure must be tested again, there is a potential savings of time and money.

The most common problems encountered are plain glass, panel inserts of insufficient thickness, and improper fit of a door in its frame. The latter problem can be significant because a fire can develop a substantial positive pressure, and the fire will work its way through otherwise innocent-looking gaps between door and frame.

One approach to solving these problems is as follows. The plain glass is replaced with approved or listed wire glass in a steel frame. The panel inserts can be upgraded by adding an additional layer of material. Gypsum wallboard is often used for this purpose. Intumescent paint applied to the edges of the door and frame will expand when exposed to fire, forming an effective seal around the edges. This seal, coupled with the generally even thermal expansion of a wood door in a wood frame, can prevent the passage of flames and other fire gases. Figure 3 below illustrates these solutions.

Because the interior construction of a door cannot be determined by a visual inspection, there is no absolute guarantee that the remaining doors are identical to the one(s) removed from the building and tested. But the same is true for doors constructed today, and reason and judgment must be applied. Doors that appear identical upon visual inspection can be weighed. If the weights are reasonably close, the doors can be assumed to be identical and therefore provide the same level of fire performance. Another approach is to fire test more than one door or to dismantle doors selected at random to see if they had been constructed in the same manner. Original building plans showing door details or other records showing that doors were purchased at one time or obtained from a single supplier can also be evidence of similar construction.

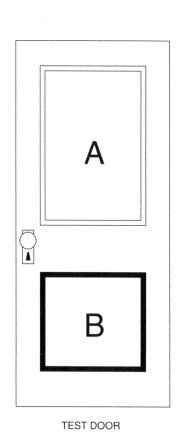
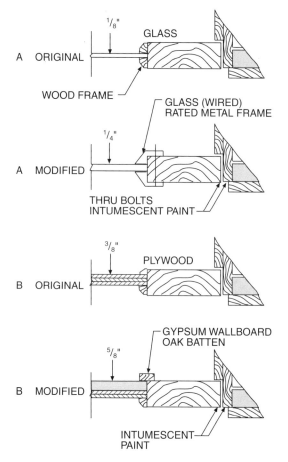

FIGURE 3
MODIFICATION DETAILS

More often though, it is what is visible to the eye that is most significant. The investigator should carefully check the condition and fit of the door and frame, and for frames out of plumb or separating from the wall. Door closers, latches, and hinges must be examined to see that they function properly and are tightly secured. If these are in order and the door and frame have passed a full-scale test, there can be a reasonable basis for allowing the existing doors to remain.

4
SUMMARY

This section summarizes the various approaches and design solutions discussed in the preceding sections of the guideline. The term "structural system" includes: frames, beams, columns, and other structural elements. "Cover" is a protective layer(s) of materials or membrane which slows the flow of heat to the structural elements. It cannot be stressed too strongly that the fire endurance of actual building elements can be greatly reduced or totally negated by removing part of the cover to allow pipes, ducts, or conduits to pass through the element. This must be repaired in the rehabilitation process.

The following approaches shall be considered equivalent.

4.1 The fire resistance of a building element can be established from the Appendix Tables. This is subject to the following limitations:

The building element in the rehabilitated building shall be constructed of the same materials with the same nominal dimensions as stated in the tables.

All penetrations in the building element or its cover for services such as electricity, plumbing, and HVAC shall be packed with noncombustible cementitious materials and so fixed that the packing material will not fall out when it loses its water of hydration.

The effects of age and wear and tear shall be repaired so that the building element is sound and the original thickness of all components, particularly covers and floor slabs, is maintained.

This approach essentially follows the approach taken by model building codes. The assembly must appear in a table either published in or accepted by the code for a given fire resistance rating to be recognized and accepted.

4.2 The fire resistance of a building element which does not explicitly appear in the Appendix Tables can be established if one or more elements of same design but different dimensions have been listed in the tables. For walls, the existing element must be thicker than the one listed. For floor/ceiling assemblies, the assembly listed in the table must have the same or less cover and the same or thinner slab constructed of the same material as the actual floor/ceiling assembly. For other structural elements, the element listed in the table must be of a similar design but with less cover thickness. The fire resistance in all instances shall be the fire resistance recommended in the table. This is subject to the following limitations:

The actual element in the rehabilitated building shall be constructed of the same materials as listed in the table. Only the following dimensions may vary from those specified: for walls, the overall thickness must exceed that specified in the table; for floor/ceiling assemblies, the thickness of the cover and the slab must be greater than, or equal to, that specified in the table; for other structural elements, the thickness of the cover must be greater than that specified in the table.

All penetrations in the building element or its cover for services such as electricity, plumbing, or HVAC shall be packed with noncombustible cementitious materials and so fixed that the packing material will not fall out when it loses its water of hydration.

The effects of age and wear and tear shall be repaired so that the building element is sound and the original thickness of all components, particularly covers and floor slabs, is maintained.

This approach is an application of the "thickness design" concept presented in Part 3.3 of the guideline. There should be many instances when a thicker building element was utilized than the one listed in the Appendix Tables. This guideline recognizes the inherent superiority of a thicker design. Note: "thickness design" for floor/ceiling assemblies and structural elements refers to cover and slab thickness rather than total thickness.

The "thickness design" concept is essentially a special case of Harmathy's Rules (specifically Rules 1 and 2). It should be recognized that the only source of data is the Appendix Tables. If other data are used, it must be in connection with the approach below.

4.3 The fire resistance of building elements can be established by applying Harmathy's Ten Rules of Fire Resistance Ratings as set forth in Part 3.2 of the guideline. This is subject to the following limitations:

The data from the tables can be utilized subject to the limitations in 4.2 above.

Test reports from recognized journals or published papers can be used to support data utilized in applying Harmathy's Rules.

Calculations utilizing recognized and well established computational techniques can be used in applying Harmathy's Rules. These include, but are not limited to, analysis of heat flow, mechanical properties, deflections, and load bearing capacity.

APPENDIX

INTRODUCTION

The fire-resistance tables that follow are a part of Resource A and provide a tabular form of assigning fire-resistance ratings to various archaic building elements and assemblies.

These tables for archaic materials and assemblies do for archaic materials what Tables 721.1(1-3) of the *International Building Code®* do for more modern building elements and assemblies. The fire resistance tables of Resource A should be used as described in the "Purpose and Procedure" that follows the table of contents for these tables.

RESOURCE A TABLE OF CONTENTS

Purpose and Procedure — A-17

Section I—Walls

1.1.1	Masonry	0 in. - 4 in. thick	A-18
1.1.2	Masonry	4 in. - 6 in. thick	A-21
1.1.3	Masonry	6 in. - 8 in. thick	A-28
1.1.4	Masonry	8 in. - 10 in. thick	A-33
1.1.5	Masonry	10 in. - 12 in. thick	A-41
1.1.6	Masonry	12 in. - 14 in. thick	A-45
1.1.7	Masonry	14 in. or more thick	A-51
1.2.1	Metal Frame	0 in. - 4 in. thick	A-54
1.2.2	Metal Frame	4 in. - 6 in. thick	A-58
1.2.3	Metal Frame	6 in. - 8 in. thick	A-60
1.2.4	Metal Frame	8 in. - 10 in. thick	A-61
1.3.1	Wood Frame	0 in. - 4 in. thick	A-62
1.3.2	Wood Frame	4 in. - 6 in. thick	A-63
1.3.3	Wood Frame	6 in. - 8 in. thick	A-71
1.4.1	Miscellaneous Materials	0 in. - 4 in. thick	A-71
1.4.2	Miscellaneous Materials	4 in. - 6 in. thick	A-72
1.5.1	Finish Ratings—Inorganic Materials	Thickness	A-73
1.5.2	Finish Ratings—Organic Materials	Thickness	A-74

Section II—Columns

2.1.1	Reinforced Concrete	Min. Dim. 0 in. - 6 in.	A-75
2.1.2	Reinforced Concrete	Min. Dim. 10 in. - 12 in.	A-76
2.1.3	Reinforced Concrete	Min. Dim. 12 in. - 14 in.	A-79
2.1.4	Reinforced Concrete	Min. Dim. 14 in. - 16 in.	A-80
2.1.5	Reinforced Concrete	Min. Dim. 16 in. - 18 in.	A-81
2.1.6	Reinforced Concrete	Min. Dim. 18 in. - 20 in.	A-83
2.1.7	Reinforced Concrete	Min. Dim. 20 in. - 22 in.	A-84
2.1.8	Hexagonal Reinforced Concrete	Diameter—12 in. - 14 in.	A-85

RESOURCE A

2.1.9	Hexagonal Reinforced Concrete	Diameter—14 in. - 16 in.	A-86
2.1.10	Hexagonal Reinforced Concrete	Diameter—16 in. - 18 in.	A-86
2.1.11	Hexagonal Reinforced Concrete	Diameter—20 in. - 22 in.	A-86
2.2	Round Cast Iron Columns	Minimum Dimension	A-87
2.3	Steel—Gypsum Encasements	Minimum Area of Solid Material	A-88
2.4	Timber	Minimum Dimension	A-89
2.5.1.1	Steel/Concrete Encasements	Minimum Dimension less than 6 in.	A-89
2.5.1.2	Steel/Concrete Encasements	Minimum Dimension 6 in. - 8 in.	A-90
2.5.1.3	Steel/Concrete Encasements	Minimum Dimension 8 in. - 10 in.	A-91
2.5.1.4	Steel/Concrete Encasements	Minimum Dimension 10 in. - 12 in.	A-93
2.5.1.5	Steel/Concrete Encasements	Minimum Dimension 12 in. - 14 in.	A-97
2.5.1.6	Steel/Concrete Encasements	Minimum Dimension 14 in. - 16 in.	A-99
2.5.1.7	Steel/Concrete Encasements	Minimum Dimension 16 in. - 18 in.	A-101
2.5.2.1	Steel/Brick and Block Encasements	Minimum Dimension 10 in. - 12 in.	A-101
2.5.2.2	Steel/Brick and Block Encasements	Minimum Dimension 12 in. - 14 in.	A-102
2.5.2.3	Steel/Brick and Block Encasements	Minimum Dimension 14 in. - 16 in.	A-102
2.5.3.1	Steel/Plaster Encasements	Minimum Dimension 6 in. - 8 in.	A-103
2.5.3.2	Steel/Plaster Encasements	Minimum Dimension 8 in. - 10 in.	A-103
2.5.4.1	Steel/Miscellaneous Encasements	Minimum Dimension 6 in. - 8 in.	A-103
2.5.4.2	Steel/Miscellaneous Encasements	Minimum Dimension 8 in. - 10 in.	A-104
2.5.4.3	Steel/Miscellaneous Encasements	Minimum Dimension 10 in. - 12 in.	A-104
2.5.4.4	Steel/Miscellaneous Encasements	Minimum Dimension 12 in. - 14 in.	A-104

Section III—Floor/Ceiling Assemblies

3.1	Reinforced Concrete	Assembly thickness	A-105
3.2	Steel Structural Elements	Membrane thickness	A-111
3.3	Wood Joist	Membrane thickness	A-117
3.4	Hollow Clay Tile with Reinforced Concrete	Membrane thickness	A-121

Section IV—Beams

4.1.1	Reinforced Concrete	Depth—10 in. - 12 in.	A-124
4.1.2	Reinforced Concrete	Depth—12 in. - 14 in.	A-127
4.1.3	Reinforced Concrete	Depth—14 in. - 16 in.	A-129
4.2.1	Reinforced Concrete/Unprotected	Depth—10 in. - 12 in.	A-130
4.2.2	Steel/Concrete Protection	Depth—10 in. - 12 in.	A-130

Section V—Doors

5.1	Resistance of Doors to Fire Exposure	Thickness	A-131

PURPOSE AND PROCEDURE

The tables and histograms which follow are to be used only within the analytical framework detailed in the main body of this guideline.

Histograms precede any table with 10 or more entries. The use and interpretation of these histograms is explained in Part 2 of the guideline. The tables are in a format similar to that found in the model building codes. The following example, taken from an entry in Table 1.1.2, best explains the table format.

1. Item Code: The item code consists of a four place series in the general form w-x-y-z in which each member of the series denotes the following:

    w = Type of building element (e.g., W=Walls; F=Floors, etc.)

    x = The building element thickness rounded down to the nearest 1-inch increment (e.g., $4^5/_8$ inches is rounded off to 4 inches)

    y = The general type of material from which the building element is constructed (e.g., M=Masonry; W=Wood, etc.)

    z = The item number of the particular building element in a given table

    The item code shown in the example W-4-M-50 denotes the following:

    W = Wall, as the building element

    4 = Wall thickness in the range of 4 inches (102 mm) to less than 5 inches (127 mm)

    M = Masonry construction

    50 = The 50th entry in Table 1.1.2

2. The specific name or heading of this column identifies the dimensions which, if varied, has the greatest impact on fire resistance. The critical dimension for walls, the example here, is thickness. It is different for other building elements (e.g., depth for beams; membrane thickness for some floor/ceiling assemblies). The table entry is the named dimension of the building element measured at the time of actual testing to within $\pm^1/_8$ inch (3.2 mm) tolerance. The thickness tabulated includes facings where facings are a part of the wall construction.

3. Construction Details: The construction details provide a brief description of the manner in which the building element was constructed.

4. Performance: This heading is subdivided into two columns. The column labeled "Load" will either list the load that the building element was subjected to during the fire test or it will contain a note number which will list the load and any other significant details. If the building element was not subjected to a load during the test, this column will contain "n/a," which means "not applicable."

    The second column under performance is labeled "Time" and denotes the actual fire endurance time observed in the fire test.

5. Reference Number: This heading is subdivided into three columns: Pre-BMS-92; BMS-92; and Post-BMS-92. The table entry under this column is the number in the Bibliography of the original source reference for the test data.

6. Notes: Notes are provided at the end of each table to allow a more detailed explanation of certain aspects of the test. In certain tables the notes given to this column have also been listed under the "Construction Details" and/or "Load" columns.

7. Rec Hours: This column lists the recommended fire endurance rating, in hours, of a building element. In some cases, the recommended fire endurance will be less than that listed under the "Time" column. In no case is the "Rec Hours" greater than given in the "Time" column.

ITEM CODE	THICKNESS	CONSTRUCTION DETAILS	PERFORMANCE		REFERENCE NUMBER			NOTES	REC. HOURS
			LOAD	TIME	PRE-BMS-92	BMS-92	POST-BMS-92		
W-4-M-50	$4^5/_8''$	Core: structural clay tile, See notes 12, 16, 21; Facings on unexposed side only, see note 18	N/A	25 min.		1		3, 4, 24	$^1/_3$

RESOURCE A

SECTION I - WALLS

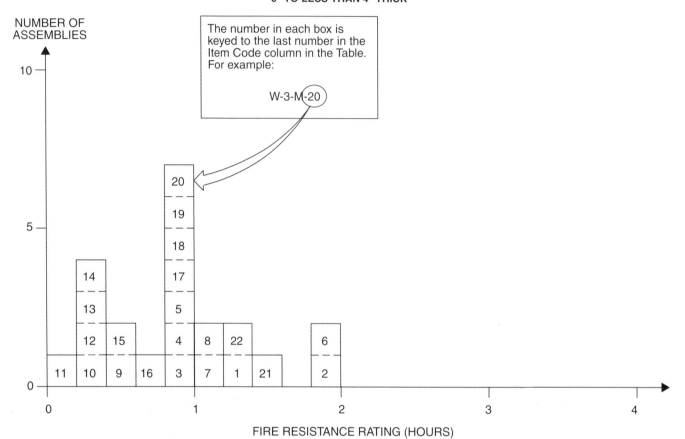

FIGURE 1.1.1
MASONRY WALLS
0″ TO LESS THAN 4″ THICK

TABLE 1.1.1
MASONRY WALLS
0″ TO LESS THAN 4″ THICK

ITEM CODE	THICKNESS	CONSTRUCTION DETAILS	PERFORMANCE		REFERENCE NUMBER			NOTES	REC. HOURS
			LOAD	TIME	PRE-BMS-92	BMS-92	POST-BMS-92		
W-2-M-1	2¼″	Solid partition; ¾″ gypsum plank- 10′ ×1′6″ ; ¾″ plus gypsum plaster each side.	N/A	1 hr. 22 min.			7	1	1¼
W-3-M-2	3″	Concrete block (18″ × 9″ × 3″) of fuel ash, portland cement and plasticizer; cement/sand mortar.	N/A	2 hrs.			7	2, 3	2
W-2-M-3	2″	Solid gypsum block wall; No facings	N/A	1 hr.		1		4	1
W-3-M-4	3″	Solid gypsum blocks, laid in 1:3 sanded gypsum mortar.	N/A	1 hr.		1		4	1
W-3-M-5	3″	Magnesium oxysulfate wood fiber blocks; 2″ thick, laid in portland cement-lime mortar; Facings: ½″ of 1:3 sanded gypsum plaster on both sides.	N/A	1 hr.		1		4	1
W-3-M-6	3″	Magnesium oxysulfate bound wood fiber blocks; 3″ thick; laid in portland cement-lime mortar; Facings: ½″ of 1:3 sanded gypsum plaster on both sides.	N/A	2 hrs.		1		4	2

(continued)

**TABLE 1.1.1—continued
MASONRY WALLS
0″ TO LESS THAN 4″ THICK**

ITEM CODE	THICKNESS	CONSTRUCTION DETAILS	PERFORMANCE		REFERENCE NUMBER			NOTES	REC. HOURS
			LOAD	TIME	PRE-BMS-92	BMS-92	POST-BMS-92		
W-3-M-7	3″	Clay tile; Ohio fire clay; single cell thick; Face plaster: $^5/_8$″ (both sides) 1:3 sanded gypsum; Design "E," Construction "A."	N/A	1 hr. 6 min.	0		2	5, 6, 7, 11, 12, 39	1
W-3-M-8	3″	Clay tile; Illinois surface clay; single cell thick; Face plaster: $^5/_8$″ (both sides) 1:3 sanded gypsum; Design "A," Construction "E."	N/A	1 hr. 1 min			2	5, 8, 9, 11, 12, 39	1
W-3-M-9	3″	Clay tile; Illinois surface clay; single cell thick; No face plaster; Design "A," Construction "C."	N/A	25 min.			2	5, 10, 11, 12, 39	$^1/_3$
W-3-M-10	$3^7/_8$″	8″ × $4^7/_8$″ glass blocks; weight 4 lbs. each; portland cement-lime mortar; horizontal mortar joints reinforced with metal lath.	N/A	15 min.		1		4	$^1/_4$
W-3-M-11	3″	Core: structural clay tile; see Notes 14, 18, 13; No facings.	N/A	10 min.		1		5, 11, 26	$^1/_6$
W-3-M-12	3″	Core: structural clay tile; see Notes 14, 19, 23; No facings.	N/A	20 min.		1		5, 11, 26	$^1/_3$
W-3-M-13	$3^5/_8$″	Core: structural clay tile; see Notes 14, 18, 23; Facings: unexposed side; see Note 20.	N/A	20 min.		1		5, 11, 26	$^1/_3$
W-3-M-14	$3^5/_8$″	Core: structural clay tile; see Notes 14, 19, 23; Facings: unexposed side only; see Note 20.	N/A	20 min.		1		5, 11, 26	$^1/_3$
W-3-M-15	$3^5/_8$″	Core: clay structural tile; see Notes 14, 18, 23; Facings: side exposed to fire; see Note 20.	N/A	30 min.		1		5, 11, 26	$^1/_2$
W-3-M-16	$3^5/_8$″	Core: clay structural tile; see Notes 14, 19, 23; Facings: side exposed to fire; see Note 20.	N/A	45 min.		1		5, 11, 26	$^3/_4$
W-2-M-17	2″	2″ thick solid gypsum blocks; see Note 27.	N/A	1 hr.		1		27	1
W-3-M-18	3″	Core: 3″ thick gypsum blocks 70% solid; see Note 2; No facings.	N/A	1 hr.		1		27	1
W-3-M-19	3″	Core: hollow concrete units; see Notes 29, 35, 36, 38; No facings.	N/A	1 hr.		1		27	1
W-3-M-20	3″	Core: hollow concrete units; see Notes 28, 35, 36, 37, 38; No facings.	N/A	1 hr.		1			1
W-3-M-21	$3^1/_2$″	Core: hollow concrete units; see Notes 28, 35, 36, 37, 38; Facings: one side; see Note 37.	N/A	$1^1/_2$ hrs.		1			$1^1/_2$
W-3-M-22	$3^1/_2$″	Core: hollow concrete units; see Notes 29, 35, 36, 38; Facings: one side, see Note 37.	N/A	$1^1/_4$ hrs.		1			$1^1/_4$

For SI: 1 inch = 25.4 mm, 1 pound per square inch = 0.00689 MPa, °C = [(°F) - 32]/1.8.

Notes:
1. Failure mode—flame thru.
2. Passed 2-hour fire test (Grade "C" fire res. - British).
3. Passed hose stream test.
4. Tested at NBS under ASA Spec. No. A2-1934. As nonload bearing partitions.
5. Tested at NBS under ASA Spec. No. 42-1934 (ASTM C19-33) except that hose stream testing where carried was run on test specimens exposed for full test duration, not for a reduced period as is contemporarily done.
6. Failure by thermal criteria—maximum temperature rise 325°F.
7. Hose stream failure.
8. Hose stream—pass.
9. Specimen removed prior to any failure occurring.
10. Failure mode—collapse.
11. For clay tile walls, unless the source or density of the clay can be positively identified or determined, it is suggested that the lowest hourly rating for the fire endurance of a clay tile partition of that thickness be followed. Identified sources of clay showing longer fire endurance can lead to longer time recommendations.

(continued)

RESOURCE A

**TABLE 1.1.1—continued
MASONRY WALLS
0″ TO LESS THAN 4″ THICK**

12. See appendix for construction and design details for clay tile walls.
13. Load: 80 psi for gross wall area.
14. One cell in wall thickness.
15. Two cells in wall thickness.
16. Double shells plus one cell in wall thickness.
17. One cell in wall thickness, cells filled with broken tile, crushed stone, slag cinders or sand mixed with mortar.
18. Dense hard-burned clay or shale tile.
19. Medium-burned clay tile.
20. Not less than $^5/_8$ inch thickness of 1:3 sanded gypsum plaster.
21. Units of not less than 30 percent solid material.
22. Units of not less than 40 percent solid material.
23. Units of not less than 50 percent solid material.
24. Units of not less than 45 percent solid material.
25. Units of not less than 60 percent solid material.
26. All tiles laid in portland cement-lime mortar.
27. Blocks laid in 1:3 sanded gypsum mortar voids in blocks not to exceed 30 percent.
28. Units of expanded slag or pumice aggregate.
29. Units of crushed limestone, blast furnace, slag, cinders and expanded clay or shale.
30. Units of calcareous sand and gravel. Coarse aggregate, 60 percent or more calcite and dolomite.
31. Units of siliceous sand and gravel. Ninety percent or more quartz, chert or flint.
32. Unit at least 49 percent solid.
33. Unit at least 62 percent solid.
34. Unit at least 65 percent solid.
35. Unit at least 73 percent solid.
36. Ratings based on one unit and one cell in wall thickness.
37. Minimum of $^1/_2$ inch—1:3 sanded gypsum plaster.
38. Nonload bearing.
39. See Clay Tile Partition Design Construction drawings, below.

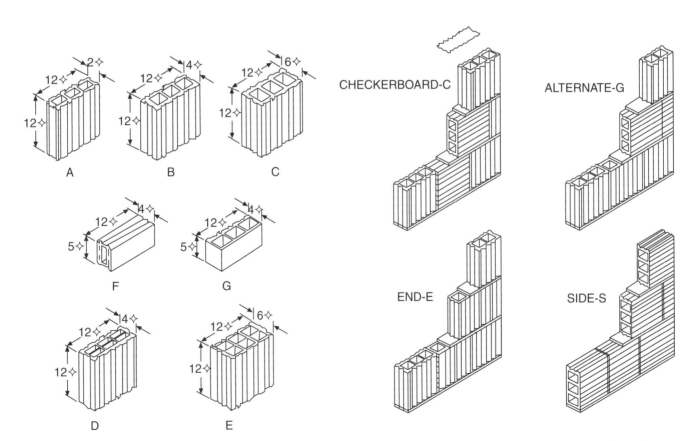

DESIGNS OF TILES USED IN FIRE-TEST PARTITIONS

THE FOUR TYPES OF CONSTRUCTION USED IN FIRE-TEST PARTITIONS

RESOURCE A

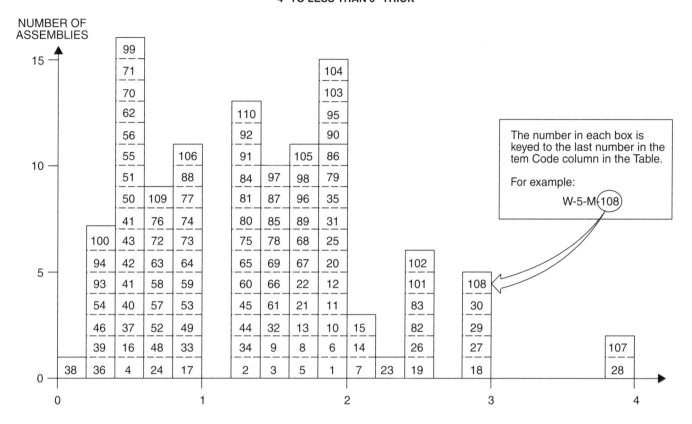

FIGURE 1.1.2
MASONRY WALLS
4″ TO LESS THAN 6″ THICK

FIRE RESISTANCE RATING (HOURS)

TABLE 1.1.2
MASONRY WALLS
4″ TO LESS THAN 6″ THICK

ITEM CODE	THICKNESS	CONSTRUCTION DETAILS	PERFORMANCE		REFERENCE NUMBER			NOTES	REC. HOURS
			LOAD	TIME	PRE-BMS-92	BMS-92	POST-BMS-92		
W-4-M-1	4″	Solid 3″ thick, gypsum blocks laid in 1:3 sanded gypsum mortar; Facings: $^1/_2$″ of 1:3 sanded gypsum plaster (both sides).	N/A	2 hrs.		1		1	2
W-4-M-2	4″	Solid clay or shale brick.	N/A	1 hr. 15 min		1		1, 2	$1^1/_4$
W-4-M-3	4″	Concrete; No facings.	N/A	1 hr. 30 min.		1		1	$1^1/_2$
W-4-M-4	4″	Clay tile; Illinois surface clay; single cell thick; No face plaster; Design "B," Construction "C."	N/A	25 min.			2	3-7, 36	$^1/_3$
W-4-M-5	4″	Solid sand-lime brick.	N/A	1 hr. 45 min.		1		1	$1^3/_4$
W-4-M-6	4″	Solid wall; 3″ thick block; $^1/_2$″ plaster each side; $17^3/_4$″ × $8^3/_4$″ × 4″ "Breeze Blocks"; portland cement/sand mortar.	N/A	1 hr. 52 min.			7	2	$1^3/_4$
W-4-M-7	4″	Concrete (4020 psi); Reinforcement: vertical $^3/_8$″; horizontal $^1/_4$″; 6″ × 6″ grid.	N/A	2 hrs. 10 min.			7	2	2
W-4-M-8	4″	Concrete wall (4340 psi crush); reinforcement $^1/_4$″ diameter rebar on 8″ centers (vertical and horizontal).	N/A	1 hr. 40 min.			7	2	$1^2/_3$

(continued)

TABLE 1.1.2—continued
MASONRY WALLS
4″ TO LESS THAN 6″ THICK

ITEM CODE	THICKNESS	CONSTRUCTION DETAILS	PERFORMANCE		REFERENCE NUMBER			NOTES	REC. HOURS
			LOAD	TIME	PRE-BMS-92	BMS-92	POST-BMS-92		
W-4-M-9	$4^3/_{16}″$	$4^3/_{16}″ \times 2^5/_8″$ cellular fletton brick (1873 psi) with $^1/_2″$ sand mortar; bricks are U-shaped yielding hollow cover (approx. 2″ × 4″) in final cross-section configuration.	N/A	1 hr. 25 min.			7	2	$1^1/_3$
W-4-M-10	$4^1/_4″$	$4^1/_4″ \times 2^1/_2″$ fletton (1831 psi) brick in $^1/_2″$ sand mortar.	N/A	1 hr. 53 min			7	2	$1^3/_4$
W-4-M-11	$4^1/_4″$	$4^1/_4″ \times 2^1/_2″$ London stock (683 psi) brick; $^1/_2″$ grout.	N/A	1 hr. 52 min.			7	2	$1^3/_4$
W-4-M-12	$4^1/_2″$	$4^1/_4″ \times 2^1/_2″$ Leicester red, wire-cut brick (4465 psi) in $^1/_2″$ sand mortar.	N/A	1 hr. 56 min.			7	6	$1^3/_4$
W-4-M-13	$4^1/_4″$	$4^1/_4″ \times 2^1/_2″$ stairfoot brick (7527 psi) $^1/_2″$ sand mortar.	N/A	1 hr. 37 min.			7	2	$1^1/_2$
W-4-M-14	$4^1/_4″$	$4^1/_4″ \times 2^1/_2″$ sand-lime brick (2603 psi) $^1/_2″$ sand mortar.	N/A	2 hrs. 6 min.			7	2	2
W-4-M-15	$4^1/_4″$	$4^1/_4″ \times 2^1/_2″$ concrete brick (2527 psi) $^1/_2″$ sand mortar.	N/A	2 hrs. 10 min.			7	2	2
W-4-M-16	$4^1/_2″$	4″ thick clay tile; Ohio fire clay; single cell thick; No plaster exposed face; $^1/_2″$ 1:2 gypsum back face; Design "F," Construction "S."	N/A	31 min.			2	3-6, 36	$^1/_2$
W-4-M-17	$4^1/_2″$	4″ thick clay tile; Ohio fire clay; single cell thick; Plaster exposed face; $^1/_2″$ 1:2 sanded gypsum; Back Face: none; Construction "S," Design "F."	80 psi	50 min.			2	3-5, 8, 36	$^3/_4$
W-4-M-18	$4^1/_2″$	Core: solid sand-lime brick; $^1/_2″$ sanded gypsum plaster facings on both sides.	80 psi	3 hrs.		1		1, 11	3
W-4-M-19	$4^1/_2″$	Core: solid sand-lime brick; $^1/_2″$ sanded gypsum plaster facings on both sides.	80 psi	2 hrs. 30 min.		1		1, 11	$2^1/_2$
W-4-M-20	$4^1/_2″$	Core: concrete brick $^1/_2″$ of 1:3 sanded gypsum plaster facings on both sides.	80 psi	2 hrs.		1		1, 11	2
W-4-M-21	$4^1/_2″$	Core: solid clay or shale brick; $^1/_2″$ thick, 1:3 sanded gypsum plaster facings on fire sides.	80 psi	1 hr. 45 min.		1		1, 2, 11	$1^3/_4$
W-4-M-22	$4^3/_4″$	4″ thick clay tile; Ohio fire clay; single cell thick; cells filled with cement and broken tile concrete; Plaster on exposed face; none on unexposed face; $^3/_4″$ 1:3 sanded gypsum; Design "G," Construction "E."	N/A	1 hr. 48 min.			2	2, 3-5, 9, 36	$1^3/_4$
W-4-M-23	$4^3/_4″$	4″ thick clay tile; Ohio fire clay; single cell thick; cells filled with cement and broken tile concrete; No plaster exposed faced; $^3/_4″$ neat gypsum plaster on unexposed face; Design "G," Construction "E."	N/A	2 hrs. 14 min.			2	2, 3-5, 9, 36	2
W-5-M-24	5″	3″ × 13″ air space; 1″ thick metal reinforced concrete facings on both sides; faces connected with wood splines.	2,250 lbs./ft.	45 min.		1		1	$^3/_4$
W-5-M-25	5″	Core: 3″ thick void filled with "nondulated" mineral wool weighing 10 lbs./ft.^{3}; 1″ thick metal reinforced concrete facings on both sides.	2,250 lbs./ft.	2 hrs.		1		1	2
W-5-M-26	5″	Core: solid clay or shale brick; $^1/_2″$ thick, 1:3 sanded gypsum plaster facings on both sides.	40 psi	2 hrs. 30 min.		1		1, 2, 11	$2^1/_2$
W-5-M-27	5″	Core: solid 4″ thick gypsum blocks, laid in 1:3 sanded gypsum mortar; $^1/_2″$ of 1:3 sanded gypsum plaster facings on both sides.	N/A	3 hrs.		1		1	3

(continued)

TABLE 1.1.2—continued
MASONRY WALLS
4″ TO LESS THAN 6″ THICK

ITEM CODE	THICKNESS	CONSTRUCTION DETAILS	PERFORMANCE		REFERENCE NUMBER			NOTES	REC. HOURS
			LOAD	TIME	PRE-BMS-92	BMS-92	POST-BMS-92		
W-5-M-28	5″	Core: 4″ thick hollow gypsum blocks with 30% voids; blocks laid in 1:3 sanded gypsum mortar; No facings.	N/A	4 hrs.		1		1	4
W-5-M-29	5″	Core: concrete brick; $^1/_2$″ of 1:3 sanded gypsum plaster facings on both sides.	160 psi	3 hrs.		1		1	3
W-5-M-30	$5^1/_4$″	4″ thick clay tile; Illinois surface clay; double cell thick; Plaster: $^5/_8$″ sanded gypsum 1:3 both faces; Design "D," Construction "S."	N/A	2 hrs. 53 min.			2	2-5, 9, 36	$2^3/_4$
W-5-M-31	$5^1/_4$″	4″ thick clay tile; New Jersey fire clay; double cell thick; Plaster: $^5/_8$″ sanded gypsum 1:3 both faces; Design "D," Construction "S."	N/A	1 hr. 52 min.			2	2-5, 9, 36	$1^3/_4$
W-5-M-32	$5^1/_4$″	4″ thick clay tile; New Jersey fire clay; single cell thick; Plaster: $^5/_8$″ sanded gypsum 1:3 both faces; Design "D," Construction "S."	N/A	1 hr. 34 min.	2		2	2-5, 9, 36	$1^1/_2$
W-5-M-33	$5^1/_4$″	4″ thick clay tile; New Jersey fire clay; single cell thick; Face plaster: $^5/_8$″ both sides; 1:3 sanded gypsum; Design "B," Construction "S."	N/A	50 min.			2	3-5, 8, 36	$^3/_4$
W-5-M-34	$5^1/_4$″	4″ thick clay tile; Ohio fire clay; single cell thick; Face plaster: $^5/_8$″ both sides; 1:3 sanded gypsum; Design "B," Construction "A."	N/A	1 hr. 19 min.			2	2-5, 9, 36	$1^1/_4$
W-5-M-35	$5^1/_4$″	4″ thick clay tile; Illinois surface clay; single cell thick; Face plaster: $^5/_8$″ both sides; 1:3 sanded gypsum; Design "B," Construction "S."	N/A	1 hr. 59 min.			2	2-5, 10, 36	$1^3/_4$
W-5-M-36	4″	Core: structural clay tile; see Notes 12, 16, 21; No facings.	N/A	15 min.		1		3, 4, 24	$^1/_4$
W-4-M-37	4″	Core: structural clay tile; see Notes 12, 17, 21; No facings.	N/A	25 min.		1		3, 4, 24	$^1/_3$
W-4-M-38	4″	Core: structural clay tile; see Notes 12, 16, 20; No facings.	N/A	10 min.		1		3, 4, 24	$^1/_6$
W-4-M-39	4″	Core: structural clay tile; see Notes 12, 17, 20; No facings.	N/A	20 min.		1		3, 4, 24	$^1/_3$
W-4-M-40	4″	Core: structural clay tile; see Notes 13, 16, 23; No facings.	N/A	30 min.		1		3, 4, 24	$^1/_2$
W-4-M-41	4″	Core: structural clay tile; see Notes 13, 17, 23; No facings.	N/A	35 min.		1		3, 4, 24	$^1/_2$
W-4-M-42	4″	Core: structural clay tile; see Notes 13, 16, 21; No facings.	N/A	25 min.		1		3, 4, 24	$^1/_3$
W-4-M-43	4″	Core: structural clay tile; see Notes 13, 17, 21; No facings.	N/A	30 min.		1		3, 4, 24	$^1/_2$
W-4-M-44	4″	Core: structural clay tile; see Notes 15, 16, 20; No facings	N/A	1 hr. 15 min.		1		3, 4, 24	$1^1/_4$
W-4-M-45	4″	Core: structural clay tile; see Notes 15, 17, 20; No facings.	N/A	1 hr. 15 min.		1		3, 4, 24	$1^1/_4$
W-4-M-46	4″	Core: structural clay tile; see Notes 14, 16, 22; No facings.	N/A	20 min.		1		3, 4, 24	$^1/_3$
W-4-M-47	4″	Core: structural clay tile; see Notes 14, 17, 22; No facings.	N/A	25 min.		1		3, 4, 24	$^1/_3$
W-4-M-48	$4^1/_4$″	Core: structural clay tile; see Notes 12, 16, 21; Facings: both sides; see Note 18.	N/A	45 min.		1		3, 4, 24	$^3/_4$

(continued)

TABLE 1.1.2—continued
MASONRY WALLS
4″ TO LESS THAN 6″ THICK

ITEM CODE	THICKNESS	CONSTRUCTION DETAILS	PERFORMANCE		REFERENCE NUMBER			NOTES	REC. HOURS
			LOAD	TIME	PRE-BMS-92	BMS-92	POST-BMS-92		
W-4-M-49	4¼″	Core: structural clay tile; see Notes 12, 17, 21; Facings: both sides; see Note 18.	N/A	1 hr.		1		3, 4, 24	1
W-4-M-50	4⅝″	Core: structural clay tile; see Notes 12, 16, 21; Facings: unexposed side only; see Note 18.	N/A	25 min.		1		3, 4, 24	⅓
W-4-M-51	4⅝″	Core: structural clay tile; see Notes 12, 17, 21; Facings: unexposed side only; see Note 18.	N/A	30 min.		1		3, 4, 24	½
W-4-M-52	4⅝″	Core: structural clay tile; see Notes 12, 16, 21; Facings: unexposed side only; see Note 18.	N/A	45 min.		1		3, 4, 24	¾
W-4-M-53	4⅝″	Core: structural clay tile; see Notes 12, 17, 21; Facings: fire side only; see Note 18.	N/A	1 hr.		1		3, 4, 24	1
W-4-M-54	4⅝″	Core: structural clay tile; see Notes 12, 16, 20; Facings: unexposed side; see Note 18.	N/A	20 min.		1		3, 4, 24	⅓
W-4-M-55	4⅝″	Core: structural clay tile; see Notes 12, 17, 20; Facings: exposed side; see Note 18.	N/A	25 min.		1		3, 4, 24	⅓
W-4-M-56	4⅝″	Core: structural clay tile; see Notes 12, 16, 20; Facings: fire side only; see Note 18.	N/A	30 min.		1		3, 4, 24	½
W-4-M-57	4⅝″	Core: structural clay tile; see Notes 12, 17, 20; Facings: fire side only; see Note 18.	N/A	45 min.		1		3, 4, 24	¾
W-4-M-58	4⅝″	Core: structural clay tile; see Notes 13, 16, 23; Facings: unexposed side only; see Note 18.	N/A	40 min.		1		3, 4, 24	⅔
W-4-M-59	4⅝″	Core: structural clay tile; see Notes 13, 17, 23; Facings: unexposed side only; see Note 18.	N/A	1 hr.		1		3, 4, 24	1
W-4-M-60	4⅝″	Core: structural clay tile; see Notes 13, 16, 23; Facings: fire side only; see Note 18.	N/A	1 hr. 15 min.		1		3, 4, 24	1¼
W-4-M-61	4⅝″	Core: structural clay tile; see Notes 13, 17, 23; Facings: fire side only; see Note 18.	N/A	1 hr. 30 min.		1		3, 4, 24	1½
W-4-M-62	4⅝″	Core: structural clay tile; see Notes 13, 16, 21; Facings: unexposed side only; see Note 18.	N/A	35 min.		1		3, 4, 24	½
W-4-M-63	4⅝″	Core: structural clay tile; see Notes 13, 17, 21; Facings: unexposed face only; see Note 18.	N/A	45 min.		1		3, 4, 24	¾
W-4-M-64	4⅝″	Core: structural clay tile; see Notes 13, 16, 23; Facings: exposed face only; see Note 18.	N/A	1 hr.		1		3, 4, 24	1
W-4-M-65	4⅝″	Core: structural clay tile; see Notes 13, 17, 21; Facings: exposed side only; see Note 18.	N/A	1 hr. 15 min.		1		3, 4, 24	1¼
W-4-M-66	4⅝″	Core: structural clay tile; see Notes 15, 17, 20; Facings: unexposed side only; see Note 18	N/A	1 hr. 30 min.		1		3, 4, 24	1½
W-4-M-67	4⅝″	Core: structural clay tile; see Notes 15, 16, 20; Facings: exposed side only; see Note 18.	N/A	1 hr. 45 min.		1		3, 4, 24	1¾
W-4-M-68	4⅝″	Core: structural clay tile; see Notes 15, 17, 20; Facings: exposed side only; see Note 18.	N/A	1 hr. 45 min.		1		3, 4, 24	1¾
W-4-M-69	4⅝″	Core: structural clay tile; see Notes 15, 16, 20; Facings: unexposed side only; see Note 18.	N/A	1 hr. 30 min.		1		3, 4, 24	1¾
W-4-M-70	4⅝″	Core: structural clay tile; see Notes 14, 16, 22; Facings: unexposed side only; see Note 18.	N/A	30 min.		1		3, 4, 24	½

(continued)

TABLE 1.1.2—continued
MASONRY WALLS
4″ TO LESS THAN 6″ THICK

ITEM CODE	THICKNESS	CONSTRUCTION DETAILS	PERFORMANCE		REFERENCE NUMBER			NOTES	REC. HOURS
			LOAD	TIME	PRE-BMS-92	BMS-92	POST-BMS-92		
W-4-M-71	4⅝″	Core: structural clay tile; see Notes 14, 17, 22; Facings: exposed side only; see Note 18.	N/A	35 min.		1		3, 4, 24	½
W-4-M-72	4⅝″	Core: structural clay tile; see Notes 14, 16, 22; Facings: fire side of wall only; see Note 18.	N/A	45 min.		1		3, 4, 24	¾
W-4-M-73	4⅝″	Core: structural clay tile; see Notes 14, 17, 22; Facings: fire side of wall only; see Note 18.	N/A	1 hr.		1		3, 4, 24	1
W-4-M-74	5¼″	Core: structural clay tile; see Notes 12, 16, 21; Facings: both sides; see Note 18.	N/A	1 hr.		1		3, 4, 24	1
W-5-M-75	5¼″	Core: structural clay tile; see Notes 12, 17, 21; Facings: both sides; see Note 18	N/A	1 hr. 15 min.		1		3, 4, 24	1¼
W-5-M-76	5¼″	Core: structural clay tile; see Notes 12, 16, 20; Facings: both sides; see Note 18.	N/A	45 min.		1		3, 4, 24	¾
W-5-M-77	5¼″	Core: structural clay tile; see Notes 12, 17, 20; Facings: both sides; see Note 18.	N/A	1 hr.		1		3, 4, 24	1
W-5-M-78	5¼″	Core: structural clay tile; see Notes 13, 16, 23; Facings: both sides of wall; see Note 18.	N/A	1 hr. 30 min.		1		3, 4, 24	1½
W-5-M-79	5¼″	Core: structural clay tile; see Notes 13, 17, 23; Facings: both sides of wall; see Note 18.	N/A	2 hrs.		1		3, 4, 24	2
W-5-M-80	5¼″	Core: structural clay tile; see Notes 13, 16, 21; Facings: both sides of wall; see Note 18.	N/A	1 hr. 15 min.		1		3, 4, 24	1¼
W-5-M-81	5¼″	Core: structural clay tile; see Notes 13, 16, 21; Facings: both sides of wall; see Note 18.	N/A	1 hr. 30 min.		1		3, 4, 24	1½
W-5-M-82	5¼″	Core: structural clay tile; see Notes 15, 16, 20; Facings: both sides; see Note 18.	N/A	2 hrs. 30 min.		1		3, 4, 24	2½
W-5-M-83	5¼″	Core: structural clay tile; see Notes 15, 17, 20; Facings: both sides; see Note 18.	N/A	2 hrs. 30 min.		1		3, 4, 24	2½
W-5-M-84	5¼″	Core: structural clay tile; see Notes 14, 16, 22; Facings: both sides of wall; see Note 18.	N/A	1 hr. 15 min.		1		3, 4, 24	1¼
W-5-M-85	5¼″	Core: structural clay tile; see Notes 14, 17, 22; Facings: both sides of wall; see Note 18.	N/A	1 hr. 30 min.		1		3, 4, 24	1½
W-4-M-86	4″	Core: 3″ thick gypsum blocks 70% solid; see Note 26; Facings: both sides; see Note 25.	N/A	2 hrs.		1			2
W-4-M-87	4″	Core: hollow concrete units; see Notes 27, 34, 35; No facings.	N/A	1 hr. 30 min.		1			1½
W-4-M-88	4″	Core: hollow concrete units; see Notes 28, 33, 35; No facings.	N/A	1 hr.		1			1
W-4-M-89	4″	Core: hollow concrete units; see Notes 28, 34, 35; Facings: both sides; see Note 25.	N/A	1 hr. 45 min.		1			1¾
W-4-M-90	4″	Core: hollow concrete units; see Notes 27, 34, 35; Facings: both sides; see Note 25.	N/A	2 hrs.		1			2
W-4-M-91	4″	Core: hollow concrete units; see Notes 27, 32, 35; No facings.	N/A	1 hr. 15 min.		1			1¼
W-4-M-92	4″	Core: hollow concrete units; see Notes 28, 34, 35; No facings.	N/A	1 hr. 15 min.		1			1¼
W-4-M-93	4″	Core: hollow concrete units; see Notes 29, 32, 35; No facings.	N/A	20 min.		1			⅓

(continued)

TABLE 1.1.2—continued
MASONRY WALLS
4″ TO LESS THAN 6″ THICK

ITEM CODE	THICKNESS	CONSTRUCTION DETAILS	PERFORMANCE		REFERENCE NUMBER			NOTES	REC. HOURS
			LOAD	TIME	PRE-BMS-92	BMS-92	POST-BMS-92		
W-4-M-94	4″	Core: hollow concrete units; see Notes 30, 34, 35; No facings.	N/A	15 min.	1				$1/4$
W-4-M-95	$4^1/_2$″	Core: hollow concrete units; see Notes 27, 34, 35; Facings: one side only; see Note 25.	N/A	2 hrs.	1				2
W-4-M-96	$4^1/_2$″	Core: hollow concrete units; see Notes 27, 32, 35; Facings: one side only; see Note 25.	N/A	1 hr. 45 min.	1				$1^3/_4$
W-4-M-97	$4^1/_2$″	Core: hollow concrete units; see Notes 28, 33, 35; Facings: one side; see Note 25.	N/A	1 hr. 30 min.	1				$1^1/_2$
W-4-M-98	$4^1/_2$″	Core: hollow concrete units; see Notes 28, 34, 35; Facings: one side only; see Note 25.	N/A	1 hr. 45 min.	1				$1^3/_4$
W-4-M-99	$4^1/_2$″	Core: hollow concrete units; see Notes 29, 32, 35; Facings: one side; see Note 25.	N/A	30 min.	1				$1/_2$
W-4-M-100	$4^1/_2$″	Core: hollow concrete units; see Notes 30, 34, 35; Facings: one side; see Note 25.	N/A	20 min.	1				$1/_3$
W-5-M-101	5″	Core: hollow concrete units; see Notes 27, 34, 35; Facings: both sides; see Note 25.	N/A	2 hrs. 30 min.	1				$2^1/_2$
W-5-M-102	5″	Core: hollow concrete units; see Notes 27, 32, 35; Facings: both sides; see Note 25.	N/A	2 hrs. 30 min.	1				$2^1/_2$
W-5-M-103	5″	Core: hollow concrete units; see Notes 28, 33, 35; Facings: both sides; see Note 25.	N/A	2 hrs.	1				2
W-5-M-104	5″	Core: hollow concrete units; see Notes 28, 31, 35; Facings: both sides; see Note 25.	N/A	2 hrs.	1				2
W-5-M-105	5″	Core: hollow concrete units; see Notes 29, 32, 35; Facings: both sides; see Note 25.	N/A	1 hr. 45 min.	1				$1^3/_4$
W-5-M-106	5″	Core: hollow concrete units; see Notes 30, 34, 35; Facings: both sides; see Note 25.	N/A	1 hr.	1				1
W-5-M-107	5″	Core: 5″ thick solid gypsum blocks; see Note 26; No facings.	N/A	4 hrs.	1				4
W-5-M-108	5″	Core: 4″ thick hollow gypsum blocks; see Note 26; Facings: both sides; see Note 25.	N/A	3 hrs.	1				3
W-5-M-109	4″	Concrete with 4″ × 4″ No. 6 welded wire mesh at wall center.	100 psi	45 min.			43	2	$3/_4$
W-4-M-110	4″	Concrete with 4″ × 4″ No. 6 welded wire mesh at wall center.	N/A	1 hr. 15 min.			43	2	$1^1/_4$

For SI: 1 inch = 25.4 mm, 1 pound per square inch = 0.00689 MPa.

Notes:
1. Tested as NBS under ASA Spec. No. A 2-1934.
2. Failure mode—maximum temperature rise.
3. Treated at NBS under ASA Spec. No. 42-1934 (ASTM C19-53) except that hose stream testing where carried out was run on test specimens exposed for full test duration, not for or reduced period as is contemporarily done.
4. For clay tile walls, unless the source the clay can be positively identified, it is suggested that the most pessimistic hour rating for the fire endurance of a clay tile partition of that thickness to be followed. Identified sources of clay showing longer fire endurance can lead to longer time recommendations.
5. See appendix for construction and design details for clay tile walls.
6. Failure mode—flame thru or crack formation showing flames.
7. Hole formed at 25 minutes; partition collapsed at 42 minutes or removal from furnace.
8. Failure mode—collapse.
9. Hose stream pass.
10. Hose stream hole formed in specimen.
11. Load: 80 psi for gross wall cross sectional area.
12. One cell in wall thickness.
13. Two cells in wall thickness.

(continued)

RESOURCE A

**TABLE 1.1.2—continued
MASONRY WALLS
4″ TO LESS THAN 6″ THICK**

14. Double cells plus one cell in wall thickness.
15. One cell in wall thickness, cells filled with broken tile, crushed stone, slag, cinders or sand mixed with mortar.
16. Dense hard-burned clay or shale tile.
17. Medium-burned clay tile.
18. Not less than $5/8$ inch thickness of 1:3 sanded gypsum plaster.
19. Units of not less than 30 percent solid material.
20. Units of not less than 40 percent solid material.
21. Units of not less than 50 percent solid material.
22. Units of not less than 45 percent solid material.
23. Units of not less than 60 percent solid material.
24. All tiles laid in portland cement-lime mortar.
25. Minimum $1/2$ inch—1:3 sanded gypsum plaster.
26. Laid in 1:3 sanded gypsum mortar. Voids in hollow units not to exceed 30 percent.
27. Units of expanded slag or pumice aggregate.
28. Units of crushed limestone, blast furnace slag, cinders and expanded clay or shale.
29. Units of calcareous sand and gravel. Coarse aggregate, 60 percent or more calcite and dolomite.
30. Units of siliceous sand and gravel. Ninety percent or more quartz, chert or flint.
31. Unit at least 49 percent solid.
32. Unit at least 62 percent solid.
33. Unit at least 65 percent solid.
34. Unit at least 73 percent solid.
35. Ratings based on one unit and one cell in wall thickness.
36. See Clay Tile Partition Design Construction drawings, below.

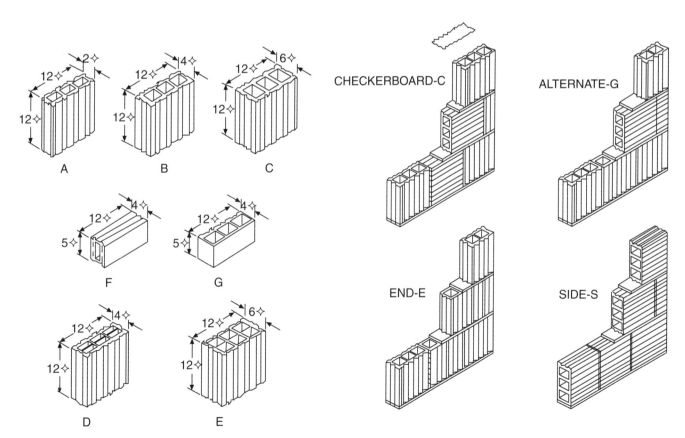

DESIGNS OF TILES USED IN FIRE-TEST PARTITIONS

THE FOUR TYPES OF CONSTRUCTION USED IN FIRE-TEST PARTITIONS

RESOURCE A

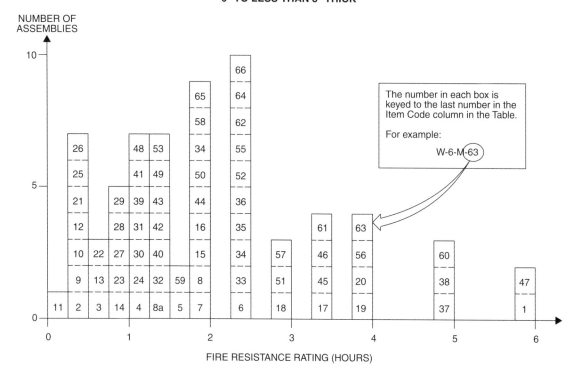

FIGURE 1.1.3
MASONRY WALLS
6″ TO LESS THAN 8″ THICK

TABLE 1.1.3
MASONRY WALLS
6″ TO LESS THAN 8″ THICK

ITEM CODE	THICKNESS	CONSTRUCTION DETAILS	PERFORMANCE		REFERENCE NUMBER			NOTES	REC. HOURS
			LOAD	TIME	PRE-BMS-92	BMS-92	POST-BMS-92		
W-6-M-1	6″	Core: 5″ thick, solid gypsum blocks laid in 1:3 sanded gypsum mortar; $^1/_2$″ of 1:3 sanded gypsum plaster facings on both sides.	N/A	6 hrs.	1				6
W-6-M-2	6″	6″ clay tile; Ohio fire clay; single cell thick; No plaster; Design "C," Construction "A."	N/A	17 min.			2	1, 3, 4, 6, 55	$^1/_4$
W-6-M-3	6″	6″ clay tile; Illinois surface clay; double cell thick; No plaster; Design "E," Construction "C."	N/A	45 min.			2	1-4, 7, 55	$^3/_4$
W-6-M-4	6″	6″ clay tile; New Jersey fire clay; double cell thick; No plaster; Design "E," Construction "S."	N/A	1 hr. 1 min.			2	1-4, 8, 55	1
W-7-M-5	$7^1/_4$″	6″ clay tile; Illinois surface clay; double cell thick; Plaster: $^5/_8$″—1:3 sanded gypsum both faces; Design "E," Construction "A."	N/A	1 hr. 41 min.			2	1-4, 55	$1^2/_3$
W-7-M-6	$7^1/_4$″	6″ clay tile; New Jersey fire clay; double cell thick; Plaster: $^5/_8$″—1:3 sanded gypsum both faces; Design "E," Construction "S."	N/A	2 hrs. 23 min.			2	1-4, 9, 55	$2^1/_3$
W-7-M-7	$7^1/_4$″	6″ clay tile; Ohio fire clay; single cell thick; Plaster: $^5/_8$″ sanded gypsum; 1:3 both faces; Design "C," Construction "A."	N/A	1 hr. 54 min.			2	1-4, 9, 55	$2^3/_4$
W-7-M-8	$7^1/_4$″	6″ clay tile; Illinois surface clay; single cell thick; Plaster: $^5/_8$″ sanded gypsum 1:3 both faces; Design "C," Construction "S."	N/A	2 hrs.			2	1, 3, 4, 9, 10, 55	2
W-7-M-8a	$7^1/_4$″	6″ clay tile; Illinois surface clay; single cell thick; Plaster: $^5/_8$″ sanded gypsum 1:3 both faces; Design "C," Construction "E."	N/A	1 hr. 23 min			2	1-4, 9, 10, 55	$1^3/_4$

(continued)

RESOURCE A

TABLE 1.1.3—continued
MASONRY WALLS
6″ TO LESS THAN 8″ THICK

ITEM CODE	THICKNESS	CONSTRUCTION DETAILS	PERFORMANCE		REFERENCE NUMBER			NOTES	REC. HOURS
			LOAD	TIME	PRE-BMS-92	BMS-92	POST-BMS-92		
W-6-M-9	6″	Core: structural clay tile; see Notes 12, 16, 20; No facings.	N/A	20 min.		1		3, 5, 24	$^1/_3$
W-6-M-10	6″	Core: structural clay tile; see Notes 12, 17, 20; No facings.	N/A	25 min.		1		3, 5, 24	$^1/_3$
W-6-M-11	6″	Core: structural clay tile; see Notes 12, 16, 19; No facings.	N/A	15 min.		1		3, 5, 24	$^1/_4$
W-6-M-12	6″	Core: structural clay tile; see Notes 12, 17, 19; No facings.	N/A	20 min.		1		3, 5, 24	$^1/_3$
W-6-M-13	6″	Core: structural clay tile; see Notes 13, 16, 22; No facings.	N/A	45 min.		1		3, 5, 24	$^3/_4$
W-6-M-14	6″	Core: structural clay tile; see Notes 13, 17, 22; No facings.	N/A	1 hr.		1		3, 5, 24	1
W-6-M-15	6″	Core: structural clay tile; see Notes 15, 17, 19; No facings.	N/A	2 hrs.		1		3, 5, 24	2
W-6-M-16	6″	Core: structural clay tile; see Notes 15, 16, 19; No facings.	N/A	2 hrs.		1		3, 5, 24	2
W-6-M-17	6″	Cored concrete masonry; see Notes 12, 34, 36, 38, 41; No facings.	80 psi	3 hrs. 30 min.		1		5, 25	$3^1/_2$
W-6-M-18	6″	Cored concrete masonry; see Notes 12, 33, 36, 38, 41; No facings.	80 psi	3 hrs.		1		5, 25	3
W-6-M-19	$6^1/_2$″	Cored concrete masonry; see Notes 12, 34, 36, 38, 41; Facings: side 1; see Note 35.	80 psi	4 hrs.		1		5, 25	4
W-6-M-20	$6^1/_2$″	Cored concrete masonry; see Notes 12, 33, 36, 38, 41; Facings: side 1; see Note 35.	80 psi	4 hrs.		1		5, 25	4
W-6-M-21	$6^5/_8$″	Core: structural clay tile; see Notes 12, 16, 20; Facings: unexposed face only; see Note 18.	N/A	30 min.		1		3, 5, 24	$^1/_2$
W-6-M-22	$6^5/_8$″	Core: structural clay tile; see Notes 12, 17, 20; Facings: unexposed face only; see Note 18.	N/A	40 min.		1		3, 5, 24	$^2/_3$
W-6-M-23	$6^5/_8$″	Core: structural clay tile; see Notes 12, 16, 20; Facings: exposed face only; see Note 18.	N/A	1 hr.		1		3, 5, 24	1
W-6-M-24	$6^5/_8$″	Core: structural clay tile; see Notes 12, 17, 20; Facings: exposed face only; see Note 18.	N/A	1 hr. 5 min.		1		3, 5, 24	1
W-6-M-25	$6^5/_8$″	Core: structural clay tile; see Notes 12, 16, 19; Facings: unexposed side only; see Note 18.	N/A	25 min.		1		3, 5, 24	$^1/_3$
W-6-M-26	$6^5/_8$″	Core: structural clay tile; see Notes 12, 7, 19; Facings: unexposed face only; see Note 18.	N/A	30 min.		1		3, 5, 24	$^1/_2$
W-6-M-27	$6^5/_8$″	Core: structural clay tile; see Notes 12, 16, 19; Facings: exposed side only; see Note 18.	N/A	1 hr.		1		3, 5, 24	1
W-6-M-28	$6^5/_8$″	Core: structural clay tile; see Notes 12, 17, 19; Facings: fire side only; see Note 18.	N/A	1 hr.		1		3, 5, 24	1
W-6-M-29	$6^5/_8$″	Core: structural clay tile; see Notes 13, 16, 22; Facings: unexposed side only; see Note 18.	N/A	1 hr.		1		3, 5, 24	1
W-6-M-30	$6^5/_8$″	Core: structural clay tile; see Notes 13, 17, 22; Facings: unexposed side only; see Note 18.	N/A	1 hr. 15 min.		1		3, 5, 24	$1^1/_4$
W-6-M-31	$6^5/_8$″	Core: structural clay tile; see Notes 13, 16, 22; Facings: fire side only; see Note 18.	N/A	1 hr. 15 min.		1		3, 5, 24	$1^1/_4$
W-6-M-32	$6^5/_8$″	Core: structural clay tile; see Notes 13, 17, 22; Facings: fire side only; see Note 18.	N/A	1 hr. 30 min.		1		3, 5, 24	$1^1/_2$

(continued)

**TABLE 1.1.3—continued
MASONRY WALLS
6″ TO LESS THAN 8″ THICK**

ITEM CODE	THICKNESS	CONSTRUCTION DETAILS	PERFORMANCE		REFERENCE NUMBER			NOTES	REC. HOURS
			LOAD	TIME	PRE-BMS-92	BMS-92	POST-BMS-92		
W-6-M-33	6⁵/₈″	Core: structural clay tile; see Notes 15, 16, 19; Facings: unexposed side only; see Note 18.	N/A	2 hrs. 30 min.		1		3, 5, 24	2¹/₂
W-6-M-34	6⁵/₈″	Core: structural clay tile; see Notes 15, 17, 19; Facings: unexposed side only; see Note 18.	N/A	2 hrs. 30 min.		1		3, 5, 24	2¹/₂
W-6-M-35	6⁵/₈″	Core: structural clay tile; see Notes 15, 16, 19; Facings: fire side only; see Note 18.	N/A	2 hrs. 30 min.		1		3, 5, 24	2¹/₂
W-6-M-36	6⁵/₈″	Core: structural clay tile; see Notes 15, 17, 19; Facings: fire side only; see Note 18.	N/A	2 hrs. 30 min.		1		3, 5, 24	2¹/₂
W-6-M-37	7″	Cored concrete masonry; see Notes 12, 34, 36, 38, 41; see Note 35 for facings on both sides.	80 psi	5 hrs.		1		5, 25	5
W-6-M-38	7″	Cored concrete masonry; see Notes 12, 33, 36, 38, 41; see Note 35 for facings.	80 psi	5 hrs.		1		5, 25	5
W-6-M-39	7¹/₄″	Core: structural clay tile; see Notes 12, 16, 20; Facings: both sides; see Note 18.	N/A	1 hr. 15 min.		1		3, 5, 24	1¹/₄
W-6-M-40	7¹/₄″	Core: structural clay tile; see Notes 12, 17, 20; Facings: both sides; see Note 18.	N/A	1 hr. 30 min.		1		3, 5, 24	1¹/₂
W-6-M-41	7¹/₄″	Core: structural clay tile; see Notes 12, 16, 19; Facings: both sides; see Note 18.	N/A	1 hr. 15 min.		1		3, 5, 24	1¹/₄
W-6-M-42	7¹/₄″	Core: structural clay tile; see Notes 12, 17, 19; Facings: both sides; see Note 18.	N/A	1 hr. 30 min.		1		3, 5, 24	1¹/₂
W-7-M-43	7¹/₄″	Core: structural clay tile; see Notes 13, 16, 22; Facings: both sides of wall; see Note 18.	N/A	1 hr. 30 min.		1		3, 5, 24	1¹/₂
W-7-M-44	7¹/₄″	Core: structural clay tile; see Notes 13, 17, 22; Facings: both sides of wall; see Note 18.	N/A	2 hrs.		1		3, 5, 24	1¹/₂
W-7-M-45	7¹/₄″	Core: structural clay tile; see Notes 15, 16, 19; Facings: both sides; see Note 18.	N/A	3 hrs. 30 min.		1		3, 5, 24	3¹/₂
W-7-M-46	7¹/₄″	Core: structural clay tile; see Notes 15, 17, 19; Facings: both sides; see Note 18.	N/A	3 hrs. 30 min.		1		3, 5, 24	3¹/₂
W-6-M-47	6″	Core: 5″ thick solid gypsum blocks; see Note 45; Facings: both sides; see Note 45.	N/A	6 hrs.		1			6
W-6-M-48	6″	Core: hollow concrete units; see Notes 47, 50, 54; No facings.	N/A	1 hr. 15 min.		1			1¹/₄
W-6-M-49	6″	Core: hollow concrete units; see Notes 46, 50, 54; No facings.	N/A	1 hr. 30 min.		1			1¹/₂
W-6-M-50	6″	Core: hollow concrete units; see Notes 46, 41, 54; No facings.	N/A	2 hrs.		1			2
W-6-M-51	6″	Core: hollow concrete units; see Notes 46, 53, 54; No facings.	N/A	3 hrs.		1			3
W-6-M-52	6″	Core: hollow concrete units; see Notes 47, 53, 54; No facings.	N/A	2 hrs. 30 min.		1			2¹/₂
W-6-M-53	6″	Core: hollow concrete units; see Notes 47, 51, 54; No facings.	N/A	1 hr. 30 min.		1			1¹/₂
W-6-M-54	6¹/₂″	Core: hollow concrete units; see Notes 46, 50, 54; Facings: one side only; see Note 35.	N/A	2 hrs.		1			2
W-6-M-55	6¹/₂″	Core: hollow concrete units; see Notes 4, 51, 54; Facings: one side; see Note 35.	N/A	2 hrs. 30 min.		1			2¹/₂
W-6-M-56	6¹/₂″	Core: hollow concrete units; see Notes 46, 53, 54; Facings: one side; see Note 35.	N/A	4 hrs.		1			4

(continued)

TABLE 1.1.3—continued
MASONRY WALLS
6″ TO LESS THAN 8″ THICK

ITEM CODE	THICKNESS	CONSTRUCTION DETAILS	PERFORMANCE		REFERENCE NUMBER			NOTES	REC. HOURS
			LOAD	TIME	PRE-BMS-92	BMS-92	POST-BMS-92		
W-6-M-57	6½″	Core: hollow concrete units; see Notes 47, 53, 54; Facings: one side; see Note 35.	N/A	3 hrs.		1			3
W-6-M-58	6½″	Core: hollow concrete units; see Notes 47, 51, 54; Facings: one side; see Note 35.	N/A	2 hrs.		1			2
W-6-M-59	6½″	Core: hollow concrete units; see Notes 47, 50, 54; Facings: one side; see Note 35.	N/A	1 hr. 45 min.		1			1¾
W-7-M-60	7″	Core: hollow concrete units; see Notes 46, 53, 54; Facings: both sides; see Note 35.	N/A	5 hrs.		1			5
W-7-M-61	7″	Core: hollow concrete units; see Notes 46, 51, 54; Facings: both sides; see Note 35.	N/A	3 hrs. 30 min.		1			3½
W-7-M-62	7″	Core: hollow concrete units; see Notes 46, 50, 54; Facings: both sides; see Note 35.	N/A	2 hrs. 30 min.		1			2½
W-7-M-63	7″	Core: hollow concrete units; see Notes 47, 53, 54; Facings: both sides; see Note 35.	N/A	4 hrs.		1			4
W-7-M-64	7″	Core: hollow concrete units; see Notes 47, 51, 54; Facings: both sides; see Note 35.	N/A	2 hrs. 30 min.		1			2½
W-7-M-65	7″	Core: hollow concrete units; see Notes 47, 50, 54; Facings: both sides; see Note 35.	N/A	2 hrs.		1			2
W-6-M-66	6″	Concrete wall with 4″ × 4″ No. 6 wire fabric (welded) near wall center for reinforcement.	N/A	2 hrs. 30 min.			43	2	2½

For SI: 1 inch = 25.4 mm, 1 pound per square inch = 0.00689 MPa.

Notes:
1. Tested at NBS under ASA Spec. No. 43-1934 (ASTM C19-53) except that hose stream testing where carried out was run on test specimens exposed for full test duration, not for a reduced period as is contemporarily done.
2. Failure by thermal criteria—maximum temperature rise.
3. For clay tile walls, unless the source or density of the clay can be positively identified or determined, it is suggested that the lowest hourly rating for the fire endurance of a clay tile partition of that thickness be followed. Identified sources of clay showing longer fire endurance can lead to longer time recommendations.
4. See Note 55 for construction and design details for clay tile walls.
5. Tested at NBS under ASA Spec. No. A2-1934.
6. Failure mode—collapse.
7. Collapsed on removal from furnace at 1 hour 9 minutes.
8. Hose stream—failed.
9. Hose stream—passed.
10. No end point met in test.
11. Wall collapsed at 1 hour 28 minutes.
12. One cell in wall thickness.
13. Two cells in wall thickness.
14. Double shells plus one cell in wall thickness.
15. One cell in wall thickness, cells filled with broken tile, crushed stone, slag, cinders or sand mixed with mortar.
16. Dense hard-burned clay or shale tile.
17. Medium-burned clay tile.
18. Not less than ⅝ inch thickness of 1:3 sanded gypsum plaster.
19. Units of not less than 30 percent solid material.
20. Units of not less than 40 percent solid material.
21. Units of not less than 50 percent solid material.
22. Units of not less than 45 percent solid material.
23. Units of not less than 60 percent solid material.
24. All tiles laid in portland cement-lime mortar.
25. Load: 80 psi for gross cross sectional area of wall.
26. Three cells in wall thickness.
27. Minimum percent of solid material in concrete units = 52.
28. Minimum percent of solid material in concrete units = 54.
29. Minimum percent of solid material in concrete units = 55.
30. Minimum percent of solid material in concrete units = 57.

(continued)

**TABLE 1.1.3—continued
MASONRY WALLS
6″ TO LESS THAN 8″ THICK**

31. Minimum percent of solid material in concrete units = 62.
32. Minimum percent of solid material in concrete units = 65.
33. Minimum percent of solid material in concrete units = 70.
34. Minimum percent of solid material in concrete units = 76.
35. Not less than $1/2$ inch of 1:3 sanded gypsum plaster.
36. Noncombustible or no members framed into wall.
37. Combustible members framed into wall.
38. One unit in wall thickness.
39. Two units in wall thickness.
40. Three units in wall thickness.
41. Concrete units made with expanded slag or pumice aggregates.
42. Concrete units made with expanded burned clay or shale, crushed limestone, air cooled slag or cinders.
43. Concrete units made with calcareous sand and gravel. Coarse aggregate, 60 percent or more calcite and dolomite.
44. Concrete units made with siliceous sand and gravel. Ninety percent or more quartz, chert or flint.
45. Laid in 1:3 sanded gypsum mortar.
46. Units of expanded slag or pumice aggregate.
47. Units of crushed limestone, blast furnace, slag, cinder and expanded clay or shale.
48. Units of calcareous sand and gravel. Coarse aggregate, 60 percent or more calcite and dolomite.
49. Units of siliceous sand and gravel. Ninety percent or more quartz, chert or flint.
50. Unit minimum 49 percent solid.
51. Unit minimum 62 percent solid.
52. Unit minimum 65 percent solid.
53. Unit minimum 73 percent solid.
54. Ratings based on one unit and one cell in wall section.
55. See Clay Tile Partition Design Construction drawings, below.

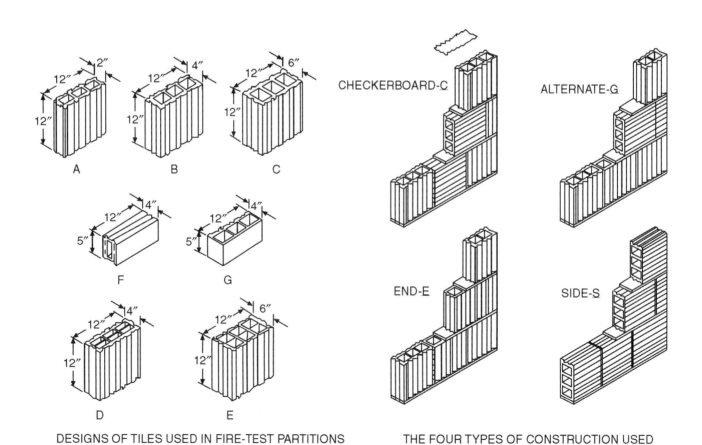

DESIGNS OF TILES USED IN FIRE-TEST PARTITIONS

THE FOUR TYPES OF CONSTRUCTION USED IN FIRE-TEST PARTITIONS

RESOURCE A

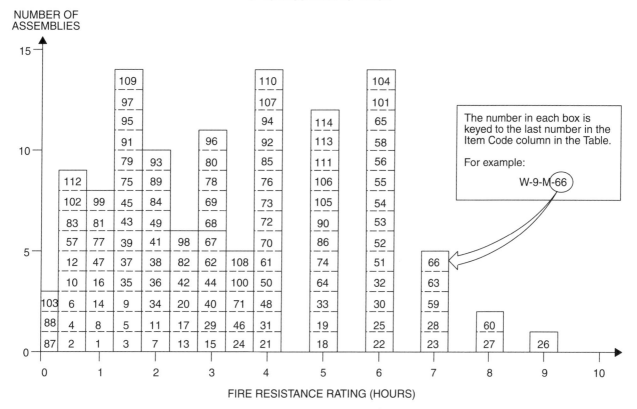

FIGURE 1.1.4
MASONRY WALLS
8″ TO LESS THAN 10″ THICK

TABLE 1.1.4
MASONRY WALLS
8″ TO LESS THAN 10″ THICK

ITEM CODE	THICKNESS	CONSTRUCTION DETAILS	PERFORMANCE		REFERENCE NUMBER			NOTES	REC. HOURS
			LOAD	TIME	PRE-BMS-92	BMS-92	POST-BMS-92		
W-8-M-1	8″	Core: clay or shale structural tile; Units in wall thickness: 1; Cells in wall thickness: 2; Minimum % solids in units: 40.	80 psi	1 hr. 15 min.		1		1, 20	$1^{1}/_{4}$
W-8-M-2	8″	Core: clay or shale structural tile; Units in wall thickness: 1; Cells in wall thickness: 2; Minimum % solids in units: 40; No facings; Result for wall with combustible members framed into interior.	80 psi	45 min.		1		1, 20	$^{3}/_{4}$
W-8-M-3	8″	Core: clay or shale structural tile; Units in wall thickness: 1; Cells in wall thickness: 2; Minimum % solids in units: 43.	80 psi	1 hr. 30 min.		1		1, 20	$1^{1}/_{2}$
W-8-M-4	8″	Core: clay or shale structural tile; Units in wall thickness: 1; Cells in wall thickness: 2; Minimum % solids in units: 43; No facings; Combustible members framed into wall.	80 psi	45 min.		1		1, 20	$^{3}/_{4}$
W-8-M-5	8″	Core: clay or shale structural tile; No facings.	See Notes	1 hr. 30 min.		1		1, 2, 5, 10, 18, 20, 21	$1^{1}/_{2}$
W-8-M-6	8″	Core: clay or shale structural tile; No facings.	See Notes	45 min.		1		1, 2, 5, 10, 19, 20, 21	$^{3}/_{4}$

(continued)

**TABLE 1.1.4—continued
MASONRY WALLS
8″ TO LESS THAN 10″ THICK**

| ITEM CODE | THICKNESS | CONSTRUCTION DETAILS | PERFORMANCE || REFERENCE NUMBER ||| NOTES | REC. HOURS |
			LOAD	TIME	PRE-BMS-92	BMS-92	POST-BMS-92		
W-8-M-7	8″	Core: clay or shale structural tile; No facings	See Notes	2 hrs.		1		1, 2, 5, 13, 18, 20, 21	2
W-8-M-8	8″	Core: clay or shale structural tile; No facings.	See Notes	1 hr. 45 min.		1		1, 2, 5, 13, 19, 20, 21	$1^1/_4$
W-8-M-9	8″	Core: clay or shale structural tile; No facings.	See Notes	1 hr. 15 min.		1		1, 2, 6, 9, 18, 20, 21	$1^3/_4$
W-8-M-10	8″	Core: clay or shale structural tile; No facings.	See Notes	45 min.		1		1, 2, 6, 9, 19, 20, 21	$^3/_4$
W-8-M-11	8″	Core: clay or shale structural tile; No facings.	See Notes	2 hrs.		1		1, 2, 6, 10, 18, 20, 21	2
W-8-M-12	8″	Core: clay or shale structural tile; No facings.	See Notes	45 min.		1		1, 2, 6, 10, 19, 20, 21	$^3/_4$
W-8-M-13	8″	Core: clay or shale structural tile; No facings.	See Notes	2 hrs. 30 min.		1		1, 3, 6, 12, 18, 20, 21	$2^1/_2$
W-8-M-14	8″	Core: clay or shale structural tile; No facings.	See Notes	1 hr.		1		1, 2, 6, 12, 19, 20, 21	1
W-8-M-15	8″	Core: clay or shale structural tile; No facings.	See Notes	3 hrs.		1		1, 2, 6, 16, 18, 20, 21	3
W-8-M-16	8″	Core: clay or shale structural tile; No facings.	See Notes	1 hr. 15 min.		1		1, 2, 6, 16, 19, 20, 21	$1^1/_4$
W-8-M-17	8″	Cored clay or shale brick; Units in wall thickness: 1; Cells in wall thickness: 1; Minimum % solids: 70; No facings.	See Notes	2 hrs. 30 min.		1		1, 44	$2^1/_2$
W-8-M-18	8″	Cored clay or shale brick; Units in wall thickness: 2; Cells in wall thickness: 2; Minimum % solids: 87; No facings.	See Notes	5 hrs.		1		1, 45	5
W-8-M-19	8″	Core: solid clay or shale brick; No facings.	See Notes	5 hrs.		1		1, 22, 45	5
W-8-M-20	8″	Core: hollow rolok of clay or shale.	See Notes	2 hrs. 30 min.		1		1, 22, 45	$2^1/_2$
W-8-M-21	8″	Core: hollow rolok bak of clay or shale; No facings.	See Notes	4 hrs.		1		1, 45	4
W-8-M-22	8″	Core: concrete brick; No facings.	See Notes	6 hrs.		1		1, 45	6
W-8-M-23	8″	Core: sand-lime brick; No facings.	See Notes	7 hrs.		1		1, 45	7
W-8-M-24	8″	Core: 4″, 40% solid clay or shale structural tile; 1 side 4″ brick facing.	See Notes	3 hrs. 30 min.		1		1, 20	$3^1/_2$
W-8-M-25	8″	Concrete wall (3220 psi); Reinforcing vertical rods 1″ from each face and 1″ diameter; horizontal rods $^5/_8$″ diameter.	22,200 lbs./ft.	6 hrs.			7		6
W-8-M-26	8″	Core: sand-line brick; $^1/_2$″ of 1:3 sanded gypsum plaster facings on one side.	See Notes	9 hrs.		1		1, 45	9
W-8-M-27	$8^1/_2$″	Core: sand-line brick; $^1/_2$″ of 1:3 sanded gypsum plaster facings on one side.	See Notes	8 hrs.		1		1, 45	8
W-8-M-28	$8^1/_2$″	Core: concrete; $^1/_2$″ of 1:3 sanded gypsum plaster facings on one side.	See Notes	7 hrs.		1		1, 45	7

(continued)

RESOURCE A

TABLE 1.1.4—continued
MASONRY WALLS
8″ TO LESS THAN 10″ THICK

ITEM CODE	THICKNESS	CONSTRUCTION DETAILS	PERFORMANCE		REFERENCE NUMBER			NOTES	REC. HOURS
			LOAD	TIME	PRE-BMS-92	BMS-92	POST-BMS-92		
W-8-M-29	8¹/₂″	Core: hollow rolok of clay or shale; ¹/₂″ of 1:3 sanded gypsum plaster facings on one side.	See Notes	3 hrs.		1		1, 45	3
W-8-M-30	8¹/₂″	Core: solid clay or shale brick ¹/₂″ thick, 1:3 sanded gypsum plaster facings on one side.	See Notes	6 hrs.		1		1, 22, 45,	6
W-8-M-31	8¹/₂″	Core: cored clay or shale brick; Units in wall thickness: 1; Cells in wall thickness: 1; Minimum % solids: 70; ¹/₂″ of 1:3 sanded gypsum plaster facings on both sides.	See Notes	4 hrs.		1		1, 44	4
W-8-M-32	8¹/₂″	Core: cored clay or shale brick; Units in wall thickness: 2; Cells in wall thickness: 2; Minimum % solids: 87; ¹/₂″ of 1:3 sanded gypsum plaster facings on one side.	See Notes	6 hrs.		1		1, 45	6
W-8-M-33	8¹/₂″	Core: hollow rolok bak of clay or shale; ¹/₂″ of 1:3 sanded gypsum plaster facings on one side.	See Notes	5 hrs.		1		1, 45	5
W-8-M-34	8⁵/₈″	Core: clay or shale structural tile; Units in wall thickness: 1; Cells in wall thickness: 2; Minimum % solids in units: 40; ⁵/₈″ of 1:3 sanded gypsum plaster facings on one side.	See Notes	2 hrs.		1		1, 20, 21	2
W-8-M-35	8⁵/₈″	Core: clay or shale structural tile; Units in wall thickness: 1; Cells in wall thickness: 2; Minimum % solids in units: 40; Exposed face: ⁵/₈″ of 1:3 sanded gypsum plaster.	See Notes	1 hr. 30 min.		1		1, 20, 21	1¹/₂
W-8-M-36	8⁵/₈″	Core: clay or shale structural tile; Units in wall thickness: 1; Cells in wall thickness: 2; Minimum % solids in units: 43; ⁵/₈″ of 1:3 sanded gypsum plaster facings on one side.	See Notes	2 hrs.		1		1, 20, 21	2
W-8-M-37	8⁵/₈″	Core: clay or shale structural tile; Units in wall thickness: 1; Cells in wall thickness: 2; Minimum % solids in units: 43; ⁵/₈″ of 1:3 sanded gypsum plaster of the exposed face only.	See Notes	1 hr. 30 min.		1		1, 20, 21	1¹/₂
W-8-M-38	8⁵/₈″	Core: clay or shale structural tile; Facings: side 1; see Note 17.	See Notes	2 hrs.		1		1, 2, 5, 10, 18, 20, 21	2
W-8-M-39	8⁵/₈″	Core: clay or shale structural tile; Facings: exposed side only; see Note 17.	See Notes	1 hr. 30 min.		1		1, 2, 5, 10, 19, 20, 21	1¹/₂
W-8-M-40	8⁵/₈″	Core: clay or shale structural tile; Facings: exposed side only; see Note 17.	See Notes	3 hrs.		1		1, 2, 5, 13, 18, 20, 21	3
W-8-M-41	8⁵/₈″	Core: clay or shale structural tile; Facings: exposed side only; see Note 17.	See Notes	2 hrs.		1		1, 2, 5, 13, 19, 20, 21	2
W-8-M-42	8⁵/₈″	Core: clay or shale structural tile; Facings: side 1; see Note 17.	See Notes	2 hrs. 30 min.		1		1, 2, 9, 18, 20, 21	2¹/₂
W-8-M-43	8⁵/₈″	Core: clay or shale structural tile; Facings: exposed side only; see Note 17.	See Notes	1 hr. 30 min.		1		1, 2, 6, 9, 19, 20, 21	1¹/₂

(continued)

TABLE 1.1.4—continued
MASONRY WALLS
8″ TO LESS THAN 10″ THICK

ITEM CODE	THICKNESS	CONSTRUCTION DETAILS	PERFORMANCE		REFERENCE NUMBER			NOTES	REC. HOURS
			LOAD	TIME	PRE-BMS-92	BMS-92	POST-BMS-92		
W-8-M-44	8⁵⁄₈″	Core: clay or shale structural tile; Facings: side 1, see Note 17; side 2, none.	See Notes	3 hrs.		1		1, 2, 10, 18, 20, 21	3
W-8-M-45	8⁵⁄₈″	Core: clay or shale structural tile; Facings: fire side only; see Note 17.	See Notes	1 hr. 30 min.		1		1, 2, 6, 10, 19, 20, 21	1¹⁄₂
W-8-M-46	8⁵⁄₈″	Core: clay or shale structural tile; Facings: side 1, see Note 17; side 2, none.	See Notes	3 hrs. 30 min.		1		1, 2, 6, 12, 18, 20, 21	3¹⁄₂
W-8-M-47	8⁵⁄₈″	Core: clay or shale structural tile; Facings: exposed side only; see Note 17.	See Notes	1 hr. 45 min.		1		1, 2, 6, 12, 19, 20, 21	1³⁄₄
W-8-M-48	8⁵⁄₈″	Core: clay or shale structural tile; Facings: side 1, see Note 17; side 2, none.	See Notes	4 hrs.		1		1, 2, 6, 16, 18, 20, 21	4
W-8-M-49	8⁵⁄₈″	Core: clay or shale structural tile; Facings: fire side only; see Note 17.	See Notes	2 hrs.		1		1, 2, 6, 16, 19, 20, 21	2
W-8-M-50	8⁵⁄₈″	Core: 4″, 40% solid clay or shale clay structural tile; 4″ brick plus ⁵⁄₈″ of 1:3 sanded gypsum plaster facings on one side.	See Notes	4 hrs.		1		1, 20	4
W-8-M-51	8³⁄₄″	8³⁄₄″ × 2¹⁄₂″ and 4″ × 2¹⁄₂″ cellular fletton (1873 psi) single and triple cell hollow brick set in ¹⁄₂″ sand mortar in alternate courses.	3.6 tons/ft.	6 hrs.			7	23, 29	6
W-8-M-52	8³⁄₄″	8³⁄₄″ thick cement brick (2527 psi) with P.C. and sand mortar.	3.6 tons/ft.	6 hrs.			7	23, 24	6
W-8-M-53	8³⁄₄″	8³⁄₄″ × 2¹⁄₂″ fletton brick (1831 psi) in ¹⁄₂″ sand mortar.	3.6 tons/ft.	6 hrs.			7	23, 24	6
W-8-M-54	8³⁄₄″	8³⁄₄″ × 2¹⁄₂″ London stock brick (683 psi) in ¹⁄₂″ P.C. - sand mortar.	7.2 tons/ft.	6 hrs.			7	23, 24	6
W-9-M-55	9″	9″ × 2¹⁄₂″ Leicester red wire-cut brick (4465 psi) in ¹⁄₂″ P.C. - sand mortar.	6.0 tons/ft.	6 hrs.			7	23, 24	6
W-9-M-56	9″	9″ × 3″ sand-lime brick (2603 psi) in ¹⁄₂″ P.C. - sand mortar.	3.6 tons/ft.	6 hrs.			7	23, 24	6
W-9-M-57	9″	2 layers 2⁷⁄₈″ fletton brick (1910 psi) with 3¹⁄₄″ air space; Cement and sand mortar.	1.5 tons/ft.	32 min.			7	23, 25	¹⁄₃
W-9-M-58	9″	9″ × 3″ stairfoot brick (7527 psi) in ¹⁄₂″ sand-cement mortar.	7.2 tons/ft.	6 hrs.			7	23, 24	6
W-9-M-59	9″	Core: solid clay or shale brick; ¹⁄₂″ thick; 1:3 sanded gypsum plaster facings on both sides.	See Notes	7 hrs.		1		1, 22, 45	7
W-9-M-60	9″	Core: concrete brick; ¹⁄₂″ of 1:3 sanded gypsum plaster facings on both sides.	See Notes	8 hrs.		1		1, 45	8
W-9-M-61	9″	Core: hollow rolok of clay or shale; ¹⁄₂″ of 1:3 sanded gypsum plaster facings on both sides.	See Notes	4 hrs.		1		1, 45	4
W-9-M-62	9″	Cored clay or shale brick; Units in wall thickness: 1; Cells in wall thickness: 1; Minimum % solids: 70; ¹⁄₂″ of 1:3 sanded gypsum plaster facings on one side.	See Notes	3 hrs.		1		1, 44	3

(continued)

TABLE 1.1.4—continued
MASONRY WALLS
8″ TO LESS THAN 10″ THICK

ITEM CODE	THICKNESS	CONSTRUCTION DETAILS	PERFORMANCE		REFERENCE NUMBER			NOTES	REC. HOURS
			LOAD	TIME	PRE-BMS-92	BMS-92	POST-BMS-92		
W-9-M-63	9″	Cored clay or shale brick; Units in wall thickness: 2; Cells in wall thickness: 2; Minimum % solids: 87; $\frac{1}{2}$″ of 1:3 sanded gypsum plaster facings on both sides.	See Notes	7 hrs.	1			1, 45	7
W-9-M-64	9-10″	Core: cavity wall of clay or shale brick; No facings.	See Notes	5 hrs.	1			1, 45	5
W-9-M-65	9-10″	Core: cavity construction of clay or shale brick; $\frac{1}{2}$″ of 1:3 sanded gypsum plaster facings on one side.	See Notes	6 hrs.	1			1, 45	6
W-9-M-66	9-10″	Core: cavity construction of clay or shale brick; $\frac{1}{2}$″ of 1:3 sanded gypsum plaster facings on both sides.	See Notes	7 hrs.	1			1, 45	7
W-9-M-67	$9\frac{1}{4}$″	Core: clay or shale structural tile; Units in wall thickness: 1; Cells in wall thickness: 2; Minimum % solids in units: 40; $\frac{5}{8}$″ of 1:3 sanded gypsum plaster facings on both sides.	See Notes	3 hrs.	1			1, 20, 21	3
W-9-M-68	$9\frac{1}{4}$″	Core: clay or shale structural tile; Units in wall thickness: 1; Cells in wall thickness: 2; Minimum % solids in units: 43; $\frac{5}{8}$″ of 1:3 sanded gypsum plaster facings on both sides.	See Notes	3 hrs.	1			1, 20, 21	3
W-9-M-69	$9\frac{1}{4}$″	Core: clay or shale structural tile; Facings: sides 1 and 2; see Note 17.	See Notes	3 hrs.	1			1, 2, 5, 10, 18, 20, 21	3
W-9-M-70	$9\frac{1}{4}$″	Core: clay or shale structural tile; Facings: sides 1 and 2; see Note 17.	See Notes	4 hrs.	1			1, 2, 5, 13, 18, 20, 21	4
W-9-M-71	$9\frac{1}{4}$″	Core: clay or shale structural tile; Facings: sides 1 and 2; see Note 17.	See Notes	3 hrs. 30 min.	1			1, 2, 6, 9, 18, 20, 21	$3\frac{1}{2}$
W-9-M-72	$9\frac{1}{4}$″	Core: clay or shale structural tile; Facings: sides 1 and 2; see Note 17.	See Notes	4 hrs.	1			1, 2, 6, 10, 18, 20, 21	4
W-9-M-73	$9\frac{1}{4}$″	Core: clay or shale structural tile; Facings: sides 1 and 2; see Note 17.	See Notes	4 hrs.	1			1, 2, 6, 12, 18, 20, 21	4
W-9-M-74	$9\frac{1}{4}$″	Core: clay or shale structural tile; Facings: sides 1 and 2; see Note 17.	See Notes	5 hrs.	1			1, 2, 6, 16, 18, 20, 21	5
W-9-M-75	8″	Cored concrete masonry; see Notes 2, 19, 26, 34, 40; No facings.	80 psi	1 hr. 30 min.	1			1, 20	$1\frac{1}{2}$
W-8-M-76	8″	Cored concrete masonry; see Notes 2, 18, 26, 34, 40; No facings	80 psi	4 hrs.	1			1, 20	4
W-8-M-77	8″	Cored concrete masonry; see Notes 2, 19, 26, 31, 40; No facings.	80 psi	1 hr. 15 min.	1			1, 20	$1\frac{1}{4}$
W-8-M-78	8″	Cored concrete masonry; see Notes 2, 18, 26, 31, 40; No facings.	80 psi	3 hrs.	1			1, 20	3
W-8-M-79	8″	Cored concrete masonry; see Notes 2, 19, 26, 36, 42; No facings.	80 psi	1 hr. 30 min.	1			1, 20	$1\frac{1}{2}$

(continued)

TABLE 1.1.4—continued
MASONRY WALLS
8″ TO LESS THAN 10″ THICK

ITEM CODE	THICKNESS	CONSTRUCTION DETAILS	PERFORMANCE		REFERENCE NUMBER			NOTES	REC. HOURS
			LOAD	TIME	PRE-BMS-92	BMS-92	POST-BMS-92		
W-8-M-80	8″	Cored concrete masonry; see Notes 2, 18, 26, 36, 41; No facings.	80 psi	3 hrs.		1		1, 20	3
W-8-M-81	8″	Cored concrete masonry; see Notes 2, 19, 26, 34, 41; No facings.	80 psi	1 hr.		1		1, 20	1
W-8-M-82	8″	Cored concrete masonry; see Notes 2, 18, 26, 34, 41; No facings.	80 psi	2 hrs. 30 min.		1		1, 20	$2^1/_2$
W-8-M-83	8″	Cored concrete masonry; see Notes 2, 19, 26, 29, 41; No facings.	80 psi	45 min.		1		1, 20	$^3/_4$
W-8-M-84	8″	Cored concrete masonry; see Notes 2, 18, 26, 29, 41; No facings.	80 psi	2 hrs.		1		1, 20	2
W-8-M-85	$8^1/_2$″	Cored concrete masonry; see Notes 3, 18, 26, 34, 41; Facings: $2^1/_4$″ brick.	80 psi	4 hrs.		1		1, 20	4
W-8-M-86	8″	Cored concrete masonry; see Notes 3, 18, 26, 34, 41; Facings: $3^3/_4$″ brick face.	80 psi	5 hrs.		1		1, 20	5
W-8-M-87	8″	Cored concrete masonry; see Notes 2, 19, 26, 30, 43; No facings.	80 psi	12 min.		1		1, 20	$^1/_5$
W-8-M-88	8″	Cored concrete masonry; see Notes 2, 18, 26, 30, 43; No facings.	80 psi	12 min.		1		1, 20	$^1/_5$
W-8-M-89	$8^1/_2$″	Cored concrete masonry; see Notes 2, 19, 26, 34, 40; Facings: fire side only; see Note 38.	80 psi	2 hrs.		1		1, 20	2
W-8-M-90	$8^1/_2$″	Cored concrete masonry; see Notes 2, 18, 26, 34, 40; Facings: side 1; see Note 38.	80 psi	5 hrs.		1		1, 20	5
W-8-M-91	$8^1/_2$″	Cored concrete masonry; see Notes 2, 19, 26, 31, 40; Facings: fire side only; see Note 38.	80 psi	1 hr. 45 min.		1		1, 20	$1^3/_4$
W-8-M-92	$8^1/_2$″	Cored concrete masonry; see Notes 2, 18, 26, 31, 40; Facings: one side; see Note 38.	80 psi	4 hrs.		1		1, 20	4
W-8-M-93	$8^1/_2$″	Cored concrete masonry; see Notes 2, 19, 26, 36, 41; Facings: fire side only; see Note 38.	80 psi	2 hrs.		1		1, 20	2
W-8-M-94	$8^1/_2$″	Cored concrete masonry; see Notes 2, 18, 26, 36, 41; Facings: fire side only; see Note 38.	80 psi	4 hrs.		1		1, 20	4
W-8-M-95	$8^1/_2$″	Cored concrete masonry; see Notes 2, 19, 26, 34, 41; Facings: fire side only; see Note 38.	80 psi	1 hr. 30 min.		1		1, 20	$1^1/_2$
W-8-M-96	$8^1/_2$″	Cored concrete masonry; see Notes 2, 18, 26, 34, 41; Facings: one side; see Note 38.	80 psi	3 hrs.				1, 20	3
W-8-M-97	$8^1/_2$″	Cored concrete masonry; see Notes 2, 19, 26, 29, 41; Facings: fire side only; see Note 38.	80 psi	1 hr. 30 min.		1		1, 20	$1^1/_2$
W-8-M-98	$8^1/_2$″	Cored concrete masonry; see Notes 2, 18, 26, 29, 41; Facings: one side; see Note 38.	80 psi	2 hrs. 30 min.		1		1, 20	$2^1/_2$
W-8-M-99	$8^1/_2$″	Cored concrete masonry; see Notes 3, 19, 23, 27, 41; No facings.	80 psi	1 hr. 15 min.		1		1, 20	$1^1/_4$

(continued)

TABLE 1.1.4—continued
MASONRY WALLS
8″ TO LESS THAN 10″ THICK

ITEM CODE	THICKNESS	CONSTRUCTION DETAILS	PERFORMANCE		REFERENCE NUMBER			NOTES	REC. HOURS
			LOAD	TIME	PRE-BMS-92	BMS-92	POST-BMS-92		
W-8-M-100	8½″	Cored concrete masonry; see Notes 3, 18, 23, 27, 41; No facings.	80 psi	3 hrs. 30 min.		1		1, 20	3½
W-8-M-101	8½″	Cored concrete masonry; see Notes 3, 18, 26, 34, 41; Facings: 3¾″ brick face; one side only; see Note 38.	80 psi	6 hrs.		1		1, 20	6
W-8-M-102	8½″	Cored concrete masonry; see Notes 2, 19, 26, 30, 43; Facings: fire side only; see Note 38.	80 psi	30 min.		1		1, 20	½
W-8-M-103	8½″	Cored concrete masonry; see Notes 2, 18, 26, 30, 43; Facings: one side only; see Note 38.	80 psi	12 min.		1		1, 20	⅕
W-8-M-104	9″	Cored concrete masonry; see Notes 2, 18, 26, 34, 40; Facings: both sides; see Note 38.	80 psi	6 hrs.		1		1, 20	6
W-8-M-105	9″	Cored concrete masonry; see Notes 2, 18, 26, 31, 40; Facings: both sides; see Note 38.	80 psi	5 hrs.		1		1, 20	5
W-8-M-106	9″	Cored concrete masonry; see Notes 2, 18, 26, 36, 41; Facings: both sides of wall; see Note 38.	80 psi	5 hrs.		1		1, 20	5
W-8-M-107	9″	Cored concrete masonry; see Notes 2, 18, 26, 34, 41; Facings: both sides; see Note 38.	80 psi	4 hrs.		1		1, 20	4
W-8-M-108	9″	Cored concrete masonry; see Notes 2, 18, 26, 29, 41; Facings: both sides; see Note 38.	80 psi	3 hrs. 30 min.		1		1, 20	3½
W-8-M-109	9″	Cored concrete masonry; see Notes 3, 19, 23, 27, 40; Facings: fire side only; see Note 38.	80 psi	1 hr. 45 min.		1		1, 20	1¾
W-8-M-110	9″	Cored concrete masonry; see Notes 3, 18, 23, 27, 41; Facings: one side only; see Note 38.	80 psi	4 hrs.		1		1, 20	4
W-8-M-111	9″	Cored concrete masonry; see Notes 3, 18, 26, 34, 41; 2¼″ brick face on one side only; see Note 38.	80 psi	5 hrs.		1		1, 20	5
W-8-M-112	9″	Cored concrete masonry; see Notes 2, 18, 26, 30, 43; Facings: both sides; see Note 38.	80 psi	30 min.		1		1, 20	½
W-9-M-113	9½″	Cored concrete masonry; see Notes 3, 18, 23, 27, 41; Facings: both sides; see Note 38.	80 psi	5 hrs.		1		1, 20	5
W-8-M-114	8″		200 psi	5 hrs.			43	22	5

For SI: 1 inch = 25.4 mm, 1 pound per square inch = 0.00689 MPa.

Notes:
1. Tested at NBS under ASA Spec. No. 43-1934 (ASTM C19-53).
2. One unit in wall thickness.
3. Two units in wall thickness.
4. Two or three units in wall thickness.
5. Two cells in wall thickness.
6. Three or four cells in wall thickness.
7. Four or five cells in wall thickness.
8. Five or six cells in wall thickness.
9. Minimum percent of solid materials in units = 40%.
10. Minimum percent of solid materials in units = 43%.
11. Minimum percent of solid materials in units = 46%.
12. Minimum percent of solid materials in units = 48%.
13. Minimum percent of solid materials in units = 49%.
14. inimum percent of solid materials in units = 45%.
15. Minimum percent of solid materials in units = 51%.
16. Minimum percent of solid materials in units = 53%.
17. Not less than ⅝ inch thickness of 1:3 sanded gypsum plaster.
18. Noncombustible or no members framed into wall.

(continued)

TABLE 1.1.4—continued
MASONRY WALLS
8″ TO LESS THAN 10″ THICK

19. Combustible members framed into wall.
20. Load: 80 psi for gross cross-sectional area of wall.
21. Portland cement-lime mortar.
22. Failure mode thermal.
23. British test.
24. Passed all criteria.
25. Failed by sudden collapse with no preceding signs of impending failure.
26. One cell in wall thickness.
27. Two cells in wall thickness.
28. Three cells in wall thickness.
29. Minimum percent of solid material in concrete units = 52.
30. Minimum percent of solid material in concrete units = 54.
31. Minimum percent of solid material in concrete units = 55.
32. Minimum percent of solid material in concrete units = 57.
33. Minimum percent of solid material in concrete units = 60.
34. Minimum percent of solid material in concrete units = 62.
35. Minimum percent of solid material in concrete units = 65.
36. Minimum percent of solid material in concrete units = 70.
37. Minimum percent of solid material in concrete units = 76.
38. Not less than $1/2$ inch of 1:3 sanded gypsum plaster.
39. Three units in wall thickness.
40. Concrete units made with expanded slag or pumice aggregates.
41. Concrete units made with expanded burned clay or shale, crushed limestone, air cooled slag or cinders.
42. Concrete units made with calcareous sand and gravel. Coarse aggregate, 60 percent or more calcite and dolomite.
43. Concrete units made with siliceous sand and gravel. Ninety percent or more quartz, chert and dolomite.
44. Load: 120 psi for gross cross-sectional area of wall.
45. Load: 160 psi for gross cross-sectional area of wall.

RESOURCE A

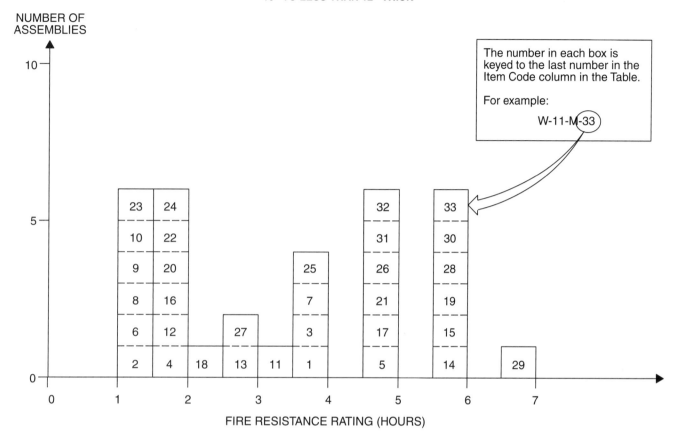

FIGURE 1.1.5
MASONRY WALLS
10″ TO LESS THAN 12″ THICK

TABLE 1.1.5
MASONRY WALLS
10″ TO LESS THAN 12″ THICK

ITEM CODE	THICKNESS	CONSTRUCTION DETAILS	PERFORMANCE		REFERENCE NUMBER			NOTES	REC. HOURS
			LOAD	TIME	PRE-BMS-92	BMS-92	POST-BMS-92		
W-10-M-1	10″	Core: two $3^3/_4''$, 40% solid clay or shale structural tiles with 2″ air space between; Facings: $^3/_4''$ portland cement plaster on stucco on both sides.	80 psi	4 hrs.		1		1, 20	4
W-10-M-2	10″	Core: cored concrete masonry, 2″ air cavity; see Notes 3, 19, 27, 34, 40; No facings.	80 psi	1 hr. 30 min.		1		1, 20	$1^1/_2$
W-10-M-3	10″	Cored concrete masonry; see Notes 3, 18, 27, 34, 40; No facings.	80 psi	4 hrs.		1		1, 20	4
W-10-M-4	10″	Cored concrete masonry; see Notes 2, 19, 26, 34, 40; No facings.	80 psi	2 hrs.		1		1, 20	2
W-10-M-5	10″	Cored concrete masonry; see Notes 2, 18, 26, 33, 40; No facings.	80 psi	5 hrs.		1		1, 20	5
W-10-M-6	10″	Cored concrete masonry; see Notes 2, 19, 26, 33, 41; No facings.	80 psi	1 hr. 30 min.		1		1, 20	$1^1/_2$
W-10-M-7	10″	Cored concrete masonry; see Notes 2, 18, 26, 33, 41; No facings.	80 psi	4 hrs.		1		1, 20	4
W-10-M-8	10″	Cored concrete masonry (cavity type 2″ air space); see Notes 3, 19, 27, 34, 42; No facings.	80 psi	1 hr. 15 min.		1		1, 20	$1^1/_4$

(continued)

RESOURCE A

**TABLE 1.1.5—continued
MASONRY WALLS
10″ TO LESS THAN 12″ THICK**

ITEM CODE	THICKNESS	CONSTRUCTION DETAILS	PERFORMANCE		REFERENCE NUMBER			NOTES	REC. HOURS
			LOAD	TIME	PRE-BMS-92	BMS-92	POST-BMS-92		
W-10-M-9	10″	Cored concrete masonry (cavity type 2″ air space); see Notes 3, 18, 27, 34, 42; No facings.	80 psi	1 hr. 15 min.		1		1, 20	$1^1/_4$
W-10-M-10	10″	Cored concrete masonry (cavity type 2″ air space); see Notes 3, 19, 27, 34, 41; No facings.	80 psi	1 hr. 15 min.		1		1, 20	$1^1/_4$
W-10-M-11	10″	Cored concrete masonry (cavity type 2″ air space); see Notes 3, 18, 27, 34, 41; No facings.	80 psi	3 hrs. 30 min.		1		1, 20	$3^1/_2$
W-10-M-12	10″	9″ thick concrete block ($11^3/_4″ \times 9″ \times 4^1/_4″$) with two 2″ thick voids included; $^3/_8″$ P.C. plaster $^1/_8″$ neat gypsum.	N/A	1 hr. 53 min.			7	23, 44	$1^3/_4$
W-10-M-13	10″	Holly clay tile block wall - $8^1/_2″$ block with two 3″ voids in each $8^1/_2″$ section; $^3/_4″$ gypsum plaster - each face.	N/A	2 hrs. 42 min.			7	23, 25	$2^1/_2$
W-10-M-14	10″	Two layers $4^1/_4″$ brick with $1^1/_2″$ air space; No ties sand cement mortar. (Fletton brick - 1910 psi).	N/A	6 hrs.			7	23, 24	6
W-10-M-15	10″	Two layers $4^1/_4″$ thick Fletton brick (1910 psi); $1^1/_2″$ air space; Ties: 18″ o.c. vertical; 3′ o.c. horizontal.	N/A	6 hrs.			7	23, 24	6
W-10-M-16	$10^1/_2″$	Cored concrete masonry; 2″ air cavity; see Notes 3, 19, 27, 34, 40; Facings: fire side only; see Note 38.	80 psi	2 hrs.		1		1, 20	2
W-10-M-17	$10^1/_2″$	Cored concrete masonry; see Notes 3, 18, 27, 34, 40; Facings: side 1 only; see Note 38.	80 psi	5 hrs.		1		1, 20	5
W-10-M-18	$10^1/_2″$	Cored concrete masonry; see Notes 2, 19, 26, 33, 40; Facings: fire side only; see Note 38.	80 psi	2 hrs. 30 min.		1		1, 20	$2^1/_2$
W-10-M-19	$10^1/_2″$	Cored concrete masonry; see Notes 2, 18, 26, 33, 40; Facings: one side; see Note 38.	80 psi	6 hrs.		1		1, 20	6
W-10-M-20	$10^1/_2″$	Cored concrete masonry; see Notes 2, 19, 26, 33, 41; Facings: fire side of wall only; see Note 38.	80 psi	2 hrs.		1		1, 20	2
W-10-M-21	$10^1/_2″$	Cored concrete masonry; see Notes 2, 18, 26, 33, 41; Facings: one side only; see Note 38.	80 psi	5 hrs.		1		1, 20	5
W-10-M-22	$10^1/_2″$	Cored concrete masonry (cavity type 2″ air space); see Notes 3,19, 27, 34, 42; Facings: fire side only; see Note 38.	80 psi	1 hr. 45 min.		1		1, 20	$1^3/_4$
W-10-M-23	$10^1/_2″$	Cored concrete masonry (cavity type 2″ air space); see Notes 3, 18, 27, 34, 42; Facings: one side only; see Note 38.	80 psi	1 hr. 15 min.		1		1, 20	$1^1/_4$
W-10-M-24	$10^1/_2″$	Cored concrete masonry (cavity type 2″ air space); see Notes 3, 19, 27, 34, 41; Facings: fire side only; see Note 38.	80 psi	2 hrs.		1		1, 20	2
W-10-M-25	$10^1/_2″$	Cored concrete masonry (cavity type 2″ air space); see Notes 3, 18, 27, 34, 41; Facings: one side only; see Note 38.	80 psi	4 hrs.		1		1, 20	4
W-10-M-26	$10^5/_8″$	Core: 8″, 40% solid tile plus 2″ furring tile; $^5/_8″$ sanded gypsum plaster between tile types; Facings: both sides $^3/_4″$ portland cement plaster or stucco.	80 psi	5 hrs.		1		1, 20	5

(continued)

TABLE 1.1.5—continued
MASONRY WALLS
10″ TO LESS THAN 12″ THICK

ITEM CODE	THICKNESS	CONSTRUCTION DETAILS	PERFORMANCE		REFERENCE NUMBER			NOTES	REC. HOURS
			LOAD	TIME	PRE-BMS-92	BMS-92	POST-BMS-92		
W-10-M-27	$10^5/_8''$	Core: 8″, 40% solid tile plus 2″ furring tile; $^5/_8''$ sanded gypsum plaster between tile types; Facings: one side $^3/_4''$ portland cement plaster or stucco.	80 psi	3 hrs. 30 min.		1		1, 20	$3^1/_2$
W-11-M-28	11″	Cored concrete masonry; see Notes 3, 18, 27, 34, 40; Facings: both sides; see Note 38.	80 psi	6 hrs.		1		1, 20	6
W-11-M-29	11″	Cored concrete masonry; see Notes 2, 18, 26, 33, 40; Facings: both sides; see Note 38.	80 psi	7 hrs.		1		1, 20	7
W-11-M-30	11″	Cored concrete masonry; see Notes 2, 18, 26, 33, 41; Facings: both sides of wall; see Note 38.	80 psi	6 hrs.		1		1, 20	6
W-11-M-31	11″	Cored concrete masonry (cavity type 2″ air space); see Notes 3, 18, 27, 34, 42; Facings: both sides; see Note 38.	80 psi	5 hrs.		1		1, 20	5
W-11-M-32	11″	Cored concrete masonry (cavity type 2″ air space); see Notes 3, 18, 27, 34, 41; Facings: both sides; see Note 38.	80 psi	5 hrs.		1		1, 20	5
W-11-M-33	11″	Two layers brick ($4^1/_2''$ Fletton, 2,428 psi) 2″ air space; galvanized ties; 18″ o.c. - horizontal; 3′ o.c. - vertical.	3 tons/ft.	6 hrs.			7	23, 24	6

For SI: 1 inch = 25.4 mm, 1 pound per square inch = 0.00689 MPa.

Notes:
1. Tested at NBS - ASA Spec. No. A2-1934.
2. One unit in wall thickness.
3. Two units in wall thickness.
4. Two or three units in wall thickness.
5. Two cells in wall thickness.
6. Three or four cells in wall thickness.
7. Four or five cells in wall thickness.
8. Five or six cells in wall thickness.
9. Minimum percent of solid materials in units = 40%.
10. Minimum percent of solid materials in units = 43%.
11. Minimum percent of solid materials in units = 46%.
12. Minimum percent of solid materials in units = 48%.
13. Minimum percent of solid materials in units = 49%.
14. Minimum percent of solid materials in units = 45%.
15. Minimum percent of solid materials in units = 51%.
16. Minimum percent of solid materials in units = 53%.
17. Not less than $^5/_8$ inch thickness of 1:3 sanded gypsum plaster.
18. Noncombustible or no members framed into wall.
19. Combustible members framed into wall.
20. Load: 80 psi for gross cross sectional area of wall.
21. Portland cement-lime mortar.
22. Failure mode—thermal.
23. British test.
24. Passed all criteria.
25. Failed by sudden collapse with no preceding signs of impending failure.
26. One cell in wall thickness.
27. Two cells in wall thickness.
28. Three cells in wall thickness.
29. Minimum percent of solid material in concrete units = 52%.
30. Minimum percent of solid material in concrete units = 54%.
31. Minimum percent of solid material in concrete units = 55%.
32. Minimum percent of solid material in concrete units = 57%.
33. Minimum percent of solid material in concrete units = 60%.
34. Minimum percent of solid material in concrete units = 62%.
35. Minimum percent of solid material in concrete units = 65%.

(continued)

TABLE 1.1.5—continued
MASONRY WALLS
10″ TO LESS THAN 12″ THICK

36. Minimum percent of solid material in concrete units = 70%.
37. Minimum percent of solid material in concrete units = 76%.
38. Not less than $^1/_2$ inch of 1:3 sanded gypsum plaster.
39. Three units in wall thickness.
40. Concrete units made with expanded slag or pumice aggregates.
41. Concrete units made with expanded burned clay or shale, crushed limestone, air cooled slag or cinders.
42. Concrete units made with calcareous sand and gravel. Coarse aggregate, 60 percent or more calcite and dolomite.

RESOURCE A

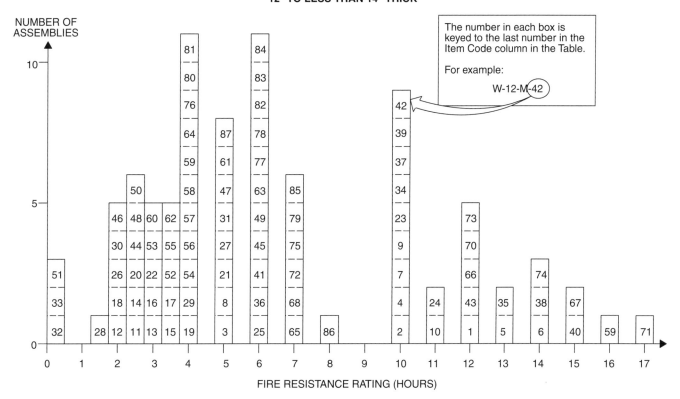

FIGURE 1.1.6
MASONRY WALLS
12″ TO LESS THAN 14″ THICK

TABLE 1.1.6
MASONRY WALLS
12″ TO LESS THAN 14″ THICK

ITEM CODE	THICKNESS	CONSTRUCTION DETAILS	PERFORMANCE		REFERENCE NUMBER			NOTES	REC. HOURS
			LOAD	TIME	PRE-BMS-92	BMS-92	POST-BMS-92		
W-12-M-1	12″	Core: solid clay or shale brick; No facings.	N/A	12 hrs.		1		1	12
W-12-M-2	12″	Core: solid clay or shale brick; No facings.	160 psi	10 hrs.		1		1, 44	10
W-12-M-3	12″	Core: hollow rolok of clay or shale; No facings.	160 psi	5 hrs.		1		1, 44	5
W-12-M-4	12″	Core: hollow rolok bak of clay or shale; No facings.	160 psi	10 hrs.		1		1, 44	10
W-12-M-5	12″	Core: concrete brick; No facings.	160 psi	13 hrs.		1		1, 44	13
W-12-M-6	12″	Core: sand-lime brick; No facings.	N/A	14 hrs.		1		1	14
W-12-M-7	12″	Core: sand-lime brick; No facings.	160 psi	10 hrs.		1		1, 44	10
W-12-M-8	12″	Cored clay or shale brick; Units in wall thickness: 1; Cells in wall thickness: 2; Minimum % solids: 70; No facings.	120 psi	5 hrs.		1		1, 45	5
W-12-M-9	12″	Cored clay or shale brick; Units in wall thickness: 3; Cells in wall thickness: 3; Minimum % solids: 87; No facings.	160 psi	10 hrs.		1		1, 44	10
W-12-M-10	12″	Cored clay or shale brick; Units in wall thickness: 3; Cells in wall thickness: 3; Minimum % solids: 87; No facings.	N/A	11 hrs.		1		1	11

(continued)

TABLE 1.1.6—continued
MASONRY WALLS
12″ TO LESS THAN 14″ THICK

ITEM CODE	THICKNESS	CONSTRUCTION DETAILS	PERFORMANCE		REFERENCE NUMBER			NOTES	REC. HOURS
			LOAD	TIME	PRE-BMS-92	BMS-92	POST-BMS-92		
W-12-M-11	12″	Core: clay or shale structural tile; see Notes 2, 6, 9, 18; No facings.	80 psi	2 hrs.		1		1, 20	2¹/₂
W-12-M-12	12″	Core: clay or shale structural tile; see Notes 2, 4, 9, 19; No facings.	80 psi	2 hrs.		1		1, 20	2
W-12-M-13	12″	Core: clay or shale structural tile; see Notes 2, 6, 14, 19; No facings.	80 psi	3 hrs.		1		1, 20	3
W-12-M-14	12″	Core: clay or shale structural tile; see Notes 2, 6, 14, 18; No facings.	80 psi	2 hrs. 30 min.		1		1, 20	2¹/₂
W-12-M-15	12″	Core: clay or shale structural tile; see Notes 2, 4, 13, 18; No facings.	80 psi	3 hrs. 30 min.		1		1, 20	3¹/₂
W-12-M-16	12″	Core: clay or shale structural tile; see Notes 2, 4, 13, 19; No facings.	80 psi	3 hrs.		1		1, 20	3
W-12-M-17	12″	Core: clay or shale structural tile; see Notes 3, 6, 9, 18; No facings.	80 psi	3 hrs. 30 min.		1		1, 20	3¹/₂
W-12-M-18	12″	Core: clay or shale structural tile; see Notes 3, 6, 9, 19; No facings.	80 psi	2 hrs.		1		1, 20	2
W-12-M-19	12″	Core: clay or shale structural tile; see Notes 3, 6, 14, 18; No facings.	80 psi	4 hrs.		1		1, 20	4
W-12-M-20	12″	Core: clay or shale structural tile; see Notes 3, 6, 14, 19; No facings.	80 psi	2 hrs. 30 min.		1		1, 20	2¹/₂
W-12-M-21	12″	Core: clay or shale structural tile; see Notes 3, 6, 16, 18; No facings.	80 psi	5 hrs.		1		1, 20	5
W-12-M-22	12″	Core: clay or shale structural tile; see Notes 3, 6, 16, 19; No facings.	80 psi	3 hrs.		1		1, 20	3
W-12-M-23	12″	Core: 8″, 70% solid clay or shale structural tile; 4″ brick facings on one side.	80 psi	10 hrs.		1		1, 20	10
W-12-M-24	12″	Core: 8″, 70% solid clay or shale structural tile; 4″ brick facings on one side.	N/A	11 hrs.		1		1	11
W-12-M-25	12″	Core: 8″, 40% solid clay or shale structural tile; 4″ brick facings on one side.	80 psi	6 hrs.		1		1, 20	6
W-12-M-26	12″	Cored concrete masonry; see Notes 1, 9, 15, 16, 20; No facings.	80 psi	2 hrs.		1		1, 20	2
W-12-M-27	12″	Cored concrete masonry; see Notes 2, 18, 26, 34, 41; No facings.	80 psi	5 hrs.		1		1, 20	5
W-12-M-28	12″	Cored concrete masonry; see Notes 2, 19, 26, 31, 41; No facings.	80 psi	1 hr. 30 min.		1		1, 20	1¹/₂
W-12-M-29	12″	Cored concrete masonry; see Notes 2, 18, 26, 31, 41; No facings.	80 psi	4 hrs.		1		1, 20	4
W-12-M-30	12″	Cored concrete masonry; see Notes 3, 19, 27, 31, 43; No facings.	80 psi	2 hrs.		1		1, 20	2
W-12-M-31	12″	Cored concrete masonry; see Notes 3, 18, 27, 31, 43; No facings.	80 psi	5 hrs.		1		1, 20	5
W-12-M-32	12″	Cored concrete masonry; see Notes 2, 19, 26, 32, 43; No facings.	80 psi	25 min.		1		1, 20	¹/₃
W-12-M-33	12″	Cored concrete masonry; see Notes 2, 18, 26, 32, 43; No facings.	80 psi	25 min.		1		1, 20	¹/₃

(continued)

**TABLE 1.1.6—continued
MASONRY WALLS
12″ TO LESS THAN 14″ THICK**

ITEM CODE	THICKNESS	CONSTRUCTION DETAILS	PERFORMANCE		REFERENCE NUMBER			NOTES	REC. HOURS
			LOAD	TIME	PRE-BMS-92	BMS-92	POST-BMS-92		
W-12-M-34	12½″	Core: solid clay or shale brick; ½″ of 1:3 sanded gypsum plaster facings on one side.	160 psi	10 hrs.		1		1, 44	10
W-12-M-35	12½″	Core: solid clay or shale brick; ½″ of 1:3 sanded gypsum plaster facings on one side.	N/A	13 hrs.		1		1	13
W-12-M-36	12½″	Core: hollow rolok of clay or shale; ½″ of 1:3 sanded gypsum plaster facings on one side.	160 psi	6 hrs.		1		1, 44	6
W-12-M-37	12½″	Core: hollow rolok bak of clay or shale; ½″ of 1:3 sanded gypsum plaster facings on one side.	160 psi	10 hrs.		1		1, 44	10
W-12-M-38	12½″	Core: concrete; ½″ of 1:3 sanded gypsum plaster facings on one side.	160 psi	14 hrs.		1		1, 44	14
W-12-M-39	12½″	Core: sand-lime brick; ½″ of 1:3 sanded gypsum plaster facings on one side.	160 psi	10 hrs.		1		1, 44	10
W-12-M-40	12½″	Core: sand-lime brick; ½″ of 1:3 sanded gypsum plaster facings on one side.	N/A	15 hrs.		1		1	15
W-12-M-41	12½″	Cored clay or shale brick; Units in wall thickness: 1; Cells in wall thickness: 2; Minimum % solids: 70; ½″ of 1:3 sanded gypsum plaster facings on one side.	120 psi	6 hrs.		1		1, 45	6
W-12-M-42	12½″	Cored clay or shale brick; Units in wall thickness: 3; Cells in wall thickness: 3; Minimum % solids: 87; ½″ of 1:3 sanded gypsum plaster facings on one side.	160 psi	10 hrs.		1		1, 44	10
W-12-M-43	12½″	Cored clay or shale brick; Units in wall thickness: 3; Cells in wall thickness: 3; Minimum % solids: 87; ½″ of 1:3 sanded gypsum plaster facings on one side.	N/A	12 hrs.		1		1	12
W-12-M-44	12½″	Cored concrete masonry; see Notes 2, 19, 26, 34, 41; Facings: fire side only; see Note 38.	80 psi	2 hrs. 30 min.		1		1, 20	2½
W-12-M-45	12½″	Cored concrete masonry; see Notes 2, 18, 26, 34, 39, 41; Facings: one side only; see Note 38.	80 psi	6 hrs.		1		1, 20	6
W-12-M-46	12½″	Cored concrete masonry; see Notes 2, 19, 26, 31, 41; Facings: fire side only; see Note 38.	80 psi	2 hrs.		1		1, 20	2
W-12-M-47	12½″	Cored concrete masonry; see Notes 2, 18, 26, 31, 41; Facings: one side of wall only; see Note 38.	80 psi	5 hrs.		1		1, 20	5
W-12-M-48	12½″	Cored concrete masonry; see Notes 3, 19, 27, 31, 43; Facings: fire side only; see Note 38.	80 psi	2 hrs. 30 min.		1		1, 20	2½
W-12-M-49	12½″	Cored concrete masonry; see Notes 3, 18, 27, 31, 43; Facings: one side only; see Note 38.	80 psi	6 hrs.		1		1, 20	6
W-12-M-50	12½″	Cored concrete masonry; see Notes 2, 19, 26, 32, 43; Facings: fire side only; see Note 38.	80 psi	2 hrs. 30 min.		1		1, 20	2½

(continued)

TABLE 1.1.6—continued
MASONRY WALLS
12″ TO LESS THAN 14″ THICK

ITEM CODE	THICKNESS	CONSTRUCTION DETAILS	PERFORMANCE		REFERENCE NUMBER			NOTES	REC. HOURS
			LOAD	TIME	PRE-BMS-92	BMS-92	POST-BMS-92		
W-12-M-51	12¹/₂″	Cored concrete masonry; see Notes 2, 18, 26, 32, 43; Facings: one side only; see Note 38.	80 psi	25 min.	1			1, 20	¹/₃
W-12-M-52	12⁵/₈″	Clay or shale structural tile; see Notes 2, 6, 9, 18; Facings: side 1, see Note 17; side 2, none.	80 psi	3 hrs. 30 min.	1			1, 20	3¹/₂
W-12-M-53	12⁵/₈″	Clay or shale structural tile; see Notes 2, 6, 9, 19; Facings: fire side only; see Note 17.	80 psi	3 hrs.	1			1, 20	3
W-12-M-54	12⁵/₈″	Clay or shale structural tile; see Notes 2, 6, 14, 19; Facings: side 1, see Note 17; side 2, none.	80 psi	4 hrs.	1			1, 20	4
W-12-M-55	12⁵/₈″	Clay or shale structural tile; see Notes 2, 6, 14, 18; Facings: exposed side only; see Note 17.	80 psi	3 hrs. 30 min.	1			1, 20	3¹/₂
W-12-M-56	12⁵/₈″	Clay or shale structural tile; see Notes 2, 4, 13, 18; Facings: side 1, see Note 17; side 2, none.	80 psi	4 hrs.	1			1, 20	4
W-12-M-57	12⁵/₈″	Clay or shale structural tile; see Notes 1, 4, 13, 19; Facings: fire side only; see Note 17.	80 psi	4 hrs.	1			1, 20	4
W-12-M-58	12⁵/₈″	Clay or shale structural tile; see Notes 3, 6, 9, 18; Facings: side 1, see Note 17; side 2, none.	80 psi	4 hrs.	1			1, 20	4
W-12-M-59	12⁵/₈″	Clay or shale structural tile; see Notes 3, 6, 9, 19; Facings: fire side only; see Note 17.	80 psi	3 hrs.	1			1, 20	3
W-12-M-60	12⁵/₈″	Clay or shale structural tile; see Notes 3, 6, 14, 18; Facings: side 1, see Note 17; side 2, none.	80 psi	5 hrs.	1			1, 20	5
W-12-M-61	12⁵/₈″	Clay or shale structural tile; see Notes 3, 6, 14, 19; Facings: fire side only; see Note 17.	80 psi	3 hrs. 30 min.	1			1, 20	3¹/₂
W-12-M-62	12⁵/₈″	Clay or shale structural tile; see Notes 3, 6, 16, 18; Facings: side 1, see Note 17; side 2, none.	80 psi	6 hrs.	1			1, 20	6
W-12-M-63	12⁵/₈″	Clay or shale structural tile; see Notes 3, 6, 16, 19; Facings: fire side only; see Note 17.	80 psi	4 hrs.	1			1, 20	4
W-12-M-64	12⁵/₈″	Core: 8″, 40% solid clay or shale structural tile; Facings: 4″ brick plus ⁵/₈″ of 1:3 sanded gypsum plaster on one side.	80 psi	7 hrs.	1			1, 20	7
W-13-M-65	13″	Core: solid clay or shale brick; ¹/₂″ of 1:3 sanded gypsum plaster facings on both sides.	160 psi	12 hrs.	1			1, 44	12
W-13-M-66	13″	Core: solid clay or shale brick; ¹/₂″ of 1:3 sanded gypsum plaster facings on both sides.	N/A	15 hrs.	1			1, 20	15
W-13-M-67	13″	Core: solid clay or shale brick; ¹/₂″ of 1:3 sanded gypsum plaster facings on both sides.	N/A	15 hrs.	1			1	15
W-13-M-68	13″	Core: hollow rolok of clay or shale; ¹/₂″ of 1:3 sanded gypsum plaster facings on both sides.	80 psi	7 hrs.	1			1, 20	7
W-13-M-69	13″	Core: concrete brick; ¹/₂″ of 1:3 sanded gypsum plaster facings on both sides.	160 psi	16 hrs.	1			1, 44	16

(continued)

RESOURCE A

TABLE 1.1.6—continued
MASONRY WALLS
12″ TO LESS THAN 14″ THICK

ITEM CODE	THICKNESS	CONSTRUCTION DETAILS	PERFORMANCE		REFERENCE NUMBER			NOTES	REC. HOURS
			LOAD	TIME	PRE-BMS-92	BMS-92	POST-BMS-92		
W-13-M-70	13″	Core: sand-lime brick; ½″ of 1:3 sanded gypsum plaster facings on both sides.	160 psi	12 hrs.		1		1, 44	12
W-13-M-71	13″	Core: sand-lime brick; ½″ of 1:3 sanded gypsum plaster facings on both sides.	N/A	17 hrs.		1		1	17
W-13-M-72	13″	Cored clay or shale brick; Units in wall thickness: 1; Cells in wall thickness: 2; Minimum % solids: 70; ½″ of 1:3 sanded gypsum plaster facings on both sides.	120 psi	7 hrs.		1		1, 45	7
W-13-M-73	13″	Cored clay or shale brick; Units in wall thickness: 3; Cells in wall thickness: 3; Minimum % solids: 87; ½″ of 1:3 sanded gypsum plaster facings on both sides.	160 psi	12 hrs.		1		1, 44	12
W-13-M-74	13″	Cored clay or shale brick; Units in wall thickness: 3; Cells in wall thickness: 2; Minimum % solids: 87; ½″ of 1:3 sanded gypsum plaster facings on both sides.	N/A	14 hrs.		1		1	14
W-13-M-75	13″	Cored concrete masonry; see Notes 18, 23, 28, 39, 41; No facings.	80 psi	7 hrs.		1		1, 20	7
W-13-M-76	13″	Cored concrete masonry; see Notes 19, 23, 28, 39, 41; No facings.	80 psi	4 hrs.		1		1, 20	4
W-13-M-77	13″	Cored concrete masonry; see Notes 3, 18, 27, 31, 43; Facings: both sides; see Note 38.	80 psi	6 hrs.		1		1, 20	6
W-13-M-78	13″	Cored concrete masonry; see Notes 2, 18, 26, 31, 41; Facings: both sides; see Note 38.	80 psi	6 hrs.		1		1, 20	6
W-13-M-79	13″	Cored concrete masonry; see Notes 2, 18, 26, 34, 41; Facings: both sides of wall; see Note 38.	80 psi	7 hrs.		1		1, 20	7
W-13-M-80	13¼″	Core: clay or shale structural tile; see Notes 2, 6, 9, 18; Facings: both sides; see Note 17.	80 psi	4 hrs.		1		1, 20	4
W-13-M-82	13¼″	Core: clay or shale structural tile; see Notes 2, 4, 13, 18; Facings: both sides; see Note 17.	80 psi	6 hrs.		1		1, 20	6
W-13-M-83	13¼″	Core: clay or shale structural tile; see Notes 3, 6, 9, 18; Facings: both sides; see Note 17.	80 psi	6 hrs.		1		1, 20	6
W-13-M-84	13¼″	Core: clay or shale structural tile; see Notes 3, 6, 14, 18; Facings: both sides; see Note 17.	80 psi	6 hrs.		1		1, 20	6
W-13-M-85	13¼″	Core: clay or shale structural tile; see Notes 3, 6, 16, 18; Facings: both sides; see Note 17.	80 psi	7 hrs.		1		1, 20	7

(continued)

TABLE 1.1.6—continued
MASONRY WALLS
12″ TO LESS THAN 14″ THICK

ITEM CODE	THICKNESS	CONSTRUCTION DETAILS	PERFORMANCE		REFERENCE NUMBER			NOTES	REC. HOURS
			LOAD	TIME	PRE-BMS-92	BMS-92	POST-BMS-92		
W-13-M-86	13$\frac{1}{2}$″	Cored concrete masonry; see Notes 18, 23, 28, 39, 41; Facings: one side only; see Note 38.	80 psi	8 hrs.		1		1, 20	8
W-13-M-87	13$\frac{1}{2}$″	Cored concrete masonry; see Notes 19, 23, 28, 39, 41; Facings: fire side only; see Note 38.	80 psi	5 hrs.		1		1, 20	5

For SI: 1 inch = 25.4 mm, 1 pound per square inch = 0.00689 MPa.

Notes:
1. Tested at NBS - ASA Spec. No. A2-1934.
2. One unit in wall thickness.
3. Two units in wall thickness.
4. Two or three units in wall thickness.
5. Two cells in wall thickness.
6. Three or four cells in wall thickness.
7. Four or five cells in wall thickness.
8. Five or six cells in wall thickness.
9. Minimum percent of solid materials in units = 40%.
10. Minimum percent of solid materials in units = 43%.
11. Minimum percent of solid materials in units = 46%.
12. Minimum percent of solid materials in units = 48%.
13. Minimum percent of solid materials in units = 49%.
14. Minimum percent of solid materials in units = 45%.
15. Minimum percent of solid materials in units = 51%.
16. Minimum percent of solid materials in units = 53%.
17. Not less than $\frac{5}{8}$ inch thickness of 1:3 sanded gypsum plaster.
18. Noncombustible or no members framed into wall.
19. Combustible members framed into wall.
20. Load: 80 psi for gross area.
21. Portland cement-lime mortar.
22. Failure mode-thermal.
23. British test.
24. Passed all criteria.
25. Failed by sudden collapse with no preceding signs of impending failure.
26. One cell in wall thickness.
27. Two cells in wall thickness.
28. Three cells in wall thickness.
29. Minimum percent of solid material in concrete units = 52%.
30. Minimum percent of solid material in concrete units = 54%.
31. Minimum percent of solid material in concrete units = 55%.
32. Minimum percent of solid material in concrete units = 57%.
33. Minimum percent of solid material in concrete units = 60%.
34. Minimum percent of solid material in concrete units = 62%.
35. Minimum percent of solid material in concrete units = 65%.
36. Minimum percent of solid material in concrete units = 70%.
37. Minimum percent of solid material in concrete units = 76%.
38. Not less than $\frac{1}{2}$ inch of 1:3 sanded gypsum plaster.
39. Three units in wall thickness.
40. Concrete units made with expanded slag or pumice aggregates.
41. Concrete units made with expanded burned clay or shale, crushed limestone, air cooled slag or cinders.
42. Concrete units made with calcareous sand and gravel. Coarse aggregate, 60 percent or more calcite and dolomite.
43. Concrete units made with siliceous sand and gravel. Ninety percent or more quartz, chert or flint.
44. Load: 160 psi of gross wall cross sectional area.
45. Load: 120 psi of gross wall cross sectional area.

RESOURCE A

**FIGURE 1.1.7
MASONRY WALLS
14″ OR MORE THICK**

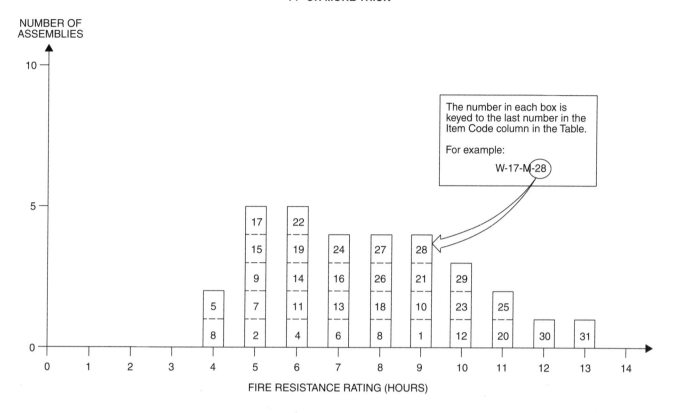

**TABLE 1.1.7
MASONRY WALLS
14″ OR MORE THICK**

ITEM CODE	THICKNESS	CONSTRUCTION DETAILS	PERFORMANCE		REFERENCE NUMBER			NOTES	REC. HOURS
			LOAD	TIME	PRE-BMS-92	BMS-92	POST-BMS-92		
W-14-M-1	14″	Core: cored masonry; see Notes 18, 28, 33, 39, 41; Facings: both sides; see Note 38.	80 psi	9 hrs.		1		1, 20	9
W-16-M-2	16″	Core: clay or shale structural tile; see Notes 4, 7, 9, 19; No facings.	80 psi	5 hrs.		1		1, 20	5
W-16-M-3	16″	Core: clay or shale structural tile; see Notes 4, 7, 9, 19; No facings.	80 psi	4 hrs.		1		1, 20	4
W-16-M-4	16″	Core: clay or shale structural tile; see Notes 4, 7, 10, 18; No facings.	80 psi	6 hrs.		1		1, 20	6
W-16-M-5	16″	Core: clay or shale structural tile; see Notes 4, 7, 10, 19; No facings.	80 psi	4 hrs.		1		1, 20	4
W-16-M-6	16″	Core: clay or shale structural tile; see Notes 4, 7, 11, 18; No facings.	80 psi	7 hrs.		1		1, 20	7
W-16-M-7	16″	Core: clay or shale structural tile; see Notes 4, 7, 11, 19; No facings.	80 psi	5 hrs.		1		1, 20	5
W-16-M-8	16″	Core: clay or shale structural tile; see Notes 4, 8, 13, 18; No facings.	80 psi	8 hrs.		1		1, 20	8
W-16-M-9	16″	Core: clay or shale structural tile; see Notes 4, 8, 13, 19; No facings.	80 psi	5 hrs.		1		1, 20	5

(continued)

TABLE 1.1.7—continued
MASONRY WALLS
14″ OR MORE THICK

ITEM CODE	THICKNESS	CONSTRUCTION DETAILS	PERFORMANCE		REFERENCE NUMBER			NOTES	REC. HOURS
			LOAD	TIME	PRE-BMS-92	BMS-92	POST-BMS-92		
W-16-M-10	16″	Core: clay or shale structural tile; see Notes 4, 8, 15, 18; No facings.	80 psi	9 hrs.		1		1, 20	9
W-16-M-11	16″	Core: clay or shale structural tile; see Notes 3, 7, 14, 18; No facings.	80 psi	6 hrs.		1		1, 20	6
W-16-M-12	16″	Core: clay or shale structural tile; see Notes 4, 8, 16, 18; No facings.	80 psi	10 hrs.		1		1, 20	10
W-16-M-13	16″	Core: clay or shale structural tile; see Notes 4, 6, 16, 19; No facings.	80 psi	7 hrs.		1		1, 20	7
W-16-M-14	16 5/8″	Core: clay or shale structural tile; see Notes 4, 7, 9, 18; Facings: side 1, see Note 17; side 2, none.	80 psi	6 hrs.		1		1, 20	6
W-16-M-15	16 5/8″	Core: clay or shale structural tile; see Notes 4, 7, 9, 19; Facings: fire side only; see Note 17.	80 psi	5 hrs.		1		1, 20	5
W-16-M-16	16 5/8″	Core: clay or shale structural tile; see Notes 4, 7, 10, 18; Facings: side 1, see Note 17; side 2, none.	80 psi	7 hrs.		1		1, 20	7
W-16-M-17	16 5/8″	Core: clay or shale structural tile; see Notes 4, 7, 10, 19; Facings: fire side only; see Note 17.	80 psi	5 hrs.		1		1, 20	5
W-16-M-18	16 5/8″	Core: clay or shale structural tile; see Notes 4, 7, 11, 18; Facings: side 1, see Note 17; side 2, none.	80 psi	5 hrs.		1		1, 20	5
W-16-M-19	16 5/8″	Core: clay or shale structural tile; see Notes 4, 7, 11, 19; Facings: fire side only; see Note 17.	80 psi	6 hrs.		1		1, 20	6
W-16-M-20	16 5/8″	Core: clay or shale structural tile; see Notes 4, 8, 13, 18; Facings: sides 1 and 2; see Note 17.	80 psi	11 hrs.		1		1, 20	11
W-16-M-21	16 5/8″	Core: clay or shale structural tile; see Notes 4, 8, 13 18; Facings: side 1, see Note 17; side 2, none.	80 psi	9 hrs.		1		1, 20	9
W-16-M-22	16 5/8″	Core: clay or shale structural tile; see Notes 4, 8, 13, 19; Facings: fire side only; see Note 17.	80 psi	6 hrs.		1		1, 20	6
W-16-M-23	16 5/8″	Core: clay or shale structural tile; see Notes 4, 8, 15, 18; Facings: side 1, see Note 17; side 2, none.	80 psi	10 hrs.		1		1, 20	10
W-16-M-24	16 5/8″	Core: clay or shale structural tile; see Notes 4, 8, 15, 19; Facings: fire side only; see Note 17.	80 psi	7 hrs.		1		1, 20	7
W-16-M-25	16 5/8″	Core: clay or shale structural tile; see Notes 4, 6, 16, 18; Facings: side 1, see Note 17; side 2, none.	80 psi	11 hrs.		1		1, 20	11
W-16-M-26	16 5/8″	Core: clay or shale structural tile; see Notes 4, 6, 16, 19; Facings: fire side only; see Note 17.	80 psi	8 hrs.		1		1, 20	8

(continued)

TABLE 1.1.7—continued
MASONRY WALLS
14" OR MORE THICK

ITEM CODE	THICKNESS	CONSTRUCTION DETAILS	PERFORMANCE		REFERENCE NUMBER			NOTES	REC. HOURS
			LOAD	TIME	PRE-BMS-92	BMS-92	POST-BMS-92		
W-17-M-27	17$\frac{1}{4}$"	Core: clay or shale structural tile; see Notes 4, 7, 9, 18; Facings: sides 1 and 2; see Note 17.	80 psi	8 hrs.	1			1, 20	8
W-17-M-28	17$\frac{1}{4}$"	Core: clay or shale structural tile; see Notes 4, 7, 10, 18; Facings: sides 1 and 2; see Note 17.	80 psi	9 hrs.	1			1, 20	9
W-17-M-29	17$\frac{1}{4}$"	Core: clay or shale structural tile; see Notes 4, 7, 11, 18; Facings: sides 1 and 2; see Note 17.	80 psi	10 hrs.	1			1, 20	10
W-17-M-30	17$\frac{1}{4}$"	Core: clay or shale structural tile; see Notes 4, 8, 15, 18; Facings: sides 1 and 2; see Note 17.	80 psi	12 hrs.	1			1, 20	12
W-17-M-31	17$\frac{1}{4}$"	Core: clay or shale structural tile; see Notes 4, 6, 16, 18; Facings: sides 1 and 2; see Note 17.	80 psi	13 hrs.	1			1, 20	13

For SI: 1 inch = 25.4 mm, 1 pound per square inch = 0.00689 MPa.

Notes:
1. Tested at NBS - ASA Spec. No. A2-1934.
2. One unit in wall thickness.
3. Two units in wall thickness.
4. Two or three units in wall thickness.
5. Two cells in wall thickness.
6. Three or four cells in wall thickness.
7. Four or five cells in wall thickness.
8. Five or six cells in wall thickness.
9. Minimum percent of solid materials in units = 40%.
10. Minimum percent of solid materials in units = 43%.
11. Minimum percent of solid materials in units = 46%.
12. Minimum percent of solid materials in units = 48%.
13. Minimum percent of solid materials in units = 49%.
14. Minimum percent of solid materials in units = 45%.
15. Minimum percent of solid materials in units = 51%.
16. Minimum percent of solid materials in units = 53%.
17. Not less than $\frac{5}{8}$ inch thickness of 1:3 sanded gypsum plaster.
18. Noncombustible or no members framed into wall.
19. Combustible members framed into wall.
20. Load: 80 psi for gross area.
21. Portland cement-lime mortar.
22. Failure mode—thermal.
23. British test.
24. Passed all criteria.
25. Failed by sudden collapse with no preceding signs of impending failure.
26. One cell in wall thickness.
27. Two cells in wall thickness.
28. Three cells in wall thickness.
29. Minimum percent of solid material in concrete units = 52%.
30. Minimum percent of solid material in concrete units = 54%.
31. Minimum percent of solid material in concrete units = 55%.
32. Minimum percent of solid material in concrete units = 57%.
33. Minimum percent of solid material in concrete units = 60%.
34. Minimum percent of solid material in concrete units = 62%.
35. Minimum percent of solid material in concrete units = 65%.
36. Minimum percent of solid material in concrete units = 70%.
37. Minimum percent of solid material in concrete units = 76%.
38. Not less than $\frac{1}{2}$ inch of 1:3 sanded gypsum plaster.
39. Three units in wall thickness.
40. Concrete units made with expanded slag or pumice aggregates.
41. Concrete units made with expanded burned clay or shale, crushed limestone, air cooled slag or cinders.
42. Concrete units made with calcareous sand and gravel. Coarse aggregate, 60 percent or more calcite and dolomite.
43. Concrete units made with siliceous sand and gravel. Ninety percent or more quartz, chert or flint.

RESOURCE A

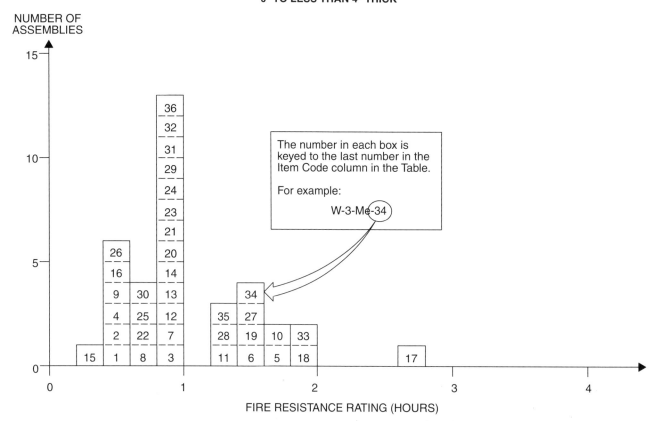

FIGURE 1.2.1
METAL FRAME WALLS
0″ TO LESS THAN 4″ THICK

TABLE 1.2.1
METAL FRAME WALLS
0″ TO LESS THAN 4″ THICK

ITEM CODE	THICKNESS	CONSTRUCTION DETAILS	PERFORMANCE		REFERENCE NUMBER			NOTES	REC. HOURS
			LOAD	TIME	PRE-BMS-92	BMS-92	POST-BMS-92		
W-3-Me-1	3″	Core: steel channels having three rows of 4″ × $^1/_8$″ staggered slots in web; core filled with heat expanded vermiculite weighing 1.5 lbs./ft.^{2} of wall area; Facings: sides 1 and 2, 18 gage steel, spot welded to core.	N/A	25 min.		1			$^1/_3$
W-3-Me-2	3″	Core: steel channels having three rows of 4″ × $^1/_8$″ staggered slots in web; core filled with heat expanded vermiculite weighing 2 lbs./ft.^{2} of wall area; Facings: sides 1 and 2, 18 gage steel, spot welded to core.	N/A	30 min.		1			$^1/_2$
W-3-Me-3	2$^1/_2$″	Solid partition: $^3/_8$″ tension rods (vertical) 3′ o.c. with metal lath; Scratch coat: cement/sand/lime plaster; Float coats: cement/sand/lime plaster; Finish coats: neat gypsum plaster.	N/A	1 hr.			7	1	1
W-2-Me-4	2″	Solid wall: steel channel per Note 1; 2″ thickness of 1:2; 1:3 portland cement on metal lath.	N/A	30 min.		1			$^1/_2$

(continued)

TABLE 1.2.1—continued
METAL FRAME WALLS
0″ TO LESS THAN 4″ THICK

ITEM CODE	THICKNESS	CONSTRUCTION DETAILS	PERFORMANCE		REFERENCE NUMBER			NOTES	REC. HOURS
			LOAD	TIME	PRE-BMS-92	BMS-92	POST-BMS-92		
W-2-Me-5	2″	Solid wall: steel channel per Note 1; 2″ thickness of neat gypsum plaster on metal lath.	N/A	1 hr. 45 min.		1			$1^3/_4$
W-2-Me-6	2″	Solid wall: steel channel per Note 1; 2″ thickness of $1:1^1/_2$; $1:1^1/_2$ gypsum plaster on metal lath.	N/A	1 hr. 30 min.		1			$1^1/_2$
W-2-Me-7	2″	Solid wall: steel channel per Note 2; 2″ thickness of 1:1; 1:1 gypsum plaster on metal lath.	N/A	1 hr.		1			1
W-2-Me-8	2″	Solid wall: steel channel per Note 1; 2″ thickness of 1:2; 1:2 gypsum plaster on metal lath.	N/A	45 min.		1			$^3/_4$
W-2-Me-9	$2^1/_4″$	Solid wall: steel channel per Note 2; $2^1/_4″$ thickness of 1:2; 1:3 portland cement on metal lath.	N/A	30 min.		1			$^1/_2$
W-2-Me-10	$2^1/_4″$	Solid wall: steel channel per Note 2; $2^1/_4″$ thickness of neat gypsum plaster on metal lath.	N/A	2 hrs.		1			2
W-2-Me-11	$2^1/_4″$	Solid wall: steel channel per Note 2; $2^1/_4″$ thickness of $1:^1/_2$; $1:^1/_2$ gypsum plaster on metal lath.	N/A	1 hr. 45 min.		1			$1^3/4$
W-2-Me-12	$2^1/_4″$	Solid wall: steel channel per Note 2; $2^1/_4″$ thickness of 1:1; 1:1 gypsum plaster on metal lath.	N/A	1 hr. 15 min.		1			$1^1/_4$
W-2-Me-13	$2^1/_4″$	Solid wall: steel channel per Note 2; $2^1/_4″$ thickness of 1:2; 1:2 gypsum plaster on metal lath.	N/A	1 hr.		1			1
W-2-Me-14	$2^1/_2″$	Solid wall: steel channel per Note 1; $2^1/_2″$ thickness of 4.5:1:7; 4.5:1:7 portland cement, sawdust and sand sprayed on wire mesh; see Note 3.	N/A	1 hr.		1			1
W-2-Me-15	$2^1/_2″$	Solid wall: steel channel per Note 2; $2^1/_2″$ thickness of 1:4; 1:4 portland cement sprayed on wire mesh; see Note 3.	N/A	20 min.		1			$^1/_3$
W-2-Me-16	$2^1/_2″$	Solid wall: steel channel per Note 2; $2^1/_2″$ thickness of 1:2; 1:3 portland cement on metal lath.	N/A	30 min.		1			$^1/_2$
W-2-Me-17	$2^1/_2″$	Solid wall: steel channel per Note 2; $2^1/_2″$ thickness of neat gypsum plaster on metal lath.	N/A	2 hrs. 30 min.		1			$2^1/_2$
W-2-Me-18	$2^1/_2″$	Solid wall: steel channel per Note 2; $2^1/_2″$ thickness of $1:^1/_2$; $1:^1/_2$ gypsum plaster on metal lath.	N/A	2 hrs.		1			2
W-2-Me-19	$2^1/_2″$	Solid wall: steel channel per Note 2; $2^1/_2″$ thickness of 1:1; 1:1 gypsum plaster on metal lath.	N/A	1 hr. 30 min.		1			$1^1/_2$

(continued)

**TABLE 1.2.1—continued
METAL FRAME WALLS
0″ TO LESS THAN 4″ THICK**

ITEM CODE	THICKNESS	CONSTRUCTION DETAILS	PERFORMANCE		REFERENCE NUMBER			NOTES	REC. HOURS
			LOAD	TIME	PRE-BMS-92	BMS-92	POST-BMS-92		
W-2-Me-20	$2^1/_2″$	Solid wall: steel channel per Note 2; $2^1/_2″$ thickness of 1:2; 1:2 gypsum plaster on metal lath.	N/A	1 hr.		1			1
W-2-Me-21	$2^1/_2″$	Solid wall: steel channel per Note 2; $2^1/_2″$ thickness of 1:2; 1:3 gypsum plaster on metal lath.	N/A	1 hr.		1			1
W-3-Me-22	3″	Core: steel channel per Note 2; 1:2; 1:2 gypsum plaster on $^3/_4″$ soft asbestos lath; plaster thickness 2″.	N/A	45 min.		1			$^3/_4$
W-3-Me-23	$3^1/_2″$	Solid wall: steel channel per Note 2; $2^1/_2″$ thickness of 1:2; 1:2 gypsum plaster on $^3/_4″$ asbestos lath.	N/A	1 hr.		1			1
W-3-Me-24	$3^1/_2″$	Solid wall: steel channel per Note 2; lath over and $1:2^1/_2$; $1:2^1/_2$ gypsum plaster on 1″ magnesium oxysulfate wood fiberboard; plaster thickness $2^1/_2″$.	N/A	1 hr.		1			1
W-3-Me-25	$3^1/_2″$	Core: steel studs; see Note 4; Facings: $^3/_4″$ thickness of $1:^1/_{30}:2$; $1:^1/_{30}:3$ portland cement and asbestos fiber plaster.	N/A	45 min.		1			$^3/_4$
W-3-Me-26	$3^1/_2″$	Core: steel studs; see Note 4; Facings: both sides $^3/_4″$ thickness of 1:2; 1:3 portland cement.	N/A	30 min.		1			$^1/_2$
W-3-Me-27	$3^1/_2″$	Core: steel studs; see Note 4; Facings: both sides $^3/_4″$ thickness of neat gypsum plaster.	N/A	1 hr. 30 min.		1			$1^1/_2$
W-3-Me-28	$3^1/_2″$	Core: steel studs; see Note 4; Facings: both sides $^3/_4″$ thickness of $1:^1/_2$; $1:^1/_2$ gypsum plaster.	N/A	1 hr. 15 min.		1			$1^1/_4$
W-3-Me-29	$3^1/_2″$	Core: steel studs; see Note 4; Facings: both sides $^3/_4″$ thickness of 1:2; 1:2 gypsum plaster.	N/A	1 hr.		1			1
W-3-Me-30	$3^1/_2″$	Core: steel studs; see Note 4; Facings: both sides $^3/_4″$ thickness of 1:2; 1:3 gypsum plaster.	N/A	45 min.		1			$^3/_4$
W-3-Me-31	$3^3/_4″$	Core: steel studs; see Note 4; Facings: both sides $^7/_8″$ thickness of $1:^1/_{30}:2$; $1:^1/_{30}:3$ portland cement and asbestos fiber plaster.	N/A	1 hr.		1			1
W-3-Me-32	$3^3/_4″$	Core: steel studs; see Note 4; Facings: both sides $^7/_8″$ thickness of 1:2; 1:3 portland cement.	N/A	45 min.		1			$^3/_4$
W-3-Me-33	$3^3/_4″$	Core: steel studs; see Note 4; Facings: both sides $^7/_8″$ thickness of neat gypsum plaster.	N/A	2 hrs.		1			2
W-3-Me-34	$3^3/_4″$	Core: steel studs; see Note 4; Facings: both sides $^7/_8″$ thickness of $1:^1/_2$; $1:^1/_2$ gypsum plaster.	N/A	1 hr. 30 min.		1			$1^1/_2$

(continued)

TABLE 1.2.1—continued
METAL FRAME WALLS
0" TO LESS THAN 4" THICK

ITEM CODE	THICKNESS	CONSTRUCTION DETAILS	PERFORMANCE		REFERENCE NUMBER			NOTES	REC. HOURS
			LOAD	TIME	PRE-BMS-92	BMS-92	POST-BMS-92		
W-3-Me-35	$3^3/_4''$	Core: steel studs; see Note 4; Facings: both sides $^7/_8''$ thickness of 1:2; 1:2 gypsum plaster.	N/A	1 hr. 15 min.	1				$1^1/_4$
W-3-Me-36	$3^3/_4''$	Core: steel; see Note 4; Facings: $^7/_8''$ thickness of 1:2; 1:3 gypsum plaster on both sides.	N/A	1 hr.	1				1

For SI: 1 inch = 25.4 mm.

Notes:
1. Failure mode—local temperature rise—back face.
2. Three-fourths inch or 1 inch channel framing—hot-rolled or strip-steel channels.
3. Reinforcement is 4-inch square mesh of No. 6 wire welded at intersections (no channels).
4. Ratings are for any usual type of nonload-bearing metal framing providing 2 inches (or more) air space.

General Note:
The construction details of the wall assemblies are as complete as the source documentation will permit. Data on the method of attachment of facings and the gauge of steel studs was provided when known. The cross-sectional area of the steel stud can be computed, thereby permitting a reasoned estimate of actual loading conditions. For load-bearing assemblies, the maximum allowable stress for the steel studs has been provided in the table "Notes." More often, it is the thermal properties of the facing materials, rather than the specific gauge of the steel, that will determine the degree of fire resistance. This is particularly true for nonbearing wall assemblies.

FIGURE 1.2.2
METAL FRAME WALLS
4″ TO LESS THAN 6″ THICK

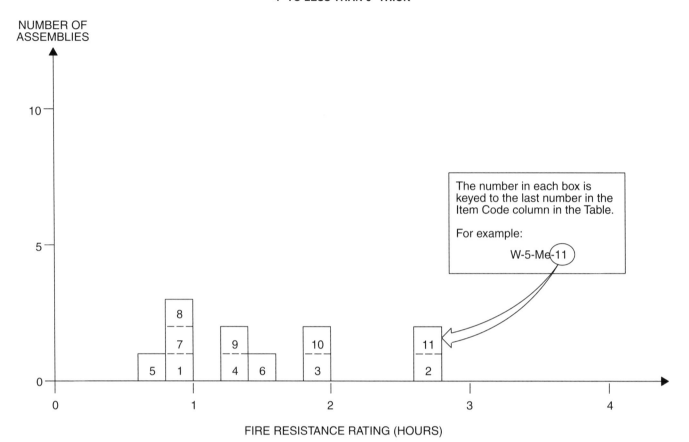

TABLE 1.2.2
METAL FRAME WALLS
4″ TO LESS THAN 6″ THICK

ITEM CODE	THICKNESS	CONSTRUCTION DETAILS	PERFORMANCE		REFERENCE NUMBER			NOTES	REC. HOURS
			LOAD	TIME	PRE-BMS-92	BMS-92	POST-BMS-92		
W-5-Me-1	5$\frac{1}{2}$″	3″ cavity with 16 ga. channel studs (3$\frac{1}{2}$″ o.c.) of $\frac{1}{2}$″ × $\frac{1}{2}$″ channel and 3″ spacer; Metal lath on ribs with plaster (three coats) $\frac{3}{4}$″ over face of lath; Plaster (each side): scratch coat, cement/lime/sand with hair; float coat, cement/lime/sand; finish coat, neat gypsum.	N/A	1 hr. 11 min.			7	1	1
W-4-Me-2	4″	Core: steel studs; see Note 2; Facings: both sides 1″ thickness of neat gypsum plaster.	N/A	2 hrs. 30 min.		1			2$\frac{1}{2}$
W-4-Me-3	4″	Core: steel studs; see Note 2; Facings: both sides 1″ thickness of 1:$\frac{1}{2}$; 1:$\frac{1}{2}$ gypsum plaster.	N/A	2 hrs.		1			2
W-4-Me-4	4″	Core: steel; see Note 2; Facings: both sides 1″ thickness of 1:2; 1:3 gypsum plaster.	N/A	1 hr. 15 min.		1			1$\frac{1}{4}$
W-4-Me-5	4$\frac{1}{2}$″	Core: lightweight steel studs 3″ in depth; Facings: both sides $\frac{3}{4}$″ thick sanded gypsum plaster, 1:2 scratch coat, 1:3 brown coat applied on metal lath.	See Note 4	45 min.		1		5	$\frac{3}{4}$

(continued)

TABLE 1.2.2—continued
METAL FRAME WALLS
4″ TO LESS THAN 6″ THICK

ITEM CODE	THICKNESS	CONSTRUCTION DETAILS	PERFORMANCE LOAD	PERFORMANCE TIME	REFERENCE NUMBER PRE-BMS-92	REFERENCE NUMBER BMS-92	REFERENCE NUMBER POST-BMS-92	NOTES	REC. HOURS
W-4-Me-6	$4\frac{1}{2}''$	Core: lightweight steel studs 3″ in depth; Facings: both sides $\frac{3}{4}''$ thick neat gypsum plaster on metal lath.	See Note 4	1 hr. 30 min.		1		5	$1\frac{1}{2}$
W-4-Me-7	$4\frac{1}{2}''$	Core: lightweight steel studs 3″ in depth; Facings: both sides $\frac{3}{4}''$ thick sanded gypsum plaster, 1:2 scratch and brown coats applied on metal lath.	See Note 4	1 hr.		1		5	1
W-4-Me-8	$4\frac{3}{4}''$	Core: lightweight steel studs 3″ in depth; Facings: both sides $\frac{7}{8}''$ thick sanded gypsum plaster, 1:2 scratch coat, 1:3 brown coat, applied on metal lath.	See Note 4	1 hr.		1		5	1
W-4-Me-9	$4\frac{3}{4}''$	Core: lightweight steel studs 3″ in depth; Facings: both sides $\frac{7}{8}''$ thick sanded gypsum plaster, 1:2 scratch and 1:3 brown coats applied on metal lath.	See Note 4	1 hr. 15 min.		1		5	$1\frac{1}{4}$
W-5-Me-10	5″	Core: lightweight steel studs 3″ in depth; Facings: both sides 1″ thick neat gypsum plaster on metal lath.	See Note 4	2 hrs.		1		5	2
W-5-Me-11	5″	Core: lightweight steel studs 3″ in depth; Facings: both sides 1″ thick neat gypsum plaster on metal lath.	See Note 4	2 hrs. 30 min.		1		5, 6	$2\frac{1}{2}$

For SI: 1 inch = 25.4 mm, 1 pound per square inch = 0.00689 MPa.

Notes:
1. Failure mode—local back face temperature rise.
2. Ratings are for any usual type of nonbearing metal framing providing a minimum 2 inches air space.
3. Facing materials secured to lightweight steel studs not less than 3 inches deep.
4. Rating based on loading to develop a maximum stress of 7270 psi for net area of each stud.
5. Spacing of steel studs must be sufficient to develop adequate rigidity in the metal-lath or gypsum-plaster base.
6. As per Note 4 but load/stud not to exceed 5120 psi.

General Note:
The construction details of the wall assemblies are as complete as the source documentation will permit. Data on the method of attachment of facings and the gauge of steel studs was provided when known. The cross sectional area of the steel stud can be computed, thereby permitting a reasoned estimate of actual loading conditions. For load-bearing assemblies, the maximum allowable stress for the steel studs has been provided in the table "Notes." More often, it is the thermal properties of the facing materials, rather than the specific gauge of the steel, that will determine the degree of fire resistance. This is particularly true for nonbearing wall assemblies.

**TABLE 1.2.3
METAL FRAME WALLS
6" TO LESS THAN 8" THICK**

ITEM CODE	THICKNESS	CONSTRUCTION DETAILS	PERFORMANCE		REFERENCE NUMBER			NOTES	REC HOURS
			LOAD	TIME	PRE-BMS-92	BMS-92	POST-BMS-92		
W-6-Me-1	$6^5/_8''$	On one side of 1" magnesium oxysulfate wood fiberboard sheathing attached to steel studs (see Notes 1 and 2), 1" air space, $3^3/_4''$ brick secured with metal ties to steel frame every fifth course; Inside facing of $^7/_8''$ 1:2 sanded gypsum plaster on metal lath secured directly to studs; Plaster side exposed to fire.	See Note 2	1 hr. 45 min.	1			1	$1^3/_4$
W-6-Me-2	$6^5/_8''$	On one side of 1" magnesium oxysulfate wood fiberboard sheathing attached to steel studs (see Notes 1 and 2), 1" air space, $3^3/_4''$ brick secured with metal ties to steel frame every fifth course; Inside facing of $^7/_8''$ 1:2 sanded gypsum plaster on metal lath secured directly to studs; Brick face exposed to fire.	See Note 2	4 hrs.	1			1	4
W-6-Me-3	$6^5/_8''$	On one side of 1" magnesium oxysulfate wood fiberboard sheathing attached to steel studs (see Notes 1 and 2), 1" air space, $3^3/_4''$ brick secured with metal ties to steel frame every fifth course; Inside facing of $^7/_8''$ vermiculite plaster on metal lath secured directly to studs; Plaster side exposed to fire.	See Note 2	2 hrs.	1			1	2

For SI: 1 inch = 25.4 mm, 1 pound per square inch = 0.00689 MPa.

Notes:

1. Lightweight steel studs (minimum 3 inches deep) used. Stud spacing dependent on loading, but in each case, spacing is to be such that adequate rigidity is provided to the metal lath plaster base.
2. Load is such that stress developed in studs is not greater than 5120 psi calculated from net stud area.

General Note:

The construction details of the wall assemblies are as complete as the source documentation will permit. Data on the method of attachment of facings and the gauge of steel studs was provided when known. The cross sectional area of the steel stud can be computed, thereby permitting a reasoned estimate of actual loading conditions. For load-bearing assemblies, the maximum allowable stress for the steel studs has been provided in the table "Notes." More often, it is the thermal properties of the facing materials, rather than the specific gauge of the steel, that will determine the degree of fire resistance. This is particularly true for nonbearing wall assemblies.

**TABLE 1.2.4
METAL FRAME WALLS
8″ TO LESS THAN 10″ THICK**

ITEM CODE	THICKNESS	CONSTRUCTION DETAILS	PERFORMANCE		REFERENCE NUMBER			NOTES	REC. HOURS
			LOAD	TIME	PRE-BMS-92	BMS-92	POST-BMS-92		
W-9-Me-1	$9^{1}/_{16}″$	On one side of $1/_{2}″$ wood fiberboard sheathing next to studs, $3/_{4}″$ air space formed with $3/_{4}″ \times 1^{5}/_{8}″$ wood strips placed over the fiberboard and secured to the studs, paper backed wire lath nailed to strips $3^{3}/_{4}″$ brick veneer held in place by filling a $3/_{4}″$ space between the brick and paper backed lath with mortar; Inside facing of $3/_{4}″$ neat gypsum plaster on metal lath attached to $5/_{16}″$ plywood strips secured to edges of steel studs; Rated as combustible because of the sheathing; See Notes 1 and 2; Plaster exposed.	See Note 2	1 hr. 45 min.		1		1	$1^{3}/_{4}$
W-9-Me-2	$9^{1}/_{16}″$	Same as above with brick exposed.	See Note 2	4 hrs.		1		1	4
W-8-Me-3	$8^{1}/_{2}″$	On one side of paper backed wire lath attached to studs and $3^{3}/_{4}″$ brick veneer held in place by filling a 1″ space between the brick and lath with mortar; Inside facing of 1″ paper-enclosed mineral wool blanket weighing 0.6 lb./ft.^{2} attached to studs, metal lath or paper backed wire lath laid over the blanket and attached to the studs, $3/_{4}″$ sanded gypsum plaster 1:2 for the scratch coat and 1:3 for the brown coat; See Notes 1 and 2; Plaster face exposed.	See Note 2	4 hrs.		1		1	4
W-8-Me-4	$8^{1}/_{2}″$	Same as above with brick exposed.	See Note 2	5 hrs.		1		1	5

For SI: 1 inch = 25.4 mm, 1 pound per square inch = 0.00689 MPa.

Notes:
1. Lightweight steel studs ≥ 3 inches in depth. Stud spacing dependent on loading, but in any case, the spacing is to be such that adequate rigidity is provided to the metal-lath plaster base.
2. Load is such that stress developed in studs is ≤ 5120 psi calculated from the net area of the stud.

General Note:
The construction details of the wall assemblies are as complete as the source documentation will permit. Data on the method of attachment of facings and the gauge of steel studs was provided when known. The cross sectional area of the steel stud can be computed, thereby permitting a reasoned estimate of actual loading conditions. For load-bearing assemblies, the maximum allowable stress for the steel studs has been provided in the table "Notes." More often, it is the thermal properties of the facing materials, rather than the specific gauge of the steel, that will determine the degree of fire resistance. This is particularly true for nonbearing wall assemblies.

**TABLE 1.3.1
WOOD FRAME WALLS
0" TO LESS THAN 4" THICK**

ITEM CODE	THICKNESS	CONSTRUCTION DETAILS	PERFORMANCE		REFERENCE NUMBER			NOTES	REC. HOURS
			LOAD	TIME	PRE-BMS-92	BMS-92	POST-BMS-92		
W-3-W-1	$3\frac{3}{4}''$	Solid wall: $2\frac{1}{4}''$ wood-wool slab core; $\frac{3}{4}''$ gypsum plaster each side.	N/A	2 hrs.			7	1, 6	2
W-3-W-2	$3\frac{7}{8}''$	2 × 4 stud wall; $\frac{3}{16}''$ thick cement asbestos board on both sides of wall.	360 psi net area	10 min.	1			2-5	$\frac{1}{6}$
W-3-W-3	$3\frac{7}{8}''$	Same as W-3-W-2 but stud cavities filled with 1 lb./ft.^{2} mineral wool batts.	360 psi net area	40 min.	1			2-5	$\frac{2}{3}$

For SI: 1 inch = 25.4 mm, 1 pound per square inch = 0.00689 MPa.

Notes:
1. Achieved "Grade C" fire resistance (British).
2. Nominal 2 × 4 wood studs of No. 1 common or better lumber set edgewise, 2 × 4 plates at top and bottom and blocking at mid height of wall.
3. All horizontal joints in facing material backed by 2 × 4 blocking in wall.
4. Load: 360 psi of net stud cross sectional area.
5. Facings secured with 6d casing nails. Nail holes predrilled and 0.02 inch to 0.03 inch smaller than nail diameter.
6. The wood-wool core is a pressed excelsior slab which possesses insulating properties similar to cellulosic insulation.

RESOURCE A

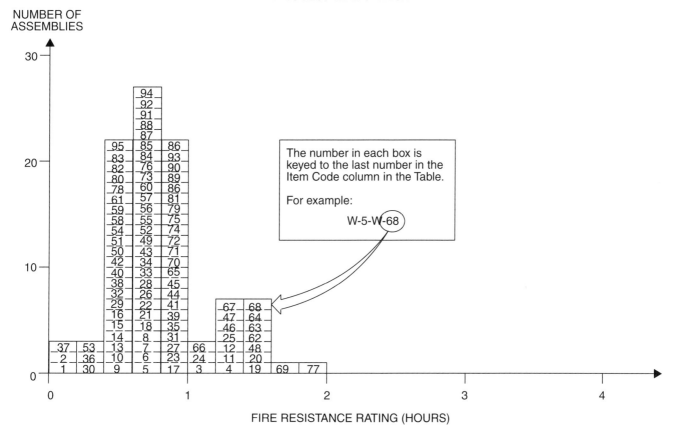

FIGURE 1.3.2
WOOD FRAME WALLS
4″ TO LESS THAN 6″ THICK

TABLE 1.3.2
WOOD FRAME WALLS
4″ TO LESS THAN 6″ THICK

ITEM CODE	THICKNESS	CONSTRUCTION DETAILS	PERFORMANCE LOAD	PERFORMANCE TIME	REFERENCE NUMBER PRE-BMS-92	REFERENCE NUMBER BMS-92	REFERENCE NUMBER POST-BMS-92	NOTES	REC. HOURS
W-4-W-1	4″	2″ × 4″ stud wall; $^3/_{16}$″ CAB; no insulation; Design A.	35 min.	10 min.			4	1-10	$^1/_6$
W-4-W-2	$4^1/_8$″	2″ × 4″ stud wall; $^3/_{16}$″ CAB; no insulation; Design A.	38 min.	9 min.			4	1-10	$^1/_6$
W-4-W-3	$4^3/_4$″	2″ × 4″ stud wall; $^3/_{16}$″ CAB and $^3/_8$″ gypsum board face (both sides); Design B.	62 min.	64 min.			4	1-10	1
W-5-W-4	5″	2″ × 4″ stud wall; $^3/_{16}$″ CAB and $^1/_2$″ gypsum board (both sides); Design B.	79 min.	Greater than 90 min.			4	1-10	1
W-4-W-5	$4^3/_4$″	2″ × 4″ stud wall; $^3/_{16}$″ CAB and $^3/_8$″ gypsum board (both sides); Design B.	45 min.	45 min.			4	1-12	—
W-5-W-6	5″	2″ × 4″ stud wall; $^3/_{16}$″ CAB and $^1/_2$″ gypsum board face (both sides); Design B.	45 min.	45 min.			4	1-10, 12, 13	—
W-4-W 7	4″	2″ × 4″ stud wall; $^3/_{16}$″ CAB face; $3^1/_2$″ mineral wool insulation; Design C.	40 min.	42 min.			4	1-10	$^2/_3$
W-4-W-8	4″	2″ × 4″ stud wall; $^3/_{16}$″ CAB face; $3^1/_2$″ mineral wool insulation; Design C.	46 min.	46 min.			4	1-10, 43	$^2/_3$
W-4-W-9	4″	2″ × 4″ stud wall; $^3/_{16}$″ CAB face; $3^1/_2$″ mineral wool insulation; Design C.	30 min.	30 min.			4	1-10, 12, 14	—

(continued)

TABLE 1.3.2—continued
WOOD FRAME WALLS
4″ TO LESS THAN 6″ THICK

ITEM CODE	THICKNESS	CONSTRUCTION DETAILS	PERFORMANCE		REFERENCE NUMBER			NOTES	REC. HOURS
			LOAD	TIME	PRE-BMS-92	BMS-92	POST-BMS-92		
W-4-W-10	$4^1/_8″$	2″ × 4″ stud wall; $^3/_{16}″$ CAB face; $3^1/_2″$ mineral wool insulation; Design C.	—	30 min.			4	1-8, 12, 14	—
W-4-W-11	$4^3/_4″$	2″ × 4″ stud wall; $^3/_{16}″$ CAB face; $^3/_8″$ gypsum strips over studs; $5^1/_2″$ mineral wool insulation; Design D.	79 min.	79 min.			4	1-10	1
W-4-W-12	$4^3/_4″$	2″ × 4″ stud wall; $^3/_{16}″$ CAB face; $^3/_8″$ gypsum strips at stud edges; $7^1/_2″$ mineral wool insulation; Design D.	82 min.	82 min.			4	1-10	1
W-4-W-13	$4^3/_4″$	2″ × 4″ stud wall; $^3/_{16}″$ CAB face; $^3/_8″$ gypsum board strips over studs; $5^1/_2″$ mineral wool insulation; Design D.	30 min.	30 min.			4	1-12	—
W-4-W-14	$4^3/_4″$	2″ × 4″ stud wall; $^3/_{16}″$ CAB face; $^3/_8″$ gypsum board strips over studs; 7″ mineral wool insulation; Design D.	30 min.	30 min.			4	1-12	—
W-5-W-15	$5^1/_2″$	2″ × 4″ stud wall; Exposed face: CAB shingles over 1″ × 6″; Unexposed face: $^1/_8″$ CAB sheet; $^7/_{16}″$ fiberboard (wood); Design E.	34 min.	—			4	1-10	$^1/_2$
W-5-W-16	$5^1/_2″$	2″ × 4″ stud wall; Exposed face: $^1/_8″$ CAB sheet; $^7/_{16}″$ fiberboard; Unexposed face: CAB shingles over 1″ × 6″; Design E.	32 min.	33 min.			4	1-10	$^1/_2$
W-5-W-17	$5^1/_2″$	2″ × 4″ stud wall; Exposed face: CAB shingles over 1″ × 6″; Unexposed face: $^1/_8″$ CAB sheet; gypsum at stud edges; $3^1/_2″$ mineral wood insulation; Design F.	51 min.	—			4	1-10	$^3/_4$
W-5-W-18	$5^1/_2″$	2″ × 4″ stud wall; Exposed face: $^1/_8″$ CAB sheet; gypsum board at stud edges; Unexposed face: CAB shingles over 1″ × 6″; $3^1/_2″$ mineral wool insulation; Design F.	42 min.	—			4	1-10	$^2/_3$
W-5-W-19	$5^5/_8″$	2″ × 4″ stud wall; Exposed face: CAB shingles over 1″ × 6″; Unexposed face: $^1/_8″$ CAB sheet; gypsum board at stud edges; $5^1/_2″$ mineral wool insulation; Design G.	74 min.	85 min.			4	1-10	1
W-5-W-20	$5^5/_8″$	2″ × 4″ stud wall; Exposed face: $^1/_8″$ CAB sheet; gypsum board at $^3/_{16}″$ stud edges; $^7/_{16}″$ fiberboard; Unexposed face: CAB shingles over 1″ × 6″; $5^1/_2″$ mineral wool insulation; Design G.	79 min.	85 min.			4	1-10	$1^1/_4$
W-5-W-21	$5^5/_8″$	2″ × 4″ stud wall; Exposed face: CAB shingles 1″ × 6″ sheathing; Unexposed face: CAB sheet; gypsum board at stud edges; $5^1/_2″$ mineral wool insulation; Design G.	38 min.	38 min.			4	1-10, 12, 14	—
W-5-W-22	$5^5/_8″$	2″ × 4″ stud wall; Exposed face: CAB sheet; gypsum board at stud edges; Unexposed face: CAB shingles 1″ × 6″ sheathing; $5^1/_2″$ mineral wool insulation; Design G.	38 min.	38 min.			4	1-12	—
W-6-W-23	6″	2″ × 4″ stud wall; 16″ o.c.; $^1/_2″$ gypsum board each side; $^1/_2″$ gypsum plaster each side.	N/A	60 min.			7	15	1

(continued)

TABLE 1.3.2—continued
WOOD FRAME WALLS
4″ TO LESS THAN 6″ THICK.

ITEM CODE	THICKNESS	CONSTRUCTION DETAILS	PERFORMANCE		REFERENCE NUMBER			NOTES	REC. HOURS
			LOAD	TIME	PRE-BMS-92	BMS-92	POST-BMS-92		
W-6-W-24	6″	2″ × 4″ stud wall; 16″ o.c.; $1/2$″ gypsum board each side; $1/2$″ gypsum plaster each side.	N/A	68 min.			7	16	1
W-6-W-25	$6^7/_8$″	2″ × 4″ stud wall; 18″ o.c.; $3/4$″ gypsum plank each side; $3/_{16}$″ gypsum plaster each side.	N/A	80 min.			7	15	$1^1/_3$
W-5-W-26	$5^1/_8$″	2″ × 4″ stud wall; 16″ o.c.; $3/8$″ gypsum board each side; $3/_{16}$″ gypsum plaster each side.	N/A	37 min.			7	15	$1/2$
W-5-W-27	$5^3/_4$″	2″ × 4″ stud wall; 16″ o.c.; $3/8$″ gypsum lath each side; $1/2$″ gypsum plaster each side.	N/A	52 min.			7	15	$3/4$
W-5-W-28	5″	2″ × 4″ stud wall; 16″ o.c.; $1/2$″ gypsum board each side.	N/A	37 min.			7	16	$1/2$
W-5-W-29	5″	2″ × 4″ stud wall; $1/2$″ fiberboard both sides 14% M.C. with F.R. paint at 35 gm./ft.^{2}.	N/A	28 min.			7	15	$1/3$
W-4-W-30	$4^3/_4$″	2″ × 4″ stud wall; Fire side: $1/2$″ (wood) fiberboard; Back side: $1/4$″ CAB; 16″ o.c.	N/A	17 min.			7	15, 16	$1/4$
W-5-W-31	$5^1/_8$″	2″ × 4″ stud wall; 16″ o.c.; $1/2$″ fiberboard insulation with $1/_{32}$″ asbestos (both sides of each board).	N/A	50 min.			7	16	$3/4$
W-4-W-32	$4^1/_4$″	2″ × 4″ stud wall; $3/8$″ thick gypsum wallboard on both faces; insulated cavities.	See Note 23	25 min.		1		17, 18, 23	$1/3$
W-4-W-33	$4^1/_2$″	2″ × 4″ stud wall; $1/2$″ thick gypsum wallboard on both faces.	See Note 17	40 min.		1		17, 23	$1/3$
W-4-W-34	$4^1/_2$″	2″ × 4″ stud wall; $1/2$″ thick gypsum wallboard on both faces; insulated cavities.	See Note 17	45 min.		1		17, 18, 23	$3/4$
W-4-W-35	$4^1/_2$″	2″ × 4″ stud wall; $1/2$″ thick gypsum wallboard on both faces; insulated cavities.	N/A	1 hr.		1		17, 18, 24	1
W-4-W-36	$4^1/_2$″	2″ × 4″ stud wall; $1/2$″ thick, 1.1 lbs./ft.^{2} wood fiberboard sheathing on both faces.	See Note 23	15 min.		1		17, 23	$1/4$
W-4-W-37	$4^1/_2$″	2″ × 4″ stud wall; $1/2$″ thick, 0.7 lb./ft.^{2} wood fiberboard sheathing on both faces.	See Note 23	10 min.		1		17, 23	$1/6$
W-4-W-38	$4^1/_2$″	2″ × 4″ stud wall; $1/2$″ thick, flameproofed 1.6 lbs./ft.^{2} wood fiberboard sheathing on both faces.	See Note 23	30 min.		1		17, 23	$1/2$
W-4-W-39	$4^1/_2$″	2″ × 4″ stud wall; $1/2$″ thick gypsum wallboard on both faces; insulated cavities.	See Note 23	1 hr.		1		17, 18, 23	1
W-4-W-40	$4^1/_2$″	2″ × 4″ stud wall; $1/2$″ thick, 1:2; 1:3 gypsum plaster on wood lath on both faces.	See Note 23	30 min.		1		17, 21, 23	$1/2$
W-4-W-41	$4^1/_2$″	2″ × 4″ stud wall; $1/2$″, 1:2; 1:3 gypsum plaster on wood lath on both faces; insulated cavities.	See Note 23	1 hr.		1		17, 18, 21, 24	1

(continued)

TABLE 1.3.2—continued
WOOD FRAME WALLS
4″ TO LESS THAN 6″ THICK

ITEM CODE	THICKNESS	CONSTRUCTION DETAILS	PERFORMANCE		REFERENCE NUMBER			NOTES	REC. HOURS
			LOAD	TIME	PRE-BMS-92	BMS-92	POST-BMS-92		
W-4-W-42	4$^1/_2$″	2″ × 4″ stud wall; $^1/_2$″, 1:5; 1:7.5 lime plaster on wood lath on both wall faces.	See Note 23	30 min.	1			17, 21, 23	$^1/_2$
W-4-W-43	4$^1/_2$″	2″ × 4″ stud wall; $^1/_2$″ thick 1:5; 1:7.5 lime plaster on wood lath on both faces; insulated cavities.	See Note 23	45 min.	1			17, 18, 21, 23	$^3/_4$
W-4-W-44	4$^5/_8$″	2″ × 4″ stud wall; $^3/_{16}$″ thick cement-asbestos over $^3/_8$″ thick gypsum board on both faces.	See Note 23	1 hr.	1			23, 25, 26, 27	1
W-4-W-45	4$^5/_8$″	2″ × 4″ stud wall; studs faced with 4″ wide strips of $^3/_8$″ thick gypsum board; $^3/_{16}$″ thick gypsum cement-asbestos board on both faces; insulated cavities.	See Note 23	1 hr.	1			23, 25, 27, 28	1
W-4-W-46	4$^5/_8$″	Same as W-4-W-45 but nonload bearing.	N/A	1 hr. 15 min.	1			24, 28	1$^1/_4$
W-4-W-47	4$^7/_8$″	2″ × 4″ stud wall; $^3/_{16}$″ thick cement-asbestos board over $^1/_2$″ thick gypsum sheathing on both faces.	See Note 23	1 hr. 15 min.	1			23, 25, 26, 27	1$^1/_4$
W-4-W-48	4$^7/_8$″	Same as W-4-W-47 but nonload bearing.	N/A	1 hr. 30 min.	1			24, 27	1$^1/_2$
W-5-W-49	5″	2″ × 4″ stud wall; Exterior face: $^3/_4$″ wood sheathing; asbestos felt 14 lbs./100 ft.^{2} and $^5/_{32}$″ cement-asbestos shingles; Interior face: 4″ wide strips of $^3/_8$″ gypsum board over studs; wall faced with $^3/_{16}$″ thick cement-asbestos board.	See Note 23	40 min.	1			18, 23, 25, 26, 29	$^2/_3$
W-5-W-50	5″	2″ × 4″ stud wall; Exterior face: as per W-5-W-49; Interior face: $^9/_{16}$″ composite board consisting of $^7/_{16}$″ thick wood fiberboard faced with $^1/_8$″ thick cement-asbestos board; Exterior side exposed to fire.	See Note 23	30 min.	1			23, 25, 26, 30	$^1/_2$
W-5-W-51	5″	Same as W-5-W-50 but interior side exposed to fire.	See Note 23	30 min.	1			23, 25, 26	$^1/_2$
W-5-W-52	5″	Same as W-5-W-49 but exterior side exposed to fire.	See Note 23	45 min.	1			18, 23, 25, 26	$^3/_4$
W-5-W-53	5″	2″ × 4″ stud wall; $^3/_4$″ thick T&G wood boards on both sides.	See Note 23	20 min.	1			17, 23	$^1/_3$
W-5-W-54	5″	Same as W-5-W-53 but with insulated cavities.	See Note 23	35 min.	1			17, 18, 23	$^1/_2$
W-5-W-55	5″	2″ × 4″ stud wall; $^3/_4$″ thick T&G wood boards on both sides with 30 lbs./100 ft.^{2} asbestos; paper, between studs and boards.	See Note 23	45 min.	1			17, 23	$^3/_4$
W-5-W-56	5″	2″ × 4″ stud wall; $^1/_2$″ thick, 1:2; 1:3 gypsum plaster on metal lath on both sides of wall.	See Note 23	45 min.	1			17, 21, 34	$^3/_4$

(continued)

**TABLE 1.3.2—continued
WOOD FRAME WALLS
4″ TO LESS THAN 6″ THICK**

ITEM CODE	THICKNESS	CONSTRUCTION DETAILS	PERFORMANCE		REFERENCE NUMBER			NOTES	REC. HOURS
			LOAD	TIME	PRE-BMS-92	BMS-92	POST-BMS-92		
W-5-W-57	5″	2″ × 4″ stud wall; $^3/_4$″ thick 2:1:8; 2:1:12 lime and Keene's cement plaster over metal lath on both sides of wall.	See Note 23	45 min.		1		17, 21, 23	$^1/_2$
W-5-W-58	5″	2″ × 4″ stud wall; $^3/_4$″ thick 2:1:8; 2:1:10 lime portland cement plaster over metal lath on both sides of wall.	See Note 23	30 min.		1		17, 21, 23	$^1/_2$
W-5-W-59	5″	2″ × 4″ stud wall; $^3/_4$″ thick 1:5; 1:7.5 lime plaster on metal lath on both sides of wall.	See Note 23	30 min.		1		17, 21, 23	$^1/_2$
W-5-W-60	5″	2″ × 4″ stud wall; $^3/_4$″ thick 1:$^1/_{30}$:2; 1:$^1/_{30}$:3 portland cement, asbestos fiber plaster on metal lath on both sides of wall.	See Note 23	45 min.		1		17, 21, 23	$^3/_4$
W-5-W-61	5″	2″ × 4″ stud wall; $^3/_4$″ thick 1:2; 1:3 portland cement plaster on metal lath on both sides of wall.	See Note 23	30 min.		1		17, 21, 23	$^1/_2$
W-5-W-62	5″	2″ × 4″ stud wall; $^3/_4$″ thick neat gypsum plaster on metal lath on both sides of wall.	N/A	1 hr. 30 min.		1		17, 22, 24	$1^1/_2$
W-5-W-63	5″	2″ × 4″ stud wall; $^3/_4$″ thick neat gypsum plaster on metal lath on both sides of wall.	See Note 23	1 hr. 30 min.		1		17, 21, 23	$1^1/_2$
W-5-W-64	5″	2″ × 4″ stud wall; $^3/_4$″ thick 1:2; 1:2 gypsum plaster on metal lath on both sides of wall; insulated cavities.	See Note 23	1 hr. 30 min.		1		17, 18, 21, 23	$1^1/_2$
W-5-W-65	5″	2″ × 4″ stud wall; same as W-5-W-64 but cavities not insulated.	See Note 23	1 hr.		1		17, 21, 23	1
W-5-W-66	5″	2″ × 4″ stud wall; $^3/_4$″ thick 1:2; 1:3 gypsum plaster on metal lath on both sides of wall; insulated cavities.	See Note 23	1 hr. 15 min.		1		17, 18, 21, 23	$1^1/_4$
W-5-W-67	$5^1/_{16}$″	Same as W-5-W-49 except cavity insulation of 1.75 lbs./ft.^{2} mineral wool bats; rating applies when either wall side exposed to fire.	See Note 23	1 hr. 15 min.		1		23, 26, 25	$1^1/_4$
W-5-W-68	$5^1/_4$″	2″ × 4″ stud wall; $^7/_8$″ thick 1:2; 1:3 gypsum plaster on metal lath on both sides of wall; insulated cavities.	See Note 23	1 hr. 30 min.		1		17, 18, 21, 23	$1^1/_2$
W-5-W-69	$5^1/_4$″	2″ × 4″ stud wall; $^7/_8$″ thick neat gypsum plaster applied on metal lath on both sides of wall.	N/A	1 hr. 45 min.		1		17, 22, 24	$1^3/_4$
W-5-W-70	$5^1/_4$″	2″ × 4″ stud wall; $^1/_2$″ thick neat gypsum plaster on $^3/_8$″ plain gypsum lath on both sides of wall.	See Note 23	1 hr.		1		17, 22, 23	1
W-5-W-71	$5^1/_4$″	2″ × 4″ stud wall; $^1/_2$″ thick of 1:2; 1:2 gypsum plaster on $^3/_8$″ thick plain gypsum lath with $1^3/_4$″ × $1^3/_4$″ metal lath pads nailed 8″ o.c. vertically and 16″ o.c. horizontally on both sides of wall.	See Note 23	1 hr.		1		17, 21, 23	1
W-5-W-72	$5^1/_4$″	2″ × 4″ stud wall; $^1/_2$″ thick of 1:2; 1:2 gypsum plaster on $^3/_8$″ perforated gypsum lath, one $^3/_4$″ diameter hole or larger per 16″ square of lath surface, on both sides of wall.	See Note 23	1 hr.		1		17, 21, 23	1

(continued)

TABLE 1.3.2—continued
WOOD FRAME WALLS
4" TO LESS THAN 6" THICK

ITEM CODE	THICKNESS	CONSTRUCTION DETAILS	PERFORMANCE		REFERENCE NUMBER			NOTES	REC. HOURS
			LOAD	TIME	PRE-BMS-92	BMS-92	POST-BMS-92		
W-5-W-73	5¼"	2" × 4" stud wall; ½" thick of 1:2; 1:2 gypsum plaster on ⅜" gypsum lath (plain, indented or perforated) on both sides of wall.	See Note 23	45 min.		1		17, 21, 23	¾
W-5-W-74	5¼"	2" × 4" stud wall; ⅞" thick of 1:2; 1:3 gypsum plaster over metal lath on both sides of wall.	See Note 23	1 hr.		1		17, 21, 23	1
W-5-W-75	5¼"	2" × 4" stud wall; ⅞" thick of 1:¹⁄₃₀:2; 1:¹⁄₃₀:3 portland cement, asbestos plaster applied over metal lath on both sides of wall.	See Note 23	1 hr.		1		17, 21, 23	1
W-5-W-76	5¼"	2" × 4" stud wall; ⅞" thick of 1:2; 1:3 portland cement plaster over metal lath on both sides of wall.	See Note 23	45 min.		1		17, 21, 23	¾
W-5-W-77	5½"	2" × 4" stud wall; 1" thick neat gypsum plaster over metal lath on both sides of wall; nonload bearing.	N/A	2 hrs.		1		17, 22, 24	2
W-5-W-78	5½"	2" × 4" stud wall; ½" thick of 1:2; 1:2 gypsum plaster on ½" thick, 0.7 lb./ft.² wood fiberboard on both sides of wall.	See Note 23	35 min.		1		17, 21, 23	½
W-4-W-79	4¾"	2" × 4" wood stud wall; ½" thick of 1:2; 1:2 gypsum plaster over wood lath on both sides of wall; mineral wool insulation.	N/A	1 hr.			43	21, 31, 35, 38	1
W-4-W-80	4¾"	Same as W-4-W-79 but uninsulated.	N/A	35 min.			43	21, 31, 35	½
W-4-W-81	4¾"	2" × 4" wood stud wall; ½" thick of 3:1:8; 3:1:12 lime, Keene's cement, sand plaster over wood lath on both sides of wall; mineral wool insulation.	N/A	1 hr.			43	21, 31, 35, 40	1
W-4-W-82	4¾"	2" × 4" wood stud wall; ½" thick of 1:6¼; 1:6¼ lime Keene's cement plaster over wood lath on both sides of wall; mineral wool insulation.	N/A	30 min.			43	21, 31, 35, 40	½
W-4-W-83	4¾"	2" × 4" wood stud wall; ½" thick of 1:5; 1:7.5 lime plaster over wood lath on both sides of wall.	N/A	30 min.			43	21, 31, 35	½
W-5-W-84	5⅛"	2" × 4" wood stud wall; ¹¹⁄₁₆" thick of 1:5; 1:7.5 lime plaster over wood lath on both sides of wall; mineral wool insulation.	N/A	45 min.			43	21, 31, 35, 39	¾
W-5-W-85	5¼"	2" × 4" wood stud wall; ¾" thick of 1:5; 1:7 lime plaster over wood lath on both sides of wall; mineral wool insulation.	N/A	40 min.			43	21, 31, 35, 40	⅔
W-5-W-86	5¼"	2" × 4" wood stud wall; ½" thick of 2:1:12 lime, Keene's cement and sand scratch coat; ½" thick 2:1:18 lime, Keene's cement and sand brown coat over wood lath on both sides of wall; mineral wool insulation.	N/A	1 hr.			43	21, 31, 35, 40	1
W-5-W-87	5¼"	2" × 4" wood stud wall; ½" thick of 1:2; 1:2 gypsum plaster over ⅜" plaster board on both sides of wall.	N/A	45 min.			43	21, 31	¾

(continued)

**TABLE 1.3.2—continued
WOOD FRAME WALLS
4″ TO LESS THAN 6″ THICK**

ITEM CODE	THICKNESS	CONSTRUCTION DETAILS	PERFORMANCE		REFERENCE NUMBER			NOTES	REC. HOURS
			LOAD	TIME	PRE-BMS-92	BMS-92	POST-BMS-92		
W-5-W-88	5¼″	2″ × 4″ wood stud wall; ½″ thick of 1:2; 1:2 gypsum plaster over ⅜″ gypsum lath on both sides of wall.	N/A	45 min.			43	21, 31	¾
W-5-W-89	5¼″	2″ × 4″ wood stud wall; ½″ thick of 1:2; 1:2 gypsum plaster over ⅜″ gypsum lath on both sides of wall.	N/A	1 hr.			43	21, 31, 33	1
W-5-W-90	5¼″	2″ × 4″ wood stud wall; ½″ thick neat plaster over ⅜″ thick gypsum lath on both sides of wall.	N/A	1 hr.			43	21, 22, 31	1
W-5-W-91	5¼″	2″ × 4″ wood stud wall; ½″ thick of 1:2; 1:2 gypsum plaster over ⅜″ thick indented gypsum lath on both sides of wall.	N/A	45 min.			43	21, 31	¾
W-5-W-92	5¼″	2″ × 4″ wood stud wall; ½″ thick of 1:2; 1:2 gypsum plaster over ⅜″ thick perforated gypsum lath on both sides of wall.	N/A	45 min.			43	21, 31, 34	¾
W-5-W-93	5¼″	2″ × 4″ wood stud wall; ½″ thick of 1:2; 1:2 gypsum plaster over ⅜″ perforated gypsum lath on both sides of wall.	N/A	1 hr.			43	21, 31	1
W-5-W-94	5¼″	2″ × 4″ wood stud wall; ½″ thick of 1:2; 1:2 gypsum plaster over ⅜″ thick perforated gypsum lath on both sides of wall.	N/A	45 min.			43	21, 31, 34	¾
W-5-W-95	5½″	2″ × 4″ wood stud wall; ½″ thick of 1:2; 1:2 gypsum plaster over ½″ thick wood fiberboard plaster base on both sides of wall.	N/A	35 min.			43	21, 31, 36	½
W-5-W-96	5¾″	2″ × 4″ wood stud wall; ½″ thick of 1:2; 1:2 gypsum plaster over ⅞″ thick flameproofed wood fiberboard on both sides of wall.	N/A	1 hr.			43	21, 31, 37	1

For SI: 1 inch = 25.4 mm, 1 foot = 305 mm, 1 pound = 0.004448 kN, 1 pound per square inch = 0.00689 MPa, 1 pound per square foot = 47.9 N/m^2.

Notes:
1. All specimens 8 feet or 8 feet 8 inches by 10 feet 4 inches, i.e. one-half of furnace size. See Note 42 for design cross section.
2. Specimens tested in tandem (two per exposure).
3. Test per ASA No. A2-1934 except where unloaded. Also, panels were of "half" size of furnace opening. Time value signifies a thermal failure time.
4. Two-inch by 4-inch studs: 16 inches on center.; where 10 feet 4 inches, blocking at 2-foot 4-inch height.
5. Facing 4 feet by 8 feet, cement-asbestos board sheets, ³⁄₁₆ inch thick.
6. Sheathing (diagonal): 25/22 inch by 5½ inch, 1 inch by 6 inches pine.
7. Facing shingles: 24 inches by 12 inches by ⁵⁄₃₂ inch where used.
8. Asbestos felt: asphalt sat between sheathing and shingles.
9. Load: 30,500 pounds or 360 psi/stud where load was tested.
10. Walls were tested beyond achievement of first test end point. A load-bearing time in excess of performance time indicates that although thermal criteria were exceeded, load-bearing ability continued.
11. Wall was rated for one hour combustible use in original source.
12. Hose steam test specimen. See table entry of similar design above for recommended rating.
13. Rated one and one-fourth hour load bearing. Rated one and one-half hour nonload bearing.
14. Failed hose stream.
15. Test terminated due to flame penetration.
16. Test terminated—local back face temperature rise.
17. Nominal 2-inch by 4-inch wood studs of No. 1 common or better lumber set edgewise. Two-inch by four-inch plates at top and bottom and blocking at mid height of wall.
18. Cavity insulation consists of rock wool bats 1.0 lb./ft.2 of filled cavity area.
19. Cavity insulation consists of glass wool bats 0.6 lb./ft.2 of filled cavity area.
20. Cavity insulation consists of blown-in forck wool 2.0 lbs./ft.2 of filled cavity area
21. Mix proportions for plastered walls as follows: first ratio indicates scratch coat mix, weight of dry plaster: dry sand; second ratio indicates brown coat mix.
22. "Neat" plaster is taken to mean unsanded wood-fiber gypsum plaster.
23. Load: 360 psi of net stud cross sectional area.
24. Rated as nonload bearing.

(continued)

RESOURCE A

**TABLE 1.3.2—continued
WOOD FRAME WALLS
4″ TO LESS THAN 6″ THICK**

25. Nominal 2-inch by 4-inch studs per Note 17, spaced at 16 inches on center.
26. Horizontal joints in facing material supported by 2-inch by 4-inch blocking within wall.
27. Facings secured with 6d casing nails. Nail holes predrilled and were 0.02 to 0.03 inch smaller than nail diameter.
28. Cavity insulation consists of mineral wool bats weighing 2 lbs./ft.^{2} of filled cavity area.
29. Interior wall face exposed to fire.
30. Exterior wall faced exposed to fire.
31. Nominal 2-inch by 4-inch studs of yellow pine or Douglas-fir spaced 16 inches on center in a single row.
32. Studs as in Note 31 except double row, with studs in rows staggered.
33. Six roofing nails with metal-lath pads around heats to each 16-inch by 48-inch lath.
34. Areas of holes less than $2^3/_4$ percent of area of lath.
35. Wood laths were nailed with either 3d or 4d nails, one nail to each bearing, and the end joining broken every seventh course.
36. One-half-inch thick fiberboard plaster base nailed with 3d or 4d common wire nails spaced 4 to 6 inches on center.
37. Seven-eighths-inch thick fiberboard plaster base nailed with 5d common wire nails spaced 4 to 6 inches on center.
38. Mineral wood bats 1.05 to 1.25 lbs./ft.^{2} with waterproofed-paper backing.
39. Blown-in mineral wool insulation, 2.2 lbs./ft.^{2}.
40. Mineral wool bats, 1.4 lbs./ft.^{2} with waterproofed-paper backing.
41. Mineral wood bats, 0.9 lb./ft.^{2}.
42. See wall design diagram, below.

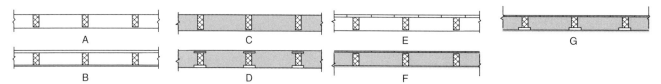

43. Duplicate specimen of W-4-W-7, tested simultaneously with W-4-W-7 in 18-foot test furnace.

TABLE 1.3.3
WOOD FRAME WALLS
6″ TO LESS THAN 8″ THICK

ITEM CODE	THICKNESS	CONSTRUCTION DETAILS	PERFORMANCE		REFERENCE NUMBER			NOTES	REC. HOURS
			LOAD	TIME	PRE-BMS-92	BMS-92	POST-BMS-92		
W-6-W-1	$6\frac{1}{4}''$	2 × 4 stud wall; $\frac{1}{2}''$ thick, 1:2; 1:2 gypsum plaster on $\frac{7}{8}''$ flameproofed wood fiberboard weighing 2.8 lbs./ft.^{2} on both sides of wall.	See Note 3	1 hr.		1		1-3	1
W-6-W-2	$6\frac{1}{2}''$	2 × 4 stud wall; $\frac{1}{2}''$ thick, 1:3; 1:3 gypsum plaster on 1″ thick magnesium oxysulfate wood fiberboard on both sides of wall.	See Note 3	45 min.		1		1-3	$\frac{3}{4}$
W-7-W-3	$7\frac{1}{4}''$	Double row of 2 × 4 studs, $\frac{1}{2}''$ thick of 1:2; 1:2 gypsum plaster applied over $\frac{3}{8}''$ thick perforated gypsum lath on both sides of wall; mineral wool insulation.	N/A	1 hr.			43	2, 4, 5	1
W-7-W-4	$7\frac{1}{2}''$	Double row of 2 × 4 studs, $\frac{5}{8}''$ thick of 1:2; 1:2 gypsum plaster applied over $\frac{3}{8}''$ thick perforated gypsum lath over laid with 2″ × 2″, 16 gage wire fabric, on both sides of wall.	N/A	1 hr. 15 min.			43	2, 4	$1\frac{1}{4}$

For SI: 1 inch = 25.4 mm, 1 pound = 0.004448 kN, 1 pound per square inch = 0.00689 MPa, 1 pound per square foot = 47.9 N/m^2.

Notes:
1. Nominal 2-inch by 4-inch wood studs of No. 1 common or better lumber set edgewise. Two-inch by 4-inch plates at top and bottom and blocking at mid height of wall.
2. Mix proportions for plastered walls as follows: first ratio indicates scratch coat mix, weight of dry plaster: dry sand; second ratio indicates brown coat mix.
3. Load: 360 psi of net stud cross sectional area.
4. Nominal 2-inch by 4-inch studs of yellow pine of Douglas-fir spaced 16 inches in a double row, with studs in rows staggered.
5. Mineral wool bats, 0.19 lb./ft.^{2}

TABLE 1.4.1
MISCELLANEOUS MATERIALS WALLS
0″ TO LESS THAN 4″ THICK

ITEM CODE	THICKNESS	CONSTRUCTION DETAILS	PERFORMANCE		REFERENCE NUMBER			NOTES	REC. HOURS
			LOAD	TIME	PRE-BMS-92	BMS-92	POST-BMS-92		
W-3-Mi-1	$3\frac{7}{8}''$	Glass brick wall: (bricks $5\frac{3}{4}'' \times 5\frac{3}{4}'' \times 3\frac{7}{8}''$) $\frac{1}{4}''$ mortar bed, cement/lime/sand; mounted in brick (9″) wall with mastic and $\frac{1}{2}''$ asbestos rope.	N/A	1 hr.			7	1, 2	1
W-3-Mi-2	3″	Core: 2″ magnesium oxysulfate wood-fiber blocks; laid in portland cement-lime mortar; Facings: on both sides; see Note 3.	N/A	1 hr.		1		3	1
W-3-Mi-3	$3\frac{7}{8}''$	Core: 8″ × $4\frac{7}{8}''$ glass blocks $3\frac{7}{8}''$ thick weighing 4 lbs. each; laid in portland cement-lime mortar; horizontal mortar joints reinforced with metal lath.	N/A	15 min.		1			$\frac{1}{4}$

For SI: 1 inch = 25.4 mm, 1 pound = 0.004448 kN.

Notes:
1. No failure reached at 1 hour.
2. These glass blocks are assumed to be solid based on other test data available for similar but hollow units which show significantly reduced fire endurance.
3. Minimum of $\frac{1}{2}$ inch of 1:3 sanded gypsum plaster required to develop this rating.

TABLE 1.4.2
MISCELLANEOUS MATERIALS WALLS
4" TO LESS THAN 6" THICK

ITEM CODE	THICKNESS	CONSTRUCTION DETAILS	PERFORMANCE		REFERENCE NUMBER			NOTES	REC. HOURS
			LOAD	TIME	PRE-BMS-92	BMS-92	POST-BMS-92		
W-4-Mi-1	4"	Core: 3" magnesium oxysulfate wood-fiber blocks; laid in portland cement mortar; Facings: both sides; see Note 1.	N/A	2 hrs.		1			2

For SI: 1 inch = 25.4 mm.

Notes:
1. One-half inch sanded gypsum plaster. Voids in hollow blocks to be not more than 30 percent.

**FIGURE 1.5.1
FINISH RATINGS—INORGANIC MATERIALS**

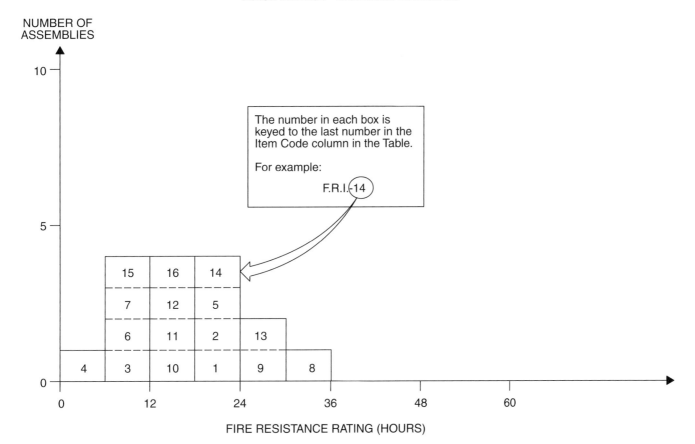

**TABLE 1.5.1
FINISH RATINGS—INORGANIC MATERIALS**

ITEM CODE	THICKNESS	CONSTRUCTION DETAILS	PERFORMANCE FINISH RATING	REFERENCE NUMBER PRE-BMS-92	BMS-92	POST-BMS-92	NOTES	REC. F.R. (MIN.)
F.R.-I-1	$9/16''$	$3/8''$ gypsum wallboard faced with $3/16''$ cement-asbestos board.	20 minutes	1			1, 2	15
F.R.-I-2	$11/16''$	$1/2''$ gypsum sheathing faced with $3/16''$ cement-asbestos board.	20 minutes	1			1, 2	20
F.R.-I-3	$3/16''$	$3/16''$ cement-asbestos board over uninsulated cavity.	10 minutes	1			1, 2	5
F.R.-I-4	$3/16''$	$3/16''$ cement-asbestos board over insulated cavities.	5 minutes	1			1, 2	5
F.R.-I-5	$3/4''$	$3/4''$ thick 1:2; 1:3 gypsum plaster over paper backed metal lath.	20 minutes	1			1, 2, 3	20
F.R.-I-6	$3/4''$	$3/4''$ thick portland cement plaster on metal lath.	10 minutes	1			1, 2	10
F.R.-I-7	$3/4''$	$3/4''$ thick 1:5; 1:7.5 lime plaster on metal lath.	10 minutes	1			1, 2	10
F.R.-I-8	$1''$	$1''$ thick neat gypsum plaster on metal lath.	35 minutes	1			1, 2, 4	35
F.R.-I-9	$3/4''$	$3/4''$ thick neat gypsum plaster on metal lath.	30 minutes	1			1, 2, 4	30

(continued)

TABLE 1.5.1—continued
FINISH RATINGS—INORGANIC MATERIALS

ITEM CODE	THICKNESS	CONSTRUCTION DETAILS	PERFORMANCE FINISH RATING	REFERENCE NUMBER PRE-BMS-92	BMS-92	POST-BMS-92	NOTES	REC. F.R. (MIN.)
F.R.-I-10	$3/4''$	$3/4''$ thick 1:2; 1:2 gypsum plaster on metal lath.	15 minutes	1			1, 2, 3	15
F.R.-I-11	$1/2''$	Same as F.R.-1-7, except $1/2''$ thick on wood lath.	15 minutes	1			1, 2, 3	15
F.R.-I-12	$1/2''$	$1/2''$ thick 1:2; 1:3 gypsum plaster on wood lath.	15 minutes	1			1, 2, 3	15
F.R.-I-13	$7/8''$	$1/2''$ thick 1:2; 1:2 gypsum plaster on $3/8''$ perforated gypsum lath.	30 minutes	1			1, 2, 3	30
F.R.-I-14	$7/8''$	$1/2''$ thick 1:2; 1:2 gypsum plaster on $3/8''$ thick plain or indented gypsum plaster.	20 minutes	1			1, 2, 3	20
F.R.-I-15	$3/8''$	$3/8''$ gypsum wallboard.	10 minutes	1			1, 2	10
F.R.-I-16	$1/2''$	$1/2''$ gypsum wallboard.	15 minutes	1			1, 2	15

For SI: 1 inch = 25.4 mm, °C = [(°F) - 32]/1.8.

Notes:
1. The finish rating is the time required to obtain an average temperature rise of 250°F, or a single point rise of 325°F, at the interface between the material being rated and the substrate being protected.
2. Tested in accordance with the Standard Specifications for Fire Tests of Building Construction and Materials, ASA No. A2-1932.
3. Mix proportions for plasters as follows: first ratio, dry weight of plaster: dry weight of sand for scratch coat; second ratio, plaster: sand for brown coat.
4. Neat plaster means unsanded wood-fiber gypsum plaster.

General Note:
The finish rating of modern building materials can be found in the current literature.

TABLE 1.5.2
FINISH RATINGS—ORGANIC MATERIALS

ITEM CODE	THICKNESS	CONSTRUCTION DETAILS	PERFORMANCE FINISH RATING	REFERENCE NUMBER PRE-BMS-92	BMS-92	POST-BMS-92	NOTES	REC. F.R. (MIN.)
F.R.-O-1	$9/16''$	$7/16''$ wood fiberboard faced with $1/8''$ cement-asbestos board.	15 minutes	1			1, 2	15
F.R.-O-2	$29/32''$	$3/4''$ wood sheathing, asbestos felt weighing 14 lbs./100 ft.^{2} and $5/32''$ cement-asbestos shingles.	20 minutes	1			1, 2	20
F.R.-O-3	$1 1/2''$	1'' thick magnesium oxysulfate wood fiberboard faced with 1:3; 1:3 gypsum plaster, $1/2''$ thick.	20 minutes	1			1, 2, 3	20
F.R.-O-4	$1/2''$	$1/2''$ thick wood fiberboard.	5 minutes	1			1, 2	5
F.R.-O-5	$1/2''$	$1/2''$ thick flameproofed wood fiberboard.	10 minutes	1			1, 2	10
F.R.-O-6	1''	$1/2''$ thick wood fiberboard faced with $1/2''$ thick 1:2; 1:2 gypsum plaster.	15 minutes	1			1, 2, 3	30
F.R.-O-7	$1 3/8''$	$7/8''$ thick flameproofed wood fiberboard faced with $1/2''$ thick 1:2; 1:2 gypsum plaster.	30 minutes	1			1, 2, 3	30
F.R.-O-8	$1 1/4''$	$1 1/4''$ thick plywood.	30 minutes			35		30

For SI: 1 inch = 25.4 mm, 1 pound = 0.004448 kN, 1 pound per square foot = 47.9 N/m^2, °C = [(°F) - 32]/1.8.

Notes:
1. The finish rating is the time required to obtain an average temperature rise of 250°F, or a single point rise of 325°F, at the interface between the material being rated and he substrate being protected.
2. Tested in accordance with the Standard Specifications for Fire Tests of Building Construction and Materials, ASA No. A2-1932.
3. Plaster ratios as follows: first ratio is for scratch coat, weight of dry plaster: weight of dry sand; second ratio is for the brown coat.

General Note:
The finish rating of thinner materials, particularly thinner woods, have not been listed because the possible effects of shrinkage, warpage and aging cannot be predicted.

SECTION II
COLUMNS

TABLE 2.1.1
REINFORCED CONCRETE COLUMNS
MINIMUM DIMENSION 0″ TO LESS THAN 6″

ITEM CODE	MINIMUM DIMENSION	CONSTRUCTION DETAILS	PERFORMANCE		REFERENCE NUMBER			NOTES	REC. HOURS
			LOAD	TIME	PRE-BMS-92	BMS-92	POST-BMS-92		
C-6-RC-1	6″	6″ × 6″ square columns; gravel aggregate concrete (4030 psi); Reinforcement: vertical, four $7/8$″ rebars; horizontal, $5/16$″ ties at 6″ pitch; Cover: 1″.	34.7 tons	62 min.			7	1, 2	1
C-6-RC-2	6″	6″ × 6″ square columns; gravel aggregate concrete (4200 psi); Reinforcement: vertical, four $1/2$″ rebars; horizontal, $5/16$″ ties at 6″ pitch; Cover: 1″.	21 tons	69 min.			7	1, 2	1

Notes:
1. Collapse.
2. British Test

FIGURE 2.1.2
REINFORCED CONCRETE COLUMNS
MINIMUM DIMENSION 10″ TO LESS THAN 12″

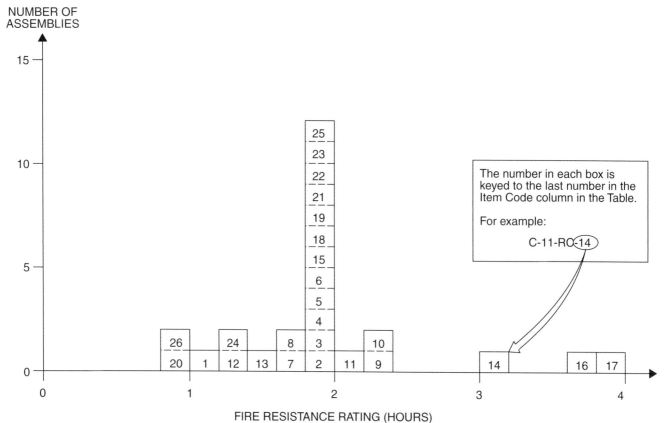

TABLE 2.1.2
REINFORCED CONCRETE COLUMNS
MINIMUM DIMENSION 10″ TO LESS THAN 12″

ITEM CODE	MINIMUM DIMENSION	CONSTRUCTION DETAILS	PERFORMANCE		REFERENCE NUMBER			NOTES	REC. HOURS
			LOAD	TIME	PRE-BMS-92	BMS-92	POST-BMS-92		
C-10-RC-1	10″	10″ square columns; aggregate concrete (4260 psi); Reinforcement: vertical, four $1\frac{1}{4}$″ rebars; horizontal, $\frac{3}{8}$″ ties at 6″ pitch; Cover: $1\frac{1}{4}$″.	92.2 tons	1 hr. 2 min.			7	1	1
C-10-RC-2	10″	10″ square columns; aggregate concrete (2325 psi); Reinforcement: vertical, four $\frac{1}{2}$″ rebars; horizontal, $\frac{5}{16}$″ ties at 6″ pitch; Cover: 1″.	46.7 tons	1 hr. 52 min.			7	1	$1\frac{3}{4}$
C-10-RC-3	10″	10″ square columns; aggregate concrete (5370 psi); Reinforcement: vertical, four $\frac{1}{2}$″ rebars; horizontal, $\frac{5}{16}$″ ties at 6″ pitch; Cover: 1″.	46.5 tons	2 hrs.			7	2, 3, 11	2
C-10-RC-4	10″	10″ square columns; aggregate concrete (5206 psi); Reinforcement: vertical, four $\frac{1}{2}$″ rebars; horizontal, $\frac{5}{16}$″ ties at 6″ pitch; Cover: 1″.	46.5 tons	2 hrs.			7	2, 7	2
C-10-RC-5	10″	10″ square columns; aggregate concrete (5674 psi); Reinforcement: vertical, four $\frac{1}{2}$″ rebars; horizontal, $\frac{5}{16}$″ ties at 6″ pitch; Cover: 1″.	46.7 tons	2 hrs.			7	1	2

(continued)

TABLE 2.1.2—continued
REINFORCED CONCRETE COLUMNS
MINIMUM DIMENSION 10″ TO LESS THAN 12″

ITEM CODE	MINIMUM DIMENSION	CONSTRUCTION DETAILS	PERFORMANCE		REFERENCE NUMBER			NOTES	REC. HOURS
			LOAD	TIME	PRE-BMS-92	BMS-92	POST-BMS-92		
C-10-RC-6	10″	10″ square columns; aggregate concrete (5150 psi); Reinforcement: vertical, four $1\frac{1}{2}$″ rebars; horizontal, $\frac{5}{16}$″ ties at 6″ pitch; Cover: 1″.	66 tons	1 hr. 43 min.			7	1	$1\frac{3}{4}$
C-10-RC-7	10″	10″ square columns; aggregate concrete (5580 psi); Reinforcement: vertical, four $\frac{1}{2}$″ rebars; horizontal, $\frac{5}{16}$″ ties at 6″ pitch; Cover: $1\frac{1}{8}$″.	62.5 tons	1 hr. 38 min.			7	1	$1\frac{1}{2}$
C-10-RC-8	10″	10″ square columns; aggregate concrete (4080 psi); Reinforcement: vertical, four $1\frac{1}{8}$″ rebars; horizontal, $\frac{5}{16}$″ ties at 6″ pitch; Cover: $1\frac{1}{8}$″.	72.8 tons	1 hr. 48 min.			7	1	$1\frac{3}{4}$
C-10-RC-9	10″	10″ square columns; aggregate concrete (2510 psi); Reinforcement: vertical, four $\frac{1}{2}$″ rebars; horizontal, $\frac{5}{16}$″ ties at 6″ pitch; Cover: 1″.	51 tons	2 hrs. 16 min.			7	1	$2\frac{1}{4}$
C-10-RC-10	10″	10″ square columns; aggregate concrete (2170 psi); Reinforcement: vertical, four $\frac{1}{2}$″ rebars; horizontal, $\frac{5}{16}$″ ties at 6″ pitch; Cover: 1″.	45 tons	2 hrs. 14 min.			7	12	$2\frac{1}{4}$
C-10-RC-11	10″	10″ square columns; gravel aggregate concrete (4015 psi); Reinforcement: vertical, four $\frac{1}{2}$″ rebars; horizontal, $\frac{5}{16}$″ ties at 6″ pitch; Cover: $1\frac{1}{8}$″.	46.5 tons	2 hrs. 6 min.			7	1	2
C-11-RC-12	11″	11″ square columns; gravel aggregate concrete (4150 psi); Reinforcement: vertical, four $1\frac{1}{4}$″ rebars; horizontal, $\frac{3}{8}$″ ties at $7\frac{1}{2}$″ pitch; Cover: $1\frac{1}{2}$″.	61 tons	1 hr. 23 min.			7	1	$1\frac{1}{4}$
C-11-RC-13	11″	11″ square columns; gravel aggregate concrete (4380 psi); Reinforcement: vertical, four $1\frac{1}{4}$″ rebars; horizontal, $\frac{3}{8}$″ ties at $7\frac{1}{2}$″ pitch; Cover: $1\frac{1}{2}$″.	61 tons	1 hr. 26 min.			7	1	$1\frac{1}{4}$
C-11-RC-14	11″	11″ square columns; gravel aggregate concrete (4140 psi); Reinforcement: vertical, four $1\frac{1}{4}$″ rebars; horizontal, $\frac{3}{8}$″ ties at $7\frac{1}{2}$″ pitch; steel mesh around reinforcement; Cover: $1\frac{1}{2}$″.	61 tons	3 hrs. 9 min.			7	1	3
C-11-RC-15	11″	11″ square columns; slag aggregate concrete (3690 psi); Reinforcement: vertical, four $1\frac{1}{4}$″ rebars; horizontal, $\frac{3}{8}$″ ties at $7\frac{1}{2}$″ pitch; Cover: $1\frac{1}{2}$″.	91 tons	2 hrs.			7	2, 3, 4, 5	2
C-11-RC-16	11″	11″ square columns; limestone aggregate concrete (5230 psi); Reinforcement: vertical, four $1\frac{1}{4}$″ rebars; horizontal, $\frac{3}{8}$″ ties at $7\frac{1}{2}$″ pitch; Cover: $1\frac{1}{2}$″.	91.5 tons	3 hrs. 41 min.			7	1	$3\frac{1}{2}$
C-11 RC-17	11″	11″ square columns; limestone aggregate concrete (5530 psi); Reinforcement: vertical, four $1\frac{1}{4}$″ rebars; horizontal, $\frac{3}{8}$″ ties at $7\frac{1}{2}$″ pitch; Cover: $1\frac{1}{2}$″.	91.5 tons	3 hrs. 47 min.			7	1	$3\frac{1}{2}$

(continued)

TABLE 2.1.2—continued
REINFORCED CONCRETE COLUMNS
MINIMUM DIMENSION 10″ TO LESS THAN 12″

ITEM CODE	MINIMUM DIMENSION	CONSTRUCTION DETAILS	PERFORMANCE LOAD	PERFORMANCE TIME	PRE-BMS-92	BMS-92	POST-BMS-92	NOTES	REC. HOURS
C-11-RC-18	11″	11″ square columns; limestone aggregate concrete (5280 psi); Reinforcement: vertical, four $1\frac{1}{4}$″ rebars; horizontal, $\frac{3}{8}$″ ties at $7\frac{1}{2}$″ pitch; Cover: $1\frac{1}{2}$″.	91.5 tons	2 hrs.			7	2, 3, 4, 6	2
C-11-RC-19	11″	11″ square columns; limestone aggregate concrete (4180 psi); Reinforcement: vertical, four $\frac{5}{8}$″ rebars; horizontal, $\frac{3}{8}$″ ties at 7″ pitch; Cover: $1\frac{1}{2}$″.	71.4 tons	2 hrs.			7	2, 7	2
C-11-RC-20	11″	11″ square columns; gravel concrete (4530 psi); Reinforcement: vertical, four $\frac{5}{8}$″ rebars; horizontal, $\frac{3}{8}$″ ties at 7″ pitch; Cover: $1\frac{1}{2}$″ with $\frac{1}{2}$″ plaster.	58.8 tons	2 hrs.			7	2, 3, 9	$1\frac{1}{4}$
C-11-RC-21	11″	11″ square columns; gravel concrete (3520 psi); Reinforcement: vertical, four $\frac{5}{8}$″ rebars; horizontal, $\frac{3}{8}$″ ties at 7″ pitch; Cover: $1\frac{1}{2}$″.	Variable	1 hr. 24 min.			7	1, 8	2
C-11-RC-22	11″	11″ square columns; aggregate concrete (3710 psi); Reinforcement: vertical, four $\frac{5}{8}$″ rebars; horizontal, $\frac{3}{8}$″ ties at 7″ pitch; Cover: $1\frac{1}{2}$″.	58.8 tons	2 hrs.			7	2, 3, 10	2
C-11-RC-23	11″	11″ square columns; aggregate concrete (3190 psi); Reinforcement: vertical, four $\frac{5}{8}$″ rebars; horizontal, $\frac{3}{8}$″ ties at 7″ pitch; Cover: $1\frac{1}{2}$″.	58.8 tons	2 hrs.			7	2, 3, 10	2
C-11-RC-24	11″	11″ square columns; aggregate concrete (4860 psi); Reinforcement: vertical, four $\frac{5}{8}$″ rebars; horizontal, $\frac{3}{8}$″ ties at 7″ pitch; Cover: $1\frac{1}{2}$″.	86.1 tons	1 hr. 20 min.			7	1	$1\frac{1}{3}$
C-11-RC-25	11″	11″ square columns; aggregate concrete (4850 psi); Reinforcement: vertical, four $\frac{5}{8}$″ rebars; horizontal, $\frac{3}{8}$″ ties at 7″ pitch; Cover: $1\frac{1}{2}$″.	58.8 tons	1 hr. 59 min.			7	1	$1\frac{3}{4}$
C-11-RC-26	11″	11″ square columns; aggregate concrete (3834 psi); Reinforcement: vertical, four $\frac{5}{8}$″ rebars; horizontal, $\frac{5}{16}$″ ties at $4\frac{1}{2}$″ pitch; Cover: $1\frac{1}{2}$″.	71.4 tons	53 min.			7	1	$\frac{3}{4}$

For SI: 1 inch = 25.4 mm, 1 pound per square inch = 0.00689 MPa, 1 ton = 8.896 kN.

Notes:
1. Failure mode—collapse.
2. Passed 2 hour fire exposure.
3. Passed hose stream test.
4. Reloaded effectively after 48 hours but collapsed at load in excess of original test load.
5. Failing load was 150 tons.
6. Failing load was 112 tons.
7. Failed during hose stream test.
8. Range of load 58.8 tons (initial) to 92 tons (92 minutes) to 60 tons (80 minutes).
9. Collapsed at 44 tons in reload after 96 hours.
10. Withstood reload after 72 hours.
11. Collapsed on reload after 48 hours.

TABLE 2.1.3
REINFORCED CONCRETE COLUMNS
MINIMUM DIMENSION 12″ TO LESS THAN 14″

ITEM CODE	MINIMUM DIMENSION	CONSTRUCTION DETAILS	PERFORMANCE LOAD	PERFORMANCE TIME	PRE-BMS-92	BMS-92	POST-BMS-92	NOTES	REC. HOURS
C-12-RC-1	12″	12″ square columns; gravel aggregate concrete (2647 psi); Reinforcement: vertical, four $5/8$″ rebars; horizontal, $5/16$″ ties at $4\frac{1}{2}$″ pitch; Cover: 2″.	78.2 tons	38 min.		1	7	1	$1/2$
C-12-RC-2	12″	Reinforced columns with $1\frac{1}{2}$″ concrete outside of reinforced steel; Gross diameter or side of column: 12″; Group I, Column A.	—	6 hrs.	1			2, 3	6
C-12-RC-3	12″	Description as per C-12-RC-2; Group I, Column B.	—	4 hrs.	1			2, 3	4
C-12-RC-4	12″	Description as per C-12-RC-2; Group II, Column A.	—	4 hrs.	1			2, 3	4
C-12-RC-5	12″	Description as per C-12-RC-2; Group II, Column B.	—	2 hrs. 30 min.	1			2, 3	$2\frac{1}{2}$
C-12-RC-6	12″	Description as per C-12-RC-2; Group III, Column A.	—	3 hrs.	1			2, 3	3
C-12-RC-7	12″	Description as per C-12-RC-2; Group III, Column B.	—	2 hrs.	1			2, 3	2
C-12-RC-8	12″	Description as per C-12-RC-2; Group IV, Column A.	—	2 hrs.	1			2, 3	2
C-12-RC-9	12″	Description as per C-12-RC-2; Group IV, Column B.	—	1 hr. 30 min.	1			2, 3	$1\frac{1}{2}$

For SI: 1 inch = 25.4 mm, 1 pound per square inch = 0.00689 MPa, 1 pound per square yard = 5.3 N/m^2.

Notes:
1. Failure mode—unspecified structural.
2. Group I: includes concrete having calcareous aggregate containing a combined total of not more than 10 percent of quartz, chert and flint for the coarse aggregate.
   Group II: includes concrete having trap-rock aggregate applied without metal ties and also concrete having cinder, sandstone or granite aggregate, if held in place with wire mesh or expanded metal having not larger than 4-inch mesh, weighing not less than 1.7 lbs./yd.^{2}, placed not more than 1 inch from the surface of the concrete.
   Group III: includes concrete having cinder, sandstone or granite aggregate tied with No. 5 gage steel wire, wound spirally over the column section on a pitch of 8 inches, or equivalent ties, and concrete having siliceous aggregates containing a combined total of 60 percent or more of quartz, chert and flint, if held in place with wire mesh or expanded metal having not larger than 4-inch mesh, weighing not less than 1.7 lbs./yd.^{2}, placed not more than 1 inch from the surface of the concrete.
   Group IV: includes concrete having siliceous aggregates containing a combined total of 60 percent or more of quartz, chert and flint, and tied with No. 5 gage steel wire wound spirally over the column section on a pitch of 8 inches, or equivalent ties.
3. Groupings of aggregates and ties are the same as for structural steel columns protected solidly with concrete, the ties to be placed over the vertical reinforcing bars and the mesh where required, to be placed within 1 inch from the surface of the column.
   Column A: working loads are assumed as carried by the area of the column inside of the lines circumscribing the reinforcing steel.
   Column B: working loads are assumed as carried by the gross area of the column.

**TABLE 2.1.4
REINFORCED CONCRETE COLUMNS
MINIMUM DIMENSION 14″ TO LESS THAN 16″**

ITEM CODE	MINIMUM DIMENSION	CONSTRUCTION DETAILS	PERFORMANCE LOAD	PERFORMANCE TIME	REFERENCE NUMBER PRE-BMS-92	REFERENCE NUMBER BMS-92	REFERENCE NUMBER POST-BMS-92	NOTES	REC. HOURS
C-14-RC-1	14″	14″ square columns; gravel aggregate concrete (4295 psi); Reinforcement: vertical four $^3/_4$″ rebars; horizontal: $^1/_4$″ ties at 9″ pitch; Cover: $1^1/_2$″	86 tons	1 hr. 22 min.			7	1	$1^1/_4$
C-14-RC-2	14″	Reinforced concrete columns with $1^1/_2$″ concrete outside reinforcing steel; Gross diameter or side of column: 12″; Group I, Column A.	—	7 hrs.		1		2, 3	7
C-14-RC-3	14″	Description as per C-14-RC-2; Group II, Column B.	—	5 hrs.		1		2, 3	5
C-14-RC-4	14″	Description as per C-14-RC-2; Group III, Column A.	—	5 hrs.		1		2, 3	5
C-14-RC-5	14″	Description as per C-14-RC-2; Group IV, Column B.	—	3 hrs. 30 min.		1		2, 3	$3^1/_2$
C-14-RC-6	14″	Description as per C-14-RC-2; Group III, Column A.	—	4 hrs.		1		2, 3	4
C-14-RC-7	14″	Description as per C-14-RC-2; Group III, Column B.	—	2 hrs. 30 min.		1		2, 3	$2^1/_2$
C-14-RC-8	14″	Description as per C-14-RC-2; Group IV, Column A.	—	2 hrs. 30 min.		1		2, 3	$2^1/_2$
C-14-RC-9	14″	Description as per C-14-RC-2; Group IV, Column B.	—	1 hr. 30 min.		1		2, 3	$1^1/_2$

For SI: 1 inch = 25.4 mm, 1 pound per square inch = 0.00689 MPa, 1 pound per square yard = 5.3 N/m^2.

Notes:
1. Failure mode—main rebars buckled between links at various points.
2. Group I: includes concrete having calcareous aggregate containing a combined total of not more than 10 percent of quartz, chert and flint for the coarse aggregate.
   Group II: includes concrete having trap-rock aggregate applied without metal ties and also concrete having cinder, sandstone or granite aggregate, if held in place with wire mesh or expanded metal having not larger than 4-inch mesh, weighing not less than 1.7 lbs./yd.^{2}, placed not more than 1 inch from the surface of the concrete.
   Group III: includes concrete having cinder, sandstone or granite aggregate tied with No. 5 gage steel wire, wound spirally over the column section on a pitch of 8 inches, or equivalent ties, and concrete having siliceous aggregates containing a combined total of 60 percent or more of quartz, chert and flint, if held in place with wire mesh or expanded metal having not larger than 4-inch mesh, weighing not less than 1.7 lbs./yd.^{2}, placed not more than 1 inch from the surface of the concrete.
   Group IV: includes concrete having siliceous aggregates containing a combined total of 60 percent or more of quartz, chert and flint, and tied with No. 5 gage steel wire wound spirally over the column section on a pitch of 8 inches, or equivalent ties.
3. Groupings of aggregates and ties are the same as for structural steel columns protected solidly with concrete, the ties to be placed over the vertical reinforcing bars and the mesh where required, to be placed within 1 inch from the surface of the column.
   Column A: working loads are assumed as carried by the area of the column inside of the lines circumscribing the reinforcing steel.
   Column B: working loads are assumed as carried by the gross area of the column.

RESOURCE A

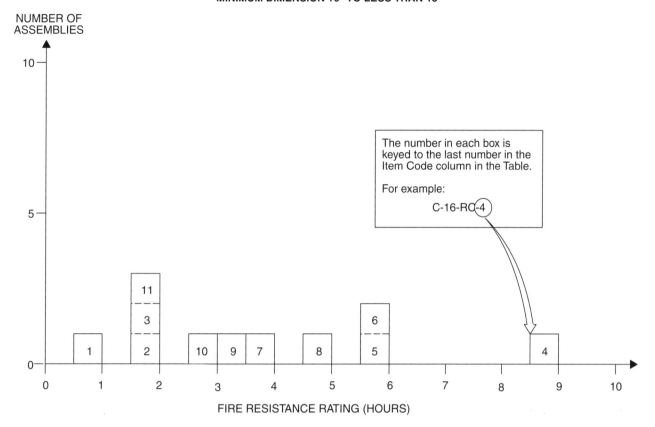

**FIGURE 2.1.5
REINFORCED CONCRETE COLUMNS
MINIMUM DIMENSION 16″ TO LESS THAN 18″**

**TABLE 2.1.5
REINFORCED CONCRETE COLUMNS
MINIMUM DIMENSION 16″ TO LESS THAN 18″**

ITEM CODE	MINIMUM DIMENSION	CONSTRUCTION DETAILS	PERFORMANCE LOAD	PERFORMANCE TIME	REFERENCE NUMBER PRE-BMS-92	REFERENCE NUMBER BMS-92	REFERENCE NUMBER POST-BMS-92	NOTES	REC. HOURS
C-16-RC-1	16″	16″ square columns; gravel aggregate concrete (4550 psi); Reinforcement: vertical, eight $1\frac{3}{8}″$ rebars; horizontal, $\frac{5}{16}″$ ties at 6″ pitch $1\frac{3}{8}″$ below column surface and $\frac{5}{16}″$ ties at 6″ pitch linking center rebars of each face forming a smaller square in column cross section.	237 tons	1 hr			7	1, 2, 3	1
C-16-RC-2	16″	16″ square columns; gravel aggregate concrete (3360 psi); Reinforcement: vertical, eight $1\frac{3}{8}″$ rebars; horizontal, $\frac{5}{16}″$ ties at 6″ pitch; Cover: $1\frac{3}{8}″$.	210 tons	2 hrs.			7	2, 4, 5, 6	2
C-16-RC-3	16″	16″ square columns; gravel aggregate concrete (3980 psi); Reinforcement: vertical, four $\frac{7}{8}″$ rebars; horizontal, $\frac{3}{8}″$ ties at 6″ pitch; Cover: 1″.	123.5 tons	2 hrs.			7	2, 4, 7	2
C-16-RC-4	16″	Reinforced concrete columns with $1\frac{1}{2}″$ concrete outside reinforcing steel; Gross diameter or side of column: 16″; Group I, Column A.	—	9 hrs.		1		8, 9	9
C-16-RC-5	16″	Description as per C-16-RC-4; Group I, Column B.	—	6 hrs.		1		8, 9	6
C-16-RC-6	16″	Description as per C-16-RC-4; Group II, Column A.	—	6 hrs.		1		8, 9	6

(continued)

**TABLE 2.1.5—continued
REINFORCED CONCRETE COLUMNS
MINIMUM DIMENSION 16″ TO LESS THAN 18**

ITEM CODE	MINIMUM DIMENSION	CONSTRUCTION DETAILS	PERFORMANCE		REFERENCE NUMBER			NOTES	REC. HOURS
			LOAD	TIME	PRE-BMS-92	BMS-92	POST-BMS-92		
C-16-RC-7	16″	Description as per C-16-RC-4; Group II, Column B.	—	4 hrs.	1			8, 9	4
C-16-RC-8	16″	Description as per C-16-RC-4; Group III, Column A.	—	5 hrs.	1			8, 9	5
C-16-RC-9	16″	Description as per C-16-RC-4; Group III, Column B.	—	3 hrs. 30 min.	1			8, 9	$3^1/_2$
C-16-RC-10	16″	Description as per C-16-RC-4; Group IV, Column A.	—	3 hrs.	1			8, 9	3
C-16-RC-11	16″	Description as per C-16-RC-4; Group IV, Column B.	—	2 hrs.	1			8, 9	2

For SI: 1 inch = 25.4 mm, 1 pound per square inch = 0.00689 MPa, 1 pound per square yard = 5.3 N/m².

Notes:
1. Column passed 1-hour fire test.
2. Column passed hose stream test.
3. No reload specified.
4. Column passed 2-hour fire test.
5. Column reloaded successfully after 24 hours.
6. Reinforcing details same as C-16-RC-1.
7. Column passed reload after 72 hours.
8. Group I: includes concrete having calcareous aggregate containing a combined total of not more than 10 percent of quartz, chert and flint for the coarse aggregate.
   Group II: includes concrete having trap-rock aggregate applied without metal ties and also concrete having cinder, sandstone or granite aggregate, if held in place with wire mesh or expanded metal having not larger than 4-inch mesh, weighing not less than 1.7 lbs./yd.², placed not more than 1 inch from the surface of the concrete.
   Group III: includes concrete having cinder, sandstone or granite aggregate tied with No. 5 gage steel wire, wound spirally over the column section on a pitch of 8 inches, or equivalent ties, and concrete having siliceous aggregates containing a combined total of 60 percent or more of quartz, chert and flint, if held in place with wire mesh or expanded metal having not larger than 4-inch mesh, weighing not less than 1.7 lbs./yd.², placed not more than 1 inch from the surface of the concrete.
   Group IV: includes concrete having siliceous aggregates containing a combined total of 60 percent or more of quartz, chert and flint, and tied with No. 5 gage steel wire wound spirally over the column section on a pitch of 8 inches, or equivalent ties.
9. Groupings of aggregates and ties are the same as for structural steel columns protected solidly with concrete, the ties to be placed over the vertical reinforcing bars and the mesh where required, to be placed within 1 inch from the surface of the column.
   Column A: working loads are assumed as carried by the area of the column inside of the lines circumscribing the reinforcing steel.
   Column B: working loads are assumed as carried by the gross area of the column.

**TABLE 2.1.6
REINFORCED CONCRETE COLUMNS
MINIMUM DIMENSION 18″ TO LESS THAN 20″**

ITEM CODE	MINIMUM DIMENSION	CONSTRUCTION DETAILS	PERFORMANCE		REFERENCE NUMBER			NOTES	REC. HOURS
			LOAD	TIME	PRE- BMS-92	BMS-92	POST- BMS-92		
C-18-RC-1	18″	Reinforced concrete columns with $1\frac{1}{2}″$ concrete outside reinforced steel; Gross diameter or side of column: 18″ ; Group I, Column A.	—	11 hrs.		1		1, 2	11
C-18-RC-2	18″	Description as per C-18-RC-1; Group I, Column B.	—	8 hrs.		1		1, 2	8
C-18-RC-3	18″	Description as per C-18-RC-1; Group II, Column A.	—	7 hrs.		1		1, 2	7
C-18-RC-4	18″	Description as per C-18-RC-1; Group II, Column B.	—	5 hrs.		1		1, 2	5
C-18-RC-5	18″	Description as per C-18-RC-1; Group III, Column A.	—	6 hrs.		1		1, 2	6
C-18-RC-6	18″	Description as per C-18-RC-1; Group III, Column B.	—	4 hrs.		1		1, 2	4
C-18-RC-7	18″	Description as per C-18-RC-1; Group IV, Column A.	—	3 hrs. 30 min.		1		1, 2	$3\frac{1}{2}$
C-18-RC-8	18″	Description as per C-18-RC-1; Group IV, Column B.	—	2 hrs. 30 min.		1		1, 2	$2\frac{1}{2}$

For SI: 1 inch = 25.4 mm, 1 pound per square yard = 5.3 N/m^2.

Notes:
1. Group I: includes concrete having calcareous aggregate containing a combined total of not more than 10 percent of quartz, chert and flint for the coarse aggregate.
  Group II: includes concrete having trap-rock aggregate applied without metal ties and also concrete having cinder, sandstone or granite aggregate, if held in place with wire mesh or expanded metal having not larger than 4-inch mesh, weighing not less than 1.7 lbs./yd.^{2}, placed not more than 1 inch from the surface of the concrete.
  Group III: includes concrete having cinder, sandstone or granite aggregate tied with No. 5 gage steel wire, wound spirally over the column section on a pitch of 8 inches, or equivalent ties, and concrete having siliceous aggregates containing a combined total of 60 percent or more of quartz, chert and flint, if held in place with wire mesh or expanded metal having not larger than 4-inch mesh, weighing not less than 1.7 lbs./yd.^{2}, placed not more than 1 inch from the surface of the concrete.
  Group IV: includes concrete having siliceous aggregates containing a combined total of 60 percent or more of quartz, chert and flint and, tied with No. 5 gage steel wire wound spirally over the column section on a pitch of 8 inches, or equivalent ties.
2. Groupings of aggregates and ties are the same as for structural steel columns protected solidly with concrete, the ties to be placed over the vertical reinforcing bars and the mesh where required, to be placed within 1 inch from the surface of the column.
  Column A: working loads are assumed as carried by the area of the column inside of the lines circumscribing the reinforcing steel.
  Column B: working loads are assumed as carried by the gross area of the column.

RESOURCE A

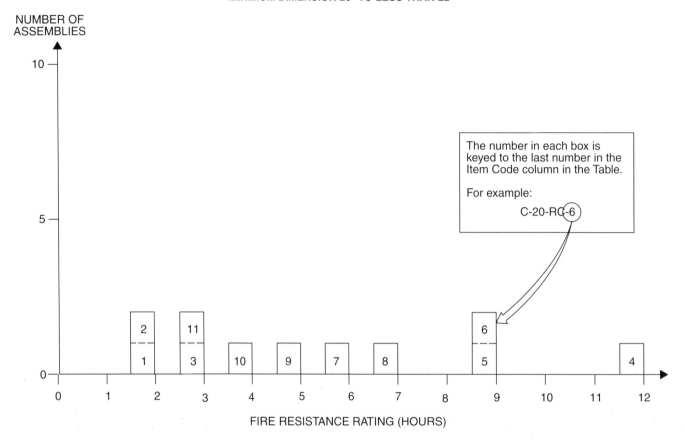

FIGURE 2.1.7
REINFORCED CONCRETE COLUMNS
MINIMUM DIMENSION 20″ TO LESS THAN 22″

TABLE 2.1.7
REINFORCED CONCRETE COLUMNS
MINIMUM DIMENSION 20″ TO LESS THAN 22″

ITEM CODE	MINIMUM DIMENSION	CONSTRUCTION DETAILS	PERFORMANCE		REFERENCE NUMBER			NOTES	REC. HOURS
			LOAD	TIME	PRE-BMS-92	BMS-92	POST-BMS-92		
C-20-RC-1	20″	20″ square columns; gravel aggregate concrete (6690 psi); Reinforcement: vertical, four $1^3/_4$″ rebars; horizontal, $^3/_8$″ wire at 6″ pitch; Cover $1^3/_4$″.	367 tons	2 hrs.			7	1, 2, 3	2
C-20-RC-2	20″	20″ square columns; gravel aggregate concrete (4330 psi); Reinforcement: vertical, four $1^3/_4$″ rebars; horizontal, $^3/_8$″ ties at 6″ pitch; Cover $1^3/_4$″.	327 tons	2 hrs.			7	1, 2, 4	2
C-20-RC-3	$20^1/_4$″	20″ square columns; gravel aggregate concrete (4230 psi); Reinforcement: vertical, four $1^1/_8$″ rebars; horizontal, $^3/_8$″ wire at 5″ pitch; Cover $1^1/_8$″.	199 tons	2 hrs. 56 min.			7	5	$2^3/_4$
C-20-RC-4	20″	Reinforced concrete columns with $1^1/_2$″ concrete outside of reinforcing steel; Gross diameter or side of column: 20″ ; Group I, Column A.	—	12 hrs.	1			6, 7	12
C-20-RC-5	20″	Description as per C-20-RC-4; Group I, Column B.	—	9 hrs.	1			6, 7	9

(continued)

TABLE 2.1.7—continued
REINFORCED CONCRETE COLUMNS
MINIMUM DIMENSION 20″ TO LESS THAN 22″

ITEM CODE	MINIMUM DIMENSION	CONSTRUCTION DETAILS	PERFORMANCE		REFERENCE NUMBER			NOTES	REC. HOURS
			LOAD	TIME	PRE-BMS-92	BMS-92	POST-BMS-92		
C-20-RC-6	20″	Description as per C-20-RC-4; Group II, Column A.	—	9 hrs.		1		6, 7	9
C-20-RC-7	20″	Description as per C-20-RC-4; Group II, Column B.	—	6 hrs		1		6, 7	6
C-20-RC-8	20″	Description as per C-20-RC-4; Group III, Column A.	—	7 hrs.		1		6, 7	7
C-20-RC-9	20″	Description as per C-20-RC-4; Group III, Column B.	—	5 hrs.		1		6, 7	5
C-20-RC-10	20″	Description as per C-20-RC-4; Group IV, Column A.	—	4 hrs.		1		6, 7	4
C-20-RC-11	20″	Description as per C-20-RC-4; Group IV, Column B.	—	3 hrs.		1		6, 7	3

For SI: 1 inch = 25.4 mm, 1 pound per square yard = 5.3 N/m^2, 1 ton = 8.896 kN.

Notes:
1. Passed 2-hour fire test.
2. Passed hose stream test.
3. Failed during reload at 300 tons.
4. Passed reload after 72 hours.
5. Failure mode—collapse.
6. Group I: includes concrete having calcareous aggregate containing a combined total of not more than 10 percent of quartz, chert and flint for the coarse aggregate.
   Group II: includes concrete having trap-rock aggregate applied without metal ties and also concrete having cinder, sandstone or granite aggregate, if held in place with wire mesh or expanded metal having not larger than 4-inch mesh, weighing not less than 1.7 lbs./yd.^{2}, placed not more than 1 inch from the surface of the concrete.
   Group III: includes concrete having cinder, sandstone or granite aggregate tied with No. 5 gage steel wire, wound spirally over the column section on a pitch of 8 inches, or equivalent ties, and concrete having siliceous aggregates containing a combined total of 60 percent or more of quartz, chert and flint, if held in place with wire mesh or expanded metal having not larger than 4-inch mesh, weighing not less than 1.7 lbs./yd.^{2}, placed not more than 1 inch from the surface of the concrete.
   Group IV: includes concrete having siliceous aggregates containing a combined total of 60 percent or more of quartz, chert and flint, and tied with No. 5 gage steel wire wound spirally over the column section on a pitch of 8 inches, or equivalent ties.
7. Groupings of aggregates and ties are the same as for structural steel columns protected solidly with concrete, the ties to be placed over the vertical reinforcing bars and the mesh where required, to be placed within 1 inch from the surface of the column.
   Column A: working loads are assumed as carried by the area of the column inside of the lines circumscribing the reinforcing steel.
   Column B: working loads are assumed as carried by the gross area of the column.

TABLE 2.1.8
HEXAGONAL REINFORCED CONCRETE COLUMNS
MINIMUM DIMENSION 12″ TO LESS THAN 14″

ITEM CODE	MINIMUM DIMENSION	CONSTRUCTION DETAILS	PERFORMANCE		REFERENCE NUMBER			NOTES	REC. HOURS
			LOAD	TIME	PRE-BMS-92	BMS-92	POST-BMS-92		
C-12-HRC-1	12″	12″ hexagonal columns; gravel aggregate concrete (4420 psi); Reinforcement: vertical, eight $1/2$″ rebars; horizontal, $5/16$″ helical winding at $1 1/2$″ pitch; Cover: $1/2$″.	88 tons	58 min.			7	1	$3/4$
C-12-HRC-2	12″	12″ hexagonal columns; gravel aggregate concrete (3460 psi); Reinforcement: vertical, eight $1/2$″ rebars; horizontal, $5/16$″ helical winding at $1 1/2$″ pitch; Cover: $1/2$″.	78.7 tons	1 hr.			7	2	1

For SI: 1 inch = 25.4 mm, 1 pound per square inch = 0.00689 MPa, 1 ton = 8.896 kN.

Notes:
1. Failure mode—collapse.
2. Test stopped at 1 hour.

TABLE 2.1.9
HEXAGONAL REINFORCED CONCRETE COLUMNS
MINIMUM DIMENSION 14″ TO LESS THAN 16″

ITEM CODE	MINIMUM DIMENSION	CONSTRUCTION DETAILS	PERFORMANCE		REFERENCE NUMBER			NOTES	REC. HOURS
			LOAD	TIME	PRE-BMS-92	BMS-92	POST-BMS-92		
C-14-HRC-1	14″	14″ hexagonal columns; gravel aggregate concrete (4970 psi); Reinforcement: vertical, eight $^1/_2$″ rebars; horizontal, $^5/_{16}$″ helical winding on 2″ pitch; Cover: $^1/_2$″.	90 tons	2 hrs.			7	1, 2, 3	2

For SI: 1 inch = 25.4 mm, 1 pound per square inch = 0.00689 MPa, 1 ton = 8.896 kN.

Notes:
1. Withstood 2-hour fire test.
2. Withstood hose stream test.
3. Withstood reload after 48 hours.

TABLE 2.1.10
HEXAGONAL REINFORCED CONCRETE COLUMNS
DIAMETER—16″ TO LESS THAN 18″

ITEM CODE	MINIMUM DIMENSION	CONSTRUCTION DETAILS	PERFORMANCE		REFERENCE NUMBER			NOTES	REC. HOURS
			LOAD	TIME	PRE-BMS-92	BMS-92	POST-BMS-92		
C-16-HRC-1	16″	16″ hexagonal columns; gravel concrete (6320 psi); Reinforcement: vertical, eight $^5/_8$″ rebars; horizontal, $^5/_{16}$″ helical winding on $^3/_4$″ pitch; Cover: $^1/_2$″.	140 tons	1 hr. 55 min.			7	1	$1^3/_4$
C-16-HRC-2	16″	16″ hexagonal columns; gravel aggregate concrete (5580 psi); Reinforcement: vertical, eight $^5/_8$″ rebars; horizontal, $^5/_{16}$″ helical winding on $1^3/_4$″ pitch; Cover: $^1/_2$″	124 tons	2 hrs.			7	2	2

For SI: 1 inch = 25.4 mm, 1 pound per square inch = 0.00689 MPa, 1 ton = 8.896 kN.

Notes:
1. Failure mode—collapse.
2. Failed on furnace removal.

TABLE 2.1.11
HEXAGONAL REINFORCED CONCRETE COLUMNS
DIAMETER—20″ TO LESS THAN 22″

ITEM CODE	MINIMUM DIMENSION	CONSTRUCTION DETAILS	PERFORMANCE		REFERENCE NUMBER			NOTES	REC. HOURS
			LOAD	TIME	PRE-BMS-92	BMS-92	POST-BMS-92		
C-20-HRC-1	20″	20″ hexagonal columns; gravel concrete (6080 psi); Reinforcement: vertical, $^3/_4$″ rebars; horizontal, $^5/_6$″ helical winding on $1^3/_4$″ pitch; Cover: $^1/_2$″.	211 tons	2 hrs.			7	1	2
C-20-HRC-2	20″	20″ hexagonal columns; gravel concrete (5080 psi); Reinforcement: vertical, $^3/_4$″ rebars; horizontal, $^5/_{16}$″ wire on $1^3/_4$″ pitch; Cover: $^1/_2$″.	184 tons	2 hrs. 15 min.			7	2, 3, 4	$2^1/_4$

For SI: 1 inch = 25.4 mm, 1 pound per square inch = 0.00689 MPa, 1 ton = 8.896 kN.

Notes:
1. Column collapsed on furnace removal.
2. Passed $2^1/_4$-hour fire test.
3. Passed hose stream test.
4. Withstood reload after 48 hours.

**TABLE 2.2
ROUND CAST IRON COLUMNS**

ITEM CODE	MINIMUM DIMENSION	CONSTRUCTION DETAILS	PERFORMANCE		REFERENCE NUMBER			NOTES	REC. HOURS
			LOAD	TIME	PRE-BMS-92	BMS-92	POST-BMS-92		
C-7-CI-1	7″ O.D.	Column: 0.6″ minimum metal thickness; unprotected.	—	30 min.	1				$^1/_2$
C-7-CI-2	7″ O.D.	Column: 0.6″ minimum metal thickness concrete filled, outside unprotected.	—	45 min.	1				$^3/_4$
C-11-CI-3	11″ O.D.	Column: 0.6″ minimum metal thickness; Protection: $1^1/_2$″ portland cement plaster on high ribbed metal lath, $^1/_2$″ broken air space.	—	3 hrs.	1				3
C-11-CI-4	11″ O.D.	Column: 0.6″ minimum metal thickness; Protection: 2″ concrete other than siliceous aggregate.	—	2 hrs. 30 min.	1				$2^1/_2$
C-12-CI-5	12.5″ O.D.	Column: 7″ O.D. 0.6″ minimum metal thickness; Protection: 2″ porous hollow tile, $^3/_4$″ mortar between tile and column, outside wire ties.	—	3 hrs.	1				3
C-7-CI-6	7.6″ O.D.	Column: 7″ I.D., $^3/_{10}$″ minimum metal thickness, concrete filled unprotected.	—	30 min.	1				$^1/_2$
C-8-CI-7	8.6″ O.D.	Column: 8″ I.D., $^3/_{10}$″ minimum metal thickness; concrete filled reinforced with four $3^1/_2$″ × $^3/_8$″ angles, in fill; unprotected outside.	—	1 hr.	1				1

For SI: 1 inch = 25.4 mm.

FIGURE 2.3
STEEL COLUMNS-GYPSUM ENCASEMENTS

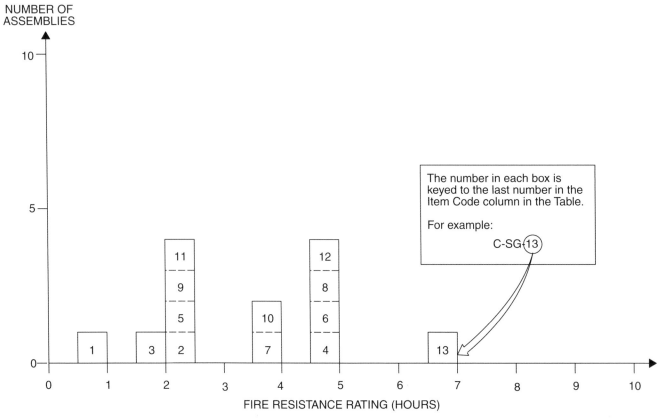

TABLE 2.3
STEEL COLUMNS—GYPSUM ENCASEMENTS

ITEM CODE	MINIMUM AREA OF SOLID MATERIAL	CONSTRUCTION DETAILS	PERFORMANCE		REFERENCE NUMBER			NOTES	REC. HOURS
			LOAD	TIME	PRE-BMS-92	BMS-92	POST-BMS-92		
C-SG-1	—	Steel protected with $3/4''$ 1:3 sanded gypsum or 1″ $1:2^1/_2$ portland cement plaster on wire or lath; one layer.	—	1 hr.		1			1
C-SG-2	—	Same as C-SG-1; two layers.	—	2 hrs. 30 min.		1			$2^1/_2$
C-SG-3	130 in.^{2}	2″ solid blocks with wire mesh in horizontal joints; 1″ mortar on flange; reentrant space filled with block and mortar.	—	2 hrs.		1			2
C-SG-4	150 in.^{2}	Same as C-130-SG-3 with $1/_2''$ sanded gypsum plaster.	—	5 hrs.		1			5
C-SG-5	130 in.^{2}	2″ solid blocks with wire mesh in horizontal joints; 1″ mortar on flange; reentrant space filled with gypsum concrete.	—	2 hrs. 30 min.		1			$2^1/_2$
C-SG-6	150 in.^{2}	Same as C-130-SG-5 with $1/_2''$ sanded gypsum plaster.	—	5 hrs.		1			5
C-SG-7	300 in.^{2}	4″ solid blocks with wire mesh in horizontal joints; 1″ mortar on flange; reentrant space filled with block and mortar.	—	4 hrs.		1			4

(continued)

TABLE 2.3—continued
STEEL COLUMNS—GYPSUM ENCASEMENTS

ITEM CODE	MINIMUM AREA OF SOLID MATERIAL	CONSTRUCTION DETAILS	PERFORMANCE		REFERENCE NUMBER			NOTES	REC. HOURS
			LOAD	TIME	PRE-BMS-92	BMS-92	POST-BMS-92		
C-SG-8	300 in.²	Same as C-300-SG-7 with reentrant space filled with gypsum concrete.	—	5 hrs.		1			5
C-SG-9	85 in.²	2″ solid blocks with cramps at horizontal joints; mortar on flange only at horizontal joints; reentrant space not filled.	—	2 hrs. 30 min.		1			2½
C-SG-10	105 in.²	Same as C-85-SG-9 with ½″ sanded gypsum plaster.	—	4 hrs.		1			4
C-SG-11	95 in.²	3″ hollow blocks with cramps at horizontal joints; mortar on flange only at horizontal joints; reentrant space not filled.	—	2 hrs. 30 min.		1			2½
C-SG-12	120 in.²	Same as C-95-SG-11 with ½″ sanded gypsum plaster.	—	5 hrs.		1			5
C-SG-13	130 in.²	2″ neat fibered gypsum reentrant space filled poured solid and reinforced with 4″ × 4″ wire mesh ½″ sanded gypsum plaster.	—	7 hrs.		1			7

For SI: 1 inch = 25.4 mm, 1 square inch = 645 mm².

TABLE 2.4
TIMBER COLUMNS MINIMUM DIMENSION

ITEM CODE	MINIMUM DIMENSION	CONSTRUCTION DETAILS	PERFORMANCE		REFERENCE NUMBER			NOTES	REC. HOURS
			LOAD	TIME	PRE-BMS-92	BMS-92	POST-BMS-92		
C-11-TC-1	11″	With unprotected steel plate cap.	—	30 min.		1		1, 2	½
C-11-TC-2	11″	With unprotected cast iron cap and pintle.	—	45 min.		1		1, 2	¾
C-11-TC-3	11″	With concrete or protected steel or cast iron cap.	—	1 hr. 15 min.		1		1, 2	1¼
C-11-TC-4	11″	With ⅜″ gypsum wallboard over column and over cast iron or steel cap.	—	1 hr. 15 min.		1		1, 2	1¼
C-11-TC-5	11″	With 1″ portland cement plaster on wire lath over column and over cast iron or steel cap; ¾″ air space.	—	2 hrs.		1		1, 2	2

For SI: 1 inch = 25.4 mm, 1 square inch = 645 mm².

Notes:
1. Minimum area: 120 square inches.
2. Type of wood: long leaf pine or Douglas fir.

TABLE 2.5.1.1
STEEL COLUMNS—CONCRETE ENCASEMENTS
MINIMUM DIMENSION LESS THAN 6″

ITEM CODE	MINIMUM DIMENSION	CONSTRUCTION DETAILS	PERFORMANCE		REFERENCE NUMBER			NOTES	REC. HOURS
			LOAD	TIME	PRE-BMS-92	BMS-92	POST-BMS-92		
C-5-SC-1	5″	5″ × 6″ outer dimensions; 4″ × 3″ × 10 lbs. "H" beam; Protection: gravel concrete (4900 psi) 6″ × 4″ - 13 SWG mesh.	12 tons	1 hr. 29 min.			7	1	1¼

For SI: 1 inch = 25.4 mm, 1 pound per square inch = 0.00689 MPa, 1 ton = 8.896 kN.

Notes:
1. Failure mode—collapse.

TABLE 2.5.1.2
STEEL COLUMNS—CONCRETE ENCASEMENTS
6″ TO LESS THAN 8″ THICK

ITEM CODE	MINIMUM DIMENSION	CONSTRUCTION DETAILS	PERFORMANCE		REFERENCE NUMBER			NOTES	REC. HOURS
			LOAD	TIME	PRE-BMS-92	BMS-92	POST-BMS-92		
C-7-SC-1	7″	7″ × 8″ column; 4″ × 3″ × 10 lbs. "H" beam; Protection: brick filled concrete (6220 psi); 6″ × 4″ mesh - 13 SWG; 1″ below column surface.	12 tons	2 hrs. 46 min.			7	1	$2^3/_4$
C-7-SC-2	7″	7″ × 8″ column; 4″ × 3″ × 10 lbs. "H" beam; Protection: gravel concrete (5140 psi); 6″ × 4″ 13 SWG mesh 1″ below surface.	12 tons	3 hrs. 1 min.			7	1	3
C-7-SC-3	7″	7″ × 8″ column; 4″ × 3″ × 10 lbs. "H" beam; Protection: concrete (4540 psi); 6″ × 4″ - 13 SWG mesh; 1″ below column surface.	12 tons	3 hrs. 9 min.			7	1	3
C-7-SC-4	7″	7″ × 8″ column; 4″ × 3″ × 10 lbs. "H" beam; Protection: gravel concrete (5520 psi); 4″ × 4″ mesh; 16 SWG.	12 tons	2 hrs. 50 min.			7	1	$2^3/_4$

For SI: 1 inch = 25.4 mm, 1 pound per square inch = 0.00689 MPa, 1 ton = 8.896 kN.

Notes:
1. Failure mode—collapse.

RESOURCE A

FIGURE 2.5.1.3
STEEL COLUMNS—CONCRETE ENCASEMENTS
MINIMUM DIMENSION 8″ TO LESS THAN 10″

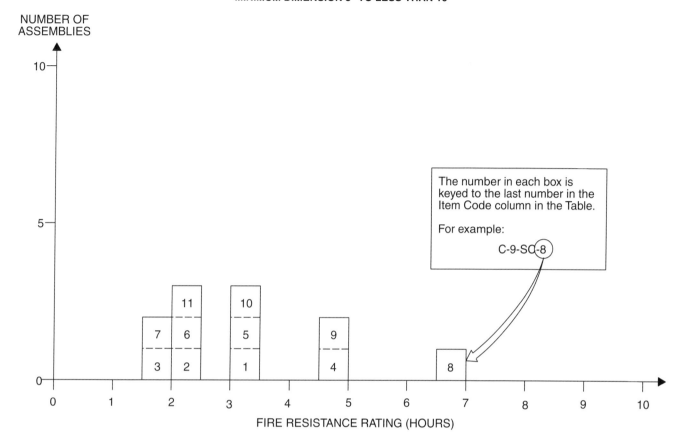

TABLE 2.5.1.3
STEEL COLUMNS—CONCRETE ENCASEMENTS
MINIMUM DIMENSION 8″ TO LESS THAN 10″

ITEM CODE	MINIMUM DIMENSION	CONSTRUCTION DETAILS	PERFORMANCE		REFERENCE NUMBER			NOTES	REC. HOURS
			LOAD	TIME	PRE-BMS-92	BMS-92	POST-BMS-92		
C-8-SC-1	8$\frac{1}{2}$″	8$\frac{1}{2}$″ × 10″ column; 6″ × 4$\frac{1}{2}$″ × 20 lbs. "H" beam; Protection: gravel concrete (5140 psi); 6″ × 4″ - 13 SWG mesh.	39 tons	3 hrs. 8 min.			7	1	3
C-8-SC-2	8″	8″ × 10″ column; 8″ × 6″ × 35 lbs. "I" beam; Protection: gravel concrete (4240 psi); 6″ × 4″ - 13 SWG mesh; $\frac{1}{2}$″ cover.	90 tons	2 hrs. 1 min.			7	1	2
C-8-SC-3	8″	8″ × 10″ concrete encased column; 8″ × 6″ × 35 lbs. "H" beam; protection: aggregate concrete (3750 psi); 4″ mesh - 16 SWG reinforcing $\frac{1}{2}$″ below column surface.	90 tons	1 hr. 58 min.			7	1	1$\frac{3}{4}$
C-8-SC-4	8″	6″ × 6″ steel column; 2″ outside protection; Group I.	—	5 hrs.	1			2	5
C-8-SC-5	8″	6″ × 6″ steel column; 2″ outside protection; Group II	—	3 hrs. 30 min.	1			2	3$\frac{1}{2}$
C-8-SC-6	8″	6″ × 6″ steel column; 2″ outside protection; Group III.	—	2 hrs. 30 min.	1			2	2$\frac{1}{2}$

(continued)

**TABLE 2.5.1.3—continued
STEEL COLUMNS—CONCRETE ENCASEMENTS
MINIMUM DIMENSION 8″ TO LESS THAN 10″**

ITEM CODE	MINIMUM DIMENSION	CONSTRUCTION DETAILS	PERFORMANCE LOAD	PERFORMANCE TIME	REFERENCE NUMBER PRE-BMS-92	REFERENCE NUMBER BMS-92	REFERENCE NUMBER POST-BMS-92	NOTES	REC. HOURS
C-8-SC-7	8″	6″ × 6″ steel column; 2″ outside protection; Group IV.	—	1 hr. 45 min.	1			2	$1^3/_4$
C-9-SC-8	9″	6″ × 6″ steel column; 3″ outside protection; Group I.	—	7 hrs.	1			2	7
C-9-SC-9	9″	6″ × 6″ steel column; 3″ outside protection; Group II.	—	5 hrs.	1			2	5
C-9-SC-10	9″	6″ × 6″ steel column; 3″ outside protection; Group III.	—	3 hrs. 30 min.	1			2	$3^1/_2$
C-9-SC-11	9″	6″ × 6″ steel column; 3″ outside protection; Group IV.	—	2 hrs. 30 min.	1			2	$2^1/_2$

For SI: 1 inch = 25.4 mm, 1 pound = 0.004448 kN, 1 pound per square inch = 0.00689 MPa, 1 pound per square yard = 5.3 N/m^2, 1 ton = 8.896 kN.

Notes:
1. Failure mode—collapse.
2. Group I: includes concrete having calcareous aggregate containing a combined total of not more than 10 percent of quartz, chert and flint for the coarse aggregate.
   Group II: includes concrete having trap-rock aggregate applied without metal ties and also concrete having cinder, sandstone or granite aggregate, if held in place with wire mesh or expanded metal having not larger than 4-inch mesh, weighing not less than 1.7 lbs./yd.^{2}, placed not more than 1 inch from the surface of the concrete.
   Group III: includes concrete having cinder, sandstone or granite aggregate tied with No. 5 gage steel wire, wound spirally over the column section on a pitch of 8 inches, or equivalent ties, and concrete having siliceous aggregates containing a combined total of 60 percent or more of quartz, chert and flint, if held in place with wire mesh or expanded metal having not larger than 4-inch mesh, weighing not less than 1.7 lbs./yd.^{2}, placed not more than 1 inch from the surface of the concrete.
   Group IV: includes concrete having siliceous aggregates containing a combined total of 60 percent or more of quartz, chert and flint, and tied with No. 5 gage steel wire wound spirally over the column section on a pitch of 8 inches, or equivalent ties.

**FIGURE 2.5.1.4
STEEL COLUMNS—CONCRETE ENCASEMENTS
MINIMUM DIMENSION 10″ TO LESS THAN 12″**

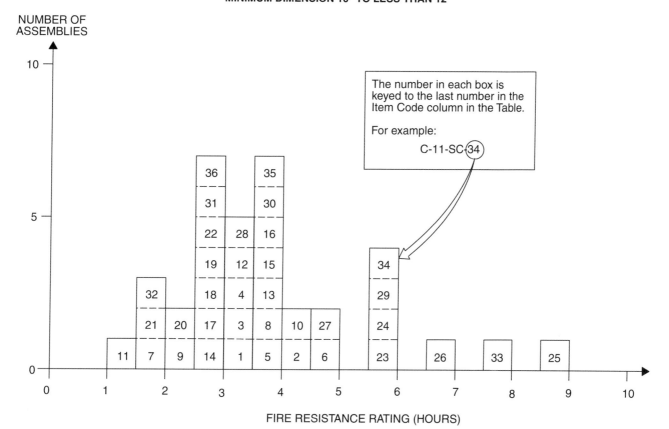

**TABLE 2.5.1.4
STEEL COLUMNS—CONCRETE ENCASEMENTS
MINIMUM DIMENSION 10″ TO LESS THAN 12″**

ITEM CODE	MINIMUM DIMENSION	CONSTRUCTION DETAILS	PERFORMANCE		REFERENCE NUMBER			NOTES	REC. HOURS
			LOAD	TIME	PRE-BMS-92	BMS-92	POST-BMS-92		
C-10-SC-1	10″	10″ × 12″ concrete encased steel column; 8″ × 6″ × 35 lbs. "H" beam; Protection: gravel aggregate concrete (3640 psi); Mesh 6″ × 4″ 13 SWG, 1″ below column surface.	90 tons	3 hrs. 7 min.			7	1,2	3
C-10-SC-2	10″	10″ × 16″ column; 8″ × 6″ × 35 lbs. "H" beam; Protection: clay brick concrete (3630 psi); 6″ × 4″ mesh; 13 SWG, 1″ below column surface.	90 tons	4 hrs. 6 min.			7	2	4
C-10-SC-3	10″	10″ × 12″ column; 8″ × 6″ × 35 lbs. "H" beam; Protection: crushed stone and sand concrete (3930 psi); 6″ × 4″ - 13 SWG mesh; 1″ below column surface.	90 tons	3 hrs. 17 min.			7	2	$3\frac{1}{4}$
C-10-SC-4	10″	10″ × 12″ column; 8″ × 6″ × 35 lbs. "H" beam; Protection: crushed basalt and sand concrete (4350 psi); 6″ × 4″ - 13 SWG mesh; 1″ below column surface.	90 tons	3 hrs. 22 min.			7	2	$3\frac{1}{3}$
C-10-SC-5	10″	10″ × 12″ column; 8″ × 6″ × 35 lbs. "H" beam; Protection: gravel aggregate concrete (5570 psi); 6″ × 4″ mesh; 13 SWG.	90 tons	3 hrs. 39 min.			7	2	$3\frac{1}{2}$
C-10-SC-6	10″	10″ × 16″ column; 8″ × 6″ × 35 lbs. "I" beam; Protection: gravel concrete (4950 psi); mesh; 6″ × 4″ 13 SWG 1″ below column surface.	90 tons	4 hrs. 32 min.			7	2	$4\frac{1}{2}$

(continued)

**TABLE 2.5.1.4—continued
STEEL COLUMNS—CONCRETE ENCASEMENTS
MINIMUM DIMENSION 10″ TO LESS THAN 12″**

ITEM CODE	MINIMUM DIMENSION	CONSTRUCTION DETAILS	PERFORMANCE		REFERENCE NUMBER			NOTES	REC. HOURS
			LOAD	TIME	PRE-BMS-92	BMS-92	POST-BMS-92		
C-10-SC-7	10″	10″ × 12″ concrete encased steel column; 8″ × 6″ × 35 lbs. "H" beam; Protection: aggregate concrete (1370 psi); 6″ × 4″ mesh; 13 SWG reinforcing 1″ below column surface.	90 tons	2 hrs.			7	3, 4	2
C-10-SC-8	10″	10″ × 12″ concrete encased steel column; 8″ × 6″ × 35 lbs. "H" column; Protection: aggregate concrete (4000 psi); 13 SWG iron wire loosely around column at 6″ pitch about 2″ beneath column surface.	86 tons	3 hrs. 36 min.			7	2	$3\frac{1}{2}$
C-10-SC-9	10″	10″ × 12″ concrete encased steel column; 8″ × 6″ × 35 lbs. "H" beam; Protection: aggregate concrete (3290 psi); 2″ cover minimum.	86 tons	2 hrs. 8 min.			7	2	2
C-10-SC-10	10″	10″ × 14″ concrete encased steel column; 8″ × 6″ × 35 lbs. "H" column; Protection: crushed brick filled concrete (5310 psi); 6″ × 4″ mesh; 13 SWG reinforcement 1″ below column surface.	90 tons	4 hrs. 28 min.			7	2	$4\frac{1}{3}$
C-10-SC-11	10″	10″ × 14″ concrete encased column; 8″ × 6″ × 35 lbs. "H" beam; Protection: aggregate concrete (342 psi); 6″ × 4″ mesh; 13 SWG reinforcement 1″ below surface.	90 tons	1 hr. 2 min.			7	2	1
C-10-SC-12	10″	10″ × 12″ concrete encased steel column; 8″ × 6″ × 35 lbs. "H" beam; Protection: aggregate concrete (4480 psi); four $\frac{3}{8}$″ vertical bars at "H" beam edges with $\frac{3}{16}$″ spacers at beam surface at 3′ pitch and $\frac{3}{16}$″ binders at 10″ pitch; 2″ concrete cover.	90 tons	3 hrs. 2 min.			7	2	3
C-10-SC-13	10″	10″ × 12″ concrete encased steel column; 8″ × 6″ × 35 lbs. "H" beam; Protection: aggregate concrete (5070 psi); 6″ × 4″ mesh; 13 SWG reinforcing at 6″ beam sides wrapped and held by wire ties across (open) 8″ beam face; reinforcements wrapped in 6″ × 4″ mesh; 13 SWG throughout; $\frac{1}{2}$″ cover to column surface.	90 tons	3 hrs. 59 min.			7	2	$3\frac{3}{4}$
C-10-SC-14	10″	10″ × 12″ concrete encased steel column; 8″ × 6″ × 35 lbs. "H" beam; Protection: aggregate concrete (4410 psi); 6″ × 4″ mesh; 13 SWG reinforcement $1\frac{1}{4}$″ below column surface; $\frac{1}{2}$″ limestone cement plaster with $\frac{3}{8}$″ gypsum plaster finish.	90 tons	2 hrs. 50 min.			7	2	$2\frac{3}{4}$
C-10-SC-15	10″	10″ × 12″ concrete encased steel column; 8″ × 6″ × 35 lbs. "H" beam; Protection: crushed clay brick filled concrete (4260 psi); 6″ × 4″ mesh; 13 SWG reinforcing 1″ below column surface.	90 tons	3 hrs. 54 min.			7	2	$3\frac{3}{4}$
C-10-SC-16	10″	10″ × 12″ concrete encased steel column; 8″ × 6″ × 35 lbs. "H" beam; Protection: limestone aggregate concrete (4350 psi); 6″ × 4″ mesh; 13 SWG reinforcing 1″ below column surface.	90 tons	3 hrs. 54 min.			7	2	$3\frac{3}{4}$

(continued)

RESOURCE A

**TABLE 2.5.1.4—continued
STEEL COLUMNS—CONCRETE ENCASEMENTS
MINIMUM DIMENSION 10″ TO LESS THAN 12″**

ITEM CODE	MINIMUM DIMENSION	CONSTRUCTION DETAILS	PERFORMANCE LOAD	PERFORMANCE TIME	PRE-BMS-92	BMS-92	POST-BMS-92	NOTES	REC. HOURS
C-10-SC-17	10″	10″ × 12″ concrete encased steel column; 8″ × 6″ × 35 lbs. "H" beam; Protection: limestone aggregate concrete (5300 psi); 6″ × 4″; 13 SWG wire mesh 1″ below column surface.	90 tons	3 hrs.			7	4, 5	3
C-10-SC-18	10″	10″ × 12″ concrete encased steel column; 8″ × 6″ × 35 lbs. "H" beam; Protection: limestone aggregate concrete (4800 psi) with 6″ × 4″; 13 SWG mesh reinforcement 1″ below surface.	90 tons	3 hrs.			7	4, 5	3
C-10-SC-19	10″	10″ × 14″ concrete encased steel column; 12″ × 8″ × 65 lbs. "H" beam; Protection: aggregate concrete (3900 psi); 4″ mesh; 16 SWG reinforcing $^1/_2$″ below column surface.	118 tons	2 hrs. 42 min.			7	2	2
C-10-SC-20	10″	10″ × 14″ concrete encased steel column; 12″ × 8″ × 65 lbs. "H" beam; Protection: aggregate concrete (4930 psi); 4″ mesh; 16 SWG reinforcing $^1/_2$″ below column surface.	177 tons	2 hrs. 8 min.			7	2	2
C-10-SC-21	$10^3/_8$″	$10^3/_8$″ × $12^3/_8$″ concrete encased steel column; 8″ × 6″ × 35 lbs. "H" beam; Protection: aggregate concrete (835 psi) with 6″ × 4″ mesh; 13 SWG reinforcing $1^3/_{16}$″ below column surface; $^3/_{16}$″ gypsum plaster finish.	90 tons	2 hrs.			7	3, 4	2
C-11-SC-22	11″	11″ × 13″ concrete encased steel column; 8″ × 6″ × 35 lbs. "H" beam; Protection: "open texture" brick filled concrete (890 psi) with 6″ × 4″ mesh; 13 SWG reinforcing $1^1/_2$″ below column surface; $^3/_8$″ lime cement plaster; $^1/_8$″ gypsum plaster finish.	90 tons	3 hrs.			7	6, 7	3
C-11-SC-23	11″	11″ × 12″ column; 4″ × 3″ × 10 lbs. "H" beam; gravel concrete (4550 psi); 6″ × 4″ - 13 SWG mesh reinforcing; 1″ below column surface.	12 tons	6 hrs.			7	7, 8	6
C-11-SC-24	11″	11″ × 12″ column; 4″ × 3″ × 10 lbs. "H" beam; Protection: gravel aggregate concrete (3830 psi); with 4″ × 4″ mesh; 16 SWG, 1″ below column surface.	16 tons	5 hrs. 32 min.			7	2	$5^1/_2$
C-10-SC-25	10″	6″ × 6″ steel column with 4″ outside protection; Group I.	—	9 hrs.		1		9	9
C-10-SC-26	10″	Description as per C-SC-25; Group II.	—	7 hrs.		1		9	7
C-10-SC-27	10″	Description as per C-10-SC-25; Group III.	—	5 hrs.		1		9	5
C-10-SC-28	10″	Description as per C-10-SC-25; Group IV.	—	3 hrs. 30 min.		1		9	$3^1/_2$
C-10-SC-29	10″	8″ × 8″ steel column with 2″ outside protection; Group I.	—	6 hrs.		1		9	6
C-10-SC-30	10″	Description as per C-10-SC-29; Group II.	—	4 hrs.		1		9	4
C-10-SC-31	10″	Description as per C-10-SC-29; Group III.	—	3 hrs.		1		9	3
C-10-SC-32	10″	Description as per C-10-SC-29; Group IV.	—	2 hrs.		1		9	2

(continued)

**TABLE 2.5.1.4—continued
STEEL COLUMNS—CONCRETE ENCASEMENTS
MINIMUM DIMENSION 10″ TO LESS THAN 12″**

ITEM CODE	MINIMUM DIMENSION	CONSTRUCTION DETAILS	PERFORMANCE		REFERENCE NUMBER			NOTES	REC. HOURS
			LOAD	TIME	PRE-BMS-92	BMS-92	POST-BMS-92		
C-11-SC-33	11″	8″ × 8″ steel column with 3″ outside protection; Group I.	—	8 hrs.	1			9	8
C-11-SC-34	11″	Description as per C-10-SC-33; Group II.	—	6 hrs.	1			9	6
C-11-SC-35	11″	Description as per C-10-SC-33; Group III.	—	4 hrs.	1			9	4
C-11-SC-36	11″	Description as per C-10-SC-33; Group IV.	—	3 hrs.	1			9	3

For SI: 1 inch = 25.4 mm, 1 pound = 0.004448 kN, 1 pound per square inch = 0.00689 MPa, 1 pound per square yard = 5.3 N/m^2, 1 ton = 8.896 kN.

Notes:
1. Tested under total restraint load to prevent expansion—minimum load 90 tons.
2. Failure mode—collapse.
3. Passed 2-hour fire test (Grade "C," British).
4. Passed hose stream test.
5. Column tested and passed 3-hour grade fire resistance (British).
6. Column passed 3-hour fire test.
7. Column collapsed during hose stream testing.
8. Column passed 6-hour fire test.
9. Group I: includes concrete having calcareous aggregate containing a combined total of not more than 10 percent of quartz, chert and flint for the coarse aggregate.
    Group II: includes concrete having trap-rock aggregate applied without metal ties and also concrete having cinder, sandstone or granite aggregate, if held in place with wire mesh or expanded metal having not larger than 4-inch mesh, weighing not less than 1.7 lbs./yd.^{2}, placed not more than 1 inch from the surface of the concrete.
    Group III: includes concrete having cinder, sandstone or granite aggregate tied with No. 5 gage steel wire, wound spirally over the column section on a pitch of 8 inches, or equivalent ties, and concrete having siliceous aggregates containing a combined total of 60 percent or more of quartz, chert and flint, if held in place with wire mesh or expanded metal having not larger than 4-inch mesh, weighing not less than 1.7 lbs./yd.^{2}, placed not more than 1 inch from the surface of the concrete.
    Group IV: includes concrete having siliceous aggregates containing a combined total of 60 percent or more of quartz, chert and flint, and tied with No. 5 gage steel wire wound spirally over the column section on a pitch of 8 inches, or equivalent ties.

RESOURCE A

FIGURE 2.5.1.5
STEEL COLUMNS—CONCRETE ENCASEMENTS
MINIMUM DIMENSION 12″ TO LESS THAN 14″

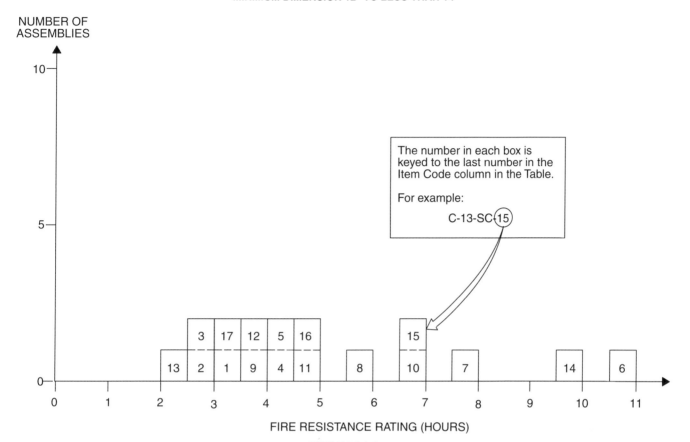

TABLE 2.5.1.5
STEEL COLUMNS—CONCRETE ENCASEMENTS
MINIMUM DIMENSION 12″ TO LESS THAN 14″

ITEM CODE	MINIMUM DIMENSION	CONSTRUCTION DETAILS	PERFORMANCE		REFERENCE NUMBER			NOTES	REC. HOURS
			LOAD	TIME	PRE-BMS-92	BMS-92	POST-BMS-92		
C-12-SC-1	12″	12″ × 14″ concrete encased steel column; 8″ × 6″ × 35 lbs. "H" beam; Protection: aggregate concrete (4150 psi) with 4″ mesh; 16 SWG reinforcing 1″ below column surface.	120 tons	3 hrs. 24 min.			7	1	$3^1/_3$
C-12-SC-2	12″	12″ × 16″ concrete encased column; 8″ × 6″ × 35 lbs. "H" beam; Protection: aggregate concrete (4300 psi) with 4″ mesh; 16 SWG reinforcing 1″ below column surface.	90 tons	2 hrs. 52 min.			7	1	$2^3/_4$
C-12-SC-3	12″	12″ × 16″ concrete encased steel column; 12″ × 8″ × 65 lbs. "H" column; Protection: gravel aggregate concrete (3550 psi) with 4″ mesh; 16 SWG reinforcement 1″ below column surface.	177 tons	2 hrs. 31 min.			7	1	$2^1/_2$
C-12-SC-4	12″	12″ × 16″ concrete encased column; 12″ × 8″ × 65 lbs. "H" beam; Protection: aggregate concrete (3450 psi) with 4″ mesh; 16 SWG reinforcement 1″ below column surface.	118 tons	4 hrs. 4 min.			7	1	4

(continued)

TABLE 2.5.1.5
STEEL COLUMNS—CONCRETE ENCASEMENTS
MINIMUM DIMENSION 12″ TO LESS THAN 14″

ITEM CODE	MINIMUM DIMENSION	CONSTRUCTION DETAILS	PERFORMANCE LOAD	PERFORMANCE TIME	REFERENCE NUMBER PRE-BMS-92	REFERENCE NUMBER BMS-92	REFERENCE NUMBER POST-BMS-92	NOTES	REC. HOURS
C-12-SC-5	12$\frac{1}{2}$″	12$\frac{1}{2}$″ × 14″ column; 6″ × 4$\frac{1}{2}$″ × 20 lbs. "H" beam; Protection: gravel aggregate concrete (3750 psi) with 4″ × 4″ mesh; 16 SWG reinforcing 1″ below column surface.	52 tons	4 hrs. 29 min.			7	1	4$\frac{1}{3}$
C-12-SC-6	12″	8″ × 8″ steel column; 2″ outside protection; Group I.	—	11 hrs.		1		2	11
C-12-SC-7	12″	Description as per C-12-SC-6; Group II.	—	8 hrs.		1		2	8
C-12-SC-8	12″	Description as per C-12-SC-6; Group III.	—	6 hrs.		1		2	6
C-12-SC-9	12″	Description as per C-12-SC-6; Group IV.	—	4 hrs.		1		2	4
C-12-SC-10	12″	10″ × 10″ steel column; 2″ outside protection; Group I.	—	7 hrs.		1		2	7
C-12-SC-11	12″	Description as per C-12-SC-10; Group II.	—	5 hrs.		1		2	5
C-12-SC-12	12″	Description as per C-12-SC-10; Group III.	—	4 hrs.		1		2	4
C-12-SC-13	12″	Description as per C-12-SC-10; Group IV.	—	2 hrs. 30 min.		1		2	2$\frac{1}{2}$
C-13-SC-14	13″	10″ × 10″ steel column; 3″ outside protection; Group I.	—	10 hrs.		1		2	10
C-13-SC-15	13″	Description as per C-12-SC-14; Group II.	—	7 hrs.		1		2	7
C-13-SC-16	13″	Description as per C-12-SC-14; Group III.	—	5 hrs.		1		2	5
C-13-SC-17	13″	Description as per C-12-SC-14; Group IV.	—	3 hrs. 30 min.		1		2	3$\frac{1}{2}$

For SI: 1 inch = 25.4 mm, 1 pound = 0.004448 kN, 1 pound per square inch = 0.00689 MPa, 1 pound per square yard = 5.3 N/m^2, 1 ton = 8.896 kN.

Notes:
1. Failure mode—collapse.
2. Group I: includes concrete having calcareous aggregate containing a combined total of not more than 10 percent of quartz, chert and flint for the coarse aggregate.
   Group II: includes concrete having trap-rock aggregate applied without metal ties and also concrete having cinder, sandstone or granite aggregate, if held in place with wire mesh or expanded metal having not larger than 4-inch mesh, weighing not less than 1.7 lbs./yd.^{2}, placed not more than 1 inch from the surface of the concrete.
   Group III: includes concrete having cinder, sandstone or granite aggregate tied with No. 5 gage steel wire, wound spirally over the column section on a pitch of 8 inches, or equivalent ties, and concrete having siliceous aggregates containing a combined total of 60 percent or more of quartz, chert and flint, if held in place with wire mesh or expanded metal having not larger than 4-inch mesh, weighing not less than 1.7 lbs./yd.^{2}, placed not more than 1 inch from the surface of the concrete.
   Group IV: includes concrete having siliceous aggregates containing a combined total of 60 percent or more of quartz, chert and flint, and tied with No. 5 gage steel wire wound spirally over the column section on a pitch of 8 inches, or equivalent ties.

FIGURE 2.5.1.6
STEEL COLUMNS—CONCRETE ENCASEMENTS
MINIMUM DIMENSION 14″ TO LESS THAN 16″

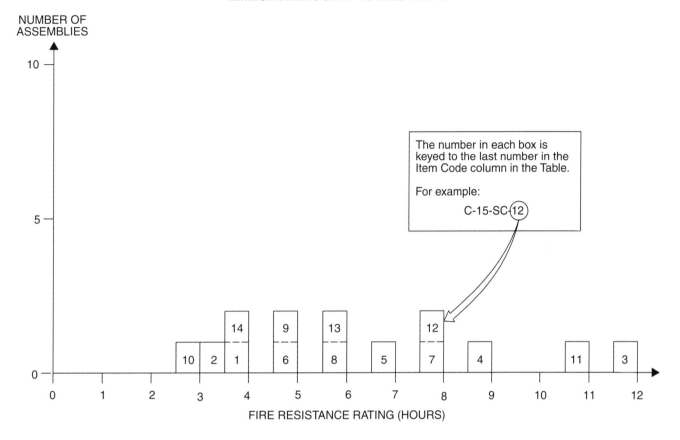

TABLE 2.5.1.6
STEEL COLUMNS—CONCRETE ENCASEMENTS
MINIMUM DIMENSION 14″ TO LESS THAN 16″

ITEM CODE	MINIMUM DIMENSION	CONSTRUCTION DETAILS	PERFORMANCE LOAD	PERFORMANCE TIME	REFERENCE NUMBER PRE-BMS-92	REFERENCE NUMBER BMS-92	REFERENCE NUMBER POST-BMS-92	NOTES	REC. HOURS
C-14-SC-1	14″	24″ × 16″ concrete encased steel column; 8″ × 6″ × 35 lbs. "H" column; Protection: aggregate concrete (4240 psi); 4″ mesh - 16 SWG reinforcing 1″ below column surface.	90 tons	3 hrs. 40 min.			7	1	3
C-14-SC-2	14″	14″ × 18″ concrete encased steel column; 12″ × 8″ × 65 lbs. "H" beam; Protection: gravel aggregate concrete (4000 psi) with 4″ - 16 SWG wire mesh reinforcement 1″ below column surface.	177 tons	3 hrs. 20 min.			7	1	3
C-14-SC-3	14″	10″ × 10″ steel column; 4″ outside protection; Group I.	—	12 hrs.		1		2	12
C-14-SC-4	14″	Description as per C-14-SC-3; Group II.	—	9 hrs.		1		2	9
C-14-SC-1	14″	24″ × 16″ concrete encased steel column; 8″ × 6″ × 35 lbs. "H" column; Protection: aggregate concrete (4240 psi); 4″ mesh - 16 SWG reinforcing 1″ below column surface.	90 tons	3 hrs. 40 min.			7	1	3

(continued)

TABLE 2.5.1.6—continued
STEEL COLUMNS—CONCRETE ENCASEMENTS
MINIMUM DIMENSION 14″ TO LESS THAN 16″

ITEM CODE	MINIMUM DIMENSION	CONSTRUCTION DETAILS	PERFORMANCE LOAD	PERFORMANCE TIME	REFERENCE NUMBER PRE-BMS-92	REFERENCE NUMBER BMS-92	REFERENCE NUMBER POST-BMS-92	NOTES	REC. HOURS
C-14-SC-2	14″	14″ × 18″ concrete encased steel column; 12″ × 8″ × 65 lbs. "H" beam; Protection: gravel aggregate concrete (4000 psi) with 4″-16 SWG wire mesh reinforcement 1″ below column surface.	177 tons	3 hrs. 20 min.			7	1	3
C-14-SC-3	14″	10″ × 10″ steel column; 4″ outside protection; Group I.	—	12 hrs.		1		2	12
C-14-SC-4	14″	Description as per C-14-SC-3; Group II.	—	9 hrs.		1		2	9
C-14-SC-5	14″	Description as per C-14-SC-3; Group III.	—	7 hrs.		1		2	7
C-14-SC-6	14″	Description as per C-14-SC-3; Group IV.	—	5 hrs.		1		2	5
C-14-SC-7	14″	12″ × 12″ steel column; 2″ outside protection; Group I.	—	8 hrs.		1		2	8
C-14-SC-8	14″	Description as per C-14-SC-7; Group II.	—	6 hrs.		1		2	6
C-14-SC-9	14″	Description as per C-14-SC-7; Group III.	—	5 hrs.		1		2	5
C-14-SC-10	14″	Description as per C-14-SC-7; Group IV	—	3 hrs.		1		2	3
C-15-SC-11	15″	12″ × 12″ steel column; 3″ outside protection; Group I.	—	11 hrs.		1		2	11
C-15-SC-12	15″	Description as per C-15-SC-11; Group II.	—	8 hrs.		1		2	8
C-15-SC-13	15″	Description as per C-15-SC-11; Group III.	—	6 hrs.		1		2	6
C-15-SC-14	15″	Description as per C-15-SC-11; Group IV.	—	4 hrs.		1		2	4

For SI: 1 inch = 25.4 mm, 1 pound = 0.004448 kN, 1 pound per square inch = 0.00689 MPa, 1 pound per square yard = 5.3 N/m^2, 1 ton = 8.896 kN.

Notes:
1. Collapse.
2. Group I: includes concrete having calcareous aggregate containing a combined total of not more than 10 percent of quartz, chert and flint for the coarse aggregate.
   Group II: includes concrete having trap-rock aggregate applied without metal ties and also concrete having cinder, sandstone or granite aggregate, if held in place with wire mesh or expanded metal having not larger than 4-inch mesh, weighing not less than 1.7 lbs./yd.^{2}, placed not more than 1 inch from the surface of the concrete.
   Group III: includes concrete having cinder, sandstone or granite aggregate tied with No. 5 gage steel wire, wound spirally over the column section on a pitch of 8 inches, or equivalent ties, and concrete having siliceous aggregates containing a combined total of 60 percent or more of quartz, chert and flint, if held in place with wire mesh or expanded metal having not larger than 4-inch mesh, weighing not less than 1.7 lbs./yd.^{2}, placed not more than 1 inch from the surface of the concrete.
   Group IV: includes concrete having siliceous aggregates containing a combined total of 60 percent or more of quartz, chert and flint, and tied with No. 5 gage steel wire wound spirally over the column section on a pitch of 8 inches, or equivalent ties.

RESOURCE A

TABLE 2.5.1.7
STEEL COLUMNS—CONCRETE ENCASEMENTS
MINIMUM DIMENSION 16″ TO LESS THAN 18″

ITEM CODE	MINIMUM DIMENSION	CONSTRUCTION DETAILS	PERFORMANCE		REFERENCE NUMBER			NOTES	REC. HOURS
			LOAD	TIME	PRE-BMS-92	BMS-92	POST-BMS-92		
C-16-SC-1	16″	12″ × 12″ steel column; 4″ outside protection; Group I.	—	14 hrs.		1		1	14
C-16-SC-2	16″	Description as per C-16-SC-1; Group II.	—	10 hrs.		1		1	10
C-16-SC-3	16″	Description as per C-16-SC-1; Group III.	—	8 hrs.		1		1	8
C-16-SC-4	16″	Description as per C-16-SC-1; Group IV.	—	5 hrs.		1		1	5

For SI: 1 inch = 25.4 mm.

Notes:

1. Group I: includes concrete having calcareous aggregate containing a combined total of not more than 10 percent of quartz, chert and flint for the coarse aggregate.

    Group II: includes concrete having trap-rock aggregate applied without metal ties and also concrete having cinder, sandstone or granite aggregate, if held in place with wire mesh or expanded metal having not larger than 4-inch mesh, weighing not less than 1.7 lbs./yd.^{2}, placed not more than 1 inch from the surface of the concrete.

    Group III: includes concrete having cinder, sandstone or granite aggregate tied with No. 5 gage steel wire, wound spirally over the column section on a pitch of 8 inches, or equivalent ties, and concrete having siliceous aggregates containing a combined total of 60 percent or more of quartz, chert and flint, if held in place with wire mesh or expanded metal having not larger than 4-inch mesh, weighing not less than 1.7 lbs./yd.^{2}, placed not more than 1 inch from the surface of the concrete.

    Group IV: includes concrete having siliceous aggregates containing a combined total of 60 percent or more of quartz, chert and flint, and tied with No. 5 gage steel wire wound spirally over the column section on a pitch of 8 inches, or equivalent ties.

TABLE 2.5.2.1
STEEL COLUMNS—BRICK AND BLOCK ENCASEMENTS
MINIMUM DIMENSION 10″ TO LESS THAN 12″

ITEM CODE	MINIMUM DIMENSION	CONSTRUCTION DETAILS	PERFORMANCE		REFERENCE NUMBER			NOTES	REC. HOURS
			LOAD	TIME	PRE-BMS-92	BMS-92	POST-BMS-92		
C-10-SB-1	10$^1/_2$″	10$^1/_2$″ × 13″ brick encased steel columns; 8″ × 6″ × 35 lbs. "H" beam; Protection. Fill of broken brick and mortar; 2″ brick on edge; joints broken in alternate courses; cement-sand grout; 13 SWG wire reinforcement in every third horizontal joint.	90 tons	3 hrs. 6 min.			7	1	3
C-10-SB-2	10$^1/_2$″	10$^1/_2$″ × 13″ brick encased steel columns; 8″ × 6″ × 35 lbs. "H" beam; Protection: 2″ brick; joints broken in alternate courses; cement-sand grout; 13 SWG iron wire reinforcement in alternate horizontal joints.	90 tons	2 hrs.			7	2, 3, 4	2
C-10-SB-3	10″	10″ × 12″ block encased columns; 8″ × 6″ × 35 lbs. "H" beam; Protection: 2″ foamed slag concrete blocks; 13 SWG wire at each horizontal joint; mortar at each joint.	90 tons	2 hrs.			7	5	2
C-10-SB-4	10$^1/_2$″	10$^1/_2$″ × 12″ block encased steel columns; 8″ × 6″ × 35 lbs. "H" beam; Protection: gravel aggregate concrete fill (unconsolidated) 2″ thick hollow clay tiles with mortar at edges.	86 tons	56 min.			7	1	$^3/_4$
C-10-SB-5	10$^1/_2$″	10$^1/_2$″ × 12″ block encased steel columns; 8″ × 6″ × 35 lbs. "H" beam; Protection: 2″ hollow clay tiles with mortar at edges.	86 tons	22 min.			7	1	$^1/_4$

For SI: 1 inch = 25.4 mm, 1 pound = 0.004448 kN, 1 ton = 8.896 kN.

Notes:
1. Failure mode—collapse.
2. Passed 2-hour fire test (Grade "C" - British).
3. Passed hose stream test.
4. Passed reload test.
5. Passed 2-hour fire exposure but collapsed immediately following hose stream test.

TABLE 2.5.2.2
STEEL COLUMNS—BRICK AND BLOCK ENCASEMENTS
MINIMUM DIMENSION 12″ TO LESS THAN 14″

ITEM CODE	MINIMUM DIMENSION	CONSTRUCTION DETAILS	PERFORMANCE		REFERENCE NUMBER			NOTES	REC. HOURS
			LOAD	TIME	PRE-BMS-92	BMS-92	POST-BMS-92		
C-12-SB-1	12″	12″ × 15″ brick encased steel columns; 8″ × 6″ × 35 lbs. "H" beam; Protection: $2^5/_8$″ thick brick; joints broken in alternate courses; cement-sand grout; fill of broken brick and mortar.	90 tons	1 hr. 49 min.			7	1	$1^3/_4$

For SI: 1 inch = 25.4 mm, 1 pound = 0.004448 kN, 1 ton = 8.896 kN.

Notes:
1. Failure mode—collapse.

TABLE 2.5.2.3
STEEL COLUMNS—BRICK AND BLOCK ENCASEMENTS
MINIMUM DIMENSION 14″ TO LESS THAN 16″

ITEM CODE	MINIMUM DIMENSION	CONSTRUCTION DETAILS	PERFORMANCE		REFERENCE NUMBER			NOTES	REC. HOURS
			LOAD	TIME	PRE-BMS-92	BMS-92	POST-BMS-92		
C-15-SB-1	15″	15″ × 17″ brick encased steel columns; 8″ × 6″ × 35 lbs. "H" beam; Protection: $4^1/_2$″ thick brick; joints broken in alternate courses; cement-sand grout; fill of broken brick and mortar.	45 tons	6 hrs.			7	1	6
C-15-SB-2	15″	15″ × 17″ brick encased steel columns; 8″ × 6″ × 35 lbs. "H" beam; Protection. Fill of broken brick and mortar; $4^1/_2$″ brick; joints broken in alternate courses; cement-sand grout.	86 tons	6 hrs.			7	2, 3, 4	6
C-15-SB-3	15″	15″ × 18″ brick encased steel columns; 8″ × 6″ × 35 lbs. "H" beam; Protection: $4^1/_2$″ brick work; joints alternating; cement-sand grout.	90 tons	4 hrs.			7	5, 6	4
C-15-SB-4	14″	14″ × 16″ block encased steel columns; 8″ × 6″ × 35 lbs. "H" beam; Protection: 4″ thick foam slag concrete blocks; 13 SWG wire reinforcement in each horizontal joint; mortar in joints.	90 tons	5 hrs. 52 min.			7	7	$4^3/_4$

For SI: 1 inch = 25.4 mm, 1 pound = 0.004448 kN, 1 ton = 8.896 kN.

Notes:
1. Only a nominal load was applied to specimen.
2. Passed 6-hour fire test (Grade "A" - British).
3. Passed (6 minute) hose stream test.
4. Reload not specified.
5. Passed 4-hour fire exposure.
6. Failed by collapse between first and second minute of hose stream exposure.
7. Mode of failure-collapse.

**TABLE 2.5.3.1
STEEL COLUMNS—PLASTER ENCASEMENTS
MINIMUM DIMENSION 6″ TO LESS THAN 8″**

ITEM CODE	MINIMUM DIMENSION	CONSTRUCTION DETAILS	PERFORMANCE		REFERENCE NUMBER			NOTES	REC. HOURS
			LOAD	TIME	PRE-BMS-92	BMS-92	POST-BMS-92		
C-7-SP-1	$7^1/_2''$	$7^1/_2'' \times 9^1/_2''$ plaster protected steel columns; $8'' \times 6'' \times 35$ lbs. "H" beam; Protection: 24 SWG wire metal lath; $1^1/_4''$ lime plaster.	90 tons	57 min.			7	1	$^3/_4$
C-7-SP-2	$7^7/_8''$	$7^7/_8'' \times 10''$ plaster protected steel columns; $8'' \times 6'' \times 35$ lbs. "H" beam; Protection: $^3/_8''$ gypsum bal wire wound with 16 SWG wire helically wound at 4″ pitch; $^1/_2''$ gypsum plaster.	90 tons	1 hr. 13 min.			7	1	1
C-7-SP-3	$7^1/_4''$	$7^1/_4'' \times 9^3/_8''$ plaster protected steel columns; $8'' \times 6'' \times 35$ lbs. "H" beam; Protection: $^3/_8''$ gypsum board; wire helically wound 16 SWG at 4″ pitch; $^1/_4''$ gypsum plaster finish.	90 tons	1 hr. 14 min.			7	1	1

Notes:
1. Failure mode—collapse.

**TABLE 2.5.3.2
STEEL COLUMNS—PLASTER ENCASEMENTS
MINIMUM DIMENSION 8″ TO LESS THAN 10″**

ITEM CODE	MINIMUM DIMENSION	CONSTRUCTION DETAILS	PERFORMANCE		REFERENCE NUMBER			NOTES	REC. HOURS
			LOAD	TIME	PRE-BMS-92	BMS-92	POST-BMS-92		
C-8-SP-1	8″	$8'' \times 10''$ plaster protected steel columns; $8'' \times 6'' \times 35$ lbs. "H" beam; Protection: 24 SWG wire lath; 1″ gypsum plaster.	86 tons	1 hr. 23 min.			7	1	$1^1/_4$
C-8-SP-2	$8^1/_2''$	$8^1/_2'' \times 10^1/_2''$ plaster protected steel columns; $8'' \times 6'' \times 35$ lbs. "H" beam; Protection: 24 SWG metal lath wrap; $1^1/_4''$ gypsum plaster.	90 tons	1 hr. 36 min.			7	1	$1^1/_2$
C-9-SP-3	9″	$9'' \times 11''$ plaster protected steel columns; $8'' \times 6'' \times 35$ lbs. "H" beam; Protection: 24 SWG metal lath wrap; $^1/_8''$ M.S. ties at 12″ pitch wire netting $1^1/_2'' \times 22$ SWG between first and second plaster coats; $1^1/_2''$ gypsum plaster.	90 tons	1 hr. 33 min.			7	1	$1^1/_2$
C-8-SP-4	$8^3/_4''$	$8^3/_4'' \times 10^3/_4''$ plaster protected steel columns; $8'' \times 6'' \times 35$ lbs. "H" beam; Protection: $^3/_4''$ gypsum board; wire wound spirally (#16 SWG) at $1^1/_2''$ pitch; $^1/_2''$ gypsum plaster.	90 tons	2 hrs.			7	2, 3, 4	2

For SI: 1 inch = 25.4 mm, 1 pound = 0.004448 kN, 1 ton = 8.896 kN.
Notes:
1. Failure mode—collapse.
2. Passed 2 hour fire exposure test (Grade "C" - British).
3. Passed hose stream test.

**TABLE 2.5.4.1
STEEL COLUMNS—MISCELLANEOUS ENCASEMENTS
MINIMUM DIMENSION 6″ TO LESS THAN 8″**

ITEM CODE	MINIMUM DIMENSION	CONSTRUCTION DETAILS	PERFORMANCE		REFERENCE NUMBER			NOTES	REC. HOURS
			LOAD	TIME	PRE-BMS-92	BMS-92	POST-BMS-92		
C-7-SM-1	$7^5/_8''$	$7^5/_8'' \times 9^1/_2''$ (asbestos plaster) protected steel columns; $8'' \times 6'' \times 35$ lbs. "H" beam; Protection: 20 gage $^1/_2''$ metal lath; $^9/_{16}''$ asbestos plaster (minimum).	90 tons	1 hr. 52 min.			7	1	$1^3/_4$

For SI: 1 inch = 25.4 mm, 1 pound = 0.004448 kN, 1 ton = 8.896 kN.
Notes:
1. Failure mode—collapse.

**TABLE 2.5.4.2
STEEL COLUMNS—MISCELLANEOUS ENCASEMENTS
MINIMUM DIMENSION 8″ TO LESS THAN 10″**

ITEM CODE	MINIMUM DIMENSION	CONSTRUCTION DETAILS	PERFORMANCE		REFERENCE NUMBER			NOTES	REC. HOURS
			LOAD	TIME	PRE-BMS-92	BMS-92	POST-BMS-92		
C-9-SM-1	$9^5/_8''$	$9^5/_8'' \times 11^3/_8''$ asbestos slab and cement plaster protected columns; 8″ × 6″ × 35 lbs. "H" beam; Protection: 1″ asbestos slab; wire wound; $^5/_8''$ plaster.	90 tons	2 hrs.			7	1, 2	2

For SI: 1 inch = 25.4 mm, 1 pound = 0.004448 kN, 1 ton = 8.896 kN.
Notes:
1. Passed 2 hour fire exposure test.
2. Collapsed during hose stream test.

**TABLE 2.5.4.3
STEEL COLUMNS—MISCELLANEOUS ENCASEMENTS
MINIMUM DIMENSION 10″ TO LESS THAN 12″**

ITEM CODE	MINIMUM DIMENSION	CONSTRUCTION DETAILS	PERFORMANCE		REFERENCE NUMBER			NOTES	REC. HOURS
			LOAD	TIME	PRE-BMS-92	BMS-92	POST-BMS-92		
C-11-SM-1	$11^1/_2''$	$11^1/_2'' \times 13^1/_2''$ wood wool and plaster protected steel columns; 8″ × 6″ × 35 lbs. "H" beam; Protection: wood-wool-cement paste as fill and to 2″ cover over beam; $^3/_4''$ gypsum plaster finish.	90 tons	2 hrs.			7	1, 2, 3	2
C-10-SM-1	10″	10″ × 12″ asbestos protected steel columns; 8″ × 6″ × 35 lbs. "H" beam; Protection: sprayed on asbestos paste to 2″ cover over column.	90 tons	4 hrs.			7	2, 3, 4	4

For SI: 1 inch = 25.4 mm, 1 pound = 0.004448 kN, 1 ton = 8.896 kN.
Notes:
1. Passed 2 hour fire exposure (Grade "C" - British).
2. Passed hose stream test.
3. Passed reload test.
4. Passed 4 hour fire exposure test.

**TABLE 2.5.4.4
STEEL COLUMNS—MISCELLANEOUS ENCASEMENTS
MINIMUM DIMENSION 12″ TO LESS THAN 14″**

ITEM CODE	MINIMUM DIMENSION	CONSTRUCTION DETAILS	PERFORMANCE		REFERENCE NUMBER			NOTES	REC. HOURS
			LOAD	TIME	PRE-BMS-92	BMS-92	POST-BMS-92		
C-12-SM-1	12″	12″ × $14^1/_4''$ cement and asbestos protected columns; 8″ × 6″ × 35 lbs. "H" beam; Protection: fill of asbestos packing pieces 1″ thick 1′3″ o.c.; cover of 2″ molded asbestos inner layer; 1″ molded asbestos outer layer; held in position by 16 SWG nichrome wire ties; wash of refractory cement on outer surface.	86 tons	4 hrs. 43 min.			7	1, 2, 3	$4^2/_3$

For SI: 1 inch = 25.4 mm, 1 pound = 0.004448 kN, 1 ton = 8.896 kN.
Notes:
1. Passed 4 hour fire exposure (Grade "B" - British).
2. Passed hose stream test.
3. Passed reload test.

RESOURCE A

SECTION III
FLOOR/CEILING ASSEMBLIES

FIGURE 3.1
FLOOR/CEILING ASSEMBLIES—REINFORCED CONCRETE

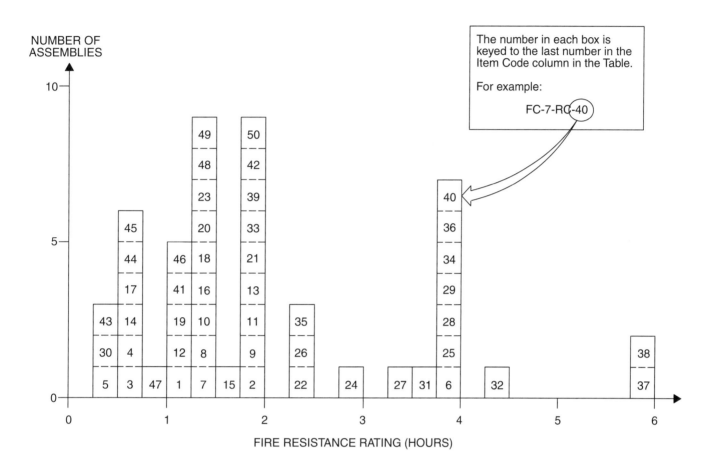

TABLE 3.1
FLOOR/CEILING ASSEMBLIES—REINFORCED CONCRETE

ITEM CODE	ASSEMBLY THICKNESS	CONSTRUCTION DETAILS	PERFORMANCE		REFERENCE NUMBER			NOTES	REC. HOURS
			LOAD	TIME	PRE-BMS-92	BMS-92	POST-BMS-92		
F/C-3-RC-1	$3^3/_4''$	$3^3/_4''$ thick floor; $3^1/_4''$ (5475 psi) concrete deck; $1/_2''$ plaster under deck; $3/_8''$ main reinforcement bars at $5^1/_2''$ pitch with $7/_8''$ concrete cover; $3/_8''$ main reinforcement bars at $4^1/_2''$ pitch perpendicular with $1/_2''$ concrete cover; 13'1" span restrained.	195 psf	24 min.			7	1, 2	$1/_3$
F/C-3-RC-2	$3^1/_4''$	$3^1/_4''$ deep (3540 psi) concrete deck; $3/_8''$ main reinforcement bars at $5^1/_2''$ pitch with $7/_8''$ cover; $3/_8''$ main reinforcement bars at $4^1/_2''$ pitch perpendicular with $1/_2''$ cover; 13'1" span restrained.	195 psf	2 hrs.			7	1, 3, 4	2

(continued)

**TABLE 3.1—continued
FLOOR/CEILING ASSEMBLIES—REINFORCED CONCRETE**

ITEM CODE	ASSEMBLY THICKNESS	CONSTRUCTION DETAILS	PERFORMANCE		REFERENCE NUMBER			NOTES	REC. HOURS
			LOAD	TIME	PRE-BMS-92	BMS-92	POST-BMS-92		
F/C-3-RC-3	3¼″	3¼″ deep (4175 psi) concrete deck; ⅜″ main reinforcement bars at 5½″ pitch with ⅞″ cover; ⅜″ main reinforcement bars at 4½″ pitch perpendicular with ½″ cover; 13′1″ span restrained.	195 psf	31 min.			7	1, 5	½
F/C-3-RC-4	3¼″	3¼″ deep (4355 psi) concrete deck; ⅜″ main reinforcement bars at 5½″ pitch with ⅞″ cover; ⅜″ main reinforcement bars at 4½″ pitch perpendicular with ½″ cover; 13′1″ span restrained.	195 psf	41 min.			7	1, 5, 6	½
F/C-3-RC-5	3¼″	3¼″ thick (3800 psi) concrete deck; ⅜″ main reinforcement bars at 5½″ pitch with ⅞″ cover; ⅜″ main reinforcement bars at 4½″ pitch perpendicular with ½″ cover; 13′1″ span restrained.	195 psf	1 hr. 5 min.			7	1, 5	1
F/C-4-RC-6	4¼″	4¼″ thick; 3¼″ (4000 psi) concrete deck; 1″ sprayed asbestos lower surface; ⅜″ main reinforcement bars at 5⅞″ pitch with ⅞″ concrete cover; ⅜″ main reinforcement bars at 4½″ pitch perpendicular with ½″ concrete cover; 13′1″ span restrained.	195 psf	4 hrs.			7	1, 7	4
F/C-4-RC-7	4″	4″ (5025 psi) concrete deck; ¼″ reinforcement bars at 7½″ pitch with ¾″ cover; ⅜″ main reinforcement bars at 3¾″ pitch perpendicular with ½″ cover; 13′1″ span restrained.	140 psf	1 hr. 16 min.			7	1, 2	1¼
F/C-4-RC-8	4″	4″ thick (4905 psi) deck; ¼″ reinforcement bars at 7½″ pitch with ⅞″ cover; ⅜″ main reinforcement bars at 3¾″ pitch perpendicular with ½″ cover; 13′1″ span restrained.	100 psf	1 hr. 23 min.			7	1, 2	1⅓
F/C-4-RC-9	4″	4″ deep (4370 psi); ¼″ reinforcement bars at 6″ pitch with ¾″ cover; ¼″ main reinforcement bars at 4″ pitch perpendicular with ½″ cover; 13′1″ span restrained.	150 psf	2 hrs.			7	1, 3	2
F/C-4-RC-10	4″	4″ thick (5140 psi) deck; ¼″ reinforcement bars at 7½″ pitch with ⅞″ cover; ⅜″ main reinforcement bars at 3¾″ pitch perpendicular with ½″ cover; 13′1″ span restrained.	140 psf	1 hr. 16 min.			7	1, 5	1¼
F/C-4-RC-11	4″	4″ thick (4000 psi) concrete deck; 3″ × 1½″ × 4 lbs. R.S.J.; 2′6″ C.R.S.; flush with top surface; 4″ × 6″ × 13 SWG mesh reinforcement 1″ from bottom of slab; 6′6″ span restrained.	150 psf	2 hrs.			7	1, 3	2

(continued)

TABLE 3.1—continued
FLOOR/CEILING ASSEMBLIES—REINFORCED CONCRETE

ITEM CODE	ASSEMBLY THICKNESS	CONSTRUCTION DETAILS	PERFORMANCE		REFERENCE NUMBER			NOTES	REC. HOURS
			LOAD	TIME	PRE-BMS-92	BMS-92	POST-BMS-92		
F/C-4-RC-12	4″	4″ deep (2380 psi) concrete deck; 3″ × 1$^1/_2$″ × 4 lbs. R.S.J.; 2′6″ C.R.S.; flush with top surface; 4″ × 6″ x 13 SWG mesh reinforcement 1″ from bottom surface; 6′6″ span restrained.	150 psf	1 hr. 3 min.			7	1, 2	1
F/C-4-RC-13	4$^1/_2$″	4$^1/_2$″ thick (5200 psi) deck; $^1/_4$″ reinforcement bars at 7$^1/_4$″ pitch with $^7/_8$″ cover; $^3/_8$″ main reinforcement bars at 3$^3/_4$″ pitch perpendicular with $^1/_2$″ cover; 13′1″ span restrained.	140 psf	2 hrs.			7	1, 3	2
F/C-4-RC-14	4$^1/_2$″	4$^1/_2$″ deep (2525 psi) concrete deck; $^1/_4$″ reinforcement bars at 7$^1/_2$″ pitch with $^7/_8$″ cover; $^3/_8$″ main reinforcement bars at 3$^3/_8$″ pitch perpendicular with $^1/_2$″ cover; 13′1″ span restrained.	150 psf	42 min.			7	1, 5	$^2/_3$
F/C-4-RC-15	4$^1/_2$″	4$^1/_2$″ deep (4830 psi) concrete deck; 1$^1/_2$″ × No. 15 gauge wire mesh; $^3/_8$″ reinforcement bars at 15″ pitch with 1″ cover; $^1/_2$″ main reinforcement bars at 6″ pitch perpendicular with $^1/_2$″ cover; 12′ span simply supported.	75 psf	1 hr. 32 min.			7	1, 8	1$^1/_2$
F/C-4-RC-16	4$^1/_2$″	4$^1/_2$″ deep (4595 psi) concrete deck; $^1/_4$″ reinforcement bars at 7$^1/_2$″ pitch with $^7/_8$″ cover; $^3/_8$″ main reinforcement bars at 3$^1/_2$″ pitch perpendicular with $^1/_2$″ cover; 12′ span simply supported.	75 psf	1 hr. 20 min.			7	1, 8	1$^1/_3$
F/C-4-RC-17	4$^1/_2$″	4$^1/_2$″ deep (3625 psi) concrete deck; $^1/_4$″ reinforcement bars at 7$^1/_2$″ pitch with $^7/_8$″ cover; $^3/_8$″ main reinforcement bars at 3$^1/_2$″ pitch perpendicular with $^1/_2$″ cover; 12′ span simply supported.	75 psf	35 min.			7	1, 8	$^1/_2$
F/C-4-RC-18	4$^1/_2$″	4$^1/_2$″ deep (4410 psi) concrete deck; $^1/_4$″ reinforcement bars at 7$^1/_2$″ pitch with $^7/_8$″ cover; $^3/_8$″ main reinforcement bars at 3$^1/_2$″ pitch perpendicular with $^1/_2$″ cover; 12′ span simply supported.	85 psf	1 hr. 27 min.			7	1, 8	1$^1/_3$
F/C-4-RC-19	4$^1/_2$″	4$^1/_2$″ deep (4850 psi) deck; $^3/_8$″ reinforcement bars at 15″ pitch with 1″ cover; $^1/_2$″ main reinforcement bars at 6″ pitch perpendicular with $^1/_2$″ cover; 12′ span simply supported.	75 psf	2 hrs. 15 min.			7	1, 9	1$^1/_4$
F/C-4-RC-20	4$^1/_2$″	4$^1/_2$″ deep (3610 psi) deck; $^1/_4$″ reinforcement bars at 7$^1/_2$″ pitch with $^7/_8$″ cover; $^3/_8$″ main reinforcement bars at 3$^1/_2$″ pitch perpendicular with $^1/_2$″ cover; 12′ span simply supported.	75 psf	1 hr. 22 min.			7	1, 8	1$^1/_3$

(continued)

TABLE 3.1—continued
FLOOR/CEILING ASSEMBLIES—REINFORCED CONCRETE

ITEM CODE	ASSEMBLY THICKNESS	CONSTRUCTION DETAILS	PERFORMANCE		REFERENCE NUMBER			NOTES	REC. HOURS
			LOAD	TIME	PRE-BMS-92	BMS-92	POST-BMS-92		
F/C-5-RC-21	5″	5″ deep; 4$^1/_2$″ (5830 psi) concrete deck; $^1/_2$″ plaster finish bottom of slab; $^1/_4$″ reinforcement bars at 7$^1/_2$″ pitch with $^7/_8$″ cover; $^3/_8$″ main reinforcement bars at 3$^1/_2$″ pitch perpendicular with $^1/_2$″ cover; 12′ span simply supported.	69 psf	2 hrs.			7	1, 3	2
F/C-5-RC-22	5″	4$^1/_2$″ (5290 psi) concrete deck; $^1/_2$″ plaster finish bottom of slab; $^1/_4$″ reinforcement bars at 7$^1/_2$″ pitch with $^7/_8$″ cover; $^3/_8$″ main reinforcement bars at 3$^1/_2$″ pitch perpendicular with $^1/_2$″ cover; 12′ span simply supported.	No load	2 hrs. 28 min.			7	1, 10, 11	2$^1/_4$
F/C-5-RC-23	5″	5″ (3020 psi) concrete deck; 3″ × 1$^1/_2$″ × 4 lbs. R.S.J.; 2′ C.R.S. with 1″ cover on bottom and top flanges; 8′ span restrained.	172 psf	1 hr. 24 min.			7	1, 2, 12	1$^1/_2$
F/C-5-RC-24	5$^1/_2$″	5″ (5180 psi) concrete deck; $^1/_2$″ retarded plaster underneath slab; $^1/_4$″ reinforcement bars at 7$^1/_2$″ pitch with 1$^3/_8$″ cover; $^3/_8$″ main reinforcement bars at 3$^1/_2$″ pitch perpendicular with 1″ cover; 12′ span simply supported.	60 psf	2 hrs. 48 min.			7	1, 10	2$^3/_4$
F/C-6-RC-25	6″	6″ deep (4800 psi) concrete deck; $^1/_4$″ reinforcement bars at 7$^1/_2$″ pitch with $^7/_8$″ cover; $^3/_8$″ main reinforcement bars at 3$^1/_2$″ pitch perpendicular with $^7/_8$″ cover; 13′1″ span restrained.	195 psf	4 hrs.			7	1, 7	4
F/C-6-RC-26	6″	6″ (4650 psi) concrete deck; $^1/_4$″ reinforcement bars at 7$^1/_2$″ pitch with $^7/_8$″ cover; $^3/_8$″ main reinforcement bars at 3$^1/_2$″ pitch perpendicular with $^1/_2$″ cover; 13′1″ span restrained.	195 psf	2 hrs. 23 min.			7	1, 2	2$^1/_4$
F/C-6-RC-27	6″	6″ deep (6050 psi) concrete deck; $^1/_4$″ reinforcement bars at 7$^1/_2$″ pitch $^7/_8$″ cover; $^3/_8$″ reinforcement bars at 3$^1/_2$″ pitch perpendicular with $^1/_2$″ cover; 13′1″ span restrained.	195 psf	3 hrs. 30 min.			7	1, 10	3$^1/_2$
F/C-6-RC-28	6″	6″ deep (5180 psi) concrete deck; $^1/_4$″ reinforcement bars at 8″ pitch $^3/_4$″ cover; $^1/_4$″ reinforcement bars at 5$^1/_2$″ pitch perpendicular with $^1/_2$″ cover; 13′1″ span restrained.	150 psf	4 hrs.			7	1, 7	4
F/C-6-RC-29	6″	6″ thick (4180 psi) concrete deck; 4″ × 3″ × 10 lbs. R.S.J.; 2′ 6″ C.R.S. with 1″ cover on both top and bottom flanges; 13′1″ span restrained.	160 psf	3 hrs. 48 min.			7	1, 10	3$^3/_4$
F/C-6-RC-30	6″	6″ thick (3720 psi) concrete deck; 4″ × 3″ × 10 lbs. R.S.J.; 2′ 6″ C.R.S. with 1″ cover on both top and bottom flanges; 12′ span simply supported.	115 psf	29 min.			7	1, 5, 13	$^1/_4$

(continued)

TABLE 3.1—continued
FLOOR/CEILING ASSEMBLIES—REINFORCED CONCRETE

ITEM CODE	ASSEMBLY THICKNESS	CONSTRUCTION DETAILS	PERFORMANCE		REFERENCE NUMBER			NOTES	REC. HOURS
			LOAD	TIME	PRE-BMS-92	BMS-92	POST-BMS-92		
F/C-6-RC-31	6″	6″ deep (3450 psi) concrete deck; 4″ × 1³/₄″ × 5 lbs. R.S.J.; 2′ 6″ C.R.S. with 1″ cover on both top and bottom flanges; 12′ span simply supported.	25 psf	3 hrs. 35 min.			7	1, 2	3¹/₂
F/C-6-RC-32	6″	6″ deep (4460 psi) concrete deck; 4″ × 1³/₄″ × 5 lbs. R.S.J.; 2′ C.R.S.; with 1″ cover on both top and bottom flanges; 12′ span simply supported.	60 psf	4 hrs. 30 min.			7	1, 10	4¹/₂
F/C-6-RC-33	6″	6″ deep (4360 psi) concrete deck; 4″ × 1³/₄″ × 5 lbs. R.S.J.; 2′ C.R.S.; with 1″ cover on both top and bottom flanges; 13′1″ span restrained.	60 psf	2 hrs.			7	1, 3	2
F/C-6-RC-34	6¹/₄″	6¹/₄″ thick; 4³/₄″ (5120 psi) concrete core; 1″ T&G board flooring; ¹/₂″ plaster undercoat; 4″ × 3″ × 10 lbs. R.S.J.; 3′ C.R.S. flush with top surface concrete; 12′ span simply supported; 2″ × 1′3″ clinker concrete insert.	100 psf	4 hrs.			7	1, 7	4
F/C-6-RC-35	6¹/₄″	4³/₄″ (3600 psi) concrete core; 1″ T&G board flooring; ¹/₂″ plaster undercoat; 4″ × 3″ × 10 lbs. R.S.J.; 3′ C.R.S.; flush with top surface concrete; 12′ span simply supported; 2″ × 1′3″ clinker concrete insert.	100 psf	2 hrs. 30 min.			7	1, 5	2¹/₂
F/C-6-RC-36	6¹/₄″	4³/₄″ (2800 psi) concrete core; 1″ T&G board flooring; ¹/₂″ plaster undercoat; 4″ × 3″ × 10 lbs. R.S.J.; 3′ C.R.S.; flush with top surface concrete; 12″ span simply supported; 2″ × 1′3″ clinker concrete insert.	80 psf	4 hrs.			7	1, 7	4
F/C-7-RC-37	7″	(3640 psi) concrete deck; ¹/₄″ reinforcement bars at 6″ pitch with 1¹/₂″ cover; ¹/₄″ reinforcement bars at 5″ pitch perpendicular with 1¹/₂″ cover; 13′1″ span restrained.	169 psf	6 hrs.			7	1, 14	6
F/C-7-RC-38	7″	(4060 psi) concrete deck; 4″ × 3″ × 10 lbs. R.S.J.; 2′ 6″ C.R.S. with 1¹/₂″ cover on both top and bottom flanges; 4″ × 6″ × 13 SWG mesh reinforcement 1¹/₂″ from bottom of slab; 13′1″ span restrained.	175 psf	6 hrs.			7	1, 14	6
F/C-7-RC-39	7¹/₄″	5³/₄″ (4010 psi) concrete core; 1″ T&G board flooring; ¹/₂″ plaster undercoat; 4″ × 3″ × 10 lbs. R.S.J.; 2′ 6″ C.R.S.; 1″ down from top surface of concrete; 12′ simply supported span; 2″ × 1′ 3″ clinker concrete insert.	95 psf	2 hrs.			7	1, 3	2
F/C-7-RC-40	7¹/₄″	5³/₄″ (3220 psi) concrete core; 1″ T&G flooring; ¹/₂″ plaster undercoat; 4″ × 3″ × 10 lbs. R.S.J.; 2′6″ C.R.S.; 1″ down from top surface of concrete; 12′ simply supported span; 2″ × 1′3″ clinker concrete insert.	95 psf	4 hrs.			7	1, 7	4

(continued)

RESOURCE A

TABLE 3.1—continued
FLOOR/CEILING ASSEMBLIES—REINFORCED CONCRETE

ITEM CODE	ASSEMBLY THICKNESS	CONSTRUCTION DETAILS	PERFORMANCE		REFERENCE NUMBER			NOTES	REC. HOURS
			LOAD	TIME	PRE-BMS-92	BMS-92	POST-BMS-92		
F/C-7-RC-41	10" (2¼" Slab)	Ribbed floor, see Note 15 for details; slab 2½" deep (3020 psi); ¼" reinforcement bars at 6" pitch with ¾" cover; beams 7½" deep × 5" wide; 24" C.R.S.; ⅝" reinforcement bars two rows ½" vertically apart with 1" cover; 13'1" span restricted.	195 psf	1 hr. 4 min.			7	1, 2, 15	1
F/C-5-RC-42	5½"	Composite ribbed concrete slab assembly; see Note 17 for details.	See Note 16	2 hrs.			43	16, 17	2
F/C-3-RC-43	3"	2500 psi concrete; ⅝" cover; fully restrained at test.	See Note 16	30 min.			43	16	½
F/C-3-RC-44	3"	2000 psi concrete; ⅝" cover; free or partial restraint at test.	See Note 16	45 min.			43	16	¾
F/C-4-RC-45	4"	2500 psi concrete; ⅝" cover; fully restrained at test.	See Note 16	40 min.			43	16	⅔
F/C-4-RC-46	4"	2000 psi concrete; ¾" cover; free or partial restraint at test.	See Note 16	1 hr. 15 min.			43	16	1¼
F/C-5-RC-47	5"	2500 psi concrete; ¾" cover; fully restrained at test.	See Note 16	1 hr.			43	16	1
F/C-5-RC-48	5"	2000 psi concrete; ¾" cover; free or partial restraint at test.	See Note 16	1 hr. 30 min.			43	16	1½
F/C-6-RC-49	6"	2500 psi concrete; 1" cover; fully restrained at test.	See Note 16	1 hr. 30 min.			43	16	1½
F/C-6-RC-50	6"	2000 psi concrete; 1" cover; free or partial restraint at test.	See Note 16	2 hrs.			43	16	2

For SI: 1 inch = 25.4 mm, 1 foot = 305 mm, 1 pound per square inch = 0.00689 MPa, 1 pound per square foot = 47.9 N/m².

Notes :
1. British test.
2. Failure mode—local back face temperature rise.
3. Tested for Grade "C" (2 hour) fire resistance
4. Collapse imminent following hose stream.
5. Failure mode—flame thru.
6. Void formed with explosive force and report.
7. Achieved Grade "B" (4 hour) fire resistance (British).
8. Failure mode—collapse.
9. Test was run to 2 hours, but specimen was partially supported by the furnace at 1¼ hours.
10. Failure mode—average back face temperature.
11. Recommended endurance for nonload bearing performance only.
12. Floor maintained load bearing ability to 2 hours at which point test was terminated.
13. Test was run to 3 hours at which time failure mode 2 (above) was reached in spite of crack formation at 29 minutes.
14. Tested for Grade "A" (6 hour) fire resistance.
15.

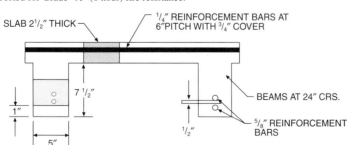

16. Load unspecified.
17. Total assembly thickness 5½ inches. Three-inch thick blocks of molded excelsior bonded with portlandcement used as inserts with 2½-inch cover (concrete) above blocks and ¾-inch gypsum plaster below. Nine-inch wide ribs containing reinforcing steel of unspecified size interrupted 20-inch wide segments of slab composite (i.e., plaster, excelsior blocks, concrete cover).

RESOURCE A

**FIGURE 3.2
FLOOR/CEILING ASSEMBLIES—STEEL STRUCTURAL ELEMENTS**

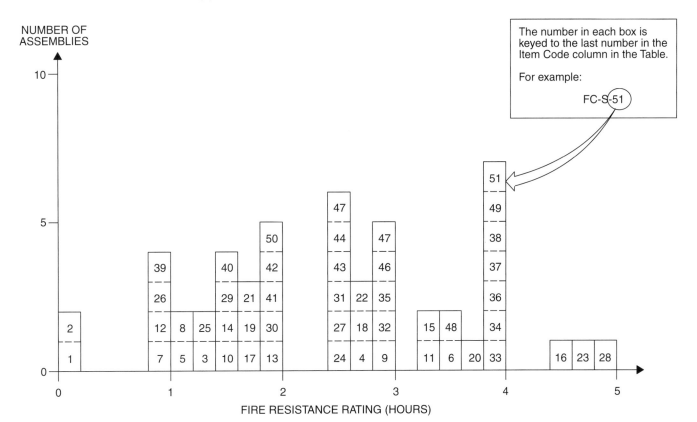

**TABLE 3.2
FLOOR/CEILING ASSEMBLIES—STEEL STRUCTURAL ELEMENTS**

ITEM CODE	MEMBRANE THICKNESS	CONSTRUCTION DETAILS	PERFORMANCE		REFERENCE NUMBER			NOTES	REC. HOURS
			LOAD	TIME	PRE-BMS-92	BMS-92	POST-BMS-92		
F/C-S-1	0″	10′ × 13′6″; S.J. 103 - 24″ o.c.; Deck: 2″ concrete; Membrane: none.	145 psf	7 min.			3	1, 2, 3, 8	0
F/C-S-2	0″	10′ × 13′6″; S.J. 103 - 24″ o.c.; Deck: 2″ concrete; Membrane: none	145 psf	7 min.			3	1, 2, 3, 8	0
F/C-S-3	$^1/_2$″	10′ × 13′ 6″; S.J. 103 - 24″ o.c.; Deck: 2″ concrete 1:2:4; Membrane: furring 12″ o.c.; Clips A, B, G; No extra reinforcement; $^1/_2$″ plaster - 1.5:2.5.	145 psf	1 hr. 15 min.			3	2, 3, 8	$1^1/_4$
F/C-S-4	$^1/_2$″	10′ × 13′ 6″; S.J. 103 - 24″ o.c.; Deck: 2″ concrete 1:2:4; Membrane: furring 16″ o.c.; Clips D, E, F, G; Diagonal wire reinforcement; $^1/_2$″ plaster - 1.5:2.5.	145 psf	2 hrs. 46 min.			3	3, 8	$2^3/_4$
F/C-S-5	$^1/_2$″	10′ × 13′6″; S.J. 103 - 24″ o.c.; Deck: 2″ concrete 1:2:4; Membrane: furring 16″ o.c.; Clips A, B, G; No extra reinforcement; $^1/_2$″ plaster - 1.5:2.5.	145 psf	1 hr. 4 min.			3	2, 3, 8	1
F/C-S-6	$^1/_2$″	10′ × 13′6″; S.J. 103 - 24″ o.c.; Deck: 2″ concrete 1:2:4; Membrane: furring 16″ o.c.; Clips D, E, F, G; Hexagonal mesh reinforcement; $^1/_2$″ plaster.	145 psf	3 hrs. 28 min.			3	2, 3, 8	$2^1/_3$

(continued)

RESOURCE A

TABLE 3.2—continued
FLOOR/CEILING ASSEMBLIES—STEEL STRUCTURAL ELEMENTS

ITEM CODE	MEMBRANE THICKNESS	CONSTRUCTION DETAILS	PERFORMANCE LOAD	PERFORMANCE TIME	REFERENCE NUMBER PRE-BMS-92	REFERENCE NUMBER BMS-92	REFERENCE NUMBER POST-BMS-92	NOTES	REC. HOURS
F/C-S-7	$1/2''$	10′ × 13′6″; S.J. 103 - 24″ o.c.; Deck: 4 lbs. rib lath; 6″ × 6″ - 10 × 10 ga. reinforcement; 2″ deck gravel concrete; Membrane: furring 16″ o.c.; Clips C, E; Reinforcement: none; $1/2''$ plaster - 1.5:2.5 mill mix.	N/A	55 min.			3	5, 8	$3/4$
F/C-S-8	$1/2''$	Spec. 9′ × 4′4″; S.J. 103 bar joists - 18″ o.c.; Deck: 4 lbs. rib lath base; 6″ × 6″ - 10 × 10 ga. reinforcement; 2″ deck 1:2:4 gravel concrete; Membrane: furring, $3/4''$ C.R.S., 16″ o.c.; Clips C, E; Reinforcement: none; $1/2''$ plaster - 1.5:2.5 mill mix.	300 psf	1 hr. 10 min.			3	2, 3, 8	1
F/C-S-9	$5/8''$	10′ × 13′6″; S.J. 103 - 24″ o.c.; Deck: 2″ concrete 1:2:4; Membrane: furring 12″ o.c.; Clips A, B, G; Extra "A" clips reinforcement; $5/8''$ plaster - 1.5:2; 1.5:3.	145 psf	3 hrs.			3	6, 8	3
F/C-S-10	$5/8''$	18′ × 13′6″; Joists, S.J. 103 - 24″ o.c.; Deck: 4 lbs. rib lath; 6″ × 6″ - 10 × 10 ga. reinforcement; 2″ deck 1:2:3.5 gravel concrete; Membrane: furring, spacing 16″ o.c.; Clips C, E; Reinforcement: none; $5/8''$ plaster - 1.5:2.5 mill mix.	145 psf	1 hr. 25 min.			3	2, 3, 8	$1 1/3$
F/C-S-11	$5/8''$	10′ × 13′6″; S.J. 103 - 24″ o.c.; Deck: 2″ concrete 1:2:4; Membrane: furring 12″ o.c.; Clips D, E, F, G; Diagonal wire reinforcement; $5/8''$ plaster - 1.5:2; 0.5:3.	145 psf	3 hrs. 15 min.			3	2, 4, 8	$3 1/4$
F/C-S-12	$5/8''$	10′ × 13′6″; Joists, S.J. 103 - 24″ o.c.; Deck: 3.4 lbs. rib lath; 6″ × 6″ - 10 × 10 ga. reinforcement; 2″ deck 1:2:4 gravel concrete; Membrane: furring 16″ o.c.; Clips D, E, F, G; Reinforcement: none; $5/8''$ plaster - 1.5:2.5.	145 psf	1 hr.			3	7, 8	1
F/C-S-13	$3/4''$	Spec. 9′ × 4′4″; S.J. 103 - 18″ o.c.; Deck: 4 lbs. rib lath; 6″ × 6″ - 10 × 10 ga. reinforcement; 2″ deck 1:2:4 gravel concrete; Membrane: furring, $3/4''$ C.R.S., 16″ o.c.; Clips C, E; Reinforcement: none; $3/4''$ plaster - 1.5:2.5 mill mix.	300 psf	1 hr. 56 min.			3	3, 8	$1 3/4$
F/C-S-14	$7/8''$	Floor finish: 1″ concrete; plate cont. weld; 4″ - 7.7 lbs. "I" beams; Ceiling: $1/4''$ rods 12″ o.c.; $7/8''$ gypsum sand plaster.	105 psf	1 hr. 35 min.			6	2, 4, 9, 10	$1 1/2$
F/C-S-15	1″	Floor finish: $1 1/2''$ L.W. concrete; $1/2''$ limestone cement; plate cont. weld; 5″ - 10 lbs. "I" beams; Ceiling: $1/4''$ rods 12″ o.c. tack welded to beams metal lath; 1″ P. C. plaster.	165 psf	3 hrs. 20 min.			6	4, 9, 11	$3 1/3$
F/C-S-16	1″	10′ × 13′6″; S.J. 103 - 24″ o.c.; Deck: 2″ concrete 1:2:4; Membrane: furring 12″ o.c.; Clips D, E, F, G; Hexagonal mesh reinforcement; 1″ thick plaster - 1.5:2; 1.5:3.	145 psf	4 hrs. 26 min.			3	2, 4, 8	$4 1/3$
F/C-S-17	1″	10′ × 13′6″; Joists - S.J. 103 - 24″ o.c.; Deck: 3.4 lbs. rib lath; 6″ × 6″ - 10 × 10 ga. reinforcement; 2″ deck 1:2:4 gravel concrete; Membrane: furring 16″ o.c.; Clips D, E, F, G; 1″ plaster.	145 psf	1 hr. 42 min.			3	2, 4, 8	$1 2/3$

(continued)

TABLE 3.2—continued
FLOOR/CEILING ASSEMBLIES—STEEL STRUCTURAL ELEMENTS

ITEM CODE	MEMBRANE THICKNESS	CONSTRUCTION DETAILS	PERFORMANCE LOAD	PERFORMANCE TIME	REFERENCE NUMBER PRE-BMS-92	REFERENCE NUMBER BMS-92	REFERENCE NUMBER POST-BMS-92	NOTES	REC. HOURS
F/C-S-18	1 1/8"	10' × 13'6" ; S. J. 103 - 24" o.c.; Deck: 2" concrete 1:2:4; Membrane: furring 12" o.c.; Clips C, E, F, G; Diagonal wire reinforcement; 1 1/8" plaster.	145 psf	2 hrs. 44 min.			3	2, 4, 8	2 2/3
F/C-S-19	1 1/8"	10' × 13'6" ; Joists - S.J. 103 - 24" o.c.; Deck: 1 1/2" gypsum concrete over; 1/2" gypsum board; Membrane: furring 12" o.c.; Clips D, E, F, G; 1 1/8" plaster - 1.5:2; 1.5:3.	145 psf	1 hr. 40 min.			3	2, 3, 8	1 2/3
F/C-S-20	1 1/8"	2 1/2" cinder concrete; 1/2" topping; plate 6" welds 12" o.c.; 5" - 18.9 lbs. "H" center; 5" - 10 lbs. "I" ends; 1" channels 18" o.c.; 1 1/8" gypsum sand plaster.	150 psf	3 hrs. 43 min.			6	2, 4, 9, 11	3 2/3
F/C-S-21	1 1/4"	10' × 13'6" ; Joists - S.J. 103 - 24" o.c.; Deck: 1 1/2" gypsum concrete over; 1/2" gypsum board base; Membrane: furring 12" o.c.; Clips D, E, F, G; 1 1/4" plaster - 1.5:2; 1.5:3.	145 psf	1 hr. 48 min.			3	2, 3, 8	1 2/3
F/C-S-22	1 1/4"	Floor finish: 1 1/2" limestone concrete; 1/2" sand cement topping; plate to beams 3 1/2"; 12" o.c. welded; 5" - 10 lbs. "I" beams; 1" channels 18" o.c.; 1 1/4" wood fiber gypsum sand plaster on metal lath.	292 psf	2 hrs. 45 min.			6	2, 4, 9, 10	2 3/4
F/C-S-23	1 1/2"	2 1/2" L.W. (gas exp.) concrete; Deck: 1/2" topping; plate 6 1/4" welds 12" o.c.; Beams: 5" - 18.9 lbs. "H" center; 5" - 10 lbs. "I" ends; Membrane: 1" channels 18" o.c.; 1 1/2" gypsum sand plaster.	150 psf	4 hrs. 42 min.			6	2, 4, 9	4 2/3
F/C-S-24	1 1/2"	Floor finish: 1 1/2" limestone concrete; 1/2" cement topping; plate 3 1/2" - 12" o.c. welded; 5" - 10 lbs. "I" beams; Ceiling: 1" channels 18" o.c.; 1 1/2" gypsum plaster.	292 psf	2 hrs. 34 min.			6	2, 4, 9, 10	2 1/2
F/C-S-25	1 1/2"	Floor finish: 1 1/2" gravel concrete on exp. metal; plate cont. weld; 4" - 7.7 lbs. "I" beams; Ceiling: 1/4" rods 12" o.c. welded to beams; 1 1/2" fiber gypsum sand plaster.	70 psf	1 hr. 24 min.			6	2, 4, 9, 10	1 1/3
F/C-S-26	2 1/2"	Floor finish: bare plate; 6 1/4" welding - 12" o.c.; 5" - 18.9 lbs. "H" girders (inner); 5" - 10 lbs "I" girders (two outer); 1" channels 18" o.c.; 2" reinforced gypsum tile; 1/2" gypsum sand plaster.	122 psf	1 hr.			6	7, 9, 11	1
F/C-S-27	2 1/2"	Floor finish: 2" gravel concrete; plate to beams 3 1/2" - 12" o.c. welded; 4" - 7.7 lbs. "I" beams; 2" gypsum ceiling tiles; 1/2" 1:3 gypsum sand plaster.	105 psf	2 hrs. 31 min.			6	2, 4, 9, 10	2 1/2
F/C-S-28	2 1/2"	Floor finish: 1 1/2" gravel concrete; 1/2" gypsum asphalt; plate continuous weld; 4" - 7.7 lbs. "I" beams; 12" - 31.8 lbs. "I" beams - girder at 5' from one end; 1" channels 18" o.c.; 2" reinforcement gypsum tile; 1/2" 1:3 gypsum sand plaster.	200 psf	4 hrs. 55 min.			6	2, 4, 9, 11	4 2/3

(continued)

TABLE 3.2—continued
FLOOR/CEILING ASSEMBLIES—STEEL STRUCTURAL ELEMENTS

ITEM CODE	MEMBRANE THICKNESS	CONSTRUCTION DETAILS	PERFORMANCE LOAD	PERFORMANCE TIME	REFERENCE NUMBER PRE-BMS-92	REFERENCE NUMBER BMS-92	REFERENCE NUMBER POST-BMS-92	NOTES	REC. HOURS
F/C-S-29	$^3/_4''$	Floor: 2" reinforced concrete or 2" precast reinforced gypsum tile; Ceiling: $^3/_4''$ portland cement-sand plaster 1:2 for scratch coat and 1:3 for brown coat with 15 lbs. hydrated lime and 3 lbs. of short asbestos fiber bag per cement or $^3/_4''$ sanded gypsum plaster 1:2 for scratch coat and 1:3 for brown coat.	See Note 12	1 hr. 30 min.		1		12, 13, 14	$1^1/_2$
F/C-S-30	$^3/_4''$	Floor: $2^1/_4''$ reinforced concrete or 2" reinforced gypsum tile; the latter with $^1/_4''$ mortar finish; Ceiling: $^3/_4''$ sanded gypsum plaster; 1:2 for scratch coat and 1:3 for brown coat.	See Note 12	2 hrs.		1		12, 13, 14	2
F/C-S-31	$^3/_4''$	Floor: $2^1/_2''$ reinforced concrete or 2" reinforced gypsum tile; the latter with $^1/_4''$ mortar finish; Ceiling: 1" neat gypsum plaster or $^3/_4''$ gypsum-vermiculite plaster, ratio of gypsum to fine vermiculite 2:1 to 3:1.	See Note 12	2 hrs. 30 min.		1		12, 13, 14	$2^1/_2$
F/C-S-32	$^3/_4''$	Floor: $2^1/_2''$ reinforced concrete or 2" reinforced gypsum tile; the latter with $^1/_2''$ mortar finish; Ceiling: 1" neat gypsum plaster or $^3/_4''$ gypsum-vermiculite plaster, ratio of gypsum to fine vermiculite 2:1 to 3:1.	See Note 12	3 hrs.		1		12, 13, 14	3
F/C-S-33	1"	Floor: $2^1/_2''$ reinforced concrete or 2" reinforced gypsum slabs; the latter with $^1/_2''$ mortar finish; Ceiling: 1" gypsum-vermiculite plaster applied on metal lath and ratio 2:1 to 3:1 gypsum to vermiculite by weight.	See Note 12	4 hrs.		1		12, 13, 14	4
F/C-S-34	$2^1/_2''$	Floor: 2" reinforced concrete or 2" precast reinforced portland cement concrete or gypsum slabs; precast slabs to be finished with $^1/_4''$ mortar top coat; Ceiling: 2" precast reinforced gypsum tile, anchored into beams with metal ties or clips and covered with $^1/_2''$ 1:3 sanded gypsum plaster.	See Note 12	4 hrs.		1		12, 13, 14	4
F/C-S-35	1"	Floor: 1:3:6 portland cement, sand and gravel concrete applied directly to the top of steel units and $1^1/_2''$ thick at top of cells, plus $^1/_2''$ $1:2^1/_2''$ cement-sand finish, total thickness at top of cells, 2"; Ceiling: 1" neat gypsum plaster, back of lath 2" or more from underside of cellular steel.	See Note 15	3 hrs.		1		15, 16, 17, 18	3
F/C-S-36	1"	Floor: same as F/C-S-35; Ceiling: 1" gypsum-vermiculite plaster (ratio of gypsum to vermiculite 2:1 to 3:1), the back of lath 2" or more from under-side of cellular steel.	See Note 15	4 hrs.		1		15, 16, 17, 18	4
F/C-S-37	1"	Floor: same as F/C-S-35; Ceiling: 1" neat gypsum plaster; back of lath 9" or more from underside of cellular steel.	See Note 15	4 hrs.		1		15, 16, 17, 18	4
F/C-S-38	1"	Floor: same as F/C-S-35; Ceiling: 1" gypsum-vermiculite plaster (ratio of gypsum to vermiculite 2:1 to 3:1), the back of lath being 9" or more from underside of cellular steel.	See Note 15	5 hrs.		1		15, 16, 17,18	5

(continued)

TABLE 3.2—continued
FLOOR/CEILING ASSEMBLIES—STEEL STRUCTURAL ELEMENTS

ITEM CODE	MEMBRANE THICKNESS	CONSTRUCTION DETAILS	PERFORMANCE LOAD	PERFORMANCE TIME	REFERENCE NUMBER PRE-BMS-92	REFERENCE NUMBER BMS-92	REFERENCE NUMBER POST-BMS-92	NOTES	REC. HOURS
F/C-S-39	$3/4''$	Floor: asbestos paper 14 lbs./100 ft.^{2} cemented to steel deck with waterproof linoleum cement, wood screeds and $7/8''$ wood floor; Ceiling: $3/4''$ sanded gypsum plaster 1:2 for scratch coat and 1:3 for brown coat.	See Note 19	1 hr.		1		19, 20, 21, 22	1
F/C-S-40	$3/4''$	Floor: $1 1/2''$, 1:2:4 portland cement concrete; Ceiling: $3/4''$ sanded gypsum plaster 1:2 for scratch coat and 1:3 for brown coat.	See Note 19	1 hr. 30 min.		1		19, 20, 21, 22	$1 1/2$
F/C-S-41	$3/4''$	Floor: 2'', 1:2:4 portland cement concrete; Ceiling: $3/4''$ sanded gypsum plaster, 1:2 for scratch coat and 1:3 for brown coat.	See Note 19	2 hrs.		1		19, 20, 21, 22	2
F/C-S-42	1''	Floor: 2'', 1:2:4 portland cement concrete; Ceiling: 1'' portland cement-sand plaster with 10 lbs. of hydrated lime for @ bag of cement 1:2 for scratch coat and $1:2 1/2''$ for brown coat.	See Note 19	2 hrs.		1		19, 20, 21, 22	2
F/C-S-43	$1 1/2''$	Floor: 2'', 1:2:4 portland cement concrete; Ceiling: $1 1/2''$, 1:2 sanded gypsum plaster on ribbed metal lath.	See Note 19	2 hrs. 30 min.		1		19, 20, 21, 22	$2 1/2$
F/C-S-44	$1 1/8''$	Floor: 2'', 1:2:4 portland cement concrete; Ceiling: $1 1/8''$, 1:1 sanded gypsum plaster.	See Note 19	2 hrs. 30 min.		1		19, 20, 21, 22	$2 1/2$
F/C-S-45	1''	Floor: $2 1/2''$, 1:2:4 portland cement concrete; Ceiling: 1'', 1:2 sanded gypsum plaster.	See Note 19	2 hrs. 30 min.		1		19, 20, 21, 22	$2 1/2$
F/C-S-46	$3/4''$	Floor: $2 1/2''$, 1:2:4 portland cement concrete; Ceiling: 1'' neat gypsum plaster or $3/4''$ gypsum-vermiculite plaster, ratio of gypsum to vermiculite 2:1 to 3:1.	See Note 19	3 hrs.		1		19, 20, 21, 22	3
F/C-S-47	$1 1/8''$	Floor: $2 1/2''$, 1:2:4 portland cement, sand and cinder concrete plus $1/2''$, $1:2 1/2''$ cement-sand finish; total thickness 3''; Ceiling: $1 1/8''$, 1:1 sanded gypsum plaster.	See Note 19	3 hrs.		1		19, 20, 21, 22	3
F/C-S-48	$1 1/8''$	Floor: $2 1/2''$, gas expanded portland cement-sand concrete plus $1/2''$, 1:2.5 cement-sand finish; total thickness 3''; Ceiling: $1 1/8''$, 1:1 sanded gypsum plaster.	See Note 19	3 hrs. 30 min.		1		19, 20, 21, 22	$3 1/2$
F/C-S-49	1''	Floor: $2 1/2''$, 1:2:4 portland cement concrete; Ceiling: 1'' gypsum-vermiculite plaster; ratio of gypsum to vermiculite 2:1 to 3:1.	See Note 19	4 hrs.		1		19, 20, 21, 22	4
F/C-S-50	$2 1/2''$	Floor: 2'', 1:2:4 portland cement concrete; Ceiling: 2'' interlocking gypsum tile supported on upper face of lower flanges of beams, $1/2''$ 1:3 sanded gypsum plaster.	See Note 19	2 hrs.		1		19, 20, 21, 22	2
F/C-S-51	$2 1/2''$	Floor: 2'', 1:2:4 portland cement concrete; Ceiling: 2'' precast metal reinforced gypsum tile, $1/2''$ 1:3 sanded gypsum plaster (tile clipped to channels which are clipped to lower flanges of beams).	See Note 19	4 hrs.		1		19, 20, 21, 22	4

For SI: 1 inch = 25.4 mm, 1 foot = 305 mm, 1 pound per square inch = 0.00689 MPa, 1 pound per square foot = 47.9 N/m^2.

Notes:
1. No protective membrane over structural steel.
2. Performance time indicates first endpoint reached only several tests were continued to points where other failures occurred.
3. Load failure.

(continued)

TABLE 3.2—continued
FLOOR/CEILING ASSEMBLIES—STEEL STRUCTURAL ELEMENTS

4. Thermal failure.
5. This is an estimated time to load bearing failure. The same joist and deck specimen was used for a later test with different membrane protection.
6. Test stopped at 3 hours to reuse specimen; no endpoint reached.
7. Test stopped at 1 hour to reuse specimen; no endpoint reached.
8. All plaster used = gypsum.
9. Specimen size - 18 feet by $13^1/_2$ inches. Floor deck - base material - $^1/_4$-inch by 18-foot steel plate welded to "I" beams.
10. "I" beams - 24 inches o.c.
11. "I" beams - 48 inches o.c.
12. Apply to open web joists, pressed steel joists or rolled steel beams, which are not stressed beyond 18,000 lbs./in.^{2} in flexure for open-web pressed or light rolled joists, and 20,000 lbs./in.^{2} for American standard or heavier rolled beams.
13. Ratio of weight of portland cement to fine and coarse aggregates combined for floor slabs shall not be less than $1:6^1/_2$.
14. Plaster for ceiling shall be applied on metal lath which shall be tied to supports to give the equivalent of single No. 18 gage steel wires 5 inches o.c.
15. Load: maximum fiber stress in steel not to exceed 16,000 psi.
16. Prefabricated units 2 feet wide with length equal to the span, composed of two pieces of No. 18 gage formed steel welded together to give four longitudinal cells.
17. Depth not less than 3 inches and distance between cells no less than 2 inches.
18. Ceiling: metal lath tied to furring channels secured to runner channels hung from cellular steel.
19. Load: rolled steel supporting beams and steel plate base shall not be stressed beyond 20,000 psi in flexure. Formed steel (with wide upper flange) construction shall not be stressed beyond 16,000 psi.
20. Some type of expanded metal or woven wire shall be embedded to prevent cracking in concrete flooring.
21. Ceiling plaster shall be metal lath wired to rods or channels which are clipped or welded to steel construction. Lath shall be no smaller than 18 gage steel wire and not more than 7 inches o.c.
22. The securing rods or channels shall be at least as effective as single $^3/_{16}$-inch rods with 1-inch of their length bent over the lower flanges of beams with the rods or channels tied to this clip with 14 gage iron wire.

RESOURCE A

FIGURE 3.3
FLOOR/CEILING ASSEMBLIES—WOOD JOIST

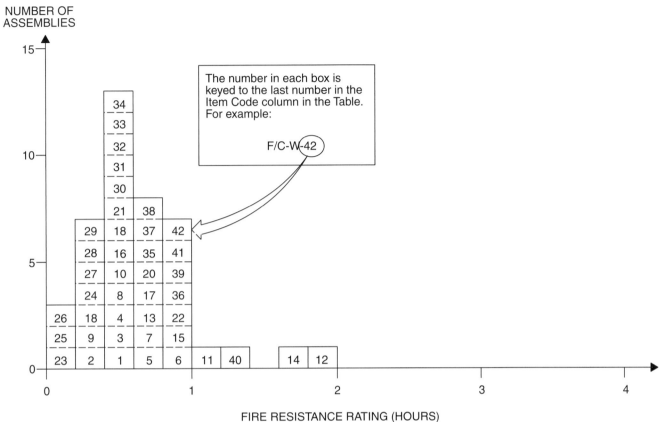

TABLE 3.3
FLOOR/CEILING ASSEMBLIES—WOOD JOIST

ITEM CODE	MEMBRANE THICKNESS	CONSTRUCTION DETAILS	PERFORMANCE LOAD	PERFORMANCE TIME	REFERENCE NUMBER PRE-BMS-92	REFERENCE NUMBER BMS-92	REFERENCE NUMBER POST-BMS-92	NOTES	REC. HOURS
F/C-W-1	$3/8''$	12' clear span - 2" × 9" wood joists; 18" o.c.; Deck: 1" T&G; Filler: 3" of ashes on $1/2''$ boards nailed to joist sides 2" from bottom; 2" air space; Membrane: $3/8''$ gypsum board.	60 psf	36 min.			7	1, 2	$1/2$
F/C-W-2	$1/2''$	12' clear span - 2" × 7" joists; 15" o.c.; Deck: 1" nominal lumber; Membrane: $1/2''$ fiber board.	60 psf	22 min.			7	1, 2, 3	$1/4$
F/C-W-3	$1/2''$	12' clear span - 2" × 7" wood joists; 16" o.c.; 2" × $1^1/_2''$ bridging at center; Deck: 1" T&G; Membrane: $1/2''$ fiber board; 2 coats "distemper" paint.	30 psf	28 min.			7	1, 3, 15	$1/3$
F/C-W-4	$3/16''$	12' clear span - 2" × 7" wood joists; 16" o.c.; 2" × $1^1/_2''$ bridging at center span; Deck: 1" nominal lumber; Membrane: $1/2''$ fiber board under $3/16''$ gypsum plaster.	30 psf	32 min.			7	1, 2	$1/2$
F/C-W-5	$5/8''$	As per previous F/C-W-4 except membrane is $5/8''$ lime plaster.	70 psf	48 min.			7	1, 2	$3/4$
F/C-W-6	$5/8''$	As per previous F/C-W-5 except membrane is $5/8''$ gypsum plaster on 22 gage $3/8''$ metal lath.	70 psf	49 min.			7	1, 2	$3/4$

(continued)

TABLE 3.3—continued
FLOOR/CEILING ASSEMBLIES—WOOD JOIST

ITEM CODE	MEMBRANE THICKNESS	CONSTRUCTION DETAILS	PERFORMANCE LOAD	PERFORMANCE TIME	REFERENCE NUMBER PRE-BMS-92	REFERENCE NUMBER BMS-92	REFERENCE NUMBER POST-BMS-92	NOTES	REC. HOURS
F/C-W-7	$1/2''$	As per previous F/C-W-6 except membrane is $1/2''$ fiber board under $1/2''$ gypsum plaster.	60 psf	43 min.			7	1, 2, 3	$2/3$
F/C-W-8	$1/2''$	As per previous F/C-W-7 except membrane is $1/2''$ gypsum board.	60 psf	33 min.			7	1, 2, 3	$1/2$
F/C-W-9	$9/16''$	12' clear span - 2" × 7" wood joists; 15" o.c.; 2" × $1 1/2''$ bridging at center; Deck: 1" nominal lumber; Membrane: $3/8''$ gypsum board; $3/16''$ gypsum plaster.	60 psf	24 min.			7	1, 2, 3	$1/3$
F/C-W-10	$5/8''$	As per F/C-W-9 except membrane is $5/8''$ gypsum plaster on wood lath.	60 psf	27 min.			7	1, 2, 3	$1/3$
F/C-W-11	$7/8''$	12' clear span - 2" × 9" wood joists; 15" o.c.; 2" × $1 1/2''$ bridging at center span; Deck: 1" T&G; Membrane: original ceiling joists have $3/8''$ plaster on wood lath; 4" metal hangers attached below joists creating 15" chases filled with mineral wool and closed with $7/8''$ plaster (gypsum) on $3/8''$ S.W.M. metal lath to form new ceiling surface.	75 psf	1 hr. 10 min.			7	1, 2	1
F/C-W-12	$7/8''$	12' clear span - 2" × 9" wood joists; 15" o.c.; 2" × $1 1/2''$ bridging at center; Deck: 1" T&G; Membrane: 3" mineral wood below joists; 3" hangers to channel below joists; $7/8''$ gypsum plaster on metal lath attached to channels.	75 psf	2 hrs.			7	1, 4	2
F/C-W-13	$7/8''$	12' clear span - 2" × 9" wood joists; 16" o.c.; 2" × $1 1/2''$ bridging at center span; Deck: 1" T&G on 1" bottoms on $3/4''$ glass wool strips on $3/4''$ gypsum board nailed to joists; Membrane: $3/4''$ glass wool strips on joists; $3/8''$ perforated gypsum lath; $1/2''$ gypsum plaster.	60 psf	41 min.			7	1, 3	$2/3$
F/C-W-14	$7/8''$	12' clear span - 2" × 9" wood joists; 15" o.c.; Deck: 1" T&G; Membrane: 3" foam concrete in cavity on $1/2''$ boards nailed to joists; wood lath nailed to 1" × $1 1/4''$ straps 14 o.c. across joists; $7/8''$ gypsum plaster.	60 psf	1 hr. 40 min.			7	1, 5	$1 2/3$
F/C-W-15	$7/8''$	12' clear span - 2" × 9" wood joists; 18" o.c.; Deck: 1" T&G; Membrane: 2" foam concrete on $1/2''$ boards nailed to joist sides 2" from joist bottom; 2" air space; 1" × $1 1/4''$ wood straps 14" o.c. across joists; $7/8''$ lime plaster on wood lath.	60 psf	53 min.			7	1, 2	$3/4$
F/C-W-16	$7/8''$	12' clear span - 2" × 9" wood joists; Deck: 1" T&G; Membrane: 3" ashes on $1/2''$ boards nailed to joist sides 2" from joist bottom; 2" air space; 1" × $1 1/4''$ wood straps 14" o.c. ; $7/8''$ gypsum plaster on wood lath.	60 psf	28 min.			7	1, 2	$1/3$
F/C-W-17	$7/8''$	As per previous F/C-W-16 but with lime plaster mix.	60 psf	41 min.			7	1, 2	$2/3$
F/C-W-18	$7/8''$	12' clear span - 2" × 9" wood joists; 18" o.c.; 2" × $1 1/2''$ bridging at center; Deck: 1" T&G; Membrane: $7/8''$ gypsum plster on wood lath.	60 psf	36 min.			7	1, 2	$1/2$
F/C-W-19	$7/8''$	As per previous F/C-W-18 except with lime plaster membrane and deck is 1" nominal boards (plain edge).	60 psf	19 min.			7	1, 2	$1/4$

(continued)

TABLE 3.3—continued
FLOOR/CEILING ASSEMBLIES—WOOD JOIST

ITEM CODE	MEMBRANE THICKNESS	CONSTRUCTION DETAILS	PERFORMANCE LOAD	PERFORMANCE TIME	REFERENCE NUMBER PRE-BMS-92	REFERENCE NUMBER BMS-92	REFERENCE NUMBER POST-BMS-92	NOTES	REC. HOURS
F/C-W-20	$7/8''$	As per F/C-W-19, except deck is 1" T&G boards.	60 psf	43 min.			7	1, 2	$2/3$
F/C-W-21	1"	12' clear span - 2" × 9" wood joists; 16" o.c.; 2" × 1 1/2" bridging at center; Deck: 1" T&G; Membrane: $3/8''$ gypsum base board; $5/8''$ gypsum plaster.	70 psf	29 min.			7	1, 2	$1/3$
F/C-W-22	$1 1/8''$	12' clear span - 2" × 9" wood joists; 16" o.c.; 2" × 2" wood bridging at center; Deck: 1" T&G; Membrane: hangers, channel with $3/8''$ gypsum baseboard affixed under $3/4''$ gypsum plaster.	60 psf	1 hr.			7	1, 2, 3	1
F/C-W-23	$3/8''$	Deck: 1" nominal lumber; Joists: 2" × 7"; 15" o.c.; Membrane: $3/8''$ plasterboard with plaster skim coat.	60 psf	$11 1/2$ min.			12	2, 6	$1/6$
F/C-W-24	$1/2''$	Deck: 1" T&G lumber; Joists: 2" × 9"; 16" o.c.; Membrane: $1/2''$ plasterboard.	60 psf	18 min.			12	2, 7	$1/4$
F/C-W-25	$1/2''$	Deck: 1" T&G lumber; Joists: 2" × 7"; 16" o.c.; Membrane: $1/2''$ fiber insulation board.	30 psf	8 min.			12	2, 8	$2/15$
F/C-W-26	$1/2''$	Deck: 1" nominal lumber; Joists: 2" × 7"; 15" o.c.; Membrane: $1/2''$ fiber insulation board.	60 psf	8 min.			12	2, 9	$2/15$
F/C-W-27	$5/8''$	Deck: 1" nominal lumber; Joists: 2" × 7"; 15" o.c.; Membrane: $5/8''$ gypsum plaster on wood lath.	60 psf	17 min.			12	2, 10	$1/4$
F/C-W-28	$5/8''$	Deck: 1" T&G lumber; Joists: 2" × 9"; 16" o.c.; Membrane: $1/2''$ fiber insulation board; $1/2''$ plaster.	60 psf	20 min.			12	2, 11	$1/3$
F/C-W-29	No Membrane	Exposed wood joists.	See Note 13	15 min.		1		1, 12, 13, 14	$1/4$
F/C-W-30	$3/8''$	Gypsum wallboard: $3/8''$ or $1/2''$ with $1 1/2''$ No. 15 gage nails with $3/16''$ heads spaced 6" centers with asbestos paper applied with paperhangers' paste and finished with casein paint.	See Note 13	25 min.		1		1, 12, 13, 14	$1/2$
F/C-W-31	$1/2''$	Gypsum wallboard: $1/2''$ with $1 3/4''$ No. 12 gage nails with $1/2''$ heads, 6" o.c., and finished with casein paint.	See Note 13	25 min.		1		1, 12, 13, 14	$1/2$
F/C-W-32	$1/2''$	Gypsum wallboard: $1/2''$ with $1 1/2''$ No. 12 gage nails with $1/2''$ heads, 18" o.c., with asbestos paper applied with paperhangers' paste and secured with $1 1/2''$ No. 15 gage nails with $3/16''$ heads and finished with casein paint; combined nail spacing 6" o.c.	See Note 13	30 min.		1		1, 12, 13, 14	$1/2$
F/C-W-33	$3/8''$	Gypsum wallboard: two layers $3/8''$ secured with $1 1/2''$ No. 15 gage nails with $3/8''$ heads, 6" o.c.	See Note 13	30 min.		1		1, 12, 13, 14	$1/2$
F/C-W-34	$1/2''$	Perforated gypsum lath: $3/8''$, plastered with $1 1/8''$ No. 13 gage nails with $5/16''$ heads, 4" o.c.; $1/2''$ sanded gypsum plaster.	See Note 13	30 min.		1		1, 12, 13, 14	$1/2$
F/C-W-35	$1/2''$	Same as F/C-W-34, except with $1 1/8''$ No. 13 gage nails with $3/8''$ heads, 4" o.c.	See Note 13	45 min.		1		1, 12, 13, 14	$3/4$

(continued)

TABLE 3.3—continued
FLOOR/CEILING ASSEMBLIES—WOOD JOIST

ITEM CODE	MEMBRANE THICKNESS	CONSTRUCTION DETAILS	PERFORMANCE LOAD	PERFORMANCE TIME	REFERENCE NUMBER PRE-BMS-92	REFERENCE NUMBER BMS-92	REFERENCE NUMBER POST-BMS-92	NOTES	REC. HOURS
F/C-W-36	$1/2''$	Perforated gypsum lath: $3/8''$, nailed with $1 1/8''$ No. 13 gage nails with $3/8''$ heads, 4" o.c.; joints covered with 3" strips of metal lath with $1 3/4''$ No. 12 nails with $1/2''$ heads, 5" o.c.; $1/2''$ sanded gypsum plaster.	See Note 13	1 hr.		1		1, 12, 13, 14	1
F/C-W-37	$1/2''$	Gypsum lath: $3/8''$ and lower layer of $3/8''$ perforated gypsum lath nailed with $1 3/4''$ No. 13 nails with $5/16''$ heads, 4" o.c.; $1/2''$ sanded gypsum plaster or $1/2''$ portland cement plaster.	See Note 13	45 min.		1		1, 12, 13, 14	$3/4$
F/C-W-38	$3/4''$	Metal lath: nailed with $1 1/4''$ No. 11 nails with $3/8''$ heads or 6d common driven 1" and bent over, 6" o.c.; $3/4''$ sanded gypsum plaster.	See Note 13	45 min.		1		1, 12, 13, 14	$3/4$
F/C-W-39	$3/4''$	Same as F/C-W-38, except nailed with $1 1/2''$ No. 11 barbed roof nails with $7/16''$ heads, 6" o.c.	See Note 13	1 hr.		1		1, 12, 13, 14	1
F/C-W-40	$3/4''$	Same as F/C-W-38, except with lath nailed to joists with additional supports for lath 27" o.c.; attached to alternate joists and consisting of two nails driven $1 1/4''$, 2" above bottom on opposite sides of the joists, one loop of No. 18 wire slipped over each nail; the ends twisted together below lath.	See Note 13	1 hr. 15 min.		1		1, 12, 13, 14	$1 1/4$
F/C-W-41	$3/4''$	Metal lath: nailed with $1 1/2''$ No. 11 barbed roof nails with $7/16''$ heads, 6 o.c., with $3/4''$ portland cement plaster for scratch coat and 1:3 for brown coat, 3 lbs. of asbestos fiber and 15 lbs. of hydrated lime/94 lbs. bag of cement.	See Note 13	1 hr.		1		1, 12, 13, 14	1
F/C-W-42	$3/4''$	Metal lath: nailed with 8d, No. $11 1/2$ gage barbed box nails, $2 1/2''$ driven, $1 1/4''$ on slant and bent over, 6" o.c.; $3/4''$ sanded gypsum plaster, 1:2 for scratch coat and 1:3 for below coat.	See Note 13	1 hr.		1		1, 12, 13, 14	1

For SI: 1 inch = 25.4 mm, 1 foot = 305 mm, 1 pound per square inch = 0.00689 MPa, 1 pound per square foot = 47.9 N/m^2.

Notes:
1. Thickness indicates thickness of first membrane protection on ceiling surface.
2. Failure mode—flame thru.
3. Failure mode—collapse.
4. No endpoint reached at termination of test.
5. Failure imminent—test terminated.
6. Joist failure—11.5 minutes; flame thru—13 minutes; collapse—24 minutes.
7. Joist failure—17 minutes; flame thru—18 minutes; collapse—33 minutes.
8. Joist failure—18 minutes; flame thru—8 minutes; collapse—30 minutes.
9. Joist failure—12 minutes; flame thru—8 minutes; collapse—22 minutes.
10. Joist failure—11 minutes; flame thru—17 minutes; collapse—27 minutes.
11. Joist failure—17 minutes; flame thru—20 minutes; collapse—43 minutes.
12. Joists: 2-inch by 10-inch southern pine or Douglas fir; No. 1 common or better. Subfloor: $3/4$-inch wood sheating diaphragm of asbestos paper, and finish of tongue-and-groove wood flooring.
13. Loadings: not more than 1,000 psi maximum fiber stress in joists.
14. Perforations in gypsum lath are to be not less than $3/4$-inch diameter with one perforation for not more than 16/in.^{2} diameter.
15. "Distemper" is a British term for a water-based paint such as white wash or calcimine.

RESOURCE A

**FIGURE 3.4
FLOOR/CEILING ASSEMBLIES—HOLLOW CLAY TILE WITH REINFORCED CONCRETE**

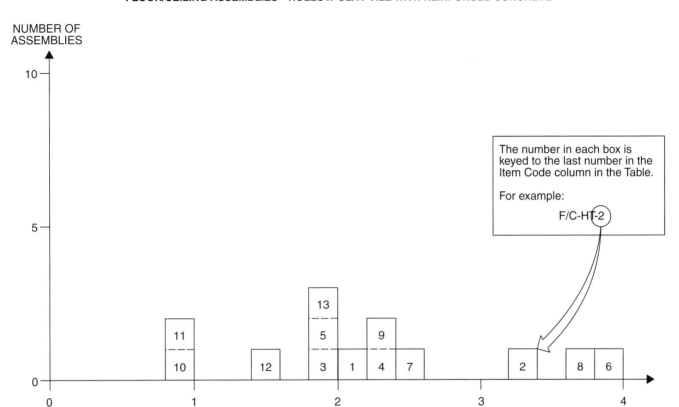

**TABLE 3.4
FLOOR/CEILING ASSEMBLIES—HOLLOW CLAY TILE WITH REINFORCED CONCRETE**

ITEM CODE	ASSEMBLY THICKNESS	CONSTRUCTION DETAILS	PERFORMANCE		REFERENCE NUMBER			NOTES	REC. HOURS
			LOAD	TIME	PRE-BMS-92	BMS-92	POST-BMS-92		
F/C-HT-1	6″	Cover: $1^1/_2$″ concrete (6080 psi); three cell hollow clay tiles, 12″ × 12″ × 4″; $3^1/_4$″ concrete between tiles including two $^1/_2$″ rebars with $^3/_4$″ concrete cover; $^1/_2$″ plaster cover, lower.	75 psf	2 hrs. 7 min.			7	1, 2, 3	2
F/C-HT-2	6″	Cover: $1^1/_2$″ concrete (5840 psi); three cell hollow clay tiles, 12″ × 12″ × 4″; $3^1/_4$″ concrete between tiles including two $^1/_2$″ rebars each with $^1/_2$″ concrete cover and $^5/_8$″ filler tiles between hollow tiles; $^1/_2$″ plaster cover, lower.	61 psf	3 hrs. 23 min.			7	3, 4, 6	$3^1/_3$
F/C-HT-3	6″	Cover: $1^1/_2$″ concrete (6280 psi); three cell hollow clay tiles, 12″ × 12″ × 4″; $3^1/_4$″ concrete between tiles including two $^1/_2$″ rebars with $^1/_2$″ cover; $^1/_2$″ plaster cover, lower.	122 psf	2 hrs.			7	1, 3, 5, 8	2
F/C-HT-4	6″	Cover: $1^1/_2$″ concrete (6280 psi); three cell hollow clay tiles, 12″ × 12″ × 4″; $3^1/_4$″ concrete between tiles including two $^1/_2$″ rebars with $^3/_4$″ cover; $^1/_2$″ plaster cover, lower.	115 psf	2 hrs. 23 min.			7	1, 3, 7	$2^1/_3$
F/C-HT-5	6″	Cover: $1^1/_2$″ concrete (6470 psi); three cell hollow clay tiles, 12″ × 12″ × 4″; $3^1/_4$″ concrete between tiles including two $^1/_2$″ rebars with $^1/_2$″ cover; $^1/_2$″ plaster cover, lower.	122 psf	2 hrs.			7	1, 3, 5, 8	2

(continued)

TABLE 3.4—continued
FLOOR/CEILING ASSEMBLIES—HOLLOW CLAY TILE WITH REINFORCED CONCRETE

ITEM CODE	ASSEMBLY THICKNESS	CONSTRUCTION DETAILS	PERFORMANCE		REFERENCE NUMBER			NOTES	REC. HOURS
			LOAD	TIME	PRE-BMS-92	BMS-92	POST-BMS-92		
F/C-HT-6	8″	Floor cover: $1^1/_2$″ gravel cement (4300 psi); three cell, 12″ × 12″ × 6″; $3^1/_2$″ space between tiles including two $^1/_2$″ rebars with 1″ cover from concrete bottom; $^1/_2$″ plaster cover, lower.	165 psf	4 hrs.			7	1, 3, 9, 10	4
F/C-HT-7	9″ (nom.)	Deck: $^7/_8$″ T&G on 2″ × $1^1/_2$″ bottoms (18″ o.c.) $1^1/_2$″ concrete cover (4600 psi); three cell hollow clay tiles, 12″ × 12″ × 4″; 3″ concrete between tiles including one $^3/_4$″ rebar $^3/_4$″ from tile bottom; $^3/_4$″ plaster cover.	95 psf	2 hrs. 26 min.			7	4, 11, 12, 13	$2^1/_3$
F/C-HT-8	9″ (nom.)	Deck: $^7/_8$″ T&G on 2″ × $1^1/_2$″ bottoms (18″ o.c.) $1^1/_2$″ concrete cover (3850 psi); three cell hollow clay tiles, 12″ × 12″ × 4″; 3″ concrete between tiles including one $^3/_4$″ rebar $^3/_4$″ from tile bottoms; $^1/_2$″ plaster cover.	95 psf	3 hrs. 28 min.			7	4, 11, 12, 13	3
F/C-HT-9	9″ (nom.)	Deck: $^7/_8$″ T&G on 2″ × $1^1/_2$″ bottoms (18″ o.c.) $1^1/_2$″ concrete cover (4200 psi); three cell hollow clay tiles, 12″ × 12″ × 4″; 3″ concrete between tiles including one $^3/_4$″ rebar $^3/_4$″ from tile bottoms; $^1/_2$″ plaster cover.	95 psf	2 hrs. 14 min.			7	3, 5, 8, 11	2
F/C-HT-10	$5^1/_2$″	Fire clay tile (4″ thick); $1^1/_2$″ concrete cover; for general details, see Note 15.	See Note 14	1 hr.			43	15	1
F/C-HT-11	8″	Fire clay tile (6″ thick); 2″ cover.	See Note 14	1 hr.			43	15	1
F/C-HT-12	$5^1/_2$″	Fire clay tile (4″ thick); $1^1/_2$″ cover; $^5/_8$″ gypsum plaster, lower.	See Note 14	1 hr. 30 min.			43	15	$1^1/_2$
F/C-HT-13	8″	Fire clay tile (6″ thick); 2″ cover; $^5/_8$″ gypsum plaster, lower.	See Note 14	2 hrs.			43	15	$1^1/_2$

For SI: 1 inch = 25.4 mm, 1 foot = 305 mm, 1 pound per square inch = 0.00689 MPa, 1 pound per square foot = 47.9 N/m².

Notes:

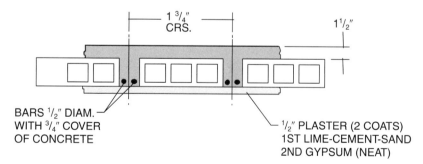

1. A generalized cross section of this floor type follows:
2. Failure mode - structural.
3. Plaster: base coat—lime-cement-sand; top coat—gypsum (neat).

(continued)

TABLE 3.4—continued
FLOOR/CEILING ASSEMBLIES-HOLLOW CLAY TILE WITH REINFORCED CONCRETE

4. Failure mode—collapse.
5. Test stopped before any endpoints were reached.
6. A generalized cross section of this floor type follows:

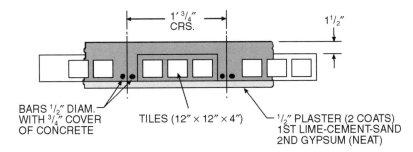

7. Failure mode—thermal—back face temperature rise.
8. Passed hose stream test.
9. Failed hose stream test.

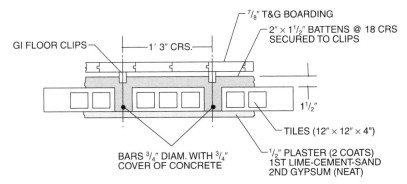

10. Test stopped at 4 hours before any endpoints were reached.
11. A generalized cross section of this floor type follows:
12. Plaster: base coat—retarded hemihydrate gypsum-sand; second coat—neat gypsum.
13. Concrete in Item 7 is P.C. based but with crushed brick aggregates while in Item 8 river sand and river gravels are used with the P.C.
14. Load - unspecified.
15. The 12-inch by 12-inch fire-clay tiles were laid end to end in rows spaced $2^1/_2$ inches or 4 inches apart. The reinforcing steel was placed between these rows and the concrete cast around them and over the tile to form the structural floor.

SECTION IV
BEAMS

TABLE 4.1.1
REINFORCED CONCRETE BEAMS
DEPTH 10″ TO LESS THAN 12″

ITEM CODE	DEPTH	CONSTRUCTION DETAILS	PERFORMANCE		REFERENCE NUMBER			NOTES	REC. HOURS
			LOAD	TIME	PRE-BMS-92	BMS-92	POST-BMS-92		
B-11-RC-1	11″	24″ wide × 11″ deep reinforced concrete "T" beam (3290 psi); Details: see Note 5 figure.	8.8 tons	4 hrs. 2 min.			7	1, 2, 14	4
B-10-RC-2	10″	24″ wide × 10″ deep reinforced concrete "T" beam (4370 psi); Details: see Note 6 figure.	8.8 tons	1 hr. 53 min.			7	1, 3	$1^3/_4$
B-10-RC-3	$10^1/_2$″	24″ wide × $10^1/_2$″ deep reinforced concrete "T" beam (4450 psi); Details: see Note 7 figure.	8.8 tons	2 hrs. 40 min.			7	1, 3	$2^2/_3$
B-11-RC-4	11″	24″ wide × 11″ deep reinforced concrete "T" beam (2400 psi); Details: see Note 8 figure.	8.8 tons	3 hrs. 32 min.			7	1, 3, 14	$3^1/_2$
B-11-RC-5	11″	24″ wide × 11″ deep reinforced concrete "T" beam (4250 psi); Details: see Note 9 figure.	8.8 tons	3 hrs. 3 min.			7	1, 3, 14	3
B-11-RC-6	11″	Concrete flange: 4″ deep × 2′ wide (4895 psi) concrete; Concrete beam: 7″ deep × $6^1/_2$″ wide beam; "I" beam reinforcement; 10″ × $4^1/_2$″ × 25 lbs. R.S.J.; 1″ cover on flanges; Flange reinforcement: $^3/_8$″ diameter bars at 6″ pitch parallel to "T"; $^1/_4$″ diameter bars perpendicular to "T"; Beam reinforcement: 4″ × 6″ wire mesh No. 13 SWG; Span: 11′ restrained; Details: see Note 10 figure.	10 tons	6 hrs.			7	1, 4	6
B-11-RC-7	11″	Concrete flange: 6″ deep × 1′ $6^1/_2$″ wide (3525 psi) concrete; Concrete beam: 5″ deep × 8″ wide precast concrete blocks $8^3/_4$″ long; "I" beam reinforcement; 7″ × 4″ × 16 lbs. R.S.J.; 2″ cover on bottom; $1^1/_2$″ cover on top; Flange reinforcement: two rows $^1/_2$″ diameter rods parallel to "T"; Beam reinforcement: $^1/_8$″ wire mesh perpendicular to 1″ ; Span: 1′ 3″ simply supported; Details: see Note 11 figure.	3.9 tons	4 hrs.			7	1, 2	4
B-11-RC-8	11″	Concrete flange: 4″ deep × 2′ wide (3525 psi) concrete; Concrete beam 7″ deep × $4^1/_2$″ wide; (scaled from drawing); "I" beam reinforcement; 10″ × $4^1/_2$″ × 25 lbs. R.S.J.; no concrete cover on bottom; Flange reinforcement: $^3/_8$″ diameter bars at 6 pitch parallel to "T"; $^1/_4$″ diameter bars perpendicular to "T"; Span: 11′ restricted.	10 tons	4 hrs.			7	1, 2, 12	4
B-11-RC-9	$11^1/_2$″	24″ wide × $11^1/_2$″ deep reinforced concrete "T" beam (4390 psi); Details: see Note 12 figure.	8.8 tons	3 hrs. 24 min.			7	1, 3	$3^1/_3$

For SI: 1 inch = 25.4 mm, 1 foot = 305 mm, 1 pound = 0.004448 kN, 1 pound per square inch = 0.00689 MPa, 1 ton = 8.896 kN.

Notes:
1. Load concentrated at mid span.
2. Achieved 4 hour performance (Class "B," British).
3. Failure mode—collapse.
4. Achieved 6 hour performance (Class "A," British).

(continued)

**TABLE 4.1.1—continued
REINFORCED CONCRETE BEAMS
DEPTH 10″ TO LESS THAN 12″**

5. 6.

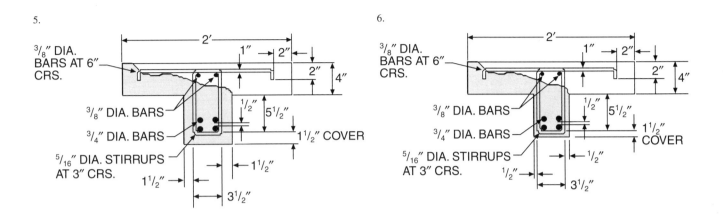

7. 8.

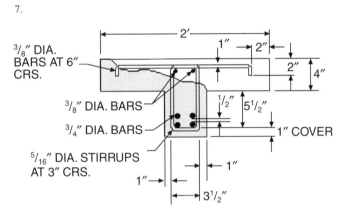
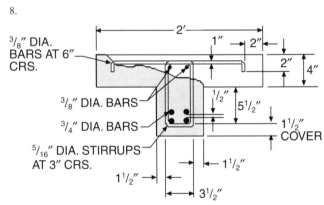

9. 10.

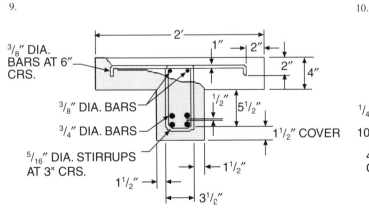
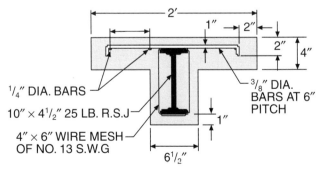

(continued)

**TABLE 4.1.1—continued
REINFORCED CONCRETE BEAMS
DEPTH 10″ TO LESS THAN 12″**

11.

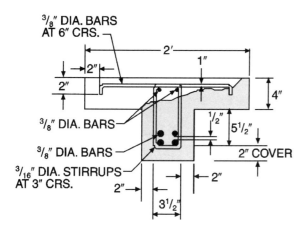

SPAN AND END CONDITIONS:-10′-3″ (CLEAR). SIMPLY SUPPORTED.

12.

13.

14. The different performances achieved by B-11-RC-1, B-11-RC-4 and B-11-RC-5 are attributable to differences in concrete aggregate compositions reported in the source document but unreported in this table. This demonstrates the significance of material composition in addition to other details.

**TABLE 4.1.2
REINFORCED CONCRETE BEAMS
DEPTH 12″ TO LESS THAN 14″**

ITEM CODE	DEPTH	CONSTRUCTION DETAILS	PERFORMANCE		REFERENCE NUMBER			NOTES	REC. HOURS
			LOAD	TIME	PRE-BMS-92	BMS-92	POST-BMS-92		
B-12-RC-1	12″	12″ × 8″ section; 4160 psi aggregate concrete; Reinforcement: 4-$^7/_8$″ rebars at corners; 1″ below each surface; $^1/_4$″ stirrups 10″ o.c.	5.5 tons	2 hrs.			7	1	2
B-12-RC-2	12″	Concrete flange: 4″ deep × 2′ wide (3045 psi) concrete at 35 days; Concrete beam: 8″ deep; "I" beam reinforcement: 10″ × 4$^1/_2$″ × 25 lbs. R.S.J.; 1″ cover on flanges; Flange reinforcement: $^3/_8$″ diameter bars at 6″ pitch parallel to "T"; $^1/_4$″ diameter bars perpendicular to "T"; Beam reinforcement: 4″ × 6″ wire mesh No. 13 SWG; Span: 10′ 3″ simply supported.	10 tons	4 hrs.			7	2, 3, 5	4
B-13-RC-3	13″	Concrete flange: 4″ deep × 2′ wide (3825 psi) concrete at 46 days; Concrete beam: 9″ deep × 8$^1/_2$″ wide; (scaled from drawing); "I" beam reinforcement: 10″ × 4$^1/_2$″ × 25 lbs. R.S.J.; 3″ cover on bottom flange; 1″ cover on top flange; Flange reinforcement: $^3/_8$″ diameter bars at 6″ pitch parallel to "T"; $^1/_4$″ diameter bars perpendicular to "T"; Beam reinforcement: 4″ × 6″ wire mesh No. 13 SWG; Span: 11′ restrained.	10 tons	6 hrs.			7	2, 3, 6, 8, 9	4
B-12-RC-4	12″	Concrete flange: 4″ deep × 2′ wide (3720 psi) concrete at 42 days; Concrete beam: 8″ deep × 8$^1/_2$″ wide; (scaled from drawing); "I" beam reinforcement: 10″ × 4$^1/_2$″ × 25 lbs. R.S.J.; 2″ cover bottom flange; 1″ cover top flange; Flange reinforcement: $^3/_8$″ diameter bars at 6″ pitch parallel to "T"; $^1/_4$″ diameter bars perpendicular to "T"; Beam reinforcement: 4″ × 6″ wire mesh No. 13 SWG; Span: 11′ restrained.	10 tons	6 hrs.			7	1, 3, 4, 7, 8, 9	4

For SI: 1 inch = 25.4 mm, 1 foot = 305 mm, 1 pound = 0.004448 kN, 1 pound per square inch = 0.00689 MPa, 1 ton = 8.896 kN.
Notes:
1. Qualified for 2 hour use. (Grade "C," British) Test included hose stream and reload at 48 hours.
2. Load concentrated at mid span.
3. British test.
4. British test—qualified for 6 hour use (Grade "A").

(continued)

**TABLE 4.1.2—continued
REINFORCED CONCRETE BEAMS
DEPTH 12″ TO LESS THAN 14″**

5.

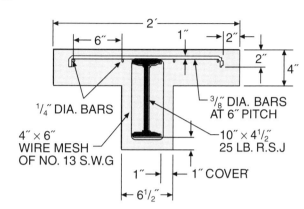

6.

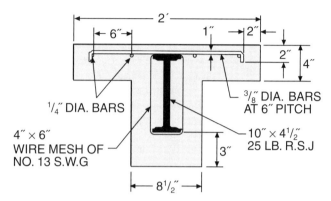

7.

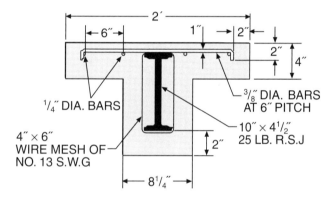

8. See Table 4.1.3, Note 5.

9. Hourly rating based upon B-12-RC-2 above.

RESOURCE A

**TABLE 4.1.3
REINFORCED CONCRETE BEAMS
DEPTH 14″ TO LESS THAN 16″**

ITEM CODE	DEPTH	CONSTRUCTION DETAILS	PERFORMANCE		REFERENCE NUMBER			NOTES	REC. HOURS
			LOAD	TIME	PRE-BMS-92	BMS-92	POST-BMS-92		
B-15-RC-1	15″	Concrete flange: 4″ deep × 2′ wide (3290 psi) concrete; Concrete beam: 10″ deep × 8½″ wide; "I" beam reinforcement: 10″ × 4½″ × 25 lbs. R.S.J.; 4″ cover on bottom flange; 1″ cover on top flange; Flange reinforcement: ⅜″ diameter bars at 6″ pitch parallel to "T"; ¼″ diameter bars perpendicular to "T"; Beam reinforcement: 4″ × 6″ wire mesh No. 13 SWG; Span: 11′ restrained.	10 tons	6 hrs.			7	1, 2, 3, 5, 6	4
B-15-RC-2	15″	Concrete flange: 4″ deep × 2′ wide (4820 psi) concrete; Concrete beam: 10″ deep × 8½″ wide; "I" beam reinforcement: 10″ × 4½″ × 25 lbs. R.S.J.; 1″ cover over wire mesh on bottom flange; 1″ cover on top flange; Flange reinforcement: ⅜″ diameter bars at 6″ pitch parallel to "T"; ¼″ diameter bars perpendicular to "T"; Beam reinforcement: 4″ × 6″ wire mesh No. 13 SWG; Span: 11′ restrained.	10 tons	6 hrs.			7	1, 2, 4, 5, 6	4

For SI: 1 inch = 25.4 mm, 1 foot = 305 mm, 1 pound = 0.004448 kN, 1 pound per square inch = 0.00689 MPa, 1 ton = 8.896 kN.

Notes:
1. Load concentrated at mid span.
2. Achieved 6 hour fire rating (Grade "A," British).
3.
4.

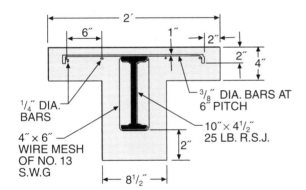
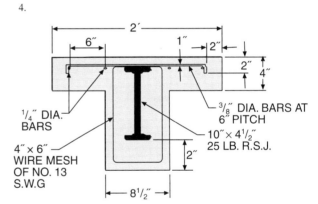

5. Section 43.147 of the 1979 edition of the *Uniform Building Code Standards* provides:
"A restrained condition in fire tests, as used in this standard, is one in which expansion at the supports of a load-carrying element resulting from the effects of the fire is resisted by forces external to the element. An unrestrained condition is one in which the load-carrying element is free to expand and rotate at its support."
"Restraint in buildings is defined as follows: Floor and roof assemblies and individual beams in buildings shall be considered restrained when the surrounding or supporting structure is capable of resisting the thermal expansion throughout the range of anticipated elevated temperatures. Construction not complying . . . is assumed to be free to rotate and expand and shall be considered as unrestrained."
"Restraint may be provided by the lateral stiffness of supports for floor and roof assemblies and intermediate beams forming part of the assembly. In order to develop restraint, connections must adequately transfer thermal thrusts to such supports. The rigidity of adjoining panels or structures shall be considered in assessing the capability of a structure to resist therm expansion."
Because it is difficult to determine whether an existing building's structural system is capable of providing the required restraint, the lower hourly ratings of a similar but unrestrained assembly have been recommended.
6. Hourly rating based upon Table 4.2.1, Item B-12-RC-2.

RESOURCE A

**TABLE 4.2.1
REINFORCED CONCRETE BEAMS—UNPROTECTED DEPTH
10″ TO LESS THAN 12″**

ITEM CODE	DEPTH	CONSTRUCTION DETAILS	PERFORMANCE		REFERENCE NUMBER			NOTES	REC. HOURS
			LOAD	TIME	PRE-BMS-92	BMS-92	POST-BMS-92		
B-SU-1	10″	10″ × 4$\frac{1}{2}$″ × 25 lbs. "I" beam.	10 tons	39 min.			7	1	$\frac{1}{3}$

For SI: 1 inch = 25.4 mm, 1 pound = 0.004448 kN, 1 ton = 8.896 kN.
Notes:
1. Concentrated at mid span.

**TABLE 4.2.2
STEEL BEAMS—CONCRETE PROTECTION DEPTH
10″ TO LESS THAN 12″**

ITEM CODE	DEPTH	CONSTRUCTION DETAILS	PERFORMANCE		REFERENCE NUMBER			NOTES	REC. HOURS
			LOAD	TIME	PRE-BMS-92	BMS-92	POST-BMS-92		
B-SC-1	10″	10″ × 8″ rectangle; aggregate concrete (4170 psi) with 1″ top cover and 2″ bottom cover; No. 13 SWG iron wire loosely wrapped at approximately 6″ pitch about 7″ × 4″ × 16 lbs. "I" beam.	3.9 tons	3 hrs. 46 min.			7	1, 2, 3	3$\frac{3}{4}$
B-SC-1	10″	10″ × 8″ rectangle; aggregate concrete (3630 psi) with 1″ top cover and 2″ bottom cover; No. 13 SWG iron wire loosely wrapped at approximately 6″ pitch about 7″ × 4″ × 16 lbs. "I" beam.	5.5 tons	5 hrs. 26 min.			7	1, 4, 5, 6, 7	3$\frac{3}{4}$

For SI: 1 inch = 25.4 mm, 1 pound = 0.004448 kN, 1 pound per square inch = 0.00689 MPa, 1 ton = 8.896 kN.
Notes:
1. Load concentrated at mid span.
2. Specimen 10-foot 3-inch clear span simply supported.
3. Passed Grade "C" fire resistance (British) including hose stream and reload.
4. Specimen 11-foot clear span—restrained.
5. Passed Grade "B" fire resistance (British) including hose stream and reload.
6. See Table 4.1.3, Note 5.
7. Hourly rating based upon B-SC-1 above.

SECTION V
DOORS

**FIGURE 5.1
RESISTANCE OF DOORS TO FIRE EXPOSURE**

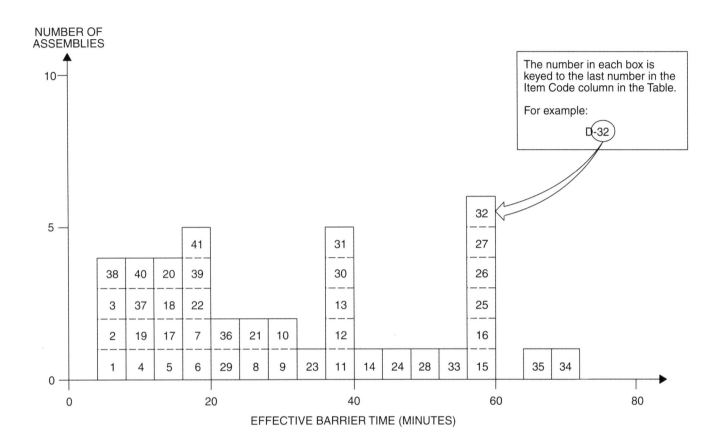

**TABLE 5.1
RESISTANCE OF DOORS TO FIRE EXPOSURE**

ITEM CODE	DOOR MINIMUM THICKNESS	CONSTRUCTION DETAILS	PERFORMANCE		REFERENCE NUMBER			NOTES	REC. (MIN.)
			EFFECTIVE BARRIER	EDGE FLAMING	PRE- BMS-92	BMS-92	POST- BMS-92		
D-1	$3/8''$	Panel door; pine perimeter ($1 3/8''$); painted (enamel).	5 min. 10 sec.	N/A			90	1, 2	5
D-2	$3/8''$	As above, with two coats U.L. listed intumescent coating.	5 min. 30 sec.	5 min.			90	1, 2, 7	5
D-3	$3/8''$	As D-1, with standard primer and flat interior paint.	5 min. 55 sec.	N/A			90	1, 3, 4	5
D-4	$2 5/8''$	As D-1, with panels covered each side with $1/2''$ plywood; edge grouted with sawdust filled plaster; door faced with $1/8''$ hardboard each side; paint see (5).	11 min. 15 sec.	3 min. 45 sec.			90	1, 2, 5, 7	10

(continued)

TABLE 5.1—continued
RESISTANCE OF DOORS TO FIRE EXPOSURE

ITEM CODE	DOOR MINIMUM THICKNESS	CONSTRUCTION DETAILS	PERFORMANCE		REFERENCE NUMBER			NOTES	REC. (MIN.)
			EFFECTIVE BARRIER	EDGE FLAMING	PRE-BMS-92	BMS-92	POST-BMS-92		
D-5	$3/8''$	As D-1, except surface protected with glass fiber reinforced intumescent fire retardant coating.	16 min.	N/A			90	1, 3, 4, 7	15
D-6	$1 5/8''$	Door detail: As D-4, except with $1/8''$ cement asbestos board facings with aluminum foil; door edges protected by sheet metal.	17 min.	10 min. 15 sec.			90	1, 3, 4	15
D-7	$1 5/8''$	Door detail with $1/8''$ hardboard cover each side as facings; glass fiber reinforced intumescent coating applied.	20 min.	N/A			90	1, 3, 4, 7	20
D-8	$1 5/8''$	Door detail same as D-4; paint was glass reinforced epoxy intumescent.	26 min.	24 min. 45 sec.			90	1, 3, 4, 6, 7	25
D-9	$1 5/8''$	Door detail same as D-4 with facings of $1/8''$ cement asbestos board.	29 min.	3 min. 15 sec.			90	1, 2	5
D-10	$1 5/8''$	As per D-9.	31 min. 30 sec.	7 min. 20 sec.			90	1, 3, 4	6
D-11	$1 5/8''$	As per D-7; painted with epoxy intumescent coating including glass fiber roving.	36 min. 25 sec.	N/A			90	1, 3, 4	35
D-12	$1 5/8''$	As per D-4 with intumescent fire retardant paint.	37 min. 30 sec.	24 min. 40 sec.			90	1, 3, 4	30
D-13	$1 1/2''$ (nom.)	As per D-4, except with 24 ga. galvanized sheet metal facings.	39 min.	39 min.			90	1, 3, 4	39
D-14	$1 5/8''$	As per D-9.	41 min. 30 sec.	17 min. 20 sec.			90	1, 3, 4, 6	20
D-15	—	Class C steel fire door.	60 min.	58 min.			90	7, 8	60
D-16	—	Class B steel fire door.	60 min.	57 min.			90	7, 8	60
D-17	$1 3/4''$	Solid core flush door; core staves laminated to facings but not each other; Birch plywood facings $1/2''$ rebate in door frame for door; $3/32''$ clearance between door and wood frame.	15 min.	13 min.			37	11	13

(continued)

TABLE 5.1—continued
RESISTANCE OF DOORS TO FIRE EXPOSURE

ITEM CODE	DOOR MINIMUM THICKNESS	CONSTRUCTION DETAILS	PERFORMANCE		REFERENCE NUMBER			NOTES	REC. (MIN.)
			EFFECTIVE BARRIER	EDGE FLAMING	PRE- BMS-92	BMS-92	POST- BMS-92		
D-18	1³/₄"	As per D-17.	14 min.	13 min.			37	11	13
D-19	1³/₄"	Door same as D-17, except with 16 ga. steel; ³/₃₂" door frame clearance.	12 min.	—			37	9, 11	10
D-20	1³/₄"	As per D-19.	16 min.	—			37	10, 11	10
D-21	1³/₄"	Doors as per D-17; intumescent paint applied to top and side edges.	26 min.	—			37	11	25
D-22	1³/₄"	Door as per D-17, except with ¹/₂" × ¹/₈" steel strip set into edges of door at top and side facing stops; matching strip on stop.	18 min.	6 min.			37	11	18
D-23	1³/₄"	Solid oak door.	36 min.	22 min.			15	13	25
D-24	1⁷/₈"	Solid oak door.	45 min.	35 min.			15	13	35
D-25	1⁷/₈"	Solid teak door.	58 min.	34 min.			15	13	35
D-26	1⁷/₈"	Solid (pitch) pine door.	57 min.	36 min.			15	13	35
D-27	1⁷/₈"	Solid deal (pine) door.	57 min.	30 min.			15	13	30
D-28	1⁷/₈"	Solid mahogany door.	49 min.	40 min.			15	13	45
D-29	1⁷/₈"	Solid poplar door.	24 min.	3 min.			15	13, 14	5
D-30	1⁷/₈"	Solid oak door.	40 min.	33 min.			15	13	35
D-31	1⁷/₈"	Solid walnut door.	40 min.	15 min.			15	13	20
D-32	2⁵/₈"	Solid Quebec pine.	60 min.	60 min.			15	13	60
D-33	2⁵/₈"	Solid pine door.	55 min.	39 min.			15	13	40
D-34	2⁵/₈"	Solid oak door.	69 min.	60 min.			15	13	60
D-35	2⁵/₈"	Solid teak door.	65 min.	17 min.			15	13	60
D-36	1¹/₂"	Solid softwood door.	23 min.	8.5 min.			15	13	10
D-37	³/₄"	Panel door.	8 min.	7.5 min.			15	13	5
D-38	⁵/₁₆"	Panel door.	5 min.	5 min.			15	13	5
D-39	³/₄"	Panel door, fire retardant treated.	17¹/₂ min.	3 min.			15	13	8
D-40	³/₄"	Panel door, fire retardant treated.	8¹/₂ min.	8¹/₂ min.			15	13	8
D-41	³/₄"	Panel door, fire retardant treated.	16³/₄ min.	11¹/₂ min.			15	13	8

For SI: 1 inch = 25.4 mm, 1 foot = 305 mm.

Notes:
1. All door frames were of standard lumber construction.
2. Wood door stop protected by asbestos millboard.
3. Wood door stop protected by sheet metal.
4. Door frame protected with sheet metal and weather strip.
5. Surface painted with intumescent coating.
6. Door edge sheet metal protected.
7. Door edge intumescent paint protected.
8. Formal steel frame and door stop.
9. Door opened into furnace at 12 feet.
10. Similar door opened into furnace at 12 feet.
11. The doors reported in these tests represent the type contemporaries used as 20 minute solid-core wood doors. The test results demonstrate the necessity of having wall anchored metal frames, minimum cleaners possible between door, frame and stops. They also indicate the utility of long throw latches and the possible use of intumescent paints to seal doors to frames in event of a fire.
12. Minimum working clearance and good latch closure are absolute necessities for effective containment for all such working door assemblies.
13. Based on British tests.
14. Failure at door-frame interface.

BIBLIOGRAPHY

1. Central Housing Committee on Research, Design, and Construction; Subcommittee on Fire Resistance Classifications, "Fire-Resistance Classifications of Building Constructions," *Building Materials and Structures,* Report BMS 92, National Bureau of Standards, Washington, Oct. 1942. (Available from NTIS No. COM-73-10974)

2. Foster, H. D., Pinkston, E. R., and Ingberg, S. H., "Fire Resistance of Structural Clay Tile Partitions," *Building Materials and Structures, Report* BMS 113, National Bureau of Standards, Washington, Oct. 1948.

3. Ryan, J. V., and Bender, E.W., "Fire Endurance of Open-Web Steel-Joist Floors with Concrete Slabs and Gypsum Ceilings," *Building Materials and Structures,* Report BMS 141, National Bureau of Standards, Washington, Aug. 1954.

4. Mitchell, N. D., "Fire Tests of Wood-Framed Walls and Partitions with Asbestos-Cement Facings," *Building Materials and Structures,* Report BMS 123, National Bureau of Standards, Washington, May 1951.

5. Robinson, H. E., Cosgrove, L. A., and Powell, F. J., "Thermal Resistance of Airspace and Fibrous Insulations Bounded by Reflective Surfaces," *Building Materials and Structures,* Report BMS 151, National Bureau of Standards, Washington, Nov. 1957.

6. Shoub, H., and Ingberg, S. H., "Fire Resistance of Steel Deck Floor Assemblies," *Building Science Series,* 11, National Bureau of Standards, Washington, Dec. 1967.

7. Davey, N., and Ashton, L. A., "Investigations on Building Fires, Part V: Fire Tests of Structural Elements," *National Building Studies,* Research Paper, No. 12, Dept. of Scientific and Industrial Research (Building Research Station), London, 1953.

8. National Board of Fire Underwriters, *Fire Resistance Ratings of Beam, Girder, and Truss Protections, Ceiling Constructions, Column Protections, Floor and Ceiling Constructions, Roof Constructions, Walls and Partitions,* New York, April 1959.

9. Mitchell, N.D., Bender, E.D., and Ryan, J.V., "Fire Resistance of Shutters for Moving-Stairway Openings," *Building Materials and Structures,* Report BMS 129, National Bureau of Standards, Washington, March 1952.

10. National Board of Fire Underwriters, *National Building Code; an Ordinance Providing for Fire Limits, and Regulations Governing the Construction, Alteration, Equipment, or Removal of Buildings or Structures,* New York, 1949.

11. Department of Scientific and Industrial Research and of the Fire Offices' Committee, Joint Committee of the Building Research Board, "Fire Gradings of Buildings, Part I: General Principles and Structural Precautions," *Post-War Building Studies,* No. 20, Ministry of Works, London, 1946.

12. Lawson, D. I., Webster, C. T., and Ashton, L. A., "Fire Endurance of Timber Beams and Floors," *National Building Studies,* Bulletin, No. 13, Dept. of Scientific and Industrial Research and Fire Offices' Committee (Joint Fire Research Organization), London, 1951.

13. Parker, T. W., Nurse, R. W., and Bessey, G. E., "Investigations on Building Fires. Part I: The Estimation of the Maximum Temperature Attained in Building Fires from Examination of the Debris, and Part II: The Visible Change in Concrete or Mortar Exposed to High Temperatures," *National Building Studies,* Technical Paper, No. 4, Dept. of Scientific and Industrial Research (Building Research Station), London, 1950.

14. Bevan, R. C., and Webster, C. T., "Investigations on Building Fires, Part III: Radiation from Building Fires," *National Building Studies,* Technical Paper, No. 5, Dept. of Scientific and Industrial Research (Building Research Station), London, 1950.

15. Webster, D. J., and Ashton, L. A., "Investigations on Building Fires, Part IV: Fire Resistance of Timber Doors," *National Building Studies,* Technical Paper, No. 6, Dept. of Scientific and Industrial Research (Building Research Station), London, 1951.

16. Kidder, F. E., *Architects' and Builders' Handbook: Data for Architects, Structural Engineers, Contractors, and Draughtsmen,* comp. by a Staff of Specialists and H. Parker, editor-in-chief, 18th ed., enl., J. Wiley, New York, 1936.

17. Parker, H., Gay, C. M., and MacGuire, J. W., *Materials and Methods of Architectural Construction,* 3rd ed., J. Wiley, New York, 1958.

18. Diets, A. G. H., *Dwelling House Construction,* The MIT Press, Cambridge, 1971.

19. Crosby, E. U., and Fiske, H. A., *Handbook of Fire Protection,* 5th ed., The Insurance Field Company, Louisville, Ky., 1914.

20. Crosby, E. U., Fiske, H. A., and Forster, H.W., *Handbook of Fire Protection,* 8th ed., R. S. Moulton, general editor, National Fire Protection Association, Boston, 1936.

21. Kidder, F. E., *Building Construction and Superintendence,* rev. and enl., by T. Nolan, W. T. Comstock, New York, 1909-1913, 2 vols.

22. National Fire Protection Association, Committee on Fire-Resistive Construction, *The Baltimore Conflagration,* 2nd ed., Chicago, 1904.

23. Przetak, L., *Standard Details for Fire-Resistive Building Construction,* McGraw-Hill Book Co., New York, 1977.

24. Hird, D., and Fischl, C. F., "Fire Hazard of Internal Linings," *National Building Studies,* Special Report, No. 22, Dept. of Scientific and Industrial Research and Fire Offices' Committee (Joint Fire Research Organization), London, 1954.

25. Menzel, C. A., *Tests of the Fire-Resistance and Strength of Walls Concrete Masonry Units,* Portland Cement Association, Chicago, 1934.

26. Hamilton, S. B., "A Short History of the Structural Fire Protection of Buildings Particularly in England," *National Building Studies,* Special Report, No. 27, Dept. of Scientific and Industrial Research (Building Research Station), London, 1958.

27. Sachs, E. O., and Marsland, E., "The Fire Resistance of Doors and Shutters being Tabulated Results of Fire Tests Conducted by the Committee," *Journal of the British Fire Prevention Committee,* No. VII, London, 1912.

28. Egan, M. D., *Concepts in Building Firesafety,* J. Wiley, New York, 1978.

29. Sachs, E. O., and Marsland, E., "The Fire Resistance of Floors being Tabulated Results of Fire Tests Conducted by the Committee," *Journal of the British Fire Prevention Committee,* No. VI, London, 1911.

30. Sachs, E. O., and Marsland, E., "The Fire Resistance of Partitions being Tabulated Results of Fire Tests Conducted by the Committee," *Journal of the British Fire Prevention Committee,* No. IX, London, 1914.

31. Ryan, J. V., and Bender, E. W., "Fire Tests of Precast Cellular Concrete Floors and Roofs," *National Bureau of Standards Monograph,* 45, Washington, April 1962.

32. Kingberg, S. H., and Foster, H. D., "Fire Resistance of Hollow Load-Bearing Wall Tile," *National Bureau of Standards* Research Paper, No. 37, (Reprint from *NBS Journal of Research,* Vol. 2) Washington, 1929.

33. Hull, W. A., and Ingberg, S. H., "Fire Resistance of Concrete Columns," *Technologic Papers of the Bureau of Standards,* No. 272, Vol. 18, Washington, 1925, pp. 635-708.

34. National Board of Fire Underwriters, *Fire Resistance Ratings of Less than One Hour,* New York, Aug. 1956.

35. Harmathy, T. Z., "Ten Rules of Fire Endurance Rating," *Fire Technology,* Vol. 1, May 1965, pp. 93-102.

36. Son, B. C., "Fire Endurance Test on a Steel Tubular Column Protected with Gypsum Board," *National Bureau of Standards,* NBSIR, 73-165, Washington, 1973.

37. Galbreath, M., "Fire Tests of Wood Door Assemblies," *Fire Study,* No. 36, Div. of Building Research, National Research Council Canada, Ottawa, May 1975.

38. Morris, W. A., "An Investigation into the Fire Resistance of Timber Doors," *Fire Research Note,* No. 855, Fire Research Station, Boreham Wood, Jan. 1971.

39. Hall, G. S., "Fire Resistance Tests of Laminated Timber Beams," *Timber Association Research Report,* WR/RR/1, High Sycombe, July 1968.

40. Goalwin, D. S., "Fire Resistance of Concrete Floors," *Building Materials and Structures,* Report BMS 134, National Bureau of Standards, Washington, Dec. 1952.

41. Mitchell, N. D., and Ryan, J. V., "Fire Tests of Steel Columns Encased with Gypsum Lath and Plaster," *Building Materials and Structures,* Report BMS 135, National Bureau of Standards, Washington, April 1953.

42. Ingberg, S. H., "Fire Tests of Brick Walls," *Building Materials and Structures,* Report BMS 143, National Bureau of Standards, Washington, Nov. 1954.

43. National Bureau of Standards, "Fire Resistance and Sound-Insulation Ratings for Walls, Partitions, and Floors," *Technical Report on Building Materials,* 44, Washington, 1944.

44. Malhotra, H. L., "Fire Resistance of Brick and Block Walls," *Fire Note,* No. 6, Ministry of Technology and Fire Offices' Committee Joint Fire Research Organization, London, HMSO, 1966.

45. Mitchell, N. D., "Fire Tests of Steel Columns Protected with Siliceous Aggregate Concrete," *Building Materials and Structures,* Report BMS 124, National Bureau of Standards, Washington, May 1951.

46. Freitag, J. K., *Fire Prevention and Fire Protection as Applied to Building Construction; a Handbook of Theory and Practice,* 2nd ed., J. Wiley, New York, 1921.

47. Ingberg, S. H., and Mitchell, N. D., "Fire Tests of Wood and Metal-Framed Partition," *Building Materials and Structures,* Report BMS 71, National Bureau of Standards, Washington, 1941.

48. Central Housing Committee on Research, Design, and Construction, Subcommittee on Definitions, "A Glossary of Housing Terms," *Building Materials and Structures,* Report BMS 91, National Bureau of Standards, Washington, Sept. 1942.

49. Crosby, E. U., Fiske, H. A., and Forster, H.W., *Handbook of Fire Protection,* 7th ed., D. Van Nostrand Co., New York 1924.

50. Bird, E. L., and Docking, S. J., *Fire in Buildings,* A. & C. Black, London, 1949.

51. American Institute of Steel Construction, Fire Resistant Construction in Modern Steel-Framed Buildings, New York, 1959.

52. Central Dockyard Laboratory, "Fire Retardant Paint Tests—a Critical Review," CDL Technical Memorandum, No. P87/73, H. M. Naval Base, Portsmouth, Dec. 1973.

53. Malhotra, H. L., "Fire Resistance of Structural Concrete Beams," *Fire Research Note,* No. 741, Fire Research Station, Borehamwood, May 1969.

54. Abrams, M. S., and Gustaferro, A. H., "Fire Tests of Poke-Thru Assemblies," *Research and Development Bulletin,* 1481-1, Portland Cement Association, Skokie, 1971.

55. Bullen, M. L., "A Note on the Relationship between Scale Fire Experiments and Standard Test Results," *Building Research Establishment Note,* N51/75, Borehamwood, May 1975.

56. The America Fore Group of Insurance Companies, Research Department, *Some Characteristic Fires in Fire Resistive Buildings,* Selected from twenty years record in the files of the N.F.P.A. "Quarterly," New York, c. 1933.

57. Spiegelhalter, F., "Guide to Design of Cavity Barriers and Fire Stops," *Current Paper,* CP 7/77, Building Research Establishment, Borehamwood, Feb. 1977.

58. Wardle, T. M. "Notes on the Fire Resistance of Heavy Timber Construction," *Information Series,* No. 53, New Zealand Forest Service, Wellington, 1966.

59. Fisher, R. W., and Smart, P. M. T., "Results of Fire Resistance Tests on Elements of Building Construction," *Building Research Establishment Report,* G R6, London, HMSO, 1975.

60. Serex, E. R., "Fire Resistance of Alta Bates Gypsum Block Non-Load Bearing Wall," Report to Alta Bates Community Hospital, *Structural Research Laboratory Report,* ES-7000, University of Calif., Berkeley, 1969.

61. Thomas, F. G., and Webster, C. T., "Investigations on Building Fires, Part VI: The Fire Resistance of Reinforced Concrete Columns," *National Building Studies,* Research Paper, No. 18, Dept. of Scientific and Industrial Research (Building Research Station), London, HMSO, 1953.

62. Building Research Establishment, "Timber Fire Doors," *Digest,* 220, Borehamwood, Nov. 1978.

63. *Massachusetts State Building Code; Recommended Provisions,* Article 22: Repairs, Alterations, Additions, and Change of Use of Existing Buildings, Boston, Oct. 23, 1978.

64. Freitag, J. K., *Architectural Engineering; with Especial Reference to High Building Construction, Including Many Examples of Prominent Office Buildings,* 2nd ed., rewritten, J. Wiley, New York, 1906.

65. Architectural Record, *Sweet's Indexed Catalogue of Building Construction for the Year 1906,* New York, 1906.

66. Dept. of Commerce, Building Code Committee, "Recommended Minimum Requirements for Fire Resistance in Buildings," *Building and Housing,* No. 14, National Bureau of Standards, Washington, 1931.

67. British Standards Institution, "Fire Tests on Building Materials and Structures," *British Standards,* 476, Pt. 1, London, 1953.

68. Löberg-Holm, K., "Glass," *The Architectural Record,* Oct. 1930, pp. 345-357.

69. Structural Clay Products Institute, "Fire Resistance," *Technical Notes on Brick and Tile Construction,* 16 rev., Washington, 1964.

70. Ramsey, C. G., and Sleeper, H. R., *Architectural Graphic Standards for Architects, Engineers, Decorators, Builders, and Draftsmen,* 3rd ed., J. Wiley, New York, 1941.

71. Underwriters' Laboratories, *Fire Protection Equipment List,* Chicago, Jan. 1957.

72. Underwriters' Laboratories, *Fire Resistance Directory; with Hourly Ratings for Beams, Columns, Floors, Roofs, Walls, and Partitions,* Chicago, Jan. 1977.

73. Mitchell, N. D., "Fire Tests of Gunite Slabs and Partitions," *Building Materials and Structures,* Report BMS 131, National Bureau of Standards, Washington, May 1952.

74. Woolson, I. H., and Miller, R. P., "Fire Tests of Floors in the United States," *Proceedings International Association for Testing Materials,* VIth Congress, New York, 1912, Section C, pp. 36-41.

75. Underwriters' Laboratories, "An Investigation of the Effects of Fire Exposure upon Hollow Concrete Building Units, Conducted for American Concrete Institute, Concrete Products Association, Portland Cement Association, Joint Submittors," *Retardant Report,* No. 1555, Chicago, May 1924.

76. Dept. of Scientific & Industrial Research and of the Fire Offices' Committee, Joint Committee of the Building Research Board, "Fire Gradings of Buildings. Part IV: Chimneys and Flues," *Post-War Building Studies,* No. 29, London, HMSO, 1952.

77. National Research Council of Canada. Associate Committee on the National Building Code, *Fire Performance Ratings,* Suppl. No. 2 to the National Building Code of Canada, Ottawa, 1965.

78. Associated Factory Mutual Fire Insurance Companies, The National Board of Fire Underwriters, and the Bureau of Standards, *Fire Tests of Building Columns; an Experimental Investigation of the Resistance of Columns, Loaded and Exposed to Fire or to Fire and Water, with Record of Characteristic Effects,* Jointly Conducted at Underwriters. Laboratories, Chicago, 1917-19.

79. Malhotra, H. L., "Effect of Age on the Fire Resistance of Reinforced Concrete Columns," *Fire Research Memorandum,* No. 1, Fire Research Station, Borehamwood, April 1970.

80. Bond, H., ed., *Research on Fire; a Description of the Facilities, Personnel and Management of Agencies Engaged in Research on Fire,* a Staff Report, National Fire Protection Association, Boston, 1957.

81. *California State Historical Building Code,* Draft, 1978.

82. Fisher, F. L., et al., "A Study of Potential Flashover Fires in Wheeler Hall and the Results from a Full Scale Fire Test of a Modified Wheeler Hall Door Assembly," *Fire Research Laboratory Report,* UCX 77-3; UCX-2480, University of Calif., Dept. of Civil Eng., Berkeley, 1977.

83. Freitag, J. K., *The Fireproofing of Steel Buildings,* 1st ed., J. Wiley, New York, 1906.

84. Gross, D., "Field Burnout Tests of Apartment Dwellings Units," *Building Science Series,* 10, National Bureau of Standards, Washington, 1967.

85. Dunlap, M. E., and Cartwright, F. P., "Standard Fire Tests for Combustible Building Materials," *Proceedings of the American Society for Testing Materials,* vol. 27, Philadelphia, 1927, pp. 534-546.

86. Menzel, C. A., "Tests of the Fire Resistance and Stability of Walls of Concrete Masonry Units," *Proceedings of the American Society for Testing Materials,* vol. 31, Philadelphia, 1931, pp. 607-660.

87. Steiner, A. J., "Method of Fire-Hazard Classification of Building Materials," *Bulletin of the American Society for Testing and Materials,* March 1943, Philadelphia, 1943, pp. 19-22.

88. Heselden, A. J. M., Smith, P. G., and Theobald, C. R., "Fires in a Large Compartment Containing Structural Steelwork; Detailed Measurements of Fire Behavior," *Fire Research Note,* No. 646, Fire Research Station, Borehamwood, Dec. 1966.

89. Ministry of Technology and Fire Offices' Committee Joint Fire Research Organization, "Fire and Structural Use of Timber in Buildings; Proceedings of the Symposium Held at the Fire Research Station, Borehamwood, Herts on 25th October, 1967," *Symposium,* No. 3, London, HMSO, 1970.

90. Shoub, H., and Gross, D., "Doors as Barriers to Fire and Smoke," *Building Science Series,* 3, National Bureau of Standards, Washington, 1966.

91. Ingberg, S. H., "The Fire Resistance of Gypsum Partitions," *Proceedings of the American Society for Testing and Materials,* vol. 25, Philadelphia, 1925, pp. 299-314.

92. Ingberg, S.H., "Influence of Mineral Composition of Aggregates on Fire Resistance of Concrete," *Proceedings of the American Society for Testing and Materials,* vol. 29, Philadelphia, 1929, pp. 824-829.

93. Ingberg, S. H., "The Fire Resistive Properties of Gypsum," *Proceedings of the American Society for Testing and Materials,* vol. 23, Philadelphia, 1923, pp. 254-256.

94. Gottschalk, F.W., "Some Factors in the Interpretation of Small-Scale Tests for Fire-Retardant Wood," *Bulletin of the American Society for Testing and Materials,* October 1945, pp. 40-43.

95. Ministry of Technology and Fire Offices' Committee Joint Fire Research Organization, "Behaviour of Structural Steel in Fire; Proceedings of the Symposium Held at the Fire Research Station Borehamwood, Herts on 24th January, 1967," *Symposium,* No. 2, London, HMSO, 1968.

96. Gustaferro, A. H., and Martin, L. D., *Design for Fire Resistance of Pre-cast Concrete,* prep. for the Prestressed Concrete Institute Fire Committee, 1st ed., Chicago, PCI, 1977.

97. "The Fire Endurance of Concrete; a Special Issue," *Concrete Construction,* vol. 18, no. 8, Aug. 1974, pp. 345-440.

98. The British Constructional Steelwork Association, "Modern Fire Protection for Structural Steelwork," *Publication,* No. FPl, London, 1961.

99. Underwriters' Laboratories, "Fire Hazard Classification of Building Materials," *Bulletin,* No. 32, Sept. 1944, Chicago, 1959.

100. Central Housing Committee on Research, Design, and Construction, Subcommittee on Building Codes, "Recommended Building Code Requirements for New Dwelling Construction with Special Reference to War Housing; Report," *Building Materials and Structures,* Report BMS 88, National Bureau of Standards, Washington, Sept. 1942.

101. De Coppet Bergh, D., *Safe Building Construction; a Treatise Giving in Simplest Forms Possible Practical and Theoretical Rules and Formulae Used in Construction of Buildings and General Instruction,* new ed., thoroughly rev. Macmillan Co., New York, 1908.

102. *Cyclopedia of Fire Prevention and Insurance; a General Reference Work on Fire and Fire Losses, Fireproof Construction, Building Inspection...,* prep. by architects, engineers, underwriters and practical insurance men. American School of Correspondence, Chicago, 1912.

103. Setchkin, N. P., and Ingberg, S. H., "Test Criterion for an Incombustible Material," *Proceedings of the American Society for Testing Materials,* vol. 45, Philadelphia, 1945, pp. 866-877.

104. Underwriters' Laboratories, "Report on Fire Hazard Classification of Various Species of Lumber," *Retardant,* 3365, Chicago, 1952.

105. Steingiser, S., "A Philosophy of Fire Testing," Journal of Fire & Flammability, vol. 3, July 1972, pp. 238-253.

106. Yuill, C. H., Bauerschlag, W. H., and Smith, H. M., "An Evaluation of the Comparative Performance of 2.4.1 Plywood and Two-Inch Lumber Roof Decking under Equivalent Fire Exposure," *Fire Protection Section, Final Report,* Project No. 717A-3-211, Southwest Research Institute, Dept. of Structural Research, San Antonio, Dec. 1962.

107. Ashton, L. A., and Smart, P.M. T., *Sponsored Fire-Resistance Tests on Structural Elements,* London, Dept. of Scientific and Industrial Research and Fire Offices. Committee, London, 1960.

108. Butcher, E. G., Chitty, T. B., and Ashton, L. A., "The Temperature Attained by Steel in Building Fires," *Fire Research Technical Paper,* No. 15, Ministry of Technology and Fire Offices. Committee, Joint Fire Research Organization, London, HMSO, 1966.

109. Dept. of the Environment and Fire Offices' Committee, Joint Fire Research Organization, "Fire-Resistance Requirements for Buildings—a New Approach; Proceedings of the Symposium Held at the Connaught Rooms, London, 28 September 1971," *Symposium,* No. 5, London, HMSO, 1973.

110. Langdon Thomas, G. J., "Roofs and Fire," *Fire Note,* No. 3, Dept. of Scientific and Industrial Research and Fire Offices' Committee, Joint Fire Research Organization, London, HMSO, 1963.

111. National Fire Protection Association and the National Board of Fire Underwriters, *Report on Fire the Edison Phonograph Works,* Thomas A. Edison, Inc., West Orange, N.J., December 9, 1914, Boston, 1915.

112. Thompson, J. P., *Fire Resistance of Reinforced Concrete Floors,* Portland Cement Association, Chicago, 1963.

113. Forest Products Laboratory, "Fire Resistance Tests of Plywood Covered Wall Panels," Information reviewed and reaffirmed, *Forest Service Report,* No. 1257, Madison, April 1961.

114. Forest Products Laboratory, "Charring Rate of Selected Woods—Transverse to Grain," Forest Service *Research Paper,* FLP 69, Madison, April 1967.

115. Bird, G. I., "Protection of Structural Steel Against Fire," *Fire Note,* No. 2, Dept. of Scientific and Industrial Research and Fire Offices' Committee, Joint Fire Research Organization, London, HMSO, 1961.

116. Robinson, W. C., *The Parker Building Fire,* Underwriters' Laboratories, Chicago, c. 1908.

117. Ferris, J. E., "Fire Hazards of Combustible Wallboards," *Commonwealth Experimental Building Station Special Report,* No. 18, Sydney, Oct. 1955.

118. Markwardt, L. J., Bruce, H. D., and Freas, A. D., "Brief Description of Some Fire-Test Methods Used for Wood and Wood Base Materials," *Forest Service Report,* No. 1976, Forest Products Laboratory, Madison, 1976.

119. Foster, H. D., Pinkston, E. R., and Ingberg, S. H., "Fire Resistance of Walls of Gravel-Aggregate Concrete Masonry Units," *Building Materials and Structures,* Report, BMS 120, National Bureau of Standards, Washington, March 1951.

120. Foster, H. D., Pinkston, E.R., and Ingberg, S. H., "Fire Resistance of Walls of Lightweight-Aggregate Concrete Masonry Units," *Building Materials and Structures,* Report BMS 117, National Bureau of Standards, Washington, May 1950.

121. Structural Clay Products Institute, "Structural Clay Tile Fireproofing," *Technical Notes on Brick & Tile Construction,* vol. 1, no. 11, San Francisco, Nov. 1950.

122. Structural Clay Products Institute, "Fire Resistance Ratings of Clay Masonry Walls—I," *Technical Notes on Brick & Tile Construction,* vol. 3, no. 12, San Francisco, Dec. 1952.

123. Structural Clay Products Institute, "Estimating the Fire Resistance of Clay Masonry Walls—II," *Technical Notes on Brick & Tile Construction,* vol. 4, no. 1, San Francisco, Jan. 1953.

124. Building Research Station, "Fire: Materials and Structures," *Digest,* No. 106, London, HMSO, 1958.

125. Mitchell, N. D., "Fire Hazard Tests with Masonry Chimneys," *NFPA Publication,* No. Q-43-7, Boston, Oct. 1949.

126. Clinton Wire Cloth Company, *Some Test Data on Fireproof Floor Construction Relating to Cinder Concrete, Terra Cotta and Gypsum,* Clinton, 1913.

127. Structural Engineers Association of Southern California, Fire Ratings Subcommittee, "Fire Ratings, a Report," part of *Annual Report,* Los Angeles, 1962, pp. 30-38.

128. Lawson, D. I., Fox, L. L., and Webster, C. T., "The Heating of Panels by Flue Pipes," *Fire Research, Special Report,* No. 1, Dept. of Scientific and Industrial Research and Fire Offices' Committee, London, HMSO, 1952.

129. Forest Products Laboratory, "Fire Resistance of Wood Construction," Excerpt from 'Wood Handbook—Basic Information on Wood as a Material of Construction with Data for its Use in Design and Specification,' *Dept. of Agriculture Handbook,* No. 72, Washington, 1955, pp. 337-350.

130. Goalwin, D. S., "Properties of Cavity Walls," *Building Materials and Structures,* Report BMS 136, National Bureau of Standards, Washington, May 1953.

131. Humphrey, R. L., "The Fire-Resistive Properties of Various Building Materials," *Geological Survey Bulletin,* 370, Washington, 1909.

132. National Lumber Manufacturers Association, "Comparative Fire Test on Wood and Steel Joists," *Technical Report,* No. 1, Washington, 1961.

133. National Lumber Manufacturers Association, "Comparative Fire Test of Timber and Steel Beams," *Technical Report,* No. 3, Washington, 1963.

134. Malhotra, H. L., and Morris, W. A., "Tests on Roof Construction Subjected to External Fire," *Fire Note,* No. 4, Dept. of Scientific and Industrial Research and Fire Offices' Committee, Joint Fire Research Organization, London, HMSO, 1963.

135. Brown, C. R., "Fire Tests of Treated and Untreated Wood Partitions," *Research Paper,* RP 1076, part of *Journal of Research of the National Bureau of Standards,* vol. 20, Washington, Feb. 1938, pp. 217-237.

136. Underwriters' Laboratories, "Report on Investigation of Fire Resistance of Wood Lath and Lime Plaster Interior Finish," *Publication,* SP. 1.230, Chicago, Nov. 1922.

137. Underwriters' Laboratories, "Report on Interior Building Construction Consisting of Metal Lath and Gypsum Plaster on Wood Supports," *Retardant,* No. 1355, Chicago, 1922.

138. Underwriters' Laboratories, "An Investigation of the Effects of Fire Exposure upon Hollow Concrete Building Units," *Retardant,* No. 1555, Chicago, May 1924.

139. Moran, T. H., "Comparative Fire Resistance Ratings of Douglas Fir Plywood," *Douglas Fir Plywood Association Laboratory Bulletin,* 57-A, Tacoma, 1957.

140. Gage Babcock & Association, "The Performance of Fire-Protective Materials under Varying Conditions of Fire Severity," Report 6924, Chicago, 1969.

141. International Conference of Building Officials, *Uniform Building Code* (1979 ed.), Whittier, CA, 1979.

142. Babrauskas, V., and Williamson, R. B., "The Historical Basis of Fire Resistance Testing, Part I and Part II," *Fire Technology,* vol. 14, no. 3 & 4, Aug. & Nov. 1978, pp. 184-194, 205, 304-316.

143. Underwriters' Laboratories, "Fire Tests of Building Construction and Materials," 8th ed., *Standard for Safety,* UL263, Chicago, 1971.

144. Hold, H. G., *Fire Protection in Buildings,* Crosby, Lockwood, London, 1913.

145. Kollbrunner, C. F., "Steel Buildings and Fire Protection in Europe," *Journal of the Structural Division,* ASCE, vol. 85, no. ST9, Proc. Paper 2264, Nov. 1959, pp. 125-149.

146. Smith, P., "Investigation and Repair of Damage to Concrete Caused by Formwork and Falsework Fire," *Journal of the American Concrete Institute,* vol. 60, Title no. 60-66, Nov. 1963, pp. 1535-1566.

147. "Repair of Fire Damage," 3 parts, *Concrete Construction,* March-May, 1972.

148. National Fire Protection Association, *National Fire Codes; a Compilation of NFPA Codes, Standards, Recommended Practices and Manuals,* 16 vols., Boston, 1978.

149. Ingberg, S. H. "Tests of Severity of Building Fires," *NFPA Quarterly,* vol. 22, no. 1, July 1928, pp. 43-61.
150. Underwriters' Laboratories, "Fire Exposure Tests of Ordinary Wood Doors," *Bulletin of Research,* no. 6, Dec. 1938, Chicago, 1942.
151. Parson, H., "The Tall Building under Test of Fire," *Red Book,* no. 17, British Fire Prevention Committee, London, 1899.
152. Sachs, E. O., "The British Fire Prevention Committee Testing Station," *Red Book,* no. 13, British Fire Prevention Committee, London, 1899.
153. Sachs, E. O., "Fire Tests with Unprotected Columns," *Red Book,* no. 11, British Fire Prevention Committee, London, 1899.
154. British Fire Prevention Committee, "Fire Tests with Floors a Floor by the Expended Metal Company," *Red Book,* no. 14, London, 1899.
155. *Engineering News,* vol. 56, Aug. 9, 1906, pp. 135-140.
156. *Engineering News,* vol. 36, Aug. 6, 1896, pp. 92-94.
157. Bauschinger, J., *Mittheilungen de Mech.-Tech. Lab. der K. Tech. Hochschule, Müchen,* vol. 12, 1885.
158. *Engineering News,* vol. 46, Dec. 26, 1901, pp. 482-486, 489-490.
159. *The American Architect and Building News,* vol. 31, March 28, 1891, pp. 195-201.
160. British Fire Prevention Committee, First International Fire Prevention Congress, *Official Congress Report,* London, 1903.
161. American Society for Testing Materials, *Standard Specifications for Fire Tests of Materials and Construction (C19-18),* Philadelphia, 1918.
162. International Organization for Standardization, *Fire Resistance Tests on Elements of Building Construction (R834),* London, 1968.
163. *Engineering Record,* vol. 35, Jan. 2, 1897, pp. 93-94; May 29, 1897, pp. 558-560; vol. 36, Sept. 18, 1897, pp. 337-340; Sept. 25, 1897, pp. 359-363; Oct. 2, 1897, pp. 382-387; Oct. 9, 1897, pp. 402-405.
164. Babrauskas, Vytenis, "Fire Endurance in Buildings," PhD Thesis. *Fire Research Group,* Report, No. UCB FRG 76-16, University of California, Berkeley, Nov. 1976.
165. The Institution of Structural Engineers and The Concrete Society, *Fire Resistance of Concrete Structures,* London, Aug. 1975.

INDEX

A

ACCESSIBILITY 107.2, 410, 605, 705,
801.1, 901.2, 906, 1006, 1012.8,
1101.2, 1105, 1105.2, 1204.1, 1205.15, 1401.2.5,
1508, Appendix B
Type A Units . . 410.6, 410.8.8, 705.1, 705.1.9, 1105.3
Type B Units 410.4, 410.4.2, 410.6, 410.7,
410.8.9, 410.9, 705.1, 705.2, 1105.4
ADDITIONS 101.2, 101.3, 104.2,
105.2.2, 106.2.3, 115.5, 202, 401.1,
401.1.1, 402, 403.4, 408.1, 410.1,
410.5, 410.8.4, 501.1, 507, Chapter 11,
1401.1, 1401.1.1, 1401.2, 1401.2.3, 1401.3,
1401.4.1, 1401.4.3, 1501.3, 1505.2
ADMINISTRATION . . . . Chapter 1
ALTERATIONS 101.2, 101.3, 101.4, 101.7,
104.2, 104.10.1, 105.1.1, 105.1.2,
105.2.2, 105.2.3, 106.2.3, 106.2.5,
108.2, 108.5, 110.2, 113.2, 115.5, 202,
301.1, 301.1.1, 301.1.2, 301.1.3, 301.2,
401.1, 401.1.1, 401.2.2, 402.4, 403,
404.1, 408.1, 410.1, 410.1.4, 410.3,
410.4.1, 410.6, 410.7, 410.8, 410.8.1,
410.8.4, 410.8.10, 410.9, 501.1, 501.1.1,
503, 504, 506, Chapter 7, Chapter 8,
Chapter 9, 1001.2.1, 1007.3.1, 1012.8,
1012.8.1, 1101.3, 1201.1, 1201.2, 1201.4,
1203.1, 1204, 1301.2, 1401.1, 1401.1.1,
1401.2, 1401.2.4, 1401.3, 1401.4.1,
1401.4.3, 1501.3, 1504.1, 1505.2,
A102.1, A105.2, A205.2, A403.2
Level 1 . . . . 104.2.1, 202, 503, 504.2,
505.2, Chapter 7, 801.2
Level 2 . . . . 106.2.3, 202, 504, 505.2, Chapter 8
Level 3 . . . . . 106.2.3, 202, 505, Chapter 9, 1007.3.1
**ALTERNATIVE MATERIALS, DESIGN,
AND METHODS OF
CONSTRUCTION** 104.11, 104.11.1, 104.11.2
APPEALS 112, 116.6
APPLICABILITY 101.4, 102, 104.2.1, 1401.2
**ARCHAIC MATERIALS AND
ASSEMBLIES** . . . . Resource A
**ARCHITECT (see REGISTERED DESIGN
PROFESSIONAL)**
AREA (see BUILDING)
AUTOMATIC SPRINKLER SYSTEM 803.2.1,
804.1.1, 804.2, 804.3, 804.4, 805.3.1.1,
805.3.1.2.1, 805.3.2, 805.5.1, 805.6, 902.2,
904.1, 1012.2.1, 1012.5.1, 1012.5.1.1,
1012.6.3, 1012.7.2, 1012.7.3, 1203.4, 1205.9,
1401.5.1, 1401.6.1, 1401.6.2, 1401.6.2.1,
1401.6.12.1, 1401.6.17, Table 1401.6.17,
1401.6.17.1, 1401.6.19, Table 1401.6.19,
Table 1407.1, 1507

B

BOARD OF APPEALS 104.8, 112
BUILDING
Area . . . . 1012.5, Table 1012.5, 1012.5.1,
1012.5.1.1, 1012.5.2, 1102, 1205.2,
1401.2.3, 1401.6.2, Table 1407.1
Dangerous 202, 1206.2
Existing 101.2, 202, 301.1, 301.2, 401.1,
401.1.1, 501.1, 507.2, 701.2, 707.1, 807.4,
908.1, 1001.3.1, 1002.1, 1007.3, 1007.3.1,
1008.1, 1008.2, 1008.3, 1008.4, 1009.1, 1010.1,
1012.1.1.1, 1012.1.1.2, 1012.5.2, 1012.8, 1101.1,
1101.2, 1101.3, 1102.1, 1102.2, 1103.1, 1103.2,
1103.3, 1103.4, 1103.5, 1104.1, 1106.1, 1201.4,
1401.1, 1401.2.1, 1401.2.3, 1401.2.4, 1401.3,
1401.4, 1401.4.1, 1401.4.3, 1401.6, 1401.6.1.1,
1501.3, 1505.1, A102.1, A201, A503.1
Height 202, 402.1, 804.2.2, 805.3.1.1,
1007.3.1, 1012.5, Table 1012.5,
1012.5.1, 1012.5.1.1, 1012.5.2, 1012.6.1,
1012.6.3 1102.1, 1401.2.3, 1401.6.1,
1401.6.1.1, 1401.6.11, Table 1407.1,
1505.1, A301.2, A506.2
Historic 202, 401.1, 408, 410.1, 410.9,
508, 601.1, 701.1, 1001.2.1, Chapter 12,
1401.1, B101.1, B101.2, B101.3, B101.4,
B101.4.2, B101.5
Relocatable 1301.1
Relocated 110.1, 301.1.2, 301.1.3,
409.1, 509, 1202.3, Chapter 13
Relocation of 101.2, 101.3, 101.4, 104.2,
301.1, 301.2, 401.1, 401.1.1, 1201.1
Underground 1002.2
Unsafe 115, 202, 602.1
**BUILDING ELEMENTS
AND MATERIALS** 602, 702, 803, 903, 1003

INDEX

C

CARPETING................105.2, 702.2
CEILING..............702.1, 801.3, 803.2.1, 803.4,
808.1, 808.3.1, 903.2.1, 1012.3,
1012.4.1, 1203.5, 1203.7, 1205.10,
1401.6.3.2, A105.3, A113.1
**CERTIFICATE OF
    OCCUPANCY**..........110, 407.2, 1001.2, 1001.4
CHANGE OF OCCUPANCY.....101.2, 101.3, 101.4,
101.5, 104.2, 104.2.1, 106.2.5,
115.5, 202, 401.1, 407, 408,
410.1, 410.4, 410.8.4, 410.9, 501.1,
506, 907.4, Chapter 10, 1201.1,
1201.2, 1201.4, 1205, 1301.2, 1401.1,
1401.2.1, 1401.2.2, 1401.2.5, 1401.4.1,
1401.4.3, 1401.6.17, B101.2, B101.3, B101.4
CODE OFFICIAL................Chapter 1, 202,
401.2.1, 404.1.1, 404.2.1, 602.1,
606.2.2.1, 804.2.5, 805.2, 907.4.1,
1001.2, 1007.3.1, 1012.4.1, 1201.2,
1201.3, 1203.2, 1203.3, 1203.11, 1203.12,
1204.1, 1205.6, 1205.7, 1205.11, 1205.12,
1205.14, 1206.1, 1206.2, 1302.7, 1401.3,
1401.3.1, 1401.4.2, 1401.4.3, 1501.6.7,
A301.1, A404.1
COMPARTMENTATION..........803.3.1, 1401.6.3,
Table 1401.6.3, Table 1401.7
COMPLIANCE METHODS....301, Chapter 4, 501.1.1
Chapter 14
  Prescriptive compliance method..301.1.1, Chapter 4
  Work area compliance method...........301.1.2
  Classification of work..................Chapter 5
  Repairs ....Chapter 6
  Alteration Level 1....................Chapter 7
  Alteration Level 2....................Chapter 8
  Alteration Level 3....................Chapter 9
  Change of occupancy ....Chapter 10
  Additions ....Chapter 11
  Historic buildings ....Chapter 12
  Relocated buildings ....Chapter 13
  Performance compliance method 301.1.3,
Chapter 14
CONFLICT 102.1, 104.10.1, 113.1, A301.2
CONSTRUCTION DOCUMENTS.......104.2, 105.3,
105.3.1, 105.4, 106,
113.4, 202, 501.2,
A105.4, A205.4, A301.1
CONSTRUCTION SAFEGUARDS..101.5, Chapter 15
CORRIDOR
  Dead-end................801.3, 805.6, 1012.4.1
  Doors....................805.5.1, 1012.4.1
  Exit access..........801.3, 803.2.1, 803.4, 803.4.1,
804.2.1, 804.2.2, 804.2.4, 804.3, 805.1,
805.3.1, 901.2, 1012.7.2, 1203.3, 1205.6
  Openings 805.5, 1012.4.1, 1012.7.2, 1205.8
  Rating..........804.1.1, 1401.6.5, Table 1401.6.5,
1401.6.5.1, Table 1401.7

D

DEFERRED SUBMITTAL........106.3.4, 106.6, 202
DEFINITIONS 202
DEMOLITION 101.6, 104.2, 106.2.5, 108.5,
113.2, 117, 1501.6, 1501.7
1502.1, 1503.1, 1504.1, 1505.2
DEPARTMENT OF BUILDING SAFETY 103,
104.10, 105.3,
109.3.7, 110.2

E

EGRESS (see MEANS OF EGRESS)
ELECTRICAL 105.1, 105.1.1, 105.2, 108.2,
108.3, 109.3.4, 202, 410.7, 607,
608.2, 705.2, 801.3, 808, 1008,
1101.2, A102.1
ELEVATOR 410.8.2, 705.1.2, 902.1.2,
1012.7.3, 1102.2, 1401.6.14,
Table 1401.6.14, Table 1401.7
**EMERGENCY ESCAPE AND
    RESCUE OPENINGS** 406.3, 702.5
EMERGENCY POWER.........805.4.5, 1401.6.15.1
ENERGY....................702.4, 707, 811, 908
**ENGINEER (see REGISTERED DESIGN
    PROFESSIONAL)**
EXISTING (see BUILDING)
EXIT 405.1.4, 803.2.1, 803.4, 805.3.1.1,
805.3.1.2, 805.3.2, 805.3.3, 805.4.1.1,
805.4.2, 805.4.3, 805.4.3.1, 805.4.4, 805.8,
805.9, 805.9.1, 805.10, 903.1, 903.3, 905.2,
905.3, 1012.7.2, Table 1012.4, 1012.4.1, 1102.2,
1203.3, 1203.6, 1203.11, 1205.6, 1205.7, 1205.12,
1205.13, 1401.6.3.1, 1401.6.6, 1401.6.10.1,
1401.6.11, 1401.6.11.1, 1401.6.12, 1401.6.13,
1401.6.15.1, Table 1401.6.15, Table 1401.7, A402

F

FIRE ESCAPES 405, 805.3.1.2, 1401.6.11
FIRE PROTECTION 106.2.2, 110.2, 202, 410.7,
603, 703, 705.2, 804, 805.5.1,
805.5.2, 904, 1004, 1012.1.2, 1012.2,
1012.7.4, 1102.3, 1501.3, 1509

INDEX

FIRE RATINGS 803.2.1, 803.3.2, 803.6, 804.1.1,
804.2.2.1, 805.3.1.1, 805.3.1.2.1,
805.5.1, 805.5.2, 1012.1.1.1, 1012.1.1.2,
1012.5.1, 1012.5.1.1, 1012.5.3, 1012.6, 1012.6.1,
1012.6.2, 1012.7.3, 1012.7.4, 1203.6, 1203.8,
1401.2.2, 1401.6.3.1, 1401.6.3.2, 1401.6.4,
1401.6.4.1, 1401.6.5, 1401.6.5.1, 1401.6.6,
1401.6.16.1, Table 1401.7, Resource A
FIRE SAFETY 101.8, 104.10, 107.2, 1101.2,
1203, 1401.5, 1401.6.1, 1401.6.2,
1401.6.3, 1401.6.4, 1401.6.5, 1401.6.6,
1401.6.7, 1401.6.8, 1401.6.9, 1401.6.14,
1401.6.16, 1401.6.17, 1401.6.18, Table 1401.7,
Table 1401.8, Table 1401.9, A102.1
FLAME SPREAD . . . . 1205.9, Resource A
FLOOD HAZARD AREA 104.10.1, 109.3.3, 202,
301.1, 402.2, 403.2, 404.5,
408.2, 601.3, 606.2.4, 701.3,
1103.5, 1201.4, 1302.6, 1401.3.3
FUEL GAS 702.4.1

G
GLASS 406, 602.3, 805.5.1, 805.5.2,
1012.7.2, 1205.8, 1401.6.10.1
GRAVITY LOADS (see STRUCTURAL LOADS/ FORCES)
GUARDS 803.5, 805.10, 1012.4.1,
1012.4.5, 1203.9, 1203.10
GUIDELINES FOR STRUCTURAL RETROFIT
Seismic. . . . . Appendix A
Wind . . . . Appendix C, Appendix D

H
HANDRAILS 805.9, 1012.4.4, 1203.9
HAZARD CATEGORIES 1007.3.1, 1012.1.3,
1012.4, Table 1012.4, 1012.5,
Table 1012.5, 1012.6, Table 1012.6
HEIGHT (see BUILDING)
HIGH-RISE BUILDING 804.2.1, 902.1,
904.1.1, 1012.5.1, 1401.6.17, 1401.6.19

I
INSPECTION 104.4, 104.6, 104.7, 105.2,
106.1, 106.3.1, 106.6, 109, 1302.7,
A105.4, A107.1, A107.4, A117.3,
A205.4, A206.2, A304.4.3, A304.5,
A406.3.4, A407.1, A503.3, A505.2
INTERIOR FINISHES 702.1, 702.2, 803.4, 903.3,
1012.1.2, 1012.3, 1203.5, 1205.9

INTERIOR TRIM 702.3

L
LIVE LOAD (see STRUCTURAL LOADS)
LOAD-BEARING ELEMENT 105.2.2, 202, A106.2

M
MAINTENANCE 101.8, 105.2, 404.1,
410.1, 410.2, 1501.6.6
MEANS OF EGRESS . . . 105.2.2, 106.2.3, 107.2, 202,
405.1.1, 405.1.2, 410.6, 418.8.10,
604, 704, 705.1, 705.1.7, 803.2.1,
803.2.3, 805, 903.1, 905, 1005, 1012.4,
1012.4.1, 1012.4.2, Table 1012.5, 1101.2,
1201.3, 1203.3, 1205.6, 1401.5, 1401.5.2,
1401.6, 1401.6.11, Table 1401.7,
Table 1401.8, Table 1401.9, 1505
MECHANICAL 105.1, 105.1.1, 105.2, 105.2.2,
108.2, 108.3, 410.7, 608, 702.4,
705.2, 809, 902.1.1, 902.2.1, 1009,
1101.2, 1401.6.7.1, 1401.6.8,
1401.6.8.1, 1401.6.10, 1401.6.10.1, A102.1
MOVED BUILDINGS (see BUILDING, RELOCATED)

O
OCCUPANCY GROUP . . . . Chapter 10
OPENINGS 106.2.4, 116.2, 803.2, 805.3.1.2.1,
805.5, 903.1 1012.4.1, 1012.6.3,
1012.7, 1102.2, 1202.3, 1203.3,
1203.10.2, 1205.3, 1205.6, 1401.6.6,
Table 1401.6.6(1), 1401.6.10.1,
Table 1401.7, 1501.6.4,

P
PERMITS 101.4.1, 101.5, 104.2, 104.2.1,
104.7, 105, 106.1, 106.2.5,
106.3.1, 106.3.2, 106.3.3, 106.6,
107.1, 107.4, 108.1, 108.2, 108.3,
108.4, 108.5, 109.1, 109.2, 109.3.9,
109.5, 109.6, 110.2, 110.3, 111.1,
113.2, 113.4, 301.1, 404.1, 706.1,
706.3, 706.3.1, A105.3
PLUMBING 105.1, 105.1.1, 105.2, 108.2,
108.3, 109.3.4, 609, 702.4, 810,
1010, 1101.2, 1501.5, A102.1,
A304.1.3, A304.3.2, A304.4.2
PRIMARY FUNCTION 202, 410.4.2, 410.5, 410.7,
705.2, 1012.8.2, 1105.1

INDEX

R

RAMPS 410.8.5, 410.8.10, 705.1.4, 705.1.7, 1205.15, B103.3, B103.4

REFERENCED STANDARDS . . . . Chapter 16

REFUGE AREAS
   Capacity. 805.10.1
   Horizontal exits 805.10.2

REGISTERED DESIGN PROFESSIONAL. 104.2.1.1, 106.1, 106.3.4, 106.6, 202, 404.2.1, 606.2.2.1, 907.4.1, 1201.2, A106.3.3.3, A206.2, A301.1, A301.2, A301.3, A304.2.2, A304.2.3, A304.4.1, A505.2, A506.2,

REHABILITATION 101.7, 202, 404.2.3, 404.3, 606.1, 606.2.2.3, 606.2.3, 804.4.3, A503.1

RELOCATABLE BUILDINGS
   (see BUILDINGS, RELOCATABLE)

RELOCATED BUILDINGS
   (see BUILDINGS, RELOCATED)

REPAIR. . . . . 101.2, 101.3, 101.4, 101.7, 104.2, 104.2.1, 104.10.1, 105.1, 105.2, 105.2.1, 105.2.2, 105.2.3, 106.2.5, 113.1, 113.2, 113.4, 115.3, 115.5, 116.1, 116.4, 116.5, 117.1, 202, 301.1, 301.1.1, 301.1.2, 301.1.3, 301.2, 401.1, 401.1.1, 401.2.2, 404, 408.1, 501.1, 502.3, Chapter 6, 1001.2.1, 1101.3, 1201.1, 1201.2, 1201.4, 1202, 1301.2, 1301.7, 1401.1, 1401.1.1, 1401.2.4, 1401.3, 1501.3, 1501.6.6, A105.2, A105.3, A106.3.3.8, A205.2, A205.3, A403.2,

RISK CATEGORY. . . . . . . Table 301.1.4.1, 301.1.4.2, Table 301.1.4.2

ROOF
   Recover 202, 706
   Repair 202, 706
   Replacement 202, 706, 707.2
   Reroofing. 202, 706
   Diaphragms 707.3.2
   Permit 707.3

S

SAFEGUARDS DURING CONSTRUCTION
   (see CONSTRUCTION SAFEGUARDS)

SAFETY PARAMETERS . . . . 1401.5.3, 1401.6, Table 1401.7

SEISMIC FORCES
   (see STRUCTURAL LOADS/FORCES)

SEISMIC LOADS
   (see STRUCTURAL LOADS/FORCES)

SEISMIC RETROFIT. . . . . Appendix A

SHAFT ENCLOSURES 803.2.2, 1012.7, 1401.6.6

SMOKE ALARMS. 804.2.2, 804.4.3, 1104

SMOKE COMPARTMENTS 403.11, 803.3, 805.10, 1401.6.20

SMOKE CONTROL . . . . 1401.6.10, Table 1401.7

SMOKE DETECTORS 608.2, 804.4, 1012.7.4

SNOW LOAD
   (see STRUCTURAL LOADS/FORCES)

SPECIAL USE AND OCCUPANCY . . . 802, 902, 1002, 1008.1, 1201.3

SPRINKLER SYSTEM
   (see AUTOMATIC SPRINKLER SYSTEM)

STAIRWAY 403.1, 405.2, 407.3, 410.8.4, 803.2.1, 803.2.2, 803.2.3, 804.1.1, 805.3.1.1, 805.3.1.2.1, 805.4.3, 805.4.3.1, 805.9.1, 805.10.1, 806.2, 808.3.5, 902.2.1, 903.1, 1012.4.1, 1012.4.2, 1012.4.4, 1012.7.2, 1012.7.3, 1012.7.4, 1102.2, 1203.3, 1203.6, 1203.9, 1205.1.3, 1205.6, 1205.11, 1205.13, 1401.6.3, 1401.6.1.10.1, 1504.1, 1505.1, 1506.1

STANDPIPE SYSTEMS . . . . . 105.2.2, 804.2.1, 804.3, 1401.6.18, Table 1401.6.18, 1401.6.18.1, Table 1401.7, 1506, 1503.2

STRUCTURAL 402.3, 402.4, 403.3, 403.4, 403.5, 404.2, 404.3, 407.4, 606, 706, 807, 907, 1007, 1103, 1206, 1302, 1401.4.1

STRUCTURAL ALTERATION–LIMITED. 301.1, 907.4.3

STRUCTURAL LOADS/FORCES
   Gravity loads . . . . 402.3, 403.3, 706.2, 807.4, 907.3 , 1007.1, 1103.2, A206.6, A506.5.1
   International Building Code-level . . . Table 301.1.4.1, 1103.3.1, 1103.3.2
   Live loads. 202, 402.3.1, 403.3.1, 404.3, 405.3, 606.2.3, 805.3.1.2.2, 1205.13, 1206.1, 1501.6.1, 1501.6.5, A108.6, A403.7
   Reduced. . . . . 301.1.4.2, Table 301.1.4.2, 606.2.2.1, 606.2.2.3, 707.2.1, 907.4.2, 907.4.3, 1007.3.1, A103, A106.3.1, A108.6, A113.3, A206.2, A402, A403.5, A403.7, A403.9, A404.1
   Seismic loads. 202, 404.2.3, 606.2.2.1, 606.2.2.3, 707.2.1, 707.3.1, 907.4.2, 907.4.3, 1007.3.1, 1103.3.1, 1302.4

Snow Loads 403.3, 606.2.3, 807.4,
1007.2, 1103.4, 1302.5
Wind Loads. 404.2.1, 404.2.3, 606.2.2.3,
707.3.2, 907.4.2, 1007.2,
1302.3, 1501.6.7
SUBSTANTIAL DAMAGE
Flood. 202, 402.2, 403.2, 404.5, 606.2.4
Structural 202, 404.2, 404.3, 404.3.1, 404.4,
606.2.1, 606.2.2, 606.2.3, 606.2.3.1
SUBSTANTIAL IMPROVEMENT . . .104.10.1, 109.3.3,
202, 402.2, 403.2,
404.5, 408.2, 601.3, 701.3,
1103.5, 1201.4, 1401.3.3
**SUBSTANTIAL STRUCTURAL
ALTERATION** 907.5.2

T

TECHNICALLY INFEASIBLE 202, 410.4.2, 410.6,
410.8.6, 410.8.11, 410.8.12,
410.9, 705.1, 705.1.6, 705.1.10,
705.1.11, 1012.8.2, 1204.1, 1205.15
TEMPORARY STRUCTURE. 107
TESTING 104.11.2, 105.2, 106.2.4,
1401.6.10.1, 1507.1, A105.4,
A205.4, A408.1, A503.3
TRAVEL DISTANCE 803.3.1, 805.3.1.1, 805.3.2,
805.4.1.1, 1401.6.13, Table 1401.7

U

UNSAFE 104.6, 105.2, 114.1, 114.3, 115,
202, 401.2.1, 602.1, 602.2,
1008.2, 1202.2, 1401.3.1
UTILITIES 111, 1503.3

V

VERTICAL OPENING PROTECTION 803.2, 903.1,
1012.7, 1401.6.6,
Table 1401.7
VIOLATIONS 101.7, 104.6, 105.2, 105.4,
105.6, 109.1, 110.1, 110.2,
110.4, 113, 114.3

W

WIND LOAD (see STRUCTURAL LOADS/FORCES)
WINDOWS
Emergency escape and
rescue openings 702.5, 406.3
Opening control devices. 702.4, 406.2
Glass replacement 406.1
Glazing. 602.3

People Helping People Build a Safer World®

ICC Plan Review Services...
Delivers the Most Detailed and Precise Plan Reviews in the Building Industry

As the association dedicated to the development of the I-Codes, no one has a better understanding of **all the International Codes** than our licensed and ICC-Certified engineers and architects. Whether you're falling behind in plan reviews or need special code expertise for highly complex projects, the I-Code experts at the International Code Council's (ICC) Plan Review Services is here to help you!

NEW! Streamline the document submittal process with our cloud-based ProjectDox, powered by Avolve plan review submittal system for easy plan review document submittal, storage and review.

To learn more about ICC Plan Review Services, visit **www.iccsafe.org/planreview** or contact: David Hunter–Director of Business Development and Outreach, Technical Services, 888-ICC-SAFE (888-422-7233), ext. 5577, dhunter@iccsafe.org

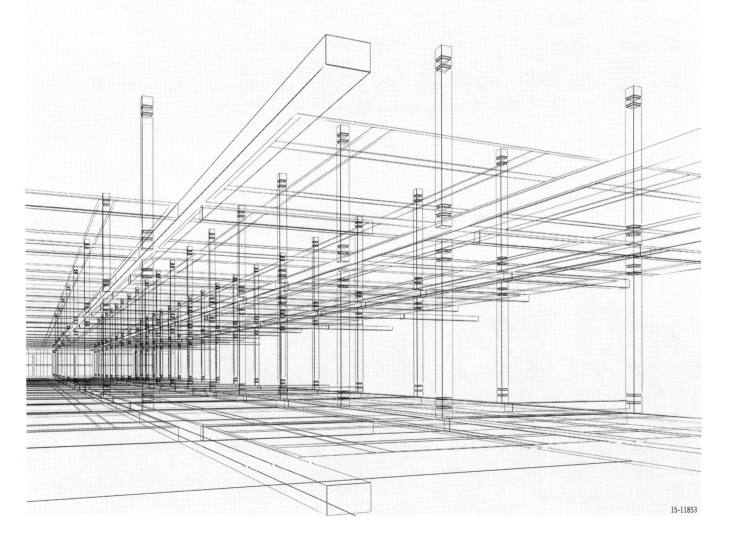

15-11853

People Helping People Build a Safer World®

Get Connected with the Best Code Resources and Member Benefits in the Industry

As a building professional, being connected to a Member-driven association that offers exclusive I-Code resources and training can play a vital role in building your career. No other building code association offers you more I-Code resources to help enhance your code knowledge and your career than the International Code Council® (ICC®).

Become an ICC Member now to take advantage of:

- Free code opinions from I-Codes experts
- Free I-Code book(s) or download to new Members*
- Discounts on I-Code resources, training and certification renewal
- Free access to employment opportunities in the ICC Career Center, including posting jobs
- Free benefits - Governmental Members: Your staff can receive free ICC benefits too*
- Great opportunities to network at ICC events
- **Savings of up to 25% on code books & Training materials**

Get connected now! Visit our Member page at www.iccsafe.org/membership or call 1-888-ICC-SAFE (422-7233), ext. 33804 to join or learn more today!

*Some restrictions apply. Speak with an ICC Member Services Representative for details.

15-11222

ICC EVALUATION SERVICE

Experts in Building Product Evaluation and Certification

Design with
CONFIDENCE

When facing new or unfamiliar materials, how do you know if they comply with building codes and standards?

- ICC-ES® Evaluation Reports are the most widely accepted and trusted technical reports for code compliance. When you specify products or materials with an ICC-ES report, you avoid delays on projects and improve your bottom line.

- ICC-ES is a subsidiary of ICC®, the publisher of the codes used throughout the U.S. and many global markets, so you can be confident in their code expertise.

- ICC-ES provides you with a free online directory of code compliant products at: **www.icc-es.org/Evaluation_Reports** and CEU courses that help you design with confidence.

INTERNATIONAL CODE COUNCIL

WWW.ICC-ES.ORG | 800-423-6587

People Helping People Build a Safer World®

2015 I-Code Essentials

This helpful series uses a straightforward, focused approach to explore code requirements with non-code language, allowing readers to gain confidence in their understanding of the material. Each book is an invaluable companion guide to the 2015 IBC, IRC, IFC or IECC for both new and experienced code users.

- Detailed full-color illustrations enhance comprehension of code provisions
- A focused, concise approach facilitates understanding of the essential code provisions
- References to corresponding code sections provide added value as companion guides to the 2015 I-Codes
- Glossary of code and construction terms clarifies their meaning in the context of the code

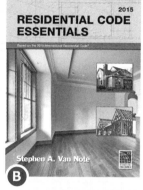

A. BUILDING CODE ESSENTIALS: BASED ON THE 2015 INTERNATIONAL BUILDING CODE
Explores those code provisions essential to understanding the application of the IBC in an easy-to-read format. The user-friendly approach of the text simplifies critical concepts so users can achieve a more complete understanding of the code's intent. Full-color illustrations, examples and simplified tables assist the reader in visualizing the code requirements. This up-to-date, step-by-step guide's topic organization reflects the intent of the IBC and facilitates understanding and application of the code provisions.
SOFT COVER #4031S15
PDF DOWNLOAD #8950P596

B. RESIDENTIAL CODE ESSENTIALS: BASED ON THE 2015 INTERNATIONAL RESIDENTIAL CODE
Explains those code provisions essential to understanding the application of the IRC to the most commonly encountered building practices. The information is presented in a user-friendly manner with an emphasis on technical accuracy and clear non-code language. Full-color illustrations, examples and simplified tables assist the reader in visualizing the code requirements.
SOFT COVER #4131S15
PDF DOWNLOAD #8950P59

C. FIRE CODE ESSENTIALS: BASED ON THE 2015 INTERNATIONAL FIRE CODE
Focuses on the essential provisions of the IFC for controlling ignition sources and fire hazards, fire department access to buildings, building uses and processes, fire protection and life safety systems, special processes and uses and hazardous materials.
SOFT COVER #4431S15
PDF DOWNLOAD #8950P59

D. ENERGY CODE ESSENTIALS: BASED ON THE 2015 INTERNATIONAL ENERGY CONSERVATION CODE®
Designed to provide a straightforward, easy-to-read companion guide to the IECC for both beginning and experienced code users. The text provides accurate information on critical energy code applications in the office and in the field for residential and commercial construction. The focused coverage examines those code provisions essential to understanding the application of the IECC to the most commonly encountered building practices and discusses the reasons behind the requirements.
SOFT COVER #4831S15
PDF DOWNLOAD #8950P651

ORDER YOUR I-CODE ESSENTIALS TODAY! 1-800-786-4452 | www.iccsafe.org/books